A TEXTBOOK OF ENGINEERING MATHEMATICS

A TEXTBOOK OF

ENGINEERING MATHEMATICS

For

B.E. II Year (*III/IV Semester*)
M.D.U., G.J.U., K.U. and DCRUST, Haryana
(Strictly According to Latest Syllabus)

By
N.P. BALI
Formerly, Principal
S.B. College, Gurgaon
Haryana

LAXMI PUBLICATIONS (P) LTD
(An ISO 9001:2008 Company)
BENGALURU • CHENNAI • COCHIN • GUWAHATI • HYDERABAD
JALANDHAR • KOLKATA • LUCKNOW • MUMBAI • RANCHI • NEW DELHI
BOSTON (USA) • ACCRA (GHANA) • NAIROBI (KENYA)

A TEXTBOOK OF ENGINEERING MATHEMATICS

Printed and bound in India
Typeset at Goswami Associates, Delhi
First Edition: 2004; Second Edition: 2005; Third Edition: 2006; Fourth Edition: 2007; Fifth Edition: 2008; Sixth Edition: 2010; Seventh Edition: 2011; Eighth Edition: 2012; Ninth Edition: 2013; Tenth Edition: 2014; Eleventh Edition: 2015, Twelfth Edition: 2016; Thirteenth Edition: 2017
ISBN 978-81-318-0493-3

PUBLISHED IN INDIA BY

Laxmi Publications (P) Ltd.
(An ISO 9001:2008 Company)
113, GOLDEN HOUSE, DARYAGANJ,
NEW DELHI - 110002, INDIA
Telephone : 91-11-4353 2500, 4353 2501
Fax : 91-11-2325 2572, 4353 2528
www.laxmipublications.com info@laxmipublications.com

Branches

Bengaluru	080-26 75 69 30	
Chennai	044-24 34 47 26,	24 35 95 07
Cochin	0484-237 70 04,	405 13 03
Guwahati	0361-254 36 69,	251 38 81
Hyderabad	040-27 55 53 83,	27 55 53 93
Jalandhar	0181-222 12 72	
Kolkata	033-22 27 43 84	
Lucknow	0522-220 99 16	
Mumbai	022-24 91 54 15,	24 92 78 69
Ranchi	0651-220 44 64	

C—
Printed at : Repro Knowledgecast Limited, Thane

CONTENTS

PREFACE TO THE THIRTEENTH EDITION

The rapid sale of the twelfth edition has encouraged me to bring in the thirteenth edition well in time. I express my sincerest thanks to the readers. I am grateful to all the professors who have patronised the book and pointed out certain missing topics for K.U.K. The needful has been done. The overwhelming response has prompted me to thoroughly revise the book again. In the present edition, many new types of questions have been added in examples as well as in exercises. A separate exercise has been provided for *Fourier Integrals*. The following topics have been added:

(*i*) De Moivre's Theorem

(*ii*) Probability Distribution of a Continuous Variable

(*iii*) Dual Simplex Method

(*iv*) Conformal Transformation (*For K.U.K. only*)

There are some changes in the syllabus also. Question 1 has been made compulsory. It is based on Short Answer Questions from all the four sections. Appendix-I contains 212 questions of this type. In the present form the book will be more useful.

I thank my student *Sandeep Yadav* for checking all answers and pointing out some printing errors.

All suggestions for the improvement of the book will be highly appreciated and gratefully acknowledged.

—Author

PREFACE TO THE FIRST EDITION

This book is a part of the original book (*with 28 chapters and covering the syllabi of engineering courses of all semesters of all the Indian Universities*) running its seventh edition and very well received by the teachers and students of all Indian Universities. The rapid sale of the seventh edition bears testimony to the overwhelming response. I thank them all for the appreciation.

The present form of the book contains only five chapters and covers the entire syllabus for the students of *third semester of M.D.U. and G.J.U.*

There is no dearth of books on Engineering Mathematics but the students find it difficult to solve most of the problems in the exercise in the absence of an adequate number of solved examples. An outstanding and distinguishing feature of the book is the large number of typical solved examples followed by well-graded problems. Many examples and problems have been selected from recent papers (2000 onwards) of various engineering examinations.

I have endeavoured to present the fundamental concepts in a comprehensive and lucid manner. I am indebted to all authors, Indian and Foreign, whose works I have frequently consulted.

All efforts have been made to keep the book free from errors. Answers to all exercises have been re-checked. All suggestions for improvement will be highly appreciated and gratefully acknowledged.

—Author

SYLLABUS

(M.D.U., Rohtak)

MAT-201-F MATHEMATICS-III

L	T	P		
3	1	-	Class Work :	50 Marks
			Theory :	100 Marks
			Total :	150 Marks
			Duration of Exam. :	3 Hours

Note: Examiner will set 9 questions in total, with two questions from each section and one question covering all sections which will be Q.1. This Q.1 is compulsory and of short answers type. Each question carries equal mark (20 marks). Students have to attempt 5 questions in total at least one question from each section.

SECTION-A

Fourier Series and Fourier Transforms: Euler's formulae, conditions for a Fourier expansion, change of interval, Fourier expansion of odd and even functions, Fourier expansion of square wave, rectangular wave, saw-toothed wave, half and full rectified wave, half range sine and cosine series.

Fourier integrals, Fourier transforms, Shifting theorem (both on time and frequency axes), Fourier transforms of derivatives, Fourier transforms of integrals, Convolution theorem, Fourier transform of Dirac-delta function.

SECTION-B

Functions of Complex Variable: Definition, Exponential function, Trigonometric and Hyperbolic functions, Logarithmic functions. Limit and Continuity of a function, Differentiability and Analyticity.

Cauchy-Riemann equations, necessary and sufficient conditions for a function to be analytic, polar form of the Cauchy-Riemann equations. Harmonic functions, application to flow problems. Integration of complex functions. Cauchy-Integral theorem and formula.

SECTION-C

Power series, radius and circle of convergence, Taylor's, Maclaurin's and Laurent's series. Zeroes and singularities of complex functions, Residues. Evaluation of real integrals using residues (around unit and semi circle only).

Probability Distributions and Hypothesis Testing: Conditional probability, Baye's theorem and its applications, expected value of a random variable. Properties and application of Binomial, Poisson and Normal distributions.

SECTION-D

Testing of a hypothesis, tests of significance for large samples, Student's t-distribution (applications only), Chi-square test of goodness of fit.

Linear Programming: Linear programming problems formulation, solving linear programming problems using (*i*) Graphical method (*ii*) Simplex method (*iii*) Dual simplex method.

SYLLABUS

(K.U., Kurukshetra)

Theory : 100 Marks
Sessional : 50 Marks
Total : 150 Marks
Duration of Exam. : 3 Hours

UNIT-I

Fourier Series: Euler's formulae. Conditions for Fourier expansions, Fourier expansion of functions having points of discontinuity, change of interval. Odd and even functions, Half-range series.

Fourier Transforms: Fourier integrals, Fourier transforms, Fourier cosine and sine transforms. Properties of Fourier transforms, Convolution theorem, Perseval's identity, Relation between Fourier and Laplace transforms, Fourier transforms of the derivatives of a function, Application to boundary value problems.

UNIT-II

Functions of a Complex Variable: Functions of a complex variable, Exponential function, Trigonometric, Hyperbolic and Logarithmic functions, limit and continuity of a function, Differentiability and analyticity.

Cauchy-Riemann Equations: Necessary and sufficient conditions for a function to be analytic, Polar Form of the Cauchy-Riemann equations, Harmonic functions, Application to flow problems, Conformal transformation, Standard transformations (Translation. Magnification and rotation, inversion and reflection, Bilinear)

UNIT-III

Probability Distributions: Probability, Baye's theorem, Discrete and Continuous probability distributions, Moment generating function. Probability generating function, Properties and applications of Binomial, Poisson and normal distributions.

UNIT-IV

Linear Programming: Linear programming problems formulation, Solution of Linear Programming Problem using Graphical method. Simplex Method, Dual-Simplex Method.

SYLLABUS

(ECE DCRUST, Murthal)

MATH - 201 MATHEMATICS-III

B.Tech. Semester-III (Common for all Branches)

L	T	P	Credits		
3	2	-	5	Class Work :	50 Marks
				Exam. :	100 Marks
				Total :	150 Marks
				Duration of Exam. :	3 Hours

PART-A

Fourier Series and Fourier Transforms: Euler's formulae, conditions for a Fourier expansion, change of interval, Fourier expansion of odd and even functions, Fourier expansion of square wave, rectangular wave, saw-toothed wave, half and full rectified wave, half range sine and cosine series.

Fourier integrals, Fourier transforms, Shifting theorem (both on time and frequency axes), Fourier transforms of derivatives, Fourier transforms of integrals, Convolution theorem, Fourier transform of Dirac-delta function.

PART-B

Functions of Complex Variable: Definition, Exponential function, Trigonometric and Hyperbolic functions, Logarithmic functions. Limit and Continuity of a function, Differentiability and Analyticity.

Cauchy-Riemann equations, necessary and sufficient conditions for a function to be analytic, polar form of the Cauchy-Riemann equations. Harmonic functions, application to flow problems. Integration of complex functions. Cauchy-Integral theorem and formula.

Power series, radius and circle of convergence, Taylor's Maclaurin's and Laurent's series. Zeroes and singularities of complex functions, Residues. Evaluation of real integrals using residues (around unit and semi circle only).

PART-C

Probability Distributions and Hypothesis Testing: Conditional probability, Baye's theorem and its applications, expected value of a random variable. Properties and application of Binomial, Poisson and Normal distributions.

Testing of a hypothesis, tests of significance for large samples, Student's t-distribution (applications only), Chi-square test of goodness of fit.

Linear Programming: Linear programming problems formulation, Solving linear programming problems using *(i)* Graphical method *(ii)* Simplex method *(iii)* Dual simplex method.

Basic Concepts

1. Greek Letters Used

α	alpha	ϕ	phi	κ	kappa	τ	tau
β	beta	ψ	psi	μ	mu	χ	chi
γ	gamma	ξ	xi	ν	nu	ω	omega
δ	delta	η	eta	π	pi	Δ	cap. delta
ε	epsilon	ζ	zeta	ρ	rho		
i	iota	λ	lambda	σ	sigma		
θ	theta	Γ	cap. gamma	Σ	cap. sigma		

2. Some Notations

$\in$	belongs to	$\Leftrightarrow$	implies and implied by
$\notin$	does not belong to	iff	if and only if
$\Rightarrow$	implies	$\cup$	union
		$\cap$	intersection

3. Useful Data

$\sqrt{2} = 1.4142$	$\frac{1}{\pi} = 0.3183$	1 rad. = 57° 17′ 45″	$\log_{10} e = 0.4343$
$\sqrt{3} = 1.732$	$e = 2.7183$	1° = 0.0174 rad.	$\log_e 2 = 0.6931$
$\pi = 3.1416$	$\frac{1}{e} = 0.3679$	$\log_e 10 = 2.3026$	$\log_e 3 = 1.0986$

4. Quadratic Equation

Roots of quadratic equation $ax^2 + bx + c = 0, \quad a \neq 0$

are $\dfrac{-b \pm \sqrt{D}}{2a}$, where $D = b^2 - 4ac$ is called discriminant

Sum of roots $= -\dfrac{b}{a}$, Product of roots $= \dfrac{c}{a}$

If $D > 0$, roots are real and distinct.

If $D = 0$, roots are equal.

If $D < 0$, roots are imaginary.

5. Progressions

(*i*) For the A.P. (Arithmetic Progression) $a, a + d, a + 2d, \ldots$

$$T_n = a + (n-1)d, \quad S_n = \frac{n}{2}[2a + (n-1)d] = \frac{n}{2}(a + l).$$

(*ii*) For the G.P. (Geometric Progression) $a, ar, ar^2, \ldots$

$$T_n = ar^{n-1}, \quad S_n = \begin{cases} \dfrac{a(1-r^n)}{1-r}, & \text{when } r \neq 1 \\ na, & \text{when } r = 1 \end{cases}$$

$$S_\infty = \frac{a}{1-r} \quad \text{provided } |r| < 1 \quad i.e., \quad -1 < r < 1$$

(*iii*) A sequence is said to be in H.P. (Harmonic Progression) if the reciprocals of its terms are in A.P.

For the H.P. $\frac{1}{a}, \frac{1}{a+d}, \frac{1}{a+2d}, \cdots, T_n = \frac{1}{a+(n-1)d}$

(*iv*) For two numbers a and b,

$$\text{A.M.} = \frac{a+b}{2}, \quad \text{G.M.} = \sqrt{ab}, \quad \text{H.M.} = \frac{2ab}{a+b}$$

(*v*) For natural numbers 1, 2, 3, ⋯, n

$$\Sigma n = \frac{n(n+1)}{2}, \Sigma n^2 = \frac{n(n+1)(2n+1)}{6}, \Sigma n^3 = \left[\frac{n(n+1)}{2}\right]^2$$

6. Permutations and Combinations

$$^nP_r = \frac{n!}{(n-r)!}, \quad ^nC_r = \frac{^nP_r}{r!} = \frac{n!}{r!(n-r)!}$$

$$^nP_n = n!, \ ^nC_r = {}^nC_{n-r}, \quad ^nC_n = {}^nC_0 = 1$$

7. Binomial Theorem

(*i*) When n is a positive integer

$$(a+b)^n = {}^nC_0\, a^n + {}^nC_1\, a^{n-1} b + {}^nC_2\, a^{n-2} b^2 + \dots + {}^nC_r\, a^{n-r} b^r + \dots + {}^nC_n\, b^n$$

$$(1+x)^n = 1 + {}^nC_1\, x + {}^nC_2\, x^2 + \dots + {}^nC_n\, x^n$$

(*ii*) When n is a negative integer or a fraction

$$(1+x)^n = 1 + nx + \frac{n(n-1)}{2!}x^2 + \frac{n(n-1)(n-2)}{3!}x^3 + \dots \infty$$

provided $|x| < 1$.

8. Logarithms

(*i*) Natural logarithm of a positive real number x is denoted by $\log_e x$ or simply $\log x$ or $\ln x$. It is the inverse of e^x.

Common logarithm of a positive real number x is denoted by $\log_{10} x$.

Relation: (*i*) $\log_{10} x = 0.4343 \log_e x$

(*ii*) $\log_a 1 = 0$, $\quad \log_a a = 1$, $\quad \log_a 0 = -\infty \quad (a > 1)$

(*iii*) $\log(mn) = \log m + \log n$, $\quad \log\left(\frac{m}{n}\right) = \log m - \log n$

$\log(m^n) = n \log m$, $\quad \log_n m \times \log_m n = 1$

9. Matrices and Determinants

(*i*) Two matrices $A = [a_{ij}]$ and $B = [b_{ij}]$ are equal if they have same order and corresponding elements are equal. *i.e.*, $a_{ij} = b_{ij}$ for all i and j.

(*ii*) If $A = [a_{ij}]$ and $B = [b_{ij}]$ are two matrices of the **same order**, then $A + B$ is defined and $A + B = [a_{ij} + b_{ij}]$, *i.e.*, add corresponding elements.

(*iii*) If $A = [a_{ij}]$ is a matrix and k is a scalar, then kA is another matrix obtained by multiplying each element of A by the scalar k. Thus, $kA = [ka_{ij}]$.

(*iv*) The product AB of two matrices A and B is defined if the number of columns in A is equal to the number of rows in B. If A is an $m \times n$ matrix and B is an $n \times p$ matrix then AB is a matrix of order $m \times p$. The $(i, j)^{th}$ element of AB is obtained by multiplying the corresponding elements of ith row of A and j^{th} column of B and adding all these products.

Matrix multiplication is not commutative, *i.e.*, AB ≠ BA in general.

If the product of two matrices is a zero matrix, it is not necessary that one of the matrices is a zero matrix.

$$(AB)C = A(BC), \quad A(B + C) = AB + AC$$

whenever both sides of equality are defined.

(*v*) For every square matrix A, there exists an identity matrix I of same order such that AI = IA = A.

(*vi*) The matrix obtained by interchanging the rows and columns of a matrix A is called the transpose of A. Transpose of A is denoted by A' or A^T.

If $A = [a_{ij}]_{m \times n}$, then $A' = [a_{ji}]_{n \times m}$

$$(A')' = A, \quad (kA)' = kA', \quad (A + B)' = A' + B',$$
$$(AB)' = B'A'.$$

(*vii*) A square matrix A is called symmetric if $A' = A$ and skew symmetric if $A' = -A$. All the diagonal elements of a skew symmetric matrix are zero.

(*viii*) A square matrix A is said to be invertible if there exists a square matrix B of same order as A, such that AB = BA = I. The matrix B is called the inverse of A and it is denoted by A^{-1}.

Thus, $\quad AA^{-1} = A^{-1} A = I$

Also, $\quad (A^{-1})^{-1} = A, \quad (AB)^{-1} = B^{-1} A^{-1}.$

(*ix*) A determinant is a function which associates each square matrix with a unique number (real or complex). The determinant of a square matrix A is denoted by | A | or det. A or Δ.

(*x*)
$$\begin{vmatrix} a & b \\ c & d \end{vmatrix} = ad - cb.$$

$$\begin{vmatrix} a_1 & a_2 & a_3 \\ b_1 & b_2 & b_3 \\ c_1 & c_2 & c_3 \end{vmatrix} = a_1\begin{vmatrix} b_2 & b_3 \\ c_2 & c_3 \end{vmatrix} - a_2\begin{vmatrix} b_1 & b_3 \\ c_1 & c_3 \end{vmatrix} + a_3\begin{vmatrix} b_1 & b_2 \\ c_1 & c_2 \end{vmatrix}$$
$$= a_1(b_2c_3 - b_3c_2) - a_2(b_1c_3 - b_3\, c_1) + a_3\,(b_1c_2 - b_2c_1)$$

$$\begin{vmatrix} a_1 & a_2 & a_3 \\ 0 & b_2 & b_3 \\ 0 & 0 & c_3 \end{vmatrix} = \begin{vmatrix} a_1 & 0 & 0 \\ b_1 & b_2 & 0 \\ c_1 & c_2 & c_3 \end{vmatrix} = a_1\, b_2c_3 = \text{product of diagonal elements.}$$

(*xi*) If A is a square matrix of order n, then $|\, kA \,| = k^n\, |\, A \,|$.

(*xii*) The value of a determinant remains unchanged if its rows and columns are interchanged, *i.e.*, $|\, A' \,| = |\, A \,|$.

(*xiii*) If any two-rows (or columns) of a determinant are interchanged, then sign of determinant changes.

(*xiv*) If any two rows (or columns) of a determinant are identical or proportional, then value of determinant is zero.

(*xv*) If each element of a row (or column) of a determinant is multiplied by a constant k, then its value gets multiplied by k. By this property, we can take out any common factor from any one row (or column) of a given determinant.

(*xvi*) If each element of a row (or column) of a determinant is the sum of m terms, then the determinant can be expressed as the sum of m determinants.

(*xvii*) If, to each element of a row (or column) of a determinant, be added equi-multiples of the corresponding elements of some other row (or column), then value of determinant remains the same.

(*xviii*) If $A = \begin{bmatrix} a_{11} & a_{12} & a_{13} \\ a_{21} & a_{22} & a_{23} \\ a_{31} & a_{32} & a_{33} \end{bmatrix}$ then adj. $A = \begin{bmatrix} A_{11} & A_{12} & A_{13} \\ A_{21} & A_{22} & A_{23} \\ A_{31} & A_{32} & A_{33} \end{bmatrix}'$, where A_{ij} is co-factor of a_{ij}.

(*xix*) For a square matrix of order n,

$$A\ (\text{adj. } A) = (\text{adj. } A) = |\ A\ |\ I$$

where I is the identity matrix of order n.

(*xx*) If A is a non-singular matrix of order n, then $|\ \text{adj. } A\ | = |\ A\ |^{n-1}$

(*xxi*) A square matrix A is invertible if and only if A is a non-singular matrix.

$$A^{-1} = \frac{1}{|A|} \text{ adj. } A$$

10. Trigonometry

(*i*)

$x°$	0°	30°	45°	60°	90°
$\sin x$	0	$\frac{1}{2}$	$\frac{1}{\sqrt{2}}$	$\frac{\sqrt{3}}{2}$	1
$\cos x$	1	$\frac{\sqrt{3}}{2}$	$\frac{1}{\sqrt{2}}$	$\frac{1}{2}$	0
$\tan x$	0	$\frac{1}{\sqrt{3}}$	1	$\sqrt{3}$	∞

(*ii*) The figure shows a unit circle with centre at origin. If $\angle XOP = x$, then $P = (\cos x, \sin x)$.

$\therefore$
$$\cos 0 = 1, \qquad \sin 0 = 0$$
$$\cos \frac{\pi}{2} = 0, \qquad \sin \frac{\pi}{2} = 1$$
$$\cos \pi = -1, \qquad \sin \pi = 0$$
$$\cos \frac{3\pi}{2} = 0, \qquad \sin \frac{3\pi}{2} = -1$$
$$\cos 2\pi = 1, \qquad \sin 2\pi = 0$$

(*iii*) Any t-ratio of $(n.\ 90° \pm x) = \pm$ same t-ratio of x, when n is even $= \pm$ co-ratio of x, when n is odd.

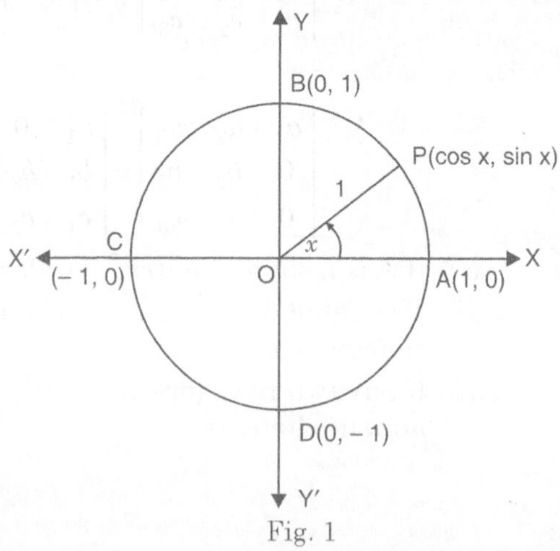

Fig. 1

where co-ratio of x is obtained by dropping co if present and adding co if absent. Thus, sin $\rightleftharpoons$ cos, tan $\rightleftharpoons$ cot, sec $\rightleftharpoons$ cosec.

The sign ± or – is decided from the quadrant in which $n.\ 90° \pm x$ lies.

(*iv*) Signs of *t*-ratios in different quadrants (Fig. 2)

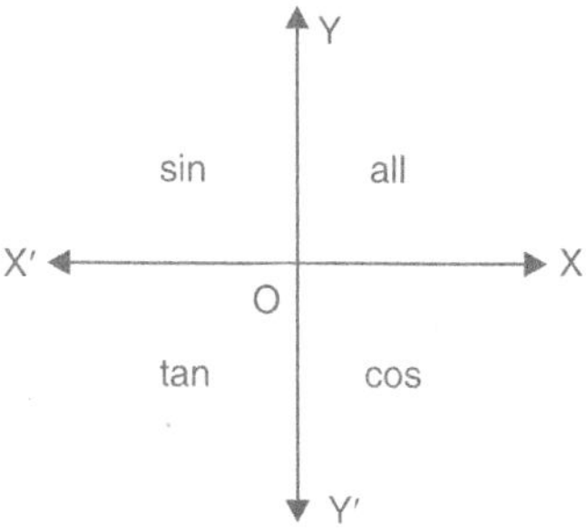

Fig. 2

(*v*) $\cos^2 x + \sin^2 x = 1$, $\sec^2 x - \tan^2 x = 1$, $\operatorname{cosec}^2 x - \cot^2 x = 1$

(*vi*)
$$\sin(x \pm y) = \sin x \cos y \pm \cos x \sin y$$
$$\cos(x \pm y) = \cos x \cos y \mp \sin x \sin y$$
$$\tan(x \pm y) = \frac{\tan x \pm \tan y}{1 \mp \tan x \tan y}$$

(*vii*)
$$\sin 2x = 2 \sin x \cos x = \frac{2\tan x}{1+\tan^2 x}$$

(*viii*)
$$\cos 2x = \cos^2 x - \sin^2 x = 2\cos^2 x - 1 = 1 - 2\sin^2 x = \frac{1-\tan^2 x}{1+\tan^2 x}$$

(*ix*)
$$\sin 3x = 3\sin x - 4\sin^3 x,\ \cos 3x = 4\cos^3 x - 3\cos x$$
$$\tan 3x = \frac{3\tan x - \tan^3 x}{1 \quad 3\tan^2 x}$$

(*x*)
$$\sin x \cos y = \frac{1}{2}[\sin(x+y) + \sin(x-y)]$$
$$\cos x \sin y = \frac{1}{2}[\sin(x+y) - \sin(x-y)]$$
$$\cos x \cos y = \frac{1}{2}[\cos(x+y) + \cos(x-y)]$$
$$\sin x \sin y = \frac{1}{2}[\cos(x-y) - \cos(x+y)]$$

(*xi*)
$$\sin x + \sin y = 2\sin\frac{x+y}{2}\cos\frac{x-y}{2}$$
$$\sin x - \sin y = 2\cos\frac{x+y}{2}\sin\frac{x-y}{2}$$
$$\cos x + \cos y = 2\cos\frac{x+y}{2}\cos\frac{x-y}{2}$$
$$\cos x - \cos y = -2\sin\frac{x+y}{2}\sin\frac{x-y}{2} = 2\sin\frac{x+y}{2}\sin\frac{y-x}{2}$$

(*xii*) $a\sin x + b\cos x = r\sin(x+\theta)$, $a\cos x + b\sin x = r\cos(x-\theta)$, where $a = r\cos\theta$, $b = r\sin\theta$

so that $r = \sqrt{a^2+b^2}$, $\theta = \tan^{-1}\left(\frac{b}{a}\right)$.

(*xiii*) In any ΔABC:

$$\frac{a}{\sin A} = \frac{b}{\sin B} = \frac{c}{\sin C} \qquad \text{(sine formula)}$$

$$\cos A = \frac{b^2 + c^2 - a^2}{2bc} \qquad \text{(cosine formula)}$$

$$a = b \cos C + c \cos B \qquad \text{(projection formula)}$$

Area of $\Delta\, ABC = \frac{1}{2}\, bc \sin A = \sqrt{s(s-a)(s-b)(s-c)}$, where $s = \frac{1}{2}(a + b + c)$

$\frac{a}{\sin A} = \frac{b}{\sin B} = \frac{c}{\sin C} = 2R$, where R is the circum-radius of Δ ABC.

$$R = \frac{abc}{4\Delta}$$

$r = \frac{\Delta}{s}$, where r is the radius of inscribed circle of Δ ABC.

11. De Moivre's Theorem: $(\cos \theta + i \sin \theta)^n = \cos n\theta + i \sin n\theta$

Euler's Theorem: $\cos \theta + i \sin \theta = e^{i\theta}$.

12. Hyperbolic Functions

$$\sinh x = \frac{e^x - e^{-x}}{2}, \cosh x = \frac{e^x + e^{-x}}{2}, \tanh x = \frac{e^x - e^{-x}}{e^x + e^{-x}}$$

$$\text{cosech } x = \frac{1}{\sinh x}, \text{sech } x = \frac{1}{\cosh x}, \coth x = \frac{1}{\tanh x}$$

$$\tanh x = \frac{\sinh x}{\cosh x}, \coth x = \frac{\cosh x}{\sinh x}, \cosh^2 x - \sinh^2 x = 1$$

$$\sin ix = i \sinh x, \cos ix = \cosh x, \tan ix = i \tanh x$$

$$\sinh^{-1} x = \log\left(x + \sqrt{x^2 + 1}\right), \cosh^{-1} x = \log\left(x + \sqrt{x^2 - 1}\right),$$

$$\tanh^{-1} x = \frac{1}{2} \log \frac{1+x}{1-x}.$$

13. Series

$$e^x = 1 + \frac{x}{1!} + \frac{x^2}{2!} + \frac{x^3}{3!} + \dots \infty$$

$$\sin x = x - \frac{x^3}{3!} + \frac{x^5}{5!} - \dots \infty, \qquad \sinh x = x + \frac{x^3}{3!} + \frac{x^5}{5!} + \dots \infty$$

$$\cos x = 1 - \frac{x^2}{2!} + \frac{x^4}{4!} - \dots \infty, \qquad \cosh x = 1 + \frac{x^2}{2!} + \frac{x^4}{4!} + \dots \infty$$

$$\log(1 + x) = x - \frac{x^2}{2} + \frac{x^3}{3} - \dots \infty, \quad \log(1 - x) = -\left(x + \frac{x^2}{2} + \frac{x^3}{3} + \dots \infty\right)$$

$$\tan^{-1} x = x - \frac{x^3}{3} + \frac{x^5}{5} - \dots \infty \qquad \text{(Gregory series)}$$

$$\tanh^{-1} x = x + \frac{x^3}{3} + \frac{x^5}{5} + \dots \infty$$

14. Calculus

(a) Standard Limits

(i) $\lim\limits_{x \to a} \dfrac{x^n - a^n}{x - a} = na^{n-1}, n \in Q$

(ii) $\lim\limits_{x \to 0} \dfrac{\sin x}{x} = 1$

(iii) $\lim\limits_{x \to 0} \dfrac{\sin^{-1} x}{x} = 1$

(iv) $\lim\limits_{x \to 0} \dfrac{\tan x}{x} = 1$

(v) $\lim\limits_{x \to 0} \dfrac{\tan^{-1} x}{x} = 1$

(vi) $\lim\limits_{x \to 0} (1 + x)^{\frac{1}{x}} = e$

(vii) $\lim\limits_{x \to 0} \dfrac{a^x - 1}{x} = \log_e a$

(viii) $\lim\limits_{x \to \infty} x^{1/x} = 1$

(ix) If $f(a) = g(a) = 0$, then

$$\lim_{x \to a} \frac{f(x)}{g(x)} = \lim_{x \to a} \frac{f'(x)}{g'(x)}$$ (L' Hospital's Rule)

(Differentiate the numerator and denominator separately)

(b) Differentiation

(i) $\dfrac{d}{dx}(uv) = u\dfrac{dv}{dx} + v\dfrac{du}{dx}$ (Product Rule)

(ii) $\dfrac{d}{dx}\left(\dfrac{u}{v}\right) = \dfrac{v\dfrac{du}{dx} - u\dfrac{dv}{dx}}{v^2}$ (Quotient Rule)

(iii) $\dfrac{dy}{dx} = \dfrac{dy}{du} \times \dfrac{du}{dx}$ (Chain Rule)

(iv) If $x = f(t)$, $y = g(t)$, then $\dfrac{dy}{dx} = \dfrac{\dfrac{dy}{dt}}{\dfrac{dx}{dt}}$

(v) $\dfrac{d}{dx}(c) = 0$

(vi) $\dfrac{d}{dx}(x^n) = nx^{n-1}$

(vii) $\dfrac{d}{dx}(e^x) = e^x$

(viii) $\dfrac{d}{dx}(a^x) = a^x \log_e a$

(ix) $\dfrac{d}{dx}(\log_e x) = \dfrac{1}{x}$

(x) $\dfrac{d}{dx}(\log_a x) = \dfrac{1}{x \log a}$

(xi) $\dfrac{d}{dx}(\sin x) = \cos x$

(xii) $\dfrac{d}{dx}(\cos x) = -\sin x$

(xiii) $\dfrac{d}{dx}(\tan x) = \sec^2 x$

(xiv) $\dfrac{d}{dx}(\cot x) = -\operatorname{cosec}^2 x$

(xv) $\dfrac{d}{dx}(\sec x) = \sec x \tan x$

(xvi) $\dfrac{d}{dx}(\operatorname{cosec} x) = -\operatorname{cosec} x \cot x$

(xvii) $\dfrac{d}{dx}(\sin^{-1} x) = \dfrac{1}{\sqrt{1 - x^2}}$

(xviii) $\dfrac{d}{dx}(\cos^{-1} x) = \dfrac{-1}{\sqrt{1 - x^2}}$

(xix) $\dfrac{d}{dx}(\tan^{-1} x) = \dfrac{1}{1+x^2}$

(xx) $\dfrac{d}{dx}(\cot^{-1} x) = \dfrac{-1}{1+x^2}$

(xxi) $\dfrac{d}{dx}(\sec^{-1} x) = \dfrac{1}{x\sqrt{x^2-1}}$

(xxii) $\dfrac{d}{dx}(\text{cosec}^{-1} x) = \dfrac{-1}{x\sqrt{x^2-1}}$

(xxiii) $\dfrac{d}{dx}(\sinh x) = \cosh x$

(xxiv) $\dfrac{d}{dx}(\cosh x) = \sinh x.$

(c) **Integration**

(i) $\int x^n\, dx = \dfrac{x^{n+1}}{n+1}\ (n \neq -1)$

(ii) $\int \dfrac{1}{x}\, dx = \log_e x$

(iii) $\int e^x\, dx = e^x$

(iv) $\int a^x\, dx = \dfrac{a^x}{\log_e a}$

(v) $\int \sin x\, dx = -\cos x$

(vi) $\int \cos x\, dx = \sin x$

(vii) $\int \tan x\, dx = -\log \cos x$

(viii) $\int \cot x\, dx = \log \sin x$

(ix) $\int \sec x \tan x\, dx = \sec x$

(x) $\int \text{cosec}\, x \cot x\, dx = -\text{cosec}\, x$

(xi) $\int \sec^2 x\, dx = \tan x$

(xii) $\int \text{cosec}^2 x\, dx = -\cot x$

(xiii) $\int \sec x\, dx = \log(\sec x + \tan x)$

(xiv) $\int \text{cosec}\, x\, dx = \log(\text{cosec}\, x - \cot x) = \log \tan \dfrac{x}{2}$

(xv) $\int \dfrac{dx}{a^2+x^2} = \dfrac{1}{a}\tan^{-1}\dfrac{x}{a}$

(xvi) $\int \dfrac{dx}{\sqrt{a^2-x^2}} = \sin^{-1}\dfrac{x}{a}$

(xvii) $\int \dfrac{dx}{a^2-x^2} = \dfrac{1}{2a}\log\dfrac{a+x}{a-x}$

(xviii) $\int \dfrac{dx}{\sqrt{a^2+x^2}} = \sinh^{-1}\dfrac{x}{a}$

(xix) $\int \dfrac{dx}{x^2-a^2} = \dfrac{1}{2a}\log\dfrac{x-a}{x+a}$

(xx) $\int \dfrac{dx}{\sqrt{x^2-a^2}} = \cosh^{-1}\dfrac{x}{a}$

(xxi) $\int \sqrt{a^2-x^2}\, dx = \dfrac{x\sqrt{a^2-x^2}}{2} + \dfrac{a^2}{2}\sin^{-1}\dfrac{x}{a}$

(xxii) $\int \sqrt{a^2+x^2}\, dx = \dfrac{x\sqrt{a^2+x^2}}{2} + \dfrac{a^2}{2}\sinh^{-1}\dfrac{x}{a}$

(xxiii) $\int \sqrt{x^2-a^2}\, dx = \dfrac{x\sqrt{x^2-a^2}}{2} - \dfrac{a^2}{2}\cosh^{-1}\dfrac{x}{a}$

(xxiv) $\int e^{ax} \sin bx\, dx = \dfrac{e^{ax}}{a^2+b^2}(a \sin bx - b \cos bx) = \dfrac{e^{ax}}{\sqrt{a^2+b^2}} \sin\left(bx - \tan^{-1}\dfrac{b}{a}\right)$

(xxv) $\int e^{ax} \cos bx\, dx = \dfrac{e^{ax}}{a^2+b^2}(a \cos bx + b \sin bx) = \dfrac{e^{ax}}{\sqrt{a^2+b^2}} \cos\left(bx - \tan^{-1}\dfrac{b}{a}\right)$

(xxvi) $\int \sinh x\, dx = \cosh x$ (xxvii) $\int \cosh x\, dx = \sinh x$

(xxviii) $\int \tanh x\, dx = \log \cosh x$ (xxix) $\int \coth x\, dx = \log \sinh x$

(xxx) $\int \text{sech}^2\, x\, dx = \tanh x$ (xxxi) $\int \text{cosech}^2\, x\, dx = -\coth x$

(xxxii) $\int_a^b f(x)\, dx = -\int_b^a f(x)\, dx$ (xxxiii) $\int_a^b f(x)\, dx = \int_a^c f(x)\, dx + \int_c^b f(x)\, dx, \quad a < c < b$

(xxxiv) $\int_0^a f(x)\, dx = \int_0^a f(a-x)\, dx$ (xxxv) $\int_a^b f(x)\, dx = \int_a^b f(a+b-x)\, dx$

(xxxvi) $\int_{-a}^a f(x)\, dx = \begin{cases} 0, & \text{if } f \text{ is an odd function} \\ 2\int_0^a f(x)\, dx, & \text{if } f \text{ is an even function} \end{cases}$

(xxxvii) $\int_0^{2a} f(x)\, dx = \begin{cases} 0, & \text{if } f(2a-x) = -f(x) \\ 2\int_0^a f(x)\, dx, & \text{if } f(2a-x) = f(x) \end{cases}$

(xxxviii) $\int_0^{\pi/2} \sin^n x\, dx = \int_0^{\pi/2} \cos^n x\, dx \quad (n > 1)$

$$= \frac{(n-1)(n-3)(n-5)\ldots}{n(n-2)(n-4)\ldots} \times \left(\frac{\pi}{2}, \text{only if } n \text{ is even}\right)$$

(xxxix) $\int_0^{\pi/2} \sin^m x \cos^n x\, dx$

$$= \frac{(m-1)(m-3)\ldots \times (n-1)(n-3)\ldots}{(m+n)(m+n-2)(m+n-4)\ldots} \times \left(\frac{\pi}{2}, \text{only if both } m \text{ and } n \text{ are even}\right).$$

SECTION-A

CHAPTER 1

Fourier Series

1.1. PERIODIC FUNCTIONS

A function $f(x)$ which satisfies the relation $f(x + T) = f(x)$ for all x is called a periodic function. The *smallest positive number* T, for which this relation holds, is called the **period** of $f(x)$.

If T is the period of $f(x)$, then $f(x) = f(x + T) = f(x + 2T) = ... = f(x + nT) = ...$

Also $f(x) = f(x - T) = f(x - 2T) = ... = f(x - nT) = ...$

$\therefore$ $f(x) = f(x \pm nT)$, where n is a positive integer.

Thus, $f(x)$ repeats itself after periods of T.

For example, sin x, cos x, sec x and cosec x are periodic functions with period 2π while tan x and cot x are periodic functions with period π. The functions sin nx and cos nx are periodic with period $\frac{2\pi}{n}$.

The sum of a number of periodic functions is also periodic. If T_1 and T_2 are the periods of $f(x)$ and $g(x)$, then the period of $a\,f(x) + b\,g(x)$ is the least common multiple of T_1 and T_2.

For example, cos x, cos $2x$, cos $3x$ are periodic functions with periods 2π, π and $\frac{2\pi}{3}$ respectively.

$\therefore$ $f(x) = \cos x + \frac{1}{2}\cos 2x + \frac{1}{3}\cos 3x$ is also periodic with period 2π, the L.C.M. of 2π, π and $\frac{2\pi}{3}$.

1.2. FOURIER SERIES

Periodic functions are of common occurrence in many physical and engineering problems; for example, in conduction of heat and mechanical vibrations. It is useful to express these functions in a series of sines and cosines. Most of the single valued functions which occur in applied mathematics can be expressed in the form $\frac{a_0}{2} + a_1 \cos x + a_2 \cos 2x + a_3 \cos 3x + ...$ $+ b_1 \sin x + b_2 \sin 2x + b_3 \sin 3x + ...$
within a desired range of values of the variable. Such a series is known as *Fourier Series*. Thus, any function $f(x)$ defined in the interval $c_1 \le x \le c_2$ can be expressed in the Fourier Series

$$f(x) = \frac{a_0}{2} + \sum_{n=1}^{\infty} (a_n \cos nx + b_n \sin nx)$$

where a_0, a_n, b_n $(n = 1, 2, 3, ...)$ are constants, called the Fourier co-efficients of $f(x)$.

Note. To determine a_0, a_n and b_n, we shall need the following results: (m and n are integers)

(*i*) $\int_c^{c+2\pi} \sin nx \, dx = -\left[\frac{\cos nx}{n}\right]_c^{c+2\pi} = 0, \quad \int_c^{c+2\pi} \cos nx \, dx = \left[\frac{\sin nx}{n}\right]_c^{c+2\pi} = 0, n \neq 0$

(*ii*) $\int_c^{c+2\pi} \sin mx \cos nx \, dx = \frac{1}{2}\int_c^{c+2\pi} [\sin (m+n)x + \sin (m-n)x] \, dx$

$$= -\frac{1}{2}\left[\frac{\cos (m+n)x}{m+n} + \frac{\cos (m-n)x}{m-n}\right]_c^{c+2\pi} = 0, m \neq n$$

(*iii*) $\int_c^{c+2\pi} \cos mx \cos nx \, dx = \frac{1}{2}\int_c^{c+2\pi} [\cos (m+n)x + \cos (m-n)x] \, dx$

$$= \frac{1}{2}\left[\frac{\sin (m+n)x}{m+n} + \frac{\sin (m-n)x}{m-n}\right]_c^{c+2\pi} = 0, m \neq n$$

(*iv*) $\int_c^{c+2\pi} \sin mx \sin nx \, dx = \frac{1}{2}\int_c^{c+2\pi} [\cos (m-n)x - \cos (m+n)x] \, dx$

$$= \frac{1}{2}\left[\frac{\sin (m-n)x}{m-n} - \frac{\sin (m+n)x}{m+n}\right]_c^{c+2\pi} = 0, m \neq n$$

(*v*) $\int_c^{c+2\pi} \cos^2 nx \, dx = \left[\frac{x}{2} + \frac{\sin 2nx}{4n}\right]_c^{c+2\pi} = \pi, \quad \int_c^{c+2\pi} \sin^2 nx \, dx = \left[\frac{x}{2} - \frac{\sin 2nx}{4n}\right]_c^{c+2\pi} = \pi, n \neq 0$

(*vi*) $\int_c^{c+2\pi} \sin nx \cos nx \, dx = \frac{1}{2}\int_c^{c+2\pi} \sin 2nx \, dx = -\frac{1}{2}\left[\frac{\cos 2nx}{2n}\right]_c^{c+2\pi} = 0, n \neq 0$

(*vii*) To integrate the product of two functions, one of which is a positive integral power of x, we apply the *generalised rule of integration by parts.* If dashes denote differentiation and suffixes denote integration w.r.t. x, the rule can be stated as follows:

$\int uv \, dx = uv_1 - u'v_2 + u''v_3 - u'''v_4 + \ldots\ldots$ where u and v are functions of x. *i.e.,* Integral of the product of two functions

= 1st function × integral of 2nd – go on differentiating 1st, integrating 2nd, signs alternately + ve and – ve.

[Simplification should be done only when the integration is over.]

For example, $\int x^3 e^{-2x} \, dx = x^3\left(\frac{e^{-2x}}{-2}\right) - 3x^2\left[\frac{e^{-2x}}{(-2)^2}\right] + 6x\left[\frac{e^{-2x}}{(-2)^3}\right] - 6\left[\frac{e^{-2x}}{(-2)^4}\right]$

$$= e^{-2x}\left[-\frac{1}{2}x^3 - \frac{3}{4}x^2 - \frac{3}{4}x - \frac{3}{8}\right] = -\frac{1}{8}e^{-2x}(4x^3 + 6x^2 + 6x + 3)$$

$$\int x^2 \cos nx \, dx = x^2\left(\frac{\sin nx}{n}\right) - 2x\left(\frac{-\cos nx}{n^2}\right) + 2\left(-\frac{\sin nx}{n^3}\right)$$

$$= \frac{x^2}{n}\sin nx + \frac{2x}{n^2}\cos nx - \frac{2}{n^3}\sin nx.$$

(*viii*) $\sin n\pi = 0$ and $\cos n\pi = (-1)^n$

$\sin\left(n + \frac{1}{2}\right)\pi = (-1)^n$ and $\cos\left(n + \frac{1}{2}\right)\pi = 0$, where n is an integer.

(*ix*) **Even and Odd Functions**

A function $f(x)$ is said to be *even* if $f(-x) = f(x)$ *e.g.,* x^2, $\cos x$, $\sin^2 x$ are even functions.

The graph of an even function is symmetrical about the y-axis.

A function $f(x)$ is said to be *odd* if $f(-x) = -f(x)$ *e.g.,* x^3, $\sin x$, $\tan^3 x$ are odd functions.

The graph of an odd function is symmetrical about the origin.

The product of two even functions or two odd functions is an even function while the product of an even function and an odd function is an odd function.

Also, $\int_{-c}^{c} f(x)\, dx = 0$, when $f(x)$ is an odd function

and $\int_{-c}^{c} f(x)\, dx = 2\int_{0}^{c} f(x)\, dx$, when $f(x)$ is an even function.

1.3. EULER'S FORMULAE

The Fourier series for the function $f(x)$ in the interval $c < x < c + 2\pi$ is given by

$$f(x) = \frac{a_0}{2} + \sum_{n=1}^{\infty} a_n \cos nx + \sum_{n=1}^{\infty} b_n \sin nx \qquad \text{...(1)}$$

In finding the co-efficients a_0, a_n and b_n, we assume that the series on the right hand side of (1) is uniformly convergent for $c < x < c + 2\pi$ and it can be integrated term by term in the given interval.

To find $\mathbf{a_0}$. Integrate both sides of (1) w.r.t. x, between the limits c to $c + 2\pi$.

$$\int_{c}^{c+2\pi} f(x)\, dx = \frac{a_0}{2}\int_{c}^{c+2\pi} dx + \int_{c}^{c+2\pi}\left(\sum_{n=1}^{\infty} a_n \cos nx\right) dx + \int_{c}^{c+2\pi}\left(\sum_{n=1}^{\infty} b_n \sin nx\right) dx$$

$$= \frac{a_0}{2}(c + 2\pi - c) + 0 + 0 \qquad \text{[By formulae } (i) \text{ above]}$$

$$= a_0 \pi$$

$$\therefore \qquad \boxed{a_0 = \frac{1}{\pi}\int_{c}^{c+2\pi} f(x)\, dx}$$

To find $\mathbf{a_n}$, multiply both sides of (1) by $\cos nx$ and integrate w.r.t. x, between the limits c to $c + 2\pi$.

$$\int_{c}^{c+2\pi} f(x) \cos nx\, dx = \frac{a_0}{2}\int_{c}^{c+2\pi} \cos nx\, dx$$

$$+ \int_{c}^{c+2\pi}\left(\sum_{n=1}^{\infty} a_n \cos nx\right) \cos nx\, dx + \int_{c}^{c+2\pi}\left(\sum_{n=1}^{\infty} b_n \sin nx\right) \cos nx\, dx$$

$$= 0 + a_n \pi + 0 \qquad \text{[By formulae } (i), (v) \text{ and } (vi)\text{]}$$

$$= a_n \pi$$

$$\therefore \qquad \boxed{a_n = \frac{1}{\pi}\int_{c}^{c+2\pi} f(x) \cos nx\, dx}$$

To find b_n, multiply both sides of (1) by sin nx and integrate w.r.t. x between the limits c to $c + 2\pi$.

$$\int_c^{c+2\pi} f(x)\sin nx\, dx = \frac{a_0}{2}\int_c^{c+2\pi} \sin nx\, dx + \int_c^{c+2\pi}\left(\sum_{n=1}^{\infty} a_n \cos nx\right)\sin nx\, dx$$

$$+ \int_c^{c+2\pi}\left(\sum_{n=1}^{\infty} b_n \sin nx\right)\sin nx\, dx$$

$$= 0 + 0 + b_n\pi \qquad \text{[By formulae } (i), (vi) \text{ and } (v)]$$

$$= b_n\pi$$

$$\therefore \qquad \boxed{b_n = \frac{1}{\pi}\int_c^{c+2\pi} f(x)\sin nx\, dx}$$

Hence $a_0 = \dfrac{1}{\pi}\displaystyle\int_c^{c+2\pi} f(x)\, dx$; $a_n = \dfrac{1}{\pi}\displaystyle\int_c^{c+2\pi} f(x)\cos nx\, dx$; and $b_n = \dfrac{1}{\pi}\displaystyle\int_c^{c+2\pi} f(x)\sin nx\, dx$ ...(I)

These values of a_0, a_n and b_n are called Euler's formulae.

Cor. 1. *If $c = 0$, the interval becomes $0 < x < 2\pi$ and the formulae I reduce to*

$$a_0 = \frac{1}{\pi}\int_0^{2\pi} f(x)\, dx;\quad a_n = \frac{1}{\pi}\int_0^{2\pi} f(x)\cos nx\, dx;\quad b_n = \frac{1}{\pi}\int_0^{2\pi} f(x)\sin nx\, dx$$

Cor. 2. *If $c = -\pi$, the interval becomes $-\pi < x < \pi$, and the formulae I reduce to*

$$a_0 = \frac{1}{\pi}\int_{-\pi}^{\pi} f(x)\, dx;\ a_n = \frac{1}{\pi}\int_{-\pi}^{\pi} f(x)\cos nx\, dx;\ b_n = \frac{1}{\pi}\int_{-\pi}^{\pi} f(x)\sin nx\, dx$$

Cor. 3. *When f(x) is an odd function* $\quad a_0 = \dfrac{1}{\pi}\displaystyle\int_{-\pi}^{\pi} f(x)\, dx = 0$

Since cos nx is an even function, therefore, $f(x)$ cos nx is an odd function.

$$a_n = \frac{1}{\pi}\int_{-\pi}^{\pi} f(x)\cos nx\, dx = 0$$

Since sin nx is an odd function, therefore, $f(x)$ sin nx is an even function.

$$\therefore \qquad b_n = \frac{1}{\pi}\int_{-\pi}^{\pi} f(x)\sin nx\, dx = \frac{2}{\pi}\int_0^{\pi} f(x)\sin nx\, dx$$

Hence, if a periodic function $f(x)$ is odd, its Fourier expansion contains only sine terms,

i.e., $$f(x) = \sum_{n=1}^{\infty} b_n \sin nx, \quad \text{where} \quad b_n = \frac{2}{\pi}\int_0^{\pi} f(x)\sin nx\, dx$$

When f(x) is an even function $\quad a_0 = \dfrac{1}{\pi}\displaystyle\int_{-\pi}^{\pi} f(x)\, dx = \dfrac{2}{\pi}\displaystyle\int_0^{\pi} f(x)\, dx$

Since cos nx is an even function, therefore, $f(x)$ cos nx is an even function.

$$\therefore \qquad a_n = \frac{1}{\pi}\int_{-\pi}^{\pi} f(x)\cos nx\, dx = \frac{2}{\pi}\int_0^{\pi} f(x)\cos nx\, dx$$

Since sin nx is an odd function, therefore, $f(x)$ sin nx is an odd function.

$$\therefore \qquad b_n = \frac{1}{\pi}\int_{-\pi}^{\pi} f(x) \sin nx\, dx = 0$$

Hence, if a periodic function $f(x)$ is even, its Fourier expansion contains only cosine terms,

i.e., $$f(x) = \frac{a_0}{2} + \sum_{n=1}^{\infty} a_n \cos nx, \text{ where } a_0 = \frac{2}{\pi}\int_0^{\pi} f(x)\, dx \text{ and } a_n = \frac{2}{\pi}\int_0^{\pi} f(x) \cos nx\, dx.$$

An Important Note. If a periodic function f is defined in the open interval (a, b), then how to find $f(a)$ and $f(b)$, the values of f at the end points?

Consider, $f(x) = x + x^2, \ -\pi < x < \pi$

where f is periodic with period 2π.

$$f(-\pi) = \frac{1}{2}\text{ (LHL + RHL)} = \frac{1}{2}\,[f(-\pi-0) + f(-\pi+0)]$$

$$= \frac{1}{2}\,[f(2\pi + (-\pi - 0)) + f(-\pi + 0)] \qquad | \because \text{ Period of } f \text{ is } 2\pi$$

$$= \frac{1}{2}\,[f(\pi - 0) + f(-\pi + 0)]$$

$$= \frac{1}{2}\,[(\pi + \pi^2) + (-\pi) + (-\pi)^2] = \pi^2$$

Clearly, $f(-\pi) \neq -\pi + \pi^2$

Similarly, $$f(\pi) = \frac{1}{2}\,[\text{ LHL + RHL}]$$

$$= \frac{1}{2}\,[f(\pi - 0) + f(\pi + 0)] = \frac{1}{2}\,[f(\pi - 0) + f(-2\pi + (\pi + 0)]$$

$| \because$ Period of f is 2π

$$= \frac{1}{2}\,[f(\pi - 0) + f(-\pi + 0)] = \frac{1}{2}\,[(\pi + \pi^2) + (-\pi + (-\pi)^2)] = \pi^2$$

Clearly, $f(\pi) \neq \pi + \pi^2$.

ILLUSTRATIVE EXAMPLES

Example 1. *Obtain the Fourier series to represent* $f(x) = \left(\frac{\pi - x}{2}\right)^2, 0 \le x \le 2\pi$.

(M.D.U. 2010, 2012, Dec. 2014; K.U.K. Jan. 2013)

Hence obtain the following relations:

(*i*) $\frac{1}{1^2} + \frac{1}{2^2} + \frac{1}{3^2} + \frac{1}{4^2} + \ldots\ldots = \sum_{n=1}^{\infty} \frac{1}{n^2} = \frac{\pi^2}{6}$ (M.D.U. Dec. 2014) (*ii*) $\frac{1}{1^2} - \frac{1}{2^2} + \frac{1}{3^2} - \frac{1}{4^2} + ... = \sum_{n=1}^{\infty} \frac{(-1)^{n-1}}{n^2} = \frac{\pi^2}{12}$ (M.D.U., 2010, Dec. 2014)

(*iii*) $\frac{1}{1^2} + \frac{1}{3^2} + \frac{1}{5^2} + ... = \sum_{n=1}^{\infty} \frac{1}{(2n-1)^2} = \frac{\pi^2}{8}$

Sol. Let $f(x) = \frac{1}{4}(\pi - x)^2 = \frac{a_0}{2} + \sum_{n=1}^{\infty} a_n \cos nx + \sum_{n=1}^{\infty} b_n \sin nx$

By Euler's formulae, we have

$$a_0 = \frac{1}{\pi}\int_0^{2\pi} f(x)\,dx = \frac{1}{\pi}\int_0^{2\pi} \frac{1}{4}(\pi - x)^2\,dx = \frac{1}{4\pi}\left[\frac{(\pi - x)^3}{-3}\right]_0^{2\pi} = -\frac{1}{12\pi}[-\pi^3 - \pi^3] = \frac{\pi^2}{6}$$

$$a_n = \frac{1}{\pi}\int_0^{2\pi} f(x)\cos nx\,dx = \frac{1}{\pi}\int_0^{2\pi} \frac{1}{4}(\pi - x)^2 \cos nx\,dx$$

$$= \frac{1}{4\pi}\left[(\pi - x)^2 \frac{\sin nx}{n} - \{-2(\pi - x)\}\left(-\frac{\cos nx}{n^2}\right) + 2\left(-\frac{\sin nx}{n^3}\right)\right]_0^{2\pi}$$

$$= \frac{1}{4\pi}\left[\left(0 + \frac{2\pi\cos 2n\pi}{n^2} + 0\right) - \left(0 - \frac{2\pi\cos 0}{n^2} + 0\right)\right] = \frac{1}{4\pi}\left[\frac{2\pi}{n^2} + \frac{2\pi}{n^2}\right] = \frac{1}{n^2}$$

$$b_n = \frac{1}{\pi}\int_0^{2\pi} f(x)\sin nx\,dx = \frac{1}{\pi}\int_0^{2\pi} \frac{1}{4}(\pi - x)^2 \sin nx\,dx$$

$$= \frac{1}{4\pi}\left[(\pi - x)^2\left(-\frac{\cos nx}{n}\right) - \{-2(\pi - x)\}\left(-\frac{\sin nx}{n^2}\right) + 2\left(\frac{\cos nx}{n^3}\right)\right]_0^{2\pi}$$

$$= \frac{1}{4\pi}\left[\left(-\frac{\pi^2\cos 2n\pi}{n} - 0 + \frac{2\cos 2n\pi}{n^3}\right) - \left(-\frac{\pi^2}{n} - 0 + \frac{2\cos 0}{n^3}\right)\right]$$

$$= \frac{1}{4\pi}\left[\left(-\frac{\pi^2}{n} + \frac{2}{n^3}\right) - \left(-\frac{\pi^2}{n} + \frac{2}{n^3}\right)\right] = 0$$

$$\therefore \quad f(x) = \frac{\pi^2}{12} + \sum_{n=1}^{\infty} \frac{\cos nx}{n^2} = \frac{\pi^2}{12} + \frac{\cos x}{1^2} + \frac{\cos 2x}{2^2} + \frac{\cos 3x}{3^2} + ... \quad ...(1)$$

Deductions

(*i*) Putting $x = 0$ in equation (1), we get

$$f(0) = \frac{\pi^2}{12} + \left(\frac{1}{1^2} + \frac{1}{2^2} + \frac{1}{3^2} + \frac{1}{4^2} + ...\right)$$

$$\Rightarrow \quad \frac{\pi^2}{4} = \frac{\pi^2}{12} + \left(\frac{1}{1^2} + \frac{1}{2^2} + \frac{1}{3^2} + \frac{1}{4^2} + ...\right)$$

$$\Rightarrow \quad \frac{1}{1^2} + \frac{1}{2^2} + \frac{1}{3^2} + \frac{1}{4^2} + ... = \frac{\pi^2}{6} \quad ...(2)$$

(*ii*) Putting $x = \pi$ in equation (1), we get

$$f(\pi) = \frac{\pi^2}{12} + \left[\left(\frac{-1}{1^2}\right) + \frac{1}{2^2} + \left(\frac{-1}{3^2}\right) + \frac{1}{4^2} + \ldots\ldots\right]$$

$$\Rightarrow \quad 0 = \frac{\pi^2}{12} - \frac{1}{1^2} + \frac{1}{2^2} - \frac{1}{3^2} + \frac{1}{4^2} + \ldots\ldots$$

$$\Rightarrow \quad \frac{1}{1^2} - \frac{1}{2^2} + \frac{1}{3^2} - \frac{1}{4^2} + \ldots\ldots = \frac{\pi^2}{12} \qquad \ldots(3)$$

(*iii*) Adding (2) and (3), we get

$$2\left(\frac{1}{1^2} + \frac{1}{3^2} + \frac{1}{5^2} + \ldots\ldots\right) = \frac{\pi^2}{6} + \frac{\pi^2}{12}$$

$$\Rightarrow \quad \frac{1}{1^2} + \frac{1}{3^2} + \frac{1}{5^2} + \ldots\ldots = \frac{1}{2}\left(\frac{\pi^2}{4}\right) = \frac{\pi^2}{8}.$$

Note. In the above example, if $f(x) = \left(\frac{\pi - x}{2}\right)^2$, $0 < x < 2\pi$, then

$$f(0) = \frac{1}{2}[f(0-0) + f(0+0)]$$

$$= \frac{1}{2}[f(2\pi - 0) + f(0+0)] \qquad [\because f \text{ is periodic with period } 2\pi]$$

$$= \frac{1}{2}\left[\left(\frac{\pi - 2\pi}{2}\right)^2 + \left(\frac{\pi - 0}{2}\right)^2\right] = \frac{1}{2}\left(\frac{\pi^2}{4} + \frac{\pi^2}{4}\right) = \frac{\pi^2}{4}.$$

Example 2. *Expand $f(x) = x \sin x$, $0 < x < 2\pi$ as a Fourier series.*

(K.U.K. 2010; M.D.U. 2010, 2013; G.B.T.U. 2010)

Sol. Let $f(x) = x \sin x = \frac{a_0}{2} + \sum_{n=1}^{\infty} a_n \cos nx + \sum_{n=1}^{\infty} b_n \sin nx$

By Euler's formulae, we have $a_0 = \frac{1}{\pi}\int_0^{2\pi} f(x)\, dx = \frac{1}{\pi}\int_0^{2\pi} x \sin x\, dx$

$$= \frac{1}{\pi}\Big[x(-\cos x) - 1.(-\sin x)\Big]_0^{2\pi} = \frac{1}{\pi}[-2\pi] = -2$$

$$a_n = \frac{1}{\pi}\int_0^{2\pi} f(x) \cos nx\, dx = \frac{1}{\pi}\int_0^{2\pi} x \sin x \cos nx\, dx = \frac{1}{2\pi}\int_0^{2\pi} x(2 \cos nx \sin x)\, dx$$

$$= \frac{1}{2\pi}\int_0^{2\pi} x[\sin (n+1)x - \sin (n-1)x]\, dx$$

$$= \frac{1}{2\pi}\left[x\left\{-\frac{\cos (n+1)x}{n+1} + \frac{\cos (n-1)x}{n-1}\right\} - 1.\left\{-\frac{\sin (n+1)x}{(n+1)^2} + \frac{\sin (n-1)x}{(n-1)^2}\right\}\right]_0^{2\pi}$$

$$= \frac{1}{2\pi}\left[2\pi\left\{-\frac{\cos 2(n+1)\pi}{n+1} + \frac{\cos 2(n-1)\pi}{n-1}\right\}\right]$$

$$= -\frac{1}{n+1} + \frac{1}{n-1} = \frac{2}{n^2-1}, n \neq 1$$

When $n = 1$, we have $a_1 = \frac{1}{\pi}\int_0^{2\pi} x \sin x \cos x \, dx = \frac{1}{2\pi}\int_0^{2\pi} x \sin 2x \, dx$

$$= \frac{1}{2\pi}\left[x\left(-\frac{\cos 2x}{2}\right) - 1.\left(-\frac{\sin 2x}{4}\right)\right]_0^{2\pi} = \frac{1}{2\pi}[-\pi] = -\frac{1}{2}$$

$$b_n = \frac{1}{\pi}\int_0^{2\pi} f(x) \sin nx \, dx = \frac{1}{\pi}\int_0^{2\pi} x \sin x \sin nx \, dx$$

$$= \frac{1}{2\pi}\int_0^{2\pi} x(2 \sin nx \sin x) \, dx = \frac{1}{2\pi}\int_0^{2\pi} x\,[\cos (n-1)\,x - \cos (n+1)\,x]\, dx$$

$$= \frac{1}{2\pi}\left[x\left\{\frac{\sin (n-1)\,x}{n-1} - \frac{\sin (n+1)\,x}{n+1}\right\} - 1.\left\{-\frac{\cos (n-1)\,x}{(n-1)^2} + \frac{\cos (n+1)\,x}{(n+1)^2}\right\}\right]_0^{2\pi}$$

$$= \frac{1}{2\pi}\left[\frac{\cos 2(n-1)\pi}{(n-1)^2} - \frac{\cos 2(n+1)\pi}{(n+1)^2} - \frac{1}{(n-1)^2} + \frac{1}{(n+1)^2}\right]$$

$$= \frac{1}{2\pi}\left[\frac{1}{(n-1)^2} - \frac{1}{(n+1)^2} - \frac{1}{(n-1)^2} + \frac{1}{(n+1)^2}\right] = 0, n \neq 1$$

When $n = 1$, we have $b_1 = \frac{1}{\pi}\int_0^{2\pi} x \sin x \sin x \, dx = \frac{1}{2\pi}\int_0^{2\pi} x(1 - \cos 2x) \, dx$

$$= \frac{1}{2\pi}\left[x\left(x - \frac{\sin 2x}{2}\right) - 1.\left(\frac{x^2}{2} + \frac{\cos 2x}{4}\right)\right]_0^{2\pi}$$

$$= \frac{1}{2\pi}\left[2\pi(2\pi) - \frac{4\pi^2}{2} - \frac{1}{4} + \frac{1}{4}\right] = \frac{1}{2\pi}(2\pi^2) = \pi$$

$$\therefore \quad f(x) = \frac{a_0}{2} + a_1 \cos x + b_1 \sin x + \sum_{n=2}^{\infty} a_n \cos nx + \sum_{n=2}^{\infty} b_n \sin nx$$

$$= -1 - \frac{1}{2}\cos x + \pi \sin x + \sum_{n=2}^{\infty} \frac{2}{n^2-1}\cos nx + 0$$

$$= -1 + \pi \sin x - \frac{1}{2}\cos x + \frac{2}{2^2-1}\cos 2x + \frac{2}{3^2-1}\cos 3x + \ldots\ldots.$$

Example 3. *Find a Fourier series to represent $x - x^2$ from $x = -\pi$ to $x = \pi$. Hence show that*

$$\frac{1}{1^2} - \frac{1}{2^2} + \frac{1}{3^2} - \frac{1}{4^2} + \ldots\ldots. = \frac{\pi^2}{12}.$$

(G.B.T.U. 2012; K.U.K. May 2013; DCRUST, Murthal May 2014)

Sol. Let $x - x^2 = \frac{a_0}{2} + \sum_{n=1}^{\infty} a_n \cos nx + \sum_{n=1}^{\infty} b_n \sin nx$

By Euler's formulae, we have $a_0 = \dfrac{1}{\pi}\displaystyle\int_{-\pi}^{\pi} (x - x^2)\, dx = \dfrac{1}{\pi}\left[\dfrac{x^2}{2} - \dfrac{x^3}{3}\right]_{-\pi}^{\pi}$

$$= \frac{1}{\pi}\left[\left(\frac{\pi^2}{2} - \frac{\pi^3}{3}\right) - \left(\frac{\pi^2}{2} + \frac{\pi^3}{3}\right)\right] = -\frac{2\pi^2}{3}$$

$$a_n = \frac{1}{\pi}\int_{-\pi}^{\pi} (x - x^2) \cos nx\, dx$$

$$= \frac{1}{\pi}\left[(x - x^2)\frac{\sin nx}{n} - (1 - 2x)\left(-\frac{\cos nx}{n^2}\right) + (-2)\left(-\frac{\sin nx}{n^3}\right)\right]_{-\pi}^{\pi}$$

$$= \frac{1}{\pi}\left[(1 - 2\pi)\frac{\cos n\pi}{n^2} - (1 + 2\pi)\frac{\cos n\pi}{n^2}\right] = \frac{1}{\pi}\left(-4\pi \,.\, \frac{\cos n\pi}{n^2}\right)$$

$$= -4\frac{(-1)^n}{n^2} \qquad [\because \cos n\pi = (-1)^n]$$

$$b_n = \frac{1}{\pi}\int_{-\pi}^{\pi} (x - x^2) \sin nx\, dx$$

$$= \frac{1}{\pi}\left[(x - x^2)\left(-\frac{\cos nx}{n}\right) - (1 - 2x)\left(-\frac{\sin nx}{n^2}\right) + (-2)\left(\frac{\cos nx}{n^3}\right)\right]_{-\pi}^{\pi}$$

$$= \frac{1}{\pi}\left[(\pi^2 - \pi)\frac{\cos n\pi}{n} - 2\frac{\cos n\pi}{n^3} + (-\pi - \pi^2)\frac{\cos n\pi}{n} + 2\frac{\cos n\pi}{n^3}\right]$$

$$= \frac{1}{\pi}\left[-2\pi \,.\, \frac{\cos n\pi}{n}\right] = -2\frac{(-1)^n}{n}$$

$$\therefore \quad x - x^2 = -\frac{\pi^2}{3} - 4\sum_{n=1}^{\infty} \frac{(-1)^n}{n^2}\cos nx - 2\sum_{n=1}^{\infty} \frac{(-1)^n}{n}\sin nx$$

$$= -\frac{\pi^2}{3} - 4\left[-\frac{\cos x}{1^2} + \frac{\cos 2x}{2^2} - \frac{\cos 3x}{3^2} + \ldots\ldots\right] - 2\left[\frac{-\sin x}{1} + \frac{\sin 2x}{2} - \frac{\sin 3x}{3} + \ldots\ldots\right]$$

$$= -\frac{\pi^2}{3} + 4\left[\frac{\cos x}{1^2} - \frac{\cos 2x}{2^2} + \frac{\cos 3x}{3^2} - \ldots\ldots\right] + 2\left[\frac{\sin x}{1} - \frac{\sin 2x}{2} + \frac{\sin 3x}{3} - \ldots\ldots\right]$$

Putting $x = 0$, we get $0 = -\dfrac{\pi^2}{3} + 4\left(\dfrac{1}{1^2} - \dfrac{1}{2^2} + \dfrac{1}{3^2} - \dfrac{1}{4^2} + \ldots\ldots\right)$

$$\Rightarrow \qquad \frac{1}{1^2} - \frac{1}{2^2} + \frac{1}{3^2} - \frac{1}{4^2} + \ldots\ldots = \frac{\pi^2}{12}.$$

Example 4. *Obtain the Fourier series for the function $f(x) = x^2$, $-\pi \le x \le \pi$. Hence show that*

(*i*) $\dfrac{1}{1^2} + \dfrac{1}{2^2} + \dfrac{1}{3^2} + \dfrac{1}{4^2} + \ldots\ldots = \displaystyle\sum_{n=1}^{\infty} \dfrac{1}{n^2} = \dfrac{\pi^2}{6}$ (K.U.K. Jan. 2014)

(*ii*) $\dfrac{1}{1^2} - \dfrac{1}{2^2} + \dfrac{1}{3^2} - \dfrac{1}{4^2} + \ldots\ldots = \displaystyle\sum_{n=1}^{\infty} \dfrac{(-1)^{n-1}}{n^2} = \dfrac{\pi^2}{12}$ (M.D.U. 2011)

(*iii*) $\dfrac{1}{1^2} + \dfrac{1}{3^2} + \dfrac{1}{5^2} + \ldots\ldots = \displaystyle\sum_{n=1}^{\infty} \dfrac{1}{(2n-1)^2} = \dfrac{\pi^2}{8}$.

Sol. Since $f(-x) = (-x)^2 = x^2 = f(x)$.

$\therefore$ $f(x)$ is an even function and hence $b_n = 0$

Let $f(x) = x^2 = \dfrac{a_0}{2} + \sum_{n=1}^{\infty} a_n \cos nx$

Then $a_0 = \dfrac{2}{\pi}\int_0^{\pi} f(x)\,dx = \dfrac{2}{\pi}\int_0^{\pi} x^2\,dx = \dfrac{2}{\pi}\left[\dfrac{x^3}{3}\right]_0^{\pi} = \dfrac{2}{3}\pi^2$

$$a_n = \frac{2}{\pi}\int_0^{\pi} f(x)\cos nx\,dx = \frac{2}{\pi}\int_0^{\pi} x^2 \cos nx\,dx$$

$$= \frac{2}{\pi}\left[x^2\left(\frac{\sin nx}{n}\right) - 2x\left(-\frac{\cos nx}{n^2}\right) + 2\left(-\frac{\sin nx}{n^3}\right)\right]_0^{\pi} = \frac{2}{\pi}\left[2\pi\,.\,\frac{\cos n\pi}{n^2}\right] = 4\frac{(-1)^n}{n^2}$$

$$\therefore \quad x^2 = \frac{\pi^2}{3} + 4\sum_{n=1}^{\infty} \frac{(-1)^n}{n^2}\cos nx$$

$$= \frac{\pi^2}{3} - 4\left(\frac{\cos x}{1^2} - \frac{\cos 2x}{2^2} + \frac{\cos 3x}{3^2} - \frac{\cos 4x}{4^2} + \ldots\ldots\right) \qquad \ldots(1)$$

Putting $x = \pi$ in (1), we get

$$\pi^2 = \frac{\pi^2}{3} - 4\left(-\frac{1}{1^2} - \frac{1}{2^2} - \frac{1}{3^2} - \frac{1}{4^2} + \ldots\ldots\right) \Rightarrow \frac{2\pi^2}{3} = 4\left(\frac{1}{1^2} + \frac{1}{2^2} + \frac{1}{3^2} + \frac{1}{4^2} + \ldots\ldots\right)$$

$$\therefore \quad \frac{1}{1^2} + \frac{1}{2^2} + \frac{1}{3^2} + \frac{1}{4^2} + \ldots\ldots = \frac{\pi^2}{6} \qquad \text{[Result } (i)\text{]}$$

Putting $x = 0$ in (1), we get $0 = \dfrac{\pi^2}{3} - 4\left(\dfrac{1}{1^2} - \dfrac{1}{2^2} + \dfrac{1}{3^2} - \dfrac{1}{4^2} + \ldots\ldots\right)$

$$\therefore \quad \frac{1}{1^2} - \frac{1}{2^2} + \frac{1}{3^2} - \frac{1}{4^2} + \ldots\ldots = \frac{\pi^2}{12} \qquad \text{[Result } (ii)\text{]}$$

Adding (*i*) and (*ii*), we get $2\left(\dfrac{1}{1^2} + \dfrac{1}{3^2} + \dfrac{1}{5^2} + \ldots\ldots\right) = \dfrac{\pi^2}{4}$

$$\therefore \quad \frac{1}{1^2} + \frac{1}{3^2} + \frac{1}{5^2} + \ldots\ldots = \frac{\pi^2}{8}. \qquad \text{[Result } (iii)\text{]}$$

Example 5. *Obtain the Fourier series for $f(x) = e^{-x}$ in the interval $0 < x < 2\pi$.*

Sol. Let $f(x) = e^{-x} = \dfrac{a_0}{2} + \sum_{n=1}^{\infty} a_n \cos nx + \sum_{n=1}^{\infty} b_n \sin nx$

Then $a_0 = \dfrac{1}{\pi}\int_0^{2\pi} f(x)\,dx = \dfrac{1}{\pi}\int_0^{2\pi} e^{-x}\,dx = \dfrac{1}{\pi}\left[-e^{-x}\right]_0^{2\pi} = \dfrac{1-e^{-2\pi}}{\pi}$

$$a_n = \frac{1}{\pi}\int_0^{2\pi} f(x)\cos nx\,dx = \frac{1}{\pi}\int_0^{2\pi} e^{-x}\cos nx\,dx$$

$$= \frac{1}{\pi}\left[\frac{e^{-x}}{1+n^2}(-\cos nx + n\sin nx)\right]_0^{2\pi}$$

$$\left[\because \int e^{ax}\cos bx\,dx = \frac{e^{ax}}{a^2+b^2}(a\cos bx + b\sin bx)\right]$$

$$= \frac{1-e^{-2\pi}}{\pi(1+n^2)}$$

$$b_n = \frac{1}{\pi}\int_0^{2\pi} f(x)\sin nx\,dx = \frac{1}{\pi}\int_0^{2\pi} e^{-x}\sin nx\,dx$$

$$= \frac{1}{\pi}\left[\frac{e^{-x}}{1+n^2}(-\sin nx - n\cos nx)\right]_0^{2\pi} = \frac{1-e^{-2\pi}}{\pi}\cdot\frac{n}{1+n^2}$$

$$\left[\because \int e^{ax}\sin bx\,dx = \frac{e^{ax}}{a^2+b^2}(a\sin bx - b\cos bx)\right]$$

$$\therefore \quad e^{-x} = \frac{1-e^{-2\pi}}{2\pi} + \frac{1-e^{-2\pi}}{\pi}\sum_{n=1}^{\infty}\frac{\cos nx}{1+n^2} + \frac{1-e^{-2\pi}}{\pi}\sum_{n=1}^{\infty}\frac{n\sin nx}{1+n^2}$$

$$= \frac{1-e^{-2\pi}}{\pi}\left[\frac{1}{2} + \left(\frac{1}{2}\cos x + \frac{1}{5}\cos 2x + \frac{1}{10}\cos 3x + \ldots\ldots\right)\right.$$

$$\left. + \left(\frac{1}{2}\sin x + \frac{2}{5}\sin 2x + \frac{3}{10}\sin 3x + \ldots\ldots\right)\right]$$

Example 6. *Find the Fourier series to represent e^{ax} in the interval $-\pi < x < \pi$. Hence derive series for $\frac{\pi}{\sinh \pi}$.*

Sol. Let $f(x) = e^{ax} = \frac{a_0}{2} + \sum_{n=1}^{\infty} a_n \cos nx + \sum_{n=1}^{\infty} b_n \sin nx$

Then $$a_0 = \frac{1}{\pi}\int_{-\pi}^{\pi} f(x)\,dx = \frac{1}{\pi}\int_{-\pi}^{\pi} e^{ax}\,dx = \frac{1}{\pi}\left[\frac{e^{ax}}{a}\right]_{-\pi}^{\pi} = \frac{1}{a\pi}(e^{a\pi} - e^{-a\pi}) = \frac{2\sinh a\pi}{a\pi}$$

$$a_n = \frac{1}{\pi}\int_{-\pi}^{\pi} f(x)\cos nx\,dx = \frac{1}{\pi}\int_{-\pi}^{\pi} e^{ax}\cos nx\,dx$$

$$= \frac{1}{\pi}\left[\frac{e^{ax}}{a^2+n^2}(a\cos nx + n\sin nx)\right]_{-\pi}^{\pi} = \frac{1}{\pi(a^2+n^2)}[ae^{a\pi}\cos n\pi - ae^{-a\pi}\cos n\pi]$$

$$= \frac{a\cos n\pi\,(e^{a\pi} - e^{-a\pi})}{\pi(a^2+n^2)} = \frac{2a(-1)^n \sinh a\pi}{\pi(a^2+n^2)}$$

Similarly, $$b_n = \frac{-2n(-1)^n \sinh a\pi}{\pi(a^2+n^2)}$$

$$\therefore \quad e^{ax} = \frac{\sinh a\pi}{a\pi} + \sum_{n=1}^{\infty}\frac{2a(-1)^n\sinh a\pi}{\pi(a^2+n^2)}\cos nx - \sum_{n=1}^{\infty}\frac{2n(-1)^n\sinh a\pi}{\pi(a^2+n^2)}\sin nx$$

$$= \frac{2\sinh a\pi}{\pi}\left[\frac{1}{2a} - a\left(\frac{\cos x}{a^2+1^2} - \frac{\cos 2x}{a^2+2^2} + \frac{\cos 3x}{a^2+3^2} - \ldots\ldots\right) + \left(\frac{\sin x}{a^2+1^2} - \frac{2\sin 2x}{a^2+2^2} + \frac{3\sin 3x}{a^2+3^2} - \ldots\ldots\right)\right]$$

Deduction. Putting $x = 0$ and $a = 1$, we get

$$1 = \frac{2\sinh\pi}{\pi}\left[\frac{1}{2} - \left(\frac{1}{1+1^2} - \frac{1}{1+2^2} + \frac{1}{1+3^2} - \frac{1}{1+4^2} + \ldots\ldots\right)\right]$$

$$\Rightarrow \qquad \frac{\pi}{\sinh\pi} = 2\left(\frac{1}{1+2^2} - \frac{1}{1+3^2} + \frac{1}{1+4^2} - \ldots\ldots\right)$$

Example 7. *Express* $f(x) = |\,x\,|, -\pi < x < \pi$, *as Fourier series.* (M.D.U. 2013, Dec. 2015)

Sol. Since $f(-x) = |-x| = |\,x\,| = f(x)$

$\therefore$ $f(x)$ is an even function and hence $b_n = 0$

Let $\qquad f(x) = |\,x\,| = \dfrac{a_0}{2} + \sum\limits_{n=1}^{\infty} a_n \cos nx$

Then, $\qquad a_0 = \dfrac{2}{\pi}\displaystyle\int_0^{\pi} f(x)\,dx = \dfrac{2}{\pi}\int_0^{\pi} |x|\,dx = \dfrac{2}{\pi}\int_0^{\pi} x\,dx = \dfrac{2}{\pi}\left[\dfrac{x^2}{2}\right]_0^{\pi} = \pi$

$$a_n = \frac{2}{\pi}\int_0^{\pi} f(x)\cos nx\,dx = \frac{2}{\pi}\int_0^{\pi} |x|\cos nx\,dx = \frac{2}{\pi}\int_0^{\pi} x\cos nx\,dx$$

$$= \frac{2}{\pi}\left[x\left(\frac{\sin nx}{n}\right) - 1.\left(-\frac{\cos nx}{n^2}\right)\right]_0^{\pi}$$

$$= \frac{2}{\pi}\left[\frac{\cos n\pi}{n^2} - \frac{1}{n^2}\right] = \frac{2}{\pi n^2}[(-1)^n - 1] = \begin{cases} 0, & \text{if } n \text{ is even} \\ \dfrac{-4}{\pi n^2}, & \text{if } n \text{ is odd} \end{cases}$$

$$\therefore \qquad |\,x\,| = \frac{\pi}{2} - \frac{4}{\pi}\left(\cos x + \frac{\cos 3x}{3^2} + \frac{\cos 5x}{5^2} + \ldots\ldots\right)$$

Note. Putting $x = 0$ in the above result, we get $\dfrac{1}{1^2} + \dfrac{1}{3^2} + \dfrac{1}{5^2} + \ldots\ldots = \dfrac{\pi^2}{8}$.

Example 8. *Expand the function* $f(x) = x\sin x$ *as a Fourier series in the interval* $-\pi \le x \le \pi$.

Deduce that $\dfrac{1}{1.3} - \dfrac{1}{3.5} + \dfrac{1}{5.7} - \dfrac{1}{7.9} + \ldots\ldots = \dfrac{\pi-2}{4}$. (M.D.U. 2012)

Sol. Since $x\sin x$ is an even function of x, $b_n = 0$

Let $\qquad f(x) = x\sin x = \dfrac{a_0}{2} + \sum\limits_{n=1}^{\infty} a_n \cos nx$

Then $\qquad a_0 = \dfrac{2}{\pi}\displaystyle\int_0^{\pi} x\sin x\,dx = \dfrac{2}{\pi}\Big[x(-\cos x) - 1.(-\sin x)\Big]_0^{\pi} = \dfrac{2}{\pi}(-\pi\cos\pi) = 2$

$$a_n = \frac{2}{\pi}\int_0^{\pi} x \sin x \cos nx\, dx = \frac{1}{\pi}\int_0^{\pi} x(2\cos nx \sin x)\, dx$$

$$= \frac{1}{\pi}\int_0^{\pi} x[\sin(n+1)x - \sin(n-1)x]\, dx$$

$$= \frac{1}{\pi}\left[x\left\{-\frac{\cos(n+1)x}{n+1} + \frac{\cos(n-1)x}{n-1}\right\} - 1.\left\{-\frac{\sin(n+1)x}{(n+1)^2} + \frac{\sin(n-1)x}{(n-1)^2}\right\}\right]_0^{\pi}$$

$$= \frac{1}{\pi}\left[\pi\left\{-\frac{\cos(n+1)\pi}{n+1} + \frac{\cos(n-1)\pi}{n-1}\right\}\right], \; n \neq 1$$

$$= \frac{\cos(n-1)\pi}{n-1} - \frac{\cos(n+1)\pi}{n+1}$$

When n is odd, $n \neq 1$, $n-1$ and $n+1$ are even

$$\therefore \quad a_n = \frac{1}{n-1} - \frac{1}{n+1} = \frac{2}{n^2-1}$$

When n is even, $n-1$ and $n+1$ are odd

$$\therefore \quad a_n = \frac{-1}{n-1} + \frac{1}{n+1} = \frac{-2}{n^2-1}$$

When $n = 1$, we have $a_1 = \dfrac{2}{\pi}\displaystyle\int_0^{\pi} x \sin x \cos x\, dx = \dfrac{1}{\pi}\int_0^{\pi} x \sin 2x\, dx$

$$= \frac{1}{\pi}\left[x\left(-\frac{\cos 2x}{2}\right) - 1.\left(-\frac{\sin 2x}{4}\right)\right]_0^{\pi} = \frac{1}{\pi}\left[-\frac{\pi\cos 2\pi}{2}\right] = -\frac{1}{2}$$

$$\therefore \quad x \sin x = 1 - \frac{1}{2}\cos x - 2\left(\frac{\cos 2x}{2^2-1} - \frac{\cos 3x}{3^2-1} + \frac{\cos 4x}{4^2-1} - \frac{\cos 5x}{5^2-1} + \ldots\ldots\right)$$

Putting $x = \dfrac{\pi}{2}$, we get $\dfrac{\pi}{2} = 1 - 2\left(\dfrac{-1}{2^2-1} + \dfrac{1}{4^2-1} - \dfrac{1}{6^2-1} + \ldots\ldots\right)$

$$\Rightarrow \quad \frac{\pi}{2} - 1 = 2\left(\frac{1}{2^2-1} - \frac{1}{4^2-1} + \frac{1}{6^2-1} - \ldots\ldots\right) \Rightarrow \frac{\pi-2}{4} = \frac{1}{1.3} - \frac{1}{3.5} + \frac{1}{5.7} - \ldots\ldots$$

Example 9. *Show that for* $-\pi < x < \pi$,

$$\sin ax = \frac{2\sin a\pi}{\pi}\left(\frac{\sin x}{1^2-a^2} - \frac{2\sin 2x}{2^2-a^2} + \frac{3\sin 3x}{3^2-a^2} - \ldots\ldots\right).$$

Sol. Since $\sin ax$ is an odd function of x, $a_0 = 0$ and $a_n = 0$.

Let $\sin ax = \displaystyle\sum_{n=1}^{\infty} b_n \sin nx$

Then $b_n = \dfrac{2}{\pi}\displaystyle\int_0^{\pi} \sin ax \sin nx\, dx = \dfrac{1}{\pi}\int_0^{\pi} [\cos(n-a)x - \cos(n+a)x]\, dx$

$$= \frac{1}{\pi}\left[\frac{\sin(n-a)x}{n-a} - \frac{\sin(n+a)x}{n+a}\right]_0^{\pi} = \frac{1}{\pi}\left[\frac{\sin(n-a)\pi}{n-a} - \frac{\sin(n+a)\pi}{n+a}\right]$$

$$= \frac{1}{\pi}\left[\frac{(-1)^n(-\sin a\pi)}{n-a} - \frac{(-1)^n \sin a\pi}{n+a}\right] = -\frac{(-1)^n \sin a\pi}{\pi}\left[\frac{1}{n-a} + \frac{1}{n+a}\right]$$

$$= (-1)^{n+1} \cdot \frac{2n \sin a\pi}{\pi(n^2 - a^2)}$$

$$\therefore \quad \sin ax = \frac{2 \sin a\pi}{\pi} \sum_{n=1}^{\infty} \frac{(-1)^{n+1}}{n^2 - a^2} \sin nx$$

$$= \frac{2 \sin a\pi}{\pi}\left(\frac{\sin x}{1^2 - a^2} - \frac{2 \sin 2x}{2^2 - a^2} + \frac{3 \sin 3x}{3^2 - a^2} - \ldots\ldots\right).$$

Example 10. *Obtain Fourier series for the function f(x) given by*

$$f(x) = 1 + \frac{2x}{\pi}, \; -\pi \le x \le 0$$

$$= 1 - \frac{2x}{\pi}, \; 0 \le x \le \pi.$$ (U.K.T.U. 2011)

Hence deduce that $\frac{1}{1^2} + \frac{1}{3^2} + \frac{1}{5^2} + \ldots\ldots = \frac{\pi^2}{8}$. (K.U.K. Jan. 2013)

Sol. When $-\pi \le x \le 0, \quad 0 \le -x \le \pi$

$$\therefore \quad f(-x) = 1 - \frac{2(-x)}{\pi} = 1 + \frac{2x}{\pi} = f(x)$$

When $0 \le x \le \pi, \quad -\pi \le -x \le 0$

$$\therefore \quad f(-x) = 1 + \frac{2(-x)}{\pi} = 1 - \frac{2x}{\pi} = f(x)$$

$\Rightarrow$ $f(x)$ is an even function of x in $[-\pi, \pi]$. This is also clear from its graph which is symmetrical above the y-axis.

$\therefore \quad b_n = 0$

Let $f(x) = \frac{a_0}{2} + \sum_{n=1}^{\infty} a_n \cos nx$

then
$$a_0 = \frac{2}{\pi}\int_0^{\pi} f(x)\, dx = \frac{2}{\pi}\int_0^{\pi}\left(1 - \frac{2x}{\pi}\right) dx = \frac{2}{\pi}\left[x - \frac{x^2}{\pi}\right]_0^{\pi} = 0$$

$$a_n = \frac{2}{\pi}\int_0^{\pi} f(x) \cos nx\, dx = \frac{2}{\pi}\int_0^{\pi}\left(1 - \frac{2x}{\pi}\right)\cos nx\, dx$$

$$= \frac{2}{\pi}\left[\left(1 - \frac{2x}{\pi}\right)\frac{\sin nx}{n} - \left(-\frac{2}{\pi}\right)\left(-\frac{\cos nx}{n^2}\right)\right]_0^{\pi}$$

$$= \frac{2}{\pi}\left[-\frac{2 \cos n\pi}{\pi n^2} + \frac{2}{\pi n^2}\right] = \frac{4}{\pi^2 n^2}[1 - (-1)^n]$$

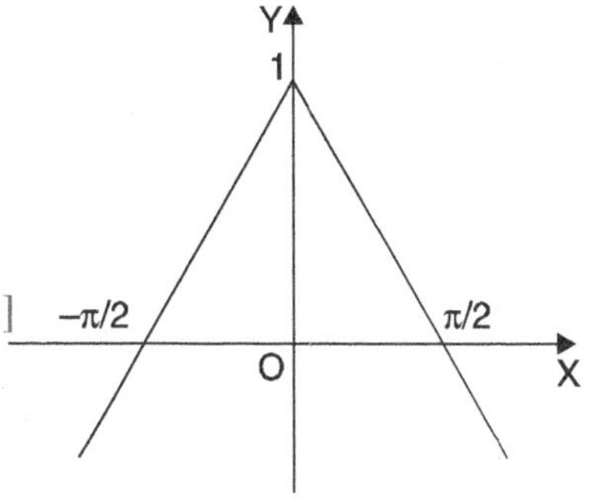

$$\therefore \quad f(x) = \frac{4}{\pi^2} \sum_{n=1}^{\infty} [1-(-1)^n] \frac{\cos nx}{n^2}$$

$$= \frac{4}{\pi^2}\left(\frac{2\cos x}{1^2} + \frac{2\cos 3x}{3^2} + \frac{2\cos 5x}{5^2} + \ldots\ldots\right)$$

$[\because \; 1-(-1)^n = 0$ when n is even$]$

$$= \frac{8}{\pi^2}\left(\frac{\cos x}{1^2} + \frac{\cos 3x}{3^2} + \frac{\cos 5x}{5^2} + \ldots\ldots\right)$$

Putting $x = 0$, we get $\dfrac{1}{1^2} + \dfrac{1}{3^2} + \dfrac{1}{5^2} + \ldots\ldots = \dfrac{\pi^2}{8}$, since $f(0) = 1$.

Example 11. *Show that for* $-\pi \le x \le \pi$,

$$\cos cx = \frac{\sin c\pi}{\pi}\left[\frac{1}{c} - \frac{2c\cos x}{c^2-1^2} + \frac{2c\cos 2x}{c^2-2^2} - \ldots\ldots\right]$$

where c is non-integral. Hence deduce that

$$\pi \operatorname{cosec}(c\pi) = \sum_{n=0}^{\infty} (-1)^n \left[\frac{1}{n+c} + \frac{1}{n+1-c}\right]. \qquad \text{(M.D.U. 2011)}$$

Sol. Since $\cos cx$ is an even function of x, $b_n = 0$

Let $$\cos cx = \frac{a_0}{2} + \sum_{n=1}^{\infty} a_n \cos nx \qquad \ldots(1)$$

Then $$a_0 = \frac{2}{\pi}\int_0^{\pi} \cos cx \, dx = \frac{2}{\pi}\left[\frac{\sin cx}{c}\right]_0^{\pi}$$

$$= \frac{2\sin c\pi}{c\pi}, \quad \text{since } c \text{ is non-integral, } \sin c\pi \ne 0$$

Also, $$a_n = \frac{2}{\pi}\int_0^{\pi} \cos cx \cos nx \, dx = \frac{1}{\pi}\int_0^{\pi} [\cos(n+c)x + \cos(n-c)x]\, dx$$

$$= \frac{1}{\pi}\left[\frac{\sin(n+c)x}{n+c} + \frac{\sin(n-c)x}{n-c}\right]_0^{\pi}$$

$$= \frac{1}{\pi}\left[\frac{\sin(n+c)\pi}{n+c} + \frac{\sin(n-c)\pi}{n-c}\right]$$

$$= \frac{1}{\pi}\left[\frac{\sin(n\pi+c\pi)}{n+c} + \frac{\sin(n\pi-c\pi)}{n-c}\right]$$

$$= \frac{1}{\pi}\left[\frac{(-1)^n \sin c\pi}{n+c} + \frac{(-1)^n \sin(-c\pi)}{n-c}\right]$$

$$= \frac{1}{\pi}\left[\frac{(-1)^n \sin c\pi}{n+c} - \frac{(-1)^n \sin c\pi}{n-c}\right] = \frac{(-1)^n \sin c\pi}{\pi}\left(\frac{1}{n+c} - \frac{1}{n-c}\right)$$

$$= \frac{(-1)^n \sin c\pi}{\pi}\left[\frac{-2c}{n^2 - c^2}\right] = \frac{(-1)^n \sin c\pi}{\pi} \cdot \frac{2c}{c^2 - n^2}$$

$\therefore$ From (1),
$$\cos cx = \frac{\sin c\pi}{c\pi} + \frac{2c \sin c\pi}{\pi} \sum_{n=1}^{\infty} \frac{(-1)^n}{c^2 - n^2} \cos nx$$

$$= \frac{\sin c\pi}{\pi}\left[\frac{1}{c} + 2c\left\{-\frac{\cos x}{c^2 - 1^2} + \frac{\cos 2x}{c^2 - 2^2} - \ldots\ldots\right\}\right]$$

$$= \frac{\sin c\pi}{\pi}\left[\frac{1}{c} - \frac{2c \cos x}{c^2 - 1^2} + \frac{2c \cos 2x}{c^2 - 2^2} - \ldots\ldots\right]$$

Deduction Put $x = 0$

$$1 = \frac{\sin c\pi}{\pi}\left[\frac{1}{c} - \frac{2c}{c^2 - 1^2} + \frac{2c}{c^2 - 2^2} - \ldots\ldots\right]$$

$$\Rightarrow \qquad \pi \operatorname{cosec}(c\pi) = \frac{1}{c} - \frac{(c+1)+(c-1)}{(c+1)(c-1)} + \frac{(c+2)+(c-2)}{(c+2)(c-2)} - \ldots\ldots$$

$$= \frac{1}{c} - \frac{1}{c-1} - \frac{1}{c+1} + \frac{1}{c-2} + \frac{1}{c+2} - \frac{1}{c-3} + \ldots\ldots$$

$$= \left(\frac{1}{c} + \frac{1}{1-c}\right) - \left(\frac{1}{c+1} + \frac{1}{2-c}\right) + \left(\frac{1}{c+2} + \frac{1}{3-c}\right) - \ldots\ldots$$

$$= \sum_{n=0}^{\infty} (-1)^n \left[\frac{1}{n+c} + \frac{1}{n+1-c}\right]$$

Hence,
$$\pi \operatorname{cosec}(c\pi) = \sum_{n=0}^{\infty} (-1)^n \left[\frac{1}{n+c} + \frac{1}{n+1-c}\right]$$

EXERCISE 1.1

1. Expand in a Fourier series the function $f(x) = x$ in the interval $0 < x < 2\pi$.
2. Express $f(x) = \frac{1}{2}(\pi - x)$ in a Fourier series in the interval $0 < x < 2\pi$.
3. Find the Fourier series for the function $f(x) = x + x^2$ in the interval $-\pi < x < \pi$.

(G.B.T.U. 2010)

Hence show that:

(*i*) $\frac{1}{1^2} - \frac{1}{2^2} + \frac{1}{3^2} - \frac{1}{4^2} + \ldots\ldots = \frac{\pi^2}{12}$ *(U.K.T.U. 2011)*

(*ii*) $1 + \frac{1}{2^2} + \frac{1}{3^2} + \frac{1}{4^2} + \ldots\ldots = \frac{\pi^2}{6}$

4. Prove that for all values of x between $-\pi$ and π, $x = 2\left[\sin x - \frac{1}{2}\sin 2x + \frac{1}{3}\sin 3x - \frac{1}{4}\sin 4x + \ldots\ldots\right]$

(M.D.U. 2011)

5. Obtain the Fourier series to represent e^x in the interval $0 < x < 2\pi$.

6. Find the Fourier series to represent e^x in the interval $-\pi < x < \pi$. *(M.D.U. 2011)*

7. Find the Fourier series to represent the function $f(x) = |\sin x|, -\pi < x < \pi$.

8. Expand $f(x) = |\cos x|$ as a Fourier series in the interval $-\pi < x < \pi$.

(M.D.U. 2010, Dec. 2013; M.T.U. 2011)

9. Prove that in the interval $-\pi < x < \pi$, $x \cos x = -\frac{1}{2}\sin x + 2\sum_{n=2}^{\infty} \frac{n(-1)^n}{n^2 - 1}\sin nx$.

10. Prove that for $-\pi < x < \pi$, $\frac{x(\pi^2 - x^2)}{12} = \frac{\sin x}{1^3} - \frac{\sin 2x}{2^3} + \frac{\sin 3x}{3^3} - \frac{\sin 4x}{4^3} + \ldots\ldots$

11. (*a*) Obtain a Fourier expansion for $\sqrt{1 - \cos x}$ in the interval $-\pi < x < \pi$.

[**Hint.** For all integral values of n, $\cos\left(n + \frac{1}{2}\right)\pi = \cos(2n+1)\frac{\pi}{2} = 0 = \cos\left(n - \frac{1}{2}\right)\pi$.

$$\sqrt{1 - \cos x} = \sqrt{2\sin^2\frac{x}{2}} = \sqrt{2}\left|\sin\frac{x}{2}\right| = \begin{cases} -\sqrt{2}\sin\frac{x}{2}, & -\pi < x \le 0 \\ \sqrt{2}\sin\frac{x}{2}, & 0 < x < \pi \end{cases}\Bigg]$$

(*b*) Obtain a Fourier series for $\sqrt{1 - \cos x}$ in the interval $(0, 2\pi)$ and hence find the value of

$$\frac{1}{1\cdot 3} + \frac{1}{3\cdot 5} + \frac{1}{5\cdot 7} + \ldots\ldots$$

12. Express $f(x) = \cos wx$, $-\pi < x < \pi$, where w is a fraction, as a Fourier series. Hence, prove that

$$\cot\theta = \frac{1}{\theta} + \frac{2\theta}{\theta^2 - \pi^2} + \frac{2\theta}{\theta^2 - 4\pi^2} + \ldots\ldots$$

13. Find the Fourier series for $f(x)$ in the interval $(-\pi, \pi)$ when

$$\begin{aligned} f(x) &= \pi + x, & -\pi < x < 0 \\ &= \pi - x, & 0 < x < \pi. \end{aligned}$$

14. Obtain a Fourier series to represent e^{-ax} from $x = -\pi$ to $x = \pi$. Hence derive series for $\frac{\pi}{\sinh \pi}$.

15. Prove that in the range $-\pi < x < \pi$, $\cosh ax = \frac{2a}{\pi}\sinh a\pi\left[\frac{1}{2a^2} + \sum_{n=1}^{\infty}\frac{(-1)^n}{n^2 + a^2}\cos nx\right]$.

16. Given $f(x) = \begin{cases} -x + 1 & \text{for } -\pi \le x \le 0 \\ x + 1 & \text{for } 0 \le x \le \pi \end{cases}$

Is the function even or odd ? Find the Fourier series for $f(x)$ and deduce the value of

$$\frac{1}{1^2} + \frac{1}{3^2} + \frac{1}{5^2} + \ldots\ldots$$

17. Find the Fourier series expansion for $f(x) = x + \frac{x^2}{4}$, $-\pi \le x \le \pi$.

18. Find the Fourier series of $f(x) = \frac{3x^2 - 6\pi x + 2\pi^2}{12}$ in the interval $(0, 2\pi)$. Hence deduce that

$\frac{\pi^2}{6} = 1 + \frac{1}{2^2} + \frac{1}{3^2} + \ldots\ldots$. *(M.T.U. 2012)*

19. Find the Fourier series to represent the function $f(x)$ given by $f(x) = \begin{cases} x, & 0 < x < \frac{\pi}{2} \\ \pi - x, & \frac{\pi}{2} < x < \pi \end{cases}$

and hence show that $1 + \frac{1}{3^2} + \frac{1}{5^2} + \ldots\ldots = \frac{\pi^2}{8}$. *(DCRUST, Murthal May 2014)*

20. Obtain the Fourier series expansion for $f(x) = \pi^2 - x^2$ in the interval $-\pi < x < \pi$. *(M.D.U. May 2014)*

21. Find the Fourier series of period 2π for the function

$$f(x) = \begin{cases} \cos x - \sin x, & -\pi < x \le 0 \\ \cos x + \sin x, & 0 \le x < \pi \end{cases}$$

22. Find the Fourier series expansion of $f(x)$ in $(-\pi, \pi)$, where $f(x) = \sin^3 x + \cos^3 x$.

23. Find the Fourier series of $f(x) = \cos^4 x$ in $(0, 2\pi)$.

Answers

1. $f(x) = \pi - 2\sum_{n=1}^{\infty} \frac{\sin nx}{n}$ **2.** $f(x) = \sum_{n=1}^{\infty} \frac{\sin nx}{n}$

3. $f(x) = \frac{\pi^2}{3} - 4\left(\frac{\cos x}{1^2} - \frac{\cos 2x}{2^2} + \frac{\cos 3x}{3^2} - \ldots\ldots\right) + 2\left(\frac{\sin x}{1} - \frac{\sin 2x}{2} + \frac{\sin 3x}{3} - \ldots\ldots\right)$

5. $e^x = \frac{e^{2\pi} - 1}{2\pi} + \frac{e^{2\pi} - 1}{\pi}\sum_{n=1}^{\infty}\left(\frac{\cos nx}{1+n^2} - \frac{n}{1+n^2}\sin nx\right)$

6. $e^x = \frac{2\sinh \pi}{\pi}$

$$\left[\frac{1}{2} - \left(\frac{1}{2}\cos x - \frac{1}{5}\cos 2x + \frac{1}{10}\cos 3x - \ldots\ldots\right) - \left(\frac{1}{2}\sin x - \frac{2}{5}\sin 2x + \frac{3}{10}\sin 3x - \ldots\ldots\right)\right]$$

7. $|\sin x| = \frac{2}{\pi} - \frac{4}{\pi}\left(\frac{\cos 2x}{3} + \frac{\cos 4x}{15} + \ldots\ldots + \frac{\cos 2nx}{4n^2 - 1} + \ldots\ldots\right)$

8. $|\cos x| = \frac{2}{\pi} + \frac{4}{\pi}\left(\frac{\cos 2x}{3} - \frac{\cos 4x}{15} + \ldots\ldots\right)$

11. (*a*) $\sqrt{1 - \cos x} = \frac{2\sqrt{2}}{\pi} - \frac{4\sqrt{2}}{\pi}\sum_{n=1}^{\infty}\frac{\cos nx}{4n^2 - 1}$ (*b*) Same as in part (*a*); $\frac{1}{2}$

12. $\cos wx = \frac{2w \sin w\pi}{\pi}\left(\frac{1}{2w^2} + \frac{\cos x}{1^2 - w^2} - \frac{\cos 2x}{2^2 - w^2} + \frac{\cos 3x}{3^2 - w^2} - \ldots\ldots\right)$

13. $f(x) = \dfrac{\pi}{2} + \dfrac{4}{\pi}\left(\dfrac{\cos x}{1^2} + \dfrac{\cos 3x}{3^2} + \dfrac{\cos 5x}{5^2} + \ldots\ldots\right)$

14. $e^{-ax} = \dfrac{2\sinh a\pi}{\pi}\left[\left(\dfrac{1}{2a} - \dfrac{a\cos x}{1^2 + a^2} + \dfrac{a\cos 2x}{2^2 + a^2} - \ldots\ldots\right) - \left(\dfrac{\sin x}{1^2 + a^2} - \dfrac{2\sin 2x}{2^2 + a^2} + \dfrac{3\sin 3x}{3^2 + a^2} - \ldots\ldots\right)\right]$

$\dfrac{\pi}{\sinh \pi} = 2\left[\dfrac{1}{2^2 + 1} - \dfrac{1}{3^2 + 1} + \dfrac{1}{4^2 + 1} - \ldots\ldots\right]$

16. Even. $f(x) = \dfrac{\pi}{2} + 1 - \dfrac{4}{\pi}\left(\cos x + \dfrac{\cos 3x}{3^2} + \dfrac{\cos 5x}{5^2} + \ldots\ldots\right); \dfrac{\pi^2}{8}$.

17. $f(x) = \dfrac{\pi^2}{12} + \sum_{n=1}^{\infty} \dfrac{(-1)^n}{n^2}\cos nx - 2\sum_{n=1}^{\infty} \dfrac{(-1)^n}{n}\sin nx$.

18. $f(x) = \sum_{n=1}^{\infty} \dfrac{\cos nx}{n^2}$.

19. $f(x) = \dfrac{4}{\pi}\left(\dfrac{\sin x}{1^2} - \dfrac{\sin 3x}{3^2} + \dfrac{\sin 5x}{5^2} - \ldots\ldots\right)$.

20. $f(x) = \dfrac{2\pi^2}{3} - 4\sum_{n=1}^{\infty} \dfrac{(-1)^n}{n^2}\cos nx$.

21. $f(x) = \dfrac{2}{\pi} + \dfrac{4}{\pi}\sum_{n=2,4,6,\ldots} \dfrac{n}{n^2 - 1}\cos nx$

22. $f(x) = \dfrac{3}{4}\cos x + \dfrac{1}{4}\cos 3x + \dfrac{3}{4}\sin x - \dfrac{1}{4}\sin 3x$.

23. $f(x) = \dfrac{3}{8} + \dfrac{1}{2}\cos 2x + \dfrac{1}{4}\cos 4x$.

1.4. DIRICHLET'S CONDITIONS *(M.D.U. May 2014, Dec. 2014, May 2015)*

The sufficient conditions for the uniform convergence of a Fourier series are called Dirichlet's conditions (after Dirichlet, a German mathematician). All the functions that normally arise in engineering problems satisfy these conditions and hence they can be expressed as a Fourier series.

Any function f(x) can be expressed as a Fourier series $\dfrac{a_0}{2} + \sum_{n=1}^{\infty} a_n \cos nx + \sum_{n=1}^{\infty} b_n \sin nx$

where a_0, a_n, b_n *are constants, provided*

(*i*) *f(x) is periodic, single valued and finite.*

(*ii*) *f(x) has a finite number of finite discontinuities in any one period.*

(*iii*) *f(x) has a finite number of maxima and minima.*

When these conditions are satisfied, the Fourier series converges to $f(x)$ at every point of continuity. At a point of discontinuity, the sum of the series is equal to the mean of the limits on the right and left *i.e.,* $\frac{1}{2}[f(x+0)+f(x-0)]$

where $f(x + 0)$ and $f(x - 0)$ denote the limit on the right and the limit on the left respectively.

1.5. FOURIER SERIES FOR DISCONTINUOUS FUNCTIONS

In Art. 1.3, we derived Euler's formulae for a_0, a_n, b_n on the assumption that $f(x)$ is continous in $(c, c + 2\pi)$. However, if $f(x)$ has finitely many points of finite discontinuity, even then it can be expressed as a Fourier series. The integrals for a_0, a_n, b_n are to be evaluated by breaking up the range of integration.

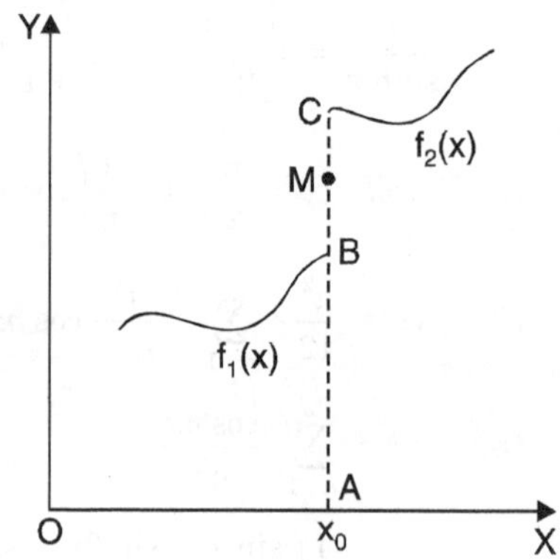

Let $f(x)$ be defined by $f(x) = f_1(x),\ c < x < x_0$

$= f_2(x),\ x_0 < x < c + 2\pi$

where x_0 is the point of finite discontinuity in the interval $(c, c + 2\pi)$.

The values of a_0, a_n, b_n are given by

$$a_0 = \frac{1}{\pi}\left[\int_c^{x_0} f_1(x)\,dx + \int_{x_0}^{c+2\pi} f_2(x)\,dx\right]$$

$$a_n = \frac{1}{\pi}\left[\int_c^{x_0} f_1(x)\cos nx\,dx + \int_{x_0}^{c+2\pi} f_2(x)\cos nx\,dx\right]$$

$$b_n = \frac{1}{\pi}\left[\int_c^{x_0} f_1(x)\sin nx\,dx + \int_{x_0}^{c+2\pi} f_2(x)\sin nx\,dx\right]$$

At $x = x_0$, there is a finite jump in the graph of the function. Both the limits $f(x_0 - 0)$ and $f(x_0 + 0)$ exist but are unequal. The sum of the Fourier series $= \frac{1}{2}[f(x_0-0)+f(x_0+0)]$ $= \frac{1}{2}$ (AB + AC) = AM, where M is the mid-point of BC.

ILLUSTRATIVE EXAMPLES

Example 1. *Find the Fourier series to represent the function f(x) given by*

$$f(x) = x \quad \text{for} \quad 0 \le x \le \pi$$

and $$= 2\pi - x \text{ for } \quad \pi \le x \le 2\pi.$$ (G.B.T.U. 2010)

Deduce that $\dfrac{1}{1^2}+\dfrac{1}{3^2}+\dfrac{1}{5^2}+\ldots\ldots=\dfrac{\pi^2}{8}$.

Sol. Let $f(x) = \dfrac{a_0}{2} + \sum_{n=1}^{\infty} a_n \cos nx + \sum_{n=1}^{\infty} b_n \sin nx$

Then $$a_0 = \frac{1}{\pi}\int_0^{2\pi} f(x)\,dx = \frac{1}{\pi}\left[\int_0^{\pi} x\,dx + \int_{\pi}^{2\pi}(2\pi - x)\,dx\right]$$

$$= \frac{1}{\pi}\left[\left|\frac{x^2}{2}\right|_0^\pi + \left|2\pi x - \frac{x^2}{2}\right|_\pi^{2\pi}\right] = \frac{1}{\pi}\left[\frac{\pi^2}{2} + (4\pi^2 - 2\pi^2) - \left(2\pi^2 - \frac{\pi^2}{2}\right)\right] = \pi$$

$$a_n = \frac{1}{\pi}\int_0^{2\pi} f(x)\cos nx\,dx = \frac{1}{\pi}\left[\int_0^\pi x\cos nx\,dx + \int_\pi^{2\pi}(2\pi - x)\cos nx\,dx\right]$$

$$= \frac{1}{\pi}\left[\left|x\,.\,\frac{\sin nx}{n} - 1.\left(-\frac{\cos nx}{n^2}\right)\right|_0^\pi + \left|(2\pi - x)\left(\frac{\sin nx}{n}\right) - (-1)\left(-\frac{\cos nx}{n^2}\right)\right|_\pi^{2\pi}\right]$$

$$= \frac{1}{\pi}\left[\left(\frac{\cos n\pi}{n^2} - \frac{1}{n^2}\right) + \left(-\frac{\cos 2n\pi}{n^2} + \frac{\cos n\pi}{n^2}\right)\right]$$

$$= \frac{1}{\pi n^2}[(-1)^n - 1 - 1 + (-1)^n] = \frac{2}{\pi n^2}\,[(-1)^n - 1] = \begin{cases} -\dfrac{4}{\pi n^2}, \text{ if } n \text{ is odd} \\ 0, \text{ if } n \text{ is even}\end{cases}$$

$$b_n = \frac{1}{\pi}\int_0^{2\pi} f(x)\sin nx\,dx = \frac{1}{\pi}\left[\int_0^\pi x\sin nx\,dx + \int_\pi^{2\pi}(2\pi - x)\sin nx\,dx\right]$$

$$= \frac{1}{\pi}\left[\left|x\left(-\frac{\cos nx}{n}\right) - 1.\left(-\frac{\sin nx}{n^2}\right)\right|_0^\pi + \left|(2\pi - x)\times\left(-\frac{\cos nx}{n}\right) - 1.\left(-\frac{\sin nx}{n^2}\right)\right|_\pi^{2\pi}\right]$$

$$= \frac{1}{\pi}\left[-\frac{\pi\cos n\pi}{n} + \frac{\pi\cos n\pi}{n}\right] = 0$$

$$\therefore \qquad f(x) = \frac{\pi}{2} - \frac{4}{\pi}\left(\frac{\cos x}{1^2} + \frac{\cos 3x}{3^2} + \frac{\cos 5x}{5^2} + \ldots\ldots\right)$$

Putting $x = 0$, we get $0 = \dfrac{\pi}{2} - \dfrac{4}{\pi}\left(\dfrac{1}{1^2} + \dfrac{1}{3^2} + \dfrac{1}{5^2} + \ldots\ldots\right) \Rightarrow \dfrac{\pi^2}{8} = \dfrac{1}{1^2} + \dfrac{1}{3^2} + \dfrac{1}{5^2} + \ldots\ldots$

Example 2. *If* $f(x) = \begin{cases} 0, & -\pi \le x \le 0 \\ \sin x, & 0 \le x \le \pi, \end{cases}$

prove that $f(x) = \dfrac{1}{\pi} + \dfrac{1}{2}\sin x - \dfrac{2}{\pi}\displaystyle\sum_{n=1}^{\infty}\frac{\cos 2nx}{4n^2 - 1}$. (M.D.U. Dec. 2015; G.B.T.U. 2012)

Hence show that

(*i*) $\dfrac{1}{1.3} + \dfrac{1}{3.5} + \dfrac{1}{5.7} + \ldots\ldots = \dfrac{1}{2}$ (*ii*) $\dfrac{1}{1.3} - \dfrac{1}{3.5} + \dfrac{1}{5.7} - \ldots\ldots = \dfrac{\pi - 2}{4}$.

Sol. Let $\quad f(x) = \dfrac{a_0}{2} + \displaystyle\sum_{n=1}^{\infty} a_n\cos nx + \sum_{n=1}^{\infty} b_n\sin nx$

Then $$a_0 = \frac{1}{\pi}\int_{-\pi}^{\pi} f(x)\,dx = \frac{1}{\pi}\left[\int_{-\pi}^{0} 0\,dx + \int_0^{\pi} \sin x\,dx\right] = \frac{2}{\pi}$$

$$a_n = \frac{1}{\pi}\int_{-\pi}^{\pi} f(x)\cos nx\,dx = \frac{1}{\pi}\left[\int_{-\pi}^{0} 0\,dx + \int_0^{\pi} \sin x \cos nx\,dx\right]$$

$$= \frac{1}{2\pi}\int_0^{\pi} 2\cos nx \sin x\,dx = \frac{1}{2\pi}\int_0^{\pi} [\sin (n+1)x - \sin (n-1)x]\,dx$$

$$= \frac{1}{2\pi}\left[-\frac{\cos (n+1)x}{n+1} + \frac{\cos (n-1)x}{n-1}\right]_0^{\pi},\ n \neq 1$$

$$= \frac{1}{2\pi}\left[-\frac{\cos (n+1)\pi}{n+1} + \frac{\cos (n-1)\pi}{n-1} + \frac{1}{n+1} - \frac{1}{n-1}\right]$$

$$= \frac{1}{2\pi}\left[-\frac{(-1)^{n+1}}{n+1} + \frac{(-1)^{n-1}}{n-1} + \frac{1}{n+1} - \frac{1}{n-1}\right]$$

$$= \begin{cases} \dfrac{1}{2\pi}\left(-\dfrac{1}{n+1} + \dfrac{1}{n-1} + \dfrac{1}{n+1} - \dfrac{1}{n-1}\right), & \text{when } n \text{ is odd} \\ \dfrac{1}{2\pi}\left(\dfrac{1}{n+1} - \dfrac{1}{n-1} + \dfrac{1}{n+1} - \dfrac{1}{n-1}\right), & \text{when } n \text{ is even} \end{cases}$$

$$= \begin{cases} 0, \text{ when } n \text{ is odd, } i.e., n = 3, 5, 7, \ldots\ldots \\ -\dfrac{2}{\pi(n^2-1)}, \text{ when } n \text{ is even} \end{cases}$$

When $n = 1$, we have $a_1 = \dfrac{1}{\pi}\displaystyle\int_0^{\pi} \sin x \cos x\,dx = \dfrac{1}{2\pi}\int_0^{\pi} \sin 2x\,dx = \dfrac{1}{2\pi}\left[-\dfrac{\cos 2x}{2}\right]_0^{\pi} = 0$

$$b_n = \frac{1}{\pi}\int_{-\pi}^{\pi} f(x)\sin nx\,dx = \frac{1}{\pi}\left[\int_{-\pi}^{0} 0\,dx + \int_0^{\pi} \sin x \sin nx\,dx\right]$$

$$= \frac{1}{2\pi}\int_0^{\pi} 2\sin nx \sin x\,dx = \frac{1}{2\pi}\int_0^{\pi} [\cos (n-1)x - \cos (n+1)x]\,dx$$

$$= \frac{1}{2\pi}\left[\frac{\sin (n-1)x}{n-1} - \frac{\sin (n+1)x}{n+1}\right]_0^{\pi} = 0,\ n \neq 1$$

When $n = 1$, we have

$$b_1 = \frac{1}{\pi}\int_0^{\pi} \sin x \sin x\,dx = \frac{1}{2\pi}\int_0^{\pi} (1 - \cos 2x)\,dx = \frac{1}{2\pi}\left[x - \frac{\sin 2x}{2}\right]_0^{\pi} = \frac{1}{2}$$

$\therefore$ $$f(x) = \frac{1}{\pi} - \frac{2}{\pi}\left[\frac{\cos 2x}{2^2-1} + \frac{\cos 4x}{4^2-1} + \frac{\cos 6x}{6^2-1} + \ldots\ldots\right] + \frac{1}{2}\sin x$$

$$= \frac{1}{\pi} + \frac{1}{2}\sin x - \frac{2}{\pi}\sum_{n=1}^{\infty} \frac{\cos 2nx}{(2n)^2 - 1} \qquad \ldots(1)$$

Putting $x = 0$ in (1), we have $0 = \dfrac{1}{\pi} - \dfrac{2}{\pi}\displaystyle\sum_{n=1}^{\infty} \dfrac{1}{4n^2-1}$

$$\Rightarrow \quad \frac{1}{2} = \sum_{n=1}^{\infty} \frac{1}{4n^2 - 1} = \sum_{n=1}^{\infty} \frac{1}{(2n-1)(2n+1)} = \frac{1}{1\cdot 3} + \frac{1}{3\cdot 5} + \frac{1}{5\cdot 7} + \ldots\ldots$$

Putting $x = \frac{\pi}{2}$ in (1), we have $1 = \frac{1}{\pi} + \frac{1}{2} - \frac{2}{\pi} \sum_{n=1}^{\infty} \frac{\cos n\pi}{4n^2 - 1}$

$$\Rightarrow \quad \frac{1}{2} - \frac{1}{\pi} = -\frac{2}{\pi} \sum_{n=1}^{\infty} \frac{(-1)^n}{4n^2 - 1}$$

$$\Rightarrow \quad \frac{\pi - 2}{4} = -\sum_{n=1}^{\infty} \frac{(-1)^n}{(2n-1)(2n+1)} = -\left(-\frac{1}{1\cdot 3} + \frac{1}{3\cdot 5} - \frac{1}{5\cdot 7} + \ldots\ldots\right)$$

$$\Rightarrow \quad \frac{1}{1\cdot 3} - \frac{1}{3\cdot 5} + \frac{1}{5\cdot 7} - \ldots\ldots = \frac{\pi - 2}{4}.$$

EXERCISE 1.2

1. Find the Fourier series to represent the function

$$f(x) = -k \quad \text{when} \quad -\pi < x < 0$$
$$= k \quad \text{when} \quad 0 < x < \pi$$

Also deduce that $\frac{\pi}{4} = 1 - \frac{1}{3} + \frac{1}{5} - \frac{1}{7} + \ldots\ldots$ *(G.B.T.U. 2010, 2013)*

2. (*a*) Develop $f(x)$ in a Fourier series in the interval $(-\pi, \pi)$ if $f(x) = 0$ when $-\pi < x < 0$
$= 1$ when $0 < x < \pi$.

Deduce that sum of the Gregory series $1 - \frac{1}{3} + \frac{1}{5} - \frac{1}{7} + \ldots\ldots$ is $\frac{\pi}{4}$.

(*b*) Find the Fourier series of the function defined by

$$f(x) = \begin{cases} 0, & -\pi \le x < 0 \\ \pi, & 0 \le x < \pi \end{cases}$$ *(M.D.U. 2011)*

3. Find the Fourier series expansion for $f(x)$, if $f(x) = -\pi, \quad -\pi < x < 0$
$= x, \quad 0 < x < \pi$

Deduce that $\frac{1}{1^2} + \frac{1}{3^2} + \frac{1}{5^2} + \ldots\ldots = \frac{\pi^2}{8}$. *(M.D.U. Dec. 2013; U.K.T.U. 2012; G.B.T.U. 2011)*

[**Hint. For the deduction, put $x = 0$ in the expansion of $f(x)$.**

$f(0-0) = -\pi$ and $f(0+0) = 0 \quad \therefore \quad f(0) = \frac{1}{2}[f(0-0) + f(0+0)] = -\frac{\pi}{2}$]

4. Find the Fourier expansion of the function defined in one period by the relations

$$f(x) = 1 \text{ for } 0 < x < \pi$$
$$= 2 \text{ for } \pi < x < 2\pi$$

and deduce that $\frac{\pi}{4} = 1 - \frac{1}{3} + \frac{1}{5} - \frac{1}{7} + \ldots\ldots$

5. Find the Fourier series of $f(x) = \begin{cases} 0, & -\pi \le x \le 0 \\ x^2, & 0 \le x \le \pi \end{cases}$

which is assumed to be periodic with period 2π. *(M.T.U. 2011)*

6. Find the Fourier series of the following function:

$$f(x) = x^2, \qquad 0 \le x \le \pi$$
$$= -x^2, \qquad -\pi \le x \le 0.$$

7. An alternating current after passing through a rectifier has the form

$$i = I_0 \sin x \quad \text{for} \quad 0 \le x \le \pi$$
$$= 0 \quad \text{for} \quad \pi \le x \le 2\pi$$

where I_0 is the maximum current and the period is 2π. Express i as a Fourier series.

8. Obtain Fourier series for the function

$$f(x) = \begin{cases} x & \text{for } -\pi < x < 0 \\ -x & \text{for } 0 < x < \pi \end{cases}$$ *(M.T.U. 2013)*

and hence show that

$$\frac{1}{1^2} + \frac{1}{3^2} + \frac{1}{5^2} + \ldots\ldots = \frac{\pi^2}{8}.$$ *(G.B.T.U. 2010)*

9. Find the Fourier series for the function

$$f(x) = \begin{cases} -1 & \text{for } -\pi < x < -\dfrac{\pi}{2} \\ 0 & \text{for } -\dfrac{\pi}{2} < x < \dfrac{\pi}{2}. \\ 1 & \text{for } \dfrac{\pi}{2} < x < \pi \end{cases}$$

Hence deduce that $\dfrac{\pi}{4} = 1 - \dfrac{1}{3} + \dfrac{1}{5} - \dfrac{1}{7} + \ldots\ldots.$ *(G.B.T.U. 2012)*

Answers

1. $f(x) = \dfrac{4k}{\pi}\left(\sin x + \dfrac{\sin 3x}{3} + \dfrac{\sin 5x}{5} + \ldots\ldots\right)$

2. (a) $f(x) = \dfrac{1}{2} + \dfrac{2}{\pi}\left(\sin x + \dfrac{\sin 3x}{3} + \dfrac{\sin 5x}{5} + \ldots\ldots\right)$

   (b) $f(x) = \dfrac{\pi}{2} + 2\left(\sin x + \dfrac{\sin 3x}{3} + \dfrac{\sin 5x}{5} + \ldots\ldots\right)$

3. $f(x) = -\dfrac{\pi}{4} - \dfrac{2}{\pi}\left(\cos x + \dfrac{\cos 3x}{3^2} + \dfrac{\cos 5x}{5^2} + \ldots\ldots\right) + \left(3 \sin x - \dfrac{\sin 2x}{2} + \sin 3x - \dfrac{\sin 4x}{4} + \ldots\ldots\right)$

4. $f(x) = \dfrac{3}{2} - \dfrac{2}{\pi}\left(\sin x + \dfrac{\sin 3x}{3} + \dfrac{\sin 5x}{5} + \ldots\ldots\right)$

5. $f(x) = \dfrac{\pi^2}{6} - 2\left(\cos x - \dfrac{\cos 2x}{2^2} + \dfrac{\cos 3x}{3^2} - \ldots\ldots\right)$

$$-\frac{1}{\pi}\left[\left(\frac{4}{1^3} - \frac{\pi^2}{1}\right)\sin x + \frac{\pi^2}{2}\sin 2x + \left(\frac{4}{3^3} - \frac{\pi^2}{3}\right)\sin 3x + \frac{\pi^2}{4}\sin 4x + \ldots\ldots\right]$$

6. $f(x) = 2\left(\pi - \dfrac{4}{\pi}\right)\sin x - \pi \sin 2x + \dfrac{2}{3}\left(\pi - \dfrac{4}{9\pi}\right)\sin 3x - \dfrac{\pi}{2}\sin 4x + \ldots\ldots$

7. $i = \dfrac{I_0}{\pi} + \dfrac{I_0}{2}\sin x - \dfrac{2I_0}{\pi}\left(\dfrac{\cos 2x}{2^2 - 1} + \dfrac{\cos 4x}{4^2 - 1} + \dfrac{\cos 6x}{6^2 - 1} + \ldots\ldots\right)$

8. $f(x) = -\frac{\pi}{2} + \frac{4}{\pi}\left(\frac{\cos x}{1^2} + \frac{\cos 3x}{3^2} + \frac{\cos 5x}{5^2} + \ldots\ldots\right)$

9. $f(x) = \frac{2}{\pi}\left(\sin x - \sin 2x + \frac{\sin 3x}{3} - \ldots\ldots\right)$

1.6. CHANGE OF INTERVAL

In many engineering problems, it is desired to expand a function in a Fourier series over an interval of length $2l$ and not 2π. In order to apply foregoing theory, this interval must be transformed into an interval of length 2π. This can be achieved by a transformation of the variable.

Consider a periodic function $f(x)$ defined in the interval $c < x < c + 2l$. To change the interval into one of length 2π, we put

$$\frac{x}{l} = \frac{z}{\pi} \quad \text{or} \quad z = \frac{\pi x}{l} \quad \text{so that when } x = c,\ z = \frac{\pi c}{l} = d \text{ (say)}$$

and when $x = c + 2l$, $\quad z = \frac{\pi(c + 2l)}{l} = \frac{\pi c}{l} + 2\pi = d + 2\pi.$

Thus the function $f(x)$ of period $2l$ in $(c, c + 2l)$ is transformed to the function $f\left(\frac{lz}{\pi}\right)$ = F(z), say, of period 2π in $(d, d + 2\pi)$ and the latter function can be expressed as the Fourier series

$$F(z) = \frac{a_0}{2} + \sum_{n=1}^{\infty} a_n \cos nz + \sum_{n=1}^{\infty} b_n \sin nz \qquad \ldots(1)$$

where $a_0 = \frac{1}{\pi}\int_d^{d+2\pi} F(z)\, dz$; $a_n = \frac{1}{\pi}\int_d^{d+2\pi} F(z) \cos nz\, dz$; and $b_n = \frac{1}{\pi}\int_d^{d+2\pi} F(z) \sin nz\, dz$ ...(2)

Now making the inverse substitution $z = \frac{\pi x}{l}$, $dz = \frac{\pi}{l}\, dx$

When $z = d$, $x = c$ and when $z = d + 2\pi$, $x = c + 2l$.

The expression (1) becomes $F(z) = F\left(\frac{\pi x}{l}\right) = f(x) = \frac{a_0}{2} + \sum_{n=1}^{\infty} a_n \cos \frac{n\pi x}{l} + \sum_{n=1}^{\infty} b_n \sin \frac{n\pi x}{l}$

and the co-efficients a_0, a_n, b_n from (2) reduce to

$$a_0 = \frac{1}{l}\int_c^{c+2l} f(x)\, dx; \quad a_n = \frac{1}{l}\int_c^{c+2l} f(x) \cos \frac{n\pi x}{l}\, dx; \quad \text{and} \quad b_n = \frac{1}{l}\int_c^{c+2l} f(x) \sin \frac{n\pi x}{l}\, dx$$

Hence the Fourier series for f(x) in the interval c < x < c + 2l is given by

$$f(x) = \frac{a_0}{2} + \sum_{n=1}^{\infty} a_n \cos \frac{n\pi x}{l} + \sum_{n=1}^{\infty} b_n \sin \frac{n\pi x}{l}$$

where $a_0 = \frac{1}{l}\int_c^{c+2l} f(x)\, dx$, $a_n = \frac{1}{l}\int_c^{c+2l} f(x) \cos \frac{n\pi x}{l}\, dx$ and $b_n = \frac{1}{l}\int_c^{c+2l} f(x) \sin \frac{n\pi x}{l}\, dx.$

Cor. 1. If we put $c = 0$, the interval becomes $0 < x < 2l$ and the above results reduce to

$$a_0 = \frac{1}{l}\int_0^{2l} f(x)\,dx; \quad a_n = \frac{1}{l}\int_0^{2l} f(x)\cos\frac{n\pi x}{l}\,dx; \quad \text{and} \quad a_n = \frac{1}{l}\int_0^{2l} f(x)\sin\frac{n\pi x}{l}\,dx.$$

Cor. 2. If we put $c = -l$, the interval becomes $-l < x < l$ and the above results reduce to

$$a_0 = \frac{1}{l}\int_{-l}^{l} f(x)\,dx, \quad a_n = \frac{1}{l}\int_{-l}^{l} f(x)\cos\frac{n\pi x}{l}\,dx \quad \text{and} \quad b_n = \frac{1}{l}\int_{-l}^{l} f(x)\sin\frac{n\pi x}{l}\,dx.$$

If f(x) is an even function, we have

$$a_0 = \frac{2}{l}\int_0^{l} f(x)\,dx, \quad a_n = \frac{2}{l}\int_0^{l} f(x)\cos\frac{n\pi x}{l}\,dx \quad \text{and} \quad b_n = 0$$

If f(x) is an odd function, we have $a_0 = 0, \quad a_n = 0$

and
$$b_n = \frac{2}{l}\int_0^{l} f(x)\sin\frac{n\pi x}{l}\,dx.$$

ILLUSTRATIVE EXAMPLES

Example 1. *Find Fourier expansion for the function $f(x) = x - x^2$, $-1 < x < 1$.*

Sol. Let $f(x) = \dfrac{a_0}{2} + \sum\limits_{n=1}^{\infty} a_n \cos n\pi x + \sum\limits_{n=1}^{\infty} b_n \sin n\pi x$

Then
$$a_0 = \int_{-1}^{1}(x - x^2)\,dx = \int_{-1}^{1} x\,dx - \int_{-1}^{1} x^2\,dx = 0 - 2\int_0^1 x^2\,dx = -2\left[\frac{x^3}{3}\right]_0^1 = -\frac{2}{3}$$

$$a_n = \int_{-1}^{1}(x - x^2)\cos n\pi x\,dx = \int_{-1}^{1} x\cos n\pi x\,dx - \int_{-1}^{1} x^2\cos n\pi x\,dx$$

$$= 0 - 2\int_0^1 x^2\cos n\pi x\,dx = -2\left[x^2 \cdot \frac{\sin n\pi x}{n\pi} - 2x\left(-\frac{\cos n\pi x}{n^2\pi^2}\right) + 2\left(-\frac{\sin n\pi x}{n^3\pi^3}\right)\right]_0^1$$

$$= -2\left[\frac{2\cos n\pi}{n^2\pi^2}\right] = \frac{-4(-1)^n}{n^2\pi^2} = \frac{4(-1)^{n+1}}{n^2\pi^2}$$

$$b_n = \int_{-1}^{1}(x - x^2)\sin n\pi x\,dx = \int_{-1}^{1} x\sin n\pi x\,dx - \int_{-1}^{1} x^2\sin n\pi x\,dx$$

$$= 2\int_0^1 x\sin n\pi x\,dx - 0 = 2\left[x\left(-\frac{\cos n\pi x}{n\pi}\right) - 1.\left(-\frac{\sin n\pi x}{n^2\pi^2}\right)\right]_0^1$$

$$= 2\left[-\frac{\cos n\pi}{n\pi}\right] = \frac{-2(-1)^n}{n\pi} = \frac{2(-1)^{n+1}}{n\pi}$$

$$\therefore \quad x - x^2 = -\frac{1}{3} + \frac{4}{\pi^2}\left(\frac{\cos \pi x}{1^2} - \frac{\cos 2\pi x}{2^2} + \frac{\cos 3\pi x}{3^2} - \ldots\right)$$
$$+ \frac{2}{\pi}\left(\frac{\sin \pi x}{1} - \frac{\sin 2\pi x}{2} + \frac{\sin 3\pi x}{3} - \ldots\right)$$

Example 2. *Find the Fourier series to represent $f(x) = x^2 - 2$, when $-2 \le x \le 2$.*

(M.D.U. 2011)

Sol. Since $f(x)$ is an even function, $b_n = 0$.

Let $\quad f(x) = x^2 - 2 = \dfrac{a_0}{2} + \sum_{n=1}^{\infty} a_n \cos \dfrac{n\pi x}{2}$

Then $\quad a_0 = \dfrac{2}{2} \int_0^2 (x^2 - 2)\, dx = \left[\dfrac{x^3}{3} - 2x \right]_0^2 = \dfrac{8}{3} - 4 = -\dfrac{4}{3}$

$$a_n = \frac{2}{2} \int_0^2 (x^2 - 2) \cos \frac{n\pi x}{2}\, dx$$

$$= \left[(x^2 - 2) . \frac{\sin \frac{n\pi x}{2}}{\frac{n\pi}{2}} - 2x \left(- \frac{\cos \frac{n\pi x}{2}}{\frac{n^2\pi^2}{4}} \right) + 2 \left(- \frac{\sin \frac{n\pi x}{2}}{\frac{n^3\pi^3}{8}} \right) \right]_0^2$$

$$= \frac{16 \cos n\pi}{n^2\pi^2} = \frac{16(-1)^n}{n^2\pi^2}$$

$\therefore \quad x^2 - 2 = -\dfrac{2}{3} - \dfrac{16}{\pi^2} \left(\cos \dfrac{\pi x}{2} - \dfrac{1}{4} \cos \pi x + \dfrac{1}{9} \cos \dfrac{3\pi x}{2} - \ldots\ldots \right).$

Example 3. *Expand $f(x) = e^{-x}$ as a Fourier series in the interval $(-l, l)$.* (M . D . U . 2010)

Sol. Let $f(x) = e^{-x} = \dfrac{a_0}{2} + \sum_{n=1}^{\infty} a_n \cos \dfrac{n\pi x}{l} + \sum_{n=1}^{\infty} b_n \sin \dfrac{n\pi x}{l}$

Then $\quad a_0 = \dfrac{1}{l} \int_{-l}^{l} e^{-x}\, dx = \dfrac{1}{l} \left[-e^{-x} \right]_{-l}^{l} = \dfrac{1}{l} (e^l - e^{-l}) = \dfrac{2 \sinh l}{l}$

$$a_n = \frac{1}{l} \int_{-l}^{l} e^{-x} \cos \frac{n\pi x}{l}\, dx = \frac{1}{l} \left[\frac{e^{-x}}{1 + \left(\frac{n\pi}{l} \right)^2} \left(-\cos \frac{n\pi x}{l} + \frac{n\pi}{l} \sin \frac{n\pi x}{l} \right) \right]_{-l}^{l}$$

$$\left[\because \int e^{ax} \cos bx\, dx = \frac{e^{ax}}{a^2 + b^2} (a \cos bx + b \sin bx) \right]$$

$$= \frac{l}{l^2 + (n\pi)^2} [-e^{-l} \cos n\pi + e^l \cos n\pi] = -\frac{2l \cos n\pi}{l^2 + (n\pi)^2} \left(\frac{e^l - e^{-l}}{2} \right) = \frac{2l\,(-1)^n \sinh l}{l^2 + (n\pi)^2}$$

$$b_n = \frac{1}{l} \int_{-l}^{l} e^{-x} \sin \frac{n\pi x}{l}\, dx$$

$$= \frac{1}{l} \left[\frac{e^{-x}}{1 + \left(\frac{n\pi}{l} \right)^2} \left(-\sin \frac{n\pi x}{l} - \frac{n\pi}{l} \cos \frac{n\pi x}{l} \right) \right]_{-l}^{l}$$

$$\left[\because \int e^{ax} \sin bx\, dx = \frac{e^{ax}}{a^2 + b^2} (a \sin bx - b \cos bx) \right]$$

$$= -\frac{1}{l^2+(n\pi)^2}\left[\frac{n\pi}{l}(e^{-l}-e^{l})\cos n\pi\right] = \frac{2n\pi\cos n\pi}{l^2+(n\pi)^2}\left(\frac{e^{l}-e^{-l}}{2}\right) = \frac{2n\pi(-1)^n\sinh l}{l^2+(n\pi)^2}$$

$$\therefore \quad e^{-x} = \sinh l\left[\frac{1}{l} - 2l\left(\frac{1}{l^2+\pi^2}\cos\frac{\pi x}{l} - \frac{1}{l^2+2^2\pi^2}\cos\frac{2\pi x}{l} + \frac{1}{l^2+3^2\pi^2}\cos\frac{3\pi x}{l} - \ldots\ldots\right)\right.$$

$$\left. - 2\pi\left(\frac{1}{l^2+\pi^2}\sin\frac{\pi x}{l} - \frac{2}{l^2+2^2\pi^2}\sin\frac{2\pi x}{l} + \frac{3}{l^2+3^2\pi^2}\sin\frac{3\pi x}{l} - \ldots\ldots\right)\right].$$

Example 4. *Obtain Fourier series for the function* $f(x) = \pi x, \quad 0 \le x \le 1$
$= \pi(2-x), \quad 1 \le x \le 2.$

(M.D.U. May 2013)

Sol. Let $f(x) = \frac{a_0}{2} + \sum_{n=1}^{\infty} a_n \cos n\pi x + \sum_{n=1}^{\infty} b_n \sin n\pi x$

Then $a_0 = \int_0^2 f(x)\,dx = \int_0^1 \pi x\,dx + \int_1^2 \pi(2-x)\,dx = \pi\left[\frac{x^2}{2}\right]_0^1 + \pi\left[2x - \frac{x^2}{2}\right]_1^2$

$$= \pi\left(\frac{1}{2}\right) + \pi\left[(4-2) - \left(2-\frac{1}{2}\right)\right] = \pi$$

$$a_n = \int_0^2 f(x)\cos n\pi x\,dx = \int_0^1 \pi x\cos n\pi x\,dx + \int_1^2 \pi(2-x)\cos n\pi x\,dx$$

$$= \left[\pi x\,.\frac{\sin n\pi x}{n\pi} - \pi\left(-\frac{\cos n\pi x}{n^2\pi^2}\right)\right]_0^1 + \left[\pi(2-x)\,.\frac{\sin n\pi x}{n\pi} - (-\pi)\left(-\frac{\cos n\pi x}{n^2\pi^2}\right)\right]_1^2$$

$$= \left[\frac{\cos n\pi}{n^2\pi} - \frac{1}{n^2\pi}\right] + \left[-\frac{\cos 2n\pi}{n^2\pi} + \frac{\cos n\pi}{n^2\pi}\right] = \frac{2}{n^2\pi}(\cos n\pi - 1) = \frac{2}{n^2\pi}[(-1)^n - 1]$$

$= 0$ or $-\frac{4}{n^2\pi}$ according as n is even or odd.

$$b_n = \int_0^2 f(x)\sin n\pi x\,dx = \int_0^1 \pi x\sin n\pi x\,dx + \int_1^2 \pi(2-x)\sin n\pi x\,dx$$

$$= \left[\pi x\left(-\frac{\cos n\pi x}{n\pi}\right) - \pi\left(-\frac{\sin n\pi x}{n^2\pi^2}\right)\right]_0^1 + \left[\pi(2-x)\left(-\frac{\cos n\pi x}{n\pi}\right) - (-\pi)\left(-\frac{\sin n\pi x}{n^2\pi^2}\right)\right]_1^2$$

$$= \left[-\frac{\cos n\pi}{n}\right] + \left[\frac{\cos n\pi}{n}\right] = 0$$

$$\therefore \quad f(x) = \frac{\pi}{2} - \frac{4}{\pi}\left(\frac{\cos \pi x}{1^2} + \frac{\cos 3\pi x}{3^2} + \frac{\cos 5\pi x}{5^2} + \ldots\ldots\right).$$

Note. Putting $x = 0$, we have $f(0) = \frac{\pi}{2} - \frac{4}{\pi}\left(\frac{1}{1^2} + \frac{1}{3^2} + \frac{1}{5^2} + \ldots\ldots\right)$

or $$0 = \frac{\pi}{2} - \frac{4}{\pi}\left(\frac{1}{1^2} + \frac{1}{3^2} + \frac{1}{5^2} + \ldots\ldots\right)$$

$$\therefore \quad \frac{1}{1^2} + \frac{1}{3^2} + \frac{1}{5^2} + \ldots\ldots = \frac{\pi^2}{8}.$$

EXERCISE 1.3

1. Find a Fourier series for $f(t) = 1 - t^2$ when $-1 \le t \le 1$.
2. Expand $f(x)$ in Fourier series in the interval $(-2, 2)$ when $f(x) = 0, \quad -2 < x < 0$
$= 1, \quad 0 < x < 2.$
3. Develop $f(x)$ in a Fourier series in the interval $(0, 2)$ if $f(x) = x, \quad 0 < x < 1$
$= 0, \quad 1 < x < 2.$
4. Find the Fourier expansion for $f(x) = \pi x$ from $x = -c$ to $x = c$.
5. Find the Fourier expansion for the function $f(x) = x - x^3$ in the interval $-1 < x < 1$.
6. (*a*) Find the Fourier series for the function given by $f(t) = t, \quad 0 < t < 1$
$= 1 - t, \quad 1 < t < 2.$

Hence deduce that $\displaystyle\sum_{n=1}^{\infty} \frac{1}{(2n-1)^2} = \frac{\pi^2}{8}$ *(G.B.T.U. 2011)*

(*b*) Find the Fourier series for the function

$$f(x) = \begin{cases} x, & -1 < x \le 0 \\ x + 2, & 0 < x < 1 \end{cases}$$

where $f(x) = f(x + 2)$

(*c*) Find the Fourier series expansion of $f(x) = 2x - x^2$ in $(0, 3)$ and hence deduce that

$$\frac{1}{1^2} - \frac{1}{2^2} + \frac{1}{3^2} - \frac{1}{4^2} + \ldots = \frac{\pi^2}{12}.$$ *(K.U.K. Dec. 2015)*

7. Find a Fourier series to represent x^2 in the interval $(-l, l)$.
8. Expand $f(x) = e^{-x}$ as a Fourier series in the interval $(-2, 2)$.
9. Expand: $f(x) = \begin{cases} \frac{1}{4} - x, & 0 < x < \frac{1}{2} \\ x - \frac{3}{4}, & \frac{1}{2} < x < 1 \\ 1, & 1 < x < \frac{3}{2} \\ x - 1, & \frac{3}{2} < x < 2 \end{cases}$

as a Fourier series.

10. Find the Fourier series for the function $f(x) = \begin{cases} 0 & \text{when } -2 < x < -1 \\ k & \text{when } -1 < x < 1 \\ 0 & \text{when } 1 < x < 2. \end{cases}$
11. A sinusoidal voltage E sin ωt is passed through a half-wave rectifier which clips the negative portion of the wave.

Expand the resulting periodic function $u(t) = \begin{cases} 0 & \text{when } -\frac{T}{2} < t < 0 \\ E \sin \omega t & \text{when } 0 < t < \frac{T}{2} \end{cases}$

and $T = \frac{2\pi}{\omega}$, in a Fourier series.

Hence deduce that $\sum_{n=1}^{\infty} \frac{1}{(2n-1)^2} = \frac{\pi^2}{8}$. *(G.B.T.U. 2011)*

12. Obtain the Fourier series expansion of

$f(x) = \left(\frac{\pi - x}{2}\right)$ for $0 < x < 2$. *(M.T.U. 2011)*

13. Find the Fourier series expansion of $f(x) = 1 + |x|$ defined in $-3 < x < 3$. *(M.T.U. 2011)*

14. Obtain the Fourier series for $f(x) = \begin{cases} -1 & \text{in } -1 < x < 0 \\ 2x & \text{in } \quad 0 < x < 1 \end{cases}$

Hence show that $1 + \frac{1}{3^2} + \frac{1}{5^2} + \cdots = \frac{\pi^2}{8}$

and $1 - \frac{1}{3} + \frac{1}{5} - \frac{1}{7} + \cdots = \frac{\pi}{4}$.

[**Hint:** For deductions, put $x = 0$ and $x = \frac{1}{2}$]

15. Obtain the Fourier series expansion of $f(x) = \begin{cases} 1 & \text{in } 0 < x < 1 \\ 2 & \text{in } 1 < x < 3 \end{cases}$.

16. Obtain the Fourier series expansion of $f(x) = x^2$ defined in the interval $0 < x < 2l$. Hence deduce that

(*i*) $\frac{1}{1^2} + \frac{1}{2^2} + \frac{1}{3^2} + \cdots = \frac{\pi^2}{6}$ (*ii*) $\frac{1}{1^2} - \frac{1}{2^2} + \frac{1}{3^2} - \cdots = \frac{\pi^2}{12}$

(*iii*) $\frac{1}{1^2} + \frac{1}{3^2} + \frac{1}{5^2} + \cdots = \frac{\pi^2}{8}$

[**Hint:** (*i*) Put $x = 0$ (*ii*) Put $x = l$ (*iii*) Add]

Answers

1. $1 - t^2 = \frac{2}{3} + \frac{4}{\pi^2}\left(\cos \pi t - \frac{\cos 2\pi t}{2^2} + \frac{\cos 3\pi t}{3^2} - \ldots\ldots\right)$

2. $f(x) = \frac{1}{2} + \frac{2}{\pi}\left(\sin \frac{\pi x}{2} + \frac{1}{3}\sin \frac{3\pi x}{2} + \frac{1}{5}\sin \frac{5\pi x}{2} + \ldots\ldots\right)$

3. $f(x) = \frac{1}{4} - \frac{2}{\pi^2}\left(\cos \pi x + \frac{\cos 3\pi x}{3^2} + \frac{\cos 5\pi x}{5^2} + \ldots\ldots\right) + \frac{1}{\pi}\left(\sin \pi x - \frac{\sin 2\pi x}{2} + \frac{\sin 3\pi x}{3} - \ldots\ldots\right)$

4. $f(x) = 2c\left[\sin\left(\frac{\pi x}{c}\right) - \frac{1}{2}\sin\left(\frac{2\pi x}{c}\right) + \frac{1}{3}\sin\left(\frac{3\pi x}{c}\right) - \ldots\ldots\right]$

5. $f(x) = \frac{12}{\pi^3}\left(\sin \pi x - \frac{\sin 2\pi x}{2^3} + \frac{\sin 3\pi x}{3^3} - \ldots\ldots\right)$

6. (a) $f(t) = -\frac{4}{\pi^2}\left(\cos \pi t + \frac{\cos 3\pi t}{3^2} + \frac{\cos 5\pi t}{5^3} + \ldots\ldots\right) + \frac{2}{\pi}\left(\sin \pi t + \frac{\sin 3\pi t}{3} + \ldots\ldots\right)$

(b) $f(x) = 1 + \frac{2}{\pi}\sum_{n=1}^{\infty} \frac{1-2(-1)^n}{n} \sin n\pi x = 1 + \frac{2}{\pi}\left(3 \sin \pi x - \frac{\sin 2\pi x}{2} + \sin 3\pi x - \frac{\sin 4\pi x}{4} + \ldots\ldots\right)$

(c) $2x - x^2 = -\sum_{n=1}^{\infty} \frac{9}{n^2\pi^2} \cos \frac{2n\pi x}{3} + \sum_{n=1}^{\infty} \frac{3}{n\pi} \sin \frac{2n\pi x}{3}$

(For deduction, put $x = \frac{3}{2}$)

7. $x^2 = \frac{l^2}{3} - \frac{4l^2}{\pi^2}\left(\frac{\cos \pi x/l}{1^2} - \frac{\cos 2\pi x/l}{2^2} + \frac{\cos 3\pi x/l}{3^2} - \frac{\cos 4\pi x/l}{4^2} + \ldots\ldots\right)$

8. $e^{-x} = \sinh 2\left[\frac{1}{2} - 4\left(\frac{1}{2^2+\pi^2}\cos\frac{\pi x}{2} - \frac{1}{2^2+2^2\pi^2}\cos \pi x + \frac{1}{2^2+3^2\pi^2}\cos\frac{3\pi x}{2} - \ldots\ldots\right)\right.$

$\left. - 2\pi\left(\frac{1}{2^2+\pi^2}\sin\frac{\pi x}{2} - \frac{2}{2^2+2^2\pi^2}\sin \pi x + \frac{3}{2^2+3^2\pi^2}\sin\frac{3\pi x}{2} - \ldots\ldots\right)\right]$

9. $f(x) = \frac{7}{16} + \frac{1}{\pi}\left(\frac{1}{\pi} - \frac{1}{2}\right)\cos \pi x - \frac{1}{\pi}\left(\frac{1}{\pi} + \frac{3}{2}\right)\sin \pi x + \frac{3}{2\pi^2}\cos 2\pi x + \frac{1}{4\pi}\sin 2\pi x + \ldots\ldots$

10. $f(x) = \frac{k}{2} + \frac{2k}{\pi}\left(\cos\frac{\pi x}{2} - \frac{1}{3}\cos\frac{3\pi x}{2} + \frac{1}{5}\cos\frac{5\pi x}{2} - \ldots\ldots\right)$

11. $u(t) = \frac{E}{\pi} + \frac{E}{2}\sin \omega t - \frac{2E}{\pi}\left(\frac{\cos 2\omega t}{1.3} + \frac{\cos 4\omega t}{3.5} + \frac{\cos 6\omega t}{5.7} + \ldots\ldots\right)$

12. $f(x) = \frac{\pi - 1}{2} + \frac{1}{\pi}\sum_{n=1}^{\infty} \frac{\sin n\pi x}{n}$.

13. $f(x) = \frac{5}{2} - \frac{12}{\pi^2}\left(\frac{\cos x}{1^2} + \frac{\cos 3x}{3^2} + \frac{\cos 5x}{5^2} + \ldots\right)$

14. $f(x) = -\frac{4}{\pi^2}\left(\cos \pi x + \frac{\cos 3\pi x}{3^2} + \frac{\cos 5\pi x}{5^2} + \ldots\right)$

$+ \frac{2}{\pi}\left(2 \sin \pi x - \frac{\sin 2\pi x}{2} + \frac{2\sin 3\pi x}{3} - \frac{\sin 4\pi x}{4} + \frac{2 \sin 5\pi x}{5} - \ldots\right)$

15. $f(x) = \frac{5}{3} - \frac{1}{\pi}\sum_{n=1}^{\infty} \frac{1}{n}\sin\frac{2n\pi}{3}\cos\frac{2n\pi x}{3} + \frac{1}{\pi}\sum_{n=1}^{\infty}\frac{1}{n}\left(\cos\frac{2n\pi}{3} - 1\right)\sin\frac{2n\pi x}{3}$

16. $f(x) = \frac{4l^2}{3} + \frac{4l^2}{\pi^2}\sum_{n=1}^{\infty}\frac{1}{n^2}\cos\frac{n\pi x}{l} - \frac{4l^2}{\pi}\sum_{n=1}^{\infty}\frac{1}{n}\sin\frac{n\pi x}{l}$

1.7. HALF RANGE SERIES

Sometimes it is required to expand a function $f(x)$ in the range $(0, \pi)$ in a Fourier series of period 2π or more generally in the range $(0, l)$ in a Fourier series of period $2l$.

If it is required to expand $f(x)$ in the interval $(0, l)$, then it is immaterial what the function may be outside the range $0 < x < l$. We are free to choose it arbitrarily in the interval $(-l, 0)$.

If we extend the function $f(x)$ by reflecting it in the y-axis so that $f(-x) = f(x)$, then the extended function is even for which $b_n = 0$. The Fourier expansion of $f(x)$ will contain only cosine terms.

If we extend the function $f(x)$ by reflecting it in the origin so that $f(-x) = -f(x)$, then the extended function is odd for which $a_0 = a_n = 0$. The Fourier expansion of $f(x)$ will contain only sine terms.

For example, consider the function

$$f(x) = x, \qquad 0 < x < \frac{\pi}{2}$$

$$= \pi - x \qquad \frac{\pi}{2} < x < \pi$$

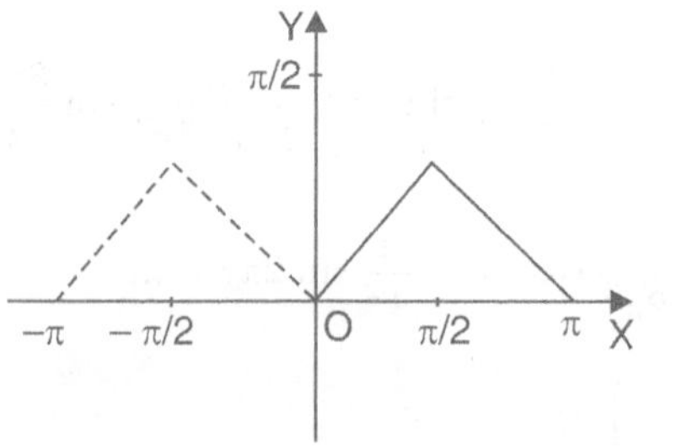

(Reflection in the y-axis)

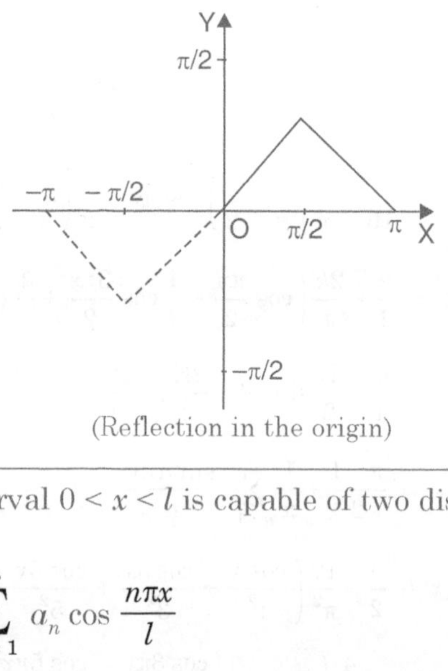

(Reflection in the origin)

Hence a function $f(x)$ defined over the interval $0 < x < l$ is capable of two distinct half-range series.

The half-range cosine series is $f(x) = \dfrac{a_0}{2} + \displaystyle\sum_{n=1}^{\infty} a_n \cos \frac{n\pi x}{l}$

where $a_0 = \dfrac{2}{l}\displaystyle\int_0^l f(x)\, dx$; $a_n = \dfrac{2}{l}\displaystyle\int_0^l f(x) \cos \frac{n\pi x}{l}\, dx$.

The half-range sine series is $f(x) = \displaystyle\sum_{n=1}^{\infty} b_n \sin \frac{n\pi x}{l}$, where $b_n = \dfrac{2}{l}\displaystyle\int_0^l f(x) \sin \frac{n\pi x}{l}\, dx$.

Cor. If the range is $0 < x < \pi$, then

(*i*) The half-range cosine series is $f(x) = \dfrac{a_0}{2} + \displaystyle\sum_{n=1}^{\infty} a_n \cos nx$

where $a_0 = \dfrac{2}{\pi}\displaystyle\int_0^\pi f(x)\, dx$; $a_n = \dfrac{2}{\pi}\displaystyle\int_0^\pi f(x) \cos nx\, dx$.

(*ii*) The half-range sine series is $f(x) = \displaystyle\sum_{n=1}^{\infty} b_n \sin nx$, where $b_n = \dfrac{2}{\pi}\displaystyle\int_0^\pi f(x) \sin nx\, dx$.

ILLUSTRATIVE EXAMPLES

Example 1. *Expand $\pi x - x^2$ in a half-range sine series in the interval $(0, \pi)$ upto the first three terms.*

Sol. Let $\pi x - x^2 = \sum_{n=1}^{\infty} b_n \sin nx$, then

$$b_n = \frac{2}{\pi}\int_0^{\pi} (\pi x - x^2)\sin nx\, dx$$

$$= \frac{2}{\pi}\left[(\pi x - x^2)\left(-\frac{\cos nx}{n}\right) - (\pi - 2x)\left(-\frac{\sin nx}{n^2}\right) + (-2)\left(\frac{\cos nx}{n^3}\right)\right]_0^{\pi}$$

$$= \frac{2}{\pi}\left[-\frac{2\cos n\pi}{n^3} + \frac{2}{n^3}\right] = \frac{4}{\pi n^3}[1 - (-1)^n]$$

$$= 0 \quad \text{or} \quad \frac{8}{\pi n^3} \text{ according as } n \text{ is even or odd.}$$

$$\therefore \quad \pi x - x^2 = \frac{8}{\pi}\left(\sin x + \frac{\sin 3x}{3^3} + \frac{\sin 5x}{5^3} + \ldots\ldots\right).$$

Example 2. *If* $f(x) = x, \quad 0 < x < \frac{\pi}{2}$

$= \pi - x, \quad \frac{\pi}{2} < x < \pi$

show that (i) $f(x) = \frac{4}{\pi}\left[\frac{\sin x}{1^2} - \frac{\sin 3x}{3^2} + \frac{\sin 5x}{5^2} - \ldots\ldots\right]$

(ii) $f(x) = \frac{\pi}{4} - \frac{2}{\pi}\left[\frac{\cos 2x}{1^2} + \frac{\cos 6x}{3^2} + \frac{\cos 10x}{5^2} + \ldots\ldots\right]$. (M.D.U. 2012)

Sol. (*i*) **For the half-range sine series**

Let $f(x) = \sum_{n=1}^{\infty} b_n \sin nx$

Then $b_n = \frac{2}{\pi}\int_0^{\pi} f(x)\sin nx\, dx = \frac{2}{\pi}\left[\int_0^{\pi/2} x\sin nx\, dx + \int_{\pi/2}^{\pi} (\pi - x)\sin nx\, dx\right]$

$$= \frac{2}{\pi}\left[x\left(-\frac{\cos nx}{n}\right) - 1.\left(-\frac{\sin nx}{n^2}\right)\right]_0^{\pi/2} + \frac{2}{\pi}\left[(\pi - x)\left(-\frac{\cos nx}{n}\right) - (-1)\left(-\frac{\sin nx}{n^2}\right)\right]_0^{\pi}$$

$$= \frac{2}{\pi}\left[-\frac{\pi}{2n}\cos\frac{n\pi}{2} + \frac{1}{n^2}\sin\frac{n\pi}{2}\right] + \frac{2}{\pi}\left[\frac{\pi}{2n}\cos\frac{n\pi}{2} + \frac{1}{n^2}\sin\frac{n\pi}{2}\right]$$

$$= \frac{2}{\pi}\left[\frac{2}{n^2}\sin\frac{n\pi}{2}\right] = \frac{4}{\pi n^2}\sin\frac{n\pi}{2}$$

When n is even, $b_n = 0$.

$$\therefore \quad f(x) = \frac{4}{\pi}\left[\frac{\sin x}{1^2} - \frac{\sin 3x}{3^2} + \frac{\sin 5x}{5^2} - \ldots\ldots\right]$$

(*ii*) **For the half-range cosine series**

Let $f(x) = \frac{a_0}{2} + \sum_{n=1}^{\infty} a_n \cos nx$

Then $a_0 = \frac{2}{\pi}\int_0^{\pi} f(x)\,dx = \frac{2}{\pi}\left[\int_0^{\pi/2} x\,dx + \int_{\pi/2}^{\pi} (\pi - x)\,dx\right]$

$$= \frac{2}{\pi}\left[\left|\frac{x^2}{2}\right|_0^{\pi/2} + \left|\pi x - \frac{x^2}{2}\right|_{\pi/2}^{\pi}\right]$$

$$= \frac{2}{\pi}\left[\frac{\pi^2}{8} + \left(\pi^2 - \frac{\pi^2}{2}\right) - \left(\frac{\pi^2}{2} - \frac{\pi^2}{8}\right)\right] = \frac{2}{\pi}\left[\frac{\pi^2}{4}\right] = \frac{\pi}{2}$$

$$a_n = \frac{2}{\pi}\int_0^{\pi} f(x)\cos nx\,dx = \frac{2}{\pi}\left[\int_0^{\pi/2} x\cos nx\,dx + \int_{\pi/2}^{\pi} (\pi - x)\cos nx\,dx\right]$$

$$= \frac{2}{\pi}\left[x\,.\frac{\sin nx}{n} - 1.\left(-\frac{\cos nx}{n^2}\right)\right]_0^{\pi/2} + \frac{2}{\pi}\left[(\pi - x)\,.\frac{\sin nx}{n} - (-1)\left(-\frac{\cos nx}{n^2}\right)\right]_{\pi/2}^{\pi}$$

$$= \frac{2}{\pi}\left[\frac{\pi}{2n}\sin\frac{n\pi}{2} + \frac{1}{n^2}\cos\frac{n\pi}{2} - \frac{1}{n^2}\right] + \frac{2}{\pi}\left[-\frac{\cos n\pi}{n^2} - \frac{\pi}{2n}\sin\frac{n\pi}{2} + \frac{1}{n^2}\cos\frac{n\pi}{2}\right]$$

$$= \frac{2}{\pi}\left[\frac{2}{n^2}\cos\frac{n\pi}{2} - \frac{\cos n\pi}{n^2} - \frac{1}{n^2}\right] = \frac{2}{\pi n^2}\left[2\cos\frac{n\pi}{2} - \cos n\pi - 1\right]$$

$\therefore \quad a_1 = 0,\ a_2 = \frac{2}{\pi\,.\,2^2}(2\cos\pi - \cos 2\pi - 1) = \frac{-2}{\pi\,.\,1^2},$

$a_3 = 0,\ a_4 = 0,\ a_5 = 0,\ a_6 = \frac{2}{\pi\,.\,6^2}(2\cos 3\pi - \cos 6\pi - 1) = \frac{-2}{\pi\,.\,3^2},$

$a_7 = 0,\ a_8 = 0,\ a_9 = 0,\ a_{10} = \frac{2}{\pi\,.\,10^2}(2\cos 5\pi - \cos 10\pi - 1) = \frac{-2}{\pi\,.\,5^2},\ \ldots\ldots$

Hence $f(x) = \frac{\pi}{4} - \frac{2}{\pi}\left[\frac{\cos 2x}{1^2} + \frac{\cos 6x}{3^2} + \frac{\cos 10x}{5^2} + \ldots\ldots\right].$

Example 3. *Find a series of cosines of multiples of x which will represent x sin x in the interval (0, π) and show that* $\frac{1}{1\cdot 3} - \frac{1}{3\cdot 5} + \frac{1}{5\cdot 7} - \ldots\ldots = \frac{\pi - 2}{4}$. (M.T.U. 2013)

Sol. Let $x\sin x = \frac{a_0}{2} + \sum_{n=1}^{\infty} a_n \cos nx$

Then $a_0 = \frac{2}{\pi}\int_0^{\pi} x\sin x\,dx = \frac{2}{\pi}\Big[x(-\cos x) - 1.(-\sin x)\Big]_0^{\pi} = \frac{2}{\pi}[-\pi\cos\pi] = 2$

$$a_n = \frac{2}{\pi}\int_0^{\pi} x\sin x\cos nx\,dx = \frac{1}{\pi}\int_0^{\pi} x\,(2\cos nx\sin x)\,dx$$

$$= \frac{1}{\pi} \int_0^{\pi} x \, [\sin (n + 1) \, x - \sin (n - 1) \, x] \, dx$$

$$= \frac{1}{\pi} \left[x \left\{ - \frac{\cos (n + 1)x}{n + 1} + \frac{\cos (n - 1)x}{n - 1} \right\} - 1. \left\{ - \frac{\sin (n + 1)x}{(n + 1)^2} - \frac{\sin (n - 1)x}{(n - 1)^2} \right\} \right]_0^{\pi}$$

$$= \frac{1}{\pi} \left[- \frac{\pi \cos (n + 1)\pi}{n + 1} + \frac{\pi \cos (n - 1)\pi}{n - 1} \right], \quad \text{when } n \neq 1$$

$$= - \frac{(-1)^{n+1}}{n + 1} + \frac{(-1)^{n-1}}{n - 1} = (-1)^{n-1} \left[\frac{1}{n - 1} - \frac{1}{n + 1} \right] = \frac{2(-1)^{n-1}}{(n - 1)(n + 1)}$$

When $n = 1$, we have

$$a_1 = \frac{2}{\pi} \int_0^{\pi} x \sin x \cos x \, dx = \frac{1}{\pi} \int_0^{\pi} x \sin 2x \, dx = \frac{1}{\pi} \left[x \left(- \frac{\cos 2x}{2} \right) - 1. \left(- \frac{\sin 2x}{2^2} \right) \right]_0^{\pi}$$

$$= \frac{1}{\pi} \left[- \frac{\pi \cos 2\pi}{2} \right] = - \frac{1}{2}$$

$$\therefore \quad x \sin x = 1 - \frac{1}{2} \cos x - 2 \left(\frac{\cos 2x}{1.3} - \frac{\cos 3x}{2.4} + \frac{\cos 4x}{3.5} - \ldots\ldots \right).$$

Putting $x = \frac{\pi}{2}$, we get $\frac{\pi}{2} = 1 - 2 \left(\frac{-1}{1.3} + \frac{1}{3.5} + \frac{-1}{5.7} - \ldots\ldots \right)$

$$\Rightarrow \qquad 1 + \frac{2}{1 \cdot 3} - \frac{2}{3 \cdot 5} + \frac{2}{5 \cdot 7} - \ldots\ldots = \frac{\pi}{2}$$

$$\Rightarrow \qquad \frac{2}{1 \cdot 3} - \frac{2}{3 \cdot 5} + \frac{2}{5 \cdot 7} - \ldots\ldots = \frac{\pi}{2} - 1$$

Hence
$$\frac{1}{1 \cdot 3} - \frac{1}{3 \cdot 5} + \frac{1}{5 \cdot 7} - \ldots\ldots = \frac{\pi - 2}{4}.$$

Example 4. *Obtain the half-range sine series for e^x in $0 < x < 1$.*

Sol. Let $e^x = \sum_{n=1}^{\infty} b_n \sin n\pi x$, (since $l = 1$)

Then $b_n = 2 \int_0^1 e^x \sin n\pi x \, dx = 2 \left[\frac{e^x}{1 + (n\pi)^2} (\sin n\pi x - n\pi \cos n\pi x) \right]_0^1$

$$\left[\because \int e^{ax} \sin bx \, dx = \frac{e^{ax}}{a^2 + b^2} (a \sin bx - b \cos bx) \right]$$

$$= 2 \left[\frac{e}{1 + (n\pi)^2} (-n\pi \cos n\pi) - \frac{1}{1 + (n\pi)^2} (-n\pi) \right]$$

$$= \frac{2}{1+n^2\pi^2} [-en\pi(-1)^n + n\pi] = \frac{2n\pi}{1+n^2\pi^2} [1 - e(-1)^n]$$

Hence $e^x = 2\pi \sum_{n=1}^{\infty} \frac{n[1-e(-1)^n]}{1+n^2\pi^2}$

$$= 2\pi \left[\frac{1+e}{1+\pi^2} \sin \pi x + \frac{2(1-e)}{1+4\pi^2} \sin 2\pi x + \frac{3(1+e)}{1+9\pi^2} \sin 3\pi x + ... \right].$$

Example 5. *Develop sin* $\left(\frac{\pi x}{l}\right)$ *in half-range cosine series in the range* $0 < x < l$.

Sol. Let $\sin\left(\frac{\pi x}{l}\right) = \frac{a_0}{2} + \sum_{n=1}^{\infty} a_n \cos \frac{n\pi x}{l}$

then $a_0 = \frac{2}{l} \int_0^l \sin \frac{\pi x}{l}\, dx = \frac{2}{l} \left[-\frac{\cos \frac{\pi x}{l}}{\frac{\pi}{l}} \right]_0^l = -\frac{2}{\pi} [\cos \pi - 1] = \frac{4}{\pi}$

$$a_n = \frac{2}{l} \int_0^l \sin \frac{\pi x}{l} \cos \frac{n\pi x}{l}\, dx$$

$$= \frac{1}{l} \int_0^l \left[\sin (n+1) \frac{\pi x}{l} - \sin (n-1) \frac{\pi x}{l} \right] dx$$

$$= \frac{1}{l} \left[-\frac{\cos (n+1) \frac{\pi x}{l}}{(n+1)\frac{\pi}{l}} + \frac{\cos (n-1) \frac{\pi x}{l}}{(n-1)\frac{\pi}{l}} \right]_0^l, \quad \text{when } n \neq 1$$

$$= \frac{1}{\pi} \left[\left\{ -\frac{\cos (n+1)\pi}{n+1} + \frac{\cos (n-1)\pi}{n-1} \right\} - \left\{ -\frac{1}{n+1} + \frac{1}{n-1} \right\} \right]$$

$$= \frac{1}{\pi} \left[-\frac{(-1)^{n+1}}{n+1} + \frac{(-1)^{n-1}}{n-1} + \frac{1}{n+1} - \frac{1}{n-1} \right]$$

When $n = 1$, we have

$$a_1 = \frac{2}{l} \int_0^l \sin \frac{\pi x}{l} \cos \frac{\pi x}{l}\, dx = \frac{1}{l} \int_0^l \sin \frac{2\pi x}{l}\, dx$$

$$= \frac{1}{l} \left[\frac{-\cos \frac{2\pi x}{l}}{\frac{2\pi}{l}} \right]_0^l = -\frac{1}{2\pi} (\cos 2\pi - \cos 0)$$

$$= -\frac{1}{2\pi} (1 - 1) = 0.$$

When n is odd, $n \neq 1$, $a_n = \frac{1}{\pi} \left[-\frac{1}{n+1} + \frac{1}{n-1} + \frac{1}{n+1} - \frac{1}{n-1} \right] = 0$, also $a_1 = 0$

When n is even, $a_n = \frac{1}{\pi} \left[\frac{1}{n+1} - \frac{1}{n-1} + \frac{1}{n+1} - \frac{1}{n-1} \right]$

$$= \frac{2}{\pi}\left[\frac{1}{n+1} - \frac{1}{n-1}\right] = -\frac{4}{\pi(n+1)(n-1)}$$

$$\therefore \qquad \sin\left(\frac{\pi x}{l}\right) = \frac{2}{\pi} - \frac{4}{\pi}\left[\frac{\cos\frac{2\pi x}{l}}{1\cdot 3} + \frac{\cos\frac{4\pi x}{l}}{3\cdot 5} + \frac{\cos\frac{6\pi x}{l}}{5\cdot 7} + \ldots\ldots\right].$$

Example 6. *Obtain a half-range cosine series for*

$$f(x) = kx \qquad \text{for} \quad 0 \le x \le \frac{l}{2}$$

$$= k(l - x) \qquad \text{for} \quad \frac{l}{2} \le x \le l.$$

Deduce the sum of the series $\frac{1}{1^2} + \frac{1}{3^2} + \frac{1}{5^2} + \ldots\ldots$ (M.D.U. Dec. 2014)

Sol. Let $f(x) = \frac{a_0}{2} + \sum_{n=1}^{\infty} a_n \cos\frac{n\pi x}{l}$

then
$$a_0 = \frac{2}{l}\int_0^l f(x)\,dx = \frac{2}{l}\left[\int_0^{l/2} kx\,dx + \int_{l/2}^{l} k(l-x)\,dx\right]$$

$$= \frac{2}{l}\left[\left|\frac{kx^2}{2}\right|_0^{l/2} + \left|k\left(lx - \frac{x^2}{2}\right)\right|_{l/2}^{l}\right]$$

$$= \frac{2}{l}\left[\frac{kl^2}{8} + k\left(l^2 - \frac{l^2}{2}\right) - k\left(\frac{l^2}{2} - \frac{l^2}{8}\right)\right] = \frac{2}{l}\left(\frac{kl^2}{4}\right) = \frac{kl}{2}$$

$$a_n = \frac{2}{l}\int_0^l f(x)\cos\frac{n\pi x}{l}\,dx = \frac{2}{l}\left[\int_0^{l/2} kx\cdot\cos\frac{n\pi x}{l}\,dx + \int_{l/2}^{l} k(l-x)\cdot\cos\frac{n\pi x}{l}\,dx\right]$$

$$= \frac{2}{l}\left[\left|kx\cdot\frac{1}{n\pi}\sin\frac{n\pi x}{l} + k\cdot\frac{l^2}{n^2\pi^2}\cos\frac{n\pi x}{l}\right|_0^{l/2}\right.$$

$$\left. + \left|k(l-x)\,.\,\frac{1}{n\pi}\sin\frac{n\pi x}{l} - k\,.\,\frac{l^2}{n^2\pi^2}\cos\frac{n\pi x}{l}\right|_{l/2}^{l}\right]$$

$$= \frac{2}{l}\left[\left|\frac{kl^2}{2n\pi}\sin\frac{n\pi}{2} + \frac{kl^2}{n^2\pi^2}\left(\cos\frac{n\pi}{2} - 1\right)\right|\right.$$

$$\left. + \left|\frac{-kl^2}{n^2\pi^2}\cos n\pi - \frac{kl^2}{2n\pi}\sin\frac{n\pi}{2} + \frac{kl^2}{n^2\pi^2}\cos\frac{n\pi}{2}\right|\right]$$

$$= \frac{2}{l}\left[\frac{2kl^2}{n^2\pi^2}\cos\frac{n\pi}{2} - \frac{kl^2}{n^2\pi^2} - \frac{kl^2}{n^2\pi^2}\cos n\pi\right] = \frac{2kl}{n^2\pi^2}\left[2\cos\frac{n\pi}{2} - 1 - \cos n\pi\right]$$

When n is odd, $\cos\frac{n\pi}{2} = 0$ and $\cos n\pi = -1 \quad \therefore \quad a_n = 0 \quad \Rightarrow \quad a_1 = a_3 = a_5 = \cdots\cdots = 0$

When n is even, $a_2 = \frac{2kl}{2^2\pi^2}[2\cos\pi - 1 - \cos 2\pi] = -\frac{8kl}{2^2\pi^2}$;

$$a_4 = \frac{2kl}{4^2\pi^2}[2\cos 2\pi - 1 - \cos 4\pi] = 0$$

$$a_6 = \frac{2kl}{6^2\pi^2}[2\cos 3\pi - 1 - \cos 6\pi]$$

$$= \frac{2kl}{6^2\pi^2}(-2-1-1) = -\frac{8kl}{6^2\pi^2} \text{ and so on.}$$

$$\therefore \quad f(x) = \frac{kl}{4} - \frac{8kl}{\pi^2}\left(\frac{1}{2^2}\cos\frac{2\pi x}{l} + \frac{1}{6^2}\cos\frac{6\pi x}{l} + \cdots\cdots\right) \quad \ldots(1)$$

Putting $x = l,\ f(l) = 0$

$\therefore$ From (1), we have $0 = \frac{kl}{4} - \frac{8kl}{\pi^2}\left(\frac{1}{2^2} + \frac{1}{6^2} + \cdots\cdots\right)$

$$\Rightarrow \quad \frac{1}{2^2} + \frac{1}{6^2} + \cdots\cdots = \frac{\pi^2}{32} \quad \Rightarrow \quad \frac{1}{2^2}\left[\frac{1}{1^2} + \frac{1}{3^2} + \cdots\cdots\right] = \frac{\pi^2}{32}$$

Hence $\frac{1}{1^2} + \frac{1}{3^2} + \cdots\cdots = \frac{\pi^2}{8}$.

EXERCISE 1.4

1. (*a*) Obtain cosine and sine series for $f(x) = x$ in the interval $0 \le x \le \pi$. Hence show that

$$\frac{1}{1^2} + \frac{1}{3^2} + \frac{1}{5^2} + \cdots\cdots = \frac{\pi^2}{8}.$$

(*b*) Prove that for $0 < x < l$

$$x = \frac{l}{2} - \frac{4l}{\pi^2}\left(\cos\frac{\pi x}{l} + \frac{1}{3^2}\cos\frac{3\pi x}{l} + \frac{1}{5^2}\cos\frac{5\pi x}{l} + \cdots\cdots\right)$$

2. Find the half-range cosine series for the function $f(x) = x^2$ in the range $0 \le x \le \pi$.

(M.D.U. Dec. 2015; K.U.K. Jan. 2014)

3. Find the half-range cosine series for the function $f(x) = (x-1)^2$ in the interval $0 < x < 1$. Hence show that

(*i*) $\frac{1}{1^2} + \frac{1}{2^2} + \frac{1}{3^2} + \frac{1}{4^2} + \cdots\cdots = \frac{\pi^2}{6}$ (*ii*) $\frac{1}{1^2} - \frac{1}{2^2} + \frac{1}{3^2} - \frac{1}{4^2} + \cdots\cdots = \frac{\pi^2}{12}$

(*iii*) $\frac{1}{1^2} + \frac{1}{3^2} + \frac{1}{5^2} + \frac{1}{7^2} + \cdots\cdots = \frac{\pi^2}{8}$.

4. (*a*) Express sin x as a cosine series in $0 < x < \pi$.

(*b*) Show that a constant function c can be expanded in an infinite series

$$\frac{4c}{\pi}\left(\sin x + \frac{\sin 3x}{3} + \frac{\sin 5x}{5} + \cdots\cdots\right) \text{ in the range } 0 < x < \pi.$$

(*c*) Find the Fourier half-range sine series of $f(x) = 1,\ 0 \le x \le 2$.

5. If

$$f(x) = \begin{cases} \frac{\pi}{3}, & 0 \le x \le \frac{\pi}{3} \\ 0, & \frac{\pi}{3} < x < \frac{2\pi}{3} \\ -\frac{\pi}{3}, & \frac{2\pi}{3} \le x \le \pi \end{cases}$$

then show that $f(x) = \frac{2}{\sqrt{3}}\left[\cos x - \frac{\cos 5x}{5} + \frac{\cos 7x}{7} - \cdots\cdots\right]$.

6. If

$$f(x) = \begin{cases} mx, & 0 \le x \le \frac{\pi}{2} \\ m(\pi - x), & \frac{\pi}{2} \le x \le \pi \end{cases}$$

then show that $f(x) = \frac{4m}{\pi}\left[\frac{\sin x}{1^2} - \frac{\sin 3x}{3^2} + \frac{\sin 5x}{5^2} - \cdots\cdots\right]$.

7. Express $f(x) = x$ as a half-range.

(*i*) sine series in $0 < x < 2$. (*M.D.U. Dec. 2013, Dec. 2014, May 2015; U.K.T.U. 2011*)

(*ii*) cosine series in $0 < x < 2$. (*K.U.K. 2010, May 2013; M.D.U. 2010*)

8. Find the Fourier sine and cosine series of

$$f(x) = \begin{cases} x, & 0 < x < \frac{\pi}{2} \\ 0, & \frac{\pi}{2} < x < \pi \end{cases}$$

9. Show that the series $\frac{1}{\pi}\sum_{n=1}^{\infty}\frac{l}{n}\sin\frac{2n\pi x}{l}$ represents $\frac{1}{2}l - x$ when $0 < x < l$.

10. Find the half-range sine series for $f(x) = \begin{cases} \frac{1}{4} - x, & 0 < x < \frac{1}{2} \\ x - \frac{3}{4}, & \frac{1}{2} < x < 1. \end{cases}$ (*U.K.T.U. 2012*)

11. Represent the following function by Fourier sine series

$$f(x) = \begin{cases} 1 & \text{when } 0 < x < \frac{l}{2} \\ 0 & \text{when } \frac{l}{2} < x < l. \end{cases}$$

12. Find the half-range sine series for the function $f(t) = t - t^2,\ 0 < t < 1$. (*M.T.U. 2011*)

13. Prove that for $0 < x < \pi$,

$$x(\pi - x) = \frac{\pi^2}{6} - \left(\frac{\cos 2x}{1^2} + \frac{\cos 4x}{2^2} + \frac{\cos 6x}{3^2} + \dots\dots\right)$$

14. Let $f(x) = \begin{cases} \omega x, & \text{when } 0 \le x \le \frac{l}{2} \\ \omega(l - x), & \text{when } \frac{l}{2} \le x \le l \end{cases}$

Show that $f(x) = \dfrac{4\omega l}{\pi^2} \displaystyle\sum_{n=0}^{\infty} \frac{(-1)^n}{(2n+1)^2} \sin \frac{(2n+1)\pi x}{l}$

Hence obtain the sum of the series

$$1 + \frac{1}{3^2} + \frac{1}{5^2} + \dots\dots$$ *(G.B.T.U. 2011)*

15. If $f(x) = \begin{cases} \sin x, & \text{for } 0 \le x < \frac{\pi}{4} \\ \cos x, & \text{for } \frac{\pi}{4} \le x \le \frac{\pi}{2} \end{cases}$ expand $f(x)$ in a series of sines. *(M.T.U. 2011)*

16. Find the half range sine series of $f(x) = lx - x^2$ in the interval $(0, l)$. Hence, deduce that

$$\frac{\pi^3}{32} = 1 - \frac{1}{3^3} + \frac{1}{5^3} - \frac{1}{7^3} + \cdots$$ *(M.T.U. 2012)*

17. Obtain the half-range sine series of the function $f(x) = x \sin x$ in $0 < x < \pi$. *(K.U.K. Dec. 2015)*

18. For the function defined by the graph OAB, find the half-range Fourier sine series.

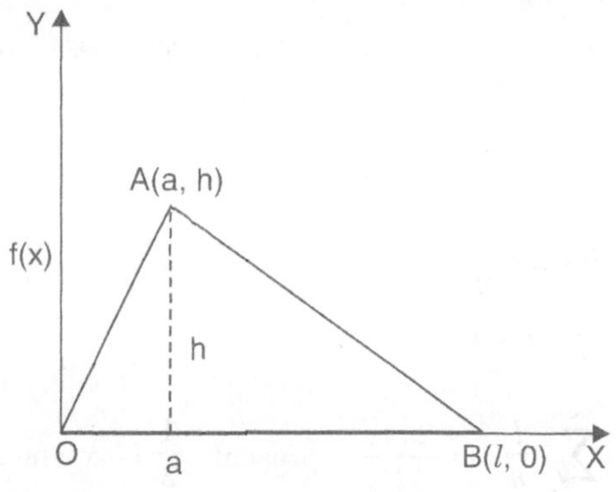

Answers

1. (a) $\frac{\pi}{2} - \frac{4}{\pi}\left(\frac{\cos x}{1^2} + \frac{\cos 3x}{3^2} + \frac{\cos 5x}{5^2} + \dots\dots\right)$; $2\left(\sin x - \frac{\sin 2x}{2} + \frac{\sin 3x}{3} - \frac{\sin 4x}{4} + \dots\dots\right)$

2. $\frac{\pi^2}{3} - 4\left[\frac{\cos x}{1^2} - \frac{\cos 2x}{2^2} + \frac{\cos 3x}{3^2} - \frac{\cos 4x}{4^2} + \dots\dots\right]$

3. $\frac{1}{3} + \frac{4}{\pi^2}\left(\frac{\cos \pi x}{1^2} + \frac{\cos 2\pi x}{2^2} + \frac{\cos 3\pi x}{3^2} + \dots\dots\right)$ **4.** (a) $\frac{2}{\pi} - \frac{4}{\pi}\left[\frac{\cos 2x}{1 \cdot 3} + \frac{\cos 4x}{3 \cdot 5} + \frac{\cos 6x}{5 \cdot 7} + \dots\dots\right]$

(c) $1 = \frac{4}{\pi}\left(\sin \frac{\pi x}{2} + \frac{1}{3} \sin \frac{3\pi x}{2} + \frac{1}{5} \sin \frac{5\pi x}{2} + \dots\dots\right)$

7. (i) $\frac{4}{\pi}\left(\sin\frac{\pi x}{2}-\frac{1}{2}\sin\frac{2\pi x}{2}+\frac{1}{3}\sin\frac{3\pi x}{2}-\frac{1}{4}\sin\frac{4\pi x}{2}+\ldots\ldots\right)$

(ii) $1-\frac{8}{\pi^2}\left[\cos\frac{\pi x}{2}+\frac{1}{3^2}\cos\frac{3\pi x}{2}+\frac{1}{5^2}\cos\frac{5\pi x}{2}+\ldots\ldots\right]$

8. (i) $f(x)=\frac{2}{\pi}\left(\frac{\sin x}{1^2}-\frac{\sin 3x}{3^2}+\frac{\sin 5x}{5^2}-\ldots\ldots\right)+\left(\frac{\sin 2x}{2}-\frac{\sin 4x}{4}+\frac{\sin 6x}{6}-\ldots\ldots\right)$

(ii) $f(x)=\frac{\pi}{8}+\frac{2}{\pi}\left[\left(\frac{\pi}{2}-1\right)\cos x-\frac{1}{2}\cos 2x-\left(\frac{\pi}{6}+\frac{1}{3^2}\right)\cos 3x+\left(\frac{\pi}{10}-\frac{1}{5^2}\right)\cos 5x\ldots\ldots\right]$

10. $f(x)=\left(\frac{1}{\pi}-\frac{4}{\pi^2}\right)\sin\pi x+\left(\frac{1}{3\pi}+\frac{4}{3^2\pi^2}\right)\sin 3\pi x+\left(\frac{1}{5\pi}-\frac{4}{5^2\pi^2}\right)\sin 5\pi x+\cdots$

11. $f(x)=\frac{2}{\pi}\left[\sin\frac{\pi x}{l}+\sin\frac{2\pi x}{l}+\frac{1}{3}\sin\frac{3\pi x}{l}+\frac{1}{5}\sin\frac{5\pi x}{l}+\ldots\ldots\right]$

12. $\frac{8}{\pi^3}\left(\frac{\sin\pi t}{1^3}+\frac{\sin 3\pi t}{3^3}+\frac{\sin 5\pi t}{5^3}+\ldots\ldots\right)$ 14. $\frac{\pi^2}{8}$

15. $\frac{4\sqrt{2}}{\pi}\left(\frac{\sin 2x}{1\cdot 3}-\frac{\sin 6x}{5\cdot 7}+\frac{\sin 10x}{9\cdot 11}\ldots\ldots\right)$

16. $f(x)=\frac{8l^2}{\pi^3}\left(\frac{1}{1^3}\sin\frac{\pi x}{l}+\frac{1}{3^3}\sin\frac{3\pi x}{l}+\frac{1}{5^3}\sin\frac{5\pi x}{l}+\cdots\right)$.

17. $f(x)=\left(\frac{\pi}{2}+\frac{2}{\pi}\right)\sin x-\frac{8}{\pi}\left[\frac{2\sin 2x}{(3\times 1)^2}+\frac{4\sin 4x}{(5\times 3)^2}+\frac{6\sin 6x}{(7\times 5)^2}+\ldots\ldots\right]$.

18. $\frac{2l^2h}{a(l-a)\pi^2}\sum_{n=1}^{\infty}\frac{1}{n^2}\sin\frac{n\pi a}{l}\sin\frac{n\pi x}{l}$.

1.8. PARSEVAL'S THEOREM ON FOURIER CONSTANTS

If the Fourier series of f(x) over an interval $c < x < c + 2l$ is given as

$$f(x)=\frac{a_0}{2}+\sum_{n=1}^{\infty}\left\{a_n\cos\frac{n\pi x}{l}+b_n\sin\frac{n\pi x}{l}\right\}$$

then
$$\frac{1}{2l}\int_c^{c+2l}[f(x)]^2\,dx=\frac{a_0^2}{4}+\frac{1}{2}\sum_{n=1}^{\infty}(a_n^2+b_n^2).$$

Proof. The Fourier series of $f(x)$ in $c < x < c + 2l$ is given as

$$f(x)=\frac{a_0}{2}+\sum_{n=1}^{\infty}\left\{a_n\cos\frac{n\pi x}{l}+b_n\sin\frac{n\pi x}{l}\right\} \quad \ldots(1)$$

where $a_0=\frac{1}{l}\int_c^{c+2l}f(x)\,dx$; $a_n=\frac{1}{l}\int_c^{c+2l}f(x)\cos\frac{n\pi x}{l}\,dx$; $b_n=\frac{1}{l}\int_c^{c+2l}f(x)\sin\frac{n\pi x}{l}\,dx$ ...(2)

Multiplying both sides of (1) by $f(x)$, we have

$$[f(x)]^2 = \frac{a_0}{2} f(x) + \sum_{n=1}^{\infty} a_n f(x) \cos \frac{n\pi x}{l} + \sum_{n=1}^{\infty} b_n f(x) \sin \frac{n\pi x}{l}$$

Integrating both sides w.r.t. x, between the limits c to $c + 2l$, we have

$$\int_c^{c+2l} [f(x)]^2\, dx = \frac{a_0}{2} \int_c^{c+2l} f(x)\, dx + \sum_{n=1}^{\infty} a_n \int_c^{c+2l} f(x) \cos \frac{n\pi x}{l}\, dx$$

$$+ \sum_{n=1}^{\infty} b_n \int_c^{c+2l} f(x) \sin \frac{n\pi x}{l}\, dx$$

$$= \frac{a_0}{2} \cdot la_0 + \sum_{n=1}^{\infty} a_n (la_n) + \sum_{n=1}^{\infty} b_n (lb_n) \qquad \text{[Using (2)]}$$

$$\Rightarrow \qquad \int_c^{c+2l} [f(x)]^2\, dx = \frac{la_0^2}{2} + l \sum_{n=1}^{\infty} (a_n^2 + b_n^2)$$

$$\therefore \qquad \frac{1}{2l} \int_c^{c+2l} [f(x)]^2\, dx = \frac{1}{2l} \left\{ \frac{la_0^2}{2} + l \sum_{n=1}^{\infty} (a_n^2 + b_n^2) \right\}$$

or

$$\mathbf{\frac{1}{2l} \int_c^{c+2l} [f(x)]^2\, dx = \frac{a_0^2}{4} + \frac{1}{2} \sum_{n=1}^{\infty} (a_n^2 + b_n^2)} \qquad \text{(Parseval's identity)}$$

Hence the proof.

Note. *Parseval's identities in different cases:*

(*i*) If $c = 0$, the interval becomes $0 < x < 2l$ and Parseval's identity reduces to

$$\frac{1}{2l} \int_0^{2l} [f(x)]^2\, dx = \frac{{a_0}^2}{4} + \frac{1}{2} \sum_{n=1}^{\infty} (a_n^2 + b_n^2)$$

If $c = -l$, the interval becomes $-l < x < l$ and Parseval's identity reduces to

$$\frac{1}{2l} \int_{-l}^{l} [f(x)]^2\, dx = \frac{{a_0}^2}{4} + \frac{1}{2} \sum_{n=1}^{\infty} (a_n^2 + b_n^2).$$

(*ii*) If $f(x)$ is an even function in $(-l, l)$ then $\frac{2}{l} \int_0^l [f(x)]^2\, dx = \frac{{a_0}^2}{2} + \sum_{n=1}^{\infty} a_n^2$

(*iii*) If $f(x)$ is an odd function in $(-l, l)$ then $\frac{2}{l} \int_0^l [f(x)]^2\, dx = \sum_{n=1}^{\infty} b_n^2$

(*iv*) If $f(x) = \frac{a_0}{2} + \sum_{n=1}^{\infty} a_n \cos \frac{n\pi x}{l}$ in $(0, l)$ then $\frac{2}{l} \int_0^l [f(x)]^2\, dx = \frac{a_0^2}{2} + \sum_{n=1}^{\infty} a_n^2$

(*v*) If $f(x) = \sum_{n=1}^{\infty} b_n \sin \frac{n\pi x}{l}$ in $(0, l)$ then $\frac{2}{l} \int_0^l [f(x)]^2\, dx = \sum_{n=1}^{\infty} b_n^2$.

ILLUSTRATIVE EXAMPLES

Example 1. *Find the Fourier sine series for unity in $0 < x < \pi$ and hence show that*

$$1 + \frac{1}{3^2} + \frac{1}{5^2} + \frac{1}{7^2} + \ldots\ldots = \frac{\pi^2}{8}.$$

Sol. We require half-range Fourier sine series for 1 in $(0, \pi)$

Let $$1 = \sum_{n=1}^{\infty} b_n \sin nx \qquad \ldots(1)$$

Then $$b_n = \frac{2}{\pi}\int_0^{\pi} (1) \sin nx \, dx = \frac{2}{\pi}\left[-\frac{\cos nx}{n}\right]_0^{\pi} = -\frac{2}{n\pi}(\cos n\pi - 1)$$

$$= \frac{2}{n\pi}[1 - (-1)^n] \qquad [\because \cos n\pi = (-1)^n]$$

Now $b_n = 0$ when n is even; and $b_n = \dfrac{4}{n\pi}$ when n is odd.

Substituting in (1), we get

$$\therefore \quad 1 = \sum_{m=1}^{\infty} \frac{4}{(2m-1)\pi} \sin (2m-1)x \quad \text{or} \quad 1 = \frac{4}{\pi}\left(\sin x + \frac{\sin 3x}{3} + \frac{\sin 5x}{5} + \ldots\ldots\right) \qquad \ldots(2)$$

Now from Parseval's theorem on Fourier constants

$$\int_c^{c+2l} [f(x)]^2 \, dx = 2l\left[\frac{{a_0}^2}{4} + \frac{l}{2}\sum_{n=1}^{\infty} ({a_n}^2 + {b_n}^2)\right] \qquad \ldots(3)$$

Applying (3) to half-range sine series for 1 in $(0, \pi)$

$$c = 0,\ 2l = \pi,\ f(x) = 1,\ a_0 = 0,\ a_n = 0, \text{ and } b_n = \frac{4}{(2m-1)\pi},\ m = 1, 2 \ldots\ldots$$

We get, $$\int_0^{\pi} (1)^2 \, dx = \pi \cdot \frac{1}{2}\sum_{m=1}^{\infty} \frac{16}{(2m-1)^2 \cdot \pi^2}$$

$$\Rightarrow \qquad \Big[x\Big]_0^{\pi} = \frac{8}{\pi}\left\{\frac{1}{1^2} + \frac{1}{3^2} + \frac{1}{5^2} + \ldots\ldots\right\} \quad \text{or} \quad \frac{\pi^2}{8} = 1 + \frac{1}{3^2} + \frac{1}{5^2} + \ldots\ldots$$

Hence the result.

Example 2. *Find Fourier series of x^2 in $(-\pi, \pi)$. Use Parseval's identity to prove that*

$$\frac{\pi^4}{90} = 1 + \frac{1}{2^4} + \frac{1}{3^4} + \ldots\ldots.$$

Sol. The Fourier series of x^2 in $(-\pi, \pi)$ is

$$x^2 = \frac{\pi^2}{3} + \sum_{n=1}^{\infty} \frac{4(-1)^n}{n^2} \cos nx \qquad \ldots(1)$$

Here $$a_0 = \frac{2\pi^2}{3},\ a_n = \frac{4(-1)^n}{n^2},\ b_n = 0,\ f(x) = x^2$$

Now by Parseval's identity from (1), we get

$$\int_{-\pi}^{\pi} (x^2)^2\, dx = 2\pi \left[\frac{\pi^4}{9} + \frac{1}{2} \sum_{n=1}^{\infty} \frac{16}{n^4} \right]$$

$$\Rightarrow \qquad \left[\frac{x^5}{5} \right]_{-\pi}^{\pi} = \frac{2\pi^5}{9} + \pi \sum_{n=1}^{\infty} \frac{16}{n^4} \quad \text{or} \quad \frac{2\pi^5}{5} - \frac{2\pi^5}{9} = \pi \sum_{n=1}^{\infty} \frac{16}{n^4}$$

or $$\frac{\pi^4}{90} = \sum_{n=1}^{\infty} \frac{1}{n^4} \quad \text{or} \quad 1 + \frac{1}{2^4} + \frac{1}{3^4} + \ldots\ldots = \frac{\pi^4}{90}.$$

EXERCISE 1.5

1. If $f(x)$ has the Fourier series expansion

$$\frac{a_0}{2} + \sum_{n=1}^{\infty} \left(a_n \cos \frac{n\pi x}{l} + b_n \sin \frac{n\pi x}{l} \right) \text{ in } a \le x \le a + 2l$$

show that $$\int_a^{a+2l} [f(x)]^2\, dx = 2l \left[\frac{a_0^2}{4} + \frac{1}{2} \sum_{n=1}^{\infty} (a_n^2 + b_n^2) \right].$$

2. If $f(x) = \dfrac{a_0}{2} + \displaystyle\sum_{n=1}^{\infty} a_n \cos \frac{n\pi x}{l}$ in $0 < x < l$, then show that

$$\int_0^l [f(x)]^2\, dx = \frac{l}{2} \left[\frac{a_0^2}{2} + \sum_{n=1}^{\infty} a_n^2 \right].$$

3. If $f(x) = \displaystyle\sum_{n=1}^{\infty} b_n \sin \frac{n\pi x}{l}$ in $(0, l)$, then show that $\displaystyle\int_0^l [f(x)]^2\, dx = \frac{l}{2} \sum_{n=1}^{\infty} b_n^2$.

4. Prove that in the range $(0, l)$, $x = \dfrac{l}{2} - \dfrac{4l}{\pi^2} \displaystyle\sum_{n=1}^{\infty} \frac{1}{(2m-1)^2} \cos \frac{(2m-1)\pi x}{l}$ and deduce that

$$\frac{1}{1^4} + \frac{1}{3^4} + \frac{1}{5^4} + \ldots\ldots = \frac{\pi^4}{96}.$$

5. Show that for $0 < x < \pi$,

$$x(\pi - x) = \frac{\pi^2}{6} - \left(\frac{\cos 2x}{1^2} + \frac{\cos 4x}{2^2} + \frac{\cos 6x}{3^2} + \ldots\ldots\right)$$

and hence evaluate $\sum_{n=1}^{\infty} \frac{1}{n^4}$.

1.9. TYPICAL WAVEFORMS

A periodic waveform is a waveform that repeats a basic pattern. It is a single-valued periodic function. Therefore it can be developed as a Fourier series.

We give below some typical waveforms usually met with in communication engineering and also the corresponding Fourier series. The student is urged to construct the Fourier series in each case.

I. Square (or Rectangular) Waveform *(M.D.U. May 2015)*

It is a periodic function of the form given below.

(*i*) $$f(x) = \begin{cases} -k & \text{for } -\pi < x < 0 \\ k & \text{for } \quad 0 < x < \pi \end{cases}, \quad f(x + 2\pi) = f(x)$$

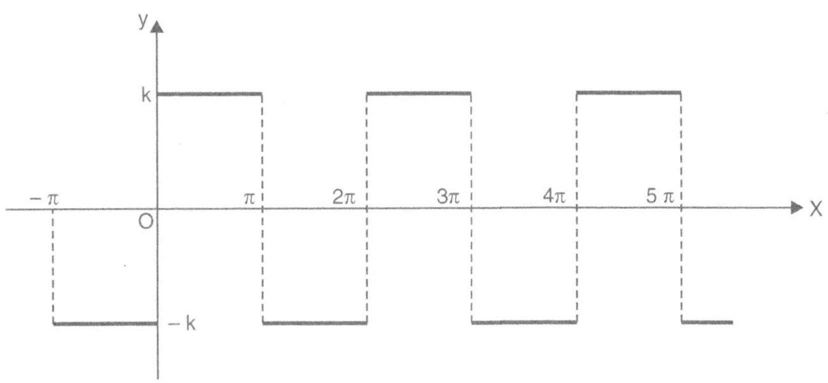

Its Fourier expansion is

$$f(x) = \frac{4k}{\pi}\left[\frac{\sin x}{1} + \frac{\sin 3x}{3} + \frac{\sin 5x}{5} + \ldots\ldots\right]$$

[See Question 1 in Exercise 1.2]

(*ii*) $$f(x) = \begin{cases} a & \text{when } 0 < x < \pi \\ 0 & \text{when } \pi < x < 2\pi \end{cases}, \quad f(x + 2\pi) = f(x)$$

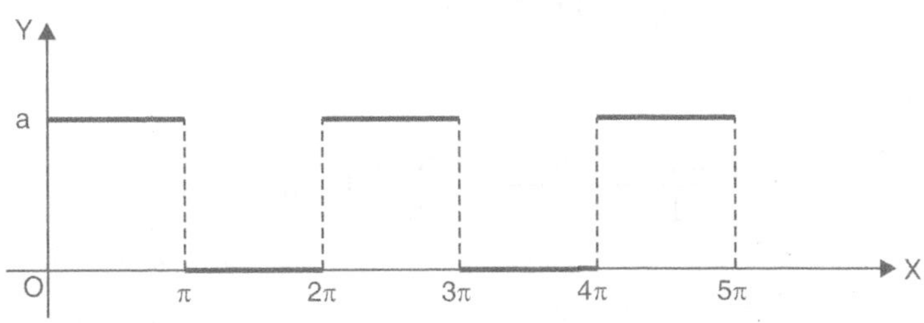

Its Fourier expansion is

$$f(x) = \frac{a}{2} + \frac{2a}{\pi}\left(\frac{\sin x}{1} + \frac{\sin 3x}{3} + \frac{\sin 5x}{5} + \ldots\ldots\right)$$

Exercise. Draw and write the name of the wave form for the following function:

$$f(x) = \begin{cases} a & \text{for } 0 < x < \pi \\ 0 & \text{for } \pi < x < 2\pi \end{cases}$$ *(M.D.U. May 2014)*

(*iii*) $$f(x) = \begin{cases} 0 & \text{for } -\pi < x < 0 \\ 1 & \text{for } \quad 0 < x < \pi \end{cases},\ f(x + 2\pi) = f(x)$$

Its Fourier expansion is

$$f(x) = \frac{1}{2} + \frac{2}{\pi}\left(\frac{\sin x}{1} + \frac{\sin 3x}{3} + \frac{\sin 5x}{5} + \ldots\ldots\right)$$

[See Question 2 in Exercise 1.2]

(*iv*) $$f(x) = \begin{cases} 1 & \text{for } \ 0 < x < \pi \\ 2 & \text{for } \pi < x < 2\pi \end{cases},\ f(x + 2\pi) = f(x)$$

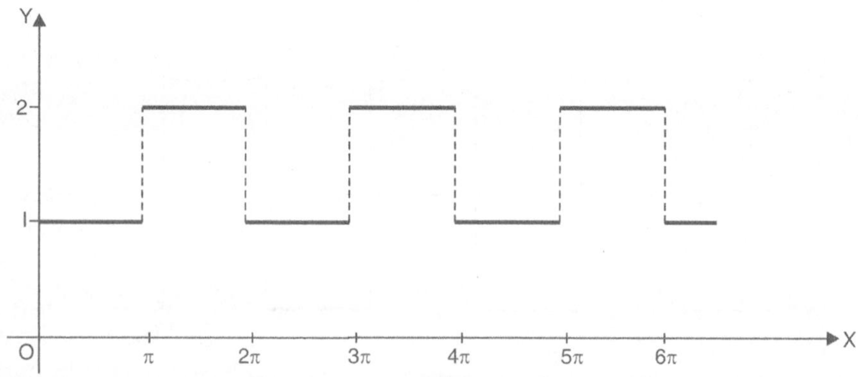

Its Fourier expansion is

$$f(x) = \frac{3}{2} - \frac{2}{\pi}\left(\frac{\sin x}{1} + \frac{\sin 3x}{3} + \frac{\sin 5x}{5} + \ldots\ldots\right)$$

(*v*) $$f(x) = \begin{cases} -\frac{\pi}{4} & \text{for } -\pi < x < 0 \\ \frac{\pi}{4} & \text{for } 0 < x < \pi \end{cases},\ f(x + 2\pi) = f(x)$$

Its Fourier expansion is

$$f(x) = \frac{\sin x}{1} + \frac{\sin 3x}{3} + \frac{\sin 5x}{5} + \ldots\ldots$$

II. Saw-toothed Waveform

It is a periodic function of the form given below.

(*i*) $f(x) = x, -\pi < x < \pi$ and $f(x + 2\pi) = f(x)$

Its Fourier expansion is

$$f(x) = 2\left(\frac{\sin x}{1} - \frac{\sin 2x}{2} + \frac{\sin 3x}{3} - \frac{\sin 4x}{4} + \ldots\ldots\right)$$

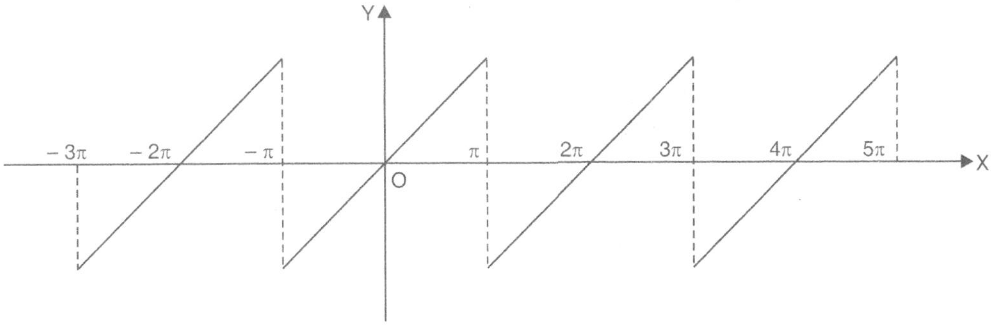

(*ii*) $f(x) = \frac{1}{T}x$ when $0 < x < T$ and $f(x + T) = f(x)$

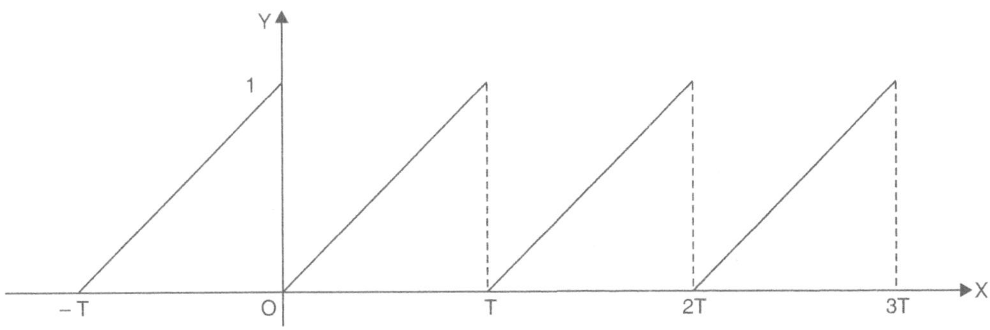

Its Fourier expansion is

$$f(x) = \frac{1}{2} - \frac{1}{\pi}\sum_{n=1}^{\infty}\frac{\sin n\omega x}{n}, \qquad \text{where } \omega = \frac{2\pi}{T}.$$

III. Modified Saw-toothed Waveform

It is a periodic function of the form given below.

$$f(x) = \begin{cases} \pi + x & \text{for } -\pi < x < 0 \\ 0 & \text{for } \quad 0 \le x < \pi \end{cases}, \; f(x + 2\pi) = f(x)$$

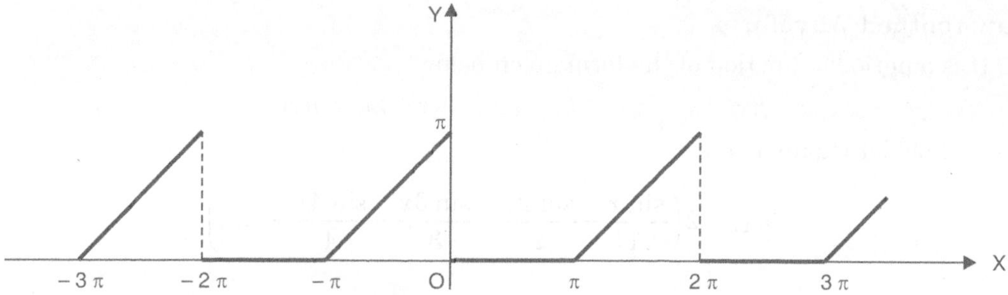

Its Fourier expansion is

$$f(x) = \frac{\pi}{4} + \frac{2}{\pi}\left(\frac{\cos x}{1^2} + \frac{\cos 3x}{3^2} + \ldots\ldots\right) - \left(\frac{\sin x}{1} + \frac{\sin 2x}{2} + \ldots\ldots\right)$$

IV. Triangular Waveform

It is a periodic function of the form given below.

(*i*)
$$f(x) = \begin{cases} 2 + x & \text{for } -2 \le x \le 0 \\ 2 - x & \text{for } \ \ 0 < x \le 2 \end{cases}, \ f(x+4) = f(x)$$

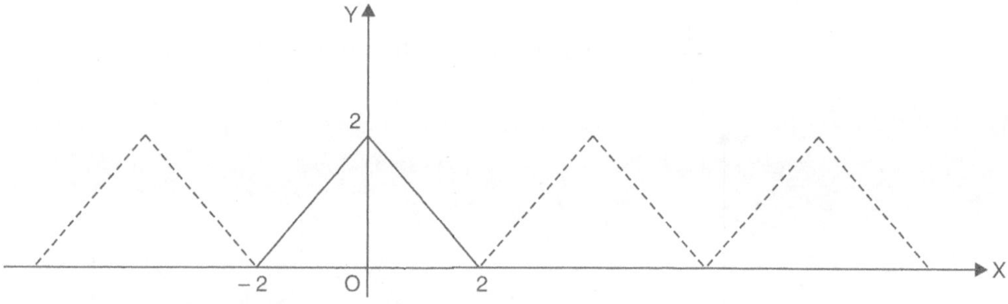

Its Fourier expansion is

$$f(x) = 1 + \frac{8}{\pi^2}\sum_{n=1}^{\infty}\frac{1}{n^2}\cos\left[(2n-1)\frac{\pi x}{2}\right]$$

(*ii*)
$$f(x) = \begin{cases} 1 + \dfrac{2x}{\pi} & \text{for } -\pi \le x \le 0 \\ 1 - \dfrac{2x}{\pi} & \text{for } \ \ 0 < x \le \pi \end{cases}, \ f(x + 2\pi) = f(x)$$

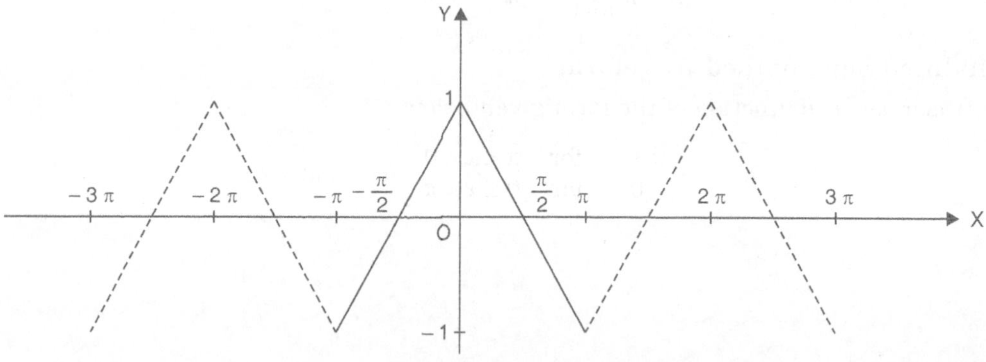

Its Fourier expansion is

$$f(x) = \frac{8}{\pi^2}\left(\frac{\cos x}{1^2} + \frac{\cos 3x}{3^2} + \frac{\cos 5x}{5^2} + \ldots\ldots\right)$$

[See Example 10 before Exercise 1.1]

V. Half Rectified Waveform

It is a periodic function of the form given below.

(*i*) $$f(x) = \begin{cases} 0 & \text{for } -\pi \le x \le 0 \\ \sin x & \text{for } \ 0 \le x \le \pi \end{cases}$$

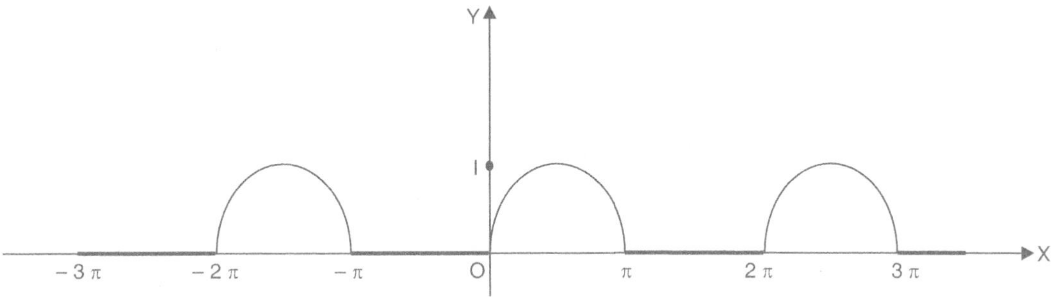

Its Fourier expansion is

$$f(x) = \frac{1}{\pi} + \frac{1}{2}\sin x - \frac{2}{\pi}\sum_{n=1}^{\infty}\frac{\cos 2nx}{4n^2 - 1}$$

(*ii*) $$f(x) = \begin{cases} I_0 \sin x & \text{for } \ 0 \le x \le \pi \\ 0 & \text{for } \pi \le x \le 2\pi \end{cases}, \ f(x + 2\pi) = f(x)$$

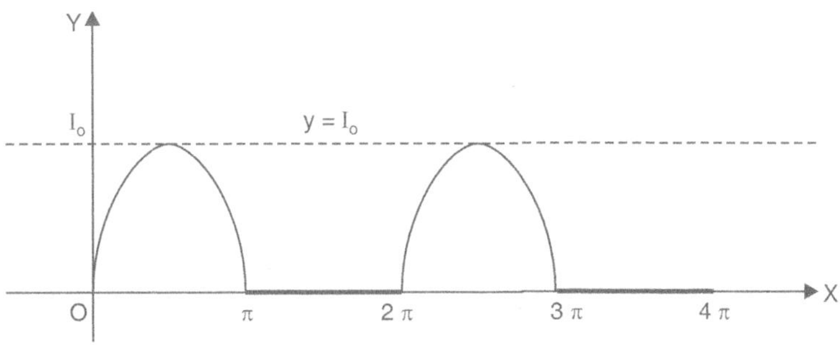

Its Fourier expansion is

$$f(x) = \frac{I_0}{\pi} + \frac{1}{2} I_0 \sin x - \frac{2I_0}{\pi}\sum_{n=1}^{\infty}\frac{\cos 2n\,x}{4n^2 - 1}$$

[See Question 6 in Exercise 1.2]

VI. Full Rectified Waveform

It is a periodic function of the form given below.

$$f(x) = a \sin x \text{ for } 0 \le x \le \pi,\ f(x + \pi) = f(x)$$

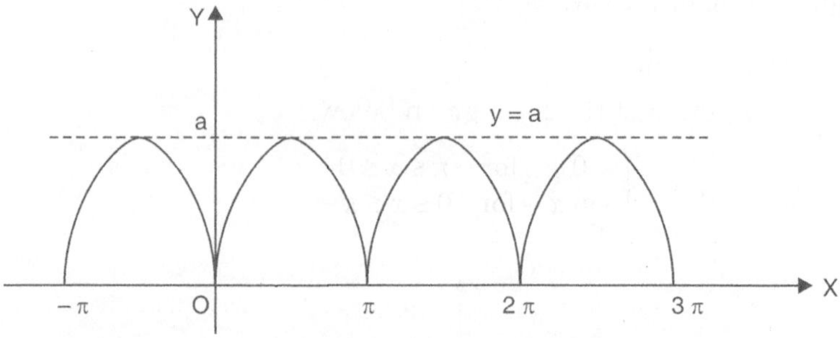

Its Fourier expansion is

$$f(x) = \frac{4a}{\pi}\left[\frac{1}{2} - \frac{\cos 2x}{1.3} - \frac{\cos 4x}{3.5} - \frac{\cos 6x}{5.7} - \ldots\ldots\right]$$

❑❑❑

CHAPTER 2

Fourier Transforms

2.1. INTEGRAL TRANSFORMS

The integral transform of a function $f(x)$ is defined by the equation

$$I\{f(x)\} = \bar{f}(s) = \int_a^b f(x)\, K(s, x)\, dx,$$

where $K(s, x)$ is a known function of s and x, called the **kernel** of the transform; s is called the **parameter** of the transform and $f(x)$ is called the **inverse transform** of $\bar{f}(s)$.

Some of the well-known transforms are given below:

(*i*) **Laplace Transform.** When $K(s, x) = e^{-sx}$, we have the Laplace transform of $f(x)$.

Thus $$L\{f(x)\} = \bar{f}(s) = \int_0^\infty f(x)\, e^{-sx}\, dx$$

(*ii*) **Fourier Transform.** When $K(s, x) = e^{isx}$, we have the Fourier transform of $f(x)$. Thus

$$F(s) = \int_{-\infty}^{\infty} f(x)\, e^{isx}\, dx$$

(*iii*) **Hankel Transform.** When $K(s, x) = x\, J_n\, (sx)$, we have the Hankel transform of $f(x)$.

Thus $$H_n(s) = \int_0^\infty f(x)\, x\, J_n\, (sx)\, dx$$

where $J_n\, (sx)$ is the Bessel function of the first kind and order n.

(*iv*) **Mellin Transform.** When $K(s, x) = x^{s-1}$, we have the Mellin transform of $f(x)$. Thus

$$M(s) = \int_0^\infty f(x)\, x^{s-1}\, dx$$

(*v*) **Fourier Sine Transform.** When $K(s, x) = \sin sx$, we have the Fourier sine transform of $f(x)$. Thus

$$F_s(s) = \int_0^\infty f(x) \sin sx\, dx$$

(*vi*) **Fourier Cosine Transform.** When $K(s, x) = \cos sx$, we have the Fourier cosine transform of $f(x)$. Thus

$$F_c(s) = \int_0^\infty f(x) \cos sx\, dx$$

We have already discussed Laplace transform and its applications to the solution of ordinary differential equations. In the present chapter, we shall discuss the Fourier integrals and Fourier transforms which are useful in solving boundary value problems arising in engineering *e.g.* conduction of heat, theory of communication, wave propagation etc. Fourier series are helpful in problems involving periodic functions. However, in many practical problems, the function is non-periodic. A suitable representation for non-periodic functions can be obtained by considering the limiting form of Fourier series when the fundamental period is made infinite. In such case, the Fourier series becomes a Fourier integral which can be expressed in terms of Fourier transform which transforms a non-periodic function.

The effect of applying an integral transform to a partial differential equation is to reduce the number of independent variables by one. The choice of a particular transform is decided by the nature of the boundary conditions and the facility with which the transform can be inverted to give $f(x)$.

2.2. FOURIER INTEGRAL THEOREM

Statement. *If*

(i) $f(x)$ satisfies Dirichlet's conditions in every interval $(-c, c)$ however large.

(ii) $\int_{-\infty}^{\infty} |f(x)|\, dx$ converges;

then
$$f(x) = \frac{1}{\pi}\int_0^{\infty}\int_{-\infty}^{\infty} f(t)\cos\lambda(t-x)\, dt\, d\lambda$$

The integral on the right hand side is called ***Fourier Integral*** *of $f(x)$.*

Proof. Consider a function $f(x)$ satisfying Dirichlet's conditions in every interval $(-c, c)$, however large. Then

$$f(x) = \frac{a_0}{2} + \sum_{n=1}^{\infty}\left(a_n\cos\frac{n\pi x}{c} + b_n\sin\frac{n\pi x}{c}\right) \qquad \ldots(1)$$

where $a_0 = \frac{1}{c}\int_{-c}^{c} f(t)\, dt$, $a_n = \frac{1}{c}\int_{-c}^{c} f(t)\cos\frac{n\pi t}{c}\, dt$ and $b_n = \frac{1}{c}\int_{-c}^{c} f(t)\sin\frac{n\pi t}{c}\, dt$

Substituting the values of a_0, a_n and b_n in (1), we get

$$f(x) = \frac{1}{2c}\int_{-c}^{c} f(t)\, dt + \frac{1}{c}\sum_{n=1}^{\infty}\int_{-c}^{c}\left[\cos\frac{n\pi x}{c}\cos\frac{n\pi t}{c} + \sin\frac{n\pi x}{c}\sin\frac{n\pi t}{c}\right] f(t)\, dt$$

$$= \frac{1}{2c}\int_{-c}^{c} f(t)\, dt + \frac{1}{c}\sum_{n=1}^{\infty}\int_{-c}^{c}\cos\frac{n\pi(t-x)}{c}\,.\, f(t)\, dt \qquad \ldots(2)$$

If we assume that the integral $\int_{-\infty}^{\infty} |f(x)|\, dx$ converges,

then
$$\lim_{c\to\infty}\left[\frac{1}{2c}\int_{-c}^{c} f(t)\, dt\right] = 0, \text{ since } \left|\frac{1}{2c}\int_{-c}^{c} f(t)\, dt\right| \le \frac{1}{2c}\int_{-\infty}^{\infty} |f(t)|\, dt$$

Putting $\frac{\pi}{c} = \Delta\lambda$, the second term in (2) becomes

$$\frac{1}{\pi}\sum_{n=1}^{\infty}\Delta\lambda\int_{-c}^{c}\cos\{n\Delta\lambda(t-x)\}\, f(t)\, dt$$

This is of the form $\sum_{n=1}^{\infty} F(n\Delta\lambda)\,\Delta\lambda$ whose limit as $\Delta\lambda \to 0$, is $\int_0^{\infty} F(\lambda)d\lambda$.

Hence as $c \to \infty$, (2) reduces to $$f(x) = \frac{1}{\pi}\int_0^{\infty}\int_{-\infty}^{\infty} f(t)\cos\lambda(t-x)\,dt\,d\lambda \quad \ldots(3)$$

which is known as **Fourier Integral** of $f(x)$. *(M.D.U. May 2015)*

Equation (3) is true at a point of continuity. At a point of discontinuity the value of the integral on the right is

$$\tfrac{1}{2}\,[f(x+0)+f(x-0)].$$

2.3. FOURIER SINE AND COSINE INTEGRALS

We know that $\cos\lambda(t-x) = \cos\lambda t\cos\lambda x + \sin\lambda t\sin\lambda x$

$\therefore$ Fourier integral of $f(x)$ can be written as

$$f(x) = \frac{1}{\pi}\int_0^{\infty}\int_{-\infty}^{\infty} f(t)\,\{\cos\lambda t\cos\lambda x + \sin\lambda t\sin\lambda x\}\,dt\,d\lambda$$

$$= \frac{1}{\pi}\int_0^{\infty}\cos\lambda x\int_{-\infty}^{\infty} f(t)\cos\lambda t\,dt\,d\lambda + \frac{1}{\pi}\int_0^{\infty}\sin\lambda x\int_{-\infty}^{\infty} f(t)\sin\lambda t\,dt\,d\lambda \quad \ldots(4)$$

When f(x) is an odd function, $f(t)\cos\lambda t$ is odd while $f(t)\sin\lambda t$ is even. Thus the first integral in (4) vanishes and, we get

$$f(x) = \frac{2}{\pi}\int_0^{\infty}\sin\lambda x\int_0^{\infty} f(t)\sin\lambda t\,dt\,d\lambda \quad \ldots(5)$$

This is called **Fourier sine integral.**

When f(x) is an even function, $f(t)\cos\lambda t$ is even while $f(t)\sin\lambda t$ is odd. Thus the second integral in (4) vanishes and, we get

$$f(x) = \frac{2}{\pi}\int_0^{\infty}\cos\lambda x\int_0^{\infty} f(t)\cos\lambda t\,dt\,d\lambda \quad \ldots(6)$$

This is called **Fourier cosine integral.**

2.4. COMPLEX FORM OF FOURIER INTEGRAL

Since $\cos\lambda(t-x)$ is an even function of λ, we have from (3)

$$f(x) = \frac{1}{2\pi}\int_{-\infty}^{\infty}\int_{-\infty}^{\infty} f(t)\cos\lambda(t-x)\,dt\,d\lambda \quad \ldots(7)$$

$$\left[\because\ 2\int_0^{a} f(x)\,dx = \int_{-a}^{a} f(x)\,dx, \text{if } f(x) \text{ is even}\right]$$

Also $\sin\lambda(t-x)$ is an odd function of λ, so that

$$0 = \frac{1}{2\pi}\int_{-\infty}^{\infty}\int_{-\infty}^{\infty} f(t)\sin\lambda\,(t-x)\,dt\,d\lambda \quad \ldots(8)$$

$$\left[\because\ \int_{-a}^{a} f(x)\,dx = 0, \text{if } f(x) \text{ is odd}\right]$$

Multiplying (8) by i and adding to (7), we get

$$f(x) = \frac{1}{2\pi}\int_{-\infty}^{\infty}\int_{-\infty}^{\infty} f(t)\,\{\cos\lambda\,(t-x) + i\sin\lambda(t-x)\}\,dt\,d\lambda$$

$$= \frac{1}{2\pi}\int_{-\infty}^{\infty}\int_{-\infty}^{\infty} f(t)\,e^{i\lambda(t-x)}\,dt\,d\lambda = \boxed{\frac{1}{2\pi}\int_{-\infty}^{\infty} e^{-i\lambda}\int_{-\infty}^{\infty} f(t)\,e^{i\lambda t}\,dt\,d\lambda}$$

which is known as the **complex form of Fourier integral.**

ILLUSTRATIVE EXAMPLES

Example 1. *Express the function* $f(x) = \begin{cases} 1 & \text{for } |x| \le 1, \\ 0 & \text{for } |x| > 1, \end{cases}$

as a Fourier integral. Hence evaluate $\displaystyle\int_0^{\infty} \frac{\sin\lambda\cos\lambda x}{\lambda}\,d\lambda$.

(M.D.U. 2012, May 2013, Dec. 2014)

Sol. The Fourier integral for $f(x)$ is

$$\frac{1}{\pi}\int_0^{\infty}\int_{-\infty}^{\infty} f(t)\cos\lambda(t-x)\,dt\,d\lambda$$

$$= \frac{1}{\pi}\int_0^{\infty}\int_{-1}^{1} \cos\lambda\,(t-x)\,dt\,d\lambda \qquad \left[\because\ f(t) = \begin{cases} 1, & -1<t<1 \\ 0, & \text{otherwise} \end{cases}\right]$$

$$= \frac{1}{\pi}\int_0^{\infty}\left[\frac{\sin\lambda(t-x)}{\lambda}\right]_{-1}^{1} d\lambda$$

$$= \frac{1}{\pi}\int_0^{\infty} \frac{\sin\lambda(1-x) - \sin\lambda(-1-x)}{\lambda}\,d\lambda$$

$$= \frac{1}{\pi}\int_0^{\infty} \frac{\sin\lambda(1+x) + \sin\lambda(1-x)}{\lambda}\,d\lambda$$

$$= \frac{2}{\pi}\int_0^{\infty} \frac{\sin\lambda\cos\lambda x}{\lambda}\,d\lambda$$

$$\therefore\quad \int_0^{\infty} \frac{\sin\lambda\cos\lambda x}{\lambda}\,d\lambda = \frac{\pi}{2}f(x) = \begin{cases} \frac{\pi}{2} & \text{for } |x| < 1 \\ 0 & \text{for } |x| > 1 \end{cases}$$

At $|x| = 1$, *i.e.*, $x = \pm 1$, $f(x)$ is discontinuous and the integral has the value

$$\frac{1}{2}\left(\frac{\pi}{2} + 0\right) = \frac{\pi}{4}.$$

Note. Putting $x = 0$, we get $\displaystyle\int_0^{\infty} \frac{\sin\lambda}{\lambda}\,d\lambda = \frac{\pi}{2}$ or $\displaystyle\int_0^{\infty} \frac{\sin x}{x}\,dx = \frac{\pi}{2}$.

Example 2. *Express* $f(x) = \begin{cases} 1 & \text{for } 0 \le x \le \pi, \\ 0 & \text{for } x > \pi, \end{cases}$ *as a Fourier sine integral and hence evaluate*

$$\int_0^{\infty} \frac{1-\cos(\pi\lambda)}{\lambda}\sin(x\lambda)\,d\lambda.$$

Sol. The Fourier sine integral for $f(x)$ is

$$\frac{2}{\pi}\int_0^\infty \sin(\lambda x)\int_0^\infty f(t)\sin(\lambda t)\,dt\,d\lambda$$

$$= \frac{2}{\pi}\int_0^\infty \sin(\lambda x)\left[\int_0^\pi f(t)\sin(\lambda t)\,dt + \int_\pi^\infty f(t)\sin(\lambda t)\,dt\right]d\lambda$$

$$= \frac{2}{\pi}\int_0^\infty \sin(\lambda x)\int_0^\pi \sin(\lambda t)\,dt\,d\lambda \qquad \text{[on substituting for } f(t)]$$

$$= \frac{2}{\pi}\int_0^\infty \sin(\lambda x)\left[-\frac{\cos(\lambda t)}{\lambda}\right]_0^\pi d\lambda$$

$$= \frac{2}{\pi}\int_0^\infty \frac{1-\cos(\lambda\pi)}{\lambda}\sin(\lambda x)\,d\lambda$$

$$\therefore \qquad f(x) = \frac{2}{\pi}\int_0^\infty \frac{1-\cos(\pi\lambda)}{\lambda}\sin(x\lambda)\,d\lambda$$

$$\Rightarrow \qquad \int_0^\infty \frac{1-\cos(\pi\lambda)}{\lambda}\sin(x\lambda)\,d\lambda = \frac{\pi}{2}f(x) = \begin{cases}\dfrac{\pi}{2} & \text{for } 0 \le x < \pi \\ 0 & \text{for } x > \pi\end{cases}$$

At $x = \pi$, which is a point of discontinuity of $f(x)$, we have

$$f(x) = \frac{1}{2}[f(\pi-0)+f(\pi+0)] = \frac{1}{2}(1+0) = \frac{1}{2}$$

$$\therefore \qquad \int_0^\infty \frac{1-\cos(\pi\lambda)}{\lambda}\sin(x\lambda)\,d\lambda = \frac{\pi}{2}\left(\frac{1}{2}\right) = \frac{\pi}{4}.$$

Example 3. *Using Fourier Integral representation, show that:*

$$\int_0^\infty \frac{\cos x\alpha + \alpha\sin x\alpha}{1+\alpha^2}\,d\alpha = \begin{cases} 0 & , \text{ if } x < 0 \\ \dfrac{\pi}{2} & , \text{ if } x = 0 \\ \pi e^{-x} & , \text{ if } x > 0\end{cases}.$$

Sol. Fourier Integral for

$$f(x) = \begin{cases} 0, & \text{if } x < 0 \\ e^{-x}, & \text{if } x > 0\end{cases}$$

is $$\frac{1}{\pi}\int_0^\infty\int_{-\infty}^\infty f(t)\cos\lambda(t-x)\,dt\,d\lambda$$

$$= \frac{1}{\pi}\int_0^\infty\left[\int_{-\infty}^0 f(t)\cos\lambda(t-x)\,dt + \int_0^\infty f(t)\cos\lambda(t-x)\,dt\right]d\lambda$$

$$= \frac{1}{\pi}\int_0^\infty\left[\int_{-\infty}^0 0\cos\lambda(t-x)\,dt + \int_0^\infty e^{-t}\cos\lambda(t-x)\,dt\right]d\lambda$$

$$= \frac{1}{\pi}\int_0^\infty\int_0^\infty e^{-t}\cos\lambda(t-x)\,dt\,d\lambda$$

$$= \frac{1}{\pi}\int_0^\infty\left[\frac{e^{-t}}{1+\lambda^2}\{-\cos\lambda(t-x)+\lambda\sin\lambda(t-x)\}\right]_0^\infty d\lambda$$

$$= \frac{1}{\pi}\int_0^\infty \frac{\cos\lambda x + \lambda\sin\lambda x}{1+\lambda^2}\,d\lambda$$

$$= \frac{1}{\pi}\int_0^\infty \frac{\cos \alpha x + \alpha \sin \alpha x}{1+\alpha^2}\, d\alpha \qquad \text{(Replacing } \lambda \text{ by } \alpha)$$

$$\Rightarrow \quad \int_0^\infty \frac{\cos x\alpha + \alpha \sin x\alpha}{1+\alpha^2}\, d\alpha = \pi f(x) = \begin{cases} 0 &, \text{ if } x < 0 \\ \pi e^{-x} &, \text{ if } x > 0 \end{cases}$$

When $x = 0$, $\displaystyle\int_0^\infty \frac{\cos x\alpha + \alpha \sin x\alpha}{1+\alpha^2}\, d\alpha = \int_0^\infty \frac{1}{1+\alpha^2}\, d\alpha = \left[\tan^{-1}\alpha\right]_0^\infty = \frac{\pi}{2}$

$$\therefore \quad \int_0^\infty \frac{\cos x\alpha + \alpha \sin x\alpha}{1+\alpha^2}\, d\alpha = \begin{cases} 0 &, \text{ if } x < 0 \\ \dfrac{\pi}{2} &, \text{ if } x = 0 \\ \pi e^{-x}, & \text{ if } x > 0 \end{cases}.$$

EXERCISE 2.1

Using Fourier integral representation, show that (1 – 8):

1. $\displaystyle\int_0^\infty \frac{\lambda \sin x\lambda}{k^2+\lambda^2}\, d\lambda = \frac{\pi}{2} e^{-kx}$, $x > 0, k > 0$ *(M.D.U. May 2015)*

2. $\displaystyle\int_0^\infty \frac{\cos x\omega}{1+\omega^2}\, d\omega = \frac{\pi}{2} e^{-x}, x \geq 0$ *(K.U.K. Jan. 2014)*

3. $\displaystyle\int_0^\infty \frac{\sin \pi\lambda \sin x\lambda}{1-\lambda^2}\, d\lambda = \begin{cases} \dfrac{\pi}{2}\sin x &, \text{ when } 0 \leq x \leq \pi \\ 0 &, \text{ when } x > \pi \end{cases}$

4. $\displaystyle\int_0^\infty \frac{\cos \lambda x}{\lambda^2+a^2}\, d\lambda = \frac{\pi}{2a} e^{-ax}, a > 0, \quad x \geq 0.$

5. $\displaystyle\int_0^\infty \left(\frac{\lambda^2+2}{\lambda^4+4}\right) \cos \lambda x\, d\lambda = \frac{\pi}{2} e^{-x}\cos x, \quad \text{if } x > 0.$

6. $\displaystyle\int_0^\infty \left(\frac{\lambda^3}{\lambda^4+4}\right) \sin \lambda x\, d\lambda = \frac{\pi}{2} e^{-x}\cos x, \quad \text{if } x > 0.$

7. $\displaystyle\int_0^\infty \frac{\cos \frac{\pi\lambda}{2} \cos \lambda x}{1-\lambda^2}\, d\lambda = \begin{cases} \dfrac{\pi}{2}\cos x &, \text{ if } |x| < \dfrac{\pi}{2} \\ 0 &, \text{ if } |x| > \dfrac{\pi}{2} \end{cases}.$

8. $\displaystyle\int_0^\infty \frac{\lambda \sin \lambda x}{(\lambda^2+\alpha^2)(\lambda^2+\beta^2)}\, d\lambda = \frac{\pi}{2}\left(\frac{e^{-\alpha x} - e^{-\beta x}}{\beta^2-\alpha^2}\right).$

9. Find Fourier sine integral representation of

$$f(x) = \begin{cases} 0, & 0 < x < 1 \\ k, & 1 < x < 2 \\ 0, & x > 2 \end{cases}$$

where k is a constant.

10. Find the Fourier integral representation for the following functions:

(*i*) $f(x) = \begin{cases} \dfrac{\pi}{2}\cos x, & |x| \leq \pi \\ 0, & |x| > \pi \end{cases}$ (*ii*) $f(x) = \begin{cases} 1, & |x| \leq a \\ 0, & |x| > a \end{cases}$

(*iii*) $f(x) = e^{-|x|}, -\infty < x < \infty.$

Answers

9. $f(x) = \dfrac{2k}{\pi}\displaystyle\int_0^\infty \left(\dfrac{\cos\lambda - \cos 2\lambda}{\lambda}\right)\sin\lambda x\, d\lambda$

10. (*i*) $f(x) = \displaystyle\int_0^\infty \dfrac{\lambda\sin\lambda\pi}{1-\lambda^2}\cos\lambda x\, d\lambda$ (*ii*) $f(x) = \dfrac{2}{\pi}\displaystyle\int_0^\infty \dfrac{\sin\lambda a\cos\lambda x}{\lambda}\, d\lambda$

(*iii*) $f(x) = \dfrac{2}{\pi}\displaystyle\int_0^\infty \dfrac{1}{1+\lambda^2}\cos\lambda x\, d\lambda.$

2.5. FOURIER TRANSFORMS AND INVERSION FORMULAE

(1) *Fourier sine transform and its inversion formula*

Fourier sine integral is $f(x) = \dfrac{2}{\pi}\displaystyle\int_0^\infty \sin\lambda x \int_0^\infty f(t)\sin\lambda t\, dt\, d\lambda$

Replacing λ by s, we get $f(x) = \dfrac{2}{\pi}\displaystyle\int_0^\infty \sin sx \int_0^\infty f(t)\sin st\, dt\, ds$

Denoting the value of the inner integral by $F_s(s)$, we have

$$f(x) = \frac{2}{\pi}\int_0^\infty F_s(s)\sin sx\, ds \qquad \ldots(1)$$

where

$$F_s(s) = \int_0^\infty f(x)\sin sx\, dx \qquad \ldots(2)$$

The function $F_s(s)$ as defined by equation (2) is known as the **Fourier sine transform** of $f(x)$ in $0 < x < \infty$.

The function $f(x)$ as defined by equation (1) is called the **inverse Fourier sine transform** of $F_s(s)$. Thus equation (1) gives the inversion formula for the Fourier sine transform.

Note. Some authors write the above formulae as:

$$F_s(s) \text{ or } F_s\{f(x)\} = \sqrt{\frac{2}{\pi}}\int_0^\infty f(x)\sin sx\, dx$$

and

$$f(x) = \sqrt{\frac{2}{\pi}}\int_0^\infty F_s(s)\sin sx\, ds$$

(2) *Fourier cosine transform and its inversion formula*

Fourier cosine integral is $f(x) = \dfrac{2}{\pi}\displaystyle\int_0^\infty \cos\lambda x \int_0^\infty f(t)\cos\lambda t\, dt\, d\lambda$

Replacing λ by s, we get $f(x) = \dfrac{2}{\pi}\displaystyle\int_0^\infty \cos sx \int_0^\infty f(t)\cos st\, dt\, ds$

Denoting the value of the inner integral by $F_c(s)$, we have

$$f(x) = \frac{2}{\pi}\int_0^\infty F_c(s)\cos sx\, ds \qquad \ldots(3)$$

where

$$F_c(s) = \int_0^\infty f(x)\cos sx\, dx \qquad \ldots(4)$$

The function $F_c(s)$ as defined by equation (4) is known as the **Fourier cosine transform** of $f(x)$ in $0 < x < \infty$.

The function $f(x)$ as defined by equation (3) is called the **inverse Fourier cosine transform** of $F_c(s)$. Thus equation (3) gives the inversion formula for the Fourier cosine transform.

Note. Some authors write the above formulae as:

$$F_c(s) \text{ or } F_c\{f(x)\} = \sqrt{\frac{2}{\pi}} \int_0^{\infty} f(x) \cos sx \, dx$$

and
$$f(x) = \sqrt{\frac{2}{\pi}} \int_0^{\infty} F_c(s) \cos sx \, ds$$

(3) *Complex Fourier transform and its inversion formula*

Complex form of Fourier integral is $f(x) = \dfrac{1}{2\pi} \displaystyle\int_{-\infty}^{\infty} e^{-i\lambda x} \int_{-\infty}^{\infty} f(t)\, e^{i\lambda t} \, dt \, d\lambda$

Replacing λ by s, we get $\quad f(x) = \dfrac{1}{2\pi} \displaystyle\int_{-\infty}^{\infty} e^{-isx} \int_{-\infty}^{\infty} f(t)\, e^{ist} \, dt \, ds$

Denoting the value of the inner integral by $F(s)$, we have

$$f(x) = \frac{1}{2\pi} \int_{-\infty}^{\infty} F(s)\, e^{-isx} \, ds \qquad \ldots(5)$$

where
$$F(s) = \int_{-\infty}^{\infty} f(x)\, e^{isx} \, dx \qquad \ldots(6)$$

The function $F(s)$ as defined by equation (6) is known as the **Fourier transform** of $f(x)$.

The function $f(x)$ as defined by equation (5) is called the **inverse Fourier transform** of $F(s)$. Thus equation (5) gives the inversion formula for the Fourier transform.

Note. Some authors write the above formulae as:

$$F(s) = \frac{1}{\sqrt{2\pi}} \int_{-\infty}^{\infty} f(x)\, e^{isx} \, dx$$

and
$$f(x) = \frac{1}{\sqrt{2\pi}} \int_{-\infty}^{\infty} F(s)\, e^{-isx} \, ds$$

(4) *Finite Fourier sine transform and its inversion formula*

The **finite Fourier sine transform** of $f(x)$ in $0 < x < c$ is defined as

$$F_s(n) = \int_0^c f(x) \sin \frac{n\pi x}{c} \, dx$$

where n is an integer.

The function $f(x)$ is then called the **inverse finite Fourier sine transform** of $F_s(n)$ and is given by

$$f(x) = \frac{2}{c} \sum_{n=1}^{\infty} F_s(n) \sin \frac{n\pi x}{c}.$$

(5) *Finite Fourier cosine transform and its inversion formula*

The **finite Fourier cosine transform** of $f(x)$ in $0 < x < c$ is defined as

$$F_c(n) = \int_0^c f(x) \cos \frac{n\pi x}{c} dx$$

where n is an integer.

The function $f(x)$ is then called the **inverse finite Fourier cosine transform** of $F_c(n)$ and is given by

$$f(x) = \frac{1}{c} F_c(0) + \frac{2}{c} \sum_{n=1}^{\infty} F_c(n) \cos \frac{n\pi x}{c}.$$

ILLUSTRATIVE EXAMPLES

Example 1. *Find the Fourier sine transform of* $e^{-|x|}$. *Hence evaluate* $\int_0^\infty \frac{x \sin mx}{1+x^2} dx$.

Sol. In the interval $(0, \infty)$, x is positive so that $e^{-|x|} = e^{-x}$. Fourier sine transform of $f(x) = e^{-x}$ is given by

$$F_s\{f(x)\} = \int_0^\infty f(x) \sin sx\, dx = \int_0^\infty e^{-x} \sin sx\, dx$$

$$= \left[\frac{e^{-x}}{1+s^2} (-\sin sx - s \cos sx) \right]_0^\infty = \frac{s}{1+s^2}$$

Using inversion formula for Fourier sine transform, we get

$$f(x) = \frac{2}{\pi} \int_0^\infty F_s\{f(x)\} \sin sx\, ds \quad \text{or} \quad e^{-x} = \frac{2}{\pi} \int_0^\infty \frac{s}{1+s^2} \sin sx\, ds$$

Replacing x by m, we have $e^{-m} = \frac{2}{\pi} \int_0^\infty \frac{s}{1+s^2} \sin ms\, ds = \frac{2}{\pi} \int_0^\infty \frac{x \sin mx}{1+x^2} dx$

Hence, $\int_0^\infty \frac{x \sin mx}{1+x^2} dx = \frac{\pi}{2} e^{-m}.$

Example 2. *Find the Fourier sine transform of*

$$f(x) = \begin{cases} x, & \text{for } 0 < x < 1 \\ 2 - x, & \text{for } 1 < x < 2. \\ 0, & \text{for } x > 2 \end{cases}$$

Sol. Fourier transform of $f(x)$ is

$$F_s\{f(x)\} = \int_0^\infty f(x) \sin sx\, dx$$

$$= \int_0^1 f(x)\sin sx\,dx + \int_1^2 f(x)\sin sx\,dx + \int_2^\infty f(x)\sin sx\,dx$$

$$= \int_0^1 x\sin sx\,dx + \int_1^2 (2-x)\sin sx\,dx + \int_2^\infty 0\,dx$$

$$= \left\{x.\frac{-\cos sx}{s} - 1.\frac{-\sin sx}{s^2}\right\}_0^1 + \left\{(2-x).\frac{-\cos sx}{s} - (-1).\frac{-\sin sx}{s^2}\right\}_1^2$$

$$= \left(-\frac{\cos s}{s} + \frac{\sin s}{s^2}\right) + \left(-\frac{\sin 2s}{s^2} + \frac{\cos s}{s} + \frac{\sin s}{s^2}\right)$$

$$= \frac{2\sin s - \sin 2s}{s^2} = \frac{2\sin s - 2\sin s\cos s}{s^2} = \frac{2\sin s\,(1-\cos s)}{s^2}.$$

Example 3. *Find the Fourier sine transform of* $\dfrac{e^{-ax}}{x}$.

Sol. Fourier sine transform of $f(x) = \dfrac{e^{-ax}}{x}$ is

$$F_s\{f(x)\} = \int_0^\infty f(x)\sin sx\,dx$$

$$= \int_0^\infty \frac{e^{-ax}}{x}\sin sx\,dx = I \qquad \text{(say)} \qquad \ldots(1)$$

Differentiating w.r.t. s, we have

$$\frac{dI}{ds} = \int_0^\infty \frac{\partial}{\partial s}\left(\frac{e^{-ax}}{x}\sin sx\right)dx$$

$$= \int_0^\infty \frac{e^{-ax}}{x}.\,x\cos sx\,dx = \int_0^\infty e^{-ax}\cos sx\,dx$$

$$= \left[\frac{e^{-ax}}{a^2+s^2}(-a\cos sx + s\sin sx)\right]_0^\infty$$

$$= 0 - \frac{1}{a^2+s^2}(-a) = \frac{a}{s^2+a^2}$$

Integrating w.r.t. s, we get

$$I = \tan^{-1}\frac{s}{a} + c \qquad \ldots(2)$$

Now, when $s = 0$, from (1), $I = 0$

$\therefore$ From (2), $0 = 0 + c \Rightarrow c = 0$

$$\therefore \qquad I = \tan^{-1}\frac{s}{a} \quad \text{or} \quad F_s\left\{\frac{e^{-ax}}{x}\right\} = \tan^{-1}\frac{s}{a}.$$

Example 4. *Find Fourier sine transform of* $\dfrac{1}{x(x^2+a^2)}$.

(M.D.U. May 2013, Dec. 2014)

Sol. Fourier sine transform of $f(x) = \dfrac{1}{x(x^2+a^2)}$ is

$$F_s\{f(x)\} = \int_0^\infty f(x)\sin sx\,dx$$

$$= \int_0^\infty \frac{\sin sx}{x(x^2+a^2)}\,dx = I \qquad \text{(say)} \quad \ldots(1)$$

Differentiating w.r.t. s, we have

$$\frac{dI}{ds} = \int_0^\infty \frac{x\cos sx}{x(x^2+a^2)}\,dx = \int_0^\infty \frac{\cos sx}{x^2+a^2}\,dx \quad \ldots(2)$$

Differentiating again w.r.t. s, we have

$$\frac{d^2I}{ds^2} = \int_0^\infty \frac{-x\sin sx}{x^2+a^2}\,dx = \int_0^\infty \frac{-x^2\sin sx}{x(x^2+a^2)}\,dx$$

$$= \int_0^\infty \frac{[a^2-(x^2+a^2)]\sin sx}{x(x^2+a^2)}\,dx = a^2\int_0^\infty \frac{\sin sx}{x(x^2+a^2)}\,dx - \int_0^\infty \frac{\sin sx}{x}\,dx$$

$$= a^2 I - \frac{\pi}{2}$$

$\Rightarrow$ $(D^2 - a^2)\,I = -\dfrac{\pi}{2}$, where $D = \dfrac{d}{ds}$

Its A.E. is $D^2 - a^2 = 0$ whence $D = \pm a$

$$C.F. = c_1e^{as} + c_2e^{-as}$$

$$P.I. = \frac{1}{D^2-a^2}\left(-\frac{\pi}{2}\right) = -\frac{\pi}{2}\cdot\frac{1}{D^2-a^2}e^{0s} = -\frac{\pi}{2}\cdot\frac{1}{-a^2} = \frac{\pi}{2a^2}$$

$\therefore$ $I = C.F. + P.I.$

$\Rightarrow$ $$I = c_1e^{as} + c_2e^{-as} + \frac{\pi}{2a^2} \quad \ldots(3)$$

$\therefore$ $$\frac{dI}{ds} = ac_1e^{as} - ac_2e^{-as} \quad \ldots(4)$$

When $s = 0$, from (1), $I = 0$

From (3), $I = c_1 + c_2 + \dfrac{\pi}{2a^2}$

$\therefore$ $$c_1 + c_2 + \frac{\pi}{2a^2} = 0 \quad \ldots(5)$$

When $s = 0$, from (2), $\frac{dI}{ds} = 1\left[\frac{1}{a}\tan^{-1}\frac{x}{a}\right]_0^\infty = \frac{\pi}{2a}$

From (4), $\frac{dI}{ds} = ac_1 - ac_2$

$$\therefore \quad ac_1 - ac_2 = \frac{\pi}{2a}$$

or $$c_1 - c_2 - \frac{\pi}{2a^2} = 0 \qquad \ldots(6)$$

Solving (5) and (6), $c_1 = 0, \quad c_2 = -\frac{\pi}{2a^2}$

$$\therefore \quad I = -\frac{\pi}{2a^2}e^{-as} + \frac{\pi}{2a^2}$$

or $$F_s\{f(x)\} = \frac{\pi}{2a^2}(1 - e^{-as}).$$

Example 5. *Find the Fourier sine transform of* $\frac{1}{x}$.

Sol. Fourier sine transform of $\frac{1}{x}$ is given by

$$F_s\left(\frac{1}{x}\right) = \int_0^\infty \frac{1}{x}\sin sx\,dx$$

Putting $sx = \theta$ *i.e.*, $x = \frac{\theta}{s}$, we have

$$F_s\left(\frac{1}{x}\right) = \int_0^\infty \frac{s}{\theta}\sin\theta\,\frac{d\theta}{s} = \int_0^\infty \frac{\sin\theta}{\theta}\,d\theta = \frac{\pi}{2}.$$

Example 6. *Find the Fourier sine and cosine transforms of* x^{n-1}, $n > 0$. (M.D.U. 2011)

Sol. We know that

$$F_s(x^{n-1}) = \int_0^\infty x^{n-1}\sin sx\,dx \qquad \ldots(1)$$

and $$F_c(x^{n-1}) = \int_0^\infty x^{n-1}\cos sx\,dx \qquad \ldots(2)$$

Now $$\Gamma(n) = \int_0^\infty e^{-t}\,t^{n-1}\,dt, \quad n > 0$$

Putting $t = ax$, $a > 0$, we have

$$\Gamma(n) = \int_0^\infty e^{-ax}(ax)^{n-1}\,a\,dx = a^n\int_0^\infty e^{-ax}\,x^{n-1}\,dx$$

$$\Rightarrow \qquad \int_0^\infty e^{-ax}\, x^{n-1}\, dx = \frac{\Gamma(n)}{a^n}$$

Putting $a = is$, we have

$$\int_0^\infty e^{-isx}\, x^{n-1}\, dx = \frac{\Gamma(n)}{(is)^n}$$

$$\Rightarrow \quad \int_0^\infty (\cos sx - i \sin sx)\, x^{n-1}\, dx = \left(\frac{i}{i^2}\right)^n \frac{\Gamma(n)}{s^n} = (-i)^n \frac{\Gamma(n)}{s^n}$$

$$= \left(\cos\frac{\pi}{2} - i\sin\frac{\pi}{2}\right)^n \frac{\Gamma(n)}{s^n}$$

$$= \left(\cos\frac{n\pi}{2} - i\sin\frac{n\pi}{2}\right) \frac{\Gamma(n)}{s^n}$$

Equating real and imaginary parts on both sides, we get

$$\int_0^\infty x^{n-1} \cos sx\, dx = \frac{\Gamma(n)}{s^n} \cos\frac{n\pi}{2}$$

and $$\int_0^\infty x^{n-1} \sin sx\, dx = \frac{\Gamma(n)}{s^n} \sin\frac{n\pi}{2}$$

$\therefore$ From (1), $F_s(x^{n-1}) = \dfrac{\Gamma(n)}{s^n} \sin\dfrac{n\pi}{2}$

and from (2), $F_c(x^{n-1}) = \dfrac{\Gamma(n)}{s^n} \cos\dfrac{n\pi}{2}$

Note. Taking $n = \dfrac{1}{2}$, we get

$$F_s\left(\frac{1}{\sqrt{x}}\right) = \frac{\Gamma\left(\frac{1}{2}\right)}{\sqrt{s}} \sin\frac{\pi}{4} = \frac{\sqrt{\pi}}{\sqrt{s}} \cdot \frac{1}{\sqrt{2}} \qquad \left[\because\ \Gamma\left(\frac{1}{2}\right) = \sqrt{\pi}\right]$$

$$= \sqrt{\frac{\pi}{2s}}$$

$$F_c\left(\frac{1}{\sqrt{x}}\right) = \frac{\Gamma\left(\frac{1}{2}\right)}{\sqrt{s}} \cos\frac{\pi}{4} = \sqrt{\frac{\pi}{2s}}.$$

Example 7. *Find the Fourier cosine transform of* e^{-x^2}.

Sol. Fourier cosine transform of e^{-x^2} is given by

$$F_c\{e^{-x^2}\} = \int_0^\infty e^{-x^2} \cos sx\, dx = \text{I (say)} \qquad \ldots(1)$$

Differentiating w.r.t. s, we have

$$\frac{dI}{ds} = -\int_0^\infty x e^{-x^2} \sin sx \, dx = \frac{1}{2}\int_0^\infty (\sin sx)(-2x\, e^{-x^2})\, dx$$

$$= \frac{1}{2}\left[\left\{\sin sx\, e^{-x^2}\right\}_0^\infty - s\int_0^\infty \cos sx\, e^{-x^2}\, dx\right]$$

(Integrating by parts)

$$= -\frac{s}{2}\int_0^\infty e^{-x^2} \cos sx\, dx = -\frac{s}{2} I$$

or $$\frac{dI}{I} = -\frac{s}{2}\, ds$$

Integrating, we have $\log I = -\dfrac{s^2}{4} + \log A$ or $I = A\, e^{-\frac{s^2}{4}}$ ...(2)

Now when $s = 0$, from (1), $I = \displaystyle\int_0^\infty e^{-x^2}\, dx = \frac{\sqrt{\pi}}{2}$

$\therefore$ From (2), $\dfrac{\sqrt{\pi}}{2} = A$

Hence $$I = F_c\{e^{-x^2}\} = \frac{\sqrt{\pi}}{2} e^{-\frac{s^2}{4}}.$$

Note. $I = \displaystyle\int_0^\infty e^{-x^2} dx$ Put $x^2 = t$ *i.e.,* $x = \sqrt{t}$

$$= \int_0^\infty \frac{e^{-t}}{2\sqrt{t}}\, dt = \frac{1}{2}\int_0^\infty t^{-1/2} e^{-t} dt = \frac{1}{2}\Gamma\left(\frac{1}{2}\right) = \frac{\sqrt{\pi}}{2}.$$

Example 8. *Find the Fourier cosine transform of*

$$f(x) = \begin{cases} \cos x, & 0 < x < a \\ 0, & x \ge a \end{cases}$$

Sol. Fourier cosine transform of $f(x)$ is

$$F_c\{f(x)\} = \int_0^\infty f(x) \cos sx\, dx = \int_0^a f(x)\cos sx\, dx + \int_a^\infty f(x) \cos sx\, dx$$

$$= \int_0^a \cos x \cos sx\, dx + \int_a^\infty 0\, dx = \frac{1}{2}\int_0^a 2\cos x \cos sx\, dx + 0$$

$$= \frac{1}{2}\int_0^a [\cos(1+s)x + \cos(1-s)x]\, dx$$

$$= \frac{1}{2}\left[\frac{\sin(1+s)x}{1+s} + \frac{\sin(1-s)x}{1-s}\right]_0^a = \frac{1}{2}\left[\frac{\sin(1+s)a}{1+s} + \frac{\sin(1-s)a}{1-s}\right].$$

Example 9. *Find the Fourier cosine transform of* $f(x) = \dfrac{1}{1+x^2}$. (M.D.U. 2011)

Hence derive Fourier sine transform of $\phi(x) = \dfrac{1}{x(1+x^2)}$.

Sol. Fourier cosine transform of $f(x) = \dfrac{1}{1+x^2}$ is

$$F_c\{f(x)\} = \int_0^\infty \frac{\cos sx}{1+x^2}\,dx = I \quad \text{(say)} \qquad \ldots(1)$$

Differentiating w.r.t. s, we have

$$\frac{dI}{ds} = \int_0^\infty \frac{-x\sin sx}{1+x^2}\,dx = \int_0^\infty \frac{-x^2\sin sx}{x(1+x^2)}\,dx$$

$$= \int_0^\infty \frac{[1-(1+x^2)]\sin sx}{x(1+x^2)}\,dx = \int_0^\infty \frac{\sin sx}{x(1+x^2)}\,dx - \int_0^\infty \frac{\sin sx}{x}\,dx$$

$$= \int_0^\infty \frac{\sin sx}{x(1+x^2)}\,dx - \frac{\pi}{2} \qquad \ldots(2)$$

Differentiating again w.r.t. s, we have

$$\frac{d^2I}{ds^2} = \int_0^\infty \frac{x\cos sx}{x(1+x^2)}\,dx = \int_0^\infty \frac{\cos sx}{1+x^2}\,dx = I$$

$\Rightarrow$ $(D^2-1)\,I = 0$, where $D = \dfrac{d}{ds}$

Its A.E. is $D^2 - 1 = 0$ whence $D = \pm 1$

$\therefore$ $I = c_1e^s + c_2e^{-s}$ ...(3)

$$\frac{dI}{ds} = c_1e^s - c_2e^{-s} \qquad \ldots(4)$$

When $s = 0$, from (1), $I = \displaystyle\int_0^\infty \frac{dx}{1+x^2} = \Big[\tan^{-1}x\Big]_0^\infty = \frac{\pi}{2}$

From (3), $I = c_1 + c_2$

$\therefore$ $c_1 + c_2 = \dfrac{\pi}{2}$ ...(5)

When $s = 0$, from (2), $\dfrac{dI}{ds} = -\dfrac{\pi}{2}$

From (4), $\dfrac{dI}{ds} = c_1 - c_2$

$\therefore$ $c_1 - c_2 = -\dfrac{\pi}{2}$ ...(6)

Solving (5) and (6), $c_1 = 0, \quad c_2 = \dfrac{\pi}{2}$

$$\therefore \qquad I = \frac{\pi}{2} e^{-s} \Rightarrow F_c\{f(x)\} = \frac{\pi}{2} e^{-s}$$

Putting the values of c_1 and c_2 in (4)

$$\frac{dI}{ds} = -\frac{\pi}{2} e^{-s}$$

$$\therefore \text{ From (2), } \quad -\frac{\pi}{2} e^{-s} = \int_0^\infty \frac{\sin sx}{x(1+x^2)} dx - \frac{\pi}{2}$$

$$\Rightarrow \qquad \int_0^\infty \frac{\sin sx}{x(1+x^2)} dx = \frac{\pi}{2} (1 - e^{-s})$$

$$\Rightarrow \qquad F_s\{\phi(x)\} = \frac{\pi}{2} (1 - e^{-s}).$$

Note. In example 9 above, we have proved that

$$\int_0^\infty \frac{\cos sx}{1+x^2} dx = \frac{\pi}{2} e^{-s}$$

Differentiating w.r.t. s, we get

$$\int_0^\infty \frac{\partial}{\partial s}\left(\frac{\cos sx}{1+x^2}\right) dx = -\frac{\pi}{2} e^{-s}$$

$$\Rightarrow \qquad \int_0^\infty \frac{-x \sin sx}{1+x^2} dx = -\frac{\pi}{2} e^{-s}$$

$$\Rightarrow \qquad \int_0^\infty \frac{x}{1+x^2} \sin sx \, dx = \frac{\pi}{2} e^{-s}$$

$$\Rightarrow \qquad F_s\{\psi(x)\} = \frac{\pi}{2} e^{-s} \qquad \text{where } \psi(x) = \frac{x}{1+x^2}$$

$\Rightarrow$ Fourier sine transform of $\dfrac{x}{1+x^2}$ is $\dfrac{\pi}{2} e^{-s}$.

Example 10. *Find the Fourier transform of*

$$f(x) = \begin{cases} 1, & |x| < a \\ 0, & |x| > a \end{cases} \qquad \text{(M.D.U. 2011)}$$

Hence evaluate

(*i*) $\displaystyle\int_{-\infty}^{\infty} \frac{\sin as \cos sx}{s} ds$ (*ii*) $\displaystyle\int_0^\infty \frac{\sin x}{x} dx = \frac{\pi}{2}$. (M.D.U. Dec. 2013)

Sol. Fourier transform of $f(x)$ is given by

$$F\{f(x)\} = \int_{-\infty}^{\infty} f(x)\, e^{isx}\, dx$$

$$= \int_{-\infty}^{-a} f(x)\, e^{isx}\, dx + \int_{-a}^{a} f(x)\, e^{isx\,dx} + \int_{a}^{\infty} f(x)\, e^{isx}\, dx$$

$$= \int_{-\infty}^{-a} 0\,.\, e^{isx}\, dx + \int_{-a}^{a} 1\,.\, e^{isx}\, dx + \int_{a}^{\infty} 0\,.\, e^{isx}\, dx$$

$$= \left[\frac{e^{isx}}{is}\right]_{-a}^{a} = \frac{e^{ias} - e^{-ias}}{is}$$

$$= \frac{2}{s}\left(\frac{e^{ias} - e^{-ias}}{2i}\right) = \frac{2\sin as}{s}, \qquad | \; s \neq 0$$

For $s = 0$, we find $F\{f(x)\} = 2a$

By inversion formula for Fourier transform, we have

$$f(x) = \frac{1}{2\pi}\int_{-\infty}^{\infty} F\{f(x)\}.\, e^{-isx}\, ds$$

$$= \frac{1}{2\pi}\int_{-\infty}^{\infty} \frac{2\sin as}{s}.\, e^{-isx}\, ds$$

$$= \frac{1}{\pi}\int_{-\infty}^{\infty} \frac{\sin as\,(\cos sx - i\sin sx)}{s}\, ds$$

$$= \frac{1}{\pi}\int_{-\infty}^{\infty} \frac{\sin as\cos sx}{s}\, ds - \frac{i}{\pi}\int_{-\infty}^{\infty} \frac{\sin as\sin sx}{s}\, ds$$

$$= \frac{1}{\pi}\int_{-\infty}^{\infty} \frac{\sin as\cos sx}{s}\, ds$$

[Second integral vanishes since the integrand is an odd function of s]

$$\Rightarrow \quad \int_{-\infty}^{\infty} \frac{\sin as\cos sx}{s}\, ds = \pi\, f(x) = \begin{cases} \pi, & |x| < a \\ 0, & |x| > a \end{cases} \qquad \ldots(1)$$

Since the integrand in (1) is an even function of s, we have

$$2\int_{0}^{\infty} \frac{\sin as\cos sx}{s}\, ds = \begin{cases} \pi, & |x| < a \\ 0, & |x| > a \end{cases}$$

$$\Rightarrow \quad \int_{0}^{\infty} \frac{\sin as\cos sx}{s}\, ds = \begin{cases} \pi/2, & |x| < a \\ 0, & |x| > a \end{cases}$$

Putting $x = 0$, $$\int_0^\infty \frac{\sin as}{s}\, ds = \begin{cases} \frac{\pi}{2}, & a > 0 \\ 0, & a < 0 \end{cases}$$

Putting $a = 1$, $$\int_0^\infty \frac{\sin s}{s}\, ds = \frac{\pi}{2}$$

$$\Rightarrow \qquad \int_0^\infty \frac{\sin x}{x}\, dx = \frac{\pi}{2}$$

Example 11. *Find the Fourier transform of* $f(x) = \begin{cases} 1 - x^2, & \text{if } |x| < 1 \\ 0, & \text{if } |x| > 1 \end{cases}$

and use it to evaluate $\int_0^\infty \left(\frac{x \cos x - \sin x}{x^3} \right) \cos \frac{x}{2}\, dx.$ (M.D.U. 2010, May 2013)

Sol. Fourier transform of $f(x)$ is given by

$$F\{f(x)\} = \int_{-\infty}^{\infty} f(x) e^{isx}\, dx$$

$$= \int_{-\infty}^{-1} f(x) e^{isx} dx + \int_{-1}^{1} f(x) e^{isx}\, dx + \int_1^\infty f(x) e^{isx}\, dx$$

$$= \int_{-\infty}^{-1} 0\, dx + \int_{-1}^{1} (1 - x^2) e^{isx}\, dx + \int_1^\infty 0\, dx$$

$$= \left[(1 - x^2) . \frac{e^{isx}}{is} - (-2x) \frac{e^{isx}}{(is)^2} + (-2) \frac{e^{isx}}{(is)^3} \right]_{-1}^{1}$$

$$= \left[-\frac{2}{s^2} (e^{is} + e^{-is}) + \frac{2}{is^3} (e^{is} - e^{-is}) \right]$$

$$= \left[-\frac{4}{s^2} \left(\frac{e^{is} + e^{-is}}{2} \right) + \frac{4}{s^3} \left(\frac{e^{is} - e^{-is}}{2i} \right) \right]$$

$$= 4 \left[-\frac{\cos s}{s^2} + \frac{\sin s}{s^3} \right] = -\frac{4}{s^3} (s \cos s - \sin s)$$

By inversion formula for Fourier transform, we have

$$f(x) = \frac{1}{2\pi} \int_{-\infty}^{\infty} F\{f(x)\} . e^{-isx}\, ds$$

$$= -\frac{2}{\pi} \int_{-\infty}^{\infty} \left(\frac{s \cos s - \sin s}{s^3} \right) (\cos sx - i \sin sx)\, ds$$

$$= -\frac{2}{\pi}\int_{-\infty}^{\infty}\left(\frac{s\cos s - \sin s}{s^3}\right)\cos sx\, ds + \frac{2i}{\pi}\int_{-\infty}^{\infty}\left(\frac{s\cos s - \sin s}{s^3}\right)\sin sx\, ds$$

$$= -\frac{4}{\pi}\int_{0}^{\infty}\left(\frac{s\cos s - \sin s}{s^3}\right)\cos sx\, ds$$

(Since the integrand in the first integral is even and that in the second integral is odd)

$$\Rightarrow \quad \int_{0}^{\infty}\left(\frac{s\cos s - \sin s}{s^3}\right)\cos sx\, ds = -\frac{\pi}{4}f(x) = \begin{cases} -\frac{\pi}{4}(1-x^2), & \text{if } |x|<1 \\ 0, & \text{if } |x|>1 \end{cases}$$

Putting $x = \frac{1}{2}$, we have $\int_{0}^{\infty}\left(\frac{s\cos s - \sin s}{s^3}\right)\cos\frac{s}{2}\, ds = -\frac{\pi}{4}\left(1-\frac{1}{4}\right) = -\frac{3\pi}{16}$

Hence $\quad \int_{0}^{\infty}\left(\frac{x\cos x - \sin x}{x^3}\right)\cos\frac{x}{2}\, dx = -\frac{3\pi}{16}$.

Example 12. *Find the inverse Fourier transform of $F(s) = e^{-|s|y}$.*

Sol. The inverse Fourier transform of $F(s) = e^{-|s|y}$ is given by

$$f(x) = \frac{1}{2\pi}\int_{-\infty}^{\infty} F(s)\, e^{-isx}\, ds - \frac{1}{2\pi}\int_{-\infty}^{\infty} e^{-|s|y} . e^{-isx}\, ds$$

$$= \frac{1}{2\pi}\left[\int_{-\infty}^{0} e^{sy} . e^{-isx}\, ds + \int_{0}^{\infty} e^{-sy} . e^{-isx}\, ds\right]$$

$$= \frac{1}{2\pi}\left[\int_{-\infty}^{0} e^{s(y-ix)}\, ds + \int_{0}^{\infty} e^{-s(y+ix)}\, ds\right]$$

$$= \frac{1}{2\pi}\left[\left\{\frac{e^{s(y-ix)}}{y-ix}\right\}_{-\infty}^{0} + \left\{\frac{e^{-s(y+ix)}}{-(y+ix)}\right\}_{0}^{\infty}\right]$$

$$= \frac{1}{2\pi}\left[\frac{1}{y-ix} + \frac{1}{y+ix}\right] = \frac{1}{2\pi}\left(\frac{2y}{y^2+x^2}\right)$$

$$= \frac{y}{\pi(y^2+x^2)}.$$

Example 13. *Solve the integral equation* $\int_{0}^{\infty} f(x)\cos px\, dx = \begin{cases} 1-p, & 0 \le p \le 1 \\ 0, & p>1 \end{cases}$

Hence deduce that $\quad \int_{0}^{\infty}\frac{\sin^2 t}{t^2}dt = \frac{\pi}{2}$. (K.U.K. May 2013)

Sol. Here $\quad \int_{0}^{\infty} f(x)\cos px\, dx = F_c(p)$

$$\therefore \quad F_c(p) = \begin{cases} 1-p, & 0 \le p \le 1 \\ 0, & p>1 \end{cases}$$

By inversion formula for Fourier cosine transform, we have

$$f(x) = \frac{2}{\pi}\int_0^\infty F_c(p)\cos px\, dp$$

$$= \frac{2}{\pi}\left[\int_0^1 F_c(p)\cos px\, dp + \int_1^\infty F_c(p)\cos px\, dp\right]$$

$$= \frac{2}{\pi}\left[\int_0^1 (1-p)\cos px\, dp + \int_1^\infty 0\, dp\right] \qquad \text{(Integrating by parts)}$$

$$= \frac{2}{\pi}\left[(1-p).\frac{\sin px}{x} - (-1).\frac{-\cos px}{x^2}\right]_0^1 = \frac{2}{\pi}\left[-\frac{\cos x}{x^2} + \frac{1}{x^2}\right] = \frac{2(1-\cos x)}{\pi x^2}.$$

Deduction. Since $\int_0^\infty f(x)\cos px\, dx = \begin{cases} 1-p, & 0 \le p \le 1 \\ 0, & p > 1 \end{cases}$, where $f(x) = \dfrac{2(1-\cos x)}{\pi x^2}$

$$\therefore \qquad \frac{2}{\pi}\int_0^\infty \left(\frac{1-\cos x}{x^2}\right)\cos px\, dx = \begin{cases} 1-p, & 0 \le p \le 1 \\ 0, & p > 1 \end{cases}$$

When $p = 0$, we have $\dfrac{2}{\pi}\displaystyle\int_0^\infty \frac{1-\cos x}{x^2} dx = 1$ or $\displaystyle\int_0^\infty \frac{2\sin^2\frac{x}{2}}{x^2} dx = \frac{\pi}{2}$

Putting $\dfrac{x}{2} = t$ so that $dx = 2dt$, we have $\displaystyle\int_0^\infty \frac{\sin^2 t}{t^2} = \frac{\pi}{2}$.

Example 14. *Find the finite Fourier sine transform of*

$$f(x) = \begin{cases} \dfrac{2k}{l}x, & 0 \le x \le \dfrac{l}{2} \\ \dfrac{2k}{l}(l-x), & \dfrac{l}{2} \le x \le l \end{cases}.$$

Sol. Finite Fourier sine transform of $f(x)$ in $0 \le x \le l$ is

$$F_s(n) = \int_0^l f(x)\sin\frac{n\pi x}{l} dx$$

$$= \int_0^{l/2} f(x)\sin\frac{n\pi x}{l} dx + \int_{l/2}^l f(x)\sin\frac{n\pi x}{l} dx$$

$$= \int_0^{l/2} \frac{2k}{l} x\sin\frac{n\pi x}{l} dx + \int_{l/2}^l \frac{2k}{l}(l-x)\sin\frac{n\pi x}{l} dx$$

$$= \frac{2k}{l}\int_0^{l/2} x\sin\frac{n\pi x}{l} dx + \frac{2k}{l}\int_{l/2}^l (l-x)\sin\frac{n\pi x}{l} dx \qquad \text{(Integrating by parts)}$$

$$= \frac{2k}{l}\left[x.\frac{-\cos\frac{n\pi x}{l}}{\frac{n\pi}{l}} - 1.\frac{-\sin\frac{n\pi x}{l}}{\left(\frac{n\pi}{l}\right)^2}\right]_0^{l/2} + \frac{2k}{l}\left[(l-x).\frac{-\cos\frac{n\pi x}{l}}{\frac{n\pi}{l}} - (-1).\frac{-\sin\frac{n\pi x}{l}}{\left(\frac{n\pi}{l}\right)^2}\right]_{l/2}^{l}$$

$$= \frac{2k}{l}\left[-\frac{l^2}{2n\pi}\cos\frac{n\pi}{2}+\frac{l^2}{n^2\pi^2}\sin\frac{n\pi}{2}\right]+\frac{2k}{l}\left[\frac{l^2}{2n\pi}\cos\frac{n\pi}{2}+\frac{l^2}{n^2\pi^2}\sin\frac{n\pi}{2}\right]$$

$$= \frac{2k}{l}\cdot\frac{2l^2}{n^2\pi^2}\sin\frac{n\pi}{2} = \frac{4kl}{n^2\pi^2}\sin\frac{n\pi}{2}.$$

Example 15. *Find the finite Fourier cosine transform of*

$$f(x) = \left(1-\frac{x}{\pi}\right)^2, \ 0 \le x < \pi.$$

Sol. Finite Fourier cosine transform of $f(x)$ in $0 \le x \le \pi$ is

$$F_c(n) = \int_0^\pi f(x)\cos\frac{n\pi x}{\pi}\,dx = \int_0^\pi \left(1-\frac{x}{\pi}\right)^2 \cos nx\,dx \qquad \text{(Integrating by parts)}$$

$$= \left[\left(1-\frac{x}{\pi}\right)^2 \cdot \frac{\sin nx}{n} - 2\left(1-\frac{x}{\pi}\right)\left(-\frac{1}{\pi}\right)\cdot\frac{-\cos nx}{n^2} + 2\left(-\frac{1}{\pi}\right)\left(-\frac{1}{\pi}\right)\cdot\frac{-\sin nx}{n^3}\right]_0^\pi$$

$$= \frac{2}{\pi n^2} \text{ when } n \ne 0.$$

If $n = 0$, then $F_c(0) = \int_0^\pi \left(1-\frac{x}{\pi}\right)^2 dx$ $\qquad [\because \cos 0x = 1]$

$$= \left[\frac{\left(1-\frac{x}{\pi}\right)^3}{3x\left(-\frac{1}{\pi}\right)}\right]_0^\pi = -\frac{\pi}{3}(0-1) = \frac{\pi}{3}$$

$$\therefore \quad F_c(n) = \begin{cases} \dfrac{2}{\pi n^2}, & \text{if } n = 1, 2, 3, \ldots \\ \dfrac{\pi}{3}, & \text{if } n = 0. \end{cases}$$

Example 16. *Find $f(x)$, if $F_s(n) = \dfrac{1-\cos n\pi}{n^2\pi^2}$, where $0 \le x \le \pi$.*

Sol. Here $F_s(n) = \dfrac{1-\cos n\pi}{n^2\pi^2}$ is the finite Fourier sine transform of $f(x)$ in $0 \le x \le \pi$.

$\therefore \qquad f(x)$ = inverse finite Fourier sine transform of $F_s(n)$

$$= \frac{2}{c}\sum_{n=1}^{\infty} F_s(n)\sin\frac{n\pi x}{c} = \frac{2}{\pi}\sum_{n=1}^{\infty}\left(\frac{1-\cos n\pi}{n^2\pi^2}\right)\sin nx \qquad (\because \text{ here } c = \pi)$$

$$= \frac{2}{\pi^3}\sum_{n=1}^{\infty}\left(\frac{1-\cos n\pi}{n^2}\right)\sin nx.$$

EXERCISE 2.2

1. Find the Fourier sine and cosine transform of the function $f(x) = e^{-x}$ and hence show that

$$\int_0^\infty \frac{x \sin mx}{1+x^2}\,dx = \frac{\pi}{2}e^{-m} \quad \text{and} \quad \int_0^\infty \frac{\cos mx}{1+x^2}\,dx = \frac{\pi}{2}e^{-m}.$$

2. Find the Fourier sine transform of

(i) $f(x) = \begin{cases} \sin x, & 0 < x < a \\ 0, & x > a \end{cases}$ (ii) $f(x) = \begin{cases} 0, & 0 \le x < a \\ x, & a \le x \le b \\ 0, & x > b \end{cases}$

3. Find the Fourier cosine transform of

(i) $f(x) = e^{-x} + e^{-2x}, x > 0$ (ii) $f(x) = e^{-\frac{x^2}{2}}$

(iii) $f(x) = \begin{cases} x, & \text{for } 0 < x < 1 \\ 2-x, & \text{for } 1 < x < 2 \\ 0, & \text{for } x > 2 \end{cases}$ (iv) $f(x) = 2e^{-5x}, + 5e^{-2x}$

(v) $f(x) = \begin{cases} x, & 0 < x < \frac{1}{2} \\ 1-x, & \frac{1}{2} < x < 1 \\ 0, & x > 1 \end{cases}$ (vi) $\left(1 - \frac{x}{\pi}\right)^2$

4. (a) Find the Fourier sine and cosine transforms of

(i) $f(x) = \begin{cases} 1, & \text{for } 0 \le x < a \\ 0, & \text{for } x \ge a \end{cases}$ (ii) $f(x) = e^{-ax}, a > 0$ *(M.D.U. 2012)*

(M.D.U. Dec. 2013, May 2014)

[**Note.** Remember the results of Q. 4.]

(b) Find the Fourier cosine transform of $f(x) = e^{-ax}$, $a > 0$ and hence find the Fourier sine transform of $x\,e^{-ax}$. *(K.U.K. Dec. 2015)*

[**Hint.** $F_c(e^{-ax}) = \dfrac{a}{s^2 + a^2} \Rightarrow \int_0^\infty e^{-ax} \cos sx\,dx = \dfrac{a}{s^2+a^2}$

Differentiating w.r.t. s, we get $\int_0^\infty \frac{\partial}{\partial s}(e^{-ax} \cos sx)\,dx = \dfrac{-2as}{(s^2+a^2)^2}$

$$\Rightarrow \int_0^\infty e^{-ax}(-x \sin sx)\,dx = \frac{-2as}{(s^2+a^2)^2} \Rightarrow \int_0^\infty (xe^{-ax}) \sin sx\,dx = \frac{2as}{(s^2+a^2)^2}$$

$$\therefore \quad F_s(x\,e^{-ax}) = \frac{2as}{(s^2+a^2)^2}$$]

5. Find the Fourier transforms of the following functions:

(i) $f(x) = \begin{cases} x, & |x| < a \\ 0, & |x| > a \end{cases}$ (ii) $f(x) = \begin{cases} x^2, & |x| < a \\ 0, & |x| > a \end{cases}$ *(K.U.K. Jan. 2013)*

(iii) $f(x) = e^{-|x|}$ (iv) $f(x) = e^{-\frac{x^2}{2}}, -\infty < x < \infty.$

(K.U.K. 2010; M.D.U. Dec. 2013)

(v) $f(x) = \begin{cases} e^{ikx}, & a < x < b \\ 0, & x < a \text{ and } x > b \end{cases}$ (vi) $f(x) = \begin{cases} 0, & x < \alpha \\ 1, & \alpha < x < \beta \\ 0, & x > \beta \end{cases}$

6. (*i*) Find the Fourier transform of $f(x) = \begin{cases} 1 & \text{for } |x| < 1 \\ 0 & \text{for } |x| > 1 \end{cases}$

Hence evaluate $\int_0^{\infty} \frac{\sin x}{x} dx$.

(*ii*) Find the Fourier transform of the function

$$f(x) = \begin{cases} 1 + \frac{x}{a}, & \text{for } -a < x < 0 \\ 1 - \frac{x}{a}, & \text{for } \quad 0 < x < a \\ 0, & \text{otherwise} \end{cases}$$

(*iii*) Find the Fourier transform of the function

$$f(t) = \begin{cases} 1\,, & \text{for } -2 < t < -1 \\ 2\,, & \text{for } -1 < t < 1 \\ 1\,, & \text{for } \quad 1 < t < 2 \\ 0\,, & \text{otherwise} \end{cases}$$

7. Using inverse Fourier sine transform, find $f(x)$ if

(*i*) $F_s(\lambda) = \frac{1}{\lambda} e^{-a\lambda}$ (*ii*) $F_s(\lambda) = \frac{\lambda}{1 + \lambda^2}$.

8. Find the finite Fourier sine and cosine transforms of the following functions:

(*i*) $f(x) = 2x,\ 0 \le x \le 4$ (*ii*) $f(x) = x(l - x),\ 0 \le x \le l$

(*M.D.U. May 2013, Dec. 2014*)

(*iii*) $f(x) = x^2,\ 0 \le x \le 2$ (*iv*) $f(x) = a\left(1 - \frac{x}{l}\right),\ 0 \le x \le l$

(*v*) $f(x) = \begin{cases} kx\,, & 0 \le x \le \frac{\pi}{2} \\ k(\pi - x)\,, & \frac{\pi}{2} \le x \le \pi \end{cases}$.

9. Find the finite Fourier sine transform of the following functions:

(*i*) $f(x) = \cos x,\ 0 \le x \le \pi$ (*ii*) $f(x) = \begin{cases} \frac{2x}{3}, & 0 \le x \le \frac{\pi}{3} \\ \frac{\pi - x}{3}, & \frac{\pi}{3} \le x \le \pi \end{cases}$.

10. If $f(x) = \sin kx$, where $0 \le x \le \pi$ and k is a positive integer, show that

$$F_s(n) = \begin{cases} 0, & \text{if } n \ne k \\ \frac{\pi}{2}, & \text{if } n = k \end{cases}.$$

11. Find the finite Fourier cosine transform of the following functions:

(*i*) $f(x) = \sin x,\ 0 \le x \le \pi$ (*ii*) $f(x) = \begin{cases} kx\,, & 0 \le x \le \frac{l}{2} \\ k(l - x)\,, & \frac{l}{2} \le x \le l \end{cases}$.

12. Find $f(x)$, if

(*i*) $F_s(n) = \dfrac{2l}{\pi n} \sin^2 \dfrac{n\pi}{4}$, $0 \le x \le l$ (*ii*) $F_s(n) = \dfrac{1}{n^2} \sin \dfrac{n\pi}{3}$, $0 \le x \le \pi$

(*iii*) $F_s(n) = \dfrac{2l^3}{n^3\pi^3} (1 - \cos n\pi)$, $0 \le x \le l$

(*iv*) $F_c(n) = \dfrac{cl}{n\pi} \sin \dfrac{n\pi a}{l}$ and $F_c(0) = ac$, where $0 \le x \le l$

(*v*) $F_c(n) = -\dfrac{l^3}{n^2\pi^2} (1 + \cos n\pi)$ and $F_c(0) = \dfrac{l^3}{6}$, where $0 \le x \le l$

(*vi*) $F_c(n) = \dfrac{\cos\left(\dfrac{2n\pi}{3}\right)}{(2n+1)^2}$, where $0 \le x \le 1$.

13. Solve the following integral equations:

(*i*) $\displaystyle\int_0^\infty f(x) \sin px \, dx = \begin{cases} 1, & 0 \le p < 1 \\ 2, & 1 \le p < 2 \\ 0, & p \ge 2 \end{cases}$ (*M.D.U. 2012*)

(*ii*) $\displaystyle\int_0^\infty f(x) \cos \lambda x \, dx = e^{-\lambda}, \lambda > 0$ (*DCRUST Murthal May 2014; M.D.U. 2011*)

(*iii*) $\displaystyle\int_0^\infty f(x) \sin \lambda x \, dx = \begin{cases} 1 - \lambda, & 0 \le \lambda \le 1 \\ 0, & \lambda \ge 1 \end{cases}$ (*iv*) $\displaystyle\int_0^\infty f(x) \sin \lambda x \, dx = \frac{\lambda}{\lambda^2 + k^2}$.

Answers

2. (*i*) $\dfrac{1}{2}\left[\dfrac{\sin(1-s)a}{1-s} - \dfrac{\sin(1+s)a}{1+s}\right]$ (*ii*) $\dfrac{a \cos sa - b \cos sb}{s} + \dfrac{\sin sb - \sin sa}{s^2}$

3. (*i*) $\dfrac{6 + 3s^2}{4 + 5s^2 + s^4}$ (*ii*) $\sqrt{\dfrac{\pi}{2}}\, e^{-\frac{s^2}{2}}$

(*iii*) $\dfrac{2 \cos s(1 - \cos s)}{s^2}$ (*iv*) $10\left(\dfrac{1}{s^2 + 25} + \dfrac{1}{s^2 + 4}\right)$

(*v*) $\dfrac{1}{s^2}\left(2 \cos \dfrac{s}{2} - 1 - \cos s\right)$ (*vi*) $\dfrac{2}{\pi s^2}$

4. (*i*) $\dfrac{1 - \cos as}{s}, \dfrac{\sin as}{s}$ (*ii*) $\dfrac{s}{s^2 + a^2}, \dfrac{a}{s^2 + a^2}$

5. (*i*) $2i\left(\dfrac{\sin as - as \cos as}{s^2}\right)$ (*ii*) $\dfrac{2}{s^3}\,[(a^2s^2 - 2) \sin as + 2as \cos as]$

(*iii*) $\dfrac{2}{1 + s^2}$ (*iv*) $\sqrt{2\pi}\, e^{-\frac{s^2}{2}}$

(*v*) $\dfrac{i}{(k+s)}\,[e^{i(k+s)a} - e^{i(k+s)b}]$ (*vi*) $\dfrac{i}{s}\,(e^{is\alpha} - e^{is\beta})$

6. (*i*) $\dfrac{2 \sin s}{s}$; $\dfrac{\pi}{2}$ (*ii*) $\dfrac{2}{as^2}(1 - \cos as)$

(*iii*) $\dfrac{2}{s} \sin s(1 + 2 \cos s)$

7. (*i*) $f(x) = \frac{2}{\pi} \tan^{-1} \frac{x}{a}$ (*ii*) $f(x) = e^{-x}$

8. (*i*) $-\frac{32}{n\pi} \cos n\pi = -\frac{32(-1)^n}{n\pi}$; $\frac{32[(-1)^n - 1]}{n^2\pi^2}$, $F_c(0) = 16$

(*ii*) $\frac{2l^3}{n^3\pi^3} [1 - (-1)^n]$; $-\frac{l^3}{n^2\pi^2} [(-1)^n + 1]$, $F_c(0) = \frac{l^3}{6}$

(*iii*) $\frac{16}{n^3\pi^3} [(-1)^n - 1] - \frac{8}{n\pi} (-1)^n$; $\frac{16}{n^2\pi^2} (-1)^n$, $F_c(0) = \frac{8}{3}$

(*iv*) $\frac{al}{n\pi}$; $\frac{al}{n^2\pi^2} [1 - (-1)^n]$, $F_c(0) = \frac{al}{2}$

(*v*) $\frac{2k}{n^2} \sin \frac{n\pi}{2}$; $\frac{k}{n^2} \left[2 \cos \frac{n\pi}{2} - (-1)^n - 1\right]$, $F_c(0) = \frac{k\pi^2}{4}$

9. (*i*) $\frac{n}{n^2 - 1} [1 + (-1)^n]$ (*ii*) $\frac{1}{n^2} \sin \frac{n\pi}{3}$

11. (*i*) $-\frac{1}{n^2 - 1} [1 + (-1)^n]$, $F_c(0) = 2$ (*ii*) $\frac{kl^2}{n^2\pi^2} \left[2 \cos \frac{n\pi}{2} - 1 - (-1)^n\right]$, $F_c(0) = \frac{kl^2}{4}$

12. (*i*) $f(x) = \frac{4}{\pi} \sum_{n-1}^{\infty} \frac{1}{n} \sin^2 \frac{n\pi}{4} \sin \frac{n\pi x}{l}$ (*ii*) $f(x) = \frac{2}{\pi} \sum_{n=1}^{\infty} \frac{1}{n^2} \sin \frac{n\pi}{3} \sin nx$

(*iii*) $f(x) = \frac{4l^2}{\pi^3} \sum_{n=1}^{\infty} \frac{1}{n^3} (1 - \cos n\pi) \sin \frac{n\pi x}{l}$ (*iv*) $f(x) = \frac{ac}{l} + \frac{2c}{\pi} \sum_{n=1}^{\infty} \frac{1}{n} \sin \frac{n\pi a}{l} \cos \frac{n\pi x}{l}$

(*v*) $f(x) = \frac{l^2}{6} - \frac{2l^2}{\pi^2} \sum_{n=1}^{\infty} \frac{1}{n^2} (1 + \cos n\pi) \cos \frac{n\pi x}{l}$

(*vi*) $f(x) = 1 + 2 \sum_{n=1}^{\infty} \frac{\cos \frac{2n\pi}{3}}{(2n + 1)^2} \cos n\pi x$

13. (*i*) $f(x) = \frac{2}{\pi x} (1 + \cos x - 2 \cos 2x)$ (*ii*) $f(x) = \frac{2}{\pi(1 + x^2)}$

(*iii*) $f(x) = \frac{2}{\pi} \left(\frac{x - \sin x}{x^2}\right)$ (*iv*) $f(x) = e^{-kx}$, $x > 0$.

2.6. PROPERTIES OF FOURIER TRANSFORMS

1. Linearity Property. *If F(s) and G(s) are Fourier transforms of f(x) and g(x) respectively, then*

$$\mathbf{F[af(x) + bg(x)] = aF(s) + bG(s)}$$

where a and b are constants.

Proof. By definition of Fourier transform, we have

$$F(s) = F\{f(x)\} = \int_{-\infty}^{\infty} f(x) \, . \, e^{isx} \, dx$$

and

$$G(s) = F\{g(x)\} = \int_{-\infty}^{\infty} g(x) \, . \, e^{isx} \, dx$$

$$\therefore \quad F[af(x) + bg(x)] = \int_{-\infty}^{\infty} [af(x) + bg(x)]\, e^{isx} dx$$

$$= a\int_{-\infty}^{\infty} f(x) . e^{isx} dx + b\int_{-\infty}^{\infty} g(x) . e^{isx}\, dx = aF(s) + bG(s).$$

Cor. (*i*) If $F_s(s)$ and $G_s(s)$ are the Fourier sine transforms of $f(x)$ and $g(x)$ respectively, then

$$F_s[af(x) + bg(x)] = aF_s(s) + bG_s(s)$$

where a and b are constants.

(*ii*) If $F_c(s)$ and $G_c(s)$ are the Fourier cosine transforms of $f(x)$ and $g(x)$ respectively, then

$$F_c[af(x) + bg(x)] = aF_c(s) + bG_c(s)$$

where a and b are constants.

2. Change of scale property (Similarity Theorem). *If F(s) is the complex Fourier transform of f(x), then*

$$\mathbf{F[f(ax)] = \frac{1}{a}F\left(\frac{s}{a}\right), a \neq 0.}$$

Proof. By definition of complex Fourier transform, we have

$$F(s) = F\{f(x)\} = \int_{-\infty}^{\infty} f(x) \cdot e^{isx}\, dx \qquad \ldots(1)$$

$$\therefore \quad F[f(ax)] = \int_{-\infty}^{\infty} f(ax) . e^{isx}\, dx \qquad \ldots(2)$$

Putting $ax = t$ *i.e.*, $x = \dfrac{t}{a}$, we have $dx = \dfrac{dt}{a}$.

When $x \to -\infty$, $t \to -\infty$ and when $x \to \infty$, $t \to \infty$

$\therefore$ From (2) we have

$$F[f(ax)] = \int_{-\infty}^{\infty} f(t)\, e^{ist/a} . \frac{dt}{a} = \frac{1}{a}\int_{-\infty}^{\infty} f(t)\, e^{i(s/a)t}\, dt = \frac{1}{a} F\left(\frac{s}{a}\right) \quad \text{[by (2)]}$$

Cor. (*i*) If $F_s(s)$ is the Fourier sine transform of $f(x)$, then

$$F_s[f(ax)] = \frac{1}{a} F_s\left(\frac{s}{a}\right).$$

(*ii*) If $F_c(s)$ is the Fourier cosine transform of $f(x)$, then

$$F_c[f(ax)] = \frac{1}{a} F_c\left(\frac{s}{a}\right).$$

3. Shifting Property. *If F(s) is the complex Fourier transform of f(x), then*

$$\mathbf{F[f(x - a)] = e^{isa}F(s).}$$

Proof. By definition of complex Fourier transform, we have

$$F(s) = F\{f(x)\} = \int_{-\infty}^{\infty} f(x) . e^{isx}\, dx$$

$$\therefore \quad F[f(x - a)] = \int_{-\infty}^{\infty} f(x - a)\, e^{isx}\, dx \qquad \ldots(1)$$

Putting $x - a = t$ *i.e.*, $x = a + t$, we have $dx = dt$

When $x \to -\infty$, $t \to -\infty$ and when $x \to \infty$, $t \to \infty$

$\therefore$ From (1), we have

$$F[f(x-a)] = \int_{-\infty}^{\infty} f(t)\, e^{is(a+t)}\, dt = \int_{-\infty}^{\infty} f(t)\,.\, e^{isa}\,.\, e^{ist}\, dt$$

$$= e^{isa} \int_{-\infty}^{\infty} f(t)\,.\, e^{ist}\, dt = e^{isa}\, F(s).$$

3. (a) Shifting on time axis. *If F(s) is the complex Fourier transform of f(t) and t_0 is any real number, then*

$$\mathbf{F[f(t-t_0)] = e^{ist_0} F(s).}$$

Proof. By definition of complex Fourier transform, we have

$$F(s) = F[f(t)] = \int_{-\infty}^{\infty} f(t)\,.\, e^{ist}\, dt$$

$\therefore$
$$F[f(t-t_0)] = \int_{-\infty}^{\infty} f(t-t_0)\, e^{ist}\, dt \qquad \ldots(1)$$

Putting $t - t_0 = T$ *i.e.,* $t = t_0 + T$, we have $dt = dT$

When $t \to -\infty$, $T \to -\infty$ and when $t \to \infty$, $T \to \infty$

$\therefore$ From (1), we have

$$F[f(t-t_0)] = \int_{-\infty}^{\infty} f(T)\, e^{is(t_0+T)}\, dT$$

$$= \int_{-\infty}^{\infty} f(T)\, e^{ist_0}\,.\, e^{isT}\, dT = e^{ist_0} \int_{-\infty}^{\infty} f(T)\, e^{isT}\, dT$$

$$= e^{ist_0}\, F(s).$$

Remark. Inverse Fourier transform of $e^{ist_0}\, F(s)$ is $f(t-t_0)$.

3. (b) Shifting on frequency axis. *If F(s) is the complex Fourier transform of f(t), and s_0 is any real number, then*

$$\mathbf{F[e^{is_0t} f(t)] = F(s+s_0).}$$

Proof. By definition of complex Fourier transform, we have

$$F(s) = F[f(t)] = \int_{-\infty}^{\infty} f(t)\,.\, e^{ist}\, dt$$

$\therefore$
$$F[e^{is_0t} f(t)] = \int_{-\infty}^{\infty} e^{is_0t} f(t)\,.\, e^{ist}\, dt = \int_{-\infty}^{\infty} f(t)\,.\, e^{i(s+s_0)t}\, dt = F(s+s_0).$$

Remark. Inverse Fourier transform of $F(s+s_0)$ is $e^{is_0t} f(t)$.

4. Modulation Theorem. *If F(s) is the complex Fourier transform of f(x), then*

$$\mathbf{F[f(x)\cos ax] = \frac{1}{2}[F(s+a) + F(s-a)].}$$

Proof. By definition of complex Fourier transform, we have

$$F(s) = F\{f(x)\} = \int_{-\infty}^{\infty} f(x) e^{isx}\, dx$$

$$\therefore \qquad F[f(x) \cos ax] = \int_{-\infty}^{\infty} f(x) \cos ax \,.\, e^{isx}\, dx$$

$$= \int_{-\infty}^{\infty} f(x) \left(\frac{e^{iax} + e^{-iax}}{2}\right) e^{isx}\, dx$$

$$= \frac{1}{2}\left[\int_{-\infty}^{\infty} f(x) e^{i(s+a)x}\, dx + \int_{-\infty}^{\infty} f(x) . e^{i(s-a)x}\, dx\right]$$

$$= \frac{1}{2}\, [F(s + a) + F(s - a)].$$

Cor. If $F_s(s)$ and $F_c(s)$ are Fourier sine and cosine transforms of $f(x)$ respectively, then

(*i*) $F_s[f(x) \cos ax] = \frac{1}{2}\, [F_s(s + a) + F_s(s - a)]$

(*ii*) $F_s[f(x) \sin ax] = \frac{1}{2}\, [F_c(s - a) - F_c(s + a)]$

(*iii*) $F_c[f(x) \cos ax] = \frac{1}{2}\, [F_c(s + a) + F_c(s - a)]$

(*iv*) $F_c[f(x) \sin ax] = \frac{1}{2}\, [F_s(s + a) - F_s(s - a)]$.

5. If $F_s(s)$ and $F_c(s)$ are Fourier sine and cosine transforms of f(x) respectively, then

(*i*) $$F_s \{x\, f(x)\} = -\frac{d}{ds} \{F_c (s)\}$$

(*ii*) $$F_c\{x\, f(x)\} = \frac{d}{ds} \{F_s (s)\}$$

Proof. (*i*) $$\frac{d}{ds} \{F_c (s)\} = \frac{d}{ds}\left\{\int_0^{\infty} f(x) \cos sx\, dx\right\}$$

$$= \int_0^{\infty} f(x)\, (- x \sin sx)\, dx$$

$$= -\int_0^{\infty} \{x\, f(x)\} \sin sx\, dx$$

$$= - F_s \{x\, f(x)\}$$

$$\Rightarrow \qquad F_s \{x\, f(x)\} = -\frac{d}{ds} \{F_c (s)\}$$

(*ii*) $$\frac{d}{ds} \{F_s (s)\} = \frac{d}{ds}\left\{\int_0^{\infty} f(x) \sin sx\, dx\right\}$$

$$= \int_0^{\infty} f(x)\, (x \cos sx)\, dx$$

$$= \int_0^{\infty} \{x\, f(x)\} \cos sx\, dx$$

$$= F_c \{x\, f(x)\}$$

$$\Rightarrow \qquad F_c \{x\, f(x)\} = \frac{d}{ds}\{F_s (s)\}.$$

Example 1. *Find the Fourier transform of* e^{-x^2}. *Hence find the Fourier transform of*

(*i*) e^{-ax^2}, $(a > 0)$ (*ii*) $e^{-\frac{x^2}{2}}$

(*iii*) $e^{-4(x-3)^2}$ (*iv*) $e^{-x^2} \cos 2x$.

Sol. Fourier transform of $f(x) = e^{-x^2}$ is given by

$$F\{f(x)\} = \int_{-\infty}^{\infty} f(x) \,.\, e^{isx}\, dx = \int_{-\infty}^{\infty} e^{-x^2} \,.\, e^{isx} dx$$

$$= \int_{-\infty}^{\infty} e^{-(x^2 - isx)}\, dx = \int_{-\infty}^{\infty} e^{-\left\{\left(x - \frac{is}{2}\right)^2 + \frac{s^2}{4}\right\}} dx$$

$$= e^{-\frac{s^2}{4}} \int_{-\infty}^{\infty} e^{-\left(x - \frac{is}{2}\right)^2} dx$$

$$= e^{-\frac{s^2}{4}} \int_{-\infty}^{\infty} e^{-z^2}\, dz \quad \text{where } z = x - \frac{is}{2}$$

$$= 2\, e^{-\frac{s^2}{4}} \int_0^{\infty} e^{-z^2}\, dz = 2\, e^{-\frac{s^2}{4}} \,.\, \frac{\sqrt{\pi}}{2}$$

$$= \sqrt{\pi}\, e^{-\frac{s^2}{4}} = F(s)$$

(*i*) $$e^{-ax^2} = e^{-(\sqrt{a}x)^2} = f(\sqrt{a}x)$$

By change of scale property, we have

$$F\{f(\sqrt{a}x)\} = \frac{1}{\sqrt{a}} F\left(\frac{s}{\sqrt{a}}\right)$$

$$\Rightarrow \qquad F(e^{-ax^2}) = \frac{1}{\sqrt{a}} \sqrt{\pi}\, e^{-\frac{1}{4}\left(\frac{s}{\sqrt{a}}\right)^2} = \sqrt{\frac{\pi}{a}}\, e^{-\frac{s^2}{4a}}$$

(*ii*) Putting $a = \frac{1}{2}$ in deduction (*i*), we have

$$F\left(e^{-\frac{x^2}{2}}\right) = \sqrt{2\pi}\, e^{-s^2/2}$$

(*iii*) $e^{-4x^2} = e^{-(2x)^2} = f(2x)$

By change of scale property, we have

$$F\{f(2x)\} = \frac{1}{2} F\left(\frac{s}{2}\right)$$

$$\Rightarrow \qquad F\{e^{-4x^2}\} = \frac{\sqrt{\pi}}{2} e^{-\frac{1}{4}\left(\frac{s}{2}\right)^2} = \frac{\sqrt{\pi}}{2} e^{-\frac{s^2}{16}}$$

By shifting property $F\{f(x-a)\} = e^{isa} F(s)$

$$\therefore \qquad F\{e^{-4(x-3)^2}\} = e^{3is} \cdot \frac{\sqrt{\pi}}{2} e^{-\frac{s^2}{16}} = \frac{\sqrt{\pi}}{2} e^{\left(3is - \frac{s^2}{16}\right)}$$

(*iv*) By modulation theorem,

$$F\{f(x) \cos ax\} = \frac{1}{2} [F(s+a) + F(s-a)]$$

$$\therefore \qquad F\{e^{-x^2} \cos 2x\} = \frac{1}{2}\left[\sqrt{\pi}\, e^{-\frac{1}{4}(s+2)^2} + \sqrt{\pi}\, e^{-\frac{1}{4}(s-2)^2}\right]$$

$$= \frac{\sqrt{\pi}}{2}\left[e^{-\frac{1}{4}(s+2)^2} + e^{-\frac{1}{4}(s-2)^2}\right].$$

Example 2. *Find the Fourier sine and cosine transform of $x\, e^{-ax}$.*

Sol. Let us first find the Fourier sine and cosine transforms of e^{-ax}.

$$F_s(e^{-ax}) = \int_0^\infty e^{-ax} \sin sx\, dx = \left[\frac{e^{-ax}}{a^2+s^2}(-a \sin sx - s \cos sx)\right]_0^\infty = \frac{s}{a^2+s^2}$$

and
$$F_c(e^{-ax}) = \int_0^\infty e^{-ax} \cos sx\, dx = \left[\frac{e^{-ax}}{a^2+s^2}(-a \cos sx + s \sin sx)\right]_0^\infty = \frac{a}{a^2+s^2}$$

$$\therefore \qquad F_s(xe^{-ax}) = -\frac{d}{ds}\{F_c(e^{-ax})\} = -\frac{d}{ds}\left(\frac{a}{a^2+s^2}\right) = \frac{2as}{(a^2+s^2)^2}$$

and
$$F_c(xe^{-ax}) = \frac{d}{ds}\{F_s(e^{-ax})\} = \frac{d}{ds}\left(\frac{s}{a^2+s^2}\right)$$

$$= \frac{(a^2+s^2).1 - s(2s)}{(a^2+s^2)^2} = \frac{a^2-s^2}{(a^2+s^2)^2}.$$

2.7. CONVOLUTION

The convolution of two functions $f(x)$ and $g(x)$ over the interval $(-\infty, \infty)$ is defined as

$$f(x) * g(x) = \int_{-\infty}^{\infty} f(u)\, g(x-u)\, du.$$

2.8. CONVOLUTION THEOREM FOR FOURIER TRANSFORMS (or Falting Theorem)

(M.D.U. 2010, 2011, 2012, May 2014; K.U.K. 2010, 2013; DCRUST Murthal May 2014)

The Fourier transform of the convolution of f(x) and g(x) is equal to the product of the Fourier transforms of f(x) and g(x), i.e.,

$$\mathbf{F[f(x) * g(x)] = F[f(x)].\ F[g(x)] = F(s)G(s).}$$

Proof. By definition of Fourier transform

$$F(s) = F[f(x)] = \int_{-\infty}^{\infty} f(x) \,.\, e^{isx}\, dx,$$

$$G(s) = F[g(x)] = \int_{-\infty}^{\infty} g(x) \,.\, e^{isx}\, dx$$

and by definition of convolution,

$$f(x) * g(x) = \int_{-\infty}^{\infty} f(u)\, g(x-u)\, du$$

$$\therefore \qquad F[f(x) * g(x)] = \int_{-\infty}^{\infty} [f(x) * g(x)]\, e^{isx} dx$$

$$= \int_{-\infty}^{\infty} \left[\int_{-\infty}^{\infty} f(u)\, g(x-u)\, du\right] e^{isx} dx$$

$$= \int_{-\infty}^{\infty} f(u) \left[\int_{-\infty}^{\infty} g(x-u)\, e^{isx}\, dx\right] du$$

(By changing the order of integration)

Putting $x - u = t$ *i.e.*, $x = u + t$, we have $dx = dt$.

When $x \to -\infty$, $t \to -\infty$ and when $x \to \infty$, $t \to \infty$

$$\therefore \qquad F[f(x) * g(x)] = \int_{-\infty}^{\infty} f(u) \left[\int_{-\infty}^{\infty} g(t)\, e^{is(u+t)}\, dt\right] du$$

$$= \int_{-\infty}^{\infty} f(u) \left[\int_{-\infty}^{\infty} g(t) \,.\, e^{isu} \,.\, e^{ist}\, dt\right] du$$

$$= \left[\int_{-\infty}^{\infty} f(u)\, e^{isu}\, du\right] \left[\int_{-\infty}^{\infty} g(t)\, e^{ist}\, dt\right]$$

$$= \left[\int_{-\infty}^{\infty} f(x)\, e^{isx}\, dx\right] \left[\int_{-\infty}^{\infty} g(x) \,.\, e^{isx}\, dx\right]$$

$$= F[f(x)]\ F[g(x)] = F(s)\ G(s).$$

Remark. The following properties of convolution can be easily proved.

(*i*) $f(x) * g(x) = g(x) * f(x)$

(*ii*) $f(x) * [g(x) * h(x)] = [f(x) * g(x)] * h(x)$

(*iii*) $f(x) * [g(x) + h(x)] = f(x) * g(x) + f(x) * h(x)$.

2.9. RELATION BETWEEN FOURIER AND LAPLACE TRANSFORMS

If $f(t) = \begin{cases} e^{-xt} g(t), & t > 0 \\ 0, & t < 0 \end{cases}$, *then* $F[f(t)] = L[g(t)]$.

Proof. By definition of Fourier transform, we have

$$F[f(t)] = \int_{-\infty}^{\infty} f(t) e^{ist}\, dt$$

$$= \int_{-\infty}^{0} 0.e^{ist}\, dt + \int_{0}^{\infty} e^{-xt} g(t) . e^{ist}\, dt \qquad \text{[By def. of } f(t)]$$

$$= \int_{0}^{\infty} e^{-(x-is)t} g(t)\, dt$$

$$= \int_{0}^{\infty} e^{-pt} g(t)\, dt, \quad \text{where } p = x - is$$

$$= L[g(t)].$$

2.10. PARSEVAL'S IDENTITY FOR FOURIER TRANSFORMS

If the Fourier transforms of f(x) and g(x) are F(s) and G(s) respectively, then

(i) $\dfrac{1}{2\pi}\displaystyle\int_{-\infty}^{\infty} F(s)\, \overline{G}(s)\, ds = \int_{-\infty}^{\infty} f(x)\, \overline{g}(x)\, dx$

(ii) $\dfrac{1}{2\pi}\displaystyle\int_{-\infty}^{\infty} |F(s)|^2\, ds = \int_{-\infty}^{\infty} |f(x)|^2\, dx$.

where bar stands for the complex conjugate.

Proof. (*i*) By definition of Fourier transform

$$F(s) = F\{f(x)\} = \int_{-\infty}^{\infty} f(x) . e^{isx}\, dx$$

and

$$G(s) = F\{g(x)\} = \int_{-\infty}^{\infty} g(x) . e^{isx}\, dx$$

By inversion formula for Fourier transform

$$f(x) = \frac{1}{2\pi}\int_{-\infty}^{\infty} F(s) . e^{-isx}\, ds$$

and

$$g(x) = \frac{1}{2\pi}\int_{-\infty}^{\infty} G(s) . e^{-isx}\, ds$$

$$\therefore \quad \int_{-\infty}^{\infty} f(x)\, \overline{g}(x)\, dx = \int_{-\infty}^{\infty} f(x) \left[\frac{1}{2\pi}\int_{-\infty}^{\infty} \overline{G}(s) . e^{isx}\, ds\right] dx$$

$$= \frac{1}{2\pi}\int_{-\infty}^{\infty} \overline{G}(s) \left[\int_{-\infty}^{\infty} f(x)\, e^{isx}\, dx\right] ds$$

(Changing the order of integration)

$$= \frac{1}{2\pi}\int_{-\infty}^{\infty} \overline{G}(s)\, [F(s)]\, ds$$

$$= \frac{1}{2\pi}\int_{-\infty}^{\infty} F(s)\,\overline{G}(s)\,ds.$$

(*ii*) By definition of Fourier transform

$$F(s) = F\{f(x)\} = \int_{-\infty}^{\infty} f(x)\,e^{isx}\,dx$$

By inversion formula for Fourier transform

$$f(x) = \frac{1}{2\pi}\int_{-\infty}^{\infty} F(s)\,\bar{e}^{isx}\,ds$$

$$\therefore \qquad \int_{-\infty}^{\infty} |f(x)|^2\,dx = \int_{-\infty}^{\infty} f(x)\,\bar{f}(x)\,dx$$

$$= \int_{-\infty}^{\infty} f(x)\left[\frac{1}{2\pi}\int_{-\infty}^{\infty} \overline{F}(s)\,e^{isx}\,ds\right]dx$$

$$= \frac{1}{2\pi}\int_{-\infty}^{\infty} \overline{F}(s)\left[\int_{-\infty}^{\infty} f(x)\,e^{isx}\,dx\right]ds$$

(Changing the order of integration)

$$= \frac{1}{2\pi}\int_{\infty}^{\infty} \overline{F}(s)\,[F(s)]\,ds$$

$$= \frac{1}{2\pi}\int_{-\infty}^{\infty} |F(s)|^2\,ds.$$

Remark. The following Parseval's identities for Fourier sine and cosine transforms can be easily proved:

(*i*) $\dfrac{2}{\pi}\displaystyle\int_0^{\infty} F_s(s)\,G_s(s)\,ds = \int_0^{\infty} f(x)\,g(x)\,dx$

(*ii*) $\dfrac{2}{\pi}\displaystyle\int_0^{\infty} [F_s(s)]^2\,ds = \int_0^{\infty} [f(x)]^2\,dx$

(*iii*) $\dfrac{2}{\pi}\displaystyle\int_0^{\infty} F_c(s)\,G_c(s)\,ds = \int_0^{\infty} f(x)\,g(x)\,dx$

(*iv*) $\dfrac{2}{\pi}\displaystyle\int_0^{\infty} [F_c(s)]^2\,ds = \int_0^{\infty} [f(x)]^2\,dx.$

ILLUSTRATIVE EXAMPLES

Example 1. *Using Parseval's identities, prove that*

(*i*) $\displaystyle\int_0^{\infty} \frac{dt}{(4+t^2)(9+t^2)} = \frac{\pi}{60}$ (*ii*) $\displaystyle\int_0^{\infty} \frac{t^2}{(4+t^2)(9+t^2)}\,dt = \frac{\pi}{10}$

(*iii*) $\displaystyle\int_0^{\infty} \frac{dt}{(4+t^2)^2} = \frac{\pi}{32}$ (*iv*) $\displaystyle\int_0^{\infty} \frac{t^2}{(9+t^2)^2}\,dt = \frac{\pi}{12}.$

Sol. Let $f(x) = e^{-2x}$ and $g(x) = e^{-3x}$

Then $$F_c(s) = \frac{2}{4+s^2}, \quad G_c(s) = \frac{3}{9+s^2}$$

$$F_s(s) = \frac{s}{4+s^2}, \quad G_s(s) = \frac{s}{9+s^2}$$

$$\left[\because f(x) = e^{-ax} \Rightarrow F_c(s) = \frac{a}{a^2+s^2}, F_s(s) = \frac{s}{a^2+s^2}\right]$$

(*i*) Using Parseval's identity for Fourier cosine transforms, *i.e.*,

$$\frac{2}{\pi}\int_0^\infty F_c(s)\, G_c(s)\, ds = \int_0^\infty f(x)\, g(x)\, dx$$

We have $$\frac{2}{\pi}\int_0^\infty \left(\frac{2}{4+s^2}\right)\left(\frac{3}{9+s^2}\right) ds = \int_0^\infty e^{-2x} \,.\, e^{-3x}\, dx$$

$$\Rightarrow \quad \frac{12}{\pi}\int_0^\infty \frac{ds}{(4+s^2)(9+s^2)} = \int_0^\infty e^{-5x}\, dx$$

$$= \left[\frac{e^{-5x}}{-5}\right]_0^\infty = -\frac{1}{5}(0-1) = \frac{1}{5}$$

$$\Rightarrow \quad \int_0^\infty \frac{ds}{(4+s^2)(9+s^2)} = \frac{\pi}{60}$$

$$\therefore \quad \int_0^\infty \frac{dt}{(4+t^2)(9+t^2)} = \frac{\pi}{60}.$$

(*ii*) Using Parseval's identity for fourier sine transforms, *i.e.*,

$$\frac{2}{\pi}\int_0^\infty F_s(s)\, G_s(s)\, ds = \int_0^\infty f(x)\, g(x)\, dx$$

We have $$\frac{2}{\pi}\int_0^\infty \left(\frac{s}{4+s^2}\right)\left(\frac{s}{9+s^2}\right) ds = \int_0^\infty e^{-2x} \,.\, e^{-3x}\, dx$$

$$\Rightarrow \quad \frac{2}{\pi}\int_0^\infty \frac{s^2}{(4+s^2)(9+s^2)}\, ds = \int_0^\infty e^{-5x}\, dx$$

$$= \left[\frac{e^{-5x}}{-5}\right]_0^\infty = -\frac{1}{5}(0-1) = \frac{1}{5}$$

$$\Rightarrow \quad \int_0^\infty \frac{s^2}{(4+s^2)(9+s^2)}\, ds = \frac{\pi}{10}$$

$$\therefore \quad \int_0^\infty \frac{t^2}{(4+t^2)(9+t^2)}\, dt = \frac{\pi}{10}.$$

(*iii*) Using Parseval's identity for Fourier cosine transform, *i.e.*,

$$\frac{2}{\pi}\int_0^\infty [F_c(s)]^2\, ds = \int_0^\infty [f(x)]^2\, dx$$

We have $$\frac{2}{\pi}\int_0^{\infty}\left(\frac{2}{4+s^2}\right)^2 ds = \int_0^{\infty}\left(e^{-2x}\right)^2 dx$$

$$\Rightarrow \quad \frac{8}{\pi}\int_0^{\infty}\frac{ds}{(4+s^2)^2} = \int_0^{\infty} e^{-4x}\, dx$$

$$= \left[\frac{e^{-4x}}{-4}\right]_0^{\infty} = -\frac{1}{4}(0-1) = \frac{1}{4}$$

$$\Rightarrow \quad \int_0^{\infty}\frac{ds}{(4+s^2)^2} = \frac{\pi}{32}$$

$$\therefore \quad \int_0^{\infty}\frac{dt}{(4+t^2)^2} = \frac{\pi}{32}.$$

(*iv*) Using Parseval's identity for Fourier sine transform, *i.e.*,

$$\frac{2}{\pi}\int_0^{\infty}[G_s(s)]^2\, ds = \int_0^{\infty}[g(x)]^2\, dx$$

We have $$\frac{2}{\pi}\int_0^{\infty}\left(\frac{s}{9+s^2}\right)^2 ds = \int_0^{\infty}(e^{-3x})^2\, dx$$

$$\Rightarrow \quad \frac{2}{\pi}\int_0^{\infty}\frac{s^2}{(9+s^2)^2}\, ds = \int_0^{\infty} e^{-6x}\, dx$$

$$= \left[\frac{e^{-6x}}{-6}\right]_0^{\infty} = -\frac{1}{6}(0-1) = \frac{1}{6}$$

$$\Rightarrow \quad \int_0^{\infty}\frac{s^2}{(9+s^2)^2}\, ds = \frac{\pi}{12}$$

$$\therefore \quad \int_0^{\infty}\frac{t^2}{(9+t^2)^2}\, dt = \frac{\pi}{12}.$$

Example 2. *Using Parseval's identity, prove that*

$$\int_0^{\infty}\frac{\sin at}{t(a^2+t^2)}\, dt = \frac{\pi}{2a^2}(1-e^{-a^2}).$$

Sol. Let $$f(x) = e^{-ax},\ a > 0 \quad \text{and} \quad g(x) = \begin{cases} 1, & 0 < x < a \\ 0, & x > a \end{cases}$$

then $$F_c(s) = \frac{a}{a^2+s^2} \text{ and } G_c(s) = \frac{\sin as}{s}.$$

Using Parseval's identity for Fourier cosine transforms, *i.e.*,

$$\frac{2}{\pi}\int_0^{\infty} F_c(s)\, G_c(s)\, ds = \int_0^{\infty} f(x)\, g(x)\, dx$$

We have $$\frac{2}{\pi}\int_0^{\infty}\left(\frac{a}{a^2+s^2}\right)\left(\frac{\sin as}{s}\right) ds = \int_0^{\infty} e^{-ax} \,.\, g(x)\, dx$$

$$\Rightarrow \quad \frac{2a}{\pi}\int_0^\infty \frac{\sin as}{s(a^2+s^2)}\,ds = \int_0^a e^{-ax}\,.\,g(x)\,dx + \int_a^\infty e^{-ax}\,.\,g(x)\,dx$$

$$= \int_0^a e^{-ax}\,.\,1\,dx + \int_a^\infty e^{-ax}\,.\,0\,dx$$

$$= \left[\frac{e^{-ax}}{-a}\right]_0^a + 0 = -\frac{1}{a}(e^{-a^2}-1) = \frac{1}{a}(1-e^{-a^2})$$

$$\Rightarrow \quad \int_0^\infty \frac{\sin as}{s(a^2+s^2)}\,ds = \frac{\pi}{2a^2}(1-e^{-a^2})$$

$$\therefore \quad \int_0^\infty \frac{\sin at}{t(a^2+t^2)}\,dt = \frac{\pi}{2a^2}(1-e^{-a^2}).$$

Example 3. *Find the Fourier transform of*

$$f(x) = \begin{cases} 1-|x|, & |x|<1 \\ 0, & |x|>1 \end{cases}$$

and hence find the value of $\int_0^\infty \frac{\sin^4 t}{t^4}\,dt$. (M.D.U. 2010, May 2014)

Sol. Fourier transform of $f(x)$ is given by

$$F\{f(x)\} = \int_{-\infty}^\infty f(x)e^{isx}\,dx$$

$$= \int_{-\infty}^{-1} f(x)e^{isx}\,dx + \int_{-1}^1 f(x)\,e^{isx}\,dx + \int_1^\infty f(x)\,e^{isx}\,dx$$

$$= \int_{-\infty}^{-1} 0\,dx + \int_{-1}^1 (1-|x|)\,e^{isx}\,dx + \int_1^\infty 0\,dx$$

$$= \int_{-1}^1 (1-|x|)(\cos sx + i\sin sx)\,dx$$

$$= \int_{-1}^1 (1-|x|)\cos sx\,dx + \int_{-1}^1 (1-|x|)(i\sin sx)\,dx$$

$$= 2\int_0^1 (1-|x|)\cos sx\,dx + 0$$

[$\because$ $(1-|x|)\cos sx$ is an even function of x where as $(1-|x|)\sin sx$ is an odd function of x]

$$= 2\int_0^1 (1-x)\cos sx\,dx \qquad [\because\ |x| = x \text{ where } x>0]$$

$$= 2\left[(1-x)\,.\,\frac{\sin sx}{s} - (-1)\left(-\frac{\cos sx}{s^2}\right)\right]_0^1$$

$$= 2\left[-\frac{\cos s}{s^2} + \frac{1}{s^2}\right] = 2\left(\frac{1-\cos s}{s^2}\right) = \mathrm{F}(s)$$

Using Parseval's identity for Fourier transform,

i.e., $$\frac{1}{2\pi}\int_{-\infty}^{\infty} |\mathrm{F}(s)|^2\, ds = \int_{-\infty}^{\infty} |f(x)|^2\, dx$$

We have

$$\frac{1}{2\pi}\int_{-\infty}^{\infty} \frac{4(1-\cos s)^2}{s^4}\, ds = \int_{-1}^{1} (1-|x|)^2\, dx$$

$$\Rightarrow \quad \frac{2}{\pi}\int_{-\infty}^{\infty} \frac{(1-\cos s)^2}{s^4}\, ds = \int_{-1}^{1} (1-|x|)^2\, dx$$

$$\Rightarrow \quad \frac{4}{\pi}\int_{0}^{\infty} \frac{(1-\cos s)^2}{s^4}\, ds = 2\int_{0}^{1} (1-|x|)^2\, dx$$

($\because$ Both integrands are even functions)

$$\Rightarrow \quad \frac{4}{\pi}\int_{0}^{\infty} \frac{(2\sin^2 s/2)^2}{s^4}\, ds = 2\int_{0}^{1} (1-x)^2\, dx$$

$$\Rightarrow \quad \frac{16}{\pi}\int_{0}^{\infty} \frac{\sin^4 s/2}{s^4}\, ds = 2\left[\frac{(1-x)^3}{-3}\right]_0^1 = \frac{2}{3}$$

Putting $\frac{s}{2} = t$ *i.e.,* $s = 2t$, we have

$$\frac{16}{\pi}\int_{0}^{\infty} \frac{\sin^4 t}{16t^4}(2dt) = \frac{2}{3}$$

$$\Rightarrow \quad \int_{0}^{\infty} \frac{\sin^4 t}{t^4}\, dt = \frac{\pi}{3}.$$

EXERCISE 2.3

1. Verify convolution theorem for $f(x) = g(x) = e^{-x^2}$. *(M.D.U. 2011)*
2. Using Parseval's identities, prove that

   (*i*) $\displaystyle\int_0^{\infty} \frac{dx}{(x^2+1)^2} = \frac{\pi}{4}$ (*ii*) $\displaystyle\int_0^{\infty} \frac{x^2}{(x^2+1)^2}\, dx = \frac{\pi}{4}$.

   [**Hint.** Use Parseval's identity for Fourier sine and cosine transforms of $f(x) = e^{-x}$]
3. Using Parseval's identity, show that $\displaystyle\int_0^{\infty} \frac{dt}{(a^2+t^2)(b^2+t^2)} = \frac{\pi}{2ab(a+b)}$. *(K.U.K. May 2013)*

   [**Hint.** Use Parseval's identity for Fourier cosine transforms of $f(x) = e^{-ax}$, $g(x) = e^{-bx}$]

4. If $f(x) = \begin{cases} 1, & |x| < a \\ 0, & |x| > a \end{cases}$ and $F(s) = \dfrac{2 \sin as}{s}$, $(s \neq 0)$, then prove that $\displaystyle\int_0^{\infty} \frac{\sin^2 ax}{x^2}\, dx = \frac{\pi a}{2}$.

5. Using Parseval's identity, prove that

(*i*) $\displaystyle\int_0^{\infty} \frac{\sin^2 x}{x^2}\, dx = \frac{\pi}{2}$ (*ii*) $\displaystyle\int_0^{\infty} \frac{1-\cos 2x}{x^2}\, dx = \pi.$

6. Using Parseval's identity, prove that

(*i*) $\displaystyle\int_0^{\infty} \left(\frac{1-\cos x}{x}\right)^2 dx = \frac{\pi}{2}$ (*ii*) $\displaystyle\int_0^{\infty} \frac{\sin^4 t}{t^2}\, dt = \frac{\pi}{4}$.

2.11. FOURIER TRANSFORMS OF THE DERIVATIVES OF A FUNCTION

The Fourier transform of the function $u(x, t)$ is given by

$$F[u(x, t)] = \int_{-\infty}^{\infty} u e^{isx}\, dx$$

(*i*) Fourier transform of $\dfrac{\partial \mathbf{u}}{\partial \mathbf{x}}$ *(M.D.U. May 2015)*

Suppose $u \to 0$ as $x \to \pm\infty$, then the Fourier transform of $\dfrac{\partial u}{\partial x}$ is given by

$$F\left[\frac{\partial u}{\partial x}\right] = \int_{-\infty}^{\infty} \frac{\partial u}{\partial x} . e^{isx}\, dx = \int_{-\infty}^{\infty} e^{isx} . \frac{\partial u}{\partial x}\, dx \qquad \text{(Integrating by parts)}$$

$$= \left[e^{isx} . u\right]_{-\infty}^{\infty} - \int_{-\infty}^{\infty} ise^{isx} . u\, dx$$

$$= 0 - is \int_{-\infty}^{\infty} u . e^{isx}\, dx \qquad [\because \ u \to 0 \text{ as } x \to \pm\infty]$$

$$= -is\, F(u)$$

Hence $\quad \mathbf{F\left[\dfrac{\partial u}{\partial x}\right] = - is\, F(u).}$

(*ii*) Fourier transform of $\dfrac{\partial^2 \mathbf{u}}{\partial \mathbf{x}^2}$ *(M.D.U. May 2015)*

Suppose u and $\dfrac{\partial u}{\partial x} \to 0$ as $x \to \pm\infty$, then the Fourier transform of $\dfrac{\partial^2 u}{\partial x^2}$ is given by

$$F\left[\frac{\partial^2 u}{\partial x^2}\right] = \int_{-\infty}^{\infty} \frac{\partial^2 u}{\partial x^2} . e^{isx}\, dx = \int_{-\infty}^{\infty} e^{isx} . \frac{\partial^2 u}{\partial x^2}\, dx$$

(Applying general rule of integration by parts)

$$= \left[e^{isx} . \frac{\partial u}{\partial x} - ise^{isx} . u\right]_{-\infty}^{\infty} + (is)^2 \int_{-\infty}^{\infty} e^{isx} . u\, dx$$

$$= 0 - s^2 \int_{-\infty}^{\infty} u.\, e^{isx}\, dx \qquad \left[\because \; u \text{ and } \frac{\partial u}{\partial x} \to 0 \text{ as } x \to \pm\infty\right]$$

$$= -\, s^2 F(u).$$

Hence $$\mathbf{F}\left[\frac{\partial^2 \mathbf{u}}{\partial \mathbf{x}^2}\right] = -\,\mathbf{s}^2\mathbf{F}(\mathbf{u}) = (-\,\mathbf{is})^2\, \mathbf{F}(\mathbf{u}).$$

(*iii*) **Fourier transform of** $\frac{\partial^n \mathbf{u}}{\partial \mathbf{x}^n}$

Suppose $u, \frac{\partial u}{\partial x}, \frac{\partial^2 u}{\partial x^2}, ..., \frac{\partial^{n-1} u}{\partial x^{n-1}} \to 0$ as $x \to \pm\infty$, then the Fourier transform of $\frac{\partial^n u}{\partial x^n}$ is given by

$$F\left[\frac{\partial^n u}{\partial x^n}\right] = \int_{-\infty}^{\infty} \frac{\partial^n u}{\partial x^n} . e^{isx}\, dx = \int_{-\infty}^{\infty} e^{isx} . \frac{\partial^n u}{\partial x^n} dx$$

(Applying general rule of integration by parts)

$$= \left[e^{isx} . \frac{\partial^{n-1} u}{\partial x^{n-1}} - ise^{isx} . \frac{\partial^{n-2} u}{\partial x^{n-2}} + (is)^2\, e^{isx} . \frac{\partial^{n-3} u}{\partial x^{n-3}} - ...... + (-\,is)^{n-1}\, u\right]_{-\infty}^{\infty}$$

$$+ (-\,is)^n \int_{-\infty}^{\infty} e^{isx} . udx$$

$$= 0 + (-\,is)^n \int_{-\infty}^{\infty} u . e^{isx}\, dx = (-\,is)^n\, F(u)$$

Hence $$\mathbf{F}\left[\frac{\partial^n \mathbf{u}}{\partial \mathbf{x}^n}\right] = (-\,\mathbf{is})^n\, \mathbf{F}(\mathbf{u}).$$

2.12. FOURIER SINE AND COSINE TRANSFORMS OF $\frac{\partial^2 \mathbf{u}}{\partial \mathbf{x}^2}$

The Fourier sine and cosine transforms of the function $u(x, t)$ are given by

$$F_s[u(x, t)] = \int_0^{\infty} u \sin sx\, dx$$

and $$F_c[u(x, t)] = \int_0^{\infty} u \cos sx\, dx .$$

(*i*) **Fourier sine transform of** $\frac{\partial^2 \mathbf{u}}{\partial \mathbf{x}^2}$

Suppose u and $\frac{\partial u}{\partial x} \to 0$ as $x \to \infty$, then the Fourier transform of $\frac{\partial^2 u}{\partial x^2}$ is given by

$$F_s\left[\frac{\partial^2 u}{\partial x^2}\right] = \int_0^{\infty} \frac{\partial^2 u}{\partial x^2} . \sin sx\, dx = \int_0^{\infty} \sin sx . \frac{\partial^2 u}{\partial x^2}\, dx \qquad \text{(Integrating by parts)}$$

$$= \left[\sin sx \, . \frac{\partial u}{\partial x}\right]_0^\infty - \int_0^\infty (s \cos sx) \, . \frac{\partial u}{\partial x} \, dx$$

$$= 0 - s \int_0^\infty (\cos sx) \frac{\partial u}{\partial x} \, dx \qquad \left[\because \quad \frac{\partial u}{\partial x} \to 0 \text{ as } x \to \infty\right]$$

(Integrating again by parts)

$$= - s \left[\{\cos sx \, . \, u\}_0^\infty - \int_0^\infty (- s \sin sx) \, . \, u \, dx\right]$$

$$= - s \left[0 - (u)_{x=0} + s \int_0^\infty u \sin sx \, dx\right] \qquad [\because \quad u \to 0 \text{ as } x \to \infty]$$

$$= s(u)_{x=0} - s^2 F_s(u)$$

Hence $\mathbf{F_s}\left[\dfrac{\partial^2 \mathbf{u}}{\partial \mathbf{x}^2}\right] = \mathbf{s(u)_{x=0} - s^2 F_s(u)}$.

(*ii*) Fourier cosine transform of $\dfrac{\partial^2 \mathbf{u}}{\partial \mathbf{x}^2}$

Suppose u and $\dfrac{\partial u}{\partial x} \to 0$ as $x \to \infty$, then the Fourier transform of $\dfrac{\partial^2 u}{\partial x^2}$ is given by

$$F_c\left[\frac{\partial^2 u}{\partial x^2}\right] = \int_0^\infty \frac{\partial^2 u}{\partial x^2} \, . \cos sx \, dx = \int_0^\infty \cos sx \, . \frac{\partial^2 u}{\partial x^2} \, dx \qquad \text{(Integrating by parts)}$$

$$= \left[\cos sx \, . \frac{\partial u}{\partial x}\right]_0^\infty - \int_0^\infty - s \sin sx \, . \frac{\partial u}{\partial x} \, dx$$

$$= \left[0 - \left(\frac{\partial u}{\partial x}\right)_{x=0}\right] + s \int_0^\infty \sin sx \frac{\partial u}{\partial x} \, dx \qquad \left[\because \quad \frac{\partial u}{\partial x} \to 0 \text{ as } x \to \infty\right]$$

$$= - \left(\frac{\partial u}{\partial x}\right)_{x=0} + s \int_0^\infty \sin sx \, . \frac{\partial u}{\partial x} \, dx \qquad \text{(Integrating again by parts)}$$

$$= - \left(\frac{\partial u}{\partial x}\right)_{x=0} + s \left[\{\sin sx . \, u\}_0^\infty - \int_0^\infty s \cos sx \, . \, u \, dx\right]$$

$$= - \left(\frac{\partial u}{\partial x}\right)_{x=0} - s^2 \int_0^\infty u \cos sx \, dx \qquad [\because \quad u \to 0 \text{ as } x \to \infty]$$

$$= - \left(\frac{\partial u}{\partial x}\right)_{x=0} - s^2 \, F_c(u)$$

Hence $\mathbf{F_c}\left[\dfrac{\partial^2 \mathbf{u}}{\partial \mathbf{x}^2}\right] = -\left(\dfrac{\partial \mathbf{u}}{\partial \mathbf{x}}\right)_{\mathbf{x=0}} - \mathbf{s^2 F_c(u)}$.

2.13. APPLICATION OF FOURIER TRANSFORMS TO BOUNDARY VALUE PROBLEMS

Fourier transforms are very useful in solving boundary value problems. We take Fourier transform of the given partial differential equation using given boundary and initial conditions. The required solution is then obtained by taking corresponding inverse transform. The choice of particular transform to be employed depends on the boundary conditions of the problem.

(*i*) *If the interval is* $-\infty < x < \infty$ and if boundary conditions are

$$u \text{ and } \frac{\partial u}{\partial x} \to 0 \text{ as } x \to \pm\infty$$

use infinite Fourier transform.

(*ii*) *If the interval is* $0 < x < \infty$ and

(*a*) boundary conditions are u and $\frac{\partial u}{\partial x} \to 0$ as $x \to \infty$ and $u(x, t) = 0$ or $f(t)$ at $x = 0$ and for all t, use Fourier sine transform.

(*b*) boundary conditions are u and $\frac{\partial u}{\partial x} \to 0$ as $x \to \infty$ and $\frac{\partial u}{\partial x} = 0$ or $f(t)$ at $x = 0$ and for all t, use Fourier cosine transform.

For the interval $0 < x < \infty$, we always assume u and $\frac{\partial u}{\partial x} \to 0$ as $x \to \infty$, even if it is not given in the problem.

(*iii*) *If the interval is* $0 < x < L$ and

(*a*) boundary conditions are $u(0, t) = u(L, t) = 0$ for all t, use finite Fourier sine transform.

(*b*) boundary conditions are $\frac{\partial u}{\partial x}(0, t) = \frac{\partial u}{\partial x}(L, t) = 0$ for all t, use finite Fourier cosine transform.

ILLUSTRATIVE EXAMPLES

Example 1. *Solve the equation* $\frac{\partial u}{\partial t} = \frac{\partial^2 u}{\partial x^2}$, $x > 0,\ t > 0$

subject to the conditions

(*i*) $u = 0$, *when* $x = 0,\ t > 0$ (*ii*) $u = \begin{cases} 1, & 0 < x < 1 \\ 0, & x \ge 1 \end{cases}$, *when* $t = 0$

and (*iii*) $u(x, t)$ *is bounded.* (M.D.U. Dec. 2015; K.U.K. Jan. 2013)

Sol. Given $$\frac{\partial u}{\partial t} = \frac{\partial^2 u}{\partial x^2},\ x > 0,\ t > 0 \qquad \ldots(1)$$

Boundary condition is $u(0, t) = 0$

Initial conditions are $u(x, 0) = \begin{cases} 1, & 0 < x < 1 \\ 0, & x \geq 1 \end{cases}$ ...(2)

and $u(x, t)$ is bounded.

Since $u(0, t)$ is given, we take Fourier sine transform of both sides of (1). Thus,

$$F_s\left(\frac{\partial u}{\partial t}\right) = F_s\left(\frac{\partial^2 u}{\partial x^2}\right)$$

$$\Rightarrow \quad \int_0^\infty \frac{\partial u}{\partial t} \sin sx\, dx = \int_0^\infty \frac{\partial^2 u}{\partial x^2} \sin sx\, dx. \text{ Integrating by parts}$$

$$\Rightarrow \quad \frac{d}{dt}\int_0^\infty u \sin sx\, dx = \left[\sin sx \,.\, \frac{\partial u}{\partial x}\right]_0^\infty - \int_0^\infty s \cos sx \,.\, \frac{\partial u}{\partial x} \,.\, dx$$

$$= 0 - s\int_0^\infty \cos sx \,.\, \frac{\partial u}{\partial x} \,.\, dx$$

$\left[\because \; u(x, t) \text{ is bounded } \therefore \; \frac{\partial u}{\partial x} \to 0 \text{ when } x \to \infty\right]$

$$= -s\left[\left\{\cos sx \,.\, u\right\}_0^\infty - \int_0^\infty - s \sin sx \,.\, u\, dx\right]$$

$$= -s\left[0 + s\int_0^\infty u \sin sx\, dx\right]$$

$[\because \text{ when } x \to \infty,\, u \to 0 \text{ and when } x = 0,\, u = u(0, t) = 0]$

$$\Rightarrow \quad \frac{d}{dt}\int_0^\infty u \sin sx\, dx = -s^2 \int_0^\infty u \sin sx\, dx$$

$$\Rightarrow \quad \frac{d\overline{u}_s}{dt} = -s^2\overline{u}_s \qquad \text{where } \overline{u}_s = \overline{u}_s(s, t) = F_s\,[u(x, t)]$$

Separating the variables,

$$\frac{d\overline{u}_s}{\overline{u}_s} = -s^2 dt$$

Integrating, $\quad \log \overline{u}_s = -s^2 t + \log c$

$$\Rightarrow \quad \log_e\left(\frac{\overline{u}_s}{c}\right) = -s^2 t \quad \text{or} \quad \overline{u}_s = ce^{-s^2 t} \qquad \text{...(3)}$$

Putting $t = 0$ in (3),

$$c = \overline{u}_s(s, 0) = F_s\,[u(x, 0)] = \int_0^\infty u(x, 0) \sin sx\, dx$$

$$= \int_0^1 u(x, 0) \sin sx\, dx + \int_1^\infty u(x, 0) \sin sx\, dx$$

$$= \int_0^1 1 . \sin sx \, dx + \int_1^\infty 0 . \sin sx \, dx \qquad \text{[from (2)]}$$

$$= \left[- \frac{\cos sx}{s} \right]_0^1 = \frac{1 - \cos s}{s}$$

$\therefore$ From (3), $\bar{u}_s(s, t) = \left(\frac{1 - \cos s}{s} \right) e^{-s^2 t}$

Taking its inverse Fourier sine transform, we get

$$u(x, t) = \frac{2}{\pi} \int_0^\infty \left(\frac{1 - \cos s}{s} \right) e^{-s^2 t} \sin sx \, ds$$

which is the required solution.

Example 2. *Solve* $\frac{\partial V}{\partial t} = K \frac{\partial^2 V}{\partial x^2}$ *for $x > 0$, $t > 0$ under the boundary conditions $V = V_0$ when $x = 0$, $t > 0$ and the initial condition $V = 0$ when $t = 0$, $x > 0$.*

Sol. Given $\frac{\partial V}{\partial t} = K \frac{\partial^2 V}{\partial x^2}, \quad x > 0, t > 0$ ...(1)

Boundary condition is $V(0, t) = V_0, t > 0$

Initial condition is $V(x, 0) = 0, x > 0$ ...(2)

Since $V(0, t)$ is given, we take Fourier sine transform of both sides of (1). Thus

$$F_s\left(\frac{\partial V}{\partial t} \right) = F_s\left(K \frac{\partial^2 V}{\partial x^2} \right)$$

$$\Rightarrow \quad \int_0^\infty \frac{\partial V}{\partial t} \sin sx \, dx = K \int_0^\infty \frac{\partial^2 V}{\partial x^2} \sin sx \, dx. \qquad \text{Integrating by parts}$$

$$\Rightarrow \quad \frac{d}{dt} \int_0^\infty V \sin sx \, dx = K \left[\left\{ \sin sx . \frac{\partial V}{\partial x} \right\}_0^\infty - \int_0^\infty s \cos sx . \frac{\partial V}{\partial x} dx \right]$$

$$= K \left[0 - s \int_0^\infty \cos sx . \frac{\partial V}{\partial x} dx \right] \qquad \left[\because \frac{\partial V}{\partial x} \to 0 \text{ when } x \to \infty \right]$$

$$= - Ks \left[\left\{ \cos sx . V \right\}_0^\infty - \int_0^\infty - s \sin sx . V dx \right]$$

$$= - Ks \left[- V_0 + s \int_0^\infty V \sin sx \, dx \right]$$

$[\because$ when $x \to \infty$, $V \to 0$ and when $x = 0$, $V = V(0, t) = V_0]$

$$\Rightarrow \quad \frac{d}{dt}\int_0^\infty \text{V} \sin sx\, dx = \text{K}s\text{V}_0 - \text{K}s^2 \int_0^\infty \text{V} \sin sx\, dx$$

$$\Rightarrow \quad \frac{d\overline{\text{V}}_s}{dt} = \text{K}s\text{V}_0 - \text{K}s^2\overline{\text{V}}_s \quad \text{where } \overline{\text{V}}_s = \overline{\text{V}}_s(s, t) = \text{F}_s[\text{V}(x, t)]$$

$$\Rightarrow \quad \frac{d\overline{\text{V}}_s}{dt} + \text{K}s^2\,\overline{\text{V}}_s = \text{K}s\text{V}_0$$

which is a linear differential equation.

$$\text{I.F.} = e^{\int \text{K}s^2 dt} = e^{\text{K}s^2 t}$$

$\therefore$ Its solution is

$$\overline{\text{V}}_s\, e^{\text{K}s^2t} = \int \text{K}s\,\text{V}_0 . e^{\text{K}s^2t}\, dt + c$$

$$= \text{K}s\text{V}_0 \cdot \frac{e^{\text{K}s^2t}}{\text{K}s^2} + c$$

or $$\overline{\text{V}}_s = \frac{\text{V}_0}{s} + ce^{-\text{K}s^2t} \qquad ...(3)$$

Putting $t = 0$ in (3), we have $\overline{\text{V}}_s\,(s, 0) = \dfrac{\text{V}_0}{s} + c$

$$\Rightarrow \quad c = -\frac{\text{V}_0}{s} + \overline{\text{V}}_s\,(s, 0) = -\frac{\text{V}_0}{s} + \text{F}_s\,[\text{V}(x, 0)]$$

$$= -\frac{\text{V}_0}{s} + \int_0^\infty 0 \sin sx\, dx \qquad \text{[from (2)]}$$

$$= -\frac{\text{V}_0}{s}$$

$\therefore$ From (3), $\overline{\text{V}}_s\,(s, t) = \dfrac{\text{V}_0}{\text{S}}\left(1 - e^{-\text{K}s^2t}\right)$

Taking its inverse Fourier sine transform, we get

$$\text{V}(x, t) = \frac{2}{\pi}\int_0^\infty \frac{\text{V}_0}{s}\left(1 - e^{-\text{K}s^2t}\right) \sin sx\, ds$$

$$= \frac{2\text{V}_0}{\pi}\left[\int_0^\infty \frac{\sin sx}{s}\, ds - \int_0^\infty \frac{e^{-\text{K}s^2t}}{s} \sin sx\, ds\right]$$

$$= \frac{2\text{V}_0}{\pi}\left[\frac{\pi}{2} - \int_0^\infty \frac{e^{-\text{K}s^2t}}{s} \sin sx\, ds\right] \qquad \left[\because \int_0^\infty \frac{\sin sx}{s}\, ds = \frac{\pi}{2}\right]$$

or $$\text{V}(x, t) = \text{V}_0\left[1 - \frac{2}{\pi}\int_0^\infty \frac{e^{-\text{K}s^2t}}{s} \sin sx\, ds\right]$$

which is the required solution.

Example 3. *The temperature u in a semi-infinite rod $0 \le x < \infty$ is determined by the differential equation* $\frac{\partial u}{\partial t} = k \frac{\partial^2 u}{\partial x^2}$ *subject to the conditions:*

(i) $u = 0$ when $t = 0$, $x \ge 0$.

(ii) $\frac{\partial u}{\partial x} = -\mu$ (a constant) when $x = 0$ and $t > 0$. (K.U.K. Dec. 2015)

Determine the temperature formula.

Sol. Given
$$\frac{\partial u}{\partial t} = k \frac{\partial^2 u}{\partial x^2} \qquad \ldots(1)$$

Boundary condition is $\frac{\partial u}{\partial x} = -\mu$ when $x = 0$, $t > 0$

Initial condition is $u(x, 0) = 0$ ...(2)

Since $\frac{\partial u}{\partial x}$ at $x = 0$ is given, we take Fourier cosine transform of both sides of (1). Thus

$$F_c\left(\frac{\partial u}{\partial t}\right) = F_c\left(k \frac{\partial^2 u}{\partial x^2}\right)$$

$$\Rightarrow \quad \int_0^\infty \frac{\partial u}{\partial t} \cos sx \, dx = k \int_0^\infty \frac{\partial^2 u}{\partial x^2} \cos sx \, dx.$$ Integrating by parts

$$\Rightarrow \quad \frac{d}{dt}\int_0^\infty u \cos sx \, dx = k\left[\left\{\cos sx \, . \frac{\partial u}{\partial x}\right\}_0^\infty - \int_0^\infty - s \sin sx \, . \frac{\partial u}{\partial x} dx\right]$$

$$= k\left[0 - (-\mu) + s \int_0^\infty \sin sx \frac{\partial u}{\partial x} dx\right]$$

$$\left[\because \frac{\partial u}{\partial x} \to 0 \text{ when } x \to \infty \text{ and } \frac{\partial u}{\partial x} = -\mu \text{ when } x = 0\right]$$

$$= k\left[\mu + s\left\{\left|\sin sx \, . u\right|_0^\infty - \int_0^\infty s \cos sx \, . u \, dx\right\}\right]$$

$$= k\left[\mu + s\{0\} - s \int_0^\infty s \cos sx \, . u \, dx\right] \qquad [\because u \to 0 \text{ when } x \to \infty]$$

$$\Rightarrow \quad \frac{d}{dt}\int_0^\infty u \cos sx \, dx = k\mu - ks^2 \int_0^\infty u \cos sx \, dx$$

$$\Rightarrow \quad \frac{d\overline{u}_c}{dt} = k\mu - ks^2 \overline{u}_c \qquad \text{where } \overline{u}_c = \overline{u}_c(s, t) = F_c[u(x, t)]$$

$$\Rightarrow \qquad \frac{d\overline{u}_c}{dt} + ks^2\overline{u}_c = k\mu$$

which is a linear differential equation.

$$\text{I.F.} = e^{\int ks^2 dt} = e^{ks^2 t}$$

$\therefore$ Its solution is

$$\overline{u}_c \,.\, e^{ks^2 t} = \int k\mu \,.\, e^{ks^2 t}\, dt + c = k\mu \,.\, \frac{e^{ks^2 t}}{ks^2} + c$$

or
$$\overline{u}_c = \frac{\mu}{s^2} + ce^{-ks^2 t} \qquad \ldots(3)$$

Putting $t = 0$ in (3), we have

$$\overline{u}_c\,(s, 0) = \frac{\mu}{s^2} + c$$

$$\Rightarrow \qquad c = -\frac{\mu}{s^2} + \overline{u}_c\,(s, 0) = -\frac{\mu}{s^2} + \mathrm{F}_c\,[u(x, 0)]$$

$$= -\frac{\mu}{s^2} + \int_0^\infty 0 \cos sx \, dx \qquad \text{[from (2)]}$$

$$= -\frac{\mu}{s^2}$$

$\therefore$ From (3), $\overline{u}_c(s, t) = \dfrac{\mu}{s^2}\left(1 - e^{-ks^2 t}\right)$

Taking its inverse Fourier cosine transform, we get

$$u(x, t) = \frac{2}{\pi}\int_0^\infty \frac{\mu}{s^2}\left(1 - e^{-ks^2 t}\right) \cos sx\, ds$$

or
$$u(x, t) = \frac{2\mu}{\pi}\int_0^\infty \frac{\cos sx}{s^2}\left(1 - e^{-ks^2 t}\right) ds$$

which is the required solution.

Example 4. *If the initial temperature of an infinite bar is given by*

$$\theta(x) = \begin{cases} \theta_0 & for\,|\,x\,| < a \\ 0 & for\,|\,x\,| > a \end{cases},$$

determine the temperature at any instant x and at any instant t.

Sol. To determine the temperature $\theta(x, t)$, we have to solve the one-dimensional heat-flow equation

$$\frac{\partial \theta}{\partial t} = c^2 \frac{\partial^2 \theta}{\partial x^2}, \quad t > 0 \qquad \ldots(1)$$

subject to the initial condition $\theta(x, 0) = \begin{cases} \theta_0 & \text{for}\,|\,x\,| < a \\ 0 & \text{for}\,|\,x\,| > a \end{cases} \qquad \ldots(2)$

Taking Fourier transform of (1), we get

$$\int_{-\infty}^{\infty} \frac{\partial \theta}{\partial t} e^{isx}\, dx = c^2 \int_{-\infty}^{\infty} \frac{\partial^2 \theta}{\partial x^2} e^{isx}\, dx$$

or
$$\frac{d}{dt}\int_{-\infty}^{\infty} \theta e^{isx}\, dx = c^2(-s^2\overline{\theta})$$

or
$$\frac{d\overline{\theta}}{dt} = -c^2 s^2 \overline{\theta} \text{ where } \overline{\theta} = \overline{\theta}(s,t) = F[\theta(x,t)] \qquad \ldots(3)$$

Now taking the Fourier transform of (2), we get

$$\overline{\theta}(s,0) = \int_{-\infty}^{\infty} \theta(x,0)\, e^{isx}\, dx = \int_{-a}^{a} \theta_0 e^{isx}\, dx = \theta_0 \left[\frac{e^{isx}}{is}\right]_{-a}^{a}$$

$$= \theta_0 \left[\frac{e^{isa} - e^{-isa}}{is}\right] = \frac{2\theta_0}{s}\left[\frac{e^{isa} - e^{-isa}}{2i}\right]$$

$$= \frac{2\theta_0 \sin sa}{s} \qquad \ldots(4)$$

From (3), $\dfrac{d\overline{\theta}}{\overline{\theta}} = -c^2 s^2 dt$

Integrating $\log \overline{\theta} = -c^2 s^2 t + \log A$ or $\overline{\theta} = Ae^{-c^2 s^2 t}$

Since $\overline{\theta} = \dfrac{2\theta_0 \sin sa}{s}$ when $t = 0$, from (4), we get $A = \dfrac{2\theta_0 \sin sa}{s}$

$$\therefore \quad \overline{\theta} = \frac{2\theta_0 \sin sa}{s}\, e^{-c^2 s^2 t}$$

Taking its inverse Fourier transform, we get

$$\theta(x,t) = \frac{1}{2\pi}\int_{-\infty}^{\infty} \frac{2\theta_0 \sin as}{s} . e^{-c^2 s^2 t} . e^{-isx}\, ds$$

$$= \frac{\theta_0}{\pi}\int_{-\infty}^{\infty} \frac{\sin as}{s} . e^{-c^2 s^2 t} (\cos xs - i \sin xs)\, ds$$

$$= \frac{\theta_0}{\pi}\left[\int_{-\infty}^{\infty} \frac{\sin as}{s} . e^{-c^2 s^2 t} \cos xs\, ds - i\int_{-\infty}^{\infty} \frac{\sin as}{s} . e^{-c^2 s^2 t} \sin xs\, ds\right]$$

$$= \frac{\theta_0}{\pi}\int_{-\infty}^{\infty} \frac{\sin as}{s} . e^{-c^2 s^2 t} \cos xs\, ds$$

(The second integral vanishes since its integrand is an odd function)

$$= \frac{2\theta_0}{\pi}\int_{0}^{\infty} \frac{\sin as}{s} . e^{-c^2 s^2 t} \cos xs\, ds = \frac{\theta_0}{\pi}\int_{0}^{\infty} \frac{e^{-c^2 s^2 t}}{s} . 2 \sin as \cos xs\, ds$$

$$= \frac{\theta_0}{\pi}\int_{0}^{\infty} e^{-c^2 s^2 t}\left(\frac{\sin (a+x)s + \sin (a-x)s}{s}\right) ds$$

which is the required solution.

Example 5. *Use the method of Fourier transform to determine the displacement $y(x, t)$ of an infinite string, given that the string is initially at rest and that the initial displacement is $f(x)$, $(-\infty < x < \infty)$.*

Sol. The equation for the vibration of the string is

$$\frac{\partial^2 y}{\partial t^2} = c^2 \frac{\partial^2 y}{\partial x^2} \qquad \text{...(1)}$$

subject to the initial conditions

$$\left(\frac{\partial y}{\partial t}\right)_{t=0} = 0 \quad \text{and} \quad y(x, 0) = f(x) \qquad \text{...(2)}$$

Taking Fourier transform of (1), we get

$$\frac{d^2 \bar{y}}{dt^2} = c^2 (-s^2 \bar{y}) \qquad \text{where} \quad \bar{y} = \text{F}[y(x, t)]$$

or

$$\frac{d^2 \bar{y}}{dt^2} + c^2 s^2 \bar{y} = 0$$

Its solution is $\bar{y} = \text{A} \cos cst + \text{B} \sin cst$ ...(3)

where A, B are constants.

Now taking Fourier transform of (2), we get

$$\frac{\partial \bar{y}}{\partial t} = 0 \quad \text{and} \quad \bar{y} = \text{F}(s) \quad \text{when } t = 0$$

Putting $t = 0$, $\bar{y} = \text{F}(s)$ in (3), we get $\text{A} = \text{F}(s)$

Also
$$\frac{\partial \bar{y}}{\partial t} = -cs\text{A} \sin cst + cs\text{B} \cos cst$$

Putting $t = 0$, $\frac{\partial \bar{y}}{\partial t} = 0$, we get $\text{B} = 0$

$$\therefore \qquad \bar{y} = \text{F}(s) \cos cst$$

Taking inverse Fourier transform, we get

$$y(x, t) = \frac{1}{2\pi} \int_{-\infty}^{\infty} \text{F}(s) \cos cst \,.\, e^{-isx} \, dx$$

$$= \frac{1}{2\pi} \int_{-\infty}^{\infty} \text{F}(s) \left[\frac{e^{icst} + e^{-icst}}{2}\right] . \, e^{-isx} \, dx$$

$$= \frac{1}{4\pi} \int_{-\infty}^{\infty} [\text{F}(s) \, e^{-is(x - ct)} + \text{F}(s) \, e^{-is(x + ct)}] \, dx$$

$$= \frac{1}{2} [f(x - ct) + f(x + ct)] \qquad \left[\because f(x) = \frac{1}{2\pi} \int_{-\infty}^{\infty} \text{F}(s) e^{-isx} \, dx\right]$$

Example 6. *Use the complex form of Fourier transform to show that*

$$u = \frac{1}{2\sqrt{\pi t}} \int_{-\infty}^{\infty} f(u) \, e^{-\frac{(x-u)^2}{4t}} \, du$$

is the solution of the boundary value problem

$$\frac{\partial u}{\partial t} = \frac{\partial^2 u}{\partial x^2}, \; -\infty < x < \infty, \; t > 0; \qquad u = f(x) \text{ when } t = 0.$$

Sol. Given $$\frac{\partial u}{\partial t} = \frac{\partial^2 u}{\partial x^2} \qquad ...(1)$$

Initial condition is $u(x, 0) = f(x)$ ...(2)

Taking Fourier transform of both sides of (1)

$$\int_{-\infty}^{\infty} \frac{\partial u}{\partial t} e^{isx}\, dx = \int_{-\infty}^{\infty} \frac{\partial^2 u}{\partial x^2} e^{isx}\, dx$$

$$\Rightarrow \quad \frac{d}{dt}\int_{-\infty}^{\infty} ue^{isx} dx = \left[e^{isx} \cdot \frac{\partial u}{\partial x}\right]_{-\infty}^{\infty} - \int_{-\infty}^{\infty} ise^{isx} \cdot \frac{\partial u}{\partial x}\, dx$$

$$= 0 - is \int_{-\infty}^{\infty} e^{isx} \cdot \frac{\partial u}{\partial x}\, dx \qquad \left[\because \frac{\partial u}{\partial x} \to 0 \text{ as } x \to \pm\infty\right]$$

$$= -is\left[\left\{e^{isx} . u\right\}_{-\infty}^{\infty} - \int_{-\infty}^{\infty} ise^{isx} . u\, dx\right]$$

$$= -is\left[0 - is\int_{-\infty}^{\infty} u\, e^{isx}\, dx\right] \qquad [\because u \to 0 \text{ as } x \to \pm\infty]$$

$$\Rightarrow \quad \frac{d}{dt}\int_{-\infty}^{\infty} ue^{isx}\, dx = i^2 s^2 \int_{-\infty}^{\infty} ue^{isx}\, dx$$

$$\Rightarrow \quad \frac{d\overline{u}}{dt} = -s^2 \overline{u}$$

$$\Rightarrow \quad \frac{d\overline{u}}{\overline{u}} = -s^2 dt$$

Integrating $\quad \log \overline{u} = -s^2 t + \log c$

$$\Rightarrow \quad \overline{u} = ce^{-s^2 t} \qquad ...(3)$$

Putting $\quad t = 0,\ \overline{u}(s, 0) = c$

$$\therefore \quad c = \overline{u}(s, 0) = \int_{-\infty}^{\infty} u(x, 0)\, e^{isx} dx$$

$$= \int_{-\infty}^{\infty} f(x)\, e^{isx} dx \qquad \text{[from (2)]}$$

$\therefore$ From (3),

$$\overline{u} = e^{-s^2 t} \int_{-\infty}^{\infty} f(x)\, e^{isx} dx$$

$$= e^{-s^2 t} \int_{-\infty}^{\infty} f(u) e^{isu} du \qquad \text{[by changing the variable } x \text{ to } u\text{]}$$

Taking its inverse Fourier transform, we get

$$u(x, t) = \frac{1}{2\pi}\int_{-\infty}^{\infty}\left[e^{-s^2 t}\int_{-\infty}^{\infty} f(u)\, e^{isu}\, du\right] e^{-isx}\, ds$$

$$= \frac{1}{2\pi}\int_{-\infty}^{\infty}\int_{-\infty}^{\infty} e^{-s^2 t} f(u)\, e^{isu} \,.\, e^{-isx}\, du\, ds$$

$$= \frac{1}{2\pi}\int_{-\infty}^{\infty}\int_{-\infty}^{\infty} e^{-s^2 t} f(u)\, e^{isu-isx}\, ds\, du$$

$$= \frac{1}{2\pi}\int_{-\infty}^{\infty} f(u)\left[\int_{-\infty}^{\infty} e^{-s^2 t + is(u-x)}\, ds\right] du$$

$$= \frac{1}{2\pi}\int_{-\infty}^{\infty} f(u)\left[\int_{-\infty}^{\infty} e^{-t\left[s^2 + is\left(\frac{x-u}{t}\right)\right]}\, ds\right] du$$

$$= \frac{1}{2\pi}\int_{-\infty}^{\infty} f(u)\left[\int_{-\infty}^{\infty} e^{-t\left[s^2 + 2s\,.\,\frac{i(x-u)}{2t} + \frac{i^2(x-u)^2}{4t^2} - \frac{i^2(x-u)^2}{4t^2}\right]}\, ds\right] du$$

$$= \frac{1}{2\pi}\int_{-\infty}^{\infty} f(u)\left[\int_{-\infty}^{\infty} e^{-t\left[s + i\frac{x-u}{2t}\right]^2 - \frac{(x-u)^2}{4t}}\, ds\right] du$$

$$= \frac{1}{2\pi}\int_{-\infty}^{\infty} f(u)\left[\int_{-\infty}^{\infty} e^{-t\left[s + i\frac{x-u}{2t}\right]^2} \,.\, e^{-\frac{(x-u)^2}{4t}}\, ds\right] du$$

$$= \frac{1}{2\pi}\int_{-\infty}^{\infty} f(u)\, e^{-\frac{(x-u)^2}{4t}}\left[\int_{-\infty}^{\infty} e^{-t\left[s + i\frac{x-u}{2t}\right]^2}\, ds\right] du$$

Put $\sqrt{t}\left(s + i\frac{x-u}{2t}\right) = y$, then $\sqrt{t}\; ds = dy$ or $ds = \frac{dy}{\sqrt{t}}$

$\therefore$
$$u(x, t) = \frac{1}{2\pi}\int_{-\infty}^{\infty} f(u) \,.\, e^{-\frac{(x-u)^2}{4t}}\left[\int_{-\infty}^{\infty} e^{-y^2}\frac{dy}{\sqrt{t}}\right] du$$

$$= \frac{1}{2\pi}\int_{-\infty}^{\infty}\frac{f(u)\, e^{-\frac{(x-u)^2}{4t}}}{\sqrt{t}}\sqrt{\pi}\, du \qquad \left[\because \int_{-\infty}^{\infty} e^{-y^2}\, dy = \sqrt{\pi}\right]$$

$$= \frac{1}{2\sqrt{\pi t}}\int_{-\infty}^{\infty} f(u)\, e^{-\frac{(x-u)^2}{4t}}\, du.$$

Example 7. *Using suitable transforms, solve the differential equation* $\frac{\partial^2 V}{\partial x^2} = \frac{\partial V}{\partial t}$, $0 \le x \le \pi$, $t \ge 0$ *where* $V(0, t) = 0 = V(\pi, t)$ *and* $V(x, 0) = V_0$ *constant.*

Sol. Given $$\frac{\partial V}{\partial t} = \frac{\partial^2 V}{\partial x^2},\ 0 \le x \le \pi,\ t \ge 0 \qquad ...(1)$$

Boundary conditions are $V(0, t) = 0 = V(\pi, t)$

Initial condition is $V(x, 0) = V_0$ ...(2)

Since the interval $0 \le x \le \pi$ is finite and $V(0, t)$ and $V(\pi, t)$ are given, we use finite Fourier sine transform.

Let $\overline{V}_s(n, t)$ denote finite Fourier sine transform of $V(x, t)$, then

$$\overline{V}_s(n, t) = \int_0^\pi V(x, t) \sin\left(\frac{n\pi x}{\pi}\right) dx = \int_0^\pi V(x, t) \sin nx\, dx$$

Taking finite Fourier sine transform of both sides of (1), we have

$$\int_0^\pi \frac{\partial V}{\partial t} \sin nx\, dx = \int_0^\pi \frac{\partial^2 V}{\partial x^2} \sin nx\, dx$$

$$\Rightarrow \quad \frac{d}{dt}\int_0^\pi V \sin nx\, dx = \left[\sin nx \,.\, \frac{\partial V}{\partial x}\right]_0^\pi - \int_0^\pi n \cos nx \,.\, \frac{\partial V}{\partial x}\, dx$$

$$= 0 - n\int_0^\pi \cos nx \,.\, \frac{\partial V}{\partial x}\, dx$$

$$= -n\left[\left\{\cos nx \,.\, V\right\}_0^\pi - \int_0^\pi -n \sin nx \,.\, V dx\right]$$

$$= -n\left[0 + n\int_0^\pi V \sin nx\, dx\right] \qquad [\because\ V = 0 \text{ at } x = 0 \text{ and } \pi]$$

$$\Rightarrow \quad \frac{d}{dt}\int_0^\pi V \sin nx\, dx = -n^2 \int_0^\pi V \sin nx\, dx$$

$$\Rightarrow \quad \frac{d\overline{V}_s}{dt} = -n^2\overline{V}_s$$

$$\Rightarrow \quad \frac{d\overline{V}_s}{\overline{V}_s} = -n^2 dt$$

Integrating, $$\log \overline{V}_s = -n^2 t + \log c$$

$$\therefore \quad \overline{V}_s = ce^{-n^2 t} \qquad ...(3)$$

Putting $t = 0$ in (3),

$$c = \overline{V}_s\,(n, 0) = \int_0^\pi V(x, 0) \sin nx\, dx$$

$$= \int_0^\pi V_0 \sin nx\, dx \qquad \text{[from (2)]}$$

$$= V_0 \left[-\frac{\cos nx}{n}\right]_0^\pi = \frac{V_0}{n}\,[1 - \cos n\pi]$$

$$= \frac{V_0}{n}\,[1 - (-1)^n]$$

$\therefore$ From (3), $\overline{V}_s\,(n, t) = \dfrac{V_0}{n}\,[1 - (-1)^n]\, e^{-n^2 t}$

Taking its inverse finite Fourier sine transform, we get

$$V(x, t) = \frac{2}{\pi} \sum_{n=1}^{\infty} \overline{V}_s\,(n, t) \sin nx$$

$$= \frac{2}{\pi} \sum_{n=1}^{\infty} \frac{V_0}{n}\,[1 - (-1)^n]\, e^{-n^2 t} \sin nx$$

$$= \frac{2V_0}{\pi} \sum_{n=1,3,5,\ldots}^{\infty} \frac{1}{n}\,(2)\, e^{-n^2 t} \sin nx$$

$[\because\ 1 - (-1)^n = 0 \text{ for } n = 2, 4, 6, \ldots]$

or
$$V(x, t) = \frac{4V_0}{\pi} \sum_{n=1,3,5,\ldots}^{\infty} \frac{1}{n}\, e^{-n^2 t} \sin nx$$

which is the required solution.

Example 8. *Using finite Fourier transform, find the solution of the wave equation* $\dfrac{\partial^2 u}{\partial t^2} = a^2 \dfrac{\partial^2 u}{\partial x^2}$ *subject to the conditions :*

$$u\,(0, t) = u\,(\pi, t) = 0,\ t > 0$$

$$u\,(x, 0) = 3 \sin x + 4 \sin 4x \quad \text{and} \quad u_t\,(x, 0) = 0 \text{ for } 0 < x < \pi.$$

Sol. Given $\dfrac{\partial^2 u}{\partial t^2} = a^2 \dfrac{\partial^2 u}{\partial x^2}$, $0 \le x \le \pi,\ t > 0$ ...(1)

and
$$u(0, t) = u(\pi, t) = 0,\ t > 0$$

$$\left.\begin{array}{l} u(x, 0) = 3 \sin x + 4 \sin 4x \\ u_t(x, 0) = 0 \end{array}\right\} \text{for } 0 < x < \pi \qquad \text{...(2)}$$

Taking finite Fourier sine transform of both sides of (1), we have

$$\int_0^{\pi} \frac{\partial^2 u}{\partial t^2} \sin nx\, dx = a^2 \int_0^{\pi} \frac{\partial^2 u}{\partial x^2} \sin nx\, dx$$

$$\Rightarrow \quad \frac{d^2}{dt^2}\int_0^{\pi} u \sin nx\, dx = a^2\left[\left\{\sin nx . \frac{\partial u}{\partial x}\right\}_0^{\pi} - \int_0^{\pi} n \cos nx . \frac{\partial u}{\partial x}\, dx\right]$$

$$= a^2\left[0 - n\int_0^{\pi} \cos nx \frac{\partial u}{\partial x}\, dx\right]$$

$$= -a^2 n\left[\left\{\cos nx . u\right\}_0^{\pi} - \int_0^{\pi} - n \sin nx . u\, dx\right]$$

$$= -a^2 n\left[0 + n\int_0^{\pi} u \sin nx\, dx\right] \qquad [\because \; u(0, t) = u(\pi, t) = 0]$$

$$= -a^2 n^2 \int_0^{\pi} u \sin nx\, dx$$

$$\Rightarrow \quad \frac{d^2 \bar{u}_s}{dt^2} + a^2 n^2\, \bar{u}_s = 0$$

where $\bar{u}_s = \bar{u}_s\,(n, t) = \int_0^{\pi} u(x, t) \sin\left(\frac{n\pi x}{\pi}\right) dx = \int_0^{\pi} u\,(x, t) \sin nx\, dx$

A.E. is $D^2 + a^2 n^2 = 0 \;\Rightarrow\; D = \pm\, ian$

$\therefore \qquad \bar{u}_s = c_1 \cos ant + c_2 \sin ant$ ...(3)

Put $t = 0$ in (3), $\bar{u}_s(n, 0) = c_1$

$$\Rightarrow \quad c_1 = \bar{u}_s(n, 0) = \int_0^{\pi} u(x, 0) \sin nx\, dx$$

$$= \int_0^{\pi} (3 \sin x + 4 \sin 4x) \sin nx\, dx \qquad \text{[from (2)]}$$

$$= 3\int_0^{\pi} \sin nx \sin x\, dx + 4\int_0^{\pi} \sin nx \sin 4x\, dx$$

$$= \frac{3}{2}\int_0^{\pi} 2 \sin nx \sin x\, dx + 2\int_0^{\pi} 2 \sin nx \sin 4x\, dx$$

$$= \frac{3}{2}\int_0^{\pi} [\cos (n-1)\, x - \cos (n+1)\, x]\, dx + 2\int_0^{\pi} [\cos(n-4)\, x - \cos (n+4)\, x]\, dx$$

$$= \frac{3}{2}\left[\frac{\sin(n-1)x}{n-1} - \frac{\sin(n+1)x}{n+1}\right]_0^{\pi} + 2\left[\frac{\sin(n-4)x}{n-4} - \frac{\sin(n+4)x}{n+4}\right]_0^{\pi}$$

$= 0$ except for $n = 1$ and $n = 4$

For $n = 1$, $\quad c_1 = \overline{u}_s(n, 0) = \int_0^{\pi} u(x, 0) \sin x \, dx$

$$= \int_0^{\pi} (3 \sin x + 4 \sin 4x) \sin x \, dx$$

$$= 3 \int_0^{\pi} \sin^2 x \, dx + 2 \int_0^{\pi} 2 \sin 4x \sin x dx$$

$$= 3 \int_0^{\pi} \frac{1 - \cos 2x}{2} dx + 2 \int_0^{\pi} (\cos 3x - \cos 5x) \, dx$$

$$= \frac{3}{2}\left[x - \frac{\sin 2x}{2}\right]_0^{\pi} + 2\left[\frac{\sin 3x}{3} - \frac{\sin 5x}{5}\right]_0^{\pi} = \frac{3}{2}\pi$$

For $n = 4$, $\quad c_1 = \overline{u}_s(n, 0) = \int_0^{\pi} u(x, 0) \sin 4x \, dx$

$$= \int_0^{\pi} (3 \sin x + 4 \sin 4x) \sin 4x \, dx$$

$$= \frac{3}{2} \int_0^{\pi} 2 \sin 4x \sin x \, dx + 4 \int_0^{\pi} \sin^2 4x \, dx$$

$$= \frac{3}{2} \int_0^{\pi} (\cos 3x - \cos 5x) dx + 4 \int_0^{\pi} \frac{1 - \cos 8x}{2} dx$$

$$= \frac{3}{2}\left[\frac{\sin 3x}{3} - \frac{\sin 5x}{5}\right]_0^{\pi} + 2\left[x - \frac{\sin 8x}{8}\right]_0^{\pi} = \pi$$

$\therefore$
$$c_1 = \begin{cases} \frac{3\pi}{2} & \text{for } n = 1 \\ 2\pi & \text{for } n = 4 \end{cases}$$

Again from (3), $\quad \frac{\partial \overline{u}_s}{\partial t} = -c_1 an \sin ant + c_2 an \cos ant$

Putting $t = 0$, $\quad \frac{\partial \overline{u}_s}{\partial t} = 0 \qquad \left[\because \text{ when } t = 0, \frac{\partial u}{\partial t} = 0 \;\therefore\; \frac{\partial \overline{u}_s}{\partial t} = 0 \text{ for } t = 0\right]$

$0 = c_2 an \Rightarrow c_2 = 0$

$\therefore$ (3) reduces to $\overline{u}_s = c_1 \cos ant$ ...(4)

where $c_1 = \frac{3\pi}{2}$ for $n = 1$ and $c_1 = 2\pi$ for $n = 4$

Taking inverse finite Fourier sine transform of (4), we have

$$u(x, t) = \frac{2}{\pi}\sum_{n=1}^{\infty} \overline{u}_s\ (n, t) \sin nx = \frac{2}{\pi}\sum_{n=1}^{\infty} c_1 \cos ant \sin nx$$

$$= \frac{2}{\pi}\left[\underset{(\text{for } n = 1)}{\frac{3\pi}{2}\cos at \sin x} + \underset{(\text{for } n = 4)}{2\pi \cos 4\, at \sin 4x}\right]$$

or $$u(x, t) = 3 \cos at \sin x + 4 \cos 4\, at \sin 4x$$

which is the required solution.

Example 9. *Using finite Fourier transform, solve* $\frac{\partial u}{\partial t} = \frac{\partial^2 u}{\partial x^2}$ *subject to the conditions:*

(a) $u_x(0, t) = u_x(6, t) = 0, \quad 0 < x < 6, t > 0$

(b) $u(x, 0) = x(6 - x), \quad 0 < x < 6.$

Sol. Since the boundary conditions are u_x (0, t) = u_x (6, 0) = 0, we take finite Fourier cosine transform.

Let $\overline{u}_c$ (n, t) denote finite Fourier cosine transform of $u(x, t)$, then

$$\overline{u}_c\ (n, t) = \int_0^6 u(x, t) \cos\left(\frac{n\pi x}{6}\right) dx \qquad [\because \quad L = 6]$$

Taking finite Fourier cosine transform of both sides of the given equation $\frac{\partial u}{\partial t} = \frac{\partial^2 u}{\partial x^2}$

we get $$\int_0^6 \frac{\partial u}{\partial t} \cos\frac{n\pi x}{6} dx = \int_0^6 \frac{\partial^2 u}{\partial x^2} \cos\frac{n\pi x}{6} dx$$

$$\Rightarrow \quad \frac{d}{dt}\int_0^6 u \cos\frac{n\pi x}{6} dx = \left[\cos\frac{n\pi x}{6} . \frac{\partial u}{\partial x}\right]_0^6 - \int_0^6 -\frac{n\pi}{6}\sin\frac{n\pi x}{6} . \frac{\partial u}{\partial x} dx$$

$$= 0 + \frac{n\pi}{6}\int_0^6 \sin\frac{n\pi x}{6} . \frac{\partial u}{\partial x} dx \left[\because \quad \frac{\partial u}{\partial x} = 0 \text{ at } x = 0 \text{ and } x = 6\right]$$

$$= \frac{n\pi}{6}\left[\left\{\sin\frac{n\pi x}{6} . u\right\}_0^6 - \int_0^6 \frac{n\pi}{6}\cos\frac{n\pi x}{6} . u dx\right]$$

$$= \frac{n\pi}{6}\left[0 - \frac{n\pi}{6}\int_0^6 u \cos\frac{n\pi x}{6} dx\right]$$

$$\Rightarrow \quad \frac{d}{dt}\int_0^6 u \cos\frac{n\pi x}{6} dx = -\frac{n^2\pi^2}{36}\int_0^6 u \cos\frac{n\pi x}{6} dx$$

$$\Rightarrow \qquad \frac{d\bar{u}_c}{dt} = -\frac{n^2\pi^2}{36}\bar{u}_c$$

$$\Rightarrow \qquad \frac{d\bar{u}_c}{\bar{u}_c} = -\frac{n^2\pi^2}{36}dt$$

Integrating, $\log \bar{u}_c = -\frac{n^2\pi^2}{36} t + \log A$ where $A = A(n)$

$$\Rightarrow \qquad \bar{u}_c = Ae^{-\frac{n^2\pi^2}{36}t} \Rightarrow \bar{u}_c(n,t) = A(n)e^{-\frac{n^2\pi^2}{36}t} \qquad \text{...(1)}$$

Putting $t = 0$, $\bar{u}_c\ (n, 0) = A(n)$

$\therefore$ $A(n) = \bar{u}_c\ (n, 0) =$ finite Fourier cosine transform of $u\ (x, 0)$

$$= \int_0^6 u(x, 0) \cos \frac{n\pi x}{6} dx = \int_0^6 x(6-x) \cos \frac{n\pi x}{6} dx \qquad \text{...(2)}$$

$$= \left[(6x - x^2) . \frac{\sin \frac{n\pi x}{6}}{\frac{n\pi}{6}} - (6 - 2x) . \frac{-\cos\frac{n\pi x}{6}}{\left(\frac{n\pi}{6}\right)^2} + (-2) . \frac{-\sin\frac{n\pi x}{6}}{\left(\frac{n\pi}{6}\right)^3}\right]_0^6$$

$$= -6 . \left(\frac{6}{n\pi}\right)^2 \cos n\pi - 6 . \left(\frac{6}{n\pi}\right)^2$$

$$= -\frac{216}{n^2\pi^2}(1 + \cos n\pi)$$

$\therefore$ From (1), $$\bar{u}_c = -\frac{216}{n^2\pi^2}(1 + \cos n\pi)\ e^{-\frac{n^2\pi^2}{36}t} \qquad \text{...(3)}$$

To find inverse finite Fourier cosine transform, we need $\bar{u}_c\ (0, t)$.

From (1), $\bar{u}_c\ (0, t) = A(0)$

From (2), $A(0) = \bar{u}_c\,(0, 0)$

$$= \int_0^6 x(6-x)dx = \left[3x^2 - \frac{x^3}{3}\right]_0^6$$

$$= 108 - 72 = 36$$

$\therefore$ $\bar{u}_c\,(0, t) = 36$

Taking inverse finite Fourier cosine transform of (3), we have

$$u(x, t) = \frac{1}{L} \bar{u}_c (0, t) + \frac{2}{L} \sum_{n=1}^{\infty} \bar{u}_c (n, t) \cos \frac{n\pi x}{L}$$

$$= \frac{1}{6} (36) + \frac{2}{6} \sum_{n=1}^{\infty} -\frac{216}{n^2\pi^2} (1 + \cos n\pi)\, e^{-\frac{n^2\pi^2}{36}t} \cos \frac{n\pi x}{6}$$

[$\because$ L = 6]

or $$u(x, t) = 6 - \frac{72}{\pi^2} \sum_{n=1}^{\infty} \frac{1+\cos n\pi}{n^2} e^{-\frac{n^2\pi^2 t}{36}} \cos \frac{n\pi x}{6}$$

which is the required solution.

EXERCISE 2.4

1. Solve the partial differential equation

$$\frac{\partial u}{\partial t} = 2\frac{\partial^2 u}{\partial x^2}, \; x > 0 \,,\; t > 0$$

subject to the following conditions:

(*a*) $u(0, t) = 0, \; t > 0$ (*b*) $u(x, 0) = e^{-x}, \; x > 0$

(*c*) u and $\frac{\partial u}{\partial x} \to 0$ as $x \to \infty$.

2. Solve the equation $\frac{\partial u}{\partial t} = \frac{\partial^2 u}{\partial x^2}$, $x > 0$, $t > 0$ where $u(x, t)$ satisfies the conditions:

(*a*) $\left(\frac{\partial u}{\partial x}\right)_{x=0} = 0, \; t > 0$ (*b*) $u(x, 0) = \begin{cases} x, & 0 \le x \le 1 \\ 0, & x > 1 \end{cases}$

(*c*) $| u(x, t) | < M$.

3. The initial temperature along the length of an infinite bar is given by

$$u(x, 0) = \begin{cases} 2, & |x| < 1 \\ 0, & |x| > 1 \end{cases}$$

If the temperature $u(x, t)$ satisfies the equation $\frac{\partial u}{\partial t} = \frac{\partial^2 u}{\partial x^2}$, $-\infty < x < \infty$, $t > 0$, find the temperature at any point of the bar at any time t.

4. Determine the distribution of temperature in the semi-infinite medium $x \ge 0$, when the end $x = 0$ is maintained at zero temperature and the initial distribution of temperature is $f(x)$.

5. (*a*) If the flow of heat is linear so that the variation of θ (temperature) with z and y may be neglected and if it is assumed that no heat is generated in the medium, then solve the differential equation

$$\frac{\partial \theta}{\partial t} = k\frac{\partial^2 \theta}{\partial x^2}$$

by using Fourier transform, where $-\infty < x < \infty$ and $\theta = f(x)$ when $t = 0$, $f(x)$ being a function of x.

(*b*) If the initial temperature of an infinite bar is given by

$$\mu(x, 0) = \begin{cases} 1, & \text{for } -c < x < c \\ 0, & \text{otherwise} \end{cases}$$

determine the temperature of the infinite bar at any point x and at any time $t > 0$.

6. Solve the equation $\dfrac{\partial u}{\partial t} = \dfrac{\partial^2 u}{\partial x^2}$, $0 < x < \pi$, $t > 0$ subject to the conditions

(*a*) $u(x, 0) = 1$, $0 < x < \pi$ (*b*) $u(0, t) = u(\pi, t) = 0$, $t > 0$

using appropriate Fourier transform.

7. Use finite Fourier transform to solve $\dfrac{\partial u}{\partial t} = \dfrac{\partial^2 u}{\partial x^2}$ given that

$$u(0, t) = 0,\ u(\pi, t) = 0,\ u(x, 0) = 2x,\ 0 < x < \pi,\ t > 0.$$

8. Use finite Fourier transform to solve

$$\frac{\partial u}{\partial t} = k\frac{\partial^2 u}{\partial x^2},\ 0 < x < 4,\ t > 0$$

subject to the conditions

(*a*) $u(x, 0) = 2x$, $0 < x < 4$ (*b*) $u(0, t) = u(4, t) = 0$, $t > 0$.

9. Solve $\dfrac{\partial u}{\partial t} = \dfrac{\partial^2 u}{\partial x^2}$, $0 < x < 6$, $t > 0$ subject to the conditions

(*a*) $u(0, t) = u(6, t) = 0$, $t > 0$ (*b*) $u(x, 0) = \begin{cases} 1, & 0 < x < 3 \\ 0, & 3 < x < 6 \end{cases}$.

10. Solve $\dfrac{\partial^2 u}{\partial t^2} = 4\dfrac{\partial^2 u}{\partial x^2}$, given

(*a*) $u(0, t) = u(\pi, t) = 0$ for $t > 0$ (*b*) $u(x, 0) = \dfrac{1}{10}\sin x + \dfrac{1}{100}\sin 4x$

(*c*) $u_t(x, 0) = 0$ for $0 < x < \pi$.

11. Solve $\dfrac{\partial u}{\partial t} = \dfrac{\partial^2 u}{\partial x^2}$, $0 < x < 6$, $t > 0$ subject to conditions $u_x(0, t) = u_x(6, t) = 0$, $u(x, 0) = 2x$.

12. Solve by using finite Fourier transform

$$\frac{\partial^2 u}{\partial t^2} = 9\frac{\partial^2 u}{\partial x^2},\ 0 \le x \le 2$$

subject to the conditions:

$$u(0, t) = u(2, t) = 0,\ u(x, 0) = x(2 - x) \text{ and } u_t(x, 0) = 0.$$

Answers

1. $u(x, t) = \dfrac{2}{\pi}\displaystyle\int_0^\infty \frac{s}{1+s^2} e^{-2s^2t} \sin sx\, ds$

2. $u(x, t) = \dfrac{2}{\pi}\displaystyle\int_0^\infty \left(\frac{s \sin s + \cos s - 1}{s^2}\right) e^{-s^2t} \cos sx ds$

3. $u(x, t) = \dfrac{2}{\pi}\displaystyle\int_0^\infty e^{-s^2t} \left(\frac{\sin (1+x)s + \sin (1-x)s}{s}\right) ds$

4. $u(x, t) = \dfrac{2}{\pi}\displaystyle\int_0^\infty \bar{f}_s (s)\, e^{-c^2s^2t} \sin sx\, ds$ where $\bar{f}_s (s) = \mathrm{F}_s [f(x)]$

5. (a) $\theta (x, t) = \dfrac{1}{2\pi}\displaystyle\int_{-\infty}^\infty \bar{f}(s)\, e^{-ks^2t - isx} ds$, where $\bar{f}(s) = \mathrm{F}[f(x)]$

(b) $\mu (x, t) = \dfrac{1}{2\pi}\displaystyle\int_{-\infty}^\infty \frac{2}{s} \sin cs e^{-k^2s^2t} . e^{-isx}\, ds = \frac{1}{\pi}\int_0^\infty \frac{e^{-k^2s^2t}}{s} [\sin (c+x)\, s + \sin (c-x)\, s]\, ds$

6. $u(x, t) = \dfrac{2}{\pi}\displaystyle\sum_{n=1}^\infty \frac{1-\cos n\pi}{n} e^{-n^2t} \sin nx$ or $\dfrac{4}{\pi}\displaystyle\sum_{n=1,3,5......}^\infty \frac{1}{n} e^{-n^2t} \sin nx$

7. $u(x, t) = 4\displaystyle\sum_{n=1}^\infty \frac{1}{n}(-1)^{n+1} e^{-n^2t} \sin nx$

8. $u(x, t) = \dfrac{16}{\pi}\displaystyle\sum_{n=1}^\infty \frac{(-1)^{n+1}}{n} e^{-\frac{kn^2\pi^2t}{16}} \sin \frac{n\pi x}{4}$

9. $u(x, t) = \dfrac{2}{\pi}\displaystyle\sum_{n=1}^\infty \frac{1}{n}\left(1-\cos\frac{n\pi}{2}\right) e^{-\frac{n^2\pi^2t}{36}} \sin\frac{n\pi x}{6}$

10. $u(x, t) = \dfrac{1}{10} \cos 2t \sin x + \dfrac{1}{100} \cos 8t \sin 4x$

11. $u(x, t) = 6 + \dfrac{24}{\pi^2}\displaystyle\sum_{n=1}^\infty \frac{(-1)^n - 1}{n^2} e^{-\frac{n^2\pi^2t}{36}} \cos\frac{n\pi x}{6}$

12. $u(x, t) = \dfrac{16}{\pi^3}\displaystyle\sum_{n=1}^\infty \frac{1-\cos n\pi}{n^3} \cos\frac{3n\pi t}{2} \sin\frac{n\pi x}{2}$.

2.14. FOURIER TRANSFORM OF AN INTEGRAL

Theorem. Let $f(t)$ be piecewise continuous on every interval $[-l, l]$ and $\int_{-\infty}^\infty |f(t)|dt$ converge. Let $\mathrm{F}[f(t)] = \mathrm{F}(s)$ and $\mathrm{F}(s)$ satisfies $\mathrm{F}(0) = 0$. Then

$$\mathrm{F}\left[\int_{-\infty}^t f(\mathrm{T})\, d\mathrm{T}\right] = \frac{1}{is}\mathrm{F}(s).$$

2.15. FOURIER TRANSFORM OF DIRAC-DELTA FUNCTION

Dirac-delta function (or unit impulse function) $\delta(t-a)$ is defined as

$$\delta(t-a) = \lim_{k \to 0} \delta_k(t-a) \text{ where}$$

$$\delta_k(t-a) = \begin{cases} 0, & \text{for } t < a \\ \dfrac{1}{k}, & \text{for } a \le t < a+k \\ 0, & \text{for } t \ge a+k \end{cases}$$

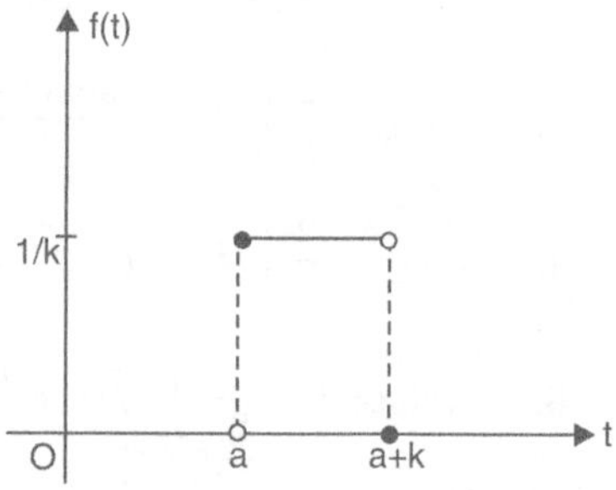

$$F\{\delta(t-a)\} = \int_{-\infty}^{\infty} \delta(t-a) e^{ist}\, dt$$

$$= \lim_{k \to 0} \int_a^{a+k} \frac{1}{k} e^{ist}\, dt$$

$$= \lim_{k \to 0} \frac{1}{k} \left[\frac{e^{ist}}{is} \right]_a^{a+k}$$

$$= \lim_{k \to 0} \frac{e^{is(a+k)} - e^{isa}}{isk} = \lim_{k \to 0} e^{isa} \left(\frac{e^{isk} - 1}{isk} \right)$$

$$= e^{isa} \times 1 \qquad \left[\because \lim_{x \to 0} \frac{e^x - 1}{x} = 1 \right]$$

$$= e^{isa}.$$

❑ ❑ ❑

SECTION-B

CHAPTER 3

Functions of a Complex Variable

3.1. (a) INTRODUCTION

A complex number z is an ordered pair (x, y) of real numbers and is written as

$$z = x + iy, \quad \text{where } i = \sqrt{-1}.$$

The real numbers x and y are called the real and imaginary parts of z. In the Argand's diagram, the complex number z is represented by the point $P(x, y)$. If (r, θ) are the polar coordinates of P, then $r = \sqrt{x^2 + y^2}$ is called the modulus of z and is denoted by $|z|$. Also $\theta = \tan^{-1} \frac{y}{x}$ is called the amplitude or argument of z and is denoted by amp. (z) or arg. (z). Every non-zero complex number z can be expressed as

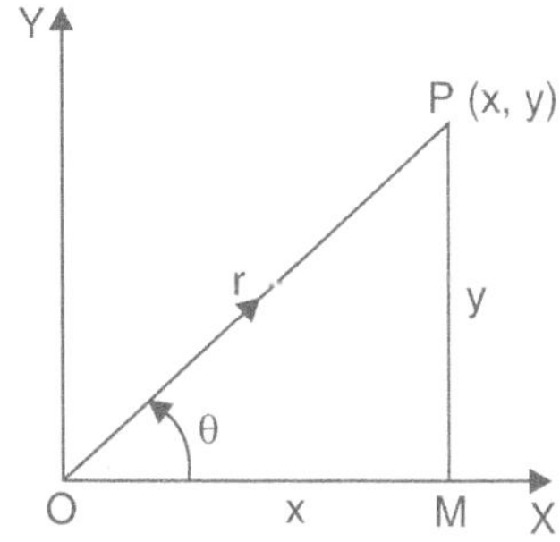

$$z = r(\cos\theta + i\sin\theta) = re^{i\theta}$$

$\cos\theta + i\sin\theta$ is briefly written as cis θ.

If $z = x + iy$, then the complex number $x - iy$ is called the conjugate of the complex number z and is denoted by $\bar{z}$.

Clearly, $\quad |\bar{z}| = |z|, |z|^2 = z\bar{z},$

$$R(z) = \frac{z + \bar{z}}{2}, \quad I(z) = \frac{z - \bar{z}}{2i}.$$

3.1. (b) DE MOIVRE'S THEOREM (STATEMENT)

(*i*) If n is an integer, positive or negative, then

$$(\cos\theta + i\sin\theta)^n = \cos n\theta + i\sin n\theta \quad \text{or} \quad (\text{cis }\theta)^n = \text{cis } n\theta$$

(*ii*) If n is a fraction, positive or negative, then one of the values of $(\cos\theta + i\sin\theta)^n$ is $\cos n\theta + i\sin n\theta$.

Remarks. (*i*) De Moivre's Theorem cannot be applied directly to $(\sin\theta + i\cos\theta)^n$. We must write

$$(\sin\theta + i\cos\theta)^n = \left[\cos\left(\frac{\pi}{2} - \theta\right) + i\sin\left(\frac{\pi}{2} - \theta\right)\right]^n$$

$$= \cos n\left(\frac{\pi}{2} - \theta\right) + i \sin n\left(\frac{\pi}{2} - \theta\right)$$

The real part must be with cos and imaginary part with sin.

(*ii*) The angle with sin and cos must be same. De Moivre's Theorem cannot be applied to $(\cos \theta + i \sin \phi)^n$.

(*iii*) $1 = \text{cis } 0, \quad -1 = \text{cis } \pi, \; i = \text{cis } \frac{\pi}{2}, \quad -i = \text{cis}\left(-\frac{\pi}{2}\right)$

(*iv*) $(\cos \theta - i \sin \theta)^n = \cos n\theta - i \sin n\theta = (\cos \theta + i \sin \theta)^{-n}$

(*v*) $\text{cis } \theta_1 \text{ cis } \theta_2 \text{ cis } \theta_3 \ldots \text{cis } \theta_n = \text{cis } (\theta_1 + \theta_2 + \theta_3 + \ldots + \theta_n)$

(*vi*) $\left(\frac{\text{cis } \theta}{\text{cis } \phi}\right) = \text{cis } (\theta - \phi)$

Example 1. *Simplify:* $\dfrac{(\cos 3\theta + i \sin 3\theta)^5 (\cos 2\theta - i \sin 2\theta)^3}{(\cos 4\theta + i \sin 4\theta)^{-9} (\cos 5\theta + i \sin 5\theta)^9}$.

Sol.
$$(\cos 3\theta + i \sin 3\theta)^5 = [(\cos \theta + i \sin \theta)^3]^5 = (\cos \theta + i \sin \theta)^{15}$$
$$(\cos 2\theta - i \sin 2\theta)^3 = [(\cos \theta + i \sin \theta)^{-2}]^3 = (\cos \theta + i \sin \theta)^{-6}$$
$$(\cos 4\theta + i \sin 4\theta)^{-9} = [(\cos \theta + i \sin \theta)^4]^{-9} = (\cos \theta + i \sin \theta)^{-36}$$
$$(\cos 5\theta + i \sin 5\theta)^9 = [(\cos \theta + i \sin \theta)^5]^9 = (\cos \theta + i \sin \theta)^{45}$$

$$\therefore \quad \text{Given expression} = \frac{(\text{cis } \theta)^{15} (\text{cis } \theta)^{-6}}{(\text{cis } \theta)^{-36} (\text{cis } \theta)^{45}} = \frac{(\text{cis } \theta)^9}{(\text{cis } \theta)^9} = 1.$$

Example 2. *Prove that:*

$$(1 + \cos \theta + i \sin \theta)^n + (1 + \cos \theta - i \sin \theta)^n = 2^{n+1} \cos^n \left(\frac{\theta}{2}\right) \cos\left(\frac{n\theta}{2}\right).$$

Sol.
$$\text{LHS} = \left(2 \cos^2 \frac{\theta}{2} + i \cdot 2 \sin \frac{\theta}{2} \cos \frac{\theta}{2}\right)^n + \left(2 \cos^2 \frac{\theta}{2} - i \cdot 2 \sin \frac{\theta}{2} \cos \frac{\theta}{2}\right)^n$$

$$= \left[2 \cos \frac{\theta}{2}\left(\cos \frac{\theta}{2} + i \sin \frac{\theta}{2}\right)\right]^n + \left[2 \cos \frac{\theta}{2}\left(\cos \frac{\theta}{2} - i \sin \frac{\theta}{2}\right)\right]^n$$

$$= 2^n \cos^n \left(\frac{\theta}{2}\right)\left[\left(\cos \frac{\theta}{2} + i \sin \frac{\theta}{2}\right)^n + \left(\cos \frac{\theta}{2} - i \sin \frac{\theta}{2}\right)^n\right]$$

$$= 2^n \cos^n \left(\frac{\theta}{2}\right)\left[\left(\cos \frac{n\theta}{2} + i \sin \frac{n\theta}{2}\right) + \left(\cos \frac{n\theta}{2} - i \sin \frac{n\theta}{2}\right)\right]$$

$$= 2^n \cos^n \left(\frac{\theta}{2}\right)\left[2 \cos\left(\frac{n\theta}{2}\right)\right]$$

$$= 2^{n+1} \cos^n \left(\frac{\theta}{2}\right) \cos\left(\frac{n\theta}{2}\right) = \text{RHS}.$$

Example 3. *If* $\sin\alpha + \sin\beta + \sin\gamma = \cos\alpha + \cos\beta + \cos\gamma = 0$, *prove that* $\sin 3\alpha + \sin 3\beta + \sin 3\gamma = 3\sin(\alpha+\beta+\gamma)$ *and* $\cos 3\alpha + \cos 3\beta + \cos 3\gamma = 3\cos(\alpha+\beta+\gamma)$

Sol. Let $a = \cos\alpha + i\sin\alpha,\ b = \cos\beta + i\sin\beta,\ c = \cos\gamma + i\sin\gamma$

then
$$a + b + c = (\cos\alpha + \cos\beta + \cos\gamma) + i(\sin\alpha + \sin\beta + \sin\gamma)$$
$$= (0) + i(0) = 0$$
$$\Rightarrow \quad a^3 + b^3 + c^3 = 3abc$$
$$\Rightarrow \quad (\cos\alpha + i\sin\alpha)^3 + (\cos\beta + i\sin\beta)^3 + (\cos\gamma + i\sin\gamma)^3$$
$$= 3(\cos\alpha + i\sin\alpha)(\cos\beta + i\sin\beta)(\cos\gamma + i\sin\gamma)$$
$$\Rightarrow \quad (\cos 3\alpha + i\sin 3\alpha) + (\cos 3\beta + i\sin 3\beta) + (\cos 3\gamma + i\sin 3\gamma)$$
$$= 3[\cos(\alpha+\beta+\gamma) + i\sin(\alpha+\beta+\gamma)]$$

Equating the real and imaginary parts on both sides, we get
$$\cos 3\alpha + \cos 3\beta + \cos 3\gamma = 3\cos(\alpha+\beta+\gamma)$$
and
$$\sin 3\alpha + \sin 3\beta + \sin 3\gamma = 3\sin(\alpha+\beta+\gamma).$$

Example 4. *If* $x + \dfrac{1}{x} = 2\cos\theta,\ y + \dfrac{1}{y} = 2\cos\phi$, *prove that one of the value of* $\dfrac{x^m}{y^n} + \dfrac{y^n}{x^m}$ *is* $2\cos(m\theta - n\phi)$.

Sol. Given
$$x + \frac{1}{x} = 2\cos\theta \Rightarrow x^2 - 2x\cos\theta + 1 = 0$$
$$\Rightarrow \quad x = \frac{2\cos\theta \pm \sqrt{4\cos^2\theta - 4}}{2} = \cos\theta \pm \sqrt{-\sin^2\theta} = \cos\theta \pm i\sin\theta$$

$\Rightarrow$ One of the values of x is $\cos\theta + i\sin\theta$.

Similarly, one of the values of y is $\cos\phi + i\sin\phi$.

$\Rightarrow$ One of the values of
$$\frac{x^m}{y^n} = \frac{(\cos\theta + i\sin\theta)^m}{(\cos\phi + i\sin\phi)^n} = \frac{\cos m\theta + i\sin m\theta}{\cos n\phi + i\sin n\phi}$$
$$= \cos(m\theta - n\phi) + i\sin(m\theta - n\phi)$$

One of the value of $\dfrac{y^n}{x^m}$ *i.e.* $\left[\dfrac{x^m}{y^n}\right]^{-1}$
$$= [\cos(m\theta - n\phi) + i\sin(m\theta - n\phi)]^{-1}$$
$$= \cos(m\theta - n\phi) - i\sin(m\theta - n\phi)$$

Hence, one of the values of $\dfrac{x^m}{y^n} + \dfrac{y^n}{x^m}$ is $2\cos(m\theta - n\phi)$.

Example 5. *If* $x_r = \cos\frac{\pi}{2^r} + i\sin\frac{\pi}{2^r}$, *prove that* $x_1 x_2 x_3 \ldots\ldots \infty = -1$.

Sol. Here, $x_r = \cos\frac{\pi}{2^r} + i\sin\frac{\pi}{2^r}$ or $x_r = \text{cis}\,\frac{\pi}{2^r}$

$$\therefore \quad x_1 x_2 x_3 \ldots\ldots \infty = \text{cis}\,\frac{\pi}{2}\,\text{cis}\,\frac{\pi}{2^2}\,\text{cis}\,\frac{\pi}{2^3} \ldots \infty$$

$$= \text{cis}\left(\frac{\pi}{2} + \frac{\pi}{2^2} + \frac{\pi}{2^3} + \ldots \infty\right)$$

$$= \text{cis}\,\frac{\frac{\pi}{2}}{1-\frac{1}{2}} \qquad \left|\because \text{ for an infinite G.P., } S = \frac{a}{1-r}\right.$$

$$= \text{cis}\,\pi = -1.$$

Example 6. *Prove that:* $(a+ib)^{m/n} + (a-ib)^{m/n} = 2(a^2+b^2)^{m/2n}\cos\left(\frac{m}{n}\tan^{-1}\frac{b}{a}\right)$.

Sol. Let $a = r\cos\theta$ and $b = r\sin\theta$

Squaring and adding

$$a^2 + b^2 = r^2 \Rightarrow r = \sqrt{a^2+b^2}$$

Dividing $\frac{b}{a} = \tan\theta \Rightarrow \theta = \tan^{-1}\frac{b}{a}$

$$\therefore \quad (a+ib)^{m/n} + (a-ib)^{m/n} = [r(\cos\theta + i\sin\theta)]^{m/n} + [r(\cos\theta - i\sin\theta)]^{m/n}$$

$$= r^{m/n}[(\cos\theta + i\sin\theta)^{m/n} + (\cos\theta - i\sin\theta)^{m/n}]$$

$$= r^{m/n}\left[\left(\cos\frac{m}{n}\theta + i\sin\frac{m}{n}\theta\right) + \left(\cos\frac{m}{n}\theta - i\sin\frac{m}{n}\theta\right)\right]$$

$$= r^{\frac{m}{n}}\left(2\cos\frac{m}{n}\theta\right) = 2(a^2+b^2)^{m/2n}\cos\left(\frac{m}{n}\tan^{-1}\frac{b}{a}\right).$$

3.1. (c) ROOTS OF A COMPLEX NUMBER

There are q and only q distinct values of $(\cos\theta + i\sin\theta)^{1/q}$**, q being a positive integer.**

By De Moivre's Theorem, we know that $\cos\frac{\theta}{q} + i\sin\frac{\theta}{q}$ is one of the values of $(\cos\theta + i\sin\theta)^{1/q}$. To find all the values of $(\cos\theta + i\sin\theta)^{1/q}$, we proceed as follows:

$$(\cos \theta + i \sin \theta)^{1/q} = [\cos (2n\pi + \theta) + i \sin (2n\pi + \theta)]^{1/q}$$

$$= \cos \frac{2n\pi + \theta}{q} + i \sin \frac{2n\pi + \theta}{q}$$

Now put $n = 0, 1, 2, \ldots, (q - 1)$ to get all the q values.

Remark. $(\cos \theta + i \sin \theta)^{p/q} = [(\cos \theta + i \sin \theta)^p]^{1/q} = (\cos p\theta + i \sin p\theta)^{1/q}$ has q and only q distinct values, p and q being integers prime to each other, *i.e.*, p and q should have no common factor greater than 1.

Thus, $(\cos \theta + i \sin \theta)^{9/12}$ does not have 12 distinct values but only 4, since $\frac{9}{12} = \frac{3}{4}$.

Example 1. *Find the five fifth roots of unity and show that their sum vanishes.*

Sol. $$(1)^{1/5} = (\cos 0 + i \sin 0)^{1/5} = [\cos (2n\pi + 0) + i \sin (2n\pi + 0)]^{1/5}$$

$$= \cos \frac{2n\pi}{5} + i \sin \frac{2n\pi}{5}$$

Putting $n = 0, 1, 2, 3, 4$, the required fifth roots of unity are $\cos 0 + i \sin 0$, $\cos \frac{2\pi}{5} + i \sin \frac{2\pi}{5}$, $\cos \frac{4\pi}{5} + i \sin \frac{4\pi}{5}$, $\cos \frac{6\pi}{5} + i \sin \frac{6\pi}{5}$, $\cos \frac{8\pi}{5} + i \sin \frac{8\pi}{5}$

i.e., $1, \cos \frac{2\pi}{5} + i \sin \frac{2\pi}{5}, \cos \frac{4\pi}{5} + i \sin \frac{4\pi}{5}, \cos \frac{4\pi}{5} - i \sin \frac{4\pi}{5}, \cos \frac{2\pi}{5} - i \sin \frac{2\pi}{5}$

i.e., $1, \cos \frac{2\pi}{5} \pm i \sin \frac{2\pi}{5}, \cos \frac{4\pi}{5} \pm i \sin \frac{4\pi}{5}$

Now, let $$\alpha = \cos \frac{2\pi}{5} + i \sin \frac{2\pi}{5}$$

then $$\alpha^2 = \left(\cos \frac{2\pi}{5} + i \sin \frac{2\pi}{5}\right)^2 = \cos \frac{4\pi}{5} + i \sin \frac{4\pi}{5}$$

$$\alpha^3 = \left(\cos \frac{2\pi}{5} + i \sin \frac{2\pi}{5}\right)^3 = \cos \frac{6\pi}{5} + i \sin \frac{6\pi}{5}$$

$$\alpha^4 = \left(\cos \frac{2\pi}{5} + i \sin \frac{2\pi}{5}\right)^4 = \cos \frac{8\pi}{5} + i \sin \frac{8\pi}{5}$$

$\therefore$ Sum of the five fifth roots of unity

$$= 1 + \alpha + \alpha^2 + \alpha^3 + \alpha^4$$

$$= \frac{1(1 - \alpha^5)}{1 - \alpha} \qquad \left|\ \text{For a G.P., } S_n = \frac{a(1 - r^n)}{1 - r}\right.$$

$$= \frac{1}{1-\alpha}\left[1-\left(\cos\frac{2\pi}{5}+i\sin\frac{2\pi}{5}\right)^5\right]$$

$$= \frac{1}{1-\alpha}\,[1-(\cos 2\pi + i\sin 2\pi)]$$

$$= \frac{1}{1-\alpha}\,[1-(1+i\cdot 0)] = \frac{0}{1-\alpha} = 0.$$

Example 2. *Find all the values of* $\left(\frac{1}{2}+\frac{\sqrt{3}i}{2}\right)^{3/4}$ *and show that their continued product is 1.*

Sol. Let $\frac{1}{2} = r\cos\theta$ and $\frac{\sqrt{3}}{2} = r\sin\theta$ so that $r = 1$ and $\theta = \frac{\pi}{3}$

$$\therefore \quad \left(\frac{1}{2}+\frac{\sqrt{3}i}{2}\right)^{3/4} = \left(\cos\frac{\pi}{3}+i\sin\frac{\pi}{3}\right)^{3/4} = \left[\left(\cos\frac{\pi}{3}+i\sin\frac{\pi}{3}\right)^3\right]^{1/4}$$

$$= (\cos\pi + i\sin\pi)^{1/4} = [\cos(2n\pi+\pi) + i\sin(2n\pi+\pi)]^{1/4}$$

$$= \cos\frac{(2n+1)\pi}{4} + i\sin\frac{(2n+1)\pi}{4}$$

Putting $n = 0, 1, 2, 3$ the four values are

$$\cos\frac{\pi}{4}+i\sin\frac{\pi}{4} = \frac{1}{\sqrt{2}}+i\cdot\frac{1}{\sqrt{2}} = \frac{1+i}{\sqrt{2}},$$

$$\cos\frac{3\pi}{4}+i\sin\frac{3\pi}{4} = -\frac{1}{\sqrt{2}}+i\cdot\frac{1}{\sqrt{2}} = \frac{-1+i}{\sqrt{2}},$$

$$\cos\frac{5\pi}{4}+i\sin\frac{5\pi}{4} = -\frac{1}{\sqrt{2}}-i\cdot\frac{1}{\sqrt{2}} = \frac{-1-i}{\sqrt{2}},$$

$$\cos\frac{7\pi}{4}+i\sin\frac{7\pi}{4} = \frac{1}{\sqrt{2}}-i\cdot\frac{1}{\sqrt{2}} = \frac{1-i}{\sqrt{2}} \quad i.e., \quad \frac{1\pm i}{\sqrt{2}}, \frac{-1\pm i}{\sqrt{2}}$$

The continued product of the four values

$$= \text{cis}\,\frac{\pi}{4}\ \text{cis}\,\frac{3\pi}{4}\ \text{cis}\,\frac{5\pi}{4}\ \text{cis}\,\frac{7\pi}{4} = \text{cis}\left(\frac{\pi}{4}+\frac{3\pi}{4}+\frac{5\pi}{4}+\frac{7\pi}{4}\right)$$

$$= \cos 4\pi + i\sin 4\pi = 1 + i\cdot 0 = 1$$

EXERCISE 3.1

1. Simplify:

(i) $\dfrac{(\cos 3\theta + i \sin 3\theta)^5 (\cos \theta - i \sin \theta)^3}{(\cos 5\theta + i \sin 5\theta)^7 (\cos 2\theta - i \sin 2\theta)^5}$ (ii) $\left(\dfrac{\cos \theta + i \sin \theta}{\sin \theta + i \cos \theta}\right)^4$

2. If $z = \cos \theta + i \sin \theta$, show that $\dfrac{z^{2n} - 1}{z^{2n} + 1} = i \tan n\theta$, n being an integer.

3. If $x = \cos \alpha + i \sin \alpha$ and $y = \cos \beta + i \sin \beta$, prove that $\dfrac{x - y}{x + y} = i \tan \dfrac{\alpha - \beta}{2}$.

4. If $a = \text{cis } \alpha$, $b = \text{cis } \beta$ and $c = \text{cis } \gamma$, prove that

(i) $\dfrac{ab}{c} + \dfrac{c}{ab} = 2 \cos (\alpha + \beta - \gamma)$ (ii) $abc + \dfrac{1}{abc} = 2 \cos (\alpha + \beta + \gamma)$

(iii) $a^p b^q c^r + \dfrac{1}{a^p b^q c^r} = 2 \cos (p\alpha + q\beta + r\gamma)$

5. If $a = \text{cis } 2\alpha$, $b = \text{cis } 2\beta$, $c = \text{cis } 2\gamma$ and $d = \text{cis } 2\delta$, prove that

(i) $\sqrt{abcd} + \dfrac{1}{\sqrt{abcd}} = 2 \cos (\alpha + \beta + \gamma + \delta)$ (ii) $\sqrt{\dfrac{ab}{cd}} + \sqrt{\dfrac{cd}{ab}} = 2 \cos (\alpha + \beta - \gamma - \delta)$

6. If $\sin \alpha + \sin \beta + \sin \gamma = \cos \alpha + \cos \beta + \cos \gamma = 0$, prove that

(i) $\cos (\beta + \gamma) + \cos (\gamma + \alpha) + \cos (\alpha + \beta) = 0$

(ii) $\sin (\beta + \gamma) + \sin (\gamma + \alpha) + \sin (\alpha + \beta) = 0$

(iii) $\cos 2\alpha + \cos 2\beta + \cos 2\gamma = 0$

(iv) $\sin 2\alpha + \sin 2\beta + \sin 2\gamma = 0$

[**Hint.** Use $\dfrac{1}{a} + \dfrac{1}{b} + \dfrac{1}{c} = a^{-1} + b^{-1} + c^{-1} = 0 \Rightarrow bc + ca + ab = 0$ for parts (i) and (ii).

Also $a + b + c = 0 \Rightarrow (a + b + c)^2 = 0 \Rightarrow a^2 + b^2 + c^2 = 0$ $(\because bc + ca + ab = 0)$

Use it for parts (iii) and (iv)]

7. If $\cos \alpha + 2 \cos \beta + 3 \cos \gamma = 0$ and $\sin \alpha + 2 \sin \beta + 3 \sin \gamma = 0$, prove that

(i) $\cos 3\alpha + 8 \cos 3\beta + 27 \cos 3\gamma = 18 \cos (\alpha + \beta + \gamma)$

(ii) $\sin 3\alpha + 8 \sin 3\beta + 27 \sin 3\gamma = 18 \sin (\alpha + \beta + \gamma)$

[**Hint.** Put $a = \cos \alpha + i \sin \alpha$, $b = 2(\cos \beta + i \sin \beta)$ and $c = 3(\cos \gamma + i \sin \gamma)$]

8. Prove that the general value of θ which satisfies the equation $(\cos \theta + i \sin \theta)(\cos 2\theta + i \sin 2\theta)$ $(\cos 3\theta + i \sin 3\theta) \ldots (\cos n\theta + i \sin n\theta) = 1$ is $\dfrac{4m\pi}{n(n+1)}$, where m is any integer.

9. If $x_r = \cos \dfrac{\pi}{3^r} + i \sin \dfrac{\pi}{3^r}$, show that $x_1 x_2 x_3 \ldots x_n = \cos \left[\dfrac{\pi}{2}\left(1 - \dfrac{1}{3^n}\right)\right] + i \sin \left[\dfrac{\pi}{2}\left(1 - \dfrac{1}{3^n}\right)\right]$

Hence show that $\lim\limits_{n \to \infty} x_1 x_2 x_3 \ldots x_n = i$.

10. If $2 \cos \theta = a + \dfrac{1}{a}$ and $2 \cos \phi = b + \dfrac{1}{b}$, prove that one of the values of $ab + \dfrac{1}{ab}$ is $2 \cos (\theta + \phi)$.

11. Prove that: (*i*) $\left(\dfrac{1 + \cos \theta + i \sin \theta}{1 + \cos \theta - i \sin \theta}\right)^n = \cos n\theta + i \sin n\theta$

(*ii*) $\left(\dfrac{1 + \sin \theta + i \cos \theta}{1 + \sin \theta - i \cos \theta}\right)^n = \cos \left(\dfrac{n\pi}{2} - n\theta\right) + i \sin \left(\dfrac{n\pi}{2} - n\theta\right)$.

12. Apply De Moivre's Theorem to find an equation whose roots are the nth powers of the roots of the equation $x^2 - 2x \cos \theta + 1 = 0$.

13. Prove that: (*i*) $(1 + i)^n + (1 - i)^n = 2^{n/2 + 1} \cos \left(\dfrac{n\pi}{4}\right)$.

(*ii*) $(\sqrt{3} + i)^n + (\sqrt{3} - i)^n = 2^{n + 1} \cos \left(\dfrac{n\pi}{6}\right)$.

(*iii*) $(1 + i\sqrt{3})^n + (1 - i\sqrt{3})^n = 2^{n + 1} \cos \left(\dfrac{n\pi}{3}\right)$.

where n is a positive integer.

14. If α and β be the roots of $x^2 - 2x + 4 = 0$, prove that $\alpha^n + \beta^n = 2^{n + 1} \cos \dfrac{n\pi}{3}$.

15. If α, β be the roots of $t^2 - 2t + 2 = 0$, then prove that

(*i*) $\dfrac{(x + \alpha)^n - (x + \beta)^n}{\alpha - \beta} = \dfrac{\sin n\phi}{\sin^n \phi}$

(*ii*) $\dfrac{(x + \alpha)^n + (x + \beta)^n}{\alpha + \beta} = \dfrac{\cos n\phi}{\sin^n \phi}$, where $x + 1 = \cot \phi$.

16. If $x = \cos \alpha + i \sin \alpha$, $y = \cos \beta + i \sin \beta$, prove that: $\dfrac{(x + y)(xy - 1)}{(x - y)(xy + 1)} = \dfrac{\sin \alpha + \sin \beta}{\sin \alpha - \sin \beta}$.

17. Find the cube roots of unity.

18. Find all the values of:

(*i*) $(1)^{1/4}$ (*ii*) $(-1)^{1/4}$ (*K.U.K. Dec. 2015*) (*iii*) $(1 + i)^{1/3}$

(*iv*) $(1 + \sqrt{-3})^{3/4}$ (*v*) $(i)^{1/4}$ (*iv*) $(-i)^{1/6}$

19. Prove that the nth roots of unity form a series in G.P. Hence show that

(*i*) their product $= (-1)^{n-1}$ (*ii*) their sum $= 0$

20. Use De Moivre's theorem to solve the equation $x^4 - x^3 + x^2 - x + 1 = 0$.

Answers

1. (*i*) $\cos 13\theta - i \sin 13\theta$ (*ii*) $\cos 8\theta + i \sin 8\theta$

12. $x^2 - 2x \cos n\theta + 1 = 0$ **17.** $1, \dfrac{-1 \pm i\sqrt{3}}{2}$

18. (*i*) $\pm 1, \pm i$ (*ii*) $\dfrac{1 \pm i}{\sqrt{2}}, \dfrac{-1 \pm i}{\sqrt{2}}$

(*iii*) $2^{1/6} \text{ cis } \frac{r\pi}{12}$, where $r = 1, 9, 17$ (*iv*) $2^{1/3} \text{ cis } \frac{r\pi}{18}$, where $r = 1, 13, 25$

(*v*) $\pm \text{ cis } \frac{\pi}{8}, \pm \text{ cis } \frac{5\pi}{8}$ (*vi*) $\pm \text{ cis } \frac{r\pi}{12}$, where $r = 3, 7, 11$

20. $\cos \frac{r\pi}{5} \pm i \sin \frac{r\pi}{5}$, where $r = 1, 3$

3.2. FUNCTION OF A COMPLEX VARIABLE

If x and y are real variables, then $z = x + iy$ is called a **complex variable.** If corresponding to each value of a complex variable $z(= x + iy)$ in a given region R, there correspond one or more values of another complex variable w $(= u + iv)$, then w is called a function of the complex variable z and is denoted by

$$w = f(z) = u + iv$$

For example, if $w = z^2$, where $z = x + iy$ and $w = f(z) = u + iv$

then $u + iv = (x + iy)^2 = (x^2 - y^2) + i(2xy)$

$\Rightarrow$ $u = x^2 - y^2$ and $v = 2xy$

Thus u and v, the real and imaginary parts of w, are functions of the real variables x and y.

$$\therefore \qquad w = f(z) = u(x, y) + iv(x, y)$$

If to each value of z there corresponds one and only one value of w, then w is called a *single-valued function* of z. If to each value of z there correspond more than one values of w, then w is called a *multi-valued function of* z.

To represent $w = f(z)$ graphically, we take two Argand diagrams: one to represent the point z and the other to represent w. The former diagram is called the xOy-plane or the z-plane and the latter uOv-plane or the w-plane.

3.3. EXPONENTIAL FUNCTION OF A COMPLEX VARIABLE

Def. The exponential function of the complex variable $z = x + iy$, where x and y are real, is defined as

$$e^z = 1 + \frac{z}{1!} + \frac{z^2}{2!} + \dots + \frac{z^n}{n!} + \dots \infty$$

where, $e = 2.718$ is the base of natural logarithms.

Replacing z by $x + iy$, we have

$$e^{x+iy} = 1 + \frac{x+iy}{1!} + \frac{(x+iy)^2}{2!} + \frac{(x+iy)^3}{3!} + \dots$$

Putting, $x = 0$, we get

$$e^{iy} = 1 + iy + \frac{(iy)^2}{2!} + \frac{(iy)^3}{3!} + \frac{(iy)^4}{4!} + \frac{(iy)^5}{5!} + \dots$$

$$= 1 + iy - \frac{y^2}{2!} - \frac{iy^3}{3!} + \frac{y^4}{4!} + \frac{iy^5}{5!} + \dots \qquad \left[\begin{matrix} \because i^2 = -1, & i^3 = -i \\ i^4 = 1, & i^5 = i \end{matrix}\right]$$

$$= \left(1 - \frac{y^2}{2!} + \frac{y^4}{4!} + \dots\right) + i\left(y - \frac{y^3}{3!} + \frac{y^5}{5!} - \dots\right) = \cos y + i \sin y$$

$\therefore \qquad e^{x+iy} = e^x \,.\, e^{iy} = e^x (\cos y + i \sin y)$

[**Note.** We have proved $e^{iy} = \cos y + i \sin y$

Changing i to $-i$ (*i.e.*, taking conjugates of both sides), $e^{-iy} = \cos y - i \sin y$.]

3.4. PERIODICITY

e^z IS A PERIODIC FUNCTION, WHERE z IS A COMPLEX VARIABLE

(*P.T.U. 2008*)

Proof. Let $\qquad z = x + iy$

then, by definition $\quad e^z = e^{x+iy} = e^x (\cos y + i \sin y) = e^x [\cos (2n\pi + y) + i \sin (2n\pi + y)]$

$$= e^x \,.\, e^{i(2n\pi + y)}$$

$$= e^{x + i(2n\pi + y)} = e^{(x+iy) + 2n\pi i} = e^{z + 2n\pi i}$$

i.e., e^z remains unchanged when z is increased by any multiple of $2\pi i$.

$\Rightarrow$ e^z is a periodic function with period $2\pi i$.

ILLUSTRATIVE EXAMPLES

Example 1. *Split up into real and imaginary parts:*

(*i*) $e^{5 + \frac{1}{2} i\pi}$ (*ii*) $e^{(5+3i)^2}$ (*iii*) e^{e^z}. (M.D.U. Dec. 2013)

Sol. (*i*) $\qquad e^{5+\frac{1}{2}i\pi} = e^5 \left[\cos \frac{\pi}{2} + i \sin \frac{\pi}{2}\right] = e^5 [0 + i \cdot 1] = ie^5$

$\therefore \qquad \text{Re}\,[e^{5+\frac{1}{2}i\pi}] = 0, \quad \text{Im}\,[e^{5+\frac{1}{2}i\pi}] = e^5.$

(*ii*) $\qquad (5 + 3i)^2 = 25 + 9i^2 + 30i = 16 + 30i \qquad [\because i^2 = -1]$

$\therefore \qquad e^{(5+3i)^2} = e^{16+30i} = e^{16} (\cos 30 + i \sin 30)$

$\therefore \qquad \text{Re}\,[e^{(5+3i)^2}] = e^{16} \cos 30, \ \text{Im}\,[e^{(5+3i)^2}] = e^{16} \sin 30.$

(*iii*)
$$e^{e^z} = e^{e^{x+iy}} = e^{e^x(\cos y + i \sin y)} = e^{(e^x \cos y) + i (e^x \sin y)}$$

$$= e^{e^x \cos y} [\cos (e^x \sin y) + i \sin (e^x \sin y)]$$

$\therefore \qquad \text{Re}(e^{e^z}) = e^{e^x \cos y} \,.\cos (e^x \sin y)$

$\qquad\quad \text{Im}\,(e^{e^z}) = e^{e^x \cos y} \,.\sin (e^x \sin y).$

Example 2. *Find all values of z which satisfy* $e^z = 1 + i$.

Sol. Since $\quad e^z = e^{x+iy} = e^x (\cos y + i \sin y)$

$\therefore \quad e^z = 1 + i \Rightarrow e^x(\cos y + i \sin y) = 1 + i$

Equating real parts $\quad e^x \cos y = 1 \quad$ ...(1)

Equating imaginary parts $\quad e^x \sin y = 1 \quad$ ...(2)

Squaring and adding (1) and (2), we get

$$e^{2x} (\cos^2 y + \sin^2 y) = 1 + 1 \quad \text{or} \quad e^{2x} = 2$$

or
$$2x = \log 2 \quad \therefore \quad x = \frac{1}{2} \log 2$$

Dividing (2) by (1), $\quad \tan y = 1 = \tan \dfrac{\pi}{4}$

$$\Rightarrow \quad y = n\pi + \frac{\pi}{4}, \quad \text{where } n \text{ is an integer.}$$

$$\therefore \quad z = x + iy = \frac{1}{2} \log 2 + i\left(n\pi + \frac{\pi}{4}\right), \quad \text{where} \quad n = 0, \pm 1, \pm 2, \ldots\ldots$$

Example 3. *Prove that* $[\sin (\alpha - \theta) + e^{-i\alpha} \sin \theta]^n = \sin^{n-1} \alpha\, [\sin (\alpha - n\theta) + e^{-i\alpha} \sin n\theta]$.

Sol.
$$\begin{aligned}
\text{LHS} &= [(\sin \alpha \cos \theta - \cos \alpha \sin \theta) + (\cos \alpha - i \sin \alpha) \sin \theta]^n \\
&= (\sin \alpha \cos \theta - i \sin \alpha \sin \theta)^n = [\sin \alpha (\cos \theta - i \sin \theta)]^n \\
&= [\sin \alpha \,.\, e^{-i\theta}]^n = \sin^n \alpha \,.\, e^{-in\theta} \\
\text{RHS} &= \sin^{n-1} \alpha\, [(\sin \alpha \cos n\theta - \cos \alpha \sin n\theta) + (\cos \alpha - i \sin \alpha) \sin n\theta] \\
&= \sin^{n-1} \alpha\, [\sin \alpha \cos n\theta - i \sin \alpha \sin n\theta] = \sin^n \alpha\, [\cos n\theta - i \sin n\theta] \\
&= \sin^n \alpha \,.\, e^{-in\theta}
\end{aligned}$$

$\therefore \quad$ LHS = RHS.

Example 4. *Given* $\dfrac{1}{\rho} = \dfrac{1}{LPi} + CPi + \dfrac{1}{R}$ *where L, P, R are real, express* ρ *in the form* $Ae^{i\theta}$ *giving the values of A and* θ.

Sol.
$$\frac{1}{\rho} = \frac{1}{\text{LP}i} + \text{CP}i + \frac{1}{\text{R}}$$

$$\Rightarrow \quad \frac{1}{\rho} = \frac{\text{R} + \text{LP}^2\text{CR}\,(-1) + \text{LP}i}{\text{LPR}i}$$

$$\Rightarrow \quad \rho = \frac{\text{LPR}i}{(\text{R} - \text{LP}^2\text{CR}) + \text{LP}i}$$

$$= \frac{\text{LPR}i}{(\text{R} - \text{LP}^2\text{CR}) + \text{LP}i} \times \frac{(\text{R} - \text{LP}^2\text{CR}) - \text{LP}i}{(\text{R} - \text{LP}^2\text{CR}) - \text{LP}i}$$

$$= \frac{L^2P^2R + iLPR(R - LP^2CR)}{(R - LP^2CR)^2 + L^2P^2}$$

$$= A(\cos\theta + i\sin\theta), \text{ say}$$

Equating real and imaginary parts, we get

$$A\cos\theta = \frac{L^2P^2R}{(R - LP^2CR)^2 + L^2P^2} \quad ...(1)$$

$$A\sin\theta = \frac{LPR(R - LP^2CR)}{(R - LP^2CR)^2 + L^2P^2} \quad ...(2)$$

Squaring and adding (1) and (2),

$$A^2 = \frac{L^4P^4R^2 + L^2P^2R^2(R - LP^2CR)^2}{[(R - LP^2CR)^2 + L^2P^2]^2}$$

$$= \frac{L^2P^2R^2[L^2P^2 + (R - LP^2CR)^2]}{[(R - LP^2CR)^2 + L^2P^2]^2}$$

$$= \frac{L^2P^2R^2}{(R - LP^2CR)^2 + L^2P^2}$$

$$\Rightarrow \qquad A = \frac{LPR}{\sqrt{(R - LP^2CR)^2 + L^2P^2}}$$

Dividing (2) by (1),

$$\tan\theta = \frac{R - LP^2CR}{LP}$$

$$\Rightarrow \qquad \theta = \tan^{-1}\left[\frac{R(1 - LP^2C)}{LP}\right]$$

Hence, $\rho = A(\cos\theta + i\sin\theta) = Ae^{i\theta}$

where $A = \dfrac{LPR}{\sqrt{(R - LP^2CR)^2 + L^2P^2}}$ and $\theta = \tan^{-1}\left[\dfrac{R(1 - LP^2C)}{LP}\right]$.

3.5. TRIGONOMETRIC FUNCTIONS OF A COMPLEX VARIABLE

1. Definitions. For all real values of x, we know that

$$e^{ix} = \cos x + i\sin x \quad \text{and} \quad e^{-ix} = \cos x - i\sin x$$

Adding and subtracting, we get $\cos x = \dfrac{e^{ix} + e^{-ix}}{2}$; $\sin x = \dfrac{e^{ix} - e^{-ix}}{2i}$

These are called Euler's Exponential values of $\sin x$ and $\cos x$, where $x \in R$.

If $z = x + iy$ then trigonometric functions of z are defined as follows:

$$\cos z = \frac{e^{iz} + e^{-iz}}{2}, \qquad \sin z = \frac{e^{iz} - e^{-iz}}{2i}, \qquad \tan z = \frac{\sin z}{\cos z} = \frac{e^{iz} - e^{-iz}}{i(e^{iz} + e^{-iz})}$$

$$\cot z = \frac{\cos z}{\sin z} = \frac{i(e^{iz} + e^{-iz})}{e^{iz} - e^{-iz}}, \qquad \sec z = \frac{1}{\cos z} = \frac{2}{e^{iz} + e^{-iz}}, \qquad \text{cosec } z = \frac{1}{\sin z} = \frac{2i}{e^{iz} - e^{-iz}}$$

2. Euler's Theorem.

For all values of θ, real or complex, $e^{i\theta} = \cos\theta + i\sin\theta$.

For all values of θ, real or complex $\cos\theta = \dfrac{e^{i\theta} + e^{-i\theta}}{2}$ and $\sin\theta = \dfrac{e^{i\theta} - e^{-i\theta}}{2i}$

$$\therefore \quad \cos\theta + i\sin\theta = \frac{e^{i\theta} + e^{-i\theta}}{2} + \frac{e^{i\theta} - e^{-i\theta}}{2} = \frac{2e^{i\theta}}{2} = e^{i\theta}.$$

Hence $e^{i\theta} = \cos\theta + i\sin\theta$ for all values of θ.

3. Periodicity of Circular Functions.

(*a*) **To prove that sin z and cos z are periodic functions with period 2π.**

We know that $\sin z = \dfrac{e^{iz} - e^{-iz}}{2i}$

If n is any integer, then

$$\sin(z + 2n\pi) = \frac{e^{i(z+2n\pi)} - e^{-i(z+2n\pi)}}{2i}$$

$$= \frac{e^{iz}.e^{2n\pi i} - e^{-iz}.e^{-2n\pi i}}{2i} = \frac{e^{iz} - e^{-iz}}{2i} \qquad [\because e^{2n\pi i} = 1 = e^{-2n\pi i}]$$

$$= \sin z$$

$\Rightarrow$ $\sin z$ remains unchanged when z is increased by any multiple of 2π.

$\therefore$ **sin z is a periodic function with period 2π.**

Similarly, $\cos z$ can be shown to be a periodic function with period 2π.

(*b*) **To prove that tan z is a periodic function with period π.**

We know that $\tan z = \dfrac{e^{iz} - e^{-iz}}{i(e^{iz} + e^{-iz})}$

If n is any integer, $\tan(z + n\pi) = \dfrac{e^{i(z+n\pi)} - e^{-i(z+n\pi)}}{i[e^{i(z+n\pi)} + e^{-i(z+n\pi)}]} = \dfrac{e^{iz}.e^{in\pi} - e^{-iz}.e^{-in\pi}}{i[e^{iz}.e^{in\pi} + e^{-iz}.e^{n\pi}]}$

Multiplying the numerator and denominator by $e^{in\pi}$

$$= \frac{e^{iz}.e^{2n\pi i} - e^{-iz}}{i[e^{iz}.e^{2n\pi i} + e^{-iz}]} = \frac{e^{iz} - e^{-iz}}{i(e^{iz} + e^{-iz})} = \tan z \qquad |\because e^{2n\pi i} = 1$$

$\Rightarrow$ $\tan z$ remains unchanged when z is increased by any multiple of π.

$\therefore$ **tan z is a periodic function with period π.**

3.6. TRIGONOMETRIC IDENTITIES

If z is a complex variable, prove that

(i) $\sin^2 z + \cos^2 z = 1$ (ii) $\sin 2z = 2 \sin z \cos z$

(iii) $\cos 2z = \cos^2 z - \sin^2 z = 2 \cos^2 z - 1 = 1 - 2 \sin^2 z$

(iv) $\tan 2z = \dfrac{2 \tan z}{1 - \tan^2 z}$ (v) $\sin(-z) = -\sin z$

(vi) $\sin 3z = 3 \sin z - 4 \sin^3 z$ (vii) $\tan 3z = \dfrac{3 \tan z - \tan^3 z}{1 - 3 \tan^2 z}$.

Proof. (i) LHS $= \sin^2 z + \cos^2 z = \left(\dfrac{e^{iz} - e^{-iz}}{2i}\right)^2 + \left(\dfrac{e^{iz} + e^{-iz}}{2}\right)^2$

$$= -\frac{1}{4}(e^{2iz} + e^{-2iz} - 2) + \frac{1}{4}(e^{2iz} + e^{-2iz} + 2) = \frac{1}{2} + \frac{1}{2} = 1 = \text{RHS}$$

(ii) $\text{RHS} = 2 \sin z \cos z = 2 . \dfrac{e^{iz} - e^{-iz}}{2i} . \dfrac{e^{iz} + e^{-iz}}{2} = \dfrac{e^{2iz} - e^{-2iz}}{2} = \sin 2z = \text{LHS}$

(iii) $\cos^2 z - \sin^2 z = \left(\dfrac{e^{iz} + e^{-iz}}{2}\right)^2 - \left(\dfrac{e^{iz} - e^{-iz}}{2i}\right)^2$

$$= \frac{1}{4}[(e^{2iz} + e^{-2iz} + 2) + (e^{2iz} + e^{-2iz} - 2)] = \frac{e^{2iz} + e^{-2iz}}{2} = \cos 2z$$

$$2 \cos^2 z - 1 = 2\left(\frac{e^{iz} + e^{-iz}}{2i}\right)^2 - 1 = \frac{1}{2}(e^{2iz} + e^{-2iz} + 2) - 1 = \frac{e^{2iz} + e^{-2iz}}{2} = \cos 2z$$

$$1 - 2 \sin^2 z = 1 - 2\left(\frac{e^{iz} - e^{-iz}}{2i}\right)^2 = 1 + \frac{1}{2}(e^{2iz} + e^{-2iz} - 2) = \frac{e^{2iz} + e^{-2iz}}{2} = \cos 2z$$

Hence the result.

(iv) $\text{RHS} = \dfrac{2 \tan z}{1 - \tan^2 z} = \dfrac{2 . \dfrac{e^{iz} - e^{-iz}}{i(e^{iz} + e^{-iz})}}{1 - \left[\dfrac{e^{iz} - e^{-iz}}{i(e^{iz} + e^{-iz})}\right]^2} = \dfrac{2(e^{iz} - e^{-iz})(e^{iz} + e^{-iz})}{i[(e^{iz} + e^{-iz})^2 + (e^{iz} - e^{-iz})^2]}$

$$= \frac{2(e^{2iz} - e^{-2iz})}{i.2(e^{2iz} + e^{-2iz})} = \frac{e^{2iz} - e^{-2iz}}{i(e^{2iz} + e^{-2iz})} = \tan 2z = \text{LHS}$$

(v) $\sin(-z) = \dfrac{e^{i(-z)} - e^{-i(-z)}}{2i} = \dfrac{e^{-iz} - e^{iz}}{2i} = -\dfrac{e^{iz} - e^{-iz}}{2i} = -\sin z.$

(vi) $\sin 3z = \dfrac{e^{3iz} - e^{-3iz}}{2i} = \dfrac{x^3 - y^3}{2i}$, where $x = e^{iz}$, $y = e^{-iz}$

$$= \frac{(x-y)^3 + 3xy(x-y)}{2i} = \frac{1}{2i}\left[(e^{iz}-e^{-iz})^3 + 3\,.\,e^{iz}\,.\,e^{-iz}(e^{iz}-e^{-iz})\right]$$

$$= \frac{1}{2i}\left[(2i\sin z)^3 + 3(2i\sin z)\right] = \frac{1}{2i}\left[-8i\sin^3 z + 6i\sin z\right] = 3\sin z - 4\sin^3 z.$$

(*vii*) RHS $= \dfrac{3\tan z - \tan^3 z}{1 - 3\tan^2 z}$

$$= \frac{3.\left[\dfrac{e^{iz}-e^{-iz}}{i(e^{iz}+e^{-iz})}\right] - \left[\dfrac{e^{iz}-e^{-iz}}{i(e^{iz}+e^{-iz})}\right]^3}{1 - 3.\left[\dfrac{e^{iz}-e^{-iz}}{i(e^{iz}+e^{-iz})}\right]^2} = \frac{3\left[\dfrac{e^{iz}-e^{-iz}}{i(e^{iz}+e^{-iz})}\right] + \dfrac{1}{i}.\left[\dfrac{e^{iz}-e^{-iz}}{e^{iz}+e^{-iz}}\right]^3}{1 + 3.\left[\dfrac{e^{iz}-e^{-iz}}{e^{iz}+e^{-iz}}\right]^2}$$

$$= \frac{3\dfrac{x}{iy} + \dfrac{1}{i}.\left(\dfrac{x}{y}\right)^3}{1 + 3\left(\dfrac{x}{y}\right)^2}, \text{ where } x = e^{iz} - e^{-iz},\ y = e^{iz} + e^{-iz}$$

$$= \frac{3xy^2 + x^3}{iy^3}\cdot\frac{y^2}{y^2 + 3x^2} = \frac{x(3y^2 + x^2)}{iy(y^2 + 3x^2)} = \frac{x(3e^{2iz} + 3e^{-2iz} + 6 + e^{2iz} + e^{-2iz} - 2)}{iy(e^{2iz} + e^{-2iz} + 2 + 3e^{2iz} + 3e^{-2iz} - 6)}$$

$$= \frac{x(4e^{2iz} + 4e^{-2iz} + 4)}{iy(4e^{2iz} + 4e^{-2iz} - 4)} = \frac{(e^{iz} - e^{-iz})(e^{2iz} + e^{-2iz} + 1)}{i(e^{iz} + e^{-iz})(e^{2iz} + e^{-2iz} - 1)}$$

$$= \frac{e^{3iz} - e^{-3iz}}{i(e^{3iz} + e^{-3iz})} \qquad \left[\because \begin{array}{l}(a-b)(a^2+b^2+ab) = a^3 - b^3\\ (a+b)(a^2+b^2-ab) = a^3 + b^3\end{array}\right.$$

$= \tan 3z.$

EXERCISE 3.2

1. If $z = x + iy$, find the real and imaginary parts of (*i*) $e^{\bar{z}}$ (*ii*) $\exp(z^2)$ (*iii*) $\dfrac{e^{i\theta}}{1 - \lambda e^{i\phi}}$.

2. Prove that:

   (*i*) $\sin(\alpha + n\theta) - e^{i\alpha}\sin n\theta = e^{-in\theta}\sin\alpha$ (*M.D.U. May 2013*)

   (*ii*) $[\sin(\alpha + \theta) - e^{i\alpha}\sin\theta]^n = \sin^n\alpha\,.\,e^{-in\theta}$

   (*iii*) $\dfrac{z^2 - 1}{z^2 + 1} = i\tan\theta$, where $z = e^{i\theta}$.

3. Prove that by a proper choice of p and q, $pe^{2i\theta} + qe^{-2i\theta}$ can be made equal to $5\cos 2\theta - 7\sin 2\theta$.

4. If z is a complex number, prove that

   (*i*) $\cos(-z) = \cos z$

   (*ii*) $\tan(-z) = -\tan z$

   (*iii*) $\cos 3z = 4\cos^3 z - 3\cos z$

   (*iv*) $\tan z = \dfrac{\sin 2z}{1 + \cos 2z}$.

5. If z_1, z_2 are complex numbers, show that

(*i*) $\sin(z_1 + z_2) = \sin z_1 \cos z_2 + \cos z_1 \sin z_2$ (*ii*) $\cos(z_1 - z_2) = \cos z_1 \cos z_2 + \sin z_1 \sin z_2$

(*iii*) $\tan(z_1 + z_2) = \dfrac{\tan z_1 + \tan z_2}{1 - \tan z_1 \tan z_2}$ (*iv*) $\sin z_1 + \sin z_2 = 2 \sin \dfrac{z_1 + z_2}{2} \cos \dfrac{z_1 - z_2}{2}$

(*v*) $\cos z_1 - \cos z_2 = 2 \sin \dfrac{z_1 + z_2}{2} \sin \dfrac{z_2 - z_1}{2}$.

6. Show that

(*i*) $\cos(\alpha + i\beta) = \dfrac{1}{2}(e^{-\beta} + e^{\beta}) \cos \alpha + \dfrac{i}{2}(e^{-\beta} - e^{\beta}) \sin \alpha$

(*ii*) $\sin(\alpha - i\beta) = \dfrac{1}{2}(e^{-\beta} + e^{\beta}) \sin \alpha + \dfrac{i}{2}(e^{-\beta} - e^{\beta}) \cos \alpha$.

7. If α and β are the imaginary cube roots of unity, prove that

$$\alpha e^{\alpha x} + \beta e^{\beta x} = -e^{-\frac{x}{2}} \left(\cos \frac{\sqrt{3}}{2} x + \sqrt{3} \sin \frac{\sqrt{3}}{2} x \right).$$

Answers

1. (*i*) $e^x \cos y, -e^x \sin y$ (*ii*) $e^{x^2 - y^2} \cos 2xy, e^{x^2 - y^2} \sin 2xy$.

(*iii*) $\dfrac{\cos\theta - \lambda\cos(\theta - \phi)}{1 - 2\lambda\cos\phi + \lambda^2}, \dfrac{\sin\theta - \lambda\sin(\theta - \phi)}{1 - 2\lambda\cos\phi + \lambda^2}$

3. $p = \dfrac{1}{2}(5 + 7i), q = \dfrac{1}{2}(5 - 7i)$.

3.7. LOGARITHMIC FUNCTION OF A COMPLEX VARIABLE

Definition. If $\omega = e^z$, where z and ω are complex numbers, then z is called a logarithm of ω to the base e. Thus $\log_e \omega = z$.

1. Prove that $\log_e \omega$ is a many-valued function.

Proof. We know that $e^{2n\pi i} = \cos 2n\pi + i \sin 2n\pi = 1$

Let $e^z = \omega$, then $e^{z + 2n\pi i} = e^z \cdot e^{2n\pi i} = e^z \cdot 1 = \omega$

$\therefore$ By definition $\log_e \omega = z + 2n\pi i$, where n is zero, or any + ve or – ve integer.

Thus if z be a logarithm of ω, so is $z + 2n\pi i$.

Hence the logarithm of a complex number has infinite values and is thus a many-value function.

Note. The value $z + 2n\pi i$ is called the *general value* of $\log_e \omega$ and is denoted by $\mathbf{Log}_e\, \omega$.

Thus $\mathbf{Log}_e\, \omega = z + 2n\pi i = 2n\pi i + \mathbf{log}_e\, \omega$

If $\omega = x + iy$, then $\mathbf{Log}(\mathbf{x} + \mathbf{iy}) = \mathbf{2n\pi i} + \mathbf{log}(\mathbf{x} + \mathbf{iy})$.

If we put n = 0, in the general value, we get the principal value of z, i.e., $\log_e \omega$.

2. Prove that $\log(-N) = \pi i + \log N$, where N is positive.

Proof. $-N = N(-1) = N(\cos \pi + i \sin \pi) = N \cdot e^{i\pi}$

$\therefore$ $\log(-N) = \log(N \cdot e^{i\pi}) = \log N + \log e^{i\pi} = \log N + \pi i$.

3. Separate $\mathrm{Log}(\alpha + i\beta)$ into real and imaginary parts.

Proof. Let $\alpha + i\beta = r(\cos \theta + i \sin \theta)$ so that $r = \sqrt{\alpha^2 + \beta^2}$, $\theta = \tan^{-1} \dfrac{\beta}{\alpha}$

$\therefore \quad \text{Log}(\alpha + i\beta) = 2n\pi i + \log(\alpha + i\beta) = 2n\pi i + \log[r(\cos\theta + i\sin\theta)]$

$$= 2n\pi i + \log(r\,e^{i\theta}) = 2n\pi i + \log r + \log e^{i\theta} = 2n\pi i + \log r + i\theta$$

$$= 2n\pi i + \log\sqrt{\alpha^2+\beta^2} + i\tan^{-1}\frac{\beta}{\alpha} = 2n\pi i + \frac{1}{2}\log(\alpha^2+\beta^2) + i\tan^{-1}\frac{\beta}{\alpha}$$

$\therefore \quad \text{Re}[\text{Log}(\alpha + i\beta)] = \frac{1}{2}\log(\alpha^2+\beta^2)$

$\text{Im}[\text{Log}(\alpha + i\beta)] = 2n\pi + \tan^{-1}\frac{\beta}{\alpha}.$

Note. Putting $n = 0$, the principal value of $\mathbf{log}\,(\alpha + \mathbf{i}\beta) = \frac{\mathbf{1}}{\mathbf{2}}\mathbf{log}\,(\alpha^2+\beta^2) + \mathbf{i}\,\mathbf{tan}^{-1}\frac{\beta}{\alpha}$.

ILLUSTRATIVE EXAMPLES

Example 1. *Prove that* $\log(1 + re^{i\theta}) = \frac{1}{2}\log(1 + 2r\cos\theta + r^2) + i\tan^{-1}\frac{r\sin\theta}{1 + r\cos\theta}$.

Deduce that $\log(1 + \cos\theta + i\sin\theta) = \log\left(2\cos\frac{\theta}{2}\right) + i\frac{\theta}{2}$.

Sol. $\log(1 + re^{i\theta}) = \log[1 + r(\cos\theta + i\sin\theta)] = \log[(1 + r\cos\theta) + i(r\sin\theta)]$

$$= \frac{1}{2}\log[(1 + r\cos\theta)^2 + (r\sin\theta)^2] + i\tan^{-1}\frac{r\sin\theta}{1 + r\cos\theta}$$

$$= \frac{1}{2}\log[1 + 2r\cos\theta + r^2\cos^2\theta + r^2\sin^2\theta] + i\tan^{-1}\frac{r\sin\theta}{1 + r\cos\theta}$$

$$= \frac{1}{2}\log[1 + 2r\cos\theta + r^2] + i\tan^{-1}\frac{r\sin\theta}{1 + r\cos\theta} \qquad \ldots(i)$$

Now $\quad \log(1 + \cos\theta + i\sin\theta) = \log(1 + e^{i\theta})$

Putting $r = 1$ in (i),

$$\log(1 + \cos\theta + i\sin\theta) = \frac{1}{2}\log(1 + 2\cos\theta + 1) + i\tan^{-1}\frac{\sin\theta}{1 + \cos\theta}$$

$$= \frac{1}{2}\log[2(1 + \cos\theta)] + i\tan^{-1}\frac{2\sin\frac{\theta}{2}\cos\frac{\theta}{2}}{2\cos^2\frac{\theta}{2}}$$

$$= \frac{1}{2}\log\left[2\,.\,2\cos^2\frac{\theta}{2}\right] + i\tan^{-1}\left(\tan\frac{\theta}{2}\right) = \frac{1}{2}\log\left[\left(2\cos\frac{\theta}{2}\right)^2\right] + i\frac{\theta}{2}$$

$$= \frac{1}{2}\,.\,2\log\left(2\cos\frac{\theta}{2}\right) + i\,.\,\frac{\theta}{2} = \log\left(2\cos\frac{\theta}{2}\right) + i\,.\,\frac{\theta}{2}.$$

Example 2. *Find the general value of log (– 3).*

Sol. $\because \quad -3 = 3(-1) = 3\,\text{cis}\,\pi = 3\,e^{i\pi}$

$\therefore \quad \text{Log}(-3) = \text{Log}(3e^{i\pi}) = 2n\pi i + \log(3\,e^{i\pi})$

$$= 2n\pi i + \log 3 + \log e^{i\pi} = 2n\pi i + \log 3 + i\pi = \log 3 + i(2n + 1)\pi.$$

Example 3. *Separate into real and imaginary parts Log (4 + 3i).* (P.T.U. 2010)

Sol. Let $4 + 3i = r(\cos\theta + i\sin\theta)$

Equating real and imaginary parts $r\cos\theta = 4$; $r\sin\theta = 3$

Squaring and adding, $r^2 = 16 + 9 = 25 \quad \therefore \quad r = 5$

Dividing, $\tan\theta = \dfrac{3}{4} \qquad \therefore \quad \theta = \tan^{-1}\dfrac{3}{4}$

$$\therefore \quad \text{Log}\,(4+3i) = \text{Log}\,[r(\cos\theta + i\sin\theta)] = \text{Log}\,(re^{i\theta}) = 2n\pi i + \log(re^{i\theta})$$

$$= 2n\pi i + \log r + \log e^{i\theta} = 2n\pi i + \log 5 + i\theta = \log 5 + 2n\pi i + i\tan^{-1}\frac{3}{4}$$

$\therefore$ Re[Log (4 + 3i)] = log 5

$$\text{Im}[\text{Log}\,(4+3i)] = \left(2n\pi + \tan^{-1}\frac{3}{4}\right).$$

Example 4. *Prove that* $\tan\left(i\log\dfrac{a-ib}{a+ib}\right) = \dfrac{2ab}{a^2-b^2}$. (M.D.U. 2012; K.U.K. May 2013)

Sol. Let $a + ib = r(\cos\theta + i\sin\theta)$

Equating real and imaginary parts $r\cos\theta = a$, $r\sin\theta = b$.

Dividing, $\tan\theta = \dfrac{b}{a}$ ...(i)

Also $a - ib = r(\cos\theta - i\sin\theta)$

$$\text{L.H.S.} = \tan\left[i\log\frac{r(\cos\theta - i\sin\theta)}{r(\cos\theta + i\sin\theta)}\right] = \tan\left[i\log\frac{e^{-i\theta}}{e^{i\theta}}\right]$$

$$= \tan[i\log e^{-2i\theta}] = \tan[i(-2i\theta)\log e]$$

$$= \tan 2\theta = \frac{2\tan\theta}{1-\tan^2\theta} = \frac{2\dfrac{b}{a}}{1-\dfrac{b^2}{a^2}} \qquad [\because \text{ of } (i)]$$

$$= \frac{2ab}{a^2-b^2}.$$

Example 5. *Express Log (Log i) in the form A + iB.*

Sol. $i = \cos\dfrac{\pi}{2} + i\sin\dfrac{\pi}{2} = e^{i\pi/2}$

$$\therefore \quad \text{Log}\, i = 2n\pi i + \log e^{i\pi/2} = 2n\pi i + i\frac{\pi}{2} = i(4n+1)\frac{\pi}{2}$$

$$\therefore \quad \text{Log}\,(\text{Log}\, i) = \text{Log}\left[i(4n+1)\frac{\pi}{2}\right] = 2m\pi i + \log\left[i(4n+1)\frac{\pi}{2}\right]$$

$$= 2m\pi i + \log i + \log(4n+1)\frac{\pi}{2}$$

$$= 2m\pi i + \log e^{i\pi/2} + \log(4n+1)\frac{\pi}{2} = 2m\pi i + i\frac{\pi}{2} + \log(4n+1)\frac{\pi}{2}$$

$$= \log(4n+1)\frac{\pi}{2} + i(4m+1)\frac{\pi}{2}.$$

3.8. THE GENERAL EXPONENTIAL FUNCTION

Definition. The general exponential function a^z is defined by the equation $a^z = e^{z \operatorname{Log} a}$, where a and z are any numbers, real or complex.

1. Prove that a^z is a many valued function.

Proof. Since $\operatorname{Log} a = 2n\pi i + \log a$

$\therefore \quad a^z = e^{z(2n\pi i + \log a)}$

Hence a^z is a many valued function and its principal value is obtained by putting $n = 0$.

2. Separate $(a + ib)^{x+iy}$ into real and imaginary parts.

Proof. $(a + ib)^{x+iy} = e^{(x+iy)\operatorname{Log}(a+ib)}$

$$= e^{(x+iy)[2n\pi i + \log(a+ib)]}$$

$$= e^{(x+iy)\left[2n\pi i + \frac{1}{2}\log(a^2+b^2) + i\tan^{-1}\frac{b}{a}\right]}$$

$$= e^{\left[\frac{x}{2}\log(a^2+b^2) - 2n\pi y - y\tan^{-1}\frac{b}{a}\right] + i\left[2n\pi x + x\tan^{-1}\frac{b}{a} + \frac{y}{2}\log(a^2+b^2)\right]}$$

$$= e^{\alpha + i\beta}$$

where, $\alpha = \frac{x}{2}\log(a^2+b^2) - 2n\pi y - y\tan^{-1}\frac{b}{a}$, $\beta = \frac{y}{2}\log(a^2+b^2) + 2n\pi x + x\tan^{-1}\frac{b}{a}$

$\Rightarrow \quad (a+ib)^{x+iy} = e^{\alpha}(\cos\beta + i\sin\beta)$

$\therefore \quad \operatorname{Re}[(a+ib)^{x+iy}] = e^{\alpha}\cos\beta$

and $\operatorname{Im}[(a+ib)^{x+iy}] = e^{\alpha}\sin\beta$

ILLUSTRATIVE EXAMPLES

Example 1. *Prove that i^i is wholly real and find its principal value. Also show that the values of i^i form a G.P.* (P.T.U. 2010)

Sol. $i^i = e^{i \operatorname{Log} i}$ [By definition]

$$= e^{i(2n\pi i + \log i)} = e^{i[2n\pi i + \log(\cos\pi/2 + i\sin\pi/2)]}$$

$$= e^{i[2n\pi i + \log e^{i\pi/2}]} = e^{i[2n\pi i + i\pi/2]} = e^{i^2(4n+1)\pi/2} = e^{-(4n+1)\pi/2}$$

which is wholly real.

The principal value of $i^i = e^{-\pi/2}$ (Putting $n = 0$)

Putting $n = 0, 1, 2, \ldots\ldots$ the values of i^i are $e^{-\pi/2}, e^{-5\pi/2}, e^{-9\pi/2}, \ldots\ldots$ which form a G.P. whose common ratio is $e^{-2\pi}$.

Example 2. *If $i^{\alpha+i\beta} = \alpha + i\beta$, prove that $\alpha^2 + \beta^2 = e^{-(4n+1)\pi\beta}$.*

Sol. $\alpha + i\beta = i^{\alpha+i\beta} = e^{(\alpha+i\beta)\operatorname{Log} i}$

$$= e^{(\alpha+i\beta)[2n\pi i + \log i]} = e^{(\alpha+i\beta)[2n\pi i + \log(\cos\pi/2 + i\sin\pi/2)]}$$

$$= e^{(\alpha+i\beta)[2n\pi i + \log e^{i\pi/2}]} = e^{(\alpha+i\beta)[2n\pi i + i\pi/2]} = e^{-\beta(4n+1)\pi/2 + i\alpha(4n+1)\pi/2}$$

$$= e^{-\beta(4n+1)\pi/2} \cdot e^{i\alpha(4n+1)\pi/2} = e^{-\beta(4n+1)\pi/2}\left[\cos(4n+1)\frac{\alpha\pi}{2} + i\sin(4n+1)\frac{\alpha\pi}{2}\right]$$

$[\because \quad e^{i\theta} = \cos\theta + i\sin\theta]$

Equating real and imaginary parts

$$\alpha = e^{-(4n+1)\beta\pi/2} \,.\, \cos (4n+1)\frac{\alpha\pi}{2}\,;\quad \beta = e^{-(4n+1)\beta\pi/2} \,.\, \sin (4n+1)\frac{\alpha\pi}{2}$$

Squaring and adding, $\alpha^2 + \beta^2 = e^{-(4n+1)\beta\pi}\left[\cos^2 (4n+1)\frac{\alpha\pi}{2} + \sin^2 (4n+1)\frac{\alpha\pi}{2}\right]$

$$= e^{-(4n+1)\beta\pi}.$$

Example 3. *Considering only the principal value, prove that the real part of*

$$(1+i\sqrt{3})^{1+i\sqrt{3}} \text{ is } 2e^{-\pi/\sqrt{3}} \cos\left(\frac{\pi}{3} + \sqrt{3}\log 2\right).$$

Sol. $(1+i\sqrt{3})^{1+i\sqrt{3}} = e^{(1+i\sqrt{3})\log(1+i\sqrt{3})} = e^{(1+i\sqrt{3})\left[\frac{1}{2}\log(1+3) + i\tan^{-1}\sqrt{3}\right]}$

$$= e^{(1+i\sqrt{3})\left[\frac{1}{2}\log 4 + i\pi/3\right]} = e^{(1+i\sqrt{3})\left[\frac{1}{2}.2\log 2 + i\pi/3\right]}$$

$$= e^{(\log 2 - \pi/\sqrt{3}) + i(\pi/3 + \sqrt{3}\log 2)} = e^{\log 2 - \pi/\sqrt{3}} \,.\, e^{i(\pi/3 + \sqrt{3}\log 2)}$$

$$= e^{\log 2}\, e^{-\pi/\sqrt{3}}\left[\cos\left(\frac{\pi}{3} + \sqrt{3}\log 2\right) + i\sin\left(\frac{\pi}{3} + \sqrt{3}\log 2\right)\right]$$

$$= 2e^{-\pi/\sqrt{3}}\left[\cos\left(\frac{\pi}{3} + \sqrt{3}\log 2\right) + i\sin\left(\frac{\pi}{3} + \sqrt{3}\log 2\right)\right] \quad [\because\ e^{\log f(x)} = f(x)]$$

$\Rightarrow$ Real part of $(1+i\sqrt{3})^{1+i\sqrt{3}}$ is $2e^{-\pi/\sqrt{3}}\cos\left(\frac{\pi}{3} + \sqrt{3}\log 2\right)$.

Example 4. *If* $i^{i^{i \ldots\ldots\ ad.inf.}} = A + iB$ *and only principal values are considered, prove that*

(*a*) $\tan\dfrac{\pi A}{2} = \dfrac{B}{A}$ (*b*) $A^2 + B^2 = e^{-\pi B}$. (M.D.U. 2010, Dec. 2013)

Sol. $i^{i^{i \ldots\ldots\ ad.inf.}} = A + iB \quad \Rightarrow \quad i^{A+iB} = A + iB$

Now $A + iB = i^{A+iB} = e^{(A+iB)\log i}$ (Taking principal values only)

$$= e^{(A+iB)\log(\cos\pi/2 + i\sin\pi/2)} = e^{(A+iB)\log(e^{i\pi/2})}$$

$$= e^{(A+iB)(i\pi/2)} = e^{-(B\pi/2) + i\,.\,(A\pi/2)}$$

$$= e^{-(B\pi/2)} \,.\, e^{i\,A\pi/2} = e^{-B\pi/2}\left(\cos\frac{A\pi}{2} + i\sin\frac{A\pi}{2}\right)$$

Equating real and imaginary parts

$$A = e^{-B(\pi/2)}\cos\frac{A\pi}{2} \quad \ldots(i)$$

$$B = e^{-B(\pi/2)}\sin\frac{A\pi}{2} \quad \ldots(ii)$$

Dividing (*ii*) by (*i*), $\tan\dfrac{A\pi}{2} = \dfrac{B}{A}$ ...(I)

Squaring and adding (*i*) and (*ii*), $A^2 + B^2 = e^{-B\pi}\left(\cos^2\dfrac{A\pi}{2} + \sin\dfrac{A\pi}{2}\right) = e^{-B\pi}$. ...(II)

Example 5. *If* $(a + ib)^p = m^{x+iy}$, *then prove that* $\frac{y}{x} = \frac{2\tan^{-1}\left(\frac{b}{a}\right)}{\log(a^2+b^2)}$ *when only principal values are considered.*

Sol. $$(a + ib)^p = m^{x+iy}$$

Taking log of both sides, $\log (a + ib)^p = \log m^{x+iy}$

or $$p \log (a + ib) = (x + iy) \log m$$

or $$p\left[\frac{1}{2}\log(a^2+b^2) + i\tan^{-1}\frac{b}{a}\right] = x \log m + iy \log m$$

(Considering only the principal values)

Equating real and imaginary parts $x \log m = \frac{1}{2} p \log (a^2 + b^2)$ ...(*i*)

$$y \log m = p \tan^{-1}\frac{b}{a} \quad ...(ii)$$

Dividing (*ii*) by (*i*), $$\frac{y}{x} = \frac{p\tan^{-1}\frac{b}{a}}{\frac{1}{2}p\log(a^2+b^2)} = \frac{2\tan^{-1}\frac{b}{a}}{\log(a^2+b^2)}.$$

Example 6. *If* $\tan \log (x + iy) = a + ib$ *and* $a^2 + b^2 \neq 1$, *then prove that*

$$\tan \log (x^2 + y^2) = \frac{2a}{1-a^2-b^2}.$$

Sol. $$\tan \log (x + iy) = a + ib \quad ...(i)$$

$\Rightarrow$ $$\tan \log (x - iy) = a - ib \quad ...(ii)$$

Now, $\tan \log (x^2 + y^2) = \tan \log (x + iy)(x - iy)$

$$= \tan [\log (x + iy) + \log (x - iy)] = \frac{\tan\log(x+iy) + \tan\log(x-iy)}{1 - \tan\log(x+iy).\tan\log(x-iy)}$$

$$= \frac{a+ib+a-ib}{1-(a+ib)(a-ib)} = \frac{2a}{1-a^2-b^2}, \text{ where } a^2 + b^2 \neq 1.$$

EXERCISE 3.3

1. Find the general value of

   (*i*) $\log (- i)$ (*ii*) $\log (1 + i)$ (*iii*) $\log(-1 + i\sqrt{3})$.

   (*P.T.U. 2012*)

2. Prove that

   (*i*) $i \log \left(\frac{x-i}{x+i}\right) = \pi - 2\tan^{-1} x$ (*ii*) $\cos\left[i\log\left(\frac{a+ib}{a-ib}\right)\right] = \frac{a^2-b^2}{a^2+b^2}$

   (*iii*) $i^i = e^{-(4n+1)\frac{\pi}{2}}$ (*P.T.U. 2010*) (*iv*) $\text{Log } i^i = -\left(2n+\frac{1}{2}\right)\pi$.

3. Show that

   (*i*) $\log (1 + i \tan \alpha) = \log \sec \alpha + i\alpha$ (*ii*) $\text{Log}_e\left(\frac{3-i}{3+i}\right) = 2i\left(n\pi - \tan^{-1}\frac{1}{3}\right)$.

(*iii*) $\log(6+8i) = \log 10 + i\tan^{-1}\dfrac{4}{3}$

(*iv*) $(\sqrt{i})^{\sqrt{i}} = e^{-\frac{\pi}{4\sqrt{2}}}\left(\cos\dfrac{\pi}{4\sqrt{2}} + i\sin\dfrac{\pi}{4\sqrt{2}}\right)$

(*v*) $\log(e^{i\alpha} + e^{i\beta}) = \log\left(2\cos\dfrac{\alpha-\beta}{2}\right) + i\left(\dfrac{\alpha+\beta}{2}\right)$

(*vi*) $\log\dfrac{(1+i)(1+i\sqrt{3})}{\sqrt{3}+i} = \dfrac{1}{2}\log 2 + i\,\dfrac{5\pi}{12}$.

4. Prove that $\sin\left[i\log\left(\dfrac{1+ie^{-i\theta}}{1-ie^{i\theta}}\right)\right]$ is wholly real.
5. Prove that $\sin(\log i^i) = -1$.
6. Prove that $\tan^{-1} x = \dfrac{1}{2i}\log\left(\dfrac{1+ix}{1-ix}\right)$.
7. If $\log\log(x+iy) = p+iq$, show that $y = x\tan[\tan q\log\sqrt{x^2+y^2}]$.
8. Prove that the principal value of

   $\dfrac{(a+ib)^{p+iq}}{(a-ib)^{p-iq}}$ is $\cos 2(p\alpha + q\log r) + i\sin 2(p\alpha + q\log r)$, where $r = \sqrt{a^2+b^2}$ and $\alpha = \tan^{-1}\dfrac{b}{a}$.
9. If $\dfrac{(1+i)^{x+iy}}{(1-i)^{x-iy}} = \alpha + i\beta$, prove that one of the values of $\tan^{-1}\dfrac{\beta}{\alpha}$ is $\dfrac{1}{2}\pi x + y\log 2$.
10. Prove that $\dfrac{(1+i)^{1-i}}{(1-i)^{1+i}} = \sin(\log 2) + i\cos(\log 2)$.
11. Prove that the real part of the principal value of $i^{\log(1+i)}$ is $e^{-\frac{\pi^2}{8}}\cos\left(\dfrac{\pi}{4}\log 2\right)$.
12. Prove that $\text{Log}_i\, i = \dfrac{4m+1}{4n+1}$, where m and n are integers.
13. If (x, y) is a point on the circle having its centre at the origin and radius a, prove that

$$\log\frac{x+iy-a}{x+iy+a} = \frac{1}{2}\log\frac{a-x}{a+x} + \frac{\pi}{2}i.$$

14. Prove that $\log\dfrac{1}{1-e^{i\theta}} = \log\left(\dfrac{1}{2}\operatorname{cosec}\dfrac{\theta}{2}\right) + i\left(\dfrac{\pi}{2} - \dfrac{\theta}{2}\right)$.
15. If $(a+ib)^{c+id}$ is wholly real and principal values are considered, prove that

$$(a+ib)^{c+id} = (a^2+b^2)^{\frac{c^2+d^2}{2c}}.$$

16. If $(a_1+ib_1)(a_2+ib_2)\ldots\ldots(a_n+ib_n) = A+iB$, prove that

   (*i*) $(a_1^2+b_1^2)(a_2^2+b_2^2)\ldots\ldots(a_n^2+b_n^2) = A^2+B^2$

   (*ii*) $\tan^{-1}\dfrac{b_1}{a_1} + \tan^{-1}\dfrac{b_2}{a_2} + \ldots\ldots + \tan^{-1}\dfrac{b_n}{a_n} = \tan^{-1}\dfrac{B}{A}$.

Answers

1. (*i*) $(4n-1)\dfrac{\pi i}{2}$, (*ii*) $\dfrac{1}{2}\log 2 + i(8n+1)\dfrac{\pi}{4}$ (*iii*) $\log 2 + \dfrac{2}{3}(3n+1)\pi i$.

3.9. HYPERBOLIC FUNCTIONS

1. **Definitions.** For all values of x, real or complex

(*i*) the quantity $\dfrac{e^x - e^{-x}}{2}$ is called *hyperbolic sine of* x and is written as *sinh* x

(*ii*) the quantity $\dfrac{e^x + e^{-x}}{2}$ is called *hyperbolic cosine of* x and is written as *cosh* x.

Thus $\quad \sinh x = \dfrac{e^x - e^{-x}}{2}; \quad \cosh x = \dfrac{e^x + e^{-x}}{2}$

The other hyperbolic functions are defined in terms of hyperbolic sine and cosine as follows:

$$\tanh x = \frac{\sinh x}{\cosh x} = \frac{e^x - e^{-x}}{e^x + e^{-x}}; \qquad \coth x = \frac{\cosh x}{\sinh x} = \frac{e^x + e^{-x}}{e^x - e^{-x}}$$

$$\text{sech } x = \frac{1}{\cosh x} = \frac{2}{e^x + e^{-x}}; \qquad \text{cosech } x = \frac{1}{\sinh x} = \frac{2}{e^x - e^{-x}}$$

Note. $\sinh 0 = \dfrac{e^0 - e^{-0}}{2} = \dfrac{1-1}{2} = 0; \qquad \cosh 0 = \dfrac{e^0 + e^{-0}}{2} = \dfrac{1+1}{2} = 1$

$\cosh x + \sinh x = \dfrac{e^x + e^{-x}}{2} + \dfrac{e^x - e^{-x}}{2} = e^x; \qquad \cosh x - \sinh x = \dfrac{e^x + e^{-x}}{2} - \dfrac{e^x - e^{-x}}{2} = e^{-x}.$

2. (*a*) **Relations between hyperbolic and trigonometric functions**

$$\cos \theta = \frac{e^{i\theta} + e^{-i\theta}}{2}; \ \sin \theta = \frac{e^{i\theta} - e^{-i\theta}}{2i}$$

Putting $\theta = ix$ in these equations, we get

$$\cos(ix) = \frac{e^{i(ix)} + e^{-i(ix)}}{2} = \frac{e^{-x} + e^x}{2} = \cosh x$$

$$\sin(ix) = \frac{e^{i(ix)} - e^{-i(ix)}}{2i} = \frac{e^{-x} - e^x}{2i} = \frac{-(e^x - e^{-x})}{2i} \qquad (K.U.K.\ 2013)$$

$$= \frac{i^2(e^x - e^{-x})}{2i} = i \cdot \frac{e^x - e^{-x}}{2} = i \sinh x$$

$$\tan(ix) = \frac{\sin(ix)}{\cos(ix)} = \frac{i \sinh x}{\cosh x} = i \tanh x$$

$$\cot(ix) = \frac{\cos(ix)}{\sin(ix)} = \frac{\cosh x}{i \sinh x} = \frac{i \cosh x}{i^2 \sinh x} = -i \coth x$$

$$\sec(ix) = \frac{1}{\cos(ix)} = \frac{1}{\cosh x} = \text{sech } x$$

$$\text{cosec}(ix) = \frac{1}{\sin(ix)} = \frac{1}{i \sinh x} = \frac{1}{i^2 \sinh x} = -i \text{ cosech } x.$$

(*b*) **By Definition,** $\sinh \theta = \dfrac{e^\theta - e^{-\theta}}{2}; \ \cosh \theta = \dfrac{e^\theta + e^{-\theta}}{2}; \ \tanh \theta = \dfrac{e^\theta - e^{-\theta}}{e^\theta + e^{-\theta}}$

Putting $\theta = ix$, we get

$$\sinh(ix) = \frac{e^{ix} - e^{-ix}}{2} = i \cdot \frac{e^{ix} - e^{-ix}}{2i} = i \sin x;$$

$$\cosh(ix) = \frac{e^{ix} - e^{-ix}}{2} = \cos x$$

$$\tanh(ix) = \frac{e^{ix} - e^{-ix}}{e^{ix} + e^{-ix}} = i \cdot \frac{\dfrac{e^{ix} - e^{-ix}}{2i}}{\dfrac{e^{ix} + e^{-ix}}{2}} = i \cdot \frac{\sin x}{\cos x} = i \tan x$$

3. Prove that hyperbolic functions are periodic and find their periods.

(*a*) We know that $\sinh x = \dfrac{e^x - e^{-x}}{2}$

$$\therefore \sinh(x + 2n\pi i) = \frac{e^{x+2n\pi i} - e^{-(x+2n\pi i)}}{2}, \text{ where } n \text{ is any integer}$$

$$= \frac{1}{2}[e^x \cdot e^{2n\pi i} - e^{-x} \cdot e^{-2n\pi i}] = \frac{1}{2}[e^x \cdot 1 - e^{-x} \cdot 1] = \frac{e^x - e^{-x}}{2} = \sinh x$$

Thus $\sinh x$ remains unchanged when x is increased by any multiple of $2\pi i$. **Hence sinh x is a periodic function and its period is 2πi.**

(*b*) $$\cosh x = \frac{e^x + e^{-x}}{2}$$

$$\cosh(x + 2n\pi i) = \frac{e^{x+2n\pi i} + e^{-(x+2n\pi i)}}{2}, \quad \text{where } n \text{ is any integer}$$

$$= \tfrac{1}{2}[e^x \cdot e^{2n\pi i} + e^{-x} \cdot e^{-2n\pi i}] = \tfrac{1}{2}[e^x \cdot 1 + e^{-x} \cdot 1] = \frac{e^x + e^{-x}}{2} = \cosh x$$

Thus $\cosh x$ remains unchanged when x is increased by any multiple of $2\pi i$. Hence **cosh x is a periodic function and its period is 2πi.**

(*c*) $$\tanh x = \frac{e^x - e^{-x}}{e^x + e^{-x}}$$

$$\tanh(x + n\pi i) = \frac{e^{x+n\pi i} - e^{-(x+n\pi i)}}{e^{x+n\pi i} + e^{-(x+n\pi i)}}, \text{ where } n \text{ is any integer}$$

$$= \frac{e^x \cdot e^{n\pi i} - e^{-x} \cdot e^{-n\pi i}}{e^x \cdot e^{n\pi i} + e^{-x} \cdot e^{-n\pi i}}$$

Multiplying the numerator and denominator by $e^{n\pi i}$

$$= \frac{e^x \cdot e^{2n\pi i} - e^{-x}}{e^x \cdot e^{2n\pi i} + e^{-x}} = \frac{e^x - e^{-x}}{e^x + e^{-x}} = \tanh x$$

$[\because \; e^{2n\pi i} = \cos 2n\pi + i \sin 2n\pi = 1]$

Thus $\tanh x$ remains unchanged when x is increased by any multiple of πi. Hence **tanh x is a periodic function and its period is πi.**

Note. cosech x, sech x and coth x being reciprocals of $\sinh x$, $\cosh x$ and $\tanh x$ respectively, are also periodic functions with periods $2\pi i$, $2\pi i$ and πi respectively.

3.10. FORMULAE OF HYPERBOLIC FUNCTIONS

1. Prove that (*a*) $\cosh^2 x - \sinh^2 x = 1$, (*b*) $\text{sech}^2 x + \tanh^2 x = 1$, (*c*) $\coth^2 x - \text{cosech}^2 x = 1$

Proof. (*a*) For all values of θ, $\cos^2 \theta + \sin^2 \theta = 1$

Putting $\theta = ix$, we get $\cos^2 (ix) + \sin^2 (ix) = 1$

or $$(\cosh x)^2 + (i \sinh x)^2 = 1 \qquad [\because \cos ix = \cosh x;\ \sin (ix) = i \sinh x]$$

or $$\cosh^2 x - \sinh^2 x = 1 \qquad [\because i^2 = -1]$$

(*b*) We know that $\cosh^2 x - \sinh^2 x = 1$

Dividing both sides by $\cosh^2 x$, we have

$$1 - \tanh^2 x = \text{sech}^2 x \Rightarrow \text{sech}^2 x + \tanh^2 x = 1$$

(*c*) We know that $\cosh^2 x - \sinh^2 x = 1$

Dividing both sides by $\sinh^2 x$, we have

$$\coth^2 x - 1 = \text{cosech}^2 x \Rightarrow \coth^2 x - \text{cosech}^2 x = 1$$

2. Prove that (*a*) $\sinh (x \pm y) = \sinh x \cosh y \pm \cosh x \sinh y$

(*b*) $\cosh (x \pm y) = \cosh x \cosh y \pm \sinh x \sinh y$

(*c*) $\tanh (x \pm y) = \dfrac{\tanh x \pm \tanh y}{1 \pm \tanh x \tanh y}$

Proof. (*a*) $$\sinh (x \pm y) = \frac{1}{i} \sin i (x \pm y) \qquad \left[\because \sinh x = \frac{1}{i} \sin ix\right]$$

$$= \frac{1}{i} (\sin ix \cos iy \pm \cos ix \sin iy)$$

$$= \frac{1}{i} (i \sinh x \cosh y \pm \cosh x \cdot i \sinh y)$$

$$[\because \sin i\theta = i \sinh \theta;\ \cos i\theta = \cosh \theta]$$

$$= \sinh x \cosh y \pm \cosh x \sinh y$$

(*b*) $$\cosh (x \pm y) = \cos i (x \pm y) \qquad [\because \cosh x = \cos ix]$$

$$= \cos ix \cos iy \mp \sin ix \sin iy = \cosh x \cosh y \mp i \sinh x \cdot i \sinh y$$

$$= \cosh x \cosh y \mp (-\sinh x \cdot \sinh y) \qquad [\because i^2 = -1]$$

$$= \cosh x \cosh y \pm \sinh x \sinh y$$

(*c*) $$\tanh (x \pm y) = \frac{\sinh (x \pm y)}{\cosh (x \pm y)} = \frac{\sinh x \cosh y \pm \cosh x \sinh y}{\cosh x \cosh y \pm \sinh x \sinh y}$$

Dividing the numerator and denominator by $\cosh x \cosh y = \dfrac{\tanh x \pm \tanh y}{1 \pm \tanh x \tanh y}$

3. Prove that (*a*) $\sinh 2x = 2 \sinh x \cosh x = \dfrac{2 \tanh x}{1 - \tanh^2 x}$

(*b*) $\cosh 2x = \cosh^2 x + \sinh^2 x = 2\cosh^2 x - 1 = 1 + 2\sinh^2 x = \dfrac{1+\tanh^2 x}{1-\tanh^2 x}$

(*c*) $\tanh 2x = \dfrac{2\tanh x}{1+\tanh^2 x}$

Proof. (*a*) We know that $\sin 2\theta = 2\sin\theta\cos\theta$

Putting $\theta = ix$, we get $\sin(2ix) = 2\sin(ix)\cos(ix)$ or $i\sinh 2x = 2\,.\,i\sinh x\,.\,\cosh x$

or $$\mathbf{\sinh 2x = 2\sinh x\cosh x}$$

Also $$\sin 2\theta = \frac{2\tan\theta}{1+\tan^2\theta}$$

Putting $\theta = ix$, we get $\sin(2ix) = \dfrac{2\tan ix}{1+\tan^2 ix} = \dfrac{2\,.\,i\tanh x}{1+(i\tanh x)^2}$

or $$i\sinh 2x = \frac{2i\tanh x}{1-\tanh^2 x} \quad \text{or} \quad \sinh 2x = \frac{2\tanh x}{1-\tanh^2 x}$$

(*b*) We know that $\cos 2\theta = \cos^2\theta - \sin^2\theta$

Putting $\theta = ix$, we get $\cos(2ix) = \cos^2(ix) - \sin^2(ix)$ or $\cosh 2x = (\cosh x)^2 - (i\sinh x)^2$

or $$\mathbf{\cosh 2x = \cosh^2 x + \sinh^2 x}$$

We know that $\cos 2\theta = 2\cos^2\theta - 1$

Putting $\theta = ix$, we get $\cos(2ix) = 2\cos^2(ix) - 1$ or $\mathbf{\cosh 2x = 2\cosh^2 x - 1}$

Cor. $$\cosh^2 x = \frac{\cosh 2x + 1}{2}$$

We know that $\cos 2\theta = 1 - 2\sin^2\theta$

Putting $\theta = ix$, we get $\cos(2ix) = 1 - 2\sin^2(ix)$

or $$\mathbf{\cosh 2x} = 1 - 2(i\sinh x)^2 = \mathbf{1 + 2\sinh^2 x}$$

Cor. $$\sinh^2 x = \frac{\cosh 2x - 1}{2}$$

We know that $$\cos 2\theta = \frac{1-\tan^2\theta}{1+\tan^2\theta}$$

Putting $\theta = ix$, we get $\cos(2ix) = \dfrac{1-\tan^2(ix)}{1+\tan^2(ix)} = \dfrac{1-(i\tanh x)^2}{1+(i\tanh x)^2}$

or $$\cosh 2x = \frac{1+\tanh^2 x}{1-\tanh^2 x}$$

(*c*) We know that $\tan 2\theta = \dfrac{2\tan\theta}{1-\tan^2\theta}$

Putting $\theta = ix$, we get

$$\tan(2ix) = \frac{2\tan(ix)}{1-\tan^2(ix)}$$

$$i\tanh 2x = \frac{2i\tanh x}{1-(i\tanh x)^2} = \frac{2i\tanh x}{1+\tanh^2 x}$$

$$\therefore \qquad \tanh 2x = \frac{2\tanh x}{1+\tanh^2 x}.$$

4. Prove that (*a*) $\sinh 3x = 3\sinh x + 4\sinh^3 x$.

(*b*) $\cosh 3x = 4\cosh^3 x - 3\cosh x$ (*c*) $\tanh 3x = \dfrac{3\tanh x + \tanh^3 x}{1+3\tanh^2 x}$

Proof. (*a*) We know that $\sin 3\theta = 3\sin\theta - 4\sin^3\theta$

Putting $\theta = ix$, we get $\sin(3ix) = 3\sin(ix) - 4\sin^3(ix)$

or $\qquad i\sinh 3x = 3i\sinh x - 4(i\sinh x)^3$

or $\qquad i\sinh 3x = 3i\sinh x + 4i\sinh^3 x \qquad [\because\ i^3 = -i]$

or $\qquad \sinh 3x = 3\sinh x + 4\sinh^3 x$

(*b*) We know that $\cos 3\theta = 4\cos^3\theta - 3\cos\theta$

Putting $\theta = ix$, we get $\cos(3ix) = 4\cos^3(ix) - 3\cos(ix)$

or $\qquad \cosh 3x = 4\cosh^3 x - 3\cosh x$

(*c*) We know that $\qquad \tan 3\theta = \dfrac{3\tan\theta - \tan^3\theta}{1-3\tan^2\theta}$

Putting $\theta = ix$, we get $\quad \tan(3ix) = \dfrac{3\tan(ix) - \tan^3(ix)}{1-3\tan^2(ix)}$

or $\qquad i\tanh 3x = \dfrac{3\,.\,i\tanh x - (i\tanh x)^3}{1-3(i\tanh x)^2}$

or $\qquad i\tanh 3x = \dfrac{3\,.\,i\tanh x + i\tanh^3 x}{1+3\tanh^2 x}$ or $\tanh 3x = \dfrac{3\tanh x + \tanh^3 x}{1+3\tanh^2 x}$.

5. Prove that

(*i*) $2\sinh A\cosh B = \sinh(A+B) + \sinh(A-B)$

(*ii*) $2\cosh A\sinh B = \sinh(A+B) - \sinh(A-B)$

(*iii*) $2\cosh A\cosh B = \cosh(A+B) + \cosh(A-B)$

(*iv*) $2\sinh A\sinh B = \cosh(A+B) - \cosh(A-B)$

Proof. We shall prove only the last result.

The first three are left as an exercise for the student.

We know that $2\sin x\sin y = \cos(x-y) - \cos(x+y)$

Putting $x = iA$; $y = iB$, we get

$$2\sin(iA)\,.\,\sin(iB) = \cos i(A-B) - \cos i(A+B)$$

or $\qquad 2\,.\,i\sinh A\,.\,i\sinh B = \cosh(A-B) - \cosh(A+B)$

or $\qquad -2\sinh A\sinh B = \cosh(A-B) - \cosh(A+B) \qquad [\because\ i^2 = -1]$

or $\qquad 2\sinh A\sinh B = \cosh(A+B) - \cosh(A-B)$.

6. Prove that

(*i*) $\sinh C + \sinh D = 2\sinh\dfrac{C+D}{2}\cosh\dfrac{C-D}{2}$

(*ii*) $\sinh C - \sinh D = 2\cosh\dfrac{C+D}{2}\sinh\dfrac{C-D}{2}$

(*iii*) $\cosh C + \cosh D = 2\cosh\frac{C+D}{2}\cosh\frac{C-D}{2}$

(*iv*) $\cosh C - \cosh D = 2\sinh\frac{C+D}{2}\sinh\frac{C-D}{2}$

Proof. We shall prove only the last result. The first three are left as an exercise for the student.

We know that $\cos x - \cos y = 2\sin\frac{x+y}{2}\sin\frac{y-x}{2}$

Putting $x = iA$ and $y = iB$, we get

$$\cos(iA) - \cos(iB) = 2\sin\left(i\frac{A+B}{2}\right)\sin\left(i\frac{B-A}{2}\right)$$

$$\Rightarrow \quad \cosh A - \cosh B = 2i\sinh\frac{A+B}{2}\,.\,i\sinh\frac{B-A}{2}$$

$$= -2\sinh\frac{A+B}{2}\sinh\frac{B-A}{2} = 2\sinh\frac{A+B}{2}\sinh\frac{A-B}{2}$$

$[\because \quad \sinh(-x) = \sinh x]$

7. Prove that

$$\tanh(x+y+z) = \frac{\tanh x + \tanh y + \tanh z + \tanh x\tanh y\tanh z}{1+\tanh x\tanh y + \tanh y\tanh z + \tanh z\tanh x}$$

Proof. We know that,

$$\tan(\alpha+\beta+\gamma) = \frac{\tan\alpha + \tan\beta + \tan\gamma - \tan\alpha\tan\beta\tan\gamma}{1-\tan\alpha\tan\beta - \tan\beta\tan\gamma - \tan\gamma\tan\alpha}$$

Putting $\alpha = ix$; $\beta = iy$; $\gamma = iz$, we get

$$\tan i(x+y+z) = \frac{\tan(ix)+\tan(iy)+\tan(iz)-\tan(ix)\tan(iy)\tan(iz)}{1-\tan(ix)\tan(iy)-\tan(iy)\tan(iz)-\tan(iz)\tan(ix)}$$

$$i\tanh(x+y+z) = \frac{i\tanh x + i\tanh y + i\tanh z - i\tanh x\,.\,i\tanh y\,.\,i\tanh z}{1-i\tanh x\,.\,i\tanh y - i\tanh y\,.\,i\tanh z - i\tanh z\,.\,i\tanh x}$$

or $$\tanh(x+y+z) = \frac{\tanh x + \tanh y + \tanh z + \tanh x\tanh y\tanh z}{1+\tanh x\tanh y + \tanh y\tanh z + \tanh z\tanh x}.$$

ILLUSTRATIVE EXAMPLES

Example 1. *Separate into real and imaginary parts*

(*a*) *sin* $(x+iy)$ (*b*) *cos* $(x+iy)$

(*c*) *tan* $(x+iy)$ (*d*) *cot* $(x+iy)$

(M.D.U. May 2013, May 2014)

(*e*) *sec* $(x+iy)$ (*f*) *cosec* $(x+iy)$.

Sol. (*a*) $\sin(x+iy) = \sin x\cos iy + \cos x\sin iy$

$= \sin x\cosh y + \cos x\,.\,i\sinh y = \sin x\cosh y + i\,.\,\cos x\sinh y$

(*b*) $\cos(x+iy) = \cos x\cos iy - \sin x\sin iy$

$= \cos x\cosh y - \sin x\,.\,i\sinh y = \cos x\cosh y - i\,.\,\sin x\sinh y$

$$(c) \qquad \tan(x+iy) = \frac{\sin(x+iy)}{\cos(x+iy)} = \frac{2\sin(x+iy)\cos(x-iy)}{2\cos(x+iy)\cos(x-iy)}$$

$$= \frac{\sin 2x + \sin 2iy}{\cos 2x + \cos 2iy} \qquad \left[\begin{array}{l} \because \; 2\sin A\cos B = \sin(A+B)+\sin(A-B) \\ \quad 2\cos A\cos B = \cos(A+B)+\cos(A-B) \end{array}\right]$$

$$= \frac{\sin 2x + i\,.\sinh 2y}{\cos 2x + \cosh 2y} = \frac{\sin 2x}{\cos 2x + \cosh 2y} + i\,.\frac{\sinh 2y}{\cos 2x + \cosh 2y}.$$

$$(d) \qquad \cot(x+iy) = \frac{\cos(x+iy)}{\sin(x+iy)} = \frac{2\cos(x+iy)\sin(x-iy)}{2\sin(x+iy)\sin(x-iy)}$$

$$= \frac{\sin 2x - \sin 2iy}{\cos 2iy - \cos 2x} \qquad \left[\begin{array}{l} \because \; 2\cos A\sin B = \sin(A+B)-\sin(A-B) \\ \quad 2\sin A\sin B = \cos(A-B)-\cos(A+B) \end{array}\right]$$

$$= \frac{\sin 2x - i\,.\sinh 2y}{\cosh 2y - \cos 2x} = \frac{\sin 2x}{\cosh 2y - \cos 2x} - i\,.\frac{\sinh 2y}{\cosh 2y - \cos 2x}$$

$$(e) \qquad \sec(x+iy) = \frac{1}{\cos(x+iy)} = \frac{2\cos(x-iy)}{2\cos(x+iy)\cos(x-iy)}$$

$$= \frac{2(\cos x\cos iy + \sin x\sin iy)}{\cos 2x + \cos 2iy} = \frac{2(\cos x\cosh y + \sin x\,.\,i\sinh y)}{\cos 2x + \cosh 2y}$$

$$= \frac{2\cos x\cosh y}{\cos 2x + \cosh 2y} + i\,.\frac{2\sin x\sinh y}{\cos 2x + \cosh 2y}$$

$$(f) \qquad \text{cosec}(x+iy) = \frac{1}{\sin(x+iy)} = \frac{2\sin(x-iy)}{2\sin(x+iy)\sin(x-iy)}$$

$$= \frac{2(\sin x\cos iy - \cos x\sin iy)}{\cos 2iy - \cos 2x} = \frac{2(\sin x\cosh y + \cos x\,.\,i\sinh y)}{\cosh 2y - \cos 2x}$$

$$= \frac{2\sin x\cosh y}{\cosh 2y - \cos 2x} - i\,.\frac{2\cos x\sinh y}{\cosh 2y - \cosh 2x}.$$

Example 2. *Separate the following into real and imaginary parts:*

(a) sinh $(x+iy)$ *(b) cosh $(x+iy)$*

(c) tanh $(x+iy)$ *(d) coth $(x+iy)$*

(e) sech $(x+iy)$ *(f) cosech $(x+iy)$.*

Sol. *(a)* $\sinh(x+iy) = \frac{1}{i}\sin i(x+iy)$ $\qquad [\because \; i\sinh\theta = \sin i\theta]$

$$= \frac{i}{i^2}\sin(ix-y) = -i(\sin ix\cos y - \cos ix\sin y)$$

$$= -i(i\sinh x\cos y - \cosh x\sin y) = \sinh x\cos y + i\cosh x\sin y$$

(b) $\cosh(x+iy) = \cos i(x+iy)$ $\qquad [\because \; \cosh\theta = \cos i\theta]$

$$= \cos(ix-y) = \cos ix\cos y + \sin ix\sin y$$

$$= \cosh x\cos y + i\sinh x\sin y$$

(c) $\tanh(x+iy) = \frac{1}{i}\tan i(x+iy)$ $\qquad [\because \; i\tanh\theta = \tan i\theta]$

$$= \frac{i}{i^2}\tan(ix-y) = -i\frac{\sin(ix-y)}{\cos(ix-y)} = -i\,.\frac{2\sin(ix-y)\cos(ix+y)}{2\cos(ix-y)\cos(ix+y)}$$

$$= -i\,.\frac{\sin 2ix - \sin 2y}{\cos 2ix + \cos 2y} = -i\,.\frac{i\sinh 2x - \sin 2y}{\cosh 2x + \cos 2y}$$

$$= \frac{\sinh 2x}{\cosh 2x + \cos 2y} + i \cdot \frac{\sin 2y}{\cosh 2x + \cos 2y}.$$

(*d*) $\coth (x + iy) = \dfrac{\cosh (x + iy)}{\sinh (x + iy)} = \dfrac{\cos i (x + iy)}{\frac{1}{i} \sin i (x + iy)} = i \cdot \dfrac{\cos (ix - y)}{\sin (ix - y)}$

$$= i \cdot \frac{2 \sin (ix + y) \cos (ix - y)}{2 \sin (ix + y) \sin (ix - y)} = i \cdot \frac{\sin 2ix + \sin 2y}{\cos 2y - \cos 2ix}$$

$$= i \cdot \frac{i \sinh 2x + \sin 2y}{\cos 2y - \cosh 2x} = \frac{-\sinh 2x}{\cos 2y - \cosh 2x} + i \cdot \frac{\sin 2y}{\cos 2y - \cosh 2x}$$

$$= \frac{\sinh 2y}{\cosh 2x - \cos 2y} - i \cdot \frac{\sin 2y}{\cosh 2x - \cos 2y}.$$

(*e*) $\text{sech} (x + iy) = \dfrac{1}{\cosh (x + iy)} = \dfrac{1}{\cos i(x + iy)}$

$$= \frac{1}{\cos (ix - y)} = \frac{2 \cos (ix + y)}{2 \cos (ix + y) \cos (ix - y)} = \frac{2(\cos ix \cos y - \sin ix \sin y)}{\cos 2ix + \cos 2y}$$

$$= \frac{2(\cosh x \cos y - i \sinh x \sin y)}{\cosh 2x + \cos 2y}$$

$$= \frac{2 \cosh x \cos y}{\cosh 2x + \cos 2y} - i \cdot \frac{2 \sinh x \sin y}{\cosh 2x + \cos 2y}.$$

(*f*) $\text{cosech} (x + iy) = \dfrac{1}{\sinh (x + iy)} = \dfrac{1}{\frac{1}{i} \sin i (x + iy)} = \dfrac{i}{\sin (ix - y)}$

$$= i \cdot \frac{2 \sin (ix + y)}{2 \sin (ix + y) \sin (ix - y)}$$

$$= i \cdot \frac{2(\sin ix \cos y + \cos ix \sin y)}{\cos 2y - \cos 2ix} = i \cdot \frac{2(i \sinh x \cos y + \cosh x \sin y)}{\cos 2y - \cosh 2x}$$

$$= -\frac{2 \sinh x \cos y}{\cos 2y - \cosh 2x} + i \cdot \frac{2 \cosh x \sin y}{\cos 2y - \cosh 2x}$$

$$= \frac{2 \sinh x \cos y}{\cosh 2x - \cos 2y} - i \cdot \frac{2 \cosh x \sin y}{\cosh 2x - \cos 2y}.$$

Example 3. *If* $u = \log \tan \left(\dfrac{\pi}{4} + \dfrac{\theta}{2}\right)$, *then prove that*

(*i*) $\tanh \dfrac{u}{2} = \tan \dfrac{\theta}{2}$ (K.U.K. Jan. 2014) (*ii*) $\cosh u = \sec \theta$. (M.D.U. 2012)

Sol. $$u = \log \tan \left(\frac{\pi}{4} + \frac{\theta}{2}\right)$$

(*i*) $$e^u = \tan \left(\frac{\pi}{4} + \frac{\theta}{2}\right)$$

$$\Rightarrow \quad e^{u/2} \cdot e^{u/2} = \frac{1+\tan\frac{\theta}{2}}{1-\tan\frac{\theta}{2}} \quad \Rightarrow \quad \frac{e^{u/2}}{e^{-u/2}} = \frac{1+\tan\frac{\theta}{2}}{1-\tan\frac{\theta}{2}}$$

By componendo and dividendo

$$\frac{e^{u/2} - e^{-u/2}}{e^{u/2} + e^{-u/2}} = \frac{\left(1+\tan\frac{\theta}{2}\right) - \left(1-\tan\frac{\theta}{2}\right)}{\left(1+\tan\frac{\theta}{2}\right) + \left(1-\tan\frac{\theta}{2}\right)} \quad \Rightarrow \quad \tanh\frac{u}{2} = \tan\frac{\theta}{2}$$

(*ii*) $$\cosh u = \frac{1+\tanh^2\frac{u}{2}}{1-\tanh^2\frac{u}{2}} = \frac{1+\tan^2\frac{\theta}{2}}{1-\tan^2\frac{\theta}{2}} \quad \text{[Using part } (i)]$$

$$= \frac{1}{\cos\theta} = \sec\theta.$$

Example 4. *If* $\sin(A + iB) = x + iy$, *prove that*

(*i*) $\dfrac{x^2}{\cosh^2 B} + \dfrac{y^2}{\sinh^2 B} = 1$ (*ii*) $x^2 \operatorname{cosec}^2 A - y^2 \sec^2 A = 1$.

Sol. $x + iy = \sin(A + iB)$

$= \sin A \cos iB + \cos A \sin iB = \sin A \cosh B + i \cos A \sinh B$

Equating real and imaginary parts on both sides

$$x = \sin A \cosh B;\ y = \cos A \sinh B \quad \ldots(i)$$

From (*i*), $\dfrac{x}{\cosh B} = \sin A;\ \dfrac{y}{\sinh B} = \cos A$

Squaring and adding, $\dfrac{x^2}{\cosh^2 B} + \dfrac{y^2}{\sinh^2 B} = \sin^2 A + \cos^2 A = 1$

Also from (*i*), $\dfrac{x}{\sin A} = \cosh B;\ \dfrac{y}{\cos A} = \sinh B$

Squaring and subtracting, $\dfrac{x^2}{\sin^2 A} - \dfrac{y^2}{\cos^2 A} = \cosh^2 B - \sinh^2 B = 1$

or $x^2 \operatorname{cosec}^2 A - y^2 \sec^2 A = 1$.

Example 5. *If* $x + iy = \cosh(u + iv)$ *show that*

(*i*) $\dfrac{x^2}{\cosh^2 u} + \dfrac{y^2}{\sinh^2 u} = 1$ (*ii*) $x^2 \sec^2 v - y^2 \operatorname{cosec}^2 v = 1$.

(P.T.U. 2010; M.D.U. 2010)

Sol. $x + iy = \cosh(u + iv) = \cos i(u + iv)$ [$\because \cosh\theta = \cos i\theta$]

$= \cos(iu - v) = \cos iu \cos v + \sin iu \sin v = \cosh u \cos v + i \sinh u \sin v$

Equating the real and imaginary parts

$$x = \cosh u \cos v;\ y = \sinh u \sin v \quad \ldots(i)$$

From (*i*), $\dfrac{x}{\cosh u} = \cos v,\ \dfrac{y}{\sinh u} = \sin v$

Squaring and adding, $\dfrac{x^2}{\cosh^2 u} + \dfrac{y^2}{\sinh^2 u} = \cos^2 v + \sin^2 v = 1$

From (*i*), $\dfrac{x}{\cos v} = \cosh u;\ \dfrac{y}{\sin v} = \sinh u$

Squaring and subtracting, $x^2 \sec^2 v - y^2 \operatorname{cosec}^2 v = \cosh^2 u - \sinh^2 u = 1.$

Example 6. *If $x + iy = \tan (A + iB)$; prove that*

(*i*) $x^2 + y^2 + 2x \cot 2A = 1$ (*ii*) $x^2 + y^2 - 2y \coth 2B + 1 = 0.$

Sol. $x + iy = \tan (A + iB)$

Changing i into $-i$, we get $x - iy = \tan (A - iB)$

Now $\tan 2A = \tan [(A + iB) + (A - iB)]$

$$= \frac{\tan (A + iB) + \tan (A - iB)}{1 - \tan (A + iB) \tan (A - iB)} = \frac{(x + iy) + (x - iy)}{1 - (x + iy)(x - iy)} = \frac{2x}{1 - (x^2 + y^2)}$$

or $\dfrac{1}{\cot 2A} = \dfrac{2x}{1 - (x^2 + y^2)}$ or $1 - (x^2 + y^2) = 2x \cot 2A$

or $x^2 + y^2 + 2x \cot 2A = 1$ ...(I)

Again $\tan (2iB) = \tan [(A + iB) - (A - iB)]$

$$= \frac{\tan (A + iB) - \tan (A - iB)}{1 + \tan (A + iB) \tan (A - iB)} = \frac{(x + iy) - (x - iy)}{1 + (x + iy)(x - iy)} = \frac{2iy}{1 + x^2 + y^2}$$

or $i \tanh 2B = \dfrac{2iy}{1 + x^2 + y^2}$ or $\dfrac{1}{\coth 2B} = \dfrac{2y}{1 + x^2 + y^2}$

or $1 + x^2 + y^2 = 2y \coth 2B$

Hence $x^2 + y^2 - 2y \coth 2B + 1 = 0$ ...(II)

Example 7. *If $a + ib = \tanh \left(v + \dfrac{i\pi}{4}\right)$, prove that $a^2 + b^2 = 1$.*

(M.D.U. May 2015, Dec. 2015)

Sol. Given $a + ib = \tanh \left(v + \dfrac{i\pi}{4}\right)$ ...(1)

Changing i to $-i$, we get

$$a - ib = \tanh \left(v - \frac{i\pi}{4}\right) \quad ...(2)$$

Multiplying (1) and (2), we have

$$a^2 - i^2b^2 = \tanh \left(v + \frac{i\pi}{4}\right) \tanh \left(v - \frac{i\pi}{4}\right)$$

$$\Rightarrow \quad a^2 + b^2 = \frac{1}{i} \tan i\left(v + \frac{i\pi}{4}\right) . \frac{1}{i} \tan i\left(v - \frac{i\pi}{4}\right)$$

$$= \frac{1}{i^2} \tan \left(iv - \frac{\pi}{4}\right) \tan \left(iv + \frac{\pi}{4}\right)$$

$$= -\frac{\sin\left(iv - \frac{\pi}{4}\right)\sin\left(iv + \frac{\pi}{4}\right)}{\cos\left(iv - \frac{\pi}{4}\right)\cos\left(iv + \frac{\pi}{4}\right)} = -\frac{\sin^2(iv) - \sin^2\frac{\pi}{4}}{\cos^2(iv) - \sin^2\frac{\pi}{4}}$$

[$\because$ $\sin(\alpha+\beta)\sin(\alpha-\beta) = \sin^2\alpha - \sin^2\beta$; $\cos(\alpha+\beta)\cos(\alpha-\beta) = \cos^2\alpha - \sin^2\beta$]

$$= -\frac{(i\sinh v)^2 - \frac{1}{2}}{\cosh^2 v - \frac{1}{2}} = -\frac{-\sinh^2 v - \frac{1}{2}}{(1+\sinh^2 v) - \frac{1}{2}}$$

$$= \frac{\sinh^2 v + \frac{1}{2}}{\sinh^2 v + \frac{1}{2}} = 1$$

Hence, $a^2 + b^2 = 1$

Example 8. *If* $\tan(\theta + i\phi) = \cos\alpha + i\sin\alpha = e^{i\alpha}$, *prove that*

$$\theta = \frac{n\pi}{2} + \frac{\pi}{4} \text{ and } \phi = \frac{1}{2}\log\tan\left(\frac{\pi}{4} + \frac{\alpha}{2}\right). \qquad \text{(M.D.U. 2011)}$$

Sol. $\tan(\theta + i\phi) = \cos\alpha + i\sin\alpha$ ...(i)

Changing i into $-i$, we get

$\tan(\theta - i\phi) = \cos\alpha - i\sin\alpha$ ...(ii)

Now $\tan 2\theta = \tan[(\theta + i\phi) + (\theta - i\phi)] = \dfrac{\tan(\theta+i\phi) + \tan(\theta - i\phi)}{1 - \tan(\theta+i\phi)\tan(\theta-i\phi)}$

$$= \frac{(\cos\alpha + i\sin\alpha) + (\cos\alpha - i\sin\alpha)}{1 - (\cos\alpha + i\sin\alpha)(\cos\alpha - i\sin\alpha)}$$

$$= \frac{2\cos\alpha}{1 - (\cos^2\alpha - i^2\sin^2\alpha)} = \frac{2\cos\alpha}{1 - (\cos^2\alpha + \sin^2\alpha)}$$

$$= \frac{2\cos\alpha}{1-1} = \frac{2\cos\alpha}{0} = \infty = \tan\frac{\pi}{2}$$

$\therefore$ $2\theta = n\pi + \dfrac{\pi}{2}$ [$\because$ $\tan\theta = \tan\alpha \Rightarrow \theta = n\pi + \alpha$]

or $\theta = \dfrac{n\pi}{2} + \dfrac{\pi}{4}$

Also $\tan 2i\phi = \tan[(\theta + i\phi) - (\theta - i\phi)]$

$$= \frac{\tan(\theta+i\phi) - \tan(\theta-i\phi)}{1 + \tan(\theta+i\phi)\tan(\theta-i\phi)} = \frac{(\cos\alpha + i\sin\alpha) - (\cos\alpha - i\sin\alpha)}{1 + (\cos\alpha + i\sin\alpha)(\cos\alpha - i\sin\alpha)}$$

$$= \frac{2i\sin\alpha}{1 + (\cos^2\alpha + \sin^2\alpha)} = \frac{2i\sin\alpha}{1+1} = i\sin\alpha$$

$i\tanh 2\phi = i\sin\alpha$ or $\tanh 2\phi = \sin\alpha$

or $$\frac{e^{2\phi}-e^{-2\phi}}{e^{2\phi}+e^{-2\phi}}=\frac{\sin\alpha}{1} \quad \text{or} \quad \frac{e^{2\phi}+e^{-2\phi}}{e^{2\phi}-e^{-2\phi}}=\frac{1}{\sin\alpha}$$

By componendo and dividendo

$$\frac{2e^{2\phi}}{2e^{-2\phi}}=\frac{1+\sin\alpha}{1-\sin\alpha} \quad \text{or} \quad e^{4\phi}=\frac{1+\sin\alpha}{1-\sin\alpha}$$

or $$e^{4\phi}=\frac{\cos^2\frac{\alpha}{2}+\sin^2\frac{\alpha}{2}+2\cos\frac{\alpha}{2}\sin\frac{\alpha}{2}}{\cos^2\frac{\alpha}{2}+\sin^2\frac{\alpha}{2}-2\cos\frac{\alpha}{2}\sin\frac{\alpha}{2}}=\left[\frac{\cos\frac{\alpha}{2}+\sin\frac{\alpha}{2}}{\cos\frac{\alpha}{2}-\sin\frac{\alpha}{2}}\right]^2$$

or $$e^{2\phi}=\frac{\cos\frac{\alpha}{2}+\sin\frac{\alpha}{2}}{\cos\frac{\alpha}{2}-\sin\frac{\alpha}{2}}=\frac{1+\tan\frac{\alpha}{2}}{1-\tan\frac{\alpha}{2}}=\tan\left(\frac{\pi}{4}+\frac{\alpha}{2}\right)$$

Taking logarithms of both sides $\log e^{2\phi}=\log\tan\left(\frac{\pi}{4}+\frac{\alpha}{2}\right)$ or $2\phi=\log\tan\left(\frac{\pi}{4}+\frac{\alpha}{2}\right)$

$\therefore$ $$\phi=\frac{1}{2}\log\tan\left(\frac{\pi}{4}+\frac{\alpha}{2}\right).$$

Example 9. *Separate into real and imaginary parts log sin $(x+iy)$.*

Sol. $\log\sin(x+iy)=\log(\sin x\cos iy+\cos x\sin iy)$

$$=\log(\sin x\cosh y+i\cos x\sinh y)$$
$$=\log(\alpha+i\beta), \text{ where } \alpha=\sin x\cosh y,\ \beta=\cos x\sinh y$$
$$=\frac{1}{2}\log(\alpha^2+\beta^2)+i\tan^{-1}\frac{\beta}{\alpha}$$
$$=\frac{1}{2}\log(\sin^2 x\cosh^2 y+\cos^2 x\sinh^2 y)+i\tan^{-1}\left(\frac{\cos x\sinh y}{\sin x\cosh y}\right)$$
$$=\frac{1}{2}\log\left[\frac{1-\cos 2x}{2}\cdot\frac{\cosh 2y+1}{2}+\frac{1+\cos 2x}{2}\cdot\frac{\cosh 2y-1}{2}\right]+i\tan^{-1}(\cot x\tanh y)$$
$$=\tfrac{1}{2}\log\left[\tfrac{1}{4}(2\cosh 2y-2\cos 2x)\right]+i\tan^{-1}(\cot x\tanh y)$$
$$=\tfrac{1}{2}\log\left[\tfrac{1}{2}(\cosh 2y-\cos 2x)\right]+i\tan^{-1}(\cot x\tanh y).$$

Example 10. *If $z=x+iy$ is a complex variable, then prove that sin z and cos z are not bounded.*

Sol. We know that

$$\sin z=\sin(x+iy)=\sin x\cosh y+i\cos x\sinh y$$

$\therefore$ $$|\sin z|^2=|\sin x\cosh y+i\cos x\sinh y|^2$$
$$=\sin^2 x\cosh^2 y+\cos^2 x\sinh^2 y$$
$$=\sin^2 x(1+\sinh^2 y)+(1-\sin^2 x)\sinh^2 y$$
$$=\sin^2 x+\sinh^2 y$$

Since x is real, $0\le\sin^2 x\le 1$. But $\sinh y=\dfrac{e^y-e^{-y}}{2}$ is not bounded.

$\Rightarrow$ $|\sin z|$ and hence $\sin z$ is not bounded.

Similarly, $|\cos z|^2 = \cos^2 x + \sinh^2 y$

$\Rightarrow$ $|\cos z|$ and hence $\cos z$ is not bounded.

Hence for a complex variable z, $\sin z$ and $\cos z$ can have any value.

Remark. $|\sin z| \le 1$ and $|\cos z| \le 1$ only when z is real.

Example 11. *Find all values of z such that*

(i) sinh z = 0 *(ii) cosh z = 0.* (M.D.U. 2010)

Sol. We know that

$$\sinh z = \sinh(x + iy) = \sinh x \cos y + i \cosh x \sin y$$

and $$\cosh z = \cosh(x + iy) = \cosh x \cos y + i \sinh x \sin y \quad \text{[See Example 2 (a) and (b)]}$$

$$\therefore \quad |\sinh z|^2 = |\sinh x \cos y + i \cosh x \sin y|^2$$
$$= \sinh^2 x \cos^2 y + \cosh^2 x \sin^2 y$$
$$= \sinh^2 x (1 - \sin^2 y) + (1 + \sinh^2 x)\sin^2 y$$
$$= \sinh^2 x + \sin^2 y$$

Now $\sinh z = 0 \Rightarrow |\sinh z| = 0$

$\Rightarrow$ $\sinh^2 x + \sin^2 y = 0$

$\Rightarrow$ $\sinh x = 0$ and $\sin y = 0$

$\Rightarrow$ $x = 0$ and $y = n\pi$, where n is any integer.

$\Rightarrow$ $z = x + iy = n\pi i$

$\therefore$ $\sinh z = 0$ only when z is purely imaginary and $z = n\pi i$.

Similarly, $|\cosh z|^2 = \sinh^2 x + \cos^2 y$

Now $\cosh z = 0 \Rightarrow |\cosh z| = 0$

$\Rightarrow$ $\sinh^2 x + \cos^2 y = 0$

$\Rightarrow$ $\sinh x = 0$ and $\cos y = 0$

$\Rightarrow$ $x = 0$ and $y = (2n + 1)\dfrac{\pi}{2}$, where n is any integer.

$\Rightarrow$ $z = x + iy = (2n + 1)\dfrac{\pi}{2}i$

$\therefore$ $\cosh z = 0$ only when z is purely imaginary and $z = (2n + 1)\dfrac{\pi}{2}i$.

Example 12. *Find all values of z such that sin z = 4.*

Sol. Let $z = x + iy$, then $\sin z = 4$

$\Rightarrow$ $\sin(x + iy) = 4$

$\Rightarrow$ $\sin x \cosh y + i \cos x \sinh y = 4$

$\Rightarrow$ $\sin x \cosh y = 4$ ...(1)

and $\cos x \sinh y = 0$ ...(2)

by comparing real and imaginary parts.

From (2), we have $\cos x = 0$ or $\sinh y = 0$

$\Rightarrow$ $x = (2n + 1)\dfrac{\pi}{2}$, where n is any integer or $y = 0$

When $y = 0$, from (1), we get

$$\sin x = 4 \qquad (\because \cosh 0 = 1)$$

which is not possible since x is real.

When $x = (2n + 1)\frac{\pi}{2}$, from (1), we get

$$\sin\left(n\pi + \frac{\pi}{2}\right)\cosh y = 4 \quad \text{or} \quad (-1)^n \sin\frac{\pi}{2}\cosh y = 4$$

or $$(-1)^n \cosh y = 4 \qquad ...(3)$$

When n is odd, $\cosh y = -4$ which is not possible since $\cosh y > 0$ for every y.

$\therefore$ n must be an even integer.

From (3) $\cosh y = 4$ or $y = \cosh^{-1} 4$

Hence
$$z = x + iy$$
$$= (2n + 1)\frac{\pi}{2} + i\cosh^{-1} 4, \quad \text{where } n \text{ is an even integer}$$

or $$z = (4k + 1)\frac{\pi}{2} + i\cosh^{-1} 4, \quad \text{where } k \text{ is any integer.}$$

3.11. INVERSE HYPERBOLIC FUNCTIONS

1. Prove that $\sinh^{-1} x = \log(x + \sqrt{x^2 + 1})$

Proof. Let $\sinh^{-1} x = y$, then $x = \sinh y$

$$\Rightarrow \quad x = \frac{e^y - e^{-y}}{2} = \frac{e^{2y} - 1}{2e^y} \quad \Rightarrow \quad e^{2y} - 2x\,e^y - 1 = 0$$

It is a quadratic in e^y

$$\therefore \quad e^y = \frac{2x \pm \sqrt{4x^2 + 4}}{2} = x \pm \sqrt{x^2 + 1}$$

Rejecting the negative sign, $e^y = x + \sqrt{x^2 + 1}$

Taking logarithms, $y = \log(x + \sqrt{x^2 + 1})$ or $\sinh^{-1} x = \log(x + \sqrt{x^2 + 1})$

2. Prove that $\cosh^{-1} x = \log(x + \sqrt{x^2 - 1})$

Proof. Let $\cosh^{-1} x = y$ then $x = \cosh y$

or $$x = \frac{e^y + e^{-y}}{2} = \frac{e^{2y} + 1}{2e^y} \quad \text{or} \quad e^{2y} - 2x\,e^y + 1 = 0$$

It is a quadratic in e^y

$$\therefore \quad e^y = \frac{2x \pm \sqrt{4x^2 - 4}}{2} = x \pm \sqrt{x^2 - 1}$$

Rejecting the negative sign $e^y = x + \sqrt{x^2 - 1}$

$$\therefore \quad y = \log(x + \sqrt{x^2 - 1}) \quad \text{or} \quad \cosh^{-1} x = \log(x + \sqrt{x^2 - 1})$$

3. Prove that $\tanh^{-1} x = \frac{1}{2}\log\frac{1 + x}{1 - x}$

Proof. Let $y = \tanh^{-1} x$, then $x = \tanh y$ or $\frac{x}{1} = \frac{e^y - e^{-y}}{e^y + e^{-y}}$ or $\frac{1}{x} = \frac{e^y + e^{-y}}{e^y - e^{-y}}$

By componendo and dividendo $\dfrac{1+x}{1-x} = \dfrac{e^y}{e^{-y}} = e^{2y}$

$$\therefore \quad 2y = \log \frac{1+x}{1-x} \quad \text{or} \quad y = \frac{1}{2} \log \frac{1+x}{1-x}$$

$$\therefore \quad \tanh^{-1} x = \frac{1}{2} \log \frac{1+x}{1-x}.$$

4. Prove that:

(*i*) $\mathbf{\sinh^{-1} x + \sinh^{-1} y = \sinh^{-1} (x\sqrt{1+y^2} + y\sqrt{1+x^2})}$

(*ii*) $\mathbf{\cosh^{-1} x + \cosh^{-1} y = \cosh^{-1} (xy + \sqrt{(x^2-1)(y^2-1)})}$

Proof. (*i*) Let $\sinh^{-1} x = u$ and $\sinh^{-1} y = v$

so that $x = \sinh u$ and $y = \sinh v$

Now $\sinh (u + v) = \sinh u \cosh v + \cosh u \sinh v$

$$= \sinh u \sqrt{1+\sinh^2 v} + \sinh v \sqrt{1+\sinh^2 u}$$

$$= x\sqrt{1+y^2} + y\sqrt{1+x^2}$$

$$\Rightarrow \quad u + v = \sinh^{-1} (x\sqrt{1+y^2} + y\sqrt{1+x^2})$$

$$\Rightarrow \quad \sinh^{-1} x + \sinh^{-1} y = \sinh^{-1} (x\sqrt{1+y^2} + y\sqrt{1+x^2})$$

(*ii*) Let, $\cosh^{-1} x = u$ and $\cosh^{-1} y = v$

so that $x = \cosh u$ and $y = \cosh v$

Now, $\cosh (u + v) = \cosh u \cosh v + \sinh u \sinh v$

$$= \cosh u \cosh v + \sqrt{\cosh^2 u - 1} \, . \sqrt{\cosh^2 v - 1}$$

$$= xy + \sqrt{(x^2-1)(y^2-1)}$$

$$\Rightarrow \quad u + v = \cosh^{-1} \left(xy + \sqrt{(x^2-1)(y^2-1)} \right)$$

$$\Rightarrow \quad \cosh^{-1} x + \cosh^{-1} y = \cosh^{-1} \left(xy + \sqrt{(x^2-1)(y^2-1)} \right)$$

Example 1. *Separate into real and imaginary parts*

(*i*) $\sin^{-1} (\cos \theta + i \sin \theta),\ 0 < \theta < \dfrac{\pi}{2}$ (P.T.U. 2012)

(*ii*) $\tan^{-1} (x + iy)$. (K.U.K. May 2013; DCRUST Murthal May 2014)

Sol. (*i*) Let $\sin^{-1} (\cos \theta + i \sin \theta) = x + iy$

$$\therefore \quad \cos \theta + i \sin \theta = \sin (x + iy) = \sin x \cos iy + \cos x \sin iy$$

$$= \sin x \cosh y + i \cos x \sinh y$$

Equating real and imaginary parts, we have

$$\cos\theta = \sin x \cosh y \quad \ldots(i)$$

and

$$\sin\theta = \cos x \sinh y \quad \ldots(ii)$$

Squaring (*i*) and (*ii*) and adding, we have

$$1 = \sin^2 x \cosh^2 y + \cos^2 x \sinh^2 y$$

$$= \sin^2 x\,(1 + \sinh^2 y) + (1 - \sin^2 x)\sinh^2 y = \sin^2 x + \sinh^2 y$$

or $\sinh^2 y = 1 - \sin^2 x = \cos^2 x$

$\therefore$ From (*ii*), $\sin^2\theta = \cos^2 x \sinh^2 y = \cos^4 x \;\Rightarrow\; \cos^2 x = \sin\theta$

$\Rightarrow$ $\cos x = \sqrt{\sin\theta}$

$\therefore$ Real part $x = \cos^{-1}\sqrt{\sin\theta}$

From (*ii*), $\sinh y = \dfrac{\sin\theta}{\cos x} = \dfrac{\sin\theta}{\sqrt{\sin\theta}} = \sqrt{\sin\theta}$

$\therefore$ $y = \sinh^{-1}(\sqrt{\sin\theta})$

$= \log[\sqrt{\sin\theta} + \sqrt{1+\sin\theta}\,]$ $\quad | \because \sinh^{-1} x = \log(x + \sqrt{x^2+1})$

$\Rightarrow$ Imaginary part $y = \log(\sqrt{\sin\theta} + \sqrt{1+\sin\theta}\,)$

(*ii*) Let $\tan^{-1}(x + iy) = u + iv \quad \ldots(i)$

then $\tan^{-1}(x - iy) = u - iv \quad \ldots(ii)$

Adding (*i*) and (*ii*), we have

$$2u = \tan^{-1}(x + iy) + \tan^{-1}(x - iy)$$

$$= \tan^{-1}\frac{(x+iy)+(x-iy)}{1-(x+iy)(x-iy)} = \tan^{-1}\frac{2x}{1-x^2-y^2}$$

$$\therefore \quad u = \frac{1}{2}\tan^{-1}\frac{2x}{1-x^2-y^2}$$

Subtracting (*ii*) from (*i*), we have

$$2iv = \tan^{-1}(x + iy) - \tan^{-1}(x - iy)$$

$$= \tan^{-1}\frac{(x+iy)-(x-iy)}{1+(x+iy)(x-iy)} = \tan^{-1}\frac{2iy}{1+x^2+y^2}$$

$$\Rightarrow \tan 2iv = \frac{2iy}{1+x^2+y^2} \;\Rightarrow\; i\tanh 2v = \frac{2iy}{1+x^2+y^2} \;\Rightarrow\; v = \frac{1}{2}\tanh^{-1}\frac{2y}{1+x^2+y^2}.$$

Example 2. *Find all values of z such that* $\sinh z = e^{i\frac{\pi}{3}}$.

Sol. Let $z = x + iy$, then

$$\sinh z = e^{i\frac{\pi}{3}} \;\Rightarrow\; \sinh(x + iy) = e^{i\frac{\pi}{3}}$$

or

$$\sinh x \cos y + i\cosh x \sin y = \cos\frac{\pi}{3} + i\sin\frac{\pi}{3} = \frac{1}{2} + i\,\frac{\sqrt{3}}{2}$$

Equating real and imaginary parts,

$$\sinh x \cos y = \frac{1}{2} \quad \Rightarrow \quad \sinh x = \frac{1}{2\cos y} \qquad \text{...(1)}$$

$$\cosh x \sin y = \frac{\sqrt{3}}{2} \quad \Rightarrow \quad \cosh x = \frac{\sqrt{3}}{2\sin y} \qquad \text{...(2)}$$

Using $\cosh^2 x - \sinh^2 x = 1$, we have

$$\frac{3}{4\sin^2 y} - \frac{1}{4\cos^2 y} = 1 \quad \text{or} \quad 3\cos^2 y - \sin^2 y = 4\sin^2 y \cos^2 y$$

or $\quad 3(1 - \sin^2 y) - \sin^2 y = 4\sin^2 y\,(1 - \sin^2 y)$

or $\quad 4\sin^4 y - 8\sin^2 y + 3 = 0$

or $\quad (2\sin^2 y - 3)\,(2\sin^2 y - 1) = 0$

$$\therefore \quad \sin^2 y = \frac{3}{2} \quad \text{or} \quad \frac{1}{2}$$

Rejecting $\sin^2 y = \frac{3}{2}$, since $\sin^2 y \le 1$ for real y, we have

$$\sin^2 y = \frac{1}{2} \quad \Rightarrow \quad \sin y = \pm\frac{1}{\sqrt{2}}$$

When $\sin y$ is negative, from (2), $\cosh x$ is also negative which is impossible.

($\because \quad \cosh x > 0$ for all x)

$$\therefore \quad \sin y = \frac{1}{\sqrt{2}} = \sin\frac{\pi}{4}$$

$\Rightarrow \quad y = n\pi + (-1)^n \dfrac{\pi}{4}$, where n is an integer, is the general value of y.

Case I. If n is even, $y = n\pi + \dfrac{\pi}{4}$

Taking $n = 2k$, $\quad \cos y = \cos\left(2k\pi + \dfrac{\pi}{4}\right) = \cos\dfrac{\pi}{4} = \dfrac{1}{\sqrt{2}}$

From (1), $\quad \sinh x = \dfrac{1}{\sqrt{2}}$

$$x = \sinh^{-1}\left(\frac{1}{\sqrt{2}}\right) = \log\left(\frac{1}{\sqrt{2}} + \sqrt{\frac{1}{2} + 1}\right) = \log\frac{\sqrt{3}+1}{\sqrt{2}}$$

$$\therefore \quad z = x + iy = \log\frac{\sqrt{3}+1}{\sqrt{2}} + i\left(n\pi + \frac{\pi}{4}\right)$$

Case II. If n is odd, $y = n\pi - \dfrac{\pi}{4}$

Taking $n = 2k + 1$, $\cos y = \cos\left[(2k+1)\pi - \dfrac{\pi}{4}\right]$

$$= \cos\left(\pi - \frac{\pi}{4}\right) = -\cos\frac{\pi}{4} = -\frac{1}{\sqrt{2}}$$

From (1), $\quad \sinh x = -\dfrac{1}{\sqrt{2}}$

$$x = \sinh^{-1}\left(-\frac{1}{\sqrt{2}}\right) = \log\left(-\frac{1}{\sqrt{2}} + \sqrt{\frac{1}{2}+1}\right) = \log\frac{\sqrt{3}-1}{\sqrt{2}}$$

$$\therefore \quad z = x + iy = \log\frac{\sqrt{3}-1}{\sqrt{2}} + i\left(n\pi - \frac{\pi}{4}\right).$$

Example 3. *Show that $\tanh^{-1}(\cos\theta) = \cosh^{-1}(\operatorname{cosec}\theta)$.*

Sol. Let $\tanh^{-1}(\cos\theta) = \phi$ then $\cos\theta = \tanh\phi$

$\Rightarrow \quad \cos^2\theta = \tanh^2\phi \quad \Rightarrow \quad 1 - \sin^2\theta = \tanh^2\phi$

$\Rightarrow \quad 1 - \tanh^2\phi = \sin^2\theta \quad \Rightarrow \quad \operatorname{sech}^2\phi = \sin^2\theta$

$\Rightarrow \quad \operatorname{sech}\phi = \sin\theta$

Taking reciprocals $\quad \cosh\phi = \operatorname{cosec}\theta$

$\Rightarrow \quad \phi = \cosh^{-1}(\operatorname{cosec}\theta)$

$\Rightarrow \quad \tanh^{-1}(\cos\theta) = \cosh^{-1}(\operatorname{cosec}\theta).$

EXERCISE 3.4

1. Prove that:

   (*i*) $(\cosh x + \sinh x)^n = \cosh nx + \sinh nx$; n being a positive integer.

   (*ii*) $\left(\dfrac{1+\tanh x}{1-\tanh x}\right)^3 = \cosh 6x + \sinh 6x.$

2. If $y = \log\tan x$, show that $\sinh ny = \dfrac{1}{2}(\tan^n x - \cot^n x)$.

3. (*a*) If $\tan y = \tan\alpha\tanh\beta$ and $\tan z = \cot\alpha\tanh\beta$, prove that $\tan(y+z) = \sinh 2\beta \operatorname{cosec} 2\alpha$.

   (*b*) Prove that:

   (*i*) $\overline{\sin z} = \sin\bar{z}$ (*ii*) $\overline{\cos z} = \cos\bar{z}$

   (*iii*) $\overline{\tan z} = \tan\bar{z}$

4. If $\tan\theta = \tanh x\cot y$ and $\tan\phi = \tanh x\tan y$, prove that $\dfrac{\sin 2\theta}{\sin 2\phi} = \dfrac{\cosh 2x + \cos 2y}{\cosh 2x - \cos 2y}$.

5. If $c\cosh(\theta + i\phi) = x + iy$, prove that

   (*i*) $x^2\operatorname{sech}^2\theta + y^2\operatorname{cosech}^2\theta = c^2$ (*ii*) $x^2\sec^2\phi - y^2\operatorname{cosec}^2\phi = c^2$.

6. If $\tan(x + iy) = A + iB$, show that $\dfrac{A}{B} = \dfrac{\sin 2x}{\sinh 2y}$.

7. If $\sin(\theta + i\phi) = \rho(\cos\alpha + i\sin\alpha)$, prove that

   (*i*) $\rho^2 = \dfrac{1}{2}(\cosh 2\phi - \cos 2\theta)$ (*ii*) $\tan\alpha = \tanh\phi\cot\theta$.

8. If $\sin(\theta + i\phi) = \cos\alpha + i\sin\alpha$, prove that $\cos^4\theta = \sin^2\alpha = \sinh^4\phi$. *(M.D.U. May 2014)*

9. If $\cos(\theta + i\phi) = \cos\alpha + i\sin\alpha$, prove that

(*i*) $\sin^2\theta = \pm\sin\alpha$ (*ii*) $\cos 2\theta + \cosh 2\phi = 2$.

10. (*a*) If $\sin(\theta + i\phi) = \tan\alpha + i\sec\alpha$, show that $\cos 2\theta \cosh 2\phi = 3$.

(*b*) If $\tanh\alpha = \cos\theta$, prove that $\alpha = \log(\operatorname{cosec}\theta + \cot\theta)$

(*c*) Prove that $\log\left\{\dfrac{\cos(x - iy)}{\cos(x + iy)}\right\} = 2i\tan^{-1}(\tan x \tanh y)$

(*d*) If $\sin x \cosh y = \cos\theta$ and $\cos x \sinh y = \sin\theta$, prove that

(*i*) $\sinh^2 y = \cos^2 x$ (*ii*) $\cos^4 x = \sin^2\theta$

11. (*a*) If $\cos(\theta + i\phi) = R(\cos\alpha + i\sin\alpha)$, prove that $e^{2\phi} = \dfrac{\sin(\theta - \alpha)}{\sin(\theta + \alpha)}$. *(K.U.K. Jan. 2014)*

(*b*) If $u = \log\tan\left(\dfrac{\pi}{4} + \dfrac{\theta}{2}\right)$, prove that $\theta = -i\log\tan\left(\dfrac{\pi}{4} + \dfrac{iu}{2}\right)$. *(K.U.K. Jan. 2014)*

12. If $\tan(\theta + i\phi) = \tan\alpha + i\sec\alpha$, show that $e^{2\phi} = \pm\cot\dfrac{\alpha}{2}$ and $2\theta = \left(n + \dfrac{1}{2}\right)\pi + \alpha$.

(M.D.U. May 2013; K.U.K. 2009, 2010)

13. If $\tan(x + iy) = \sin(u + iv)$, prove that $\dfrac{\sin 2x}{\sinh 2y} = \dfrac{\tan u}{\tanh v}$.

14. If $\tan(x + iy) = \cosh(\alpha + i\beta)$, prove that $\tanh\alpha\tan\beta = \operatorname{cosec} 2x \sinh 2y$.

15. If $C\tan(x + iy) = A + iB$, prove that $\tan 2x = \dfrac{2CA}{C^2 - A^2 - B^2}$. *(M.D.U. 2011)*

16. (*a*) Prove that $(1 + \cosh x + \sinh x)^n = 2^n \cosh^n \dfrac{x}{2}\left(\cosh\dfrac{nx}{2} + \sinh\dfrac{nx}{2}\right)$.

(*b*) Prove that $\sinh^{-1}(\tan\theta) = \log\tan\left(\dfrac{\pi}{4} + \dfrac{\theta}{2}\right)$

17. If $\cosh x = \sec\theta$, prove that

(*i*) $\tanh^2\dfrac{x}{2} = \tan^2\dfrac{\theta}{2}$ (*ii*) $x = \log\tan\left(\dfrac{\pi}{4} + \dfrac{\theta}{2}\right)$.

18. If $\tan\dfrac{x}{2} = \tanh\dfrac{u}{2}$, prove that

(*i*) $\cos x \cosh u = 1$ (*ii*) $\tan x = \sinh u$

(*iii*) $u = \log\tan\left(\dfrac{\pi}{4} + \dfrac{x}{2}\right)$.

19. If $x = 2\cos\alpha\cosh\beta$, $y = 2\sin\alpha\sinh\beta$, prove that $\sec(\alpha + i\beta) + \sec(\alpha - i\beta) = \dfrac{4x}{x^2 + y^2}$.

20. If $\sin[\log(A + iB)] = x + iy$, show that $\dfrac{x^2}{\sin^2 u} - \dfrac{y^2}{\cos^2 u} = 1$, where $A^2 + B^2 = e^{2u}$.

(M.D.U. 2011)

21. Separate into real and imaginary parts:

(*i*) $e^{\cosh(x+iy)}$ (*ii*) $\sin^2(x + iy)$ (*iii*) $\log\cos(x + iy)$.

22. If $\tan(x + iy) = \theta + i\phi$, prove that $\theta^2 + \phi^2 = \dfrac{\cosh^2 y - \cos^2 x}{\cosh^2 y - \sin^2 x}$.

23. Prove that: $\tan \dfrac{u + iv}{2} = \dfrac{\sin u + i \sinh v}{\cos u + \cosh v}$.

24. If $x + iy = \cos(u + iv)$, show that

(*i*) $(1 + x)^2 + y^2 = (\cosh v + \cos u)^2$ (*ii*) $(1 - x)^2 + y^2 = (\cosh v - \cos u)^2$.

25. (*a*) If $\cos^{-1}(u + iv) = \alpha + i\beta$, prove that $\cos^2\alpha$ and $\cosh^2\beta$ are the roots of the equation

$$x^2 - (1 + u^2 + v^2)\,x + u^2 = 0.$$

(*b*) If $\sin^{-1}(u + iv) = \alpha + i\beta$, prove that $\sin^2\alpha$ and $\cosh^2\beta$ are the roots of the equation $x^2 - x(1 + u^2 + v^2) + u^2 = 0$.

26. Prove that $\tan^{-1}(e^{i\theta}) = \dfrac{n\pi}{2} + \dfrac{\pi}{4} - \dfrac{i}{2}\log\tan\left(\dfrac{\pi}{4} - \dfrac{\theta}{2}\right)$. (*M.D.U. Dec. 2014; K.U.K. Dec. 2015*)

27. Find $\tanh x$ if $5\sinh x - \cosh x = 5$.

[**Hint.** Divide both sides by $\cosh x$, square, replace $\text{sech}^2 x$ by $(1 - \tanh^2 x)$ and solve for $\tanh x$.]

28. If $\cos^{-1}(x + iy) = \alpha + i\beta$, show that

(*i*) $x^2 \sec^2\alpha - y^2 \text{cosec}^2\alpha = 1$ (*ii*) $x^2 \text{sech}^2\beta + y^2 \text{cosech}^2\beta = 1$.

29. Find all the roots of the equation:

(*i*) $\cos z = 2$ (*ii*) $\tanh z + 2 = 0$

(*iii*) $\sin z = \cosh 4$ (*iv*) $\sinh z = i$.

30. If $|\cos(u + iv)| = 1$, show that $\sin^2 u = \sinh^2 v$.

31. If $\log\sin(\theta + i\phi) = \alpha + i\beta$, prove that

(*i*) $2\cos 2\theta = e^{2\phi} + e^{-2\phi} - 4e^{2\alpha}$ (*ii*) $\cos(\theta - \beta) = e^{2\phi}\cos(\theta + \beta)$.

32. Solve $\tan z = e^{i\alpha}$, where α is real.

33. If $\cosh^{-1}(x + iy) + \cosh^{-1}(x - iy) = \cosh^{-1} a$, prove that $2(a - 1)x^2 + 2(a + 1)y^2 = a^2 - 1$.

Answers

21. (*i*) $e^{\cosh x \cos y}[\cos(\sinh x \sin y) + i \sin(\sinh x \sin y)]$

(*ii*) $\dfrac{1}{2}[(1 - \cos 2x \cosh 2y) + i \sin 2x \sinh 2y]$

(*iii*) $\dfrac{1}{2}\log\left[\dfrac{1}{2}(\cos 2x + \cosh 2y)\right] - i\tan^{-1}(\tan x \tanh y)$

27. $\dfrac{4}{5}, -\dfrac{3}{5}$

29. (*i*) $2n\pi + i\log(2 + \sqrt{3})$ (*ii*) $-\dfrac{1}{2}\log 3 + i\left(n + \dfrac{1}{2}\right)\pi$

(*iii*) $n\pi + (-1)^n\left(\dfrac{\pi}{2} - 4i\right)$ (*iv*) $i(4n + 1)\dfrac{\pi}{2}$

32. $z = \dfrac{n\pi}{2} + \dfrac{\pi}{4} + \dfrac{i}{2}\log\tan\left(\dfrac{\pi}{4} + \dfrac{\alpha}{2}\right)$ [**Hint.** See Example 7.]

3.12. LIMIT OF A FUNCTION

Let $f(z)$ be a single-valued function of z defined in a neighbourhood of $z = z_0$, then $f(z)$ is said to have the limit l as z approaches z_0 (along any path, straight or curved) if given an arbitrary real number $\varepsilon > 0$, however small, there exists a real number $\delta > 0$ such that

$$| f(z) - l | < \varepsilon \quad \text{whenever} \quad 0 < | z - z_0 | < \delta$$

and we write $\lim\limits_{z \to z_0} f(z) = l$

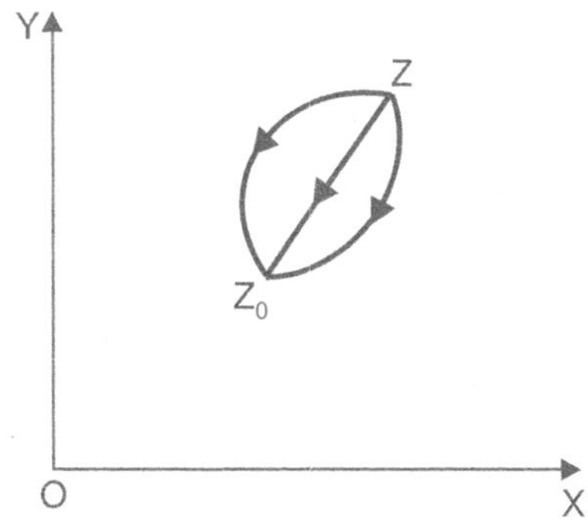

Remark 1. δ usually depends on ε.

Remark 2. In real variables, $x \to x_0$ implies that x approaches x_0 along the number line, either from left or from right. In complex variables, $z \to z_0$ implies that z approaches z_0 along any path, straight or curved. The limit must be independent of the manner in which z approaches z_0. If we get two different limits as $z \to z_0$ along two different paths then limit does not exist.

3.13. THEOREMS ON LIMITS

1. If $\lim\limits_{z \to z_0} f(z)$ exists, then it is unique.

2. If $f(z) = u(x, y) + iv(x, y)$, where $z = x + iy$ and $z_0 = x_0 + iy_0$, then $\lim\limits_{z \to z_0} f(z) = l = u_0 + iv_0$ if and only if

$$\lim_{\substack{x \to x_0 \\ y \to y_0}} u(x, y) = u_0 \quad \text{and} \quad \lim_{\substack{x \to x_0 \\ y \to y_0}} v(x, y) = v_0.$$

3. If $\lim\limits_{z \to z_0} f(z) = l$ and c is a constant, real or complex, then $\lim\limits_{z \to z_0} cf(z) = c \lim\limits_{z \to z_0} f(z) = cl$.

4. If $\lim\limits_{z \to z_0} f(z) = l_1$ and $\lim\limits_{z \to z_0} g(z) = l_2$, then

(*i*) $\lim\limits_{z \to z_0} [f(z) + g(z)] = \lim\limits_{z \to z_0} f(z) + \lim\limits_{z \to 0} g(z) = l_1 + l_2$

(*ii*) $\lim\limits_{z \to z_0} [f(z) - g(z)] = \lim\limits_{z \to z_0} f(z) - \lim\limits_{z \to z_0} g(z) = l_1 - l_2$

(*iii*) $\lim\limits_{z \to z_0} [f(z)g(z)] = \lim\limits_{z \to z_0} f(z) . \lim\limits_{z \to z_0} g(z) = l_1 l_2$

(*iv*) $\lim\limits_{z \to z_0} \dfrac{1}{g(z)} = \dfrac{1}{\lim\limits_{z \to z_0} g(z)} = \dfrac{1}{l_2}$ provided $l_2 \neq 0$

(*v*) $\lim\limits_{z \to z_0} \dfrac{f(z)}{g(z)} = \dfrac{\lim\limits_{z \to z_0} f(z)}{\lim\limits_{z \to z_0} g(z)} = \dfrac{l_1}{l_2}$ provided $l_2 \neq 0$.

ILLUSTRATIVE EXAMPLES

Example 1. *Prove that* $\lim_{z \to 0} \frac{\bar{z}}{z}$ *does not exist.* (M.D.U. May 2015)

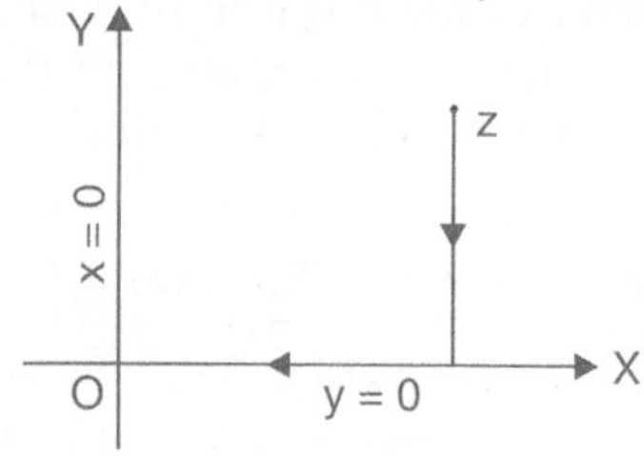

Sol. If the limit exists, then it must be independent of the manner in which z approaches 0.

Consider the path $y \to 0$ followed by $x \to 0$, we get

$$\lim_{z \to 0} \frac{\bar{z}}{z} = \lim_{x \to 0} \left[\lim_{y \to 0} \frac{x - iy}{x + iy} \right] = \lim_{x \to 0} \frac{x}{x} = 1$$

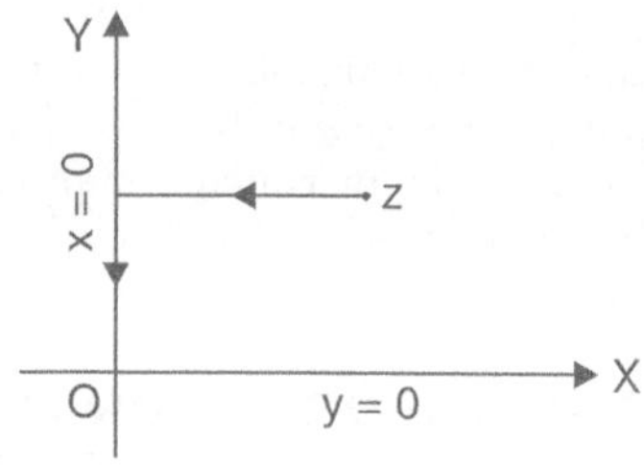

Now consider the path $x \to 0$ followed by $y \to 0$, we get

$$\lim_{z \to 0} \frac{\bar{z}}{z} = \lim_{y \to 0} \left[\lim_{x \to 0} \frac{x - iy}{x + iy} \right] = \lim_{y \to 0} \frac{-iy}{iy}$$

$$= -1$$

As $z \to 0$ along two different paths, we get different limits. Hence the limit does not exist.

Example 2. *Show that* $\lim_{z \to 0} \frac{(Re\, z - Im\, z)^2}{|z|^2}$ *does not exist.*

Sol. Let $z \to 0$ along the path $y = mx$, then

$$\lim_{z \to 0} \frac{(\text{Re } z - \text{Im } z)^2}{|z|^2} = \lim_{z \to 0} \frac{(x - y)^2}{x^2 + y^2} = \lim_{x \to 0} \frac{(x - mx)^2}{x^2 + m^2 x^2}$$

$$= \lim_{x \to 0} \frac{(1 - m)^2 x^2}{(1 + m^2) x^2} = \frac{(1 - m)^2}{1 + m^2}$$

which depends on m. For different values of m, we have different paths and different limits. Hence the limit does not exist.

Example 3. *Prove that* $\lim_{z \to 1+i} \left(\frac{z - 1 - i}{z^2 - 2z + 2} \right)^2 = -\frac{1}{4}$.

Sol. $\lim_{z \to 1+i} \left(\frac{z - 1 - i}{z^2 - 2z + 2} \right)^2 = \lim_{z \to 1+i} \left[\frac{z - 1 - i}{(z - 1)^2 - i^2} \right]^2 = \lim_{z \to 1+i} \left[\frac{z - 1 - i}{(z - 1 + i)(z - 1 - i)} \right]^2$

$$= \lim_{z \to 1+i} \frac{1}{(z - 1 + i)^2} = \frac{1}{\lim_{z \to 1+i} (z - 1 + i)^2}$$

$$= \frac{1}{(1 + i - 1 + i)^2} = \frac{1}{4i^2} = -\frac{1}{4}.$$

EXERCISE 3.5

Show that the following limits do not exist:

1. $\lim_{z \to 0} \dfrac{z}{|z|}$

2. $\lim_{z \to 0} \dfrac{z^2}{|z|^2}$

3. $\lim_{z \to 0} \left(\dfrac{z}{\bar{z}}\right)$ (M.D.U. May 2014)

4. $\lim_{z \to 0} \dfrac{\text{Re } z}{|z|}$

5. $\lim_{z \to 0} \dfrac{\text{Im}(z)^2}{|z|^2}$

Find the following limits:

6. $\lim_{z \to 1-i} (z^2 + 4z - 7)$

7. $\lim_{z \to -2i} \dfrac{(2z+3)(z-1)}{z^2 - 2z + 4}$

8. $\lim_{z \to -i} (2x^3 - 5y^3 i)$

9. $\lim_{z \to 1+i} \dfrac{z^3 - 1}{z^2 - 1}$

10. $\lim_{z \to 1+i} \dfrac{z^2 - z + 1 - i}{z^2 - 2z + 2}$

11. $\lim_{z \to 1-i} (z^2 - \bar{z}^2)$

12. $\lim_{z \to \infty} \dfrac{iz^3 + iz - 1}{(2z+3i)(z-i)^2}$.

Answers

6. $-3-6i$ 7. $-\dfrac{1}{2}+\dfrac{11}{4}i$ 8. $5i$

9. $\dfrac{7}{5}+\dfrac{4}{5}i$ 10. $1-\dfrac{1}{2}i$ 11. $-4i$ 12. $\dfrac{i}{2}$.

3.14. CONTINUITY OF A FUNCTION

Let $f(z)$ be a single-valued function of z defined in a neighbourhood of $z = z_0$, then $f(z)$ is said to continuous at $z = z_0$ if given an arbitrary real number $\varepsilon > 0$, however small, there exists a real number $\delta > 0$ such that

$$| f(z) - f(z_0) | < \varepsilon \quad \text{whenever} \quad | z - z_0 | < \delta$$

In other words, $f(z)$ is continuous at $z = z_0$ if $\lim_{z \to z_0} f(z) = f(z_0)$.

Thus three conditions must be satisfied so that $f(z)$ is continuous at $z = z_0$

1. $f(z)$ is defined at z_0 *i.e.,* $f(z_0)$ exists
2. $\lim_{z \to z_0} f(z)$ exists
3. $\lim_{z \to z_0} f(z) = f(z_0)$.

If any of the above conditions is not satisfied then $f(z)$ is said to be *discontinuous* at $z = z_0$.

3.15. REMOVABLE DISCONTINUITY

If $f(z_0)$ exists and $\lim_{z \to z_0} f(z) = l$ exists but $f(z_0) \neq l$, then $z = z_0$ is called a point of *removable discontinuity*.

By redefining $f(z)$ at $z = z_0$ such that $f(z_0) = l$, the function can be made continuous. If $\lim_{z \to z_0} f(z)$ does not exist, discontinuity cannot be removed. Existence of limit is the necessary condition for removable discontinuity.

3.16. CONTINUITY IN A REGION

A function $f(z)$ is said to be *continuous in a region* if it is continuous at all points of the region.

Remark 1. To examine the continuity of $f(z)$ at $z = \infty$, we put $z = \frac{1}{w}$ and examine the continuity of $f\left(\frac{1}{w}\right)$ at $w = 0$.

Remark 2. If $f(z)$ and $g(z)$ are continuous at $z = z_0$, then so also are the functions $f(z) + g(z)$, $f(z) - g(z)$, $f(z)\, g(z)$ and $\frac{f(z)}{g(z)}$, where $g(z_0) \neq 0$.

Remark 3. If $f(z)$ is continuous in a region, then the real and imaginary parts of $f(z)$ are also continuous in the region.

If $f(z) = u(x, y) + iv\,(x, y)$, then $f(z)$ is continuous if and only if $u(x, y)$ and $v(x, y)$ are separately continuous functions of x and y.

ILLUSTRATIVE EXAMPLES

Example 1. *Show that the function f(z) defined by*

$$f(z) = \begin{cases} \dfrac{Re(z^2)}{|z|^2}, & z \neq 0 \\ 0, & z = 0 \end{cases}$$

is not continuous at z = 0.

Sol. Given $f(0) = 0$

Let $z \to 0$ along the path $y = mx$, then

$$\lim_{z \to 0} f(z) = \lim_{z \to 0} \frac{\text{Re}(z^2)}{|z|^2} = \lim_{z \to 0} \frac{x^2 - y^2}{x^2 + y^2}$$

$$[\because \ z^2 = (x + iy)^2 = (x^2 - y^2) + 2ixy]$$

$$= \lim_{x \to 0} \frac{x^2 - m^2x^2}{x^2 + m^2x^2} = \lim_{x \to 0} \frac{(1 - m^2)x^2}{(1 + m^2)x^2} = \frac{1 - m^2}{1 + m^2}$$

which depends on m. For different values of m, we have different paths and different limits. Hence the limit does not exist and the function is not continuous at $z = 0$.

Example 2. *Find the value of $f(z_0)$ so that the function*

$$f(z) = \frac{z^2 - 2z + 2}{z^2 + 2i} \text{ is continuous at } z_0 = 1 - i.$$

Sol.
$$\lim_{z \to z_0} f(z) = \lim_{z \to 1-i} \frac{z^2 - 2z + 2}{z^2 + 2i} = \lim_{z \to 1-i} \frac{(z - 1)^2 + 1}{z^2 - (1 - i)^2}$$

$$= \lim_{z \to 1-i} \frac{(z-1)^2 - i^2}{z^2 - (1-i)^2} = \lim_{z \to 1-i} \frac{(z-1+i)(z-1-i)}{(z+1-i)(z-1+i)}$$

$$= \lim_{z \to 1-i} \frac{z-1-i}{z+1-i} = \frac{\lim_{z \to 1-i} (z-1-i)}{\lim_{z \to 1-i} (z+1-i)}$$

$$= \frac{1-i-1-i}{1-i+1-i} = \frac{-2i}{2(1-i)} = \frac{-i}{1-i} \times \frac{1+i}{1+i} = \frac{1-i}{2}$$

Now $f(z)$ will be continuous at z_0 if

$$f(z_0) = \lim_{z \to z_0} f(z) = \frac{1-i}{2}.$$

EXERCISE 3.6

1. Examine the continuity of the following functions:

(i) $f(z) = \begin{cases} \dfrac{z^2+1}{z+i}, & z \neq -i \\ 0, & z = -i \end{cases}$ at $z = -i$

(ii) $f(z) = \begin{cases} \dfrac{z^3 - iz^2 + z - i}{z-i}, & z \neq i \\ 0, & z = i \end{cases}$ at $z = i$

(iii) $f(z) = \begin{cases} \dfrac{z^2+4}{z-2i}, & z \neq 2i \\ 2+3i, & z = 2i \end{cases}$ at $z = 2i$.

2. Show that the following functions are continuous for all z.

(i) $\sin z$ (ii) e^z.

[**Hint:** Express $f(z)$ as $u(x, y) + iv(x, y)$ and show that $u(x, y)$ and $v(x, y)$ are continuous for all real values of x and y.]

3. Show that the function

$$f(z) = \begin{cases} \dfrac{\text{Im}(z)}{|z|}, & z \neq 0 \\ 0, & z = 0 \end{cases}$$

is not continuous at $z = 0$.

4. Find the value of $f(i)$ so that the function

$$f(z) = \frac{iz^3 - 1}{z - i} \text{ is continuous at } z = i.$$

Answers

1. (i) Not continuous (ii) Continuous (iii) Not continuous 4. $-3i$.

3.17. DIFFERENTIABILITY

Let $w = f(z)$ be a single-valued function of the complex variable $z(= x + iy)$.

Let $w + \Delta w = f(z + \Delta z)$, then

$$\Delta w = f(z + \Delta z) - f(z)$$

and

$$\frac{\Delta w}{\Delta z} = \frac{f(z + \Delta z) - f(z)}{\Delta z}$$

$$\lim_{\Delta z \to 0} \frac{\Delta w}{\Delta z} = \lim_{\Delta z \to 0} \frac{f(z + \Delta z) - f(z)}{\Delta z},$$

if it exists, is called the derivative of $f(z)$ and is denoted by $f'(z)$.

Thus, $$\frac{dw}{dz} = f'(z) = \lim_{\Delta z \to 0} \frac{f(z + \Delta z) - f(z)}{\Delta z}$$

The function $f(z)$ is said to be differentiable at $z = z_0$ if

$$\lim_{z \to z_0} \frac{f(z) - f(z_0)}{z - z_0} \text{ exists.}$$

This limit is called the derivative of $f(z)$ at $z = z_0$ and is denoted by $f'(z_0)$.

Thus, $$f'(z_0) = \lim_{z \to z_0} \frac{f(z) - f(z_0)}{z - z_0}$$

Equivalently, by putting $z - z_0 = \Delta z$, we have

$$f'(z_0) = \lim_{\Delta z \to 0} \frac{f(z_0 + \Delta z) - f(z_0)}{\Delta z}.$$

3.18. ANALYTIC FUNCTION

(K.U.K. 2010; M.D.U. 2012, Dec., 2013, Dec. 2014, Dec. 2015)

If a single-valued function $f(z)$ possesses a unique derivative at every point of a region R, then $f(z)$ is called an **analytic function** or a **regular function** or a **holomorphic function** of z in R.

A point where the function ceases to be analytic is called a singular point.

3.19. NECESSARY AND SUFFICIENT CONDITIONS FOR f(z) TO BE ANALYTIC

The necessary and sufficient conditions for the function

(M.D.U. Dec., 2013, Dec. 2014, Dec. 2015; M.T.U. 2012)

$$w = f(z) = u(x, y) + iv(x, y)$$

to be analytic in a region R, are

(i) $\frac{\partial u}{\partial x}, \frac{\partial u}{\partial y}, \frac{\partial v}{\partial x}, \frac{\partial v}{\partial y}$ *are continuous functions of x and y in the region R.*

(ii) $\frac{\partial u}{\partial x} = \frac{\partial v}{\partial y}, \frac{\partial u}{\partial y} = -\frac{\partial v}{\partial x}$.

The conditions in (ii) are known as **Cauchy-Riemann equations** *or briefly C-R equations.* (K.U.K. 2010; M.D.U. 2012, May 2015)

Proof. *(a)* **Necessary Condition.** Let $w = f(z) = u(x, y) + iv(x, y)$ be analytic in a region R, then $\frac{dw}{dz} = f'(z)$ exists uniquely at every point of that region.

Let δx and δy be the increments in x and y respectively. Let δu, δv and δz be the corresponding increments in u, v and z respectively. Then,

$$f'(z) = \lim_{\delta z \to 0} \frac{f(z + \delta z) - f(z)}{\delta z} = \lim_{\delta z \to 0} \frac{(u + \delta u) + i(v + \delta v) - (u + iv)}{\delta z}$$

$$= \lim_{\delta z \to 0} \left(\frac{\delta u}{\delta z} + i \frac{\delta v}{\delta z} \right) \qquad ...(1)$$

Since the function $w = f(z)$ is analytic in the region R, the limit (1) must exist independent of the manner in which $\delta z \to 0$, *i.e.*, along whichever path δx and $\delta y \to 0$.

First, let $\delta z \to 0$ along a line parallel to x-axis so that $\delta y = 0$ and $\delta z = \delta x$.

[since $z = x + iy$, $z + \delta z = (x + \delta x) + i(y + \delta y)$ and $\delta z = \delta x + i\delta y$]

$\therefore$ From (1), $$f'(z) = \lim_{\delta x \to 0} \left(\frac{\delta u}{\delta x} + i \frac{\delta v}{\delta x} \right) = \frac{\partial u}{\partial x} + i \frac{\partial v}{\partial x} \qquad ...(2)$$

Now, let $\delta z \to 0$ along a line parallel to y-axis so that $\delta x = 0$ and $\delta z = i\,\delta y$.

$\therefore$ From (1), $$f'(z) = \lim_{\delta y \to 0} \left(\frac{\delta u}{i\,\delta y} + i \frac{\delta v}{i\,\delta y} \right) = \frac{1}{i} \frac{\partial u}{\partial y} + \frac{\partial v}{\partial y}$$

$$= \frac{\partial v}{\partial y} - i \frac{\partial u}{\partial y} \qquad ...(3) \qquad \left[\because \ \frac{1}{i} = -i \right]$$

From (2) and (3), we have $\dfrac{\partial u}{\partial x} + i \dfrac{\partial v}{\partial x} = \dfrac{\partial v}{\partial y} - i \dfrac{\partial u}{\partial y}$

Equating the real and imaginary parts, $\dfrac{\partial u}{\partial x} = \dfrac{\partial v}{\partial y}$ and $\dfrac{\partial u}{\partial y} = -\dfrac{\partial v}{\partial x}$.

Hence the necessary condition for f(z) to be analytic is that the C-R equations must be satisfied.

(*b*) **Sufficient Condition.** Let $f(z) = u + iv$ be a single-valued function possessing partial derivatives $\dfrac{\partial u}{\partial x}, \dfrac{\partial u}{\partial y}, \dfrac{\partial v}{\partial x}, \dfrac{\partial v}{\partial y}$ at each point of a region R and satisfying C-R equations.

i.e., $$\frac{\partial u}{\partial x} = \frac{\partial v}{\partial y} \quad \text{and} \quad \frac{\partial u}{\partial y} = -\frac{\partial v}{\partial x}.$$

We shall show that $f(z)$ is analytic, *i.e.*, $f'(z)$ exists at every point of the region R.

By Taylor's theorem for functions of two variables, we have, on omitting second and higher degree terms of δx and δy

$$f(z + \delta z) = u(x + \delta x, y + \delta y) + iv(x + \delta x, y + \delta y)$$

$$= \left[u(x, y) + \left(\frac{\partial u}{\partial x} \delta x + \frac{\partial u}{\partial y} \delta y \right) \right] + i \left[v(x, y) + \left(\frac{\partial v}{\partial x} \delta x + \frac{\partial v}{\partial y} \delta y \right) \right]$$

$$= [u(x, y) + iv(x, y)] + \left(\frac{\partial u}{\partial x} + i \frac{\partial v}{\partial x} \right) \delta x + \left(\frac{\partial u}{\partial y} + i \frac{\partial v}{\partial y} \right) \delta y$$

$$= f(z) + \left(\frac{\partial u}{\partial x} + i \frac{\partial v}{\partial x} \right) \delta x + \left(\frac{\partial u}{\partial y} + i \frac{\partial v}{\partial y} \right) \delta y$$

or $$f(z + \delta z) - f(z) = \left(\frac{\partial u}{\partial x} + i \frac{\partial v}{\partial x} \right) \delta x + \left(\frac{\partial u}{\partial y} + i \frac{\partial v}{\partial y} \right) \delta y$$

$$= \left(\frac{\partial u}{\partial x} + i \frac{\partial v}{\partial x} \right) \delta x + \left(-\frac{\partial v}{\partial x} + i \frac{\partial u}{\partial x} \right) \delta y \qquad \text{[Using C-R equations]}$$

$$= \left(\frac{\partial u}{\partial x} + i \frac{\partial v}{\partial x} \right) \delta x + \left(\frac{\partial u}{\partial x} + i \frac{\partial v}{\partial x} \right) i\delta y \qquad [\because \ -1 = i^2]$$

$$= \left(\frac{\partial u}{\partial x} + i\frac{\partial v}{\partial x}\right)(\delta x + i\delta y) = \left(\frac{\partial u}{\partial x} + i\frac{\partial v}{\partial x}\right)\delta z \qquad [\because \ \delta x + i\delta y = \delta z]$$

$$\Rightarrow \qquad \frac{f(z+\delta z) - f(z)}{\delta z} = \frac{\partial u}{\partial x} + i\frac{\partial v}{\partial x}$$

$$\therefore \qquad f'(z) = \lim_{\delta z \to 0} \frac{f(z+\delta z) - f(z)}{\delta z} = \frac{\partial u}{\partial x} + i\frac{\partial v}{\partial x}$$

Thus $f'(z)$ exists, because $\frac{\partial u}{\partial x}, \frac{\partial v}{\partial x}$ exist.

Hence $f(z)$ is analytic.

Note 1. The real and imaginary parts of an analytic function are called **conjugate functions.** Thus, if $f(z) = u(x, y) + iv(x, y)$ is an analytic function, then $u(x, y)$ and $v(x, y)$ are conjugate functions. The relation between two conjugate functions is given by C-R equations.

Note 2. When a function $f(z)$ is known to be analytic, it can be differentiated in the ordinary way as if z is a real variable.

Thus, $\quad f(z) = z^2 \Rightarrow f'(z) = 2z$

$f(z) = \sin z \Rightarrow f'(z) = \cos z$ etc.

3.20. CAUCHY-RIEMANN EQUATIONS IN POLAR COORDINATES

(M.D.U. 2012)

Let (r, θ) be the polar coordinates of the point whose cartesian coordinates are (x, y), then

$$x = r\cos\theta, \ y = r\sin\theta,$$

$$z = x + iy = r(\cos\theta + i\sin\theta) = re^{i\theta}$$

$$\therefore \qquad u + iv = f(z) = f(re^{i\theta}) \qquad \text{...(1)}$$

Differentiating (1) partially w.r.t. r, we have

$$\frac{\partial u}{\partial r} + i\frac{\partial v}{\partial r} = f'(re^{i\theta}) \cdot e^{i\theta} \qquad \text{...(2)}$$

Differentiating (1) partially w.r.t. θ, we have

$$\frac{\partial u}{\partial \theta} + i\frac{\partial v}{\partial \theta} = f'(re^{i\theta}) \cdot ire^{i\theta} = ir\left(\frac{\partial u}{\partial r} + i\frac{\partial v}{\partial r}\right) \qquad \text{[Using (2)]}$$

$$= -r\frac{\partial v}{\partial r} + ir\frac{\partial u}{\partial r}$$

Equating real and imaginary parts, we get

$$\frac{\partial u}{\partial \theta} = -r\frac{\partial v}{\partial r} \quad \text{and} \quad \frac{\partial v}{\partial \theta} = r\frac{\partial u}{\partial r}$$

or $\quad \frac{\partial u}{\partial r} = \frac{1}{r}\frac{\partial v}{\partial \theta}$ and $\frac{\partial v}{\partial r} = -\frac{1}{r}\frac{\partial u}{\partial \theta}$, which is the polar form of C-R equations.

3.21. HARMONIC FUNCTIONS

(G.B.T.U. 2012)

Any solution of the Laplace's equation, $\frac{\partial^2 \phi}{\partial x^2} + \frac{\partial^2 \phi}{\partial y^2} = 0$ is called a harmonic function.

Let $f(z) = u + iv$ be analytic in some region of the z-plane, then u and v satisfy C-R equations.

$$\therefore \qquad \frac{\partial u}{\partial x} = \frac{\partial v}{\partial y} \qquad \text{...(1)}$$

and
$$\frac{\partial u}{\partial y} = -\frac{\partial v}{\partial x} \qquad \text{...(2)}$$

Differentiating (1) partially w.r.t. x and (2) w.r.t. y, we get

$$\frac{\partial^2 u}{\partial x^2} = \frac{\partial^2 v}{\partial x \partial y} \qquad \text{...(3)}$$

and
$$\frac{\partial^2 u}{\partial y^2} = -\frac{\partial^2 v}{\partial y \partial x} \qquad \text{...(4)}$$

Assuming $\frac{\partial^2 v}{\partial x \partial y} = \frac{\partial^2 v}{\partial y \partial x}$ and adding (3) and (4), we get

$$\frac{\partial^2 u}{\partial x^2} + \frac{\partial^2 u}{\partial y^2} = 0 \qquad \text{...(5)}$$

Now, differentiating (1) partially w.r.t. y and (2) w.r.t. x, we get

$$\frac{\partial^2 u}{\partial y \partial x} = \frac{\partial^2 v}{\partial y^2} \qquad \text{...(6)}$$

and
$$\frac{\partial^2 u}{\partial x \partial y} = -\frac{\partial^2 v}{\partial x^2} \qquad \text{...(7)}$$

Assuming $\frac{\partial^2 u}{\partial y \partial x} = \frac{\partial^2 u}{\partial x \partial y}$ and subtracting (7) from (6), we get

$$\frac{\partial^2 v}{\partial x^2} + \frac{\partial^2 v}{\partial y^2} = 0 \qquad \text{...(8)}$$

Equations (5) and (8) show that the real and imaginary parts u and v of an analytic function satisfy the Laplace's equation.

Hence, u and v are known as harmonic functions.

3.22. ORTHOGONAL SYSTEM

Every analytic function $f(z) = u + iv$ defines two families of curves $u(x, y) = c_1$ and $v(x, y) = c_2$, which form an orthogonal system. (M.T.U. 2012)

Consider the two families of curves

$$u(x, y) = c_1 \qquad \text{...(1)}$$

and
$$v(x, y) = c_2 \qquad \text{...(2)}$$

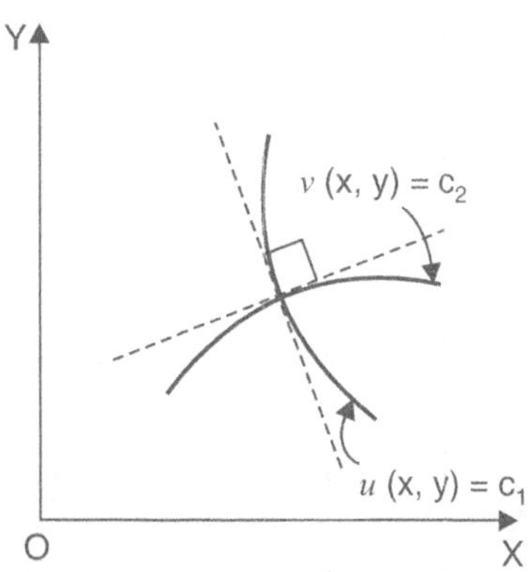

Differentiating (1) w.r.t. x, we get

$$\frac{\partial u}{\partial x} + \frac{\partial u}{\partial y} \cdot \frac{dy}{dx} = 0 \quad \text{or} \quad \frac{dy}{dx} = -\frac{\dfrac{\partial u}{\partial x}}{\dfrac{\partial u}{\partial y}} = m_1 \text{ (say)}$$

Similarly, from (2), we get
$$\frac{dy}{dx} = -\frac{\dfrac{\partial v}{\partial x}}{\dfrac{\partial v}{\partial y}} = m_2 \text{ (say)}$$

$$\therefore \qquad m_1 m_2 = \frac{\dfrac{\partial u}{\partial x} \cdot \dfrac{\partial v}{\partial x}}{\dfrac{\partial u}{\partial y} \cdot \dfrac{\partial v}{\partial y}} \qquad \ldots(3)$$

Since $f(z)$ is analytic, u and v satisfy C-R equations

i.e.,
$$\frac{\partial u}{\partial x} = \frac{\partial v}{\partial y} \quad \text{and} \quad \frac{\partial u}{\partial y} = -\frac{\partial v}{\partial x}$$

$\therefore$ From (3),
$$m_1 m_2 = \frac{\dfrac{\partial v}{\partial y} \cdot \dfrac{\partial v}{\partial x}}{-\dfrac{\partial v}{\partial x} \cdot \dfrac{\partial v}{\partial y}} = -1$$

Thus the product of the slopes of the curves (1) and (2) is – 1. Hence the curves intersect at right angles, *i.e.,* they form an orthogonal system.

3.23. APPLICATION OF ANALYTIC FUNCTIONS TO FLOW PROBLEMS

Since the real and imaginary parts of an analytic function satisfy the Laplace's equation in two variables, these conjugate functions provide solutions to a number of field and flow problems.

For example, consider the two dimensional irrotational motion of an incompressible fluid, in planes parallel to xy-plane.

Let V be the velocity of a fluid particle, then it can be expressed as

$$V = v_x \hat{i} + v_y \hat{j} \qquad \ldots(1)$$

Since the motion is irrotational, there exists a scalar function $\phi(x, y)$, such that

$$V = \nabla\phi(x, y) = \frac{\partial \phi}{\partial x} \hat{i} + \frac{\partial \phi}{\partial y} \hat{j} \qquad \ldots(2)$$

From (1) and (2), we have $\quad v_x = \dfrac{\partial \phi}{\partial x} \quad \text{and} \quad v_y = \dfrac{\partial \phi}{\partial y} \qquad \ldots(3)$

The scalar function $\phi(x, y)$, which gives the velocity components, is called the **velocity potential function** or simply the **velocity potential.**

Also the fluid being incompressible, div V = 0

$$\Rightarrow \qquad \left(\hat{i}\frac{\partial}{\partial x} + \hat{j}\frac{\partial}{\partial y}\right) \cdot (v_x \hat{i} + v_y \hat{j}) = 0$$

$$\Rightarrow \qquad \frac{\partial v_x}{\partial x} + \frac{\partial v_y}{\partial y} = 0 \qquad \ldots(4)$$

Substituting the values of v_x and v_y from (3) in (4), we get

$$\frac{\partial}{\partial x}\left(\frac{\partial \phi}{\partial x}\right) + \frac{\partial}{\partial y}\left(\frac{\partial \phi}{\partial y}\right) = 0 \quad \text{or} \quad \frac{\partial^2 \phi}{\partial x^2} + \frac{\partial^2 \phi}{\partial y^2} = 0$$

Thus the function ϕ is harmonic and can be treated as real part of an analytic function

$$w = f(z) = \phi(x, y) + i\,\psi(x, y)$$

For interpretation of conjugate function $\psi(x, y)$, the slope at any point of the curve $\psi(x, y) = c'$ is given by

$$\frac{dy}{dx} = -\frac{\dfrac{\partial \psi}{\partial x}}{\dfrac{\partial \psi}{\partial y}} = \frac{\dfrac{\partial \phi}{\partial y}}{\dfrac{\partial \phi}{\partial x}} \qquad \text{[By C-R equations]}$$

$$= \frac{v_y}{v_x} \qquad \text{[By (3)]}$$

This shows that the resultant velocity $\sqrt{v_x^2 + v_y^2}$ of the fluid particle is along the tangent to the curve $\psi(x, y) = c'$ *i.e.*, the fluid particles move along this curve. Such curves are known as *stream lines* and $\psi(x, y)$ is called the *stream function*. The curves represented by $\phi(x, y) = c$ are called *equipotential lines*.

Since $\phi(x, y)$ and $\psi(x, y)$ are conjugate functions of analytic function $w = f(z)$, the equipotential lines $\phi(x, y) = c$ and the stream lines $\psi(x, y) = c'$, intersect each other orthogonally.

Now,
$$\frac{dw}{dz} = \frac{\partial \phi}{\partial x} + i\frac{\partial \psi}{\partial x}$$

$$= \frac{\partial \phi}{\partial x} - i\frac{\partial \phi}{\partial y} \qquad \text{[By C-R equations]}$$

$$= v_x - iv_y \qquad \text{[By (3)]}$$

$\therefore$ The magnitude of resultant velocity $= \left|\dfrac{dw}{dz}\right| = \sqrt{{v_x}^2 + {v_y}^2}$

The function $w = f(z)$ which fully represents the flow pattern is called the *complex potential*.

In the study of electrostatics and gravitational fields, the curves $\phi(x, y) = c$ and $\psi(x, y) = c'$ are called *equipotential lines* and *lines of force* respectively. In heat flow problems, the curves $\phi(x, y) = c$ and $\psi(x, y) = c'$ are known as *isothermals* and *heat flow lines* respectively.

ILLUSTRATIVE EXAMPLES

Example 1. *Find the values of A and B such that the function*

$$f(z) = x^2 + Ay^2 - 2xy + i(Bx^2 - y^2 + 2xy)$$

is analytic. Also find $f'(z)$. (M.D.U. May 2014; U.K.T.U. 2011)

Sol. Given $f(z) = (x^2 + Ay^2 - 2xy) + i(Bx^2 - y^2 + 2xy)$

Comparing with $f(z) = u(x, y) + iv(x, y)$, we have

$$u(x, y) = x^2 + Ay^2 - 2xy \quad \text{and} \quad v(x, y) = Bx^2 - y^2 + 2xy$$

$\therefore$
$$\frac{\partial u}{\partial x} = 2x - 2y \quad \text{and} \quad \frac{\partial u}{\partial y} = 2Ay - 2x$$

$$\frac{\partial v}{\partial x} = 2Bx + 2y \quad \text{and} \quad \frac{\partial v}{\partial y} = -2y + 2x$$

For $f(z)$ to be analytic, it must satisfy Cauchy-Riemann equations

$$\frac{\partial u}{\partial x} = \frac{\partial v}{\partial y} \quad \text{and} \quad \frac{\partial u}{\partial y} = -\frac{\partial v}{\partial x}$$

Now $\dfrac{\partial u}{\partial x} = \dfrac{\partial v}{\partial y} \Rightarrow 2x - 2y = -2y + 2x$ which is true.

Also $\dfrac{\partial u}{\partial y} = -\dfrac{\partial v}{\partial x} \Rightarrow 2Ay - 2x = -2Bx - 2y$

Comparing co-efficients of x and y, we have

$$-2 = -2B \quad \text{and} \quad 2A = -2$$

$$\Rightarrow \quad A = -1,\ B = 1$$

Now
$$f'(z) = \frac{\partial u}{\partial x} + i\frac{\partial v}{\partial x} = (2x - 2y) + i\,(2Bx + 2y)$$
$$= (2x - 2y) + i\,(2x + 2y) \qquad (\because \; B = 1)$$
$$= 2(x + iy) + 2ix - 2y$$
$$= 2(x + iy) + 2ix + i^2 \cdot 2y$$
$$= 2(x + iy) + 2i\,(x + i\,y) = 2z + 2iz = 2(1 + i)z.$$

Example 2. *Find p such that the function $f(z)$ expressed in polar coordinates as $f(z) = r^2 \cos 2\theta + ir^2 \sin p\theta$ is analytic.*

Sol. Let $f(z) = u + iv$, then $u = r^2 \cos 2\theta$, $v = r^2 \sin p\theta$

$$\frac{\partial u}{\partial r} = 2r \cos 2\theta, \quad \frac{\partial v}{\partial r} = 2r \sin p\theta$$

$$\frac{\partial u}{\partial \theta} = -2r^2 \sin 2\theta, \quad \frac{\partial v}{\partial \theta} = pr^2 \cos p\theta$$

For $f(z)$ to be analytic, $\dfrac{\partial u}{\partial r} = \dfrac{1}{r}\dfrac{\partial v}{\partial \theta}$ and $\dfrac{\partial v}{\partial r} = -\dfrac{1}{r}\dfrac{\partial u}{\partial \theta}$

$$\therefore \quad 2r \cos 2\theta = pr \cos p\theta \quad \text{and} \quad 2r \sin p\theta = 2r \sin 2\theta$$

Both these equations are satisfied if $p = 2$.

Note. For a function $f(z)$ to be analytic, the first order partial derivatives of u and v must be continuous in addition to C-R equations.

Example 3. *Show that the function $f(z) = \sqrt{|xy|}$ is not analytic at the origin, even though Cauchy-Riemann equations are satisfied there at.*

(G.B.T.U. 2011; K.U.K. May 2013; DCRUST Murthal May 2014)

Sol. Let $f(z) = u(x, y) + iv(x, y) = \sqrt{|xy|}$, then $u(x, y) = \sqrt{|xy|}$, $v(x, y) = 0$

At the origin (0, 0), we have

$$\frac{\partial u}{\partial x} = \lim_{x \to 0} \frac{u(x, 0) - u(0, 0)}{x} = \lim_{x \to 0} \frac{0 - 0}{x} = 0$$

$$\frac{\partial u}{\partial y} = \lim_{y \to 0} \frac{u(0, y) - u(0, 0)}{y} = \lim_{y \to 0} \frac{0 - 0}{y} = 0$$

$$\frac{\partial v}{\partial x} = \lim_{x \to 0} \frac{v(x, 0) - v(0, 0)}{x} = \lim_{x \to 0} \frac{0 - 0}{x} = 0$$

$$\frac{\partial v}{\partial y} = \lim_{y \to 0} \frac{v(0, y) - v(0, 0)}{y} = \lim_{y \to 0} \frac{0 - 0}{y} = 0$$

Clearly, $\dfrac{\partial u}{\partial x} = \dfrac{\partial v}{\partial y}, \dfrac{\partial u}{\partial y} = -\dfrac{\partial v}{\partial x}$

Hence C-R equations are satisfied at the origin.

Now, $$f'(0) = \lim_{z\to 0} \frac{f(z)-f(0)}{z} = \lim_{z\to 0} \frac{\sqrt{|xy|}-0}{x+iy}$$

If $z \to 0$ along the line $y = mx$, we get

$$f'(0) = \lim_{x\to 0} \frac{\sqrt{|mx^2|}}{x(1+im)} = \lim_{x\to 0} \frac{\sqrt{|m|}}{1+im}$$

Now this limit is not unique since it depends on m. Therefore, $f'(0)$ does not exist.

Hence the function $f(z)$ is not analytic at the origin.

Example 4. *Prove that the function $f(z)$ defined by*

$$f(z) = \frac{x^3(1+i)-y^3(1-i)}{x^2+y^2},\ z \neq 0 \text{ and } f(0) = 0$$

is continuous and the Cauchy-Riemann equations are satisfied at the origin, yet $f'(0)$ does not exist. (M.D.U. 2011)

Sol. Here $$f(z) = \frac{(x^3-y^3)+i(x^3+y^3)}{x^2+y^2},\ z \neq 0$$

Let $$f(z) = u + iv = \frac{x^3-y^3}{x^2+y^2} + i\frac{x^3+y^3}{x^2+y^2}, \quad \text{then} \quad u = \frac{x^3-y^3}{x^2+y^2},\ v = \frac{x^3+y^3}{x^2+y^2}$$

Since $z \neq 0 \Rightarrow x^2 + y^2 \neq 0$

$\therefore$ u and v are rational functions of x and y with non-zero denominators. Thus, u, v and hence $f(z)$ *are continuous functions when $z \neq 0$*. To test them for continuity at $z = 0$, on changing u, v to polar co-ordinates by putting $x = r\cos\theta$, $y = r\sin\theta$, we get

$$u = r(\cos^3\theta - \sin^3\theta) \text{ and } v = r(\cos^3\theta + \sin^3\theta)$$

When $z \to 0$, $r \to 0$

$$\therefore \quad \lim_{z\to 0} u = \lim_{r\to 0} r(\cos^3\theta - \sin^3\theta) = 0$$

Similarly, $$\lim_{z\to 0} v = 0$$

$$\therefore \quad \lim_{z\to 0} f(z) = 0 = f(0)$$

$\Rightarrow$ **f(z) is continuous at z = 0.**

Hence, $f(z)$ is continuous for all values of z.

At the origin (0, 0), we have

$$\frac{\partial u}{\partial x} = \lim_{x\to 0} \frac{u(x,0)-u(0,0)}{x} = \lim_{x\to 0} \frac{x-0}{x} = 1$$

$$\frac{\partial u}{\partial y} = \lim_{y\to 0} \frac{u(0,y)-u(0,0)}{y} = \lim_{y\to 0} \frac{-y-0}{y} = -1$$

$$\frac{\partial v}{\partial x} = \lim_{x\to 0} \frac{v(x,0)-v(0,0)}{x} = \lim_{x\to 0} \frac{x-0}{x} = 1$$

$$\frac{\partial v}{\partial y} = \lim_{y\to 0} \frac{v(0,y)-v(0,0)}{y} = \lim_{y\to 0} \frac{y-0}{y} = 1$$

$$\therefore \qquad \frac{\partial u}{\partial x} = \frac{\partial v}{\partial y} \quad \text{and} \quad \frac{\partial u}{\partial y} = -\frac{\partial v}{\partial x}$$

Hence C-R equations are satisfied at the origin.

Now, $$f'(0) = \lim_{z \to 0} \frac{f(z) - f(0)}{z} = \lim_{z \to 0} \frac{(x^3 - y^3) + i(x^3 + y^3) - 0}{(x^2 + y^2)(x + iy)}$$

Let $z \to 0$ along the line $y = x$, then

$$f'(0) = \lim_{x \to 0} \frac{0 + 2ix^3}{2x^3(1+i)} = \frac{i}{1+i} = \frac{i(1-i)}{2} = \frac{1+i}{2} \qquad \ldots(1)$$

Also, let $z \to 0$ along the x-axis (*i.e.*, $y = 0$), then

$$f'(0) = \lim_{x \to 0} \frac{x^3 + ix^3}{x^3} = 1 + i \qquad \ldots(2)$$

Since the limits (1) and (2) are different, $f'(0)$ does not exist.

Example 5. *If* $f(z) = \begin{cases} \dfrac{x^3 y(y - ix)}{x^6 + y^2}, & z \neq 0 \\ 0, & z = 0 \end{cases}$, *prove that* $\dfrac{f(z) - f(0)}{z} \to 0$ *as* $z \to 0$ *along any radius vector but not as* $z \to 0$ *in any manner and also that* $f(z)$ *is not analytic at* $z = 0$.

(M.D.U. Dec. 2015; G.B.T.U. 2013; U.K.T.U. 2010)

Sol. Here $$\frac{f(z) - f(0)}{z} = \left[\frac{x^3 y(y - ix)}{x^6 + y^2} - 0\right] \div (x + iy)$$

$$= \frac{-ix^3 y(x + iy)}{x^6 + y^2} \cdot \frac{1}{x + iy} = \frac{-ix^3 y}{x^6 + y^2}$$

Let $z \to 0$ along radius vector $y = mx$, then

$$\lim_{z \to 0} \frac{f(z) - f(0)}{z} = \lim_{x \to 0} \frac{-ix^3(mx)}{x^6 + m^2 x^2} = \lim_{x \to 0} \frac{-imx^2}{x^4 + m^2} = 0$$

Hence $\dfrac{f(z) - f(0)}{z} \to 0$ as $z \to 0$ along any radius vector.

Now, let $z \to 0$ along the curve $y = x^3$, then

$$\lim_{z \to 0} \frac{f(z) - f(0)}{z} = \lim_{x \to 0} \frac{-ix^3 \cdot x^3}{x^6 + x^6} = \frac{-i}{2} \neq 0$$

Thus $\dfrac{f(z) - f(0)}{z}$ does not tend to zero as $z \to 0$ in any manner.

Since $\lim\limits_{z \to 0} \dfrac{f(z) - f(0)}{z}$ is not independent of the path along which z approaches 0, the limit does not exist. Hence $f(z)$ is not analytic at $z = 0$.

Example 6. *Prove that the function sinh z is analytic and find its derivative.*

(U.K.T.U. 2010)

Sol. Here $f(z) = u + iv = \sinh z = \sinh (x + iy) = \sinh x \cos y + i \cosh x \sin y$

$\therefore$ $u = \sinh x \cos y$ and $v = \cosh x \sin y$

$$\frac{\partial u}{\partial x} = \cosh x \cos y, \quad \frac{\partial u}{\partial y} = -\sinh x \sin y$$

$$\frac{\partial v}{\partial x} = \sinh x \sin y, \quad \frac{\partial v}{\partial y} = \cosh x \cos y$$

$$\therefore \quad \frac{\partial u}{\partial x} = \frac{\partial v}{\partial y} \quad \text{and} \quad \frac{\partial u}{\partial y} = -\frac{\partial v}{\partial x}$$

Thus C-R equations are satisfied.

Since $\sinh x$, $\cosh x$, $\sin y$ and $\cos y$ are continuous functions, $\frac{\partial u}{\partial x}, \frac{\partial u}{\partial y}, \frac{\partial v}{\partial x}$ and $\frac{\partial v}{\partial y}$ are also continuous functions satisfying C-R equations.

Hence $f(z)$ is analytic every where.

Now,
$$f'(z) = \frac{\partial u}{\partial x} + i\frac{\partial v}{\partial x}$$
$$= \cosh x \cos y + i \sinh x \sin y = \cosh (x + iy) = \cosh z.$$

Example 7. *Determine the analytic function whose real part is $e^{2x}(x \cos 2y - y \sin 2y)$.*

(M.D.U. 2011, May 2013)

Sol. Let $f(z) = u + iv$ be the analytic function, where $u = e^{2x}(x \cos 2y - y \sin 2y)$

$$\therefore \quad \frac{\partial u}{\partial x} = 2e^{2x}(x \cos 2y - y \sin 2y) + e^{2x} \cos 2y$$
$$= e^{2x}(2x \cos 2y - 2y \sin 2y + \cos 2y) \quad \text{...(1)}$$
$$\frac{\partial u}{\partial y} = e^{2x}(-2x \sin 2y - \sin 2y - 2y \cos 2y)$$
$$= -e^{2x}(2x \sin 2y + \sin 2y + 2y \cos 2y) \quad \text{...(2)}$$

Since $f(z)$ is analytic, u and v must satisfy C-R equations

$$\frac{\partial u}{\partial x} = \frac{\partial v}{\partial y} \quad \text{and} \quad \frac{\partial u}{\partial y} = -\frac{\partial v}{\partial x}$$

Now
$$\frac{\partial v}{\partial y} = \frac{\partial u}{\partial x} = e^{2x}(2x \cos 2y - 2y \sin 2y + \cos 2y)$$

Integrating w.r.t. y, treating x as constant, we get

$$v = e^{2x}\left[2x \cdot \frac{\sin 2y}{2} - \left\{2y\left(-\frac{\cos 2y}{2}\right) - (2)\left(-\frac{\sin 2y}{4}\right)\right\} + \frac{\sin 2y}{2}\right] + \phi(x)$$
$$= e^{2x}(x \sin 2y + y \cos 2y) + \phi(x) \quad \text{...(3)}$$

where $\phi(x)$ is an arbitrary function of x.

$$\therefore \quad \frac{\partial v}{\partial x} = 2e^{2x}(x \sin 2y + y \cos 2y) + e^{2x}(\sin 2y) + \phi'(x)$$
$$= e^{2x}(2x \sin 2y + \sin 2y + 2y \cos 2y) + \phi'(x)$$

But $\qquad \dfrac{\partial v}{\partial x} = -\dfrac{\partial u}{\partial y} = e^{2x}(2x \sin 2y + \sin 2y + 2y \cos 2y)$ [From (2)]

$\therefore \qquad \phi'(x) = 0$

$\Rightarrow \qquad \phi(x) = c$, an arbitrary constant.

$\therefore$ From (3), $\quad v = e^{2x}(x \sin 2y + y \cos 2y) + c$

$$\begin{aligned} f(z) = u + iv &= e^{2x}(x \cos 2y - y \sin 2y) + ie^{2x}(x \sin 2y + y \cos 2y) + ic \\ &= e^{2x}[(x + iy)\cos 2y + i(x + iy)\sin 2y] + ic \\ &= (x + iy)\, e^{2x}(\cos 2y + i \sin 2y) + ic \\ &= ze^{2x} . e^{2iy} + ic = ze^{2(x+iy)} + ic\ ze^{2z} + ic. \end{aligned}$$

(Milne-Thomson's Method)

This method determines the analytic function $f(z)$ when u or v is given.

Since $\qquad z = x + iy,\ \bar{z} = x - iy \quad$ so that $x = \dfrac{z + \bar{z}}{2},\ y = \dfrac{z - \bar{z}}{2i}$

Let $\qquad f(z) = u(x, y) + iv(x, y)$ ...(1)

$$= u\left(\frac{z + \bar{z}}{2}, \frac{z - \bar{z}}{2i}\right) + iv\left(\frac{z + \bar{z}}{2}, \frac{z - \bar{z}}{2i}\right)$$

Considering this as an identity in the two independent variables z, $\bar{z}$ and putting $\bar{z} = z$, we get

$$f(z) = u(z, 0) + iv(z, 0)$$

which is the same as (1) if we replace x by z and y by 0.

Thus to express any function in terms of z, replace x by z and y by 0.

Now $\qquad f(z) = u + iv$

$\Rightarrow \qquad f'(z) = \dfrac{\partial u}{\partial x} + i\dfrac{\partial v}{\partial x} = \dfrac{\partial u}{\partial x} - i\dfrac{\partial u}{\partial y}$ [C-R equations]

$$= e^{2x}(2x \cos 2y - 2y \sin 2y + \cos 2y) + ie^{2x}(2x \sin 2y + \sin 2y + 2y \cos 2y)$$

On replacing x by z and y by 0 on R.H.S., we get $f'(z) = e^{2z}(2z + 1)$

Integrating w.r.t. z, we have $f(z) = (2z + 1)\dfrac{e^{2z}}{2} - 2.\dfrac{e^{2z}}{4} + ic = ze^{2z} + ic$,

taking the constant of integration as imaginary since u does not contain any constant.

Example 8. *Determine the analytic function $w = u + iv$, if $v = \log(x^2 + y^2) + x - 2y$.*

(M.D.U. Dec. 2013)

Sol. Here $\qquad v = \log(x^2 + y^2) + x - 2y$

$$\therefore \qquad \frac{\partial v}{\partial x} = \frac{2x}{x^2 + y^2} + 1$$

$$\frac{\partial v}{\partial y} = \frac{2y}{x^2 + y^2} - 2$$

Since $w = u + iv$

$$\frac{dw}{dz} = \frac{\partial u}{\partial x} + i\frac{\partial v}{\partial x} = \frac{\partial v}{\partial y} + i\frac{\partial v}{\partial x} \qquad \text{[C-R equations]}$$

$$= \left(\frac{2y}{x^2+y^2} - 2\right) + i\left(\frac{2x}{x^2+y^2} + 1\right)$$

Replacing x by z and y by 0, we get $\dfrac{dw}{dz} = -2 + i\left(\dfrac{2}{z} + 1\right) = (i-2) + \dfrac{2i}{z}$

Integrating w.r.t. z, we have $w = (i-2)z + 2i \log z + c.$

Example 9. *Show that the function $u = e^{-2xy} \sin (x^2 - y^2)$ is harmonic. Find the conjugate function v and express $u + iv$ as an analytic function of z.*

Sol. Here $u = e^{-2xy} \sin (x^2 - y^2)$

$$\therefore \quad \frac{\partial u}{\partial x} = -2y\, e^{-2xy} \sin (x^2 - y^2) + 2x\, e^{-2xy} \cos (x^2 - y^2)$$

$$\frac{\partial^2 u}{\partial x^2} = 4y^2 e^{-2xy} \sin (x^2 - y^2) - 4xy\, e^{-2xy} \cos (x^2 - y^2)$$
$$+ 2e^{-2xy} \cos (x^2 - y^2) - 4xy\, e^{-2xy} \cos (x^2 - y^2) - 4x^2 e^{-2xy} \sin (x^2 - y^2) \quad ...(1)$$

$$\frac{\partial u}{\partial y} = -2x\, e^{-2xy} \sin (x^2 - y^2) - 2y\, e^{-2xy} \cos (x^2 - y^2)$$

$$\frac{\partial^2 u}{\partial y^2} = 4x^2 e^{-2xy} \sin (x^2 - y^2) + 4xy\, e^{-2xy} \cos (x^2 - y^2)$$
$$- 2e^{-2xy} \cos (x^2 - y^2) + 4xy\, e^{-2xy} \cos (x^2 - y^2) - 4y^2 e^{-2xy} \sin (x^2 - y^2) \quad ...(2)$$

Adding (1) and (2), we get

$$\frac{\partial^2 u}{\partial x^2} + \frac{\partial^2 u}{\partial y^2} = 0 \text{ which proves that } u \text{ is harmonic.}$$

Now, let $f(z) = u + iv$

then
$$f'(z) = \frac{\partial u}{\partial x} + i\frac{\partial v}{\partial x} = \frac{\partial u}{\partial x} - i\frac{\partial u}{\partial y} \qquad \text{[C-R equations]}$$
$$= [-2y\, e^{-2xy} \sin (x^2 - y^2) + 2x\, e^{-2xy} \cos (x^2 - y^2)]$$
$$+ i\, [2x\, e^{-2xy} \sin (x^2 - y^2) + 2y\, e^{-2xy} \cos (x^2 - y^2)]$$

Replacing x by z and y by 0, we get

$$f'(z) = 2z \cos z^2 + 2iz \sin z^2 = 2z (\cos z^2 + i \sin z^2) = 2z\, e^{iz^2}$$

Integrating w.r.t. z, we have

$$f(z) = -i\, e^{iz^2} + ic$$

which expresses $u + iv$ as an analytic function of z.

Since
$$u + iv = -ie^{iz^2} + ic = -ie^{i(x+iy)^2} + ic$$
$$= -ie^{i(x^2-y^2+2ixy)} + ic = -ie^{-2xy} \,.\, e^{i(x^2-y^2)} + ic$$
$$= -ie^{-2xy} [\cos (x^2 - y^2) + i \sin (x^2 - y^2)] + ic$$
$$= e^{-2xy} \sin (x^2 - y^2) + i\, [-e^{-2xy} \cos (x^2 - y^2)] + ic$$

$$\therefore \quad v = -e^{-2xy} \cos (x^2 - y^2) + c.$$

Example 10. *An electrostatic field in the xy-plane is given by the potential function $\phi = 3x^2 y - y^3$, find the stream function.*

Sol. Let $\psi(x, y)$ be a stream function.

By C-R equations $$\frac{\partial \psi}{\partial x} = -\frac{\partial \phi}{\partial y} = -3x^2 + 3y^2 \quad ...(1)$$

$$\frac{\partial \psi}{\partial y} = \frac{\partial \phi}{\partial x} = 6xy \quad ...(2)$$

Integrating (1) w.r.t. x, treating y as constant, we get

$$\psi = -x^3 + 3xy^2 + F(y)$$

so that $$\frac{\partial \psi}{\partial y} = 6xy + F'(y) \quad ...(3)$$

From (2) and (3), $6xy + F'(y) = 6xy$ or $F'(y) = 0$ $\therefore$ $F(y) = c$

Hence $$\psi = -x^3 + 3xy^2 + c.$$

Example 11. *If $u - v = (x - y)(x^2 + 4xy + y^2)$ and $f(z) = u + iv$ is an analytic function of $z = x + iy$, find $f(z)$ in terms of z.*

Sol. We have $$u - v = (x - y)(x^2 + 4xy + y^2)$$
$$= x^3 + 3x^2y - 3xy^2 - y^3$$

$\therefore$ $$\frac{\partial u}{\partial x} - \frac{\partial v}{\partial x} = 3x^2 + 6xy - 3y^2 \quad ...(1)$$

and $$\frac{\partial u}{\partial y} - \frac{\partial v}{\partial y} = 3x^2 - 6xy - 3y^2$$

or $$-\frac{\partial v}{\partial x} - \frac{\partial u}{\partial x} = 3x^2 - 6xy - 3y^2 \quad ...(2)$$

Subtracting (2) from (1)

$$2\frac{\partial u}{\partial x} = 12xy \quad \text{or} \quad \frac{\partial u}{\partial x} = 6xy$$

Adding (1) and (2)

$$-2\frac{\partial v}{\partial x} = 6x^2 - 6y^2 \quad \text{or} \quad \frac{\partial v}{\partial x} = 3y^2 - 3x^2$$

Thus $$f'(z) = \frac{\partial u}{\partial x} + i\frac{\partial v}{\partial x} = 6xy + i(3y^2 - 3x^2)$$

Replacing x by z and y by 0, we get

$$f'(z) = -3iz^2$$

Integrating $$f(z) = -iz^3 + c.$$

Example 12. *If $u + v = \dfrac{2 \sin 2x}{e^{2y} + e^{-2y} - 2\cos 2x}$ and $f(z) = u + iv$ is an analytic function of $z = x + iy$, find $f(z)$ in terms of z.*

Sol. Given $$f(z) = u + iv \quad ...(1)$$

$\Rightarrow$ $$i\,f(z) = iu - v \quad ...(2)$$

Adding (1) and (2), we get

$$(1 + i)\, f(z) = (u - v) + i(u + v)$$

or $$F(z) = U + iV$$

where $$F(z) = (1 + i)\, f(z),\ U = u - v \quad \text{and} \quad V = u + v$$

$$\therefore \quad V = u + v = \frac{2 \sin 2x}{e^{2y} + e^{-2y} - 2\cos 2x} \qquad \text{(Given)}$$

$$= \frac{2 \sin 2x}{2 \cosh 2y - 2 \cos 2x} = \frac{\sin 2x}{\cosh 2y - \cos 2x}$$

Now, we proceed to find F(z) whose imaginary part is given.

$$\frac{\partial V}{\partial x} = \frac{(\cosh 2y - \cos 2x)(2 \cos 2x) - \sin 2x\,(2 \sin 2x)}{(\cosh 2y - \cos 2x)^2}$$

$$= \frac{2 \cos 2x \cosh 2y - 2(\cos^2 2x + \sin^2 2x)}{(\cosh 2y - \cos 2x)^2}$$

$$= \frac{2 \cos 2x \cosh 2y - 2}{(\cosh 2y - \cos 2x)^2}$$

$$\frac{\partial V}{\partial y} = \frac{-2 \sin 2x \sinh 2y}{(\cosh 2y - \cos 2x)^2}$$

Since $f(z)$ is analytic, so is F(z).

$$\therefore \quad F'(z) = \frac{\partial U}{\partial x} + i\frac{\partial V}{\partial x}$$

$$= \frac{\partial V}{\partial y} + i\frac{\partial V}{\partial x} \quad \text{by C-R equations}$$

$$= \frac{-2 \sin 2x \sinh 2y}{(\cosh 2y - \cos 2x)^2} + i\,\frac{2(\cos 2x \cosh 2y - 1)}{(\cosh 2y - \cos 2x)^2}$$

By Milne-Thomson's method, replacing x by z and y by 0 on R.H.S., we get

$$F'(z) = i\,\frac{2(\cos 2z - 1)}{(1 - \cos 2z)^2} = \frac{-2i}{1 - \cos 2z} = \frac{-2i}{2 \sin^2 z}$$

$$= -i \operatorname{cosec}^2 z$$

Integrating w.r.t.z, we have

$$F(z) = i \cot z + c$$

$$\Rightarrow \quad (1 + i)\, f(z) = i \cot z + c$$

$$\Rightarrow \quad f(z) = \frac{i}{1+i} \cot z + \frac{c}{1+i}$$

$$= \frac{i(1-i)}{1-i^2} \cot z + \frac{c}{1+i}$$

$$\therefore \quad f(z) = \frac{1}{2}(1 + i) \cot z + C, \text{ where } C = \frac{c}{1+i}.$$

Example 13. *Find the analytic function f(z) = u + iv, given* $v = \left(r - \frac{1}{r}\right) \sin\theta, r \neq 0.$

Sol. Using C-R equations in polar coordinates, we have

$$r\frac{\partial u}{\partial r} = \frac{\partial v}{\partial \theta} = \left(r - \frac{1}{r}\right)\cos\theta \qquad \ldots(1)$$

$$-\frac{1}{r}\frac{\partial u}{\partial \theta} = \frac{\partial v}{\partial r} = \left(1 + \frac{1}{r^2}\right)\sin\theta \qquad \ldots(2)$$

Now (1) gives $\frac{\partial u}{\partial r} = \left(1 - \frac{1}{r^2}\right)\cos\theta$

Integrating w.r.t. r

$$u = \left(r + \frac{1}{r}\right)\cos\theta + \phi(\theta), \quad \text{where } \phi(\theta) \text{ is an arbitrary function of } \theta.$$

$$\Rightarrow \qquad \frac{\partial u}{\partial \theta} = -\left(r + \frac{1}{r}\right)\sin\theta + \phi'(\theta)$$

Also, from (2) $\frac{\partial u}{\partial \theta} = -\left(r + \frac{1}{r}\right)\sin\theta$

$$\therefore \qquad -\left(r + \frac{1}{r}\right)\sin\theta + \phi'(\theta) = -\left(r + \frac{1}{r}\right)\sin\theta$$

$$\Rightarrow \qquad \phi'(\theta) = 0 \text{ or } \phi(\theta) = c$$

$$\therefore \qquad u = \left(r + \frac{1}{r}\right)\cos\theta + c$$

Hence $f(z) = u + iv$

$$= \left(r + \frac{1}{r}\right)\cos\theta + c + i\left(r - \frac{1}{r}\right)\sin\theta.$$

Example 14. *If f(z) is an analytic function with constant modulus, show that f(z) is constant.* (K.U.K. Jan. 2013)

Sol. Let $f(z) = u + iv$ be an analytic function.

Since $|f(z)|$ = constant = c (say), $(c \neq 0)$

$$\therefore \qquad |f(z)|^2 = u^2 + v^2 = c^2 \qquad \ldots(1)$$

Differentiating (1) partially w.r.t. x and y respectively, we have

$$2u\frac{\partial u}{\partial x} + 2v\frac{\partial v}{\partial x} = 0 \quad \text{or} \quad u\frac{\partial u}{\partial x} + v\frac{\partial v}{\partial x} = 0 \qquad \ldots(2)$$

and

$$2u\frac{\partial u}{\partial y} + 2v\frac{\partial v}{\partial y} = 0 \quad \text{or} \quad u\frac{\partial u}{\partial y} + v\frac{\partial v}{\partial y} = 0 \qquad \ldots(3)$$

Using Cauchy-Riemann equations

$$\frac{\partial u}{\partial x} = \frac{\partial v}{\partial y} \quad \text{and} \quad \frac{\partial u}{\partial y} = -\frac{\partial v}{\partial x}$$

equation (3) becomes $-u\frac{\partial v}{\partial x} + v\frac{\partial u}{\partial x} = 0 \qquad \ldots(4)$

Squaring and adding (2) and (4), we have

$$(u^2 + v^2)\left[\left(\frac{\partial u}{\partial x}\right)^2 + \left(\frac{\partial v}{\partial x}\right)^2\right] = 0$$

or
$$c^2\left[\left(\frac{\partial u}{\partial x}\right)^2 + \left(\frac{\partial v}{\partial x}\right)^2\right] = 0$$

or
$$\left(\frac{\partial u}{\partial x}\right)^2 + \left(\frac{\partial v}{\partial x}\right)^2 = 0 \qquad [\because \; c \neq 0]$$

or
$$| f'(z) |^2 = 0 \qquad \left[\text{Since } f'(z) = \frac{\partial u}{\partial x} + i\frac{\partial v}{\partial x}\right]$$

$\therefore \quad f'(z) = 0 \quad \Rightarrow \quad f(z) = \text{constant}.$

Example 15. *If f(z) is a regular function of z, prove that*

$$\left(\frac{\partial^2}{\partial x^2} + \frac{\partial^2}{\partial y^2}\right) | f(z) |^2 = 4 \, | f'(z) |^2.$$ (M.D.U. 2011; K.U.K. Jan. 2014)

Sol. Let $f(z) = u + iv$ so that $| f(z) | = \sqrt{u^2 + v^2}$

or
$$| f(z) |^2 = u^2 + v^2 = \phi(x, y) \text{ (say)}$$

$\therefore$
$$\frac{\partial \phi}{\partial x} = 2u\frac{\partial u}{\partial x} + 2v\frac{\partial v}{\partial x}$$

$$\frac{\partial^2 \phi}{\partial x^2} = 2\left[u\frac{\partial^2 u}{\partial x^2} + \left(\frac{\partial u}{\partial x}\right)^2 + v\frac{\partial^2 v}{\partial x^2} + \left(\frac{\partial v}{\partial x}\right)^2\right]$$

Similarly,
$$\frac{\partial^2 \phi}{\partial y^2} = 2\left[u\frac{\partial^2 u}{\partial y^2} + \left(\frac{\partial u}{\partial y}\right)^2 + v\frac{\partial^2 v}{\partial y^2} + \left(\frac{\partial v}{\partial y}\right)^2\right]$$

Adding, we get

$$\frac{\partial^2 \phi}{\partial x^2} + \frac{\partial^2 \phi}{\partial y^2} = 2\left[u\left(\frac{\partial^2 u}{\partial x^2} + \frac{\partial^2 u}{\partial y^2}\right) + \left(\frac{\partial u}{\partial x}\right)^2 + \left(\frac{\partial u}{\partial y}\right)^2 + v\left(\frac{\partial^2 v}{\partial x^2} + \frac{\partial^2 v}{\partial y^2}\right) + \left(\frac{\partial v}{\partial x}\right)^2 + \left(\frac{\partial v}{\partial y}\right)^2\right] \quad ...(1)$$

Since $f(z) = u + iv$ is a regular function of z, u and v satisfy C-R equations and Laplace's equation.

$\therefore$
$$\frac{\partial u}{\partial x} = \frac{\partial v}{\partial y}, \frac{\partial u}{\partial y} = -\frac{\partial v}{\partial x} \quad \text{and} \quad \frac{\partial^2 u}{\partial x^2} + \frac{\partial^2 u}{\partial y^2} = 0 = \frac{\partial^2 v}{\partial x^2} + \frac{\partial^2 v}{\partial y^2}$$

$\therefore$ From (1), we get

$$\frac{\partial^2 \phi}{\partial x^2} + \frac{\partial^2 \phi}{\partial y^2} = 2\left[0 + \left(\frac{\partial u}{\partial x}\right)^2 + \left(\frac{\partial v}{\partial x}\right)^2 + 0 + \left(\frac{\partial v}{\partial x}\right)^2 + \left(\frac{\partial u}{\partial x}\right)^2\right]$$

$$= 4\left[\left(\frac{\partial u}{\partial x}\right)^2 + \left(\frac{\partial v}{\partial x}\right)^2\right] \quad ...(2)$$

Now $f(z) = u + iv$

$\therefore$ $$f'(z) = \frac{\partial u}{\partial x} + i\frac{\partial v}{\partial x} \quad \text{and} \quad |f'(z)|^2 = \left(\frac{\partial u}{\partial x}\right)^2 + \left(\frac{\partial v}{\partial x}\right)^2$$

From (2), we get $\left(\frac{\partial^2}{\partial x^2} + \frac{\partial^2}{\partial y^2}\right)\phi = 4\,|f'(z)|^2$ or $\left(\frac{\partial^2}{\partial x^2} + \frac{\partial^2}{\partial y^2}\right)|f(z)|^2 = 4\,|f'(z)|^2$.

Example 16. *If f(z) is a holomorphic function of z, show that*

$$\left\{\frac{\partial}{\partial x}|f(z)|\right\}^2 + \left\{\frac{\partial}{\partial y}|f(z)|\right\}^2 = |f'(z)|^2. \qquad \text{(M.D.U. 2010)}$$

Sol. Let $f(z) = u + iv$

then $$|f(z)| = (u^2 + v^2)^{1/2} = \phi(x, y) \text{ (say)}$$

$$\frac{\partial \phi}{\partial x} = \frac{1}{2}(u^2 + v^2)^{-1/2}\left(2u\frac{\partial u}{\partial x} + 2v\frac{\partial v}{\partial x}\right)$$

$$= \frac{1}{\sqrt{u^2 + v^2}}\left(u\frac{\partial u}{\partial x} + v\frac{\partial v}{\partial x}\right)$$

Similarly, $$\frac{\partial \phi}{\partial y} = \frac{1}{\sqrt{u^2 + v^2}}\left(u\frac{\partial u}{\partial y} + v\frac{\partial v}{\partial y}\right)$$

$$= \frac{1}{\sqrt{u^2 + v^2}}\left(-u\frac{\partial v}{\partial x} + v\frac{\partial u}{\partial x}\right) \qquad \text{[Using C-R equations]}$$

$\therefore$ $$\left(\frac{\partial \phi}{\partial x}\right)^2 + \left(\frac{\partial \phi}{\partial y}\right)^2 = \frac{1}{u^2 + v^2}\left[\left(u\frac{\partial u}{\partial x} + v\frac{\partial v}{\partial x}\right)^2 + \left(-u\frac{\partial v}{\partial x} + v\frac{\partial u}{\partial x}\right)^2\right]$$

$$= \frac{1}{u^2 + v^2}\left[(u^2 + v^2)\left(\frac{\partial u}{\partial x}\right)^2 + (u^2 + v^2)\left(\frac{\partial v}{\partial x}\right)^2\right] = \left(\frac{\partial u}{\partial x}\right)^2 + \left(\frac{\partial v}{\partial x}\right)^2$$

$\Rightarrow$ $$\left\{\frac{\partial}{\partial x}|f(z)|\right\}^2 + \left\{\frac{\partial}{\partial y}|f(z)|\right\}^2 = \left(\frac{\partial u}{\partial x}\right)^2 + \left(\frac{\partial v}{\partial x}\right)^2 \qquad \text{...(1)}$$

Also $f(z) = u + iv$

$\Rightarrow$ $$f'(z) = \frac{\partial u}{\partial x} + i\frac{\partial v}{\partial x}$$

$\therefore$ $$|f'(z)|^2 = \left(\frac{\partial u}{\partial x}\right)^2 + \left(\frac{\partial v}{\partial x}\right)^2 \qquad \text{...(2)}$$

From (1) and (2), we have

$$\left\{\frac{\partial}{\partial x}|f(z)|\right\}^2 + \left\{\frac{\partial}{\partial y}|f(z)|\right\}^2 = |f'(z)|^2.$$

Example 17. *If $z = x + iy$, show that* $\left(\dfrac{\partial^2}{\partial x^2} + \dfrac{\partial^2}{\partial y^2}\right) = 4\dfrac{\partial^2}{\partial z \partial \bar{z}}$.

Sol. We have $z = x + iy \quad \therefore \quad \bar{z} = x - iy$

$$\Rightarrow \qquad x = \frac{1}{2}(z + \bar{z}),\ y = \frac{1}{2i}(z - \bar{z}) = -\frac{i}{2}(z - \bar{z})$$

$$\therefore \qquad \frac{\partial x}{\partial z} = \frac{1}{2}, \frac{\partial x}{\partial \bar{z}} = \frac{1}{2}, \frac{\partial y}{\partial z} = -\frac{i}{2}, \frac{\partial y}{\partial \bar{z}} = \frac{i}{2}$$

Now
$$\frac{\partial}{\partial z} = \frac{\partial}{\partial x}\cdot\frac{\partial x}{\partial z} + \frac{\partial}{\partial y}\cdot\frac{\partial y}{\partial z} = \frac{1}{2}\left(\frac{\partial}{\partial x} - i\frac{\partial}{\partial y}\right)$$

and
$$\frac{\partial}{\partial \bar{z}} = \frac{\partial}{\partial x}\cdot\frac{\partial x}{\partial \bar{z}} + \frac{\partial}{\partial y}\cdot\frac{\partial y}{\partial \bar{z}} = \frac{1}{2}\left(\frac{\partial}{\partial x} + i\frac{\partial}{\partial y}\right)$$

$$\therefore \qquad \frac{\partial^2}{\partial z \partial \bar{z}} = \frac{\partial}{\partial z}\left(\frac{\partial}{\partial \bar{z}}\right) = \frac{1}{2}\left(\frac{\partial}{\partial x} - i\frac{\partial}{\partial y}\right).\frac{1}{2}\left(\frac{\partial}{\partial x} + i\frac{\partial}{\partial y}\right) = \frac{1}{4}\left(\frac{\partial^2}{\partial x^2} + \frac{\partial^2}{\partial y^2}\right)$$

Hence
$$\left(\frac{\partial^2}{\partial x^2} + \frac{\partial^2}{\partial y^2}\right) = 4\frac{\partial^2}{\partial z \partial \bar{z}}.$$

Note. Remember the result of Example 17.

Example 18. *If $f(z)$ is an analytic function of z, prove that*

$$\left(\frac{\partial^2}{\partial x^2} + \frac{\partial^2}{\partial y^2}\right) \log |f'(z)| = 0.$$

Sol. We know that $\left(\dfrac{\partial^2}{\partial x^2} + \dfrac{\partial^2}{\partial y^2}\right) = 4\dfrac{\partial^2}{\partial z \partial \bar{z}}$

Also
$$\log |f'(z)| = \frac{1}{2}\log |f'(z)|^2$$
$$= \frac{1}{2}\log [f'(z)\,.\,f'(\bar{z})] = \frac{1}{2}[\log f'(z) + \log f'(\bar{z})]$$

$$\therefore \qquad \left(\frac{\partial^2}{\partial x^2} + \frac{\partial^2}{\partial y^2}\right)\log |f'(z)| = 2\frac{\partial^2}{\partial z \partial \bar{z}}[\log f'(z) + \log f'(\bar{z})]$$
$$= 2\frac{\partial}{\partial z}\left[\frac{f''(\bar{z})}{f'(\bar{z})}\right] = 0.$$

EXERCISE 3.7

1. (*a*) Determine a, b, c, d so that the function $f(z) = (x^2 + axy + by^2) + i(cx^2 + dxy + y^2)$ is analytic.

   (*b*) Determine p such that the function $f(z) = \dfrac{1}{2}\log(x^2 + y^2) + i\tan^{-1}\dfrac{px}{y}$ be an analytic function.

   (*M.T.U. 2012*)

2. (*a*) Show that $f(z) = xy + iy$ is everywhere continuous but not analytic.

(*b*) Show that $f(z) = z + 2\bar{z}$ is not analytic anywhere in the complex plane.

(*c*) If $w = \log z$, find $\dfrac{dw}{dz}$ and determine where w is non-analytic.

3. Show that the polar form of Cauchy-Riemann equation are $\dfrac{\partial u}{\partial r} = \dfrac{1}{r}\dfrac{\partial v}{\partial \theta}, \dfrac{\partial v}{\partial r} = -\dfrac{1}{r}\dfrac{\partial u}{\partial \theta}$.

Deduce that $\dfrac{\partial^2 u}{\partial r^2} + \dfrac{1}{r}\dfrac{\partial u}{\partial r} + \dfrac{1}{r^2}\dfrac{\partial^2 u}{\partial \theta^2} = 0.$ (*M.D.U. 2010, 2012*)

4. (*a*) Show that $f(z) = \begin{cases} xy^2(x+iy) \div (x^2+y^4), & z \neq 0 \\ 0, & z = 0 \end{cases}$

is not analytic at $z = 0$, although C – R equations are satisfied at the origin.

(*b*) For the function $f(z)$ defined by

$$f(z) = \begin{cases} \dfrac{\bar{z}^2}{z}, & z \neq 0 \\ 0, & z = 0 \end{cases}$$

show that the C-R equations are satisfied at (0, 0) but the function is not differentiable at (0, 0). (*P.T.U. 2010*)

5. Examine the nature of the function

$$f(z) = \begin{cases} \dfrac{x^2y^5(x+iy)}{x^4+y^{10}}, & z \neq 0 \\ 0, & z = 0 \end{cases}$$

in a region including the origin.

6. Determine which of the following functions are analytic:

(*i*) e^z (*ii*) $\sin z$ (*iii*) $\cosh z$

(*iv*) $\dfrac{1}{z}$ (*v*) $\dfrac{x+iy}{x^2+y^2}$ or $\dfrac{z}{|z|^2}$ (*vi*) $\dfrac{1}{2}\log(x^2+y^2) + i\tan^{-1}\dfrac{y}{x}$.

7. (*a*) Show that $f(z) = |z|^2$ is continuous everywhere but not analytic at any point other than the origin.

(*b*) Show that the function $f(z) = z\,|z|$ is not analytic anywhere. (*U.K.T.U. 2010*)

8. Show that $u + iv = \dfrac{x - iy}{x - iy + a}$, where $a \neq 0$, is not an analytic function of $z = x + iy$, whereas $u - iv$ is such a function.

9. Determine the analytic function whose real part is

(*i*) $x^3 - 3xy^2 + 3x^2 - 3y^2 + 1$ (*K.U.K. 2013*)

(*ii*) $\log\sqrt{x^2+y^2}$

(*iii*) $e^x(x\cos y - y\sin y)$ (*iv*) $\cos x\cosh y$

(*v*) $e^{-x}(x \sin y - y \cos y)$ (*M.T.U. 2012; G.B.T.U. 2011*) (*vi*) $\dfrac{\sin 2x}{\cosh 2y - \cos 2x}$ (*M.D.U., 2010*)

(*vii*) $x \sin x \cosh y - y \cos x \sinh y$ (*viii*) $e^x [(x^2 - y^2) \cos y - 2xy \sin y]$.

(*K.U.K. Dec. 2015; M.D.U. 2010*)

10. Find the regular function whose imaginary part is

(*i*) $\dfrac{x-y}{x^2+y^2}$ (*M.D.U. 2011*) (*ii*) $\cos x \cosh y$ (*iii*) $\sinh x \cos y$

(*iv*) $e^{-x}(x \sin y - y \cos y)$ (*K.U.K. Jan. 2013*) (*v*) $e^{-x}(x \cos y + y \sin y)$. (*M.D.U. Dec. 2014*)

(*vi*) $\dfrac{x}{x^2+y^2} + \cosh x \cos y$ (*DCRUST Murthal, May 2014*)

11. Find the real part of the regular function whose imaginary part is $\dfrac{2 \sin x \sin y}{\cos 2x + \cosh 2y}$.

(*M.D.U. Dec. 2014*)

12. (*a*) Prove that $u = x^2 - y^2 - 2xy - 2x + 3y$ is harmonic. Find a function v such that $f(z) = u + iv$ is analytic. Also express $f(z)$ in terms of z.

(*b*) Show that the function $v(x, y) = \ln(x^2 + y^2) + x - 2y$ is harmonic. Find its conjugate harmonic function $u(x, y)$ and the corresponding analytic function $f(z)$. (*G.B.T.U. 2012*)

13. An electrostatic field in the xy-plane is given by the potential function $\phi = x^2 - y^2$, find the stream function.

14. If $w = \phi + i\psi$ represents the complex potential for an electric field and $\psi = x^2 - y^2 + \dfrac{x}{x^2+y^2}$ determine the function ϕ. (*M.D.U. May 2014; K.U.K. 2010*)

15. If the potential function is $\log(x^2 + y^2)$, find the flux function and the complex potential function.

(*M.D.U. 2011*)

16. In a two dimensional fluid flow, the stream function is $\psi = \tan^{-1}\left(\dfrac{y}{x}\right)$, find the velocity potential ϕ.

17. If $f(z) = u + iv$ is an analytic function, find $f(z)$ if

(*i*) $u - v = e^x(\cos y - \sin y)$ (*K.U.K. 2010*)

(*ii*) $u + v = \dfrac{x}{x^2+y^2}$, when $f(1) = 1$

(*iii*) $u - v = \dfrac{\cos x + \sin x - e^{-y}}{2(\cos x - \cosh y)}$, when $f\left(\dfrac{\pi}{2}\right) = 0$.

18. If $f(z)$ is an analytic function of z, prove that

(*i*) $\left(\dfrac{\partial^2}{\partial x^2} + \dfrac{\partial^2}{\partial y^2}\right) | \mathrm{R}\, f(z) |^2 = 2 \, | f'(z) |^2$ (*G.B.T.U. 2012*)

(*ii*) $\nabla^2 | f(z) |^2 = 4 | f'(z) |^2$ or $\left(\dfrac{\partial^2}{\partial x^2} + \dfrac{\partial^2}{\partial y^2}\right) | f(z) |^2 = 4 | f'(z) |^2$ (*M.D.U. Dec. 2014*)

19. Prove that $\psi = \log[(x-1)^2 + (y-2)^2]$ is harmonic in every region which does not include the point (1, 2). Find a function ϕ such that $\phi + i\psi$ is an analytic function of the complex variable $z = x + iy$. Express $\phi + i\psi$ as a function of z.

20. (*a*) Find the analytic function $f(z) = u(r, \theta) + iv(r, \theta)$ such that $v(r, \theta) = r^2 \cos 2\theta - r \cos \theta + 2$.

(*b*) Find the analytic function $f(z) = u + iv$, given $u = a(1 + \cos \theta)$.

21. Prove that $u = x^2 - y^2$ and $v = \dfrac{y}{x^2 + y^2}$ are harmonic functions of (x, y) but are not harmonic conjugates.

Answers

1. (*a*) $a = 2, b = -1, c = -1, d = 2$ (*b*) $p = -1$

2. (*c*) $\dfrac{1}{z}$, at $z = 0$

5. Not differentiable at the origin and hence not analytic in the region.

6. (*i*), (*ii*) and (*iii*); (*iv*), (*v*) and (*vi*) except when $z = 0$

9. (*i*) $z^3 + 3z^2 + 1 + ic$ (*ii*) $\log z + ic$ (*iii*) $ze^z + ic$

(*iv*) $\cos z + ic$ (*v*) $i(ze^{-z} + c)$ (*vi*) $\cot z + ic$

(*vii*) $z \sin z + ic$ (*viii*) $z^2e^z + ic$.

10. (*i*) $\dfrac{1+i}{z} + c$ (*ii*) $i \cos z + c$ (*iii*) $i \sinh z + c$

(*iv*) $-ze^{-z} + c$ (*v*) $ize^{-z} + c$ (*vi*) $i\left(\dfrac{1}{z} + \cosh z\right) + c$

11. $\dfrac{2 \cos x \cosh y}{\cos 2x + \cosh 2y} + c$.

12. (*a*) $v = x^2 - y^2 + 2xy - 2y - 3x + c$, $f(z) = (1 + i)z^2 - (2 + 3i)z + ic$

(*b*) $u(x, y) = -2x - y - 2\tan^{-1}\dfrac{y}{x} + c$; $f(z)\,(i - 2)\,z + 2i \log z + c$

13. $\psi = 2xy + c$ **14.** $-2xy + \dfrac{y}{x^2 + y^2} + c$ **15.** $2\tan^{-1}\left(\dfrac{y}{x}\right) + c$, $2 \log z + ic$

16. $\dfrac{1}{2}\log(x^2 + y^2) + c$ **17.** (*i*) $e^z + c$ (*ii*) $\dfrac{1}{2}\left[\dfrac{1+i}{z} + (1-i)\right]$ (*iii*) $\dfrac{1}{2}\left(1 - \cot\dfrac{z}{2}\right)$

19. $-2\tan^{-1}\left(\dfrac{y-2}{x-1}\right)$, $2i \log(z - 1 - 2i)$

20. (*a*) $i(r^2e^{2i\theta} - re^{i\theta} + 2) + c$ (*b*) $a(1 + \cos\theta + i \sin\theta \log r + \phi(\theta)$.

3.24. COMPLEX INTEGRATION

Let $f(z)$ be a continuous function of the complex variable $z = x + iy$ defined at all points of a curve C having end points A and B. Divide the curve C into n parts at the points

$$A = P_0(z_0),\ P_1(z_1),\ \ldots\ldots,\ P_i(z_i),\ \ldots\ldots,\ P_n(z_n) = B.$$

Let $\delta z_i = z_i - z_{i-1}$ and ξ_i be any point on the arc $P_{i-1} P_i$. Then the limit of the sum

$$\sum_{i=1}^{n} f(\xi_i)\, \delta z_i$$

as $n \to \infty$ and each $\delta z_i \to 0$, if it exists, is called the **line integral of f(z)** along the curve C. It is denoted by

$$\int_C f(z)\, dz \qquad \textit{(M.D.U. May 2013)}$$

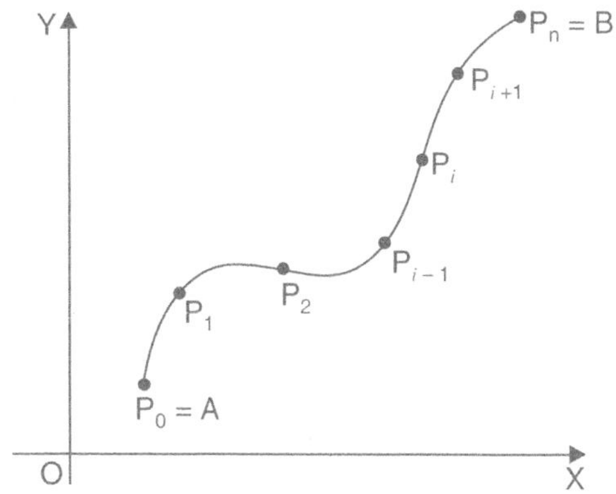

In case the points P_0 and P_n coincide so that C is a closed curve, then this integral is called **contour integral** and is denoted by $\oint_C f(z)\, dz$.

If $f(z) = u(x, y) + iv(x, y)$, then since $dz = dx + idy$, we have

$$\int_C f(z)\, dz = \int_C (u + iv)(dx + idy)$$

$$= \int_C (udx - vdy) + i \int_C (vdx + udy)$$

which shows that the evaluation of the line integral of a complex function can be reduced to the evaluation of two line integrals of real functions.

Moreover, the value of the integral depends on the path of integration unless the integrand is analytic.

When the same path of integration is used in each integral, then

$$\int_a^b f(z)\, dz = -\int_b^a f(z)\, dz$$

If c is a point on the arc joining a and b, then

$$\int_a^b f(z)\, dz = \int_a^c f(z)\, dz + \int_c^b f(z)\, dz.$$

ILLUSTRATIVE EXAMPLES

Example 1. *Evaluate* $\int_0^{1+i} (x - y + ix^2)\, dz$.

(*a*) *along the straight line from* $z = 0$ *to* $z = 1 + i$

(*b*) *along the real axis from* $z = 0$ *to* $z = 1$ *and then along a line parallel to imaginary axis from* $z = 1$ *to* $z = 1 + i$.

(*c*) *along the imaginary axis from* $z = 0$ *to* $z = i$ *and then along a line parallel to real axis from* $z = i$ *to* $z = 1 + i$.

Sol. (*a*) Along the straight line OP joining O($z = 0$) and P($z = 1 + i$), $y = x$, $dy = dx$ and x varies from 0 to 1.

$$\therefore \quad \int_0^{1+i} (x - y + ix^2)\, dz = \int_0^{1+i} (x - y + ix^2)(dx + idy)$$

$$= \int_0^1 (x - x + ix^2)(dx + idx) = \int_0^1 ix^2(1 + i)\, dx$$

$$= (-1 + i) \int_0^1 x^2\, dx = (-1 + i) \left[\frac{x^3}{3}\right]_0^1$$

$$= (-1 + i) \left(\frac{1}{3}\right) = -\frac{1}{3} + \frac{1}{3} i.$$

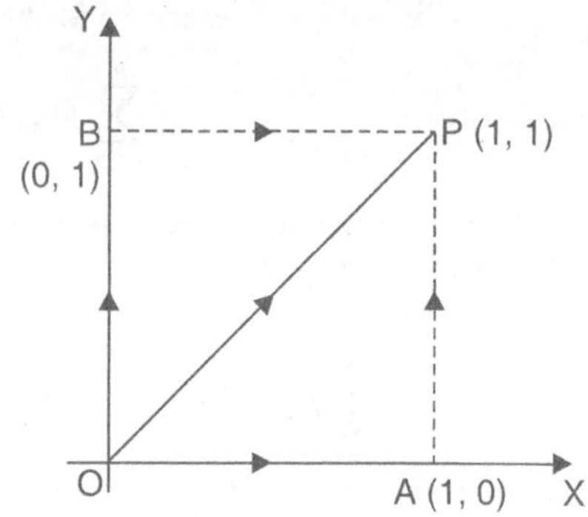

(*b*) Along the path OAP, where A is $z = 1$

$$\int_0^{1+i} (x - y + ix^2) dz = \int_{OA} (x - y + ix^2) dz + \int_{AP} (x - y + ix^2)\, dz \qquad ...(1)$$

Now along OA, $y = 0$, $dz = dx$ and x varies from 0 to 1.

$$\therefore \quad \int_{OA} (x - y + ix^2) dz = \int_0^1 (x + ix^2) dx = \left[\frac{x^2}{2} + i\frac{x^3}{3}\right]_0^1 = \frac{1}{2} + \frac{1}{3} i$$

Also, along AP, $x = 1$, $dz = idy$ and y varies from 0 to 1

$$\therefore \quad \int_{AP} (x - y + ix^2)\, dz = \int_0^1 (1 - y + i)\, idy = \left[(-1 + i)y - i\frac{y^2}{2}\right]_0^1 = -1 + i - i = -1 + \frac{1}{2} i$$

Hence from (1), $\displaystyle\int_0^{1+i} (x - y + ix^2)\, dz = \left(\frac{1}{2} + \frac{1}{3} i\right) + \left(-1 + \frac{1}{2} i\right) = -\frac{1}{2} + \frac{5}{6} i.$

(*c*) Along the path OBP, where B is $z = i$

$$\int_0^{1+i} (x - y + ix^2)\, dz = \int_{OB} (x - y + ix^2) dz + \int_{BP} (x - y + ix^2)\, dz \qquad ...(2)$$

Now along OB, $x = 0$, $dz = idy$ and y varies from 0 to 1

$$\therefore \quad \int_{OB} (x - y + ix^2) dz = \int_0^1 (-y)\, idy = -i\left[\frac{y^2}{2}\right]_0^1 = -\frac{1}{2} i$$

Also, along BP, $y = 1$, $dz = dx$ and x varies from 0 to 1

$$\therefore \quad \int_{BP} (x - y + ix^2) dz = \int_0^1 (x - 1 + ix^2)\, dx = \left[\frac{x^2}{2} - x + i\frac{x^3}{3}\right]_0^1 = -\frac{1}{2} + \frac{1}{3} i$$

Hence from (2), $\displaystyle\int_0^{1+i} (x - y + ix^2)\, dz = -\frac{1}{2} i + \left(-\frac{1}{2} + \frac{1}{3} i\right) = -\frac{1}{2} - \frac{1}{6} i.$

Note. The values of the integral are different along the three different paths.

Example 2. *Evaluate* $\displaystyle\int_0^{1+i} (x^2 - iy) dz$ *along the paths*

(*a*) $y = x$ (*b*) $y = x^2$. (G.B.T.U. 2010)

Sol. (*a*) Along the line $y = x$,

$$dy = dx \text{ so that } dz = dx + idx = (1 + i)\, dx$$

$$\therefore \quad \int_0^{1+i} (x^2 - iy)\, dz = \int_0^1 (x^2 - ix)(1 + i)\, dx$$

$$= (1 + i) \left[\frac{x^3}{3} - i\frac{x^2}{2} \right]_0^1 = (1 + i)\left(\frac{1}{3} - \frac{1}{2} i \right)$$

$$= \frac{(1 + i)(2 - 3i)}{6} = \frac{5}{6} - \frac{1}{6} i.$$

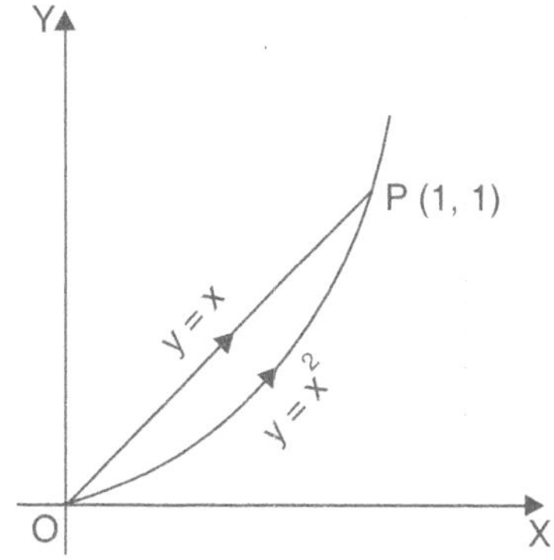

(*b*) Along the parabola $y = x^2$, $dy = 2x\, dx$ so that

$$dz = dx + 2ix\, dx = (1 + 2ix)\, dx$$

and x varies from 0 to 1.

$$\therefore \quad \int_0^{1+i} (x^2 - iy)\, dz = \int_0^1 (x^2 - ix^2)(1 + 2ix)\, dx$$

$$= (1 - i) \int_0^1 x^2(1 + 2ix)\, dx = (1 - i)\left[\frac{x^3}{3} + i\frac{x^4}{2} \right]_0^1$$

$$= (1 - i)\left(\frac{1}{3} + \frac{1}{2} i \right) = \frac{(1 - i)(2 + 3i)}{6} = \frac{5}{6} + \frac{1}{6} i.$$

Example 3. *Evaluate* $\int_0^{2+i} (\bar{z})^2\, dz$, *along*

(*a*) *the real axis to 2 and then vertically to* $2 + i$. (U.K.T.U. 2011)

(*b*) *along the line* $2y = x$.

Sol. $(\bar{z})^2 = (x - iy)^2 = (x^2 - y^2) - 2ixy$

(*a*) Along the path OAP, where A is (2, 0) and P is (2, 1).

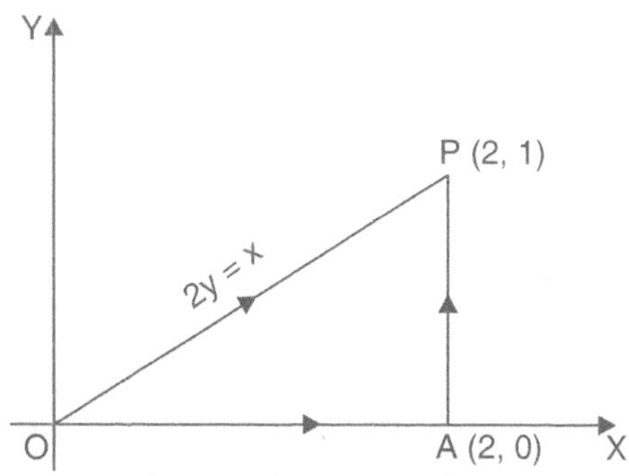

$$\int_0^{2+i} (\bar{z})^2\, dz$$

$$= \int_{OA} (x^2 - y^2 - 2ixy)\, dz + \int_{AP} (x^2 - y^2 - 2ixy)\, dz \quad \ldots(1)$$

Now, along OA, $y = 0$, $dz = dx$ and x varies from 0 to 2

$$\therefore \quad \int_{OA} (x^2 - y^2 - 2ixy)\, dz = \int_0^2 x^2\, dx = \left[\frac{x^3}{3} \right]_0^2 = \frac{8}{3}$$

Also, along AP, $x = 2$, $dz = idy$ and y varies from 0 to 1

$$\therefore \quad \int_{AP} (x^2 - y^2 - 2ixy)\, dz = \int_0^1 (4 - y^2 - 4iy)\, idy$$

$$= \left[4iy - i\frac{y^3}{3} + 2y^2 \right]_0^1 = 4i - \frac{1}{3} i + 2 = 2 + \frac{11}{3} i$$

Hence from (1), we have $\displaystyle\int_0^{2+i} (\bar{z})^2\, dz = \frac{8}{3} + 2 + \frac{11}{3} i = \frac{14}{3} + \frac{11}{3} i.$

(*b*) Along the line OP, $2y = x$, $dx = 2dy$ so that $dz = 2dy + i\,dy = (2 + i)\,dy$ and y varies from 0 to 1.

$$\therefore \quad \int_0^{2+i} (\bar{z})^2\, dz = \int_0^{2+i} (x^2 - y^2 - 2ixy)\, dz = \int_0^1 (4y^2 - y^2 - 4iy^2)(2 + i)\, dy$$

$$= (2 + i)(3 - 4i) \int_0^1 y^2\, dy = (10 - 5i)\left[\frac{y^3}{3}\right]_0^1 = \frac{10}{3} - \frac{5}{3}i.$$

Example 4. *Integrate* $f(z) = x^2 + ixy$ *from* $A(1, 1)$ *to* $B(2, 4)$ *along the curve* $x = t$, $y = t^2$.

Sol. Equations of the path of integration are $x = t$, $y = t^2$

$$\therefore \quad dx = dt, \quad dy = 2t\, dt$$

At A(1, 1), $t = 1$ and at B(2, 4), $t = 2$

$$\therefore \quad \int_A^B f(z)\, dz = \int_A^B (x^2 + ixy)(dx + idy) = \int_1^2 (t^2 + it^3)(dt + 2it\, dt)$$

$$= \int_1^2 (t^2 - 2t^4)\, dt + i\int_1^2 3t^3\, dt = \left[\frac{t^3}{3} - \frac{2t^5}{5}\right]_1^2 + i\left[\frac{3t^4}{4}\right]_1^2$$

$$= \left(\frac{8}{3} - \frac{64}{5}\right) - \left(\frac{1}{3} - \frac{2}{5}\right) + i\left(12 - \frac{3}{4}\right) = -\frac{151}{15} + \frac{45}{4}i.$$

Example 5. *Prove that*

(*i*) $\oint_C \frac{dz}{z - a} = 2\pi i$

(*ii*) $\oint_C (z - a)^n\, dz = 0$ *[n is an integer* $\neq -1$*], where C is the circle* $|\, z - a \,| = r$.

(G.B.T.U. 2011)

Sol. The equation of the circle C is

$$|\, z - a \,| = r \quad \text{or} \quad z - a = re^{i\theta}$$

where θ varies from 0 to 2π as z describes C once in the anti-clockwise direction.

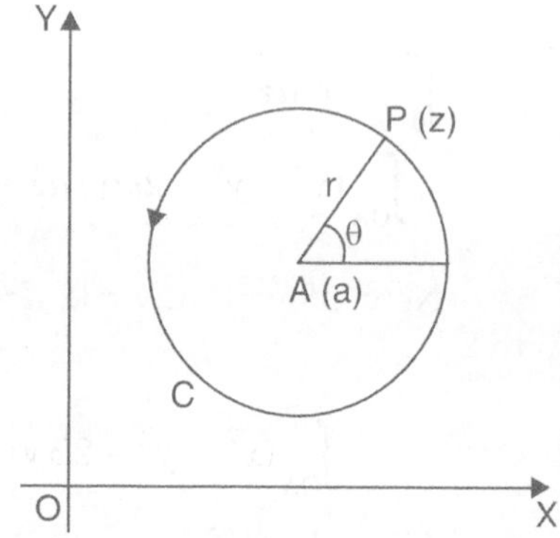

Also $\quad dz = ire^{i\theta}\, d\theta.$

$$(i) \quad \oint_C \frac{dz}{z - a} = \int_0^{2\pi} \frac{ire^{i\theta}\, d\theta}{re^{i\theta}} = i\int_0^{2\pi} d\theta = 2\pi i$$

$$(ii) \quad \oint_C (z - a)^n\, dz = \int_0^{2\pi} r^n e^{ni\theta} \,.\, ire^{i\theta}\, d\theta$$

$$= ir^{n+1} \int_0^{2\pi} e^{i(n+1)\theta}\, d\theta$$

$$= ir^{n+1} \left[\frac{e^{i(n+1)\theta}}{i(n+1)}\right]_0^{2\pi} \qquad [\because \; n \neq -1]$$

$$= \frac{r^{n+1}}{n+1}\, [e^{i2(n+1)\pi} - 1]$$

$$= 0. \qquad [\because \; e^{i2(n+1)\pi} = \cos 2(n+1)\pi + i \sin 2(n+1)\pi = 1 + i0 = 1]$$

Example 6. *Evaluate* $\int_C (z^2 + 3z + 2)\, dz$ *where C is the arc of the cycloid* $x = a(\theta + \sin\theta)$, $y = a(1 - \cos\theta)$ *between the points* $(0, 0)$ *and* $(\pi a, 2a)$.

Sol. The function $f(z) = z^2 + 3z + 2$ is a polynomial and therefore analytic in z-plane. Hence the line integral of $f(z)$ between the points $O(0, 0)$ and $A(\pi a, 2a)$ is independent of the path joining these points. Let us choose the path of integration as:

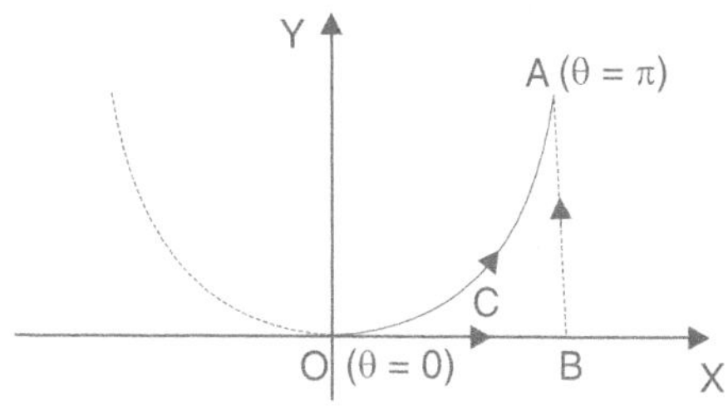

(*i*) From $O(0, 0)$ to $B(\pi a, 0)$ along the real axis; followed by

(*ii*) From $B(\pi a, 0)$ to $A(\pi a, 2a)$ vertically

$$\therefore \quad \int_C f(z)\, dz = \int_{OB} f(z)\, dz + \int_{BA} f(z)\, dz \quad \ldots(i)$$

Now along OB, $y = 0$

$$\therefore \quad z = x + iy = x, \quad dz = dx$$

and x varies from 0 to πa.

$$f(z) = z^2 + 3z + 2 = x^2 + 3x + 2$$

$$\therefore \quad \int_{OB} f(z)\, dz = \int_0^{\pi a} (x^2 + 3x + 2)\, dx = \left[\frac{x^3}{3} + \frac{3x^2}{2} + 2x\right]_0^{\pi a}$$

$$= \frac{\pi^3 a^3}{3} + \frac{3\pi^2 a^2}{2} + 2\pi a = \frac{\pi a}{6}(2\pi^2 a^2 + 9\pi a + 12)$$

Also, along BA, $x = \pi a$

$$\therefore \quad z = x + iy = \pi a + iy,$$

$$dz = i\,dy \text{ and } y \text{ varies from 0 to } 2a.$$

$$f(z) = z^2 + 3z + 2 = (\pi a + iy)^2 + 3(\pi a + iy) + 2$$

$$\therefore \quad \int_{BA} f(z)\, dz = \int_0^{2a} [(\pi a + iy)^2 + 3(\pi a + iy) + 2]\, i\,dy$$

$$= \left[\left\{\frac{(\pi a + iy)^3}{3i} + \frac{3(\pi a + iy)^2}{2i} + 2y\right\} i\right]_0^{2a}$$

$$= \frac{1}{3}(\pi a + i2a)^3 + \frac{3}{2}(\pi a + i2a)^2 + 2i\,.\,2a - \frac{1}{3}\pi^3 a^3 - \frac{3}{2}\pi^2 a^2$$

$$= \frac{a^3}{3}(\pi + 2i)^3 + \frac{3a^2}{2}(\pi + 2i)^2 + 4ia - \frac{1}{3}\pi^3 a^3 - \frac{3}{2}\pi^2 a^2$$

$\therefore$ From (1), we have

$$\int_C (z^2 + 3z + 2)\, dz = \frac{\pi a}{6}(2\pi^2 a^2 + 9\pi a + 12) + \frac{a^3}{3}(\pi + 2i)^3 + \frac{3a^2}{2}(\pi + 2i)^2 + 4ia - \frac{1}{3}\pi^3 a^3 - \frac{3}{2}\pi^2 a^2$$

$$= \frac{1}{3}\pi^3 a^3 + \frac{3}{2}\pi^2 a^2 + 2\pi a + \frac{a^3}{3}(\pi + 2i)^3 + \frac{3a^2}{2}(\pi + 2a)^2 + 4ia - \frac{1}{3}\pi^3 a^3 - \frac{3}{2}\pi^2 a^2$$

$$= \frac{a^3}{3}(\pi + 2i)^3 + \frac{3a^2}{2}(\pi + 2i)^2 + 2a(\pi + 2i)$$

EXERCISE 3.8

1. Evaluate $\int_0^{3+i} z^2\, dz$, along

   (*a*) the line $y = \frac{x}{3}$ (*b*) the real axis to 3 and then vertically to $3 + i$

   (*c*) the parabola $x = 3y^2$.

2. Evaluate $\int_{1-i}^{2+i} (2x + iy + 1)\, dz$, along

   (*a*) the straight line joining $(1 - i)$ to $(2 + i)$. (*b*) the curve $x = t + 1$, $y = 2t^2 - 1$.

3. Evaluate $\int_0^{4+2i} \bar{z}\, dz$, along the curve given by $z = t^2 + it$.

4. Evaluate $\oint_C |z|^2\, dz$, around the square with vertices at (0, 0), (1, 0), (1, 1) and (0, 1).

5. Show that $\oint_C (z + 1)\, dz = 0$, where C is the boundary of the square whose vertices are at the points $z = 0$, $z = 1$, $z = 1 + i$ and $z = i$. *(M.D.U. May 2015)*

6. Evaluate $\int_C (y - x - 3x^2 i)\, dz$, where C is the straight line from $z = 0$ to $z = 1 + i$.

7. (*a*) Evaluate $\int_C (z - z^2)\, dz$, where C is the upper half of the circle $|z| = 1$. What is the value of this integral if C is the lower half of above circle ?

   (*b*) Evaluate $\int_C (z - z^3)\, dz$, where C is the upper half of the circle $|z| = 1$. *(M.D.U. 2011)*

   (*c*) Evaluate $\int_C (z - z^2)\, dz$ where C is the upper half of the circle $|z - 2| = 3$. What is the value of the integral if C is the lower half of the above given circle ? *(M.T.U. 2013)*

8. Prove that $\int_C \frac{1}{z}\, dz = -\pi i$ or πi according as C is the semi-circular arc $|z| = 1$ from -1 to 1 above or below the real axis. *(M.D.U. 2013)*

9. Show that for every path between the limits,

$$\int_{-2}^{-2+i} (2 + z)^2\, dz = -\frac{i}{3}$$

10. Evaluate $\int_{1-i}^{2+3i} (z^2 + z)\, dz$ along the line joining the points $(1, -1)$ and $(2, 3)$.

11. Evaluate $\int_C |z|\,dz$, where C is the contour

(*a*) straight line from $z = -i$ to $z = i$.

(*b*) left half of the unit circle $|z| = 1$ from $z = -i$ to $z = i$.

(*c*) circle given by $|z+1| = 1$ described in the clockwise sense.

12. Evaluate $\oint_C \log z\, dz$, where C is the unit circle $|z| = 1$.

Answers

1. (*a*) $6 + \frac{26}{3}i$ (*b*) $6 + \frac{26}{3}i$ (*c*) $6 + \frac{26}{3}i$

2. (*a*) $4 + 8i$ (*b*) $4 + \frac{25}{3}i$ **3.** $10 - \frac{8}{3}i$ **4.** $-1 + i$

6. $1 - i$ **7.** (*a*) $\frac{2}{3}, -\frac{2}{3}$ (*b*) 0 (*c*) 30, − 30

10. $\frac{1}{6}(-103 + 64i)$ **11.** (*a*) i (*b*) $2i$ (*c*) $\frac{8}{3}i$

12. $2\pi i$

[**Hint.** On the unit circle $|z| = 1$, $x^2 + y^2 = 1$ and $z = e^{i\theta}$ $\oint_C \log z\, dz$

$$= \oint_C \log(x + iy)\,dz = \oint_C \left[\frac{1}{2}\log(x^2 + y^2) + i\tan^{-1}\frac{y}{x}\right]dz$$

$$= i\int_0^{2\pi} \tan^{-1}(\tan\theta)ie^{i\theta}\,d\theta = -\int_0^{2\pi} \theta e^{i\theta}\,d\theta\Bigg]$$

3.25. SIMPLY AND MULTIPLY CONNECTED REGIONS

A curve is called *simple closed curve* if it does not cross itself (Fig. 1). A curve which crosses itself is called a *multiple curve* (Fig. 2).

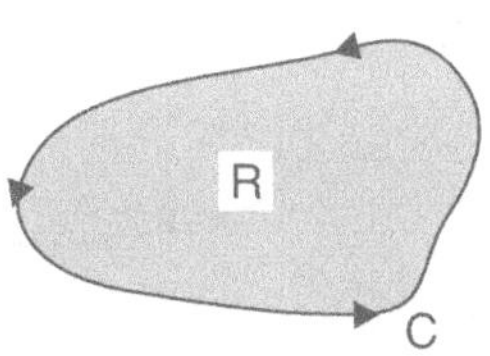

Fig.1

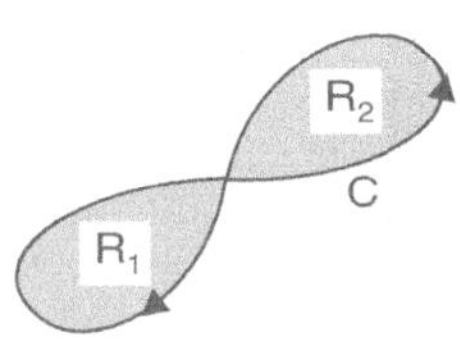

Fig. 2

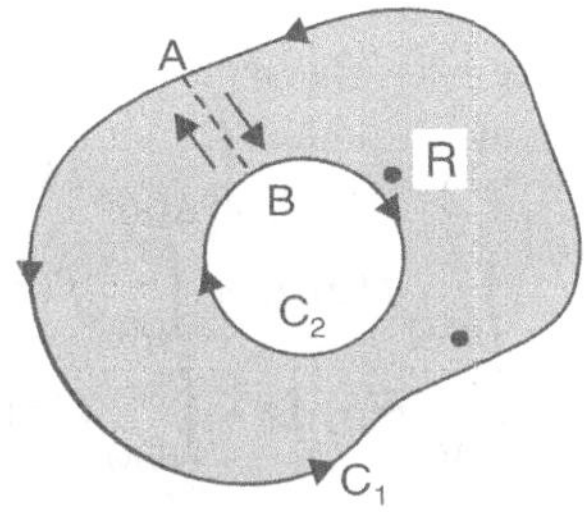

Fig. 3

A region is called *simply connected* if every closed curve in the region encloses points of the region only, *i.e.*, every closed curve lying in it can be contracted indefinitely without passing out of it. A region which is not simply connected is called a *multiply connected* region. In plain terms, a simply connected region is one which has no holes. Fig. 3 shows a multiply connected region R enclosed between two separate curves C_1 and C_2. (There can be more than two separate curves). We can convert a multiply connected region into a simply connected one, by giving it one or more cuts (*e.g.*, along the dotted line AB).

3.26. CAUCHY'S INTEGRAL THEOREM

(M.D.U. 2010, May 2015, Dec. 2015; K.U.K. 2013; M.T.U. 2013)

Statement. *If f(z) is an analytic function and f ′(z) is continuous at each point within and on a simple closed curve C, then*

$$\oint_C f(z)\,dz = 0.$$

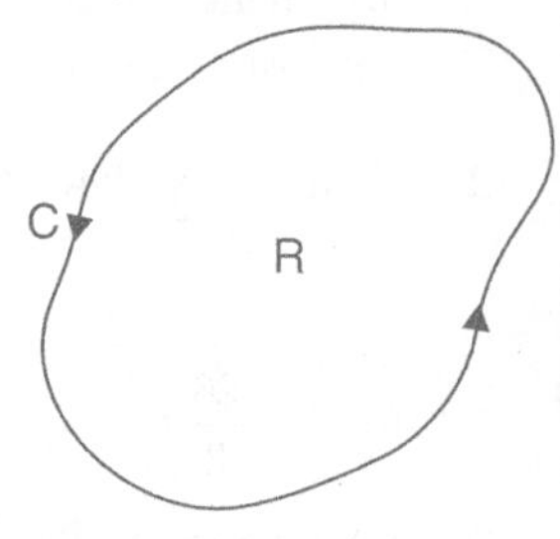

Proof. Let R be the region bounded by the curve C.

Let $f(z) = u(x, y) + iv(x, y)$, then

$$\oint_C f(z)\,dz = \oint_C (u + iv)(dx + idy)$$

$$= \oint_C (udx - vdy) + i\oint_C (vdx + udy) \quad ...(1)$$

Since $f'(z)$ is continuous, the partial derivatives $\frac{\partial u}{\partial x}, \frac{\partial u}{\partial y}, \frac{\partial v}{\partial x}, \frac{\partial v}{\partial y}$ are also continuous in R. Hence by Green's Theorem, we have

$$\oint_C f(z)\,dz = \iint_R \left(-\frac{\partial v}{\partial x} - \frac{\partial u}{\partial y}\right) dx\,dy + i\iint_R \left(\frac{\partial u}{\partial x} - \frac{\partial v}{\partial y}\right) dx\,dy \quad ...(2)$$

Now $f(z)$ being analytic at each point of the region R, by Cauchy-Riemann equations, we have

$$\frac{\partial u}{\partial x} = \frac{\partial v}{\partial y} \quad \text{and} \quad \frac{\partial u}{\partial y} = -\frac{\partial v}{\partial x}$$

Thus, the two double integrals in (2) vanish.

Hence, $\oint_C f(z)\,dz = 0.$

Cor. 1. *If f(z) is analytic in a region R and P and Q are two points in R, then* $\int_P^Q f(z)\,dz$ *is independent of the path joining P and Q and lying entirely in R.*

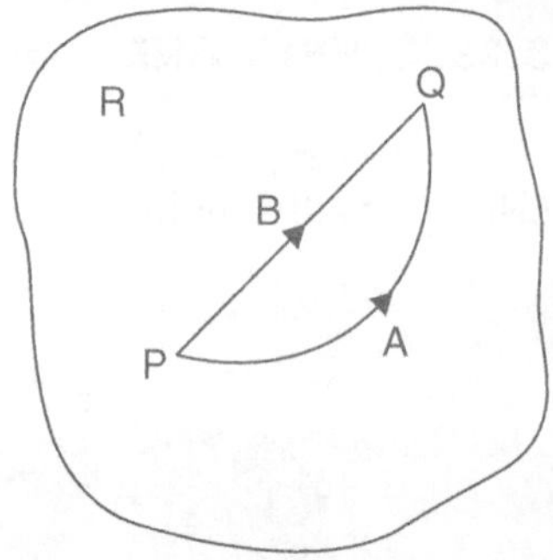

Let PAQ and PBQ be any two paths joining P and Q.

By Cauchy's theorem,

$$\int_{PAQBP} f(z)\,dz = 0$$

$$\Rightarrow \quad \int_{PAQ} f(z)\,dz + \int_{QBP} f(z)\,dz = 0$$

$$\Rightarrow \quad \int_{PAQ} f(z)\,dz - \int_{PBQ} f(z)\,dz = 0$$

Hence $$\int_{PAQ} f(z)\,dz = \int_{PBQ} f(z)\,dz.$$

Cor. 2. *If f(z) is analytic in the region bounded by two simple closed curves C and C_1, then*

$$\oint_C f(z)\,dz = \oint_{C_1} f(z)\,dz$$

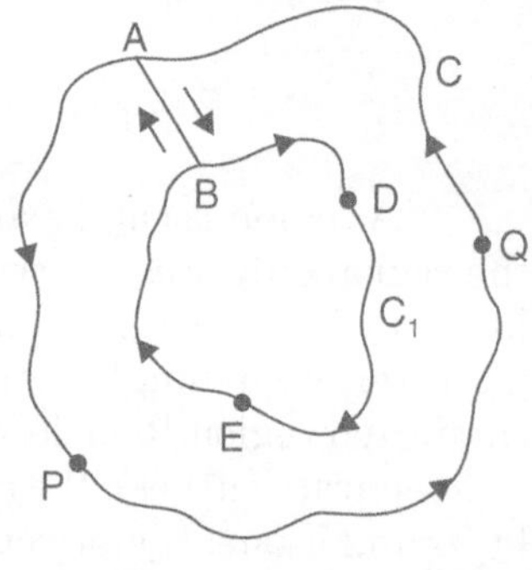

Let AB be a cross-cut joining the curves C and C_1, then the doubly connected region becomes simply connected.

By Cauchy's theorem, $\int_{APQABDEBA} f(z)\,dz = 0$

$$\Rightarrow \int_{\text{APQA}} f(z)\,dz + \int_{\text{AB}} f(z)\,dz + \int_{\text{BDEB}} f(z)\,dz + \int_{\text{BA}} f(z)\,dz = 0$$

$$\Rightarrow \oint_{\text{C}} f(z)\,dz + \int_{\text{AB}} f(z)\,dz - \int_{\text{BEDB}} f(z)\,dz - \int_{\text{AB}} f(z)\,dz = 0$$

$$\Rightarrow \oint_{\text{C}} f(z)\,dz - \oint_{\text{C}_1} f(z)\,dz = 0$$

(integrals around a closed curve are taken positive when the curve is traversed in counter-clockwise direction)

Hence $$\oint_{\text{C}} f(z)\,dz = \oint_{\text{C}_1} f(z)\,dz$$

The theorem can be extended.

If a closed curve C contains non-intersecting closed curves $C_1, C_2, \ldots\ldots, C_n$, then by introducing cross-cuts, it can be shown that

$$\oint_{\text{C}} f(z)dz = \oint_{\text{C}_1} f(z)dz + \oint_{\text{C}_2} f(z)dz + \cdots + \oint_{\text{C}_n} f(z)dz\,.$$

3.27. CAUCHY'S INTEGRAL FORMULA

(G.B.T.U. 2011, 2013; M.T.U. 2012)

Statement. *If f(z) is analytic within and on a closed curve C and a is any point within C, then*

$$f(a) = \frac{1}{2\pi i}\oint_{C} \frac{f(z)}{z-a}\,dz\,.$$

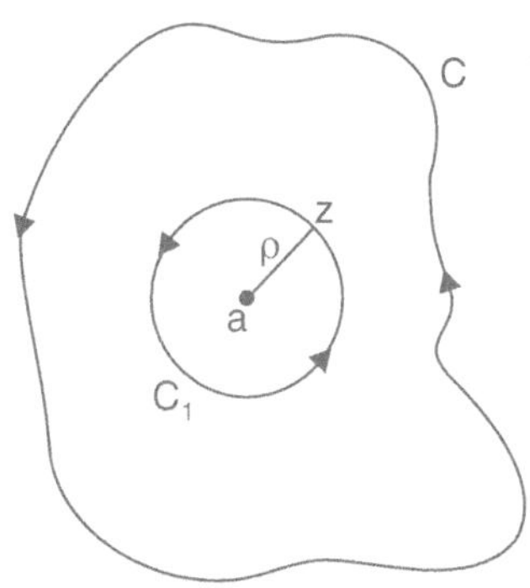

Proof. Consider the function $\frac{f(z)}{z-a}$, which is analytic at every point within C except at $z = a$.

Draw a circle C_1 with a as centre and radius ρ such that C_1 lies entirely inside C.

Thus $\frac{f(z)}{z-a}$ is analytic in the region between C and C_1.

∴ By Cauchy's theorem, we have

$$\oint_{\text{C}} \frac{f(z)}{z-a}\,dz = \oint_{\text{C}_1} \frac{f(z)}{z-a}\,dz \qquad \ldots(1)$$

Now, the equation of circle C_1 is

$$|\, z - a\,| = \rho \quad \text{or} \quad z - a = \rho e^{i\theta}$$

so that $$dz = i\rho e^{i\theta}\,d\theta$$

$$\oint_{\text{C}_1} \frac{f(z)}{z-a}\,dz = \int_0^{2\pi} \frac{f(a+\rho e^{i\theta})}{\rho e^{i\theta}}\,.\,i\rho e^{i\theta}\,d\theta = i\int_0^{2\pi} f(a+\rho e^{i\theta})\,d\theta$$

Hence by (1), we have $$\oint_{\text{C}} \frac{f(z)}{z-a}\,dz = i\int_0^{2\pi} f(a+\rho e^{i\theta})\,d\theta \qquad \ldots(2)$$

In the limiting form, as the circle C_1 shrinks to the point a, *i.e.*, $\rho \to 0$, then from (2),

$$\oint_{\text{C}} \frac{f(z)}{z-a}\,dz = i\int_0^{2\pi} f(a)d\theta = if(a)\int_0^{2\pi} d\theta = 2\pi i f(a)$$

Hence $f(a) = \dfrac{1}{2\pi i}\oint_C \dfrac{f(z)}{z-a}\,dz$ which is the required Cauchy's integral formula.

Cor. By Cauchy's integral formula, we have

$$f(a) = \frac{1}{2\pi i}\oint_C \frac{f(z)}{z-a} \qquad ...(1)$$

where a is any point within C and may be treated as a parameter.

Differentiating both sides of (1) w.r.t. a,

$$f'(a) = \frac{1}{2\pi i}\oint_C \frac{\partial}{\partial a}\left[\frac{f(z)}{z-a}\right]dz = \frac{1}{2\pi i}\oint_C \frac{f(z)}{(z-a)^2}\,dz$$

Similarly, $$f''(a) = \frac{2!}{2\pi i}\oint_C \frac{f(z)}{(z-a)^3}\,dz$$

$$f'''(a) = \frac{3!}{2\pi i}\oint_C \frac{f(z)}{(z-a)^4}\,dz$$

and in general, $$f^n(a) = \frac{n!}{2\pi i}\oint_C \frac{f(z)}{(z-a)^{n+1}}\,dz.$$

Thus, *if a function of a complex variable has a first derivative in a simply connected region, all its higher derivatives exist in that region. This property is not exhibited by the functions of real variables.*

ILLUSTRATIVE EXAMPLES

Example 1. *Evaluate* $\oint_C (x^2 - y^2 + 2ixy)\,dz$, *where C is the contour* $|\,z\,| = 1$.

Sol. $f(z) = x^2 - y^2 + 2ixy = (x + iy)^2 = z^2$ is analytic everywhere within and on $|\,z\,| = 1$.

$\therefore$ By Cauchy's integral theorem, $\oint_C f(z)\,dz = 0$.

Example 2. *Evaluate* $\int_C (3z^2 + 4z + 1)\,dz$, *where C is the arc of the cycloid* $x = a(\theta - \sin\theta)$, $y = a(1 - \cos\theta)$ *between* $(0, 0)$ *and* $(2\pi a, 0)$.

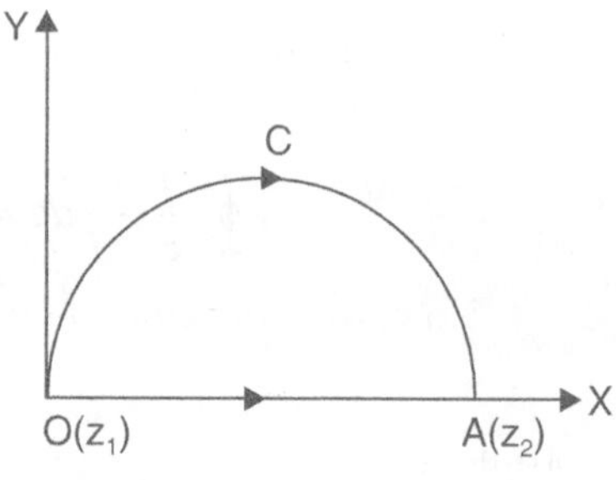

Sol. Here, $f(z) = 3z^2 + 4z + 1$ is analytic everywhere so that the integral is independent of the path of integration and depends only on the end points $z_1 = 0 + i0$ and $z_2 = 2\pi a + i0$.

$$\therefore \quad \int_C (3z^2 + 4z + 1)\,dz = \int_0^{2\pi a} (3z^2 + 4z + 1)\,dz$$

$$= \Big[z^3 + 2z^2 + z\Big]_0^{2\pi a} = 8\pi^3 a^3 + 8\pi^2 a^2 + 2\pi a = 2\pi a\,(4\pi^2 a^2 + 4\pi a + 1).$$

Example 3. *Evaluate* $\oint_C \dfrac{e^{-z}}{z+1}\,dz$, *where C is the circle*

(*a*) $|\,z\,| = 2$ (P.T.U. 2009) (*b*) $|\,z\,| = \frac{1}{2}$. (M.D.U. 2012, Dec. 2014)

Sol. $f(z) = e^{-z}$ is an analytic function.

(*a*) The point $a = -1$ lies inside the circle $|z| = 2$.

$\therefore$ By Cauchy's integral formula,

$$\frac{1}{2\pi i}\oint_C \frac{f(z)}{z-(-1)}\,dz = f(-1) \Rightarrow \oint_C \frac{e^{-z}}{z+1}\,dz = 2\pi i e.$$

(*b*) The point $a = -1$ lies outside the circle $|z| = \frac{1}{2}$.

$\therefore$ The function $\dfrac{e^{-z}}{z+1}$ is analytic within and on C.

By Cauchy's integral theorem, we have $\oint_C \dfrac{e^{-z}}{z+1}dz = 0$.

Example 4. *Evaluate* $\oint_C \dfrac{2z^2+5}{(z+2)^3(z^2+4)}dz$, *where C is the square with vertices at* $1+i,\ 2+i,\ 2+2i,\ 1+2i$.

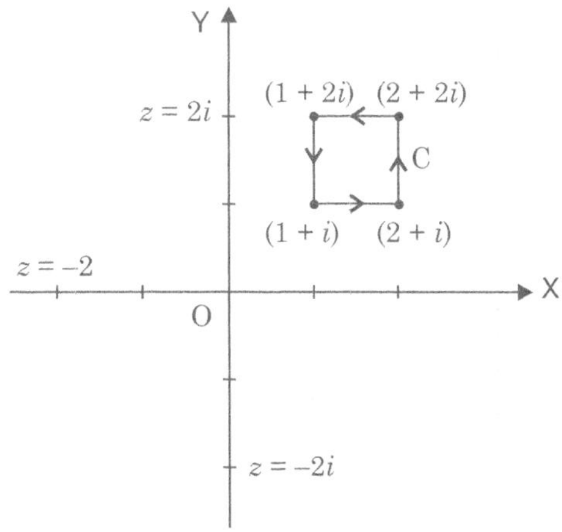

Sol. Here $f(z) = \dfrac{2z^2+5}{(z+2)^3(z^2+4)}$

Singularities are given by

$$(z+2)^3(z^2+4) = 0$$

$\Rightarrow$ $z = -2$ (order 3),

$z = \pm 2i$

All the singularities lie outside the square.

$\therefore$ $f(z)$ is analytic within and on C.

Hence by Cauchy's integral theorem, we have

$$\oint_C \frac{2z^2+5}{(z+2)^3(z^2+4)}dz = 0.$$

Example 5. *Verify Cauchy's integral theorem by integrating* e^{iz} *along the boundary of the triangle with vertices at the points* $1+i$, $-1+i$ *and* $-1-i$.

(G.B.T.U. 2012, 2013; M.T.U. 2012; M.D.U. May 2014)

Sol. Let $I = \oint_C e^{iz}\,dz$, where C is the boundary of the triangle with vertices at the points $1+i$, $-1+i$ and $-1-i$. The integrand $f(z) = e^{iz}$ is analytic for all z and $f'(z)$ is continuous within C. Hence, by Cauchy's integral theorem,

$$I = 0$$

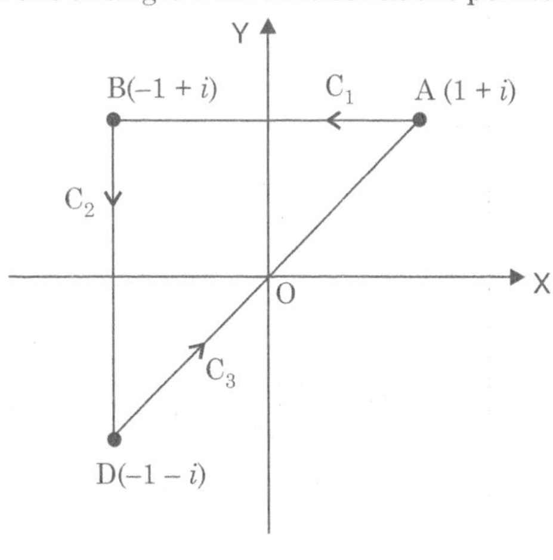

Verification. The boundary C of triangle ABD consists of three lines, C_1, C_2 and C_3 so that

$$I = \oint_C e^{iz}\,dz$$

$$= \int_{C_1} e^{iz}\,dz + \int_{C_2} e^{iz}\,dz + \int_{C_3} e^{iz}\,dz$$

$$= I_1 + I_2 + I_3 \qquad \ldots(1)$$

Along AB: $y = 1$ so that, $z = x + i$ and $dz = dx$

$$\therefore \quad I_1 = \int_{C_1} e^{iz}\, dz = \int_1^{-1} e^{i(x+i)}\, dx = \int_1^{-1} e^{-1+ix}\, dx$$

$$= \int_1^{-1} e^{-1} . e^{ix}\, dx = e^{-1}\left[\frac{e^{ix}}{i}\right]_1^{-1}$$

$$= e^{-1}\left(\frac{e^{-i} - e^{i}}{i}\right) = \frac{e^{-1-i} - e^{-1+i}}{i}$$

Along BD: $x = -1$ so that $z = -1 + iy$ and $dz = i\,dy$

$$\therefore \quad I_2 = \int_{C_2} e^{iz}\, dz = \int_1^{-1} e^{i(-1+iy)}\, i\, dy = i\int_1^{-1} e^{-i-y}\, dy$$

$$= i\int_1^{-1} e^{-i} . e^{-y}\, dy = ie^{-i}\left[\frac{e^{-y}}{-1}\right]_1^{-1} = -ie^{-i}(e^1 - e^{-1})$$

$$= \frac{1}{i} e^{-i}(e^1 - e^{-1}) = \frac{e^{1-i} - e^{-1-i}}{i}$$

Along DA: $y = x$ so that $z = x + ix = (1 + i)x$ and $dz = (1 + i)\, dx$

$$\therefore \quad I_3 = \int_{C_3} e^{iz}\, dz = \int_{-1}^{1} e^{i(1+i)x}\, (1+i)\, dx$$

$$= (1+i) \cdot \left[\frac{e^{i(1+i)x}}{i(1+i)}\right]_{-1}^{1} = \left[\frac{e^{(-1+i)x}}{i}\right]_{-1}^{1} = \frac{e^{-1+i} - e^{1-i}}{i}$$

$$\therefore \quad I = I_1 + I_2 + I_3 = \frac{1}{i}\left(e^{-1-i} - e^{-1+i} + e^{1-i} - e^{-1-i} + e^{-1+i} - e^{1-i}\right)$$

$$\Rightarrow \quad I = 0$$

Hence, Cauchy's integral theorem is verified.

Example 6. *Evaluate* $\oint_C \frac{\cos \pi z^2}{(z-1)(z-2)}\, dz$, *where C is the circle* $|z| = 3$.

Sol. The integrand has singularities where

$(z - 1)(z - 2) = 0$ *i.e.,* at $z = 1$ and $z = 2$.

Both these points lie within the circle $|z| = 3$

$f(z) = \cos \pi z^2$ is an analytic function.

$$\text{Also} \quad \frac{1}{(z-1)(z-2)} = -\frac{1}{z-1} + \frac{1}{z-2}$$

$$\therefore \quad \oint_C \frac{\cos \pi z^2}{(z-1)(z-2)}\, dz = -\oint_C \frac{\cos \pi z^2}{z-1}\, dz + \oint_C \frac{\cos \pi z^2}{z-2}\, dz \quad \text{...(1)}$$

$$= -2\pi i f(1) + 2\pi i f(2), \text{ where } f(z) = \cos \pi z^2$$
$$= -2\pi i \cos \pi + 2\pi i \cos 4\pi = 2\pi i + 2\pi i = 4\pi i.$$

Example 7. *Evaluate* $\oint_C \frac{3z^2 + z}{z^2 - 1}\, dz$*, where C is the circle* $|z - 1| = 1$.

(M.D.U. Dec. 2015)

Sol. The integrand has singularities, where $z^2 - 1 = 0$ *i.e.,* at $z = 1$ and $z = -1$.

The circle $|z - 1| = 1$ has centre at $z = 1$ and radius 1 and includes the point $z = 1$.

$f(z) = 3z^2 + z$ in an analytic function.

Also $$\frac{1}{z^2 - 1} = \frac{1}{(z-1)(z+1)} = \frac{1}{2}\left(\frac{1}{z-1} - \frac{1}{z+1}\right)$$

$$\therefore \quad \oint_C \frac{3z^2 + z}{z^2 - 1}\, dz = \frac{1}{2}\oint_C \frac{3z^2 + z}{z - 1}\, dz - \frac{1}{2}\oint_C \frac{3z^2 + z}{z + 1}\, dz \quad \text{...(1)}$$

By Cauchy's integral formula,

$$\oint_C \frac{3z^2 + z}{z - 1}\, dz = 2\pi i\, f(1), \quad \text{where } f(z) = 3z^2 + z$$
$$= 8\pi i$$

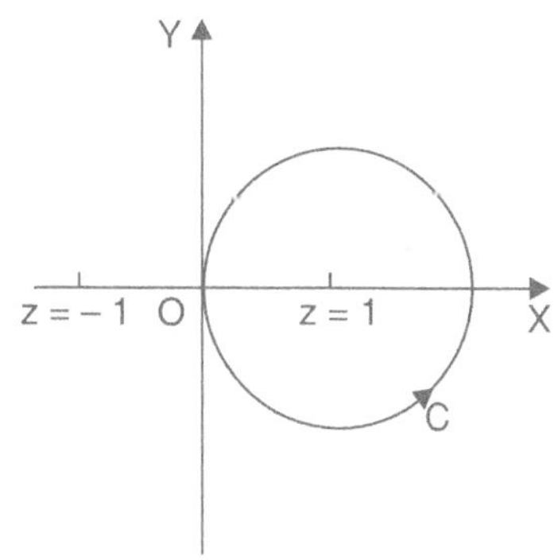

By Cauchy's integral theorem,

$$\oint_C \frac{3z^2 + z}{z + 1}\, dz = 0$$

$\therefore$ From (1), we have

$$\oint_C \frac{3z^2 + z}{z^2 - 1}\, dz = 4\pi i.$$

Example 8. *Use Cauchy's integral formula to evaluate*

$\oint_C \frac{e^{2z}}{(z+1)^4}\, dz$*, where C is the circle* $|z| = 2$. (U.K.T.U. 2011)

Sol. The integrand has a singularity at $z = -1$ which lies within the circle $|z| = 2$.

Now $$f^n(a) = \frac{n!}{2\pi i}\oint_C \frac{f(z)}{(z-a)^{n+1}}\, dz \quad \text{...(1)}$$

Here $a = -1$, $n + 1 = 4$ *i.e.,* $n = 3$, $f(z) = e^{2z}$

$$\therefore \quad f^n(z) = 2^n e^{2z},\; f'''(a) = f'''(-1) = 2^3 e^{-2}$$

From (1), we have $f'''(-1) = \frac{3!}{2\pi i}\oint_C \frac{e^{2z}}{(z+1)^4}\, dz \Rightarrow 8e^{-2} = \frac{3}{\pi i}\oint_C \frac{e^{2z}}{(z+1)^4}\, dz$

Hence $$\oint_C \frac{e^{2z}}{(z+1)^4}\, dz = \frac{8\pi i}{3e^2}.$$

Example 9. *If* $f(\xi) = \oint_C \frac{3z^2 + 7z + 1}{z - \xi} dz$, *where C is the circle* $x^2 + y^2 = 4$, *find the values of* $f(3)$, $f'(1 - i)$ *and* $f''(1 + i)$. (M.D.U. 2011)

Sol. The given circle C is $x^2 + y^2 = 4$ or $|z| = 2$.

The point $z = 3$ lies outside the circle $|z| = 2$ while $z = 1 - i$ and $z = 1 + i$ lie inside the circle.

Now, $f(3) = \oint_C \frac{3z^2 + 7z + 1}{z - 3} dz$ and $\frac{3z^2 + 7z + 1}{z - 3}$ is analytic within and on C.

$\therefore$ By Cauchy's integral theorem, we have

$$\oint_C \frac{3z^2 + 7z + 1}{z - 3} dz = 0 \Rightarrow f(3) = 0$$

Now let $\phi(z) = 3z^2 + 7z + 1$ which is analytic everywhere.

$\therefore$ By Cauchy's integral formula, we have

$$\phi(\xi) = \frac{1}{2\pi i} \oint_C \frac{\phi(z)}{z - \xi} dz, \quad \text{where } \xi \text{ is a point within C.}$$

$$\Rightarrow \quad 2\pi i \phi(\xi) = \oint_C \frac{3z^2 + 7z + 1}{z - \xi} dz = f(\xi)$$

$$\Rightarrow \quad f(\xi) = 2\pi i(3\xi^2 + 7\xi + 1) \quad \Rightarrow \quad f'(\xi) = 2\pi i(6\xi + 7)$$

$$\therefore \quad f'(1 - i) = 2\pi i\,[6(1 - i) + 7] = 2\pi i(13 - 6i) = 2\pi(6 + 13i)$$

Also $\quad f''(\xi) = 12\pi i$

$$\therefore \quad f''(1 + i) = 12\pi i.$$

EXERCISE 3.9

1. Verify Cauchy's theorem for the integral of z^3 taken over the boundary of the
   (*i*) rectangle with vertices $-1, 1, 1 + i, -1 + i$,
   (*ii*) triangle with vertices (1, 2), (1, 4), (3, 2).
2. Verify Cauchy's theorem for the function $f(z) = 3z^2 + iz - 4$ along the perimeter of square with vertices $1 \pm i$, $-1 \pm i$. (*G.B.T.U. 2011*)
3. (*a*) Evaluate $\oint_C \frac{dz}{(z - a)^n}$, $n = 2, 3, 4, ...$ where C is a closed curve containing the point $z = a$.
   (*b*) Evaluate $\oint_C \frac{z^2 - z + 1}{z - 1} dz$, where C is the circle $|z| = \frac{1}{2}$. (*M.D.U. 2011*)
4. Evaluate $\oint_C \frac{z^2 + 5}{z - 3} dz$, where C is the circle
   (*a*) $|z| = 4$ (*b*) $|z| = 1$.
5. Evaluate $\oint_C \frac{e^z}{z - 2} dz$, where C is the circle,
   (*a*) $|z| = 3$ (*b*) $|z| = 1$.

6. Evaluate $\oint_C \frac{3z^2 + 7z + 1}{z + 1}\, dz$, where C is the circle,

(a) $|z| = 1.5$ (b) $|z + i| = 1$ (c) $|z| = \frac{1}{2}$.

7. Evaluate $\oint_C \frac{\sin 3z}{z + \frac{\pi}{2}}\, dz$, where C is the circle $|z| = 5$.

8. Evaluate $\oint_C \frac{z^3 + z + 1}{z^2 - 3z + 2}\, dz$, where C is the ellipse $4x^2 + 9y^2 = 1$.

9. Evaluate $\oint_C \frac{\cos \pi z}{z^2 - 1}\, dz$ around a rectangle with vertices

(a) $2 \pm i, -2 \pm i$ (b) $-i, 2 - i, 2 + i, i$.

10. (i) Evaluate $\oint_C \frac{\sin^2 z}{\left(z - \frac{\pi}{6}\right)^3}\, dz$, where C is $|z| = 1$.

(ii) Evaluate $\oint_C \frac{\sin^6 z}{\left(z - \frac{\pi}{6}\right)^3}\, dz$, where C is the circle $|z| = 1$. (*M.D.U. 2013, May 2015*)

(iii) Evaluate $\oint_C \frac{z^3 - z}{(z - 2)^3}\, dz$, where C : $|z| = 3$

11. Evaluate $\oint_C \frac{z\, dz}{(z - 1)(z - 3)}$, where C is the circle

(a) $|z| = 3.5$ (b) $|z| = 3/2$.

12. Evaluate $\oint_C \frac{z + 4}{z^2 + 2z + 5}\, dz$, where C is the circle $|z| = 1$.

13. Evaluate (i) $\oint_C \frac{\sin \pi z^2 + \cos \pi z^2}{(z - 1)(z - 2)}\, dz$, where C is the circle $|z| = 3$.

(*M.D.U. Dec., 2014; G.B.T.U. 2010; P.T.U. May 2010*)

(ii) $\oint_C \frac{\sin \pi z^2 + \cos \pi z^2}{(z - 1)(z - 3)}\, dz$, where C is the circle $|z| = 2$ (*M.T.U. 2013*)

14. Evaluate $\oint_C \frac{e^z}{(z - 1)(z - 4)}\, dz$, where C is the circle $|z| = 2$.

15. Evaluate $\oint_C \frac{e^z}{z(z + 1)}\, dz$, where C is the circle $|z| = \frac{1}{4}$.

16. Evaluate $\oint_C \frac{\cos z}{z - \pi}\, dz$, where C is the circle $|z - 1| = 3$.

17. Let $P(z) = a + bz + cz^2$ and $\oint_C \frac{P(z)}{z}\, dz = \oint_C \frac{P(z)}{z^2}\, dz = \oint_C \frac{P(z)}{z^3}\, dz = 2\pi i$

where C is the circle $|z| = 1$. Evaluate P(z).

18. Evaluate $\oint_C \frac{e^{-z}}{z^2}\,dz$, where C is the circle $|z| = 1$.

19. (*a*) Evaluate $\oint_C \frac{e^{-2z}\,dz}{(z+1)^3}$, where C is the circle $|z| = 2$.

(*b*) Evaluate $\oint_C \frac{e^z}{(z-1)(z-2)^2}\,dz$, where C is the circle $|z| = 3$

20. If $f(\xi) = \oint_C \frac{4z^2 + z + 5}{z - \xi}\,dz$, where C is the ellipse $\frac{x^2}{4} + \frac{y^2}{9} = 1$, find $f(1)$, $f(i)$, $f'(-1)$ and $f''(-i)$.

21. Evaluate, using Cauchy's integral formula:

(*i*) $\int_C \frac{e^{2z}}{(z-1)(z-2)}\,dz$, where C is the circle $|z| = 3$. (*P.T.U. 2010*)

(*ii*) $\int_C \frac{z}{z^2 - 3z + 2}\,dz$, where C is $|z - 2| = \frac{1}{2}$.

(*iii*) $\int_C \frac{e^z\,dz}{(z+1)^2}$, where C is $|z - 1| = 3$.

(*iv*) $\int_C \frac{e^z}{(z^2 + \pi^2)^2}\,dz$, where C is $|z| = 4$. (*M.D.U. 2011*)

(*v*) $\int_C \frac{z^3 - 2z + 1}{(z - i)^2}\,dz$, where C is $|z| = 2$.

(*vi*) $\int_C \frac{\log z}{(z-1)^3}\,dz$, where C is $|z - 1| = \frac{1}{2}$. (*M.D.U. 2010*)

(*vii*) $\int_C \frac{\tan z}{z}\,dz$, where C is the circle $|z| = 1.5$.

(*viii*) $\oint_C \frac{e^{3z}}{(z - \ln 2)^4}\,dz$, where C is the square with vertices at $\pm 1, \pm i$.

(*ix*) $\int_C \frac{4 - 3z}{z(z-1)(z-2)}\,dz$, where C is the circle $|z| = 3/2$. (*M.D.U. 2012*)

Answers

3. (*a*) 0 (*b*) 0 **4.** (*a*) $28\pi i$ (*b*) 0 **5.** (*a*) $2\pi i\, e^2$ (*b*) 0

6. (*a*) $-6\pi i$ (*b*) 0 (*c*) 0 **7.** $2\pi i$ **8.** 0 **9.** (*a*) 0 (*b*) $-\pi i$

10. (*i*) πi (*ii*) $\frac{21}{16}\pi i$ (*iii*) $12\,\pi i$ **11.** (*a*) $2\pi i$ (*b*) $-\pi i$

12. 0 **13.** (*i*) $4\pi i$ (*ii*) πi **14.** $-\frac{2}{3}\,\pi i e$

15. $2\pi i$ **16.** $-2\pi i$ **17.** $1 + z + z^2$ **18.** $-2\pi i$

19. (*a*) $4\pi i e^2$ (*b*) $2\pi i e$ **20.** $20\pi i$, $2\pi(i-1)$, $-14\pi i$, $16\pi i$

21. (*i*) $2\pi i(e^4 - e^2)$ (*ii*) $4\pi i$ (*iii*) $\frac{2\pi i}{e}$ (*iv*) $\frac{i}{\pi}$

(*v*) $-10\pi i$ (*vi*) $-\pi i$ (*vii*) 0

(*viii*) $72\pi i$ [$\ln 2 = 0.6931$] (*ix*) $2\pi i$.

CONFORMAL MAPPING*

(For K.U.K. Only)

3.28. TRANSFORMATION OR MAPPING

We know that the real function $y = f(x)$ can be represented graphically by a curve in the xy-plane. Also, the real function $z = f(x, y)$ can be represented by a surface in three dimensional space. However, this method or graphical representation fails in the case of complex functions because a complex function $w = f(z)$ *i.e.*, $u + iv = f(x + iy)$ involves four real variables, two independent variables x, y and two dependent variables u, v. Thus a four dimensional region is required to represent it graphically in the cartesian fashion. As it is not possible, we choose, two complex planes and call them z-plane and w-plane. In the z-plane, we plot the point $z = x + iy$ and in the w-plane, we plot the corresponding point $w = u + iv$. Thus the function $w = f(z)$ defines a correspondence between points of these two planes. If the point z describes some curve C in the z-plane, the point w will move along a corresponding curve C in the w-plane, since to each (x, y) there corresponds a point (u, v). *The function $w = f(z)$ thus defines a mapping or transformation of the z-plane into the w-plane.*

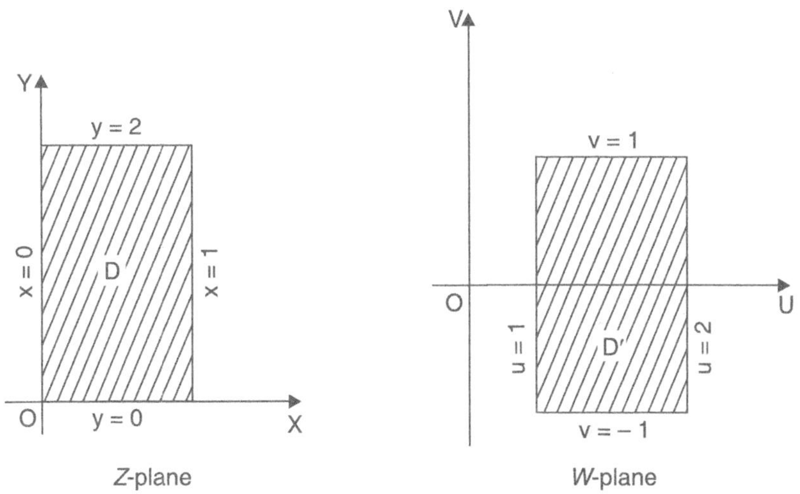

For example, consider the transformation $w = z + (1 - i)$. Let us determine the region D′ of the w-plane corresponding to the rectangular region D in the z-plane bounded by $x = 0$, $y = 0$, $x = 1$ and $y = 2$.

Since $w = z + (1 - i)$, we have

$$u + iv = (x + iy) + (1 - i) = (x + 1) + i(y - 1)$$

Thus $\qquad u = x + 1 \quad \text{and} \quad v = y - 1$

Hence the lines $x = 0$, $y = 0$, $x = 1$ and $y = 2$ in the z-plane are mapped onto the lines $u = 1$, $v = -1$, $u = 2$ and $v = 1$ in the w-plane. The regions D and D′ are shown shaded in the figure.

*This part is **not included in** the syllabus of **M.D.U.,** Rohtak.

3.29. CONFORMAL TRANSFORMATION OR GEOMETRICAL REPRESENTATION OF w = f(z) *(G.B.T.U. 2010)*

Suppose two curves C_1, C_2 in the z-plane intersect at the point P and the corresponding curves C_1', C_2' in the w-plane intersect at P′ under the transformation $w = f(z)$. *If the angle of intersection of the curves at P is the same as the angle of intersection of the curves at P′, both in magnitude and sense, then the transformation is said to be conformal at P.*

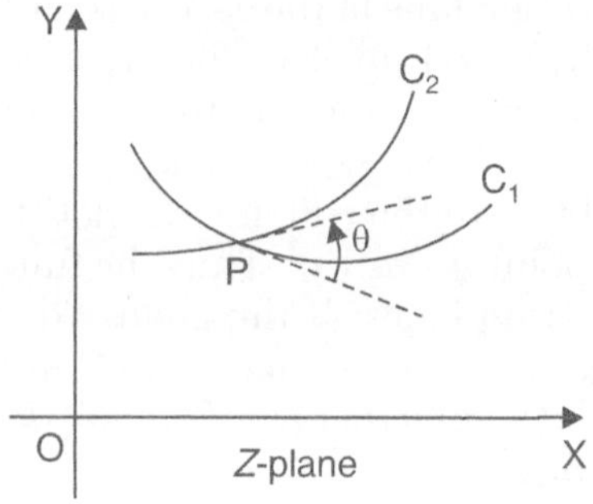

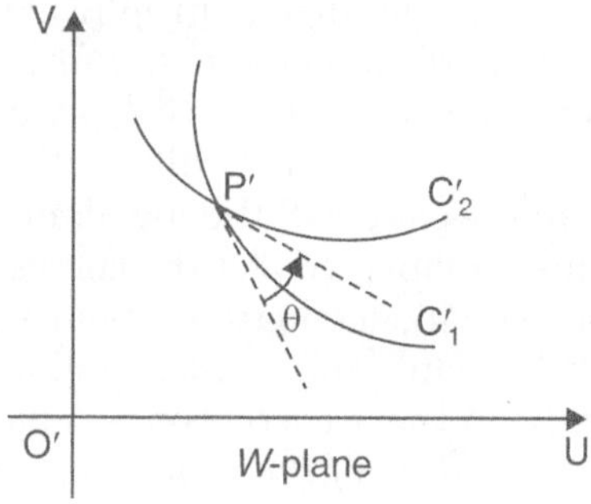

Definition. A transformation which preserves angles both in magnitude and sense between every pair of curves through a point is said to be conformal at the point.

The conditions under which the transformation $w = f(z)$ is conformal are given by the following theorem.

3.30. THEOREM *(G.B.T.U. 2010)*

If $f(z)$ is analytic and $f'(z) \neq 0$ in a region R of the z-plane, then the mapping $w = f(z)$ is conformal at all points of R.

Proof. Let P(z) be a point in the region R of the z-plane and P′(w) the corresponding point in the region R′ of the w-plane. Suppose P moves on a curve C and P′ moves on the corresponding curve C′. Let Q($z + \delta z$) be a neighbouring point on C and Q′ ($w + \delta w$) the corresponding point on C′ so that $\overrightarrow{\mathbf{PQ}} = \delta z$ and $\overrightarrow{\mathbf{P'Q'}} = \delta w$.

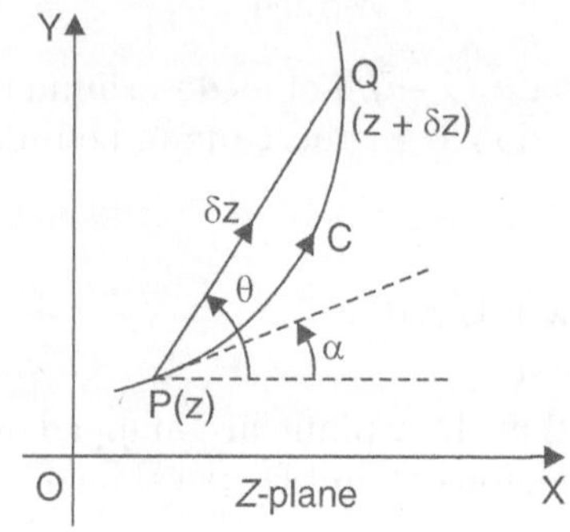

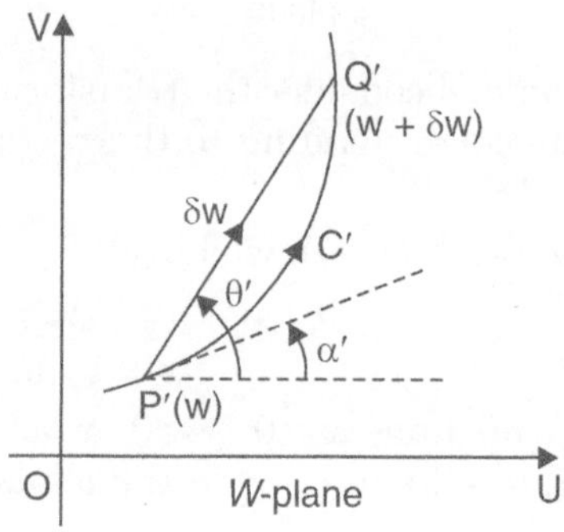

Then δz is a complex number whose modulus r is the length PQ and amplitude θ is the angle which PQ makes with the x-axis.

$\therefore \qquad \delta z = re^{i\theta}$

Similarly, $\delta w = r'e^{i\theta'}$ where r' is the modulus and θ' is the amplitude of δw.

$$\therefore \qquad \frac{\delta w}{\delta z} = \frac{r'}{r} e^{i(\theta'-\theta)}$$

Let the tangent to C at P make an angle α with x-axis and the tangent to C′ at P′ make an angle α' with u-axis, then as $\delta z \to 0$, $\theta \to \alpha$ and $\theta' \to \alpha'$.

$$\therefore \qquad f'(z) = \frac{dw}{dz} = \underset{\delta z \to 0}{\text{Lt}} \frac{\delta w}{\delta z} = \underset{\delta z \to 0}{\text{Lt}} \left(\frac{r'}{r}\right) e^{i(\theta'-\theta)} \qquad \ldots(1)$$

Since $f'(z) \neq 0$, let $f'(z) = \rho e^{i\phi}$, then $\rho = |\, f'(z)\,|$ and ϕ = amplitude of $f'(z)$.

$\therefore$ From (1), $$\rho e^{i\phi} = \underset{\delta z \to 0}{\text{Lt}} \left(\frac{r'}{r}\right) e^{i(\theta'-\theta)}$$

Thus $$\rho = \underset{\delta z \to 0}{\text{Lt}} \frac{r'}{r} \qquad \ldots(2)$$

and $$\phi = \underset{\delta z \to 0}{\text{Lt}} (\theta' - \theta) = \alpha' - \alpha \qquad \ldots(3)$$

Now let C_1 be another curve through P in the z-plane and C_1' the corresponding curve through P′ in the w-plane. If the tangent to C_1 at P makes an angle β with x-axis and the tangent to C_1' at P′ makes an angle β' with u-axis, then as in (3),

$$\phi = \beta' - \beta \qquad \ldots(4)$$

From (3) and (4), $\alpha' - \alpha = \beta' - \beta$ or $\beta - \alpha = \beta' - \alpha' = \gamma$

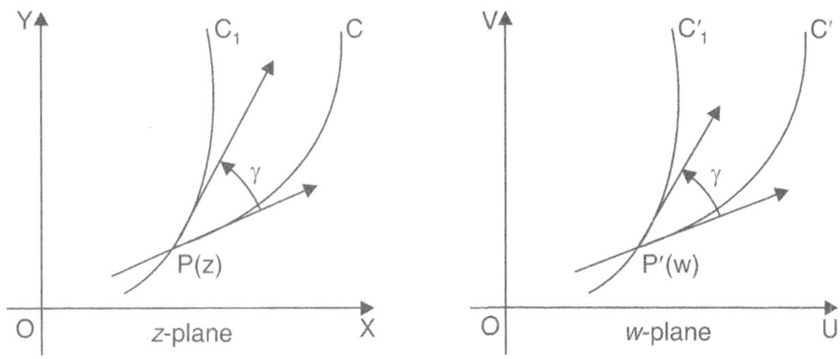

Thus angle between the curves before and after the mapping is preserved in magnitude and sense. Hence the mapping by the analytic function $w = f(z)$ is conformal at each point where $f'(z) \neq 0$.

Note 1. A point at which $f'(z) = 0$ is called a **critical point** of the transformation.

Note 2. From (2) $$\rho = \underset{\delta z \to 0}{\text{Lt}} \frac{r'}{r}$$

It follows that under the conformal transformation $w = f(z)$, the lengths of arcs through P are magnified in the ratio ρ: 1, where $\rho = |\, f'(z)\,|$. Thus an infinitesimal length in the z-plane is magnified by the factor $|\, f'(z)\,|$ in the w-plane and consequently infinitesimal areas in the z-plane are magnified by the factor $|\, f'(z)\,|$ in the ω-plane.

Note 3. From (3), $\alpha' = \alpha + \phi$ shows that the tangent to the curve C at P is rotated through an angle ϕ under the given transformation.

Note 4. A harmonic function remains harmonic under a conformal transformation.

3.31. COEFFICIENT OF MAGNIFICATION

Coefficient of magnification for the conformal transformation $w = f(z)$ at $z = \alpha + i\beta$ is given by $| f'(\alpha + i\beta) |$ where dash represents derivative.

3.32. ANGLE OF ROTATION

Angle of rotation for the conformal transformation $w = f(z)$ at $z = \alpha + i\beta$ is given by amp. $[f'(\alpha + i\beta)]$.

Example 1. *For the conformal transformation $w = z^2$, show that*

(*i*) *the coefficient of magnification at $z = 1 + i$ is $2\sqrt{2}$.*

(*ii*) *The angle of rotation at $z = 1 + i$ is $\frac{\pi}{4}$.*

Sol. Here $f(z) = w = z^2$

$\therefore$ $f'(z) = 2z$ and $f'(1 + i) = 2 + 2i$.

The coefficient of magnification at $z = 1 + i$ is $| f'(1 + i) | = \sqrt{4 + 4} = 2\sqrt{2}$.

The angle of rotation at $z = 1 + i$ is amp. $[f'(1 + i)] = \tan^{-1} \frac{2}{2} = \frac{\pi}{4}$.

3.33. SOME STANDARD TRANSFORMATIONS

1. Translation: w = z + c, where c is a complex constant

Let $z = x + iy$, $c = a + ib$ and $w = u + iv$

then the transformation becomes $u + iv = (x + iy) + (a + ib) = (x + a) + i(y + b)$

so that $u = x + a$ and $v = y + b$

Thus the transformation is a mere translation of the axes and preserves the shape and size.

For example, the rectangle OMPN in z-plane is transformed to rectangle O′M′P′N′ in the w-plane under the transformation $w = z + (1 + 2i)$.

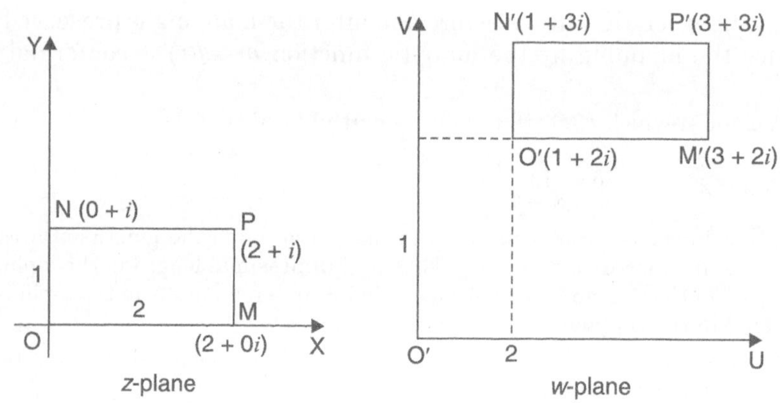

2. Rotation and Magnification: w = cz, where c is complex constant.

Let $c = \rho e^{i\alpha}$ $z = re^{i\theta}$, and $w = Re^{i\phi}$

then the transformation becomes $Re^{i\phi} = \rho\, re^{i(\theta + \alpha)}$

$\therefore$ $R = \rho r$ and $\phi = \theta + \alpha$

Thus the transformation maps a point (r, θ) in the z-plane into a point P' $(\rho r, \theta + \alpha)$ in the w-plane. Hence the transformation consists of **magnification of the radius vector of P** by $\rho = | c |$ and its **rotation through an** angle α = amp (c).

Thus under this transformation figure in w-plane is similar to the figure in z-plane (magnified by $| c |$) but rotated through an angle α.

Note 1. If $\alpha > 0$ then rotation is anticlock wise and if $\alpha < 0$ then rotation is clockwise.

Note 2. In $w = cz$; if C is real then $\alpha = 0$, then this transformation is only that of magnification (no rotation)

$\because$ in this case the two figures in z-plane and w-plane are similarly situated about their respective origins but figure in w-plane is $|c|$ times figure in z plane. Such mapping is called Magnification.

For example the transformation $w = (1 + i)\, z$ maps the square OMPN bounded by $x = 0$, $y = 0$, $x = 1$, $y = 1$ in z plane to the square OM′P′N′ w-plane.

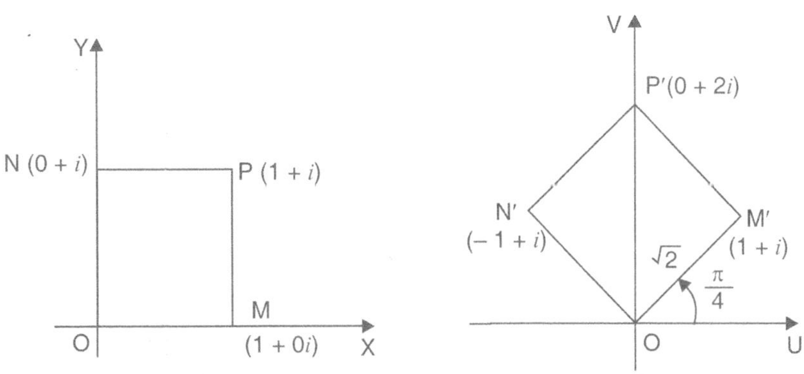

z-plane *w*-plane

Here $u + iv = (1 + i)(x + iy) = (x - y) + i\,(x + y)$

$\therefore$ $u = x - y,\ v = x + y$

$\therefore$ $x = 0$ maps into $u = -y$, $v = y$ *i.e.,* $v = -u$

$y = 0$ maps into $u = x$, $v = x$ *i.e.,* $v = u$

$x = 1$ maps into $u = 1 - y$, $v = 1 + y$ *i.e.,* $u + v = 2$

$y = 1$ maps into $u = x - 1$, $v = x + 1$ *i.e.,* $v - u = 2$

$\therefore$ Square in z plane is mapped into square in the w plane bounded by $u = v$, $u = -v$, $u + v = 2$, $v - u = 2$

Verification: Here $i = 1 + i$ $\therefore$ $| c | = \sqrt{2}$

Each side of the square in w-plane is $| c |$ times (*i.e.,* $\sqrt{2}$) the side of the square is z-plane also amp of $c = \tan^{-1}\dfrac{1}{1} = \dfrac{\pi}{4} =$ and the sides of the square are rotated through an angle $\dfrac{\pi}{4}$.

This transformation is rotation as well as magnification.

Consider another example in which c is real.

Example: *The transformation $w = 2z$ maps the triangular region OAB bound by the lines $x = 0$, $y = 0$, $x + y = 1$ into a similar triangle OA′B′ in w- plane.*

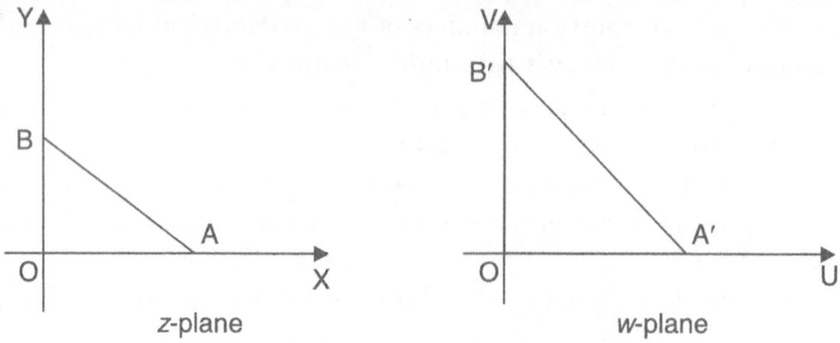

Here $w = 2z$ $\therefore$ $u + iv = 2(x + iy)$

$\therefore$ $u = 2x$ $v = 2y$

$x = 0$ maps into $u = 0$

$y = 0$ maps into $v = 0$

$$x + y = 1 \text{ maps into } \frac{u}{2} + \frac{x}{2} = 1 \quad i.e., \quad u + v = 2$$

Varifiction: The two figures are similar but figure in w-plane is 2 times the figure is z-plane.

This transformation is only Magnification

3. Inversion: $\mathbf{w = \frac{1}{z}}$

Let $z = re^{i\theta}$ and $w = \text{R}e^{i\phi}$

then the transformation becomes $\text{R}e^{i\phi} = \frac{1}{r}e^{-i\theta}$ so that $\text{R} = \frac{1}{r}$ and $\phi = -\theta$.

Thus under the transformation $w = \frac{1}{z}$, a point $\text{P}(r, \theta)$ in z-plane is mapped into the point $\text{P}\left(\frac{1}{r}, -\theta\right)$.

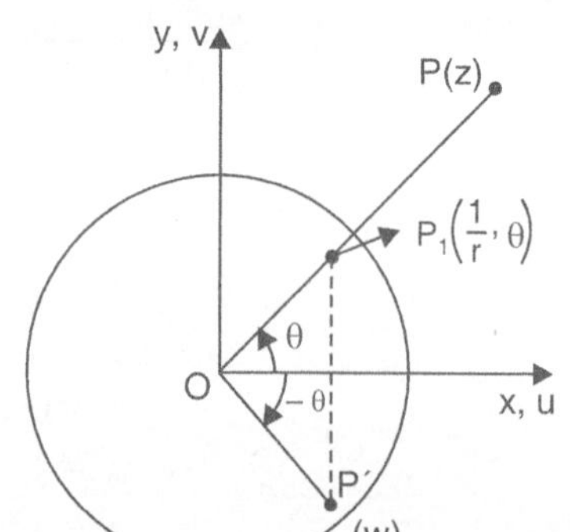

Consider the w-plane superposed on the z-plane. If P is (r, θ) and P_1 is $\left(\frac{1}{r}, \theta\right)$, then

$\text{OP}_1 = \frac{1}{r} = \frac{1}{\text{OP}}$ *i.e.,* $\text{OP} \cdot \text{OP}_1 = 1$ so that P_1 is reverse of P w.r.t. the unit circle with centre O.

[*The inverse of a point P w.r.t. a circle having centre O and radius k is defined as the point Q on OP such that OP, OQ = k^2*]

The reflection P′ of P_1 in the real axis represents $w = \frac{1}{z}$. Thus, the transformation $w = \frac{1}{z}$ is an **inversion of z w.r.t. the unit circle $|z| = 1$ followed by reflection of the inverse into the real axis.**

Obviously, the transformation $w = \frac{1}{z}$ **maps the interior of the unit circle $|z| = 1$ into the exterior of the unit circle $|w| = 1$ and the exterior of $|z| = 1$ into the interior of $|w| = 1$.**

However, the origin $z = 0$ is mapped to the point $w = \infty$, called the *point at infinity*.

Note. *This transformation $w = \frac{1}{z}$ maps a circle in z-plane to a circle in w-plane or to a straight line if the circle in z-plane passes through the origin.*

The general equation of any circle in the z-plane is

$$x^2 + y^2 + 2gx + 2fy + c = 0 \qquad \ldots(1)$$

Let $w = u + iv = \frac{1}{z}$, then $z = \frac{1}{w}$ or $x + iy = \frac{1}{u+iv} = \frac{u-iv}{u^2+v^2}$

so that
$$x = \frac{u}{u^2+v^2} \quad \text{and} \quad y = \frac{-v}{u^2+v^2}$$

Substituting the values of x and y in (1), we get

$$\frac{u^2}{(u^2+v^2)^2} + \frac{v^2}{(u^2+v^2)^2} + \frac{2gu}{u^2+v^2} - \frac{2fv}{u^2+v^2} + c = 0$$

or
$$\frac{u^2+v^2}{(u^2+v^2)^2} + \frac{2gu}{u^2+v^2} - \frac{2fv}{u^2+v^2} + c = 0$$

or
$$\frac{1}{u^2+v^2} + \frac{2gu}{u^2+v^2} - \frac{2fv}{u^2+v^2} + c = 0$$

or
$$c(u^2+v^2) + 2gu - 2fv + 1 = 0 \qquad \ldots(2)$$

If $c \neq 0$, the circle (1) does not pass through the origin and equation (2) represents a circle in the w-plane.

If $c = 0$, the circle (1) passes through the origin and equation (2) reduces to $2gu - 2fv + 1 = 0$ which is a straight line in the w-plane.

Regarding a straight line as a circle of infinite radius, we can say that the transformation $w = \frac{1}{z}$ maps circles into circles.

4. Bilinear Transformation:

A transformation of the form
$$w = \frac{az+b}{cz+d} \qquad \ldots(1)$$
where a, b, c, d are complex constants and $ad - bc \neq 0$ is called a **bilinear** or **Mobius transformation.**

The transformation given by (1) is conformal, since

$$\frac{dw}{dz} = \frac{ad - bc}{(cz + d)^2} \neq 0$$

The inverse mapping of (1) is $z = \dfrac{-dw + b}{cw - a}$ ...(2)

which is also a bilinear transformation.

The transformation (1) can be written as

$$cwz + wd - az - b = 0$$

which is linear both in w and z and hence the name bilinear transformation.

From (1), we observe that each point in the z-plane except the point $z = -\dfrac{d}{c}$ maps into a unique point in the w-plane. Similarly, from (2), we observe that each point in the w-plane except the point $w = \dfrac{a}{c}$ maps into a unique point in the z-plane. Considering the two exceptional points as points at infinity in respective planes, we can say that there is one to one correspondence between all points in the two plane.

Every bilinear transformation $w = \dfrac{az + b}{cz + d}$, $ad - bc \neq 0$ is the combination of basic transformations

(*i*) translation: $w = z + c$ (*ii*) rotation and magnification: $w = cz$

(*iii*) inversion: $w = \dfrac{1}{z}$

By actual division, we have $w = \dfrac{a}{c} + \dfrac{bc - ad}{c^2} \cdot \dfrac{1}{z + \dfrac{d}{c}}$

Taking $w_1 = z + \dfrac{d}{c}$, $w_2 = \dfrac{1}{w_1}$, $w_3 = \dfrac{bc - ad}{c^2} w_2$, we get $w = \dfrac{a}{c} + w_3$

Thus, by these transformations, we successively pass from z-plane to w_1-plane, from w_1-plane to w_2-plane, from w_2-plane to w_3-plane and finally from w_3-plane to w-plane.

Since each of these auxiliary transformations maps circles into circles, hence a bilinear transformation also maps circles into circles.

Note 1. Cross Section: If four complex numbers z_1, z_2, z_3, z_4, are taken in order then $\dfrac{(z_1 - z_2)(z_3 - z_4)}{(z_2 - z_3)(z_4 - z_1)}$ is called the cross-section of z_1, z_2, z_3, z_4.

Note 2. *The cross-ratio is invariant under a bilinear transformation.*

Thus if w_1, w_2, w_3 and w_4 are the respective images of four distinct points z_1, z_2, z_3 and z_4, then

$$\frac{(w_1 - w_3)(w_2 - w_4)}{(w_1 - w_4)(w_2 - w_3)} = \frac{(z_1 - z_3)(z_2 - z_4)}{(z_1 - z_4)(z_2 - z_3)}.$$

Note 3. In the bilinear transformation $w = \dfrac{az + b}{cz + d}$, $ad - bc \neq 0$ dividing the numerator and denominator of the right hand side by one of the four constants, we observe that there are only three independent constants. Hence *three independent conditions are required to determine a bilinear transformation.*

ILLUSTRATIVE EXAMPLES

Example 1. *What is the region of w-plane into which the rectangular region is the z-plane bounded by the lines $x = 0$, $y = 0$, $x = 1$, $y = 2$ is mapped under the transformation $w = z + (2 - i)$?*

Sol. The given transformation is

$$w = z + (2 - i) \qquad i.e., \; u + iv = x + iy + 2 - i$$

$$\therefore \qquad u = x + 2 \qquad\qquad v = y - 1$$

$x = 0$ maps into $u = 2$

$y = 0$ maps into $v = -1$

$x = 1$ maps into $u = 3$

$y = 2$ maps into $v = 1$

So the mapped region PQRS is also a rectangle bounded by $u = 2$, $v = -1$, $u = 3$, $v = 1$ (shown in the figure below)

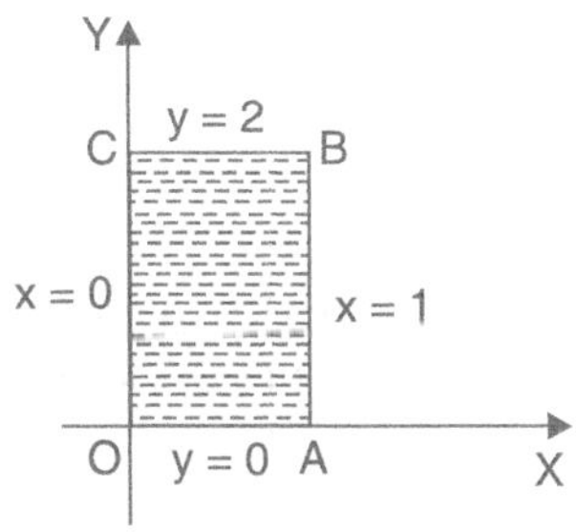

z-plane

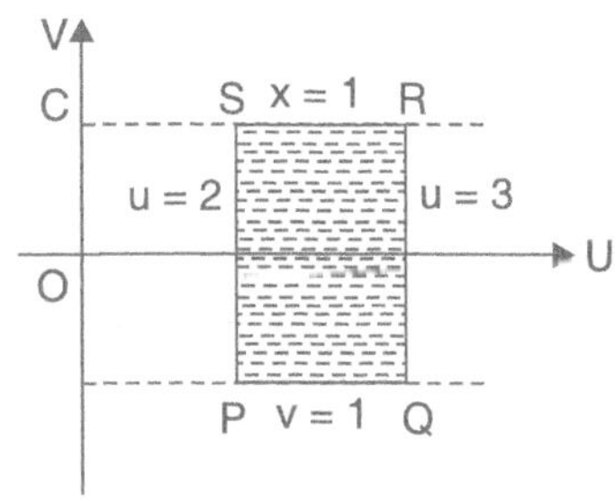

w-plane

Example 2. *Consider the transformation $w = e^{\frac{i\pi}{4}}$ and determine the region in w-plane corresponding to the triangular region bounded by the lines $x = 0$, $y = 0$ and $x + y = 1$ in the z-plane.*

Sol. The given transformation is

$$w = e^{\frac{i\pi}{4}} z = \left(\cos\frac{\pi}{4} + i\sin\frac{\pi}{4}\right)(x + iy)$$

i.e.,
$$u + iv = \frac{1}{\sqrt{2}}(1 + i)(x + iy) = \frac{1}{\sqrt{2}}[(x - y) + i(x + y)]$$

$$\therefore \qquad u = \frac{1}{\sqrt{2}}(x - y); \quad v = \frac{1}{\sqrt{2}}(x + y)$$

$$x = 0 \text{ maps into } \; u = \frac{1}{\sqrt{2}}(-y), v = \frac{1}{\sqrt{2}}y; \qquad \therefore \quad v = -u$$

$$y = 0 \text{ maps into } \; u = \frac{1}{\sqrt{2}}x, v = \frac{1}{\sqrt{2}}x \qquad \therefore \quad v = u$$

$$x + y = 1 \text{ maps into } v = \frac{1}{\sqrt{2}}.1 \qquad \therefore \quad v = \frac{1}{\sqrt{2}}$$

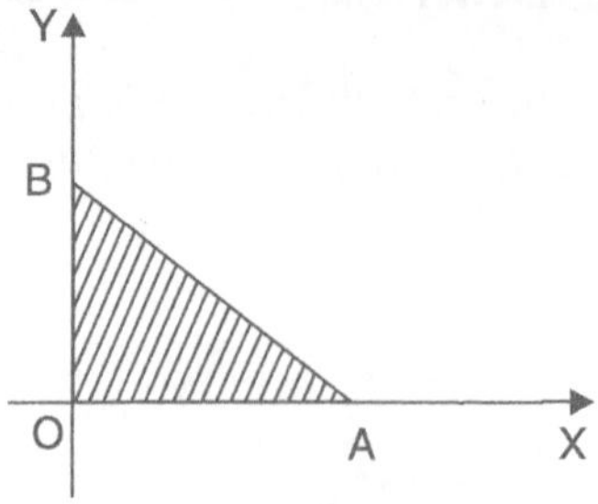

z-plane

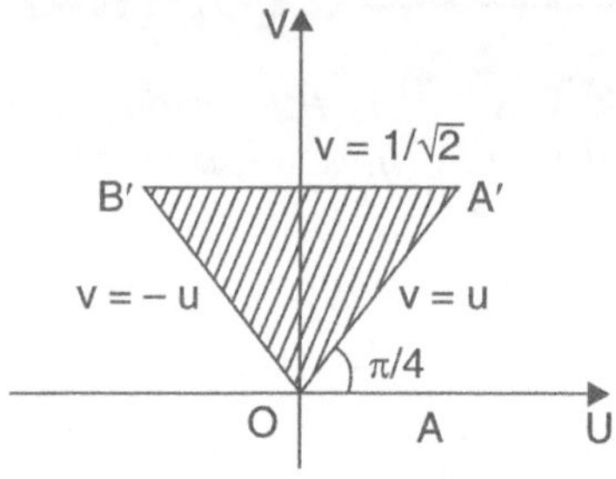

w-plane

$\therefore$ The transformation $w = e^{\frac{i\pi}{4}} z$ transforms the triangle OAB in z-plane into a triangle OA′B′ in w-plane rotated through an angle $\frac{\pi}{4}$.

Example 3. *Under the transformation* $w = \frac{1}{z}$, *find the image of the following curves*

(i) $| z - 2i | = 2$. (K.U.K. Jan. 2013)

(ii) $y - x + 1 = 0$.

Sol. The given transformation is

$$w = \frac{1}{z} \quad \text{or} \quad z = \frac{1}{w} \quad \text{or} \quad x + iy = \frac{1}{u + iv} = \frac{u - iv}{u^2 + v^2}$$

so that
$$x = \frac{u}{u^2 + v^2} \quad \text{and} \quad y = -\frac{v}{u^2 + v^2} \qquad \ldots(1)$$

(*i*) The given curve is $| z - 2i | = 2$ or $| x + i(y - 2) | = 2$

or
$$x^2 + (y - 2)^2 = 2 \quad \text{or} \quad x^2 + y^2 - 4y = 0 \qquad \ldots(2)$$

which is a circle in the z-plane with centre (0, 2) and radius 2.

Substituting the values of x and y from (1) in (2), we get

$$\frac{u^2}{(u^2 + v^2)^2} + \frac{v^2}{(u^2 + v^2)^2} + \frac{4v}{u^2 + v^2} = 0 \quad \text{or} \quad \frac{u^2 + v^2}{(u^2 + v^2)^2} + \frac{4v}{u^2 + v^2} = 0$$

or
$$\frac{1}{u^2 + v^2} + \frac{4v}{u^2 + v^2} = 0$$

or $1 + 4v = 0$, a straight line which is the required image of the given curve.

(*ii*) $y - x + 1 = 0$ maps into

$$-\frac{v}{u^2 + v^2} - \frac{u}{u^2 + v^2} + 1 = 0 \text{ i.e., } -u - v + u^2 + v^2 = 0$$

or $u^2 + v^2 - u - v = 0$ which is a circle with centre at $\left(\frac{1}{2}, \frac{1}{2}\right)$ and radius $\frac{1}{\sqrt{2}}$.

Example 4. *Find the image of infinite strip* $\frac{1}{4} \le y \le \frac{1}{2}$ *under the transformation* $w = \frac{1}{z}$. *Also show the regions graphically.*

Sol. The given transformation is

$$w = \frac{1}{z} \quad \text{or} \quad z = \frac{1}{w} \quad \text{or} \quad x + iy = \frac{1}{u+iv} = \frac{u-iv}{u^2+v^2}$$

so that
$$x = \frac{u}{u^2+v^2} \quad \text{and} \quad y = \frac{-v}{u^2+v^2}$$

If
$$y = \frac{1}{4}, \text{ then } \frac{-v}{u^2+v^2} = \frac{1}{4}$$

or
$$u^2 + v^2 + 4v = 0 \quad \text{or} \quad u^2 + (v+2)^2 = 4$$

If $y = \frac{1}{2}$, then $\frac{-v}{u^2+v^2} = \frac{1}{2}$ or $u^2 + v^2 + 2v = 0$ or $u^2 + (v+1)^2 = 1$

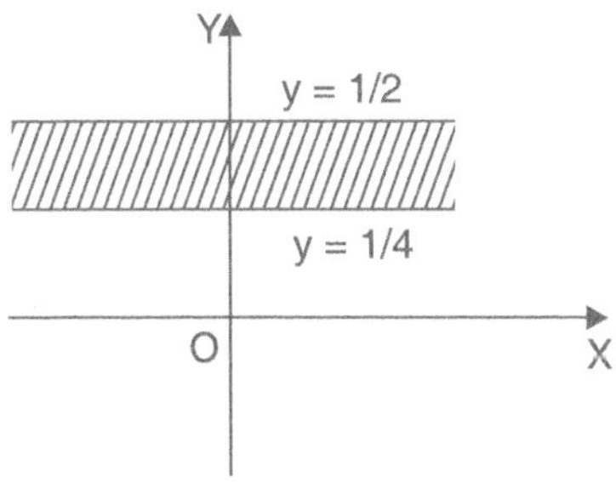

Z-plane

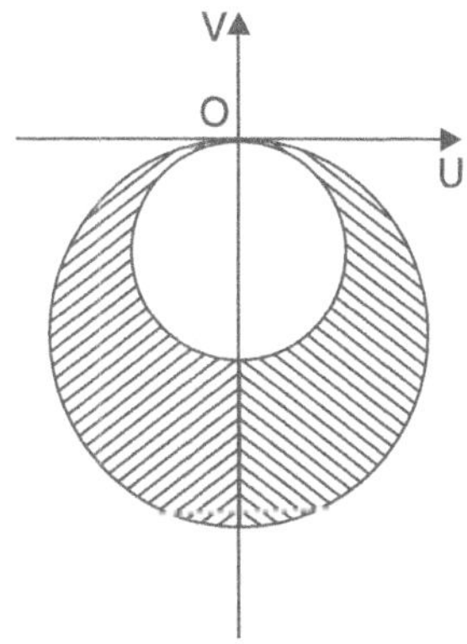

W-plane

Hence the infinite strip $\frac{1}{4} \le y \le \frac{1}{2}$ is transformed into the region between the two circles

$u^2 + (v+2)^2 = 4$, centre $(0, -2)$, radius 2

and $u^2 + (v+1)^2 = 1$, centre $(0, -1)$, radius 1.

Example 5. *Show that the image of the hyperbola $x^2 - y^2 = 1$ under the transformation $w = \frac{1}{z}$ is the Lemniscate $\rho^2 = \cos 2\phi$.* (K.U.K. Dec. 2015)

Sol. Given transformation is $w = \frac{1}{z}$

$$w = \rho e^{i\phi} \qquad \text{Let } z = re^{i\theta}$$

$$\rho e^{i\phi} = \frac{1}{re^{i\theta}} = \frac{1}{r} e^{-i\theta} \qquad \therefore \quad \rho = \frac{1}{r}, \phi = -\theta$$

Equation of the hyperbola is

$$x^2 - y^2 = 1 \quad i.e., \quad r^2 \cos^2 \theta - r^2 \sin^2 \theta = 1$$

or
$$r^2 \cos 2\theta = 1$$

or
$$\frac{1}{\rho^2} \cos 2(-\phi) = 1 \quad \text{or} \quad \rho^2 = \cos 2\phi.$$

Example 6. *Show that the transformation $w = \frac{2z+3}{z-4}$ maps the circle $x^2 + y^2 - 4x = 0$ onto the straight line $4u + 3 = 0$.*

Sol. The given transformation is $w = \dfrac{2z+3}{z-4}$

The inverse transformation is $z = \dfrac{4w+3}{w-2}$ ...(1)

Now the equation $x^2 + y^2 - 4x = 0$ can be written as $z\bar{z} - 2(z + \bar{z}) = 0$

Substituting for z and $\bar{z}$ from (1), we get

$$\frac{4w+3}{w-2} \cdot \frac{4\bar{w}+3}{\bar{w}-2} - 2\left(\frac{4w+3}{w-2} + \frac{4\bar{w}+3}{\bar{w}-2}\right) = 0$$

or $16w\bar{w} + 12w + 12\bar{w} + 9 - 2\,(4w\bar{w} + 3\bar{w} - 8w - 6 + 4w\bar{w} + 3w - 8\bar{w} - 6) = 0$

or $22(w + \bar{w}) + 33 = 0$ or $22(2u) + 33 = 0$ or $4u + 3 = 0.$

Example 7. *Show that* $w = \dfrac{i-z}{i+z}$ *maps the real axis of the z-plane into the circle* $|w| = 1$ *and the half plane* $y > 0$ *into the interior of the unit circle* $|w| = 1$ *in the w-plane.*

Sol. Given transformation is $w = \dfrac{i-z}{i+z}$

$$|w| = 1 \Rightarrow \left|\frac{i-z}{i+z}\right| = 1 \quad \text{or} \quad |i - z| = |i + z|$$

or $|i - x - iy| = |i + x + iy|$ or $|-x + i(1 - y)| = |x + i\,(1 + y)|$

or $\sqrt{x^2 + (1-y)^2} = \sqrt{x^2 + (1+y)^2}$ or $(1 - y)^2 = (1 + y)^2$

or $4y = 0$ or $y = 0$

which is the equation of real axis in z-plane

Hence the real axis of z-plane is mapped into $|w| = 1$

Now the interior of $|w| = 1$ means

$$|w| < 1 \quad i.e., \quad \left|\frac{i-z}{i+z}\right| < 1 \quad i.e., \quad |i - z| < |i + z|$$

or $x^2 + (1 - y)^2 < x^2 + (1 + y)^2$ or $-4y < 0$

or $y > 0$

Hence the half plane $y > 0$ is mapped into the interior of the circle $|w| = 1$.

Example 8. *Show that the transformation* $w = z^2$ *maps the circle* $|z - 1| = 1$ *into the cardioid* $\rho = 2(1 + \cos\phi)$, *where* $w = \rho e^{i\phi}$ *in the w-plane.*

Sol. Let $z = re^{i\theta}$, then $w = z^2 \Rightarrow \rho e^{i\phi} = r^2 e^{2i\theta}$

so that $\rho = r^2$ and $\phi = 2\theta$ ...(1)

Now the equation of circle in the z-plane is

$|z - 1| = 1$ or $|x + iy - 1| = 1$

or $|(x - 1) + iy| = 1$ or $(x - 1)^2 + y^2 = 1$

or $x^2 + y^2 - 2x = 0$ or $r^2 - 2r\cos\theta = 0$ $[\because\ z = x + iy = re^{i\theta} = r(\cos\theta + i\sin\theta)]$

or $r = 2\cos\theta$ or $r^2 = 4\cos^2\theta$

or $r^2 = 2(1 + \cos 2\theta)$ or $\rho = 2(1 + \cos\phi)$ [Using (1)]

$\therefore$ The circle $|z - 1| = 1$ in z-plane transforms into the cardioid $\rho = 2(1 + \cos\phi)$ in w-plane.

Example 9. *Determine the region of the w-plane into which the first quadrant of z-plane is mapped by the transformation $w = z^2$.*

Sol. Proceeding as in example 5, $\phi = 2\theta$

For the first quadrant in z-plane, $0 < \theta < \frac{\pi}{2} \quad \Rightarrow \quad 0 < \phi < \pi$

Hence the first quadrant of z-plane is mapped into the upper half of w-plane.

Example 10. *Determine the region of the w-plane into which the region $\frac{1}{2} \leq x \leq 1$ and $\frac{1}{2} \leq y \leq 1$ is mapped by the transformation $w = z^2$.*

Sol. The given transformation is $\quad w = z^2 \quad$ or $\quad u + iv = (x + iy)^2 = (x^2 - y^2) + 2ixy$

so that $\quad u = x^2 - y^2 \quad$ and $\quad v = 2xy$

When $x = \frac{1}{4}$, we have $\quad u = \frac{1}{4} - y^2 \quad$ and $\quad v = y \quad$ so that $v^2 = -\left(u - \frac{1}{4}\right)$

[a left-handed parabola with vertex $(\frac{1}{4}, 0)$ and latus rectum 1]

When $x = 1$, we have $u = 1 - y^2 \quad$ and $\quad v = 2y \quad$ so that $v^2 = -4(u - 1)$

[a left-handed parabola with vertex (1, 0) and latus rectum 4]

When $y = \frac{1}{2}$, we have $u = x^2 - \frac{1}{4}$ and $v = x \quad$ so that $v^2 = u + \frac{1}{4}$

[a right-handed parabola with vertex $(-\frac{1}{4}, 0)$ and latus rectum 1]

When $y = 1$, we have $u = x^2 - 1$ and $v = 2x \quad$ so that $v^2 = 4(u + 1)$

[a right-handed parabola with vertex (– 1, 0) and latus rectum 4]

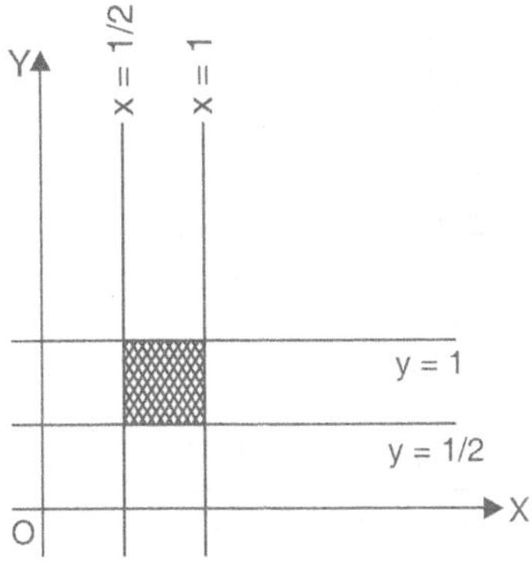

Z-plane

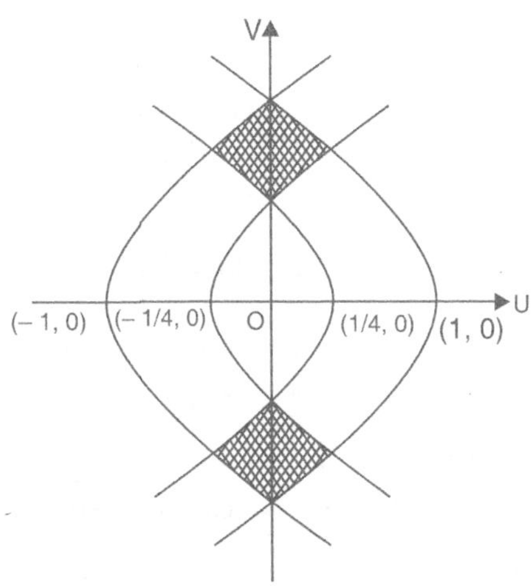

W-plane

Thus the rectangular region bounded by the lines $x = \frac{1}{2}$, $x = 1$ and $y = \frac{1}{2}$, $y = 1$ maps into the region bounded by the parabolas

$$v^2 = (u - \frac{1}{4}),\ v^2 = -4(u - 1) \quad \text{and} \quad v^2 = u + \frac{1}{4},\ v^2 = 4(u + 1).$$

Example 11. *Show that the transformation $w = z + \frac{1}{z}$ maps the circle $|z| = c$ into the ellipse $u = \left(c + \frac{1}{c}\right)\cos\theta$, $v = \left(c - \frac{1}{c}\right)\sin\theta$. Discuss the case when $c = 1$.*

Sol. A point on the circle $|z| = c$ can be written as $z = ce^{i\theta}$

$$\therefore \qquad w = z + \frac{1}{z} \text{ becomes } u + iv = ce^{i\theta} + \frac{1}{c}e^{-i\theta}$$

or
$$u + iv = c(\cos\theta + i\sin\theta) + \frac{1}{c}(\cos\theta - i\sin\theta) = \left(c + \frac{1}{c}\right)\cos\theta + i\left(c - \frac{1}{c}\right)\sin\theta$$

$$\therefore \qquad u = \left(c + \frac{1}{c}\right)\cos\theta \quad \text{and} \quad v = \left(c - \frac{1}{c}\right)\sin\theta$$

which are the parametric equations of an ellipse.

When $c = 1$, we have $u = 2\cos\theta$ and $v = 0$

Since $|\cos\theta| \le 1$, we get $-2 \le u \le 2$ and $v = 0$

$\therefore$ The transformation gives a segment of the u-axis of length 4.

Example 12. *Show that the transformation $w = z + \frac{1}{z}$ converts the straight line $\arg z = \alpha$ $\left(|\alpha| < \frac{\pi}{2}\right)$ into a branch of the hyperbola of eccentricity $\sec\alpha$.*

Sol. Let $z = re^{i\theta}$ and $w = u + iv$

$$\therefore \qquad u + iv = re^{i\theta} + \frac{1}{r}e^{-i\theta} = r(\cos\theta + i\sin\theta) + \frac{1}{r}(\cos\theta - i\sin\theta)$$

$$= \left(r + \frac{1}{r}\right)\cos\theta + i\left(r - \frac{1}{r}\right)\sin\theta$$

$$u = \left(r + \frac{1}{r}\right)\cos\theta,\ v = \left(r - \frac{1}{r}\right)\sin\theta$$

$$\arg z = \alpha \Rightarrow \theta = \alpha$$

$$\therefore \qquad u = \left(r + \frac{1}{r}\right)\cos\alpha,\ v = \left(r - \frac{1}{r}\right)\sin\alpha$$

Eliminate r;
$$r + \frac{1}{r} = \frac{u}{\cos\alpha},\ r - \frac{1}{r} = \frac{v}{\sin\alpha}$$

$$\left(r + \frac{1}{r}\right)^2 - \left(r - \frac{1}{r}\right)^2 = \frac{u^2}{\cos^2\alpha} - \frac{v^2}{\sin^2\alpha},\quad 4 = \frac{u^2}{\cos^2\alpha} - \frac{v^2}{\sin^2\alpha}$$

or
$$\frac{u^2}{4\cos^2\alpha} - \frac{v^2}{4\sin^2\alpha} = 1 \text{ where } |\alpha| < \frac{\pi}{2}$$
$\therefore$ this equation represents a branch of hyperbola in w-plane

Here $\qquad a^2 = 4\cos^2\alpha,\ b^2 = 4\sin^2\alpha$

Eccentricity e of the hyperbola is given by

$$b^2 = a^2 (e^2 - 1)$$

$$4 \sin^2 \alpha = 4 \cos^2 \alpha (e^2 - 1) \quad \therefore \quad e^2 - 1 = \tan^2 \alpha$$

$$\therefore \quad e^2 = 1 + \tan^2 \alpha = \sec^2 \alpha$$

$$\therefore \quad e = \sec \alpha.$$

$\therefore$ Eccentricity of the hyperbola = sec α

Example 13. *Discuss the transformation $w = e^z$ and show that it transforms the region between the real axis and a line parallel to real axis at $y = \pi$ into the upper half of the w-plane.*

Sol. Let $w = Re^{i\phi}$, then the given transformation becomes

$$Re^{i\phi} = e^{x+iy} = e^x . e^{iy} \text{ so that } \quad R = e^x \text{ and } \phi = y$$

The real axis *i.e.*, $y = 0$ maps into the positive u-axis $\phi = 0$ in the w-plane.

The line $y = \pi$ maps into the negative u-axis $\phi = \pi$ in the w-plane.

Thus the region between the lines $y = 0$ and $y = \pi$ maps into the upper half of w-plane.

Note. The region between the lines $y = 0$ and $y = -\pi$ maps into the lower half of w-plane.

The region between the lines $y = c$ and $y = c + 2\pi$ maps into the whole of the w-plane, since e^z is periodic with period $2\pi i$.

The imaginary axis $x = 0$ maps into a unit circle $R = e^0 = 1$ in the w-plane.

Example 14. *Show that the transformation $w = e^z$ is always conformal. Under the mapping, find the images of the regions.*

(*i*) *the line segment $0 < y < A$, $A < 2\pi$, $x < 0$*

(*ii*) *the rectangle bounded by the lines $x = 0$, $y = 0$, $x = 1$ and $y = \pi$*

(*iii*) *The rectangular region bounded by the lines* $a \le x \le b ; c \le y \le d$.

Sol.
$$w = f(z) = e^z$$

$$f'(w) = e^z \neq 0 \text{ for any } z$$

$\therefore$ The transformation is conformable in the region of w *i.e.*, for all values of z

Let $\quad z = x + iy \quad$ and $\quad w\text{-}Re^{i\phi}$

$$\therefore \quad Re^{i\phi} = e^{x+iy} = e^x (e^{iy})$$

$$\therefore \quad R = e^x, \phi = y$$

(*i*) For $0 < y < A$, $A < 2\pi$, $x < 0$ we have $0 < \phi < A$ *i.e.*, $0 < \arg w < A < 2\pi$

and $x < 0 \Rightarrow R < 1 \Rightarrow 0 < |w| < 1$

The image curves are given in Fig. (*i*).

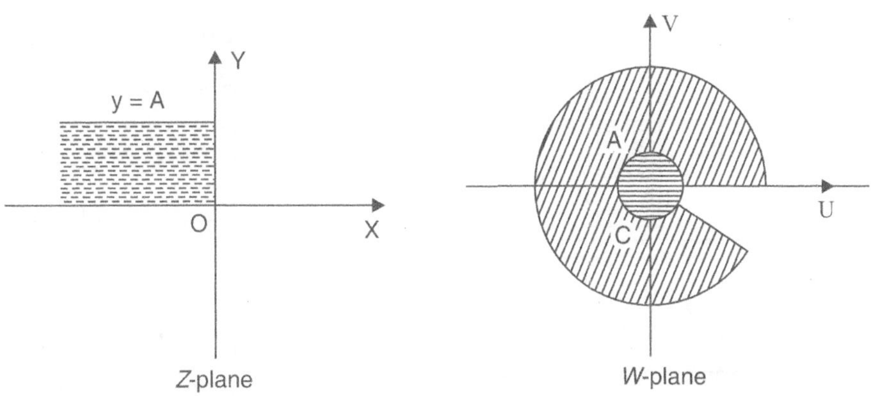

Fig. (*i*)

Image is the interior of the portion of the circle | R | = 1 whose angle of rotation is A < 2π.

(*ii*) The rectangle formed by $x = 0$, $y = 0$, $x = 1$ and $y = \pi$

$$x = 0 \quad \Rightarrow \quad R = 1$$
$$y = 0 \quad \Rightarrow \quad \phi = 0$$
$$x = 1 \quad \Rightarrow \quad R = e$$
$$y = \pi \quad \Rightarrow \quad \phi = \pi$$

If we take $w = u + iv$, then $u + iv = e^{x+iy} = e^x (\cos y + i \sin y)$

$\therefore \qquad u = e^x \cos y,\ v = e^x \sin y$

$$x = 0, y = 0 \quad \Rightarrow \quad u = 1, v = 0$$
$$x = 1, y = 0 \quad \Rightarrow \quad u = e, v = 0$$
$$x = 1, y = \pi \quad \Rightarrow \quad u = -e, v = 0$$
$$x = 0, y = \pi \quad \Rightarrow \quad u = -1, v = 0.$$

The image curves are given in Fig. (*ii*),

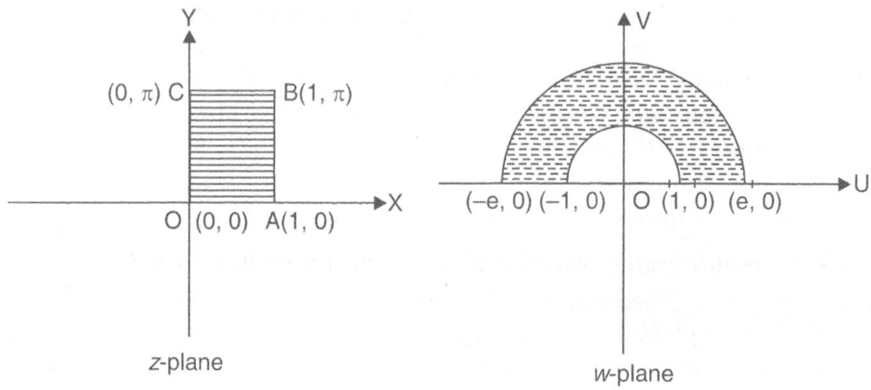

Fig. (*ii*)

Image is the region included between two semicircles |R| = 1 and |R| = e

(*iii*) The rectangle bounded by the lines $a \le x \le b$, $c \le y \le d$

$$a \le x \le b \quad \Rightarrow \quad e^a \le e^x \le e^b \Rightarrow e^a < R < e^b$$
$$c \le y \le d \quad \Rightarrow \quad c \le \phi \le d$$

∴ The image is shown in Fig. (*iii*)

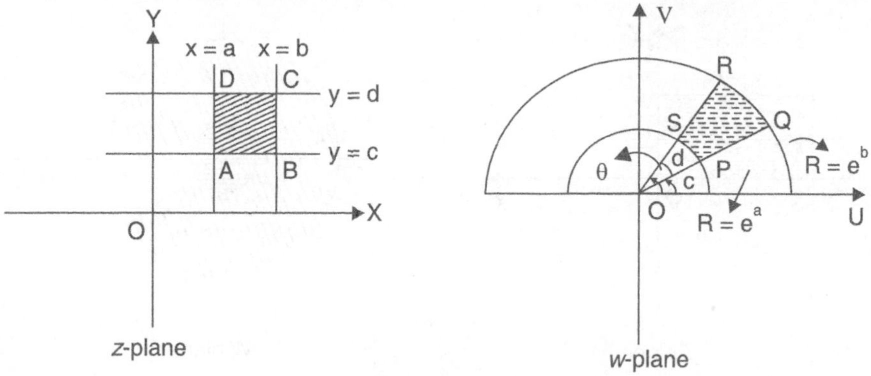

Fig. (*iii*)

The image is the region PQRS.

Example 15. *Discuss the transformation $w = \sin z$.*

Sol. The given transformation is

$$w = u + iv = \sin (x + iy)$$

or $$u + iv = \sin x \cosh y + i \cos x \sinh y$$

So that $$u = \sin x \cosh y$$

and $$v = \cos x \sinh y \quad \ldots(1)$$

Eliminating y from equation (1), we get

$$\frac{u^2}{\sin^2 x} - \frac{v^2}{\cos^2 x} = 1 \quad [\because \cosh^2 y - \sinh^2 y = 1]$$

Thus the straight lines $x = c$ in the z-plane are mapped into confocal hyperbolas in the w-plane.

Eliminating x from equation (1), we get $\dfrac{u^2}{\cosh^2 y} + \dfrac{v^2}{\sinh^2 y} = 1$

Thus the straight lines $y = c$ in the z-plane are mapped into confocal ellipses.

The lines $x = 0$ and $y = 0$ map into the lines $u = 0$ and $v = 0$ respectively in the w-plane.

Example 16. *Show that transformation $w = \cosh z$ maps the lines parallel to x-axis and lines parallel to y-axis into confocal central conics.*

Sol. Given transformation is $w = \cosh z$

i.e., $$u + iv = \cosh (x + iy) = \cosh x \cos y + i \sinh x \sin y$$

$$\therefore \quad \left.\begin{aligned} u &= \cosh x \cos y \\ v &= \sinh x \sin y \end{aligned}\right] \quad \ldots(1)$$

Lines parallel to x-axis are given by $y = k$

$$u = \cosh x \cos k$$
$$v = \sinh x \sin k$$

$$\therefore \quad \cosh x = \frac{u}{\cos k}$$

$$\sinh x = \frac{v}{\sin k}$$

Squaring and subtracting, we get

$$\frac{u^2}{\cos^2 k} - \frac{v^2}{\sin^2 k} = 1, \text{ which is a hyperbola in } w\text{-plane.}$$

Here $$a^2 = \cos^2 k,\ b^2 = \sin^2 k$$

$\therefore$ Eccentricity of the curve is given by $b^2 = a^2 (e^2 - 1)$ ...(2)

$$\sin^2 k = \cos^2 k\, (e^2 - 1) \text{ or } e^2 = 1 + \tan^2 k = \sec^2 k$$

$$\therefore \quad e = \sec k$$

Foci of (2) are $(\pm\, ae, 0) = (\pm \cos k \,.\, \sec k, 0) = (\pm\, 1, 0)$

Centre of (2) is $(0, 0)$

Now lines paralled to y- axis are $x = k'$

$\therefore$ From (1), $$u = \cosh k' \cos y$$
$$v = \sinh k' \sin y$$

$$\frac{u}{\cosh k'} = \cos x$$

$$\frac{v}{\sinh k'} = \sin x$$

Squaring and adding $\dfrac{u^2}{\cosh^2 k'} + \dfrac{v^2}{\sinh^2 k'} = 1$ which is an ellipse ...(3)

Here $a^2 = \cosh^2 k'$, $b^2 = \sinh^2 k'$

Eccentricity of ellipse is given by $b^2 = a^2 (1 - e^2)$ *i.e.*, $\sinh^2 k' = \cosh^2 k' (1 - e^2)$ or

$$e^2 = 1 - \tanh^2 k' = \text{sech}^2\, k'$$

$\therefore$ e sech k'

$\therefore$ Foci of (3) are $(\pm\, ae, 0) = (\pm \cosh k' \,\text{sech}\, k', 0) = (\pm\, 1, 0)$

Centre of (3) is (0, 0)

$\therefore$ The centres and foci of both the conics are same.

Hence lines || to x-axis and y-axis maps into confocal central conics.

Example 17. *The bilinear transformation* $w = \dfrac{az+b}{cz+d}$ *transforms the circle* $\arg \dfrac{z - z_1}{z - z_2} = \lambda$ *into similar circle of* $\arg \dfrac{w - w_1}{w - w_2}$ = *constant where* w_1, w_2 *corresponds to* z_1, z_2 *respectively.*

Sol. Here $$w = \frac{az+b}{cz+d}$$

$$\therefore \quad w_1 = \frac{az_1+b}{cz_1+d}; \; w_2 = \frac{az_2+b}{cz_2+d}$$

So that $$\frac{w-w_1}{w-w_2} = \frac{\dfrac{az+b}{cz+d} - \dfrac{az_1+b}{cz_1+d}}{\dfrac{az+b}{cz+d} - \dfrac{az_2+b}{cz_2+d}} = \frac{cz_2+d}{cz_1+d} \cdot \frac{z-z_1}{z-z_2}$$

$$\therefore \quad \frac{w-w_1}{w-w_2} = \mu \frac{z-z_1}{z-z_2} \text{ where } \mu \text{ is complex constant and} = \frac{cz_2+d}{cz_1+d}$$

$$\arg \frac{w-w_1}{w-w_2} = \arg \mu \left(\frac{z-z_1}{z-z_2}\right)$$

$$= \arg \mu + \arg \frac{z-z_1}{z-z_2}$$

$$= \arg \mu + \lambda$$

$= k$ (say) where k is real because argument of a complex number is always real.

Now, $\arg \dfrac{w - w_1}{w - w_2} = k$; represents a circle in w-plane passing through the points w_1 and w_2 corresponding to z_1 and z_2 in the z-plane.

Example 18. *Find the bilinear transformation which maps the points $z = 1, i, -1$ into the points $w = i, 0, -i$.* (K.U.K. May 2013)

Hence find the image of $|z| < 1$.

Sol. Let the required bilinear transformation be $w = \dfrac{az+b}{cz+d}$ ...(1)

Substituting the corresponding values of w and z in (1), we get

$$i = \frac{a+b}{c+d}, \qquad 0 = \frac{ai+b}{ci+d}, \quad -i = \frac{-a+b}{-c+d}$$

These equations can be written as

$$(a + b) - i(c + d) = 0 \qquad \text{...(2)}$$

$$b + ia = 0 \qquad \text{...(3)}$$

and $$(-a + b) + i(-c + d) = 0 \qquad \text{...(4)}$$

From (3), $b = -ia$

Adding (2) and (4), $2b - 2ic = 0$ or $c = \dfrac{b}{i} = -a$ [Using (3)]

Subtracting (4) from (2), $2a - 2id = 0$ or $d = \dfrac{a}{i} = -ia$

Substituting for b, c, d in (1), we get $w = \dfrac{az - ia}{-az - ia}$ or $w = \dfrac{i - z}{i + z}$ ...(5)

which is the required bilinear transformation.

Now, from (5), $z = i\,\dfrac{1-w}{1+w}$

$\therefore$ $|z| < 1$ is mapped into the region

$$\left| i\,\frac{1-w}{1+w} \right| < 1 \quad \text{or} \quad \frac{|i||1-w|}{|1+w|} < 1$$

or $$|1 - w| < |1 + w| \qquad | \because \ |i| = 1$$

or $$|1 - u - iv| < |1 + u + iv| \quad \text{or} \quad (1-u)^2 + v^2 < (1+u)^2 + v^2$$

or $$u > 0.$$

Hence the interior of the circle $x^2 + y^2 = 1$ in the z-plane is mapped into the entire half of the w-plane to the right of the imaginary axis.

EXERCISE 3.10

1. Find the image of the circle $| z | = 2$ under the transformation $w = z + 3 + 2i$.
2. Find the image of the triangle with vertices at i, $1 + i$, $1 - i$ in the z-plane under the transformation
$$w = 3z + 4 - 2i.$$
3. Determine the region in the w-plane in which the rectangle bounded by the lines $x = 0, y = 0, x = 2$ and $y = 3$ is mapped under the transformation $w = \sqrt{2}\, e^{i\frac{\pi}{4}} . z$
4. Consider the transformation $w = 2z$ and determine the region D′ of the w-plane into which the triangular region D enclosed by the lines $x = 0$, $y = 0$, $x + y = 1$ in the z- plane is mapped under this transformation .
5. Find the image of the semi-infinite strip $x > 0$, $0 < y < 2$, under the transformation $w = iz + 1$. Show the region graphically.
6. (a) Find the image of the circle $| z - 3 | = 5$ under the mapping $w = \dfrac{1}{z}$.

   (b) Show that under the transformation $w = \dfrac{1}{z}$, the circle $x^2 + y^2 - 6x = 0$, is transformed into a straight line in the w-plane.
7. Show that the map of the real axis of the z-plane on the w-plane by the transformation $w = \dfrac{1}{z + i}$ is a circle and find its centre and radius.
8. Prove that $w = \dfrac{z}{1 - z}$ maps the upper half of the z-plane into the upper half of the w-plane.
9. Under the transformation $w = \dfrac{z - i}{1 - iz}$, find the map of the circle $|z| = 1$ in the w- plane.
10. Show that the transformation $w = i\dfrac{1 - z}{1 + z}$, transforms the circle $|z| = 1$ into the real axis of w-plane and the interior of the circle $|z| < 1$ into the upper half of the w-plane.
11. Show that under the transformation $w = \dfrac{z - i}{z + i}$, real axis in the z-plane is mapped into the circle $| w | = 1$, What portion of the z-plane corresponds to the interior of the circle ?
12. Under the transformation $w = z^2$, obtain the map in the w-plane of the square with vertices (0, 0), (2, 0), (2, 2), (0, 2) in the z-plane.
13. Find the images of the straight lines $x = 0$, $y = 0$, $x = 1$ and $y = 1$ under the transformation $w = z^2$.
14. Determine the region of the w-plane in to which the triangle formed by $x = 1$, $y = 1$ and $x + y = 1$ is mapped under the transformation $w = z^2$.
15. Show that the transformation $w = \sqrt{z}$ maps the domain in the z-plane to the right of the line $x = a$ into the interior of a hyperbola in the w-plane.
16. Show that the map of the circle $|z| = 2$ under the transformation $w + 2i = z + \dfrac{1}{z}$ is an ellipse, find its axes and centre.

17. Show that the transformation $w = z + \dfrac{a^2 - b^2}{4z}$ transforms the circle $|z| = \dfrac{1}{2}(a + b)$ in the z-plane into an ellipse of semi axes a, b in the w-plane.

18. Find the bilinear transformation which maps:

(a) the points $z = 1, -i, -1$ onto the points $w = i, 0, -i$.

(b) the points $z = 1, i, -1$ onto the points $w = 0, 1, \infty$.

(c) the points $z = 0, -i, -1$ into the points $w = i, 1, 0$. Find the image of the line $y = mx$ under this transformation.

(d) the points $z = -1, i, 1$ of the z-plane into $w = 1, i, -1$ of the w-plane respectively.

(K.U.K. Jan. 2013)

19. Show that the condition for transformation $w = \dfrac{az + b}{cz + d}$ to make the circle $|w| = 1$, correspond to a straight line in the z-plane is $|a| = |c|$.

20. Find the bilinear transformation which maps the points $z = 1, i, -1$ onto the points $w = 2, i, -2$ respectively. Also find the fixed points of the transformation.

Answers

1. $(u - 3)^2 + (v - 2)^2 = 4$
2. A triangle with vertices $(4, -1)$, $(7, 1)$ and $(7, -3)$.
3. Rectangular region bounded by the lines $v = -u$, $v = u$, $u + v = 4$, $v - u = 6$.
4. A triangular region bounded by the lines $u = 0$, $v = 0$, $u + v = 2$.
5. $-1 < u < 1$, $v > 0$
6. $\left|w + \dfrac{3}{16}\right| = \dfrac{5}{16}$
7. $\left(0, -\dfrac{1}{2}\right); \dfrac{1}{2}$
9. $v = 0$
11. $y > 0$
12. $(0, 0)$, $(4, 0)$, $(0, 8)$, $(-4, 0)$; and isosceles triangle.
13. Negative real axis, positive real axis, $v^2 = 4(1 - u)$, $v^2 = 4(1 + u)$.
14. Region bounded by the parabolas $v^2 = 4(1 \pm u)$, $u^2 = 1 - 2v$
16. $\dfrac{5}{2}, \dfrac{3}{2}; (0, -2)$
18. (a) $w = \dfrac{z + i}{z - i}$ (b) $w = \dfrac{i(1 - z)}{1 + z}$

(c) $w = \dfrac{i(1 + z)}{1 - z}$; $m(u^2 + v^2) + 2u - m = 0$ (d) $w = -\dfrac{1}{z}$

20. $w = \dfrac{2i - 6z}{iz - 3}$; For fixed points $w = z$ gives

$$iz^2 + 3z - 2i = 0 \quad \text{or} \quad z^2 - 3iz - 2 = 0$$

or $(z - i)(z - 2i) = 0 \quad \therefore \quad z = i, 2i$

❑❑❑

SECTION-C

CHAPTER 4

Power Series and Contour Integration

4.1. SERIES OF COMPLEX TERMS

A series of the form $(a_1 + ib_1) + (a_2 + ib_2) + ... + (a_n + ib_n) + ...$
where $a_1, b_1, a_2, b_2, ...$ are real numbers, is called a series of complex terms and can be expressed as

$$\sum_{n=1}^{\infty} a_n + i \sum_{n=1}^{\infty} b_n \qquad ...(1)$$

The series (1) is convergent if Σa_n and Σb_n both are convergent.

If the series Σa_n and Σb_n converge to a and b respectively, then the series (1) is said to converge to $a + ib$.

The necessary (not sufficient) condition for convergence of series (1) is $\lim_{n \to \infty} (a_n + ib_n) = 0$.

The series (1) is said to be **absolutely convergent** if the series
$| a_1 + ib_1 | + | a_2 + ib_2 | + ... + | a_n + ib_n | + ...$ is convergent.

Since $| a_n | \leq | a_n + ib_n |$ and $| b_n | \leq | a_n + ib_n |$, we conclude that an absolutely convergent series is convergent. The converse is not true.

If the series of functions $u_1(z) + u_2(z) + ... + u_n(z) + ... = \sum_{n=1}^{\infty} u_n(z) \qquad ...(2)$

converges to S(z) and $S_n(z)$ is its nth partial sum, then the series (2) is said to be **uniformly convergent** in a region R, if given any positive number ε, there exists a positive number N, depending on ε but not on z, such that for every z in R,

$$| S(z) - S_n(z) | < \varepsilon \text{ for all } n > N.$$

A series of the form $\quad a_0 + a_1(z - a) + a_2(z - a)^2 + \cdots + a_n(z - a)^n + ...$

$$= \sum_{n=0}^{\infty} a_n (z - a)^n \qquad ...(3)$$

is called a **power series** in $(z - a)$.

If the series (3) is convergent for all values of z satisfying $| z - a | < R$, *i.e.*, for all values of z lying inside the circle with centre at $z = a$ and radius R, then this circle is called the **circle of convergence** and R is called the **radius of convergence.** *(M.D.U. May 2015)*

The power series (3) is uniformly convergent in a region R if there exists a series of positive terms ΣM_n such that $| a_n(z - a)^n | \leq M_n$ for all values of z in the region R.

This is called **Wierstrass M-test** for uniform convergence of a power series.

4.2. TAYLOR'S SERIES

If f(z) is analytic inside a circle C with centre at a, then for all z inside C,

$$f(z) = f(a) + (z - a) f'(a) + \frac{(z-a)^2}{2!} f''(a) + \cdots + \frac{(z-a)^n}{n!} f^n(a) + \ldots$$

Proof. Let z be any point inside the circle C. Draw a circle C_1 with centre at a and radius smaller than that of C such that z is an interior point of C_1. Let w be any point on C_1, then

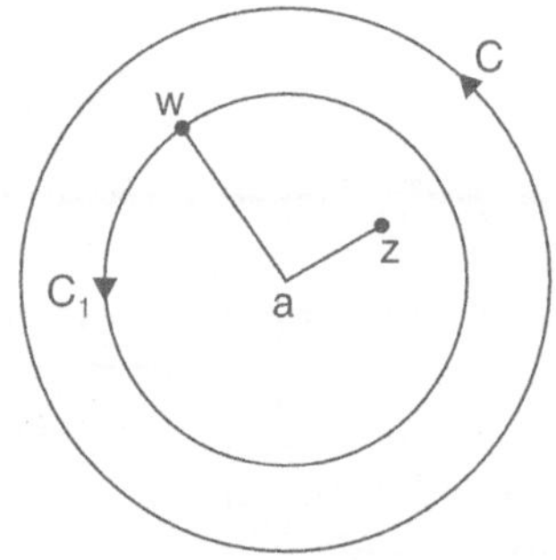

$$|z - a| < |w - a| \quad i.e., \left|\frac{z-a}{w-a}\right| < 1$$

Now, $$\frac{1}{w-z} = \frac{1}{(w-a)-(z-a)} = \frac{1}{w-a}\left[1 - \frac{z-a}{w-a}\right]^{-1}$$

Expanding R.H.S. by binomial theorem as $\left|\frac{z-a}{w-a}\right| < 1$, we get

$$\frac{1}{w-z} = \frac{1}{w-a}\left[1 + \frac{z-a}{w-a} + \left(\frac{z-a}{w-a}\right)^2 + \ldots + \left(\frac{z-a}{w-a}\right)^n + \ldots\right] \quad \ldots(1)$$

This series converges uniformly since $\left|\frac{z-a}{w-a}\right| < 1$. Multiplying both sides of (1) by $\frac{1}{2\pi i} f(w)$ and integrating term by term w.r.t. w, over C_1, we get

$$\frac{1}{2\pi i}\oint_{C_1} \frac{f(w)}{w-z}\, dw = \frac{1}{2\pi i}\oint_{C_1} \frac{f(w)}{w-a}\, dw + \frac{z-a}{2\pi i}\oint_{C_1} \frac{f(w)}{(w-a)^2}\, dw + \frac{(z-a)^2}{2\pi i}\oint_{C_1} \frac{f(w)}{(w-a)^3}\, dw +$$

$$\ldots\ldots + \frac{(z-a)^n}{2\pi i}\oint_{C_1} \frac{f(w)}{(w-a)^{n+1}}\, dw + \ldots \quad \ldots(2)$$

Since $f(w)$ is analytic on and inside C_1, by Cauchy's formula,

$$\frac{1}{2\pi i}\oint_{C_1} \frac{f(w)}{w-z}\, dw = f(z)$$

Also $$f^n(a) = \frac{n!}{2\pi i}\oint_{C_1} \frac{f(w)}{(w-a)^{n+1}}\, dw \quad i.e., \quad \frac{1}{2\pi i}\oint_{C_1} \frac{f(w)}{(w-a)^{n+1}}\, dw = \frac{f^n(a)}{n!}$$

$\therefore$ From (2), we have

$$f(z) = f(a) + (z-a) f'(a) + \frac{(z-a)^2}{2!} f''(a) + \cdots + \frac{(z-a)^n}{n!} f^n(a) + \cdots \quad \ldots(3)$$

which is the required *Taylor's series* for $f(z)$ about $z = a$.

Cor. 1. Putting $z = a + h$ in (3), we get

$$f(a+h) = f(a) + hf'(a) + \frac{h^2}{2!} f''(a) + \cdots\cdots + \frac{h^n}{n!} f^n(a) + \cdots$$

Cor. 2. If $a = 0$, the series (3) becomes

$$f(z) = f(0) + zf'(0) + \frac{z^2}{2!} f''(0) + \cdots\cdots + \frac{z^n}{n!} f^n(0) + \cdots$$

This series is called **Maclaurin's series.**

4.3. LAURENT'S SERIES

If $f(z)$ is analytic inside and on the boundary of the ring shaped region R bounded by two concentric circles C_1 and C_2 of radii r_1 and r_2 $(r_1 > r_2)$ respectively having centre at a, then for all z in R,

$$f(z) = a_0 + a_1(z-a) + a_2(z-a)^2 + \cdots + a_{-1}(z-a)^{-1} + a_{-2}(z-a)^{-2} + \cdots$$

where $$a_n = \frac{1}{2\pi i} \oint_{C_1} \frac{f(w)}{(w-a)^{n+1}} dw;\ n = 0, 1, 2, \ldots$$

and $$a_{-n} = \frac{1}{2\pi i} \oint_{C_2} \frac{f(w)}{(w-a)^{-n+1}} dw;\ n = 1, 2, 3, \ldots$$

Proof. Let z be any point in the region R, then by Cauchy's integral formula for double connected region, we have

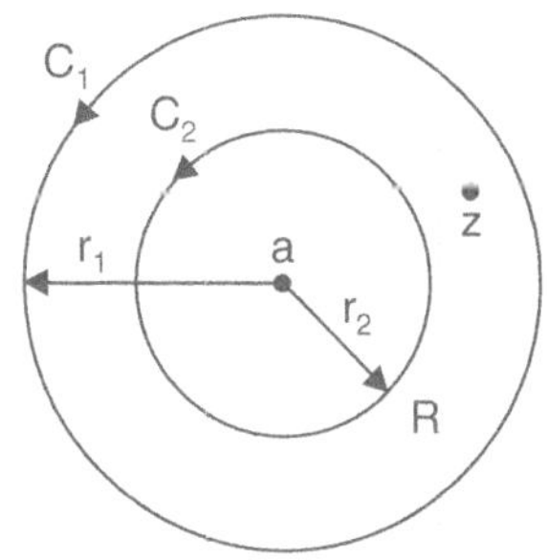

$$f(z) = \frac{1}{2\pi i} \oint_{C_1} \frac{f(w)}{w-z} dw - \frac{1}{2\pi i} \oint_{C_2} \frac{f(w)}{w-z} dw \quad \ldots(1)$$

For the first integral in (1), w lies on C_2

$$\therefore \quad |z-a| < |w-a| \quad i.e., \quad \left|\frac{z-a}{w-a}\right| < 1$$

Now, $$\frac{1}{w-a} = \frac{1}{(w-a)-(z-a)} = \frac{1}{w-a}\left(1 - \frac{z-a}{w-a}\right)^{-1}$$

$$= \frac{1}{w-a}\left[1 + \frac{z-a}{w-a} + \left(\frac{z-a}{w-a}\right)^2 + \cdots\cdots\right]$$

Multiplying both sides by $\frac{1}{2\pi i} f(w)$ and integrating term by term w.r.t. w, along the circle C_1, we get

$$\frac{1}{2\pi i} \oint_{C_1} \frac{f(w)}{w-z} dw = \frac{1}{2\pi i} \oint_{C_1} \frac{f(w)}{w-a} dw + \frac{z-a}{2\pi i} \oint_{C_1} \frac{f(w)}{(w-a)^2} dw + \frac{(z-a)^2}{2\pi i} \oint_{C_1} \frac{f(w)}{(w-a)^3} dw + \cdots\cdots$$

$$= a_0 + a_1(z-a) + a_2(z-a)^2 + \cdots\cdots$$

$$= \sum_{n=0}^{\infty} a_n (z-a)^n \quad \ldots(2) \quad \left[\because\ a_n = \frac{1}{2\pi i} \oint_{C_1} \frac{f(w)}{(w-a)^{n+1}} dw, n = 0, 1, 2, \ldots\ldots\right]$$

For the second integral in (1), w lies on C_2

$$\therefore \quad |w-a| < |z-a| \quad i.e., \quad \left|\frac{w-a}{z-a}\right| < 1$$

Now $$\frac{1}{w-z}=\frac{1}{(w-a)-(z-a)}=-\frac{1}{z-a}\left(1-\frac{w-a}{z-a}\right)^{-1}$$
$$=-\frac{1}{z-a}\left[1+\frac{w-a}{z-a}+\left(\frac{w-a}{z-a}\right)^2+\ldots\ldots\right]$$

Multiplying both sides by $\frac{1}{2\pi i}f(w)$ and integrating term by term w.r.t. w, along the circle C_2, we get

$$-\frac{1}{2\pi i}\oint_{C_2}\frac{f(w)}{w-z}\,dw=\frac{1}{z-a}\cdot\frac{1}{2\pi i}\oint_{C_2}f(w)\,dw+\frac{1}{(z-a)^2}\cdot\frac{1}{2\pi i}\oint_{C_2}(w-a)\,f(w)\,dw$$
$$+\frac{1}{(z-a)^3}\cdot\frac{1}{2\pi i}\oint_{C_2}(w-a)^2 f(w)\,dw+\ldots\ldots$$
$$=a_{-1}(z-a)^{-1}+a_{-2}(z-a)^{-2}+a_{-3}(z-a)^{-3}+\ldots\ldots$$
$$=\sum_{n=1}^{\infty}a_{-n}(z-a)^{-n}\quad\ldots(3)\quad\left[\because\ a_{-n}=\frac{1}{2\pi i}\oint_{C_2}\frac{f(w)}{(w-a)^{-n+1}}dw, n=1,2,3,\ldots\ldots\right]$$

Substituting from (2) and (3) in (1), we get

$$f(z)=\sum_{n=0}^{\infty}a_n(z-a)^n+\sum_{n=1}^{\infty}a_{-n}(z-a)^{-n}$$

Note 1. In the statement of Laurent's series, $a_n=\frac{1}{2\pi i}\oint_{C_1}\frac{f(w)}{(w-a)^{n+1}}dw\neq\frac{f^n(a)}{n!}$ because $f(z)$ is not given to be analytic inside C_1.

Note 2. In case $f(z)$ is analytic inside C_1, then $a_{-n}=0$ and $a_n=\frac{1}{2\pi i}\oint_{C_1}\frac{f(w)}{(w-a)^{n+1}}dw=\frac{f^n(a)}{n!}$ and Laurent's series reduces to Taylor's series.

Note 3. If C is any simple closed curve which lies in the ring-shaped region R and encloses the circle C_1, then

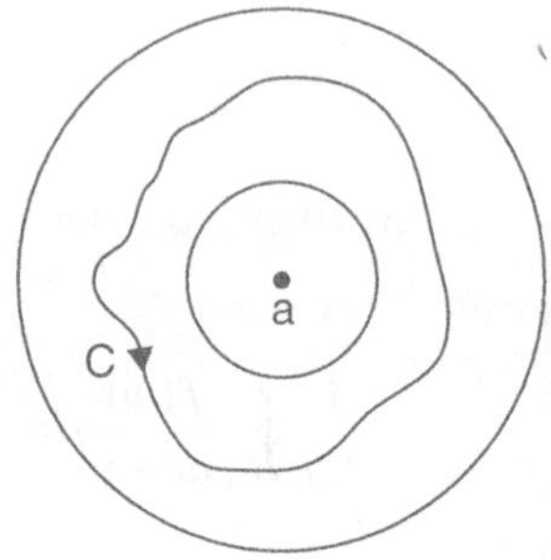

$$\oint_{C_1}\frac{f(w)}{(w-a)^{n+1}}dw=\oint_C\frac{f(w)}{(w-a)^{n+1}}dw$$

and $$\oint_{C_2}\frac{f(w)}{(w-a)^{-n+1}}dw=\oint_C\frac{f(w)}{(w-a)^{-n+1}}dw$$

$\therefore$ Laurent's series can be written as

$$f(z)=\sum_{n=-\infty}^{\infty}a_n(z-a)^n,\quad\text{where } a_n=\frac{1}{2\pi i}\oint_C\frac{f(w)}{(w-a)^{n+1}}dw.$$

Note 4. The process of finding the coefficient a_n by complex integration is complicated. In practice, we expand the function $f(z)$ by binomial theorem or by some other method to obtain Taylor's or Laurent's series.

ILLUSTRATIVE EXAMPLES

Example 1. *Expand* $\frac{1}{z^2 - 3z + 2}$ *in the region*

(*a*) $|z| < 1$ (*b*) $1 < |z| < 2$ (K.U.K. 2013; G.B.T.U. 2010)

(*c*) $|z| > 2$ (*d*) $0 < |z - 1| < 1$.

(K.U.K. 2013; G.B.T.U. 2010)

Sol. Here $f(z) = \frac{1}{z^2 - 3z + 2} = \frac{1}{(z-1)(z-2)} = \frac{1}{z-2} - \frac{1}{z-1}$ (Partial Fractions)

(*a*) If $|z| < 1$, then $\left|\frac{z}{2}\right| < 1$

$$\therefore \quad f(z) = \frac{1}{-2\left(1 - \frac{z}{2}\right)} + \frac{1}{1-z}$$

[Arranged suitably to make the binomial expansions valid]

$$= -\frac{1}{2}\left(1 - \frac{z}{2}\right)^{-1} + (1 - z)^{-1}$$

$$= -\frac{1}{2}\left(1 + \frac{z}{2} + \frac{z^2}{4} + \frac{z^3}{8} + \ldots\ldots\right) + (1 + z + z^2 + z^3 + \ldots\ldots)$$

(By binomial theorem)

$$= \frac{1}{2} + \frac{3}{4}z + \frac{7}{8}z^2 + \frac{15}{16}z^3 + \ldots\ldots, \text{ which is a Taylor's series.}$$

(*b*) If $1 < |z| < 2$, then $\left|\frac{1}{z}\right| < 1$ and $\left|\frac{z}{2}\right| < 1$

$$\therefore \quad f(z) = \frac{1}{-2\left(1 - \frac{z}{2}\right)} - \frac{1}{z\left(1 - \frac{1}{z}\right)} = -\frac{1}{2}\left(1 - \frac{z}{2}\right)^{-1} - \frac{1}{z}\left(1 - \frac{1}{z}\right)^{-1}$$

$$= -\frac{1}{2}\left(1 + \frac{z}{2} + \frac{z^2}{4} + \frac{z^3}{8} + \ldots\ldots\right) - \frac{1}{z}\left(1 + \frac{1}{z} + \frac{1}{z^2} + \frac{1}{z^3} + \ldots\ldots\right)$$

(By binomial theorem)

$$= -\frac{1}{2}\left(1 + \frac{z}{2} + \frac{z^2}{4} + \ldots\ldots\right) - (z^{-1} + z^{-2} + z^{-3} + \ldots\ldots)$$

which is a Laurent's series.

(*c*) If $|z| > 2$, then $\left|\frac{2}{z}\right| < 1$ and $\left|\frac{1}{z}\right| < 1$

$$\therefore \quad f(z) = \frac{1}{z\left(1-\frac{2}{z}\right)} - \frac{1}{z\left(1-\frac{1}{z}\right)} = \frac{1}{z}\left(1-\frac{2}{z}\right)^{-1} - \frac{1}{z}\left(1-\frac{1}{z}\right)^{-1}$$

$$= \frac{1}{z}\left(1+\frac{2}{z}+\frac{4}{z^2}+\frac{8}{z^3}+\ldots\ldots\right) - \frac{1}{z}\left(1+\frac{1}{z}+\frac{1}{z^2}+\frac{1}{z^3}+\ldots\ldots\right)$$

(By binomial theorem)

$= z^{-2} + 3z^{-3} + 7z^{-4} + \cdots\cdots$ which is a Laurent's series.

(*d*) For $0 < |z-1| < 1$, we have

$$f(z) = \frac{1}{(z-1)-1} - \frac{1}{z-1} = -\frac{1}{1-(z-1)} - \frac{1}{z-1} = -(z-1)^{-1} - [1-(z-1)]^{-1}$$

$$= -(z-1)^{-1} - [1 + (z-1) + (z-1)^2 + (z-1)^3 + \cdots\cdots].$$

Example 2. *Find the series expansion of* $f(z) = \dfrac{z^2-1}{z^2+5z+6}$ *about* $z = 0$ *in the region*

(*i*) $|z| < 2$ (*ii*) $2 < |z| < 3$. (U.K.T.U. 2011)

Sol. Here $f(z) = \dfrac{z^2-1}{z^2+5z+6} = 1 + \dfrac{3}{z+2} - \dfrac{8}{z+3}$ (Partial Fractions)

(*i*) When $|z| < 2$, we have $\left|\dfrac{z}{2}\right| < 1$ and hence $\left|\dfrac{z}{3}\right| < 1$.

$$\therefore \quad f(z) = 1 + \frac{3}{2\left(1+\frac{z}{2}\right)} - \frac{8}{3\left(1+\frac{z}{3}\right)}$$

$$= 1 + \frac{3}{2}\left(1+\frac{z}{2}\right)^{-1} - \frac{8}{3}\left(1+\frac{z}{3}\right)^{-1}$$

$$= 1 + \frac{3}{2}\left[1-\frac{z}{2}+\left(\frac{z}{2}\right)^2-\left(\frac{z}{2}\right)^3+\cdots\right] - \frac{8}{3}\left[1-\frac{z}{3}+\left(\frac{z}{3}\right)^2-\left(\frac{z}{3}\right)^3+\cdots\right]$$

(By binomial theorem)

$$= 1 + \frac{3}{2}\sum_{n=0}^{\infty}(-1)^n\left(\frac{z}{2}\right)^n - \frac{8}{3}\sum_{n=0}^{\infty}(-1)^n\left(\frac{z}{3}\right)^n$$

which is a Taylor's series.

(*ii*) When $2 < |z| < 3$, we have $\left|\dfrac{2}{z}\right| < 1$ and $\left|\dfrac{z}{3}\right| < 1$.

$$\therefore \quad f(z) = 1 + \frac{3}{z\left(1+\frac{2}{z}\right)} - \frac{8}{3\left(1+\frac{z}{3}\right)}$$

$$= 1 + \frac{3}{z}\left(1+\frac{2}{z}\right)^{-1} - \frac{8}{3}\left(1+\frac{z}{3}\right)^{-1}$$

$$= 1 + \frac{3}{z}\left[1 - \frac{2}{z} + \left(\frac{2}{z}\right)^2 - \left(\frac{2}{z}\right)^3 + \ldots\right] - \frac{8}{3}\left[1 - \frac{z}{3} + \left(\frac{z}{3}\right)^2 - \left(\frac{z}{3}\right)^3 + \ldots\right]$$

$$= 1 + \frac{3}{z}\sum_{n=0}^{\infty} (-1)^n \left(\frac{2}{z}\right)^n - \frac{8}{3}\sum_{n=0}^{\infty} (-1)^n \left(\frac{z}{3}\right)^n$$

which is a Laurent's series.

Example 3. *Expand* $\dfrac{1}{(z^2+1)(z^2+2)}$ *as a Laurent's series valid for*

(*i*) $0 < | z | < 1$ (*ii*) $1 < | z | < \sqrt{2}$ (*iii*) $| z | > \sqrt{2}$.

(M.D.U. 2010)

Sol. Here $$f(z) = \frac{1}{(z^2+1)(z^2+2)} = \frac{1}{z^2+1} - \frac{1}{z^2+2}$$ (Partial Fractions)

(*i*) When $0 < | z | < 1$, we have $z^2 < 1$ and hence $z^2 < 2$ so that $\dfrac{z^2}{2} < 1$.

$$\therefore \quad f(z) = \frac{1}{1+z^2} - \frac{1}{2\left(1+\dfrac{z^2}{2}\right)} = (1+z^2)^{-1} - \frac{1}{2}\left(1+\frac{z^2}{2}\right)^{-1}$$

$$= (1 - z^2 + z^4 - z^6 + z^8 - \ldots) - \frac{1}{2}\left[1 - \frac{z^2}{2} + \left(\frac{z^2}{2}\right)^2 - \left(\frac{z^2}{2}\right)^3 + \ldots\right]$$

$$= \left(1 - \frac{1}{2}\right) - \left(1 - \frac{1}{2^2}\right)z^2 + \left(1 - \frac{1}{2^3}\right)z^4 - \left(1 - \frac{1}{2^4}\right)z^6 + \ldots$$

$$= \sum_{n=1}^{\infty} (-1)^{n-1}\left(1 - \frac{1}{2^n}\right)z^{2n-2}$$

(*ii*) When $1 < | z | < \sqrt{2}$, we have $z^2 > 1$ and $z^2 < 2$ so that $\dfrac{1}{z^2} < 1$ and $\dfrac{z^2}{2} < 1$.

$$\therefore \quad f(z) = \frac{1}{z^2\left(1+\dfrac{1}{z^2}\right)} - \frac{1}{2\left(1+\dfrac{z^2}{2}\right)} = \frac{1}{z^2}\left(1+\frac{1}{z^2}\right)^{-1} - \frac{1}{2}\left(1+\frac{z^2}{2}\right)^{-1}$$

$$= \frac{1}{z^2}\left[1 - \frac{1}{z^2} + \frac{1}{z^4} - \frac{1}{z^6} + \ldots\right] - \frac{1}{2}\left[1 - \frac{z^2}{2} + \left(\frac{z^2}{2}\right)^2 - \left(\frac{z^2}{2}\right)^3 + \ldots\right]$$

$$= \left[\frac{1}{z^2} - \left(\frac{1}{z^2}\right)^2 + \left(\frac{1}{z^2}\right)^3 - \left(\frac{1}{z^2}\right)^4 + \ldots\right] - \frac{1}{2} + \left[\frac{z^2}{2^2} - \frac{(z^2)^2}{2^3} + \frac{(z^2)^3}{2^4} - \ldots\right]$$

$$= -\frac{1}{2} + \sum_{n=1}^{\infty} (-1)^{n+1} \left(\frac{1}{z^2}\right)^n + \sum_{n=1}^{\infty} (-1)^{n+1} \frac{(z^2)^n}{2^{n+1}}$$

$$= -\frac{1}{2} - \sum_{n=1}^{\infty} (-1)^n \left[\frac{1}{z^{2n}} + \frac{z^{2n}}{2^{n+1}}\right]$$

(*iii*) When $|z| > \sqrt{2}$, we have $z^2 > 2$ and hence $z^2 > 1$ so that $\frac{1}{z^2} < 1$ and $\frac{2}{z^2} < 1$.

$$\therefore \qquad f(z) = \frac{1}{z^2\left(1+\frac{1}{z^2}\right)} - \frac{1}{z^2\left(1+\frac{2}{z^2}\right)} = \frac{1}{z^2}\left(1+\frac{1}{z^2}\right)^{-1} - \frac{1}{z^2}\left(1+\frac{2}{z^2}\right)^{-1}$$

$$= \frac{1}{z^2}\left[1 - \frac{1}{z^2} + \frac{1}{z^4} - \frac{1}{z^6} + \ldots\right] - \frac{1}{z^2}\left[1 - \frac{2}{z^2} + \left(\frac{2}{z^2}\right)^2 - \left(\frac{2}{z^2}\right)^3 + \ldots\right]$$

$$= \left(\frac{1}{z^2} - \frac{1}{z^4} + \frac{1}{z^6} - \frac{1}{z^8} + \ldots\right) - \left(\frac{1}{z^2} - \frac{2}{z^4} + \frac{2^2}{z^6} - \frac{2^3}{z^8} + \ldots\right)$$

$$= \frac{2-1}{z^4} - \frac{2^2-1}{z^6} + \frac{2^3-1}{z^8} - \ldots = \sum_{n=1}^{\infty} (-1)^{n+1} \cdot \frac{2^n - 1}{(z^2)^{n+1}}.$$

Example 4. *Show that when* $|z+1| < 1$, $z^{-2} = 1 + \sum_{n=1}^{\infty} (n+1)(z+1)^n$.

(M.D.U. 2011)

Sol. $f(z) = z^{-2} = \frac{1}{z^2} = \frac{1}{[(z+1)-1]^2} = \frac{1}{[1-(z+1)]^2} = [1-(z+1)]^{-2}$

$= 1 + 2(z+1) + 3(z+1)^2 + 4(z+1)^3 + \ldots\ldots$

[By binomial theorem, since $|z+1| < 1$]

$$= 1 + \sum_{n=1}^{\infty} (n+1)(z+1)^n.$$

Example 5. *Expand* $\cos z$ *in a Taylor's series about* $z = \frac{\pi}{4}$.

Sol. Here $f(z) = \cos z$, $f'(z) = -\sin z$, $f''(z) = -\cos z$, $f'''(z) = \sin z$, ...

$$\therefore \qquad f\left(\frac{\pi}{4}\right) = \frac{1}{\sqrt{2}}, \quad f'\left(\frac{\pi}{4}\right) = -\frac{1}{\sqrt{2}}, \quad f''\left(\frac{\pi}{4}\right) = -\frac{1}{\sqrt{2}}, \quad f'''\left(\frac{\pi}{4}\right) = \frac{1}{\sqrt{2}}, \ldots\ldots$$

Hence, $\cos z = f(z) = f\left[\frac{\pi}{4} + \left(z - \frac{\pi}{4}\right)\right]$

$$= f\left(\frac{\pi}{4}\right) + \left(z - \frac{\pi}{4}\right) f'\left(\frac{\pi}{4}\right) + \frac{\left(z-\frac{\pi}{4}\right)^2}{2!} f''\left(\frac{\pi}{4}\right) + \frac{\left(z-\frac{\pi}{4}\right)^3}{3!} f'''\left(\frac{\pi}{4}\right) + \ldots\ldots$$

$$= \frac{1}{\sqrt{2}}\left[1 - \left(z - \frac{\pi}{4}\right) - \frac{1}{2!}\left(z - \frac{\pi}{4}\right)^2 + \frac{1}{3!}\left(z - \frac{\pi}{4}\right)^3 + \ldots\ldots\right]$$

Example 6. *Expand the function* $\dfrac{\sin z}{z-\pi}$ *about* $z = \pi$. (M.D.U. Dec. 2013)

Sol. Putting $z - \pi = t$, we have

$$\frac{\sin z}{z-\pi} = \frac{\sin(\pi+t)}{t} = \frac{-\sin t}{t}$$

$$= -\frac{1}{t}\left(t - \frac{t^3}{3!} + \frac{t^5}{5!} - \ldots\ldots\right) = -1 + \frac{t^2}{3!} - \frac{t^4}{5!} + \ldots\ldots = -1 + \frac{(z-\pi)^2}{3!} - \frac{(z-\pi)^4}{5!} + \ldots\ldots$$

Example 7. *Expand* $f(z) = \dfrac{z}{(z+1)(z+2)}$ *about* $z = -2$.

Sol. To expand $f(z)$ about $z = -2$, *i.e.*, in powers of $z + 2$, we put $z + 2 = t$.

$$\therefore \qquad f(z) = \frac{z}{(z+1)(z+2)} = \frac{t-2}{(t-1)t} = \frac{2-t}{t(1-t)} = \frac{2-t}{t}(1-t)^{-1}$$

$$= \frac{2-t}{t}(1 + t + t^2 + t^3 + \ldots\ldots) \qquad \text{for } 0 < |t| < 1$$

$$= \frac{1}{t}(2 + t + t^2 + t^3 + \ldots\ldots) = \frac{2}{t} + 1 + t + t^2 + \ldots\ldots$$

$$= \frac{2}{z+2} + 1 + (z+2) + (z+2)^2 + \ldots\ldots \qquad \text{for } 0 < |z+2| < 1$$

which is the required Laurent's series.

Example 8. *Expand* $\dfrac{e^{2z}}{(z-1)^3}$ *about the singularity* $z = 1$ *in Laurent's series.*

(M.D.U. May 2013)

Sol. To expand $f(z) = \dfrac{e^{2z}}{(z-1)^3}$ about $z = 1$, *i.e.*, in powers of $z - 1$, we put $z - 1 = t$ or $z = t + 1$

$$\therefore \qquad f(z) = \frac{e^{2(t+1)}}{t^3} = \frac{e^2}{t^3} \cdot e^{2t}$$

$$= \frac{e^2}{t^3}\left[1 + 2t + \frac{(2t)^2}{2!} + \frac{(2t)^3}{3!} + \frac{(2t)^4}{4!} + \ldots\ldots\right]$$

$$= e^2\left[\frac{1}{t^3} + \frac{2}{t^2} + \frac{2}{t} + \frac{4}{3} + \frac{2}{3}t + \ldots\ldots\right]$$

$$= e^2\left[\frac{1}{(z-1)^3} + \frac{2}{(z-1)^2} + \frac{2}{z-1} + \frac{4}{3} + \frac{2}{3}(z-1) + \ldots\ldots\right]$$

$$= e^2\left[(z-1)^{-3} + 2(z-1)^{-2} + 2(z-1)^{-1} + \frac{4}{3} + \frac{2}{3}(z-1) + \ldots\ldots\right].$$

Example 9. *Find Taylor's expansion of*

(i) $\dfrac{1}{(z+1)^2}$ *about the point* $z = -i$ *(ii)* $\dfrac{2z^3+1}{z^2+z}$ *about the point* $z = i$.

(M.D.U. 2011)

Sol. *(i)* To expand $f(z) = \dfrac{1}{(z+1)^2}$ about $z = -i$, *i.e.*, in powers of $z + i$, we put $z + i = t$

or $z = t - i$.

$$\therefore \quad f(z) = \frac{1}{(t-i+1)^2} = \frac{1}{[t+(1-i)]^2} = \frac{1}{\left[(1-i)\left(1+\dfrac{t}{1-i}\right)\right]^2}$$

$$= \frac{1}{(1-i)^2}\left(1+\frac{t}{1-i}\right)^{-2}$$

$$= \frac{1}{-2i}\left[1 - \frac{2t}{1-i} + \frac{3t^2}{(1-i)^2} - \frac{4t^3}{(1-i)^3} + \ldots\right]$$

$$= \frac{i}{2}\sum_{n=1}^{\infty}(-1)^{n-1} \cdot \frac{nt^{n-1}}{(1-i)^{n-1}}$$

$$= \frac{i}{2}\sum_{n=1}^{\infty}(-1)^{n-1} \cdot \frac{n(z+i)^{n-1}}{(1-i)^{n-1}}.$$

(ii) To expand $f(z) = \dfrac{2z^3+1}{z^2+z} = 2z - 2 + \dfrac{2z+1}{z(z+1)}$

$$= 2z - 2 + \frac{1}{z} + \frac{1}{z+1} \qquad \text{[By partial fractions]}$$

About $z = i$, *i.e.*, in powers of $z - i$, we put $z - i = t$ or $z = t + i$

$$\therefore \quad f(z) = 2(t+i) - 2 + \frac{1}{t+i} + \frac{1}{t+i+1} \qquad \ldots(1)$$

Now, $$\frac{1}{t+i} = \frac{1}{i\left(1+\dfrac{t}{i}\right)} = \frac{1}{i}\left(1+\frac{t}{i}\right)^{-1}$$

$$= \frac{1}{i}\left(1 - \frac{t}{i} + \frac{t^2}{i^2} - \frac{t^3}{i^3} + \ldots\right) = \frac{1}{i} - \frac{t}{i^2} + \frac{t^2}{i^3} - \frac{t^3}{i^4} + \ldots$$

$$= \sum_{n=1}^{\infty} (-1)^{n-1} \cdot \frac{t^{n-1}}{i^n}$$

Also, $$\frac{1}{t+i+1} = \frac{1}{(1+i)\left(1+\dfrac{t}{1+i}\right)} = \frac{1}{1+i}\left(1+\frac{t}{1+i}\right)^{-1}$$

$$= \frac{1}{1+i}\left[1 - \frac{t}{1+i} + \frac{t^2}{(1+i)^2} - \frac{t^3}{(1+i)^3} + \ldots\right]$$

$$= \frac{1}{1+i} - \frac{t}{(1+i)^2} + \frac{t^2}{(1+i)^3} - \frac{t^3}{(1+i)^4} + \ldots$$

$$= \sum_{n=1}^{\infty} (-1)^{n-1} \cdot \frac{t^{n-1}}{(1+i)^n}$$

Substituting in (1), we have

$$f(z) = -2 + 2i + 2t + \sum_{n=1}^{\infty} (-1)^{n-1}\left[\frac{1}{i^n} + \frac{1}{(1+i)^n}\right] t^{n-1}$$

$$= -2(1-i) + 2(z-i) + \sum_{n=1}^{\infty} (-1)^{n-1}\left[\frac{1}{i^n} + \frac{1}{(1+i)^n}\right](z-i)^{n-1}.$$

EXERCISE 4.1

Expand the following functions as a Taylor's series:

1. $\log(1+z)$ about $z = 0$.

2. $\sin z$ about $z = \dfrac{\pi}{4}$.

3. $\dfrac{z}{(z+1)(z+2)}$ about $z = 2$.

Expand the following functions in Laurent's series or Taylor's series:

4. $\dfrac{1}{z-2}$, for $|z| > 2$.

5. $\dfrac{1}{z^2 - 4z + 3}$, for $1 < |z| < 3$.

6. $\dfrac{1}{z(z-1)(z-2)}$, for $|z| > 2$.

7. $\dfrac{1-\cos z}{z^3}$, about $z = 0$.

(M.D.U. 2012)

8. $\dfrac{e^z}{(z-1)^2}$, about $z = 1$.

9. $\dfrac{1}{z(z-1)(z-2)}$, for $|z-1| < 1$

10. $\dfrac{1}{z(e^z - 1)}$, for $0 < |z| < 2\pi$.

11. $\dfrac{z^2 - 1}{(z+2)(z+3)}$, when *(M.D.U. Dec. 2014, May 2015)*

(*i*) $|z| < 2$ (*ii*) $2 < |z| < 3$ (*iii*) $|z| > 3$

12. $\dfrac{1}{(z+1)(z+3)}$, when

(*i*) $|z| < 1$ (*ii*) $1 < |z| < 3$
(*iii*) $|z| > 3$ (*iv*) $1 < |z+1| < 2$.

13. $\dfrac{7z - 2}{z(z+1)(z-2)}$, when

(*i*) $0 < |z+1| < 1$ (*ii*) $1 < |z+1| < 3$
(*iii*) $|z+1| > 3$ *(U.K.T.U. 2010; M.D.U. 2010)*

14. $\dfrac{(z-2)(z+2)}{(z+1)(z+4)}$ in the region

(*i*) $|z| < 1$ (*ii*) $1 < |z| < 4$ (*iii*) $|z| > 4$

15. Find the Laurent's series expansion valid in the region $0 < |z-1| < 1$ for the function

$$f(z) = \frac{2z+1}{z^3 + z^2 - 2z}.$$ *(M.D.U. May 2014)*

16. Obtain the Taylor's or Laurent's series which represents the function $f(z) = \dfrac{1}{(1+z^2)(z+2)}$, when

(*i*) $1 < |z| < 2$ (*ii*) $|z| > 2$ *(DCRUST Murthal, May 2014)*

Answers

1. $z - \dfrac{z^2}{2} + \dfrac{z^3}{3} - \dfrac{z^4}{4} + \ldots\ldots$

2. $\dfrac{1}{\sqrt{2}}\left[1 + \left(z - \dfrac{\pi}{4}\right) - \dfrac{1}{2!}\left(z - \dfrac{\pi}{4}\right)^2 - \dfrac{1}{3!}\left(z - \dfrac{\pi}{4}\right)^3 + \ldots\ldots\right]$

3. $\left(\dfrac{1}{2} - \dfrac{1}{3}\right) - \left(\dfrac{1}{2^3} - \dfrac{1}{3^2}\right)(z-2) + \left(\dfrac{1}{2^5} - \dfrac{1}{3^3}\right)(z-2)^2 - \ldots\ldots = \sum\limits_{n=1}^{\infty} (-1)^{n-1}\left(\dfrac{1}{2^{2n-1}} - \dfrac{1}{3^n}\right)(z-2)^{n-1}$

4. $\dfrac{1}{z} + \dfrac{2}{z^2} + \dfrac{2^2}{z^3} + \dfrac{2^3}{z^4} + \ldots\ldots$

5. $-\dfrac{1}{6} - \dfrac{1}{6}\sum\limits_{n=1}^{\infty} \dfrac{z^n}{3^n} - \dfrac{1}{2}\sum\limits_{n=1}^{\infty} z^{-n}$

6. $z^{-3} + 3z^{-4} + 7z^{-5} + \ldots\ldots + (2^n - 1)z^{-(n+2)} + \ldots\ldots$

7. $\sum\limits_{n=1}^{\infty} (-1)^{n-1} \cdot \dfrac{z^{2n-3}}{(2n)!}$

8. $e\left[(z-1)^{-2} + (z-1)^{-1} + \dfrac{1}{2!} + \dfrac{1}{3!}(z-1) + \dfrac{1}{4!}(z-1)^2 + \ldots\ldots\right]$

9. $-\dfrac{1}{z-1} - \sum\limits_{n=1}^{\infty} (z-1)^{2n-1}$.

10. $z^{-2} - \dfrac{1}{2}z^{-1} + \dfrac{1}{12} - \dfrac{z^2}{720} + \ldots\ldots$

11. (*i*) $1 + \frac{3}{2}\sum_{n=0}^{\infty} (-1)^n \left(\frac{z}{2}\right)^n - \frac{8}{3}\sum_{n=0}^{\infty} (-1)^n \left(\frac{z}{3}\right)^n$ (*ii*) $1 + \frac{3}{z}\sum_{n=0}^{\infty} (-1)^n \left(\frac{2}{z}\right)^n - \frac{8}{3}\sum_{n=0}^{\infty} (-1)^n \left(\frac{z}{3}\right)^n$

(*iii*) $1 + \frac{3}{z}\sum_{n=0}^{\infty} (-1)^n \left(\frac{2}{z}\right)^n - \frac{8}{z}\sum_{n=0}^{\infty} (-1)^n \left(\frac{3}{z}\right)^n$

12. (*i*) $\frac{1}{2}\left[\sum_{n=0}^{\infty} (-1)^n z^n - \frac{1}{3}\sum_{n=0}^{\infty} (-1)^n \left(\frac{z}{3}\right)^n\right]$ (*ii*) $\frac{1}{2}\left[\frac{1}{z}\sum_{n=0}^{\infty} (-1)^n \left(\frac{1}{z}\right)^n - \frac{1}{3}\sum_{n=0}^{\infty} (-1)^n \left(\frac{z}{3}\right)^n\right]$

(*iii*) $\frac{1}{2}\left[\frac{1}{z}\sum_{n=0}^{\infty} (-1)^n \left(\frac{1}{z}\right)^n - \frac{1}{z}\sum_{n=0}^{\infty} (-1)^n \left(\frac{3}{z}\right)^n\right]$ (*iv*) $\frac{1}{2(z+1)}\sum_{n=0}^{\infty} (-1)^n \left(\frac{z+1}{2}\right)^n$.

13. (*i*) $-\frac{3}{z+1} - \sum_{n=0}^{\infty} (z+1)^n - \frac{2}{3}\sum_{n=0}^{\infty} \left(\frac{z+1}{3}\right)^n$ (*ii*) $\frac{1}{z+1}\sum_{n=0}^{\infty} \left(\frac{1}{z+1}\right)^n - \frac{3}{z+1} - \frac{2}{3}\sum_{n=0}^{\infty} \left(\frac{z+1}{3}\right)^n$

(*iii*) $\frac{1}{z+1}\sum_{n=0}^{\infty} \left(\frac{1}{z+1}\right)^n - \frac{3}{z+1} + \frac{2}{z+1}\sum_{n=0}^{\infty} \left(\frac{3}{z+1}\right)^n$

14. (*i*) $-1 + \sum_{n=1}^{\infty} (-1)^{n+1} (1 + 4^{-n}) z^n$ (*ii*) $\sum_{n=1}^{\infty} \frac{(-1)^{n+1}}{4^n} z^n + \sum_{n=1}^{\infty} (-1)^n z^{-n}$

(*iii*) $1 + \sum_{n=1}^{\infty} (-1)^n (1 + 4^n) z^{-n}$

15. $-\frac{1}{2}\sum_{n=0}^{\infty} (-1)^n (z-1)^n - \frac{1}{6}\sum_{n=0}^{\infty} (-1)^n \left(\frac{z-1}{3}\right)^n + (z-1)^{-1}$.

16. (*i*) $\frac{1}{10}\sum_{n=0}^{\infty} (-1)^n \left(\frac{z}{2}\right)^n + \frac{2-z}{5z^2}\sum_{n=0}^{\infty} (-1)^n \left(\frac{1}{z^2}\right)^n$ (*ii*) $\frac{1}{5z}\sum_{n=0}^{\infty} (-1)^n \left(\frac{2}{z}\right)^n - \frac{z-2}{5z^2}\sum_{n=0}^{\infty} (-1)^n \left(\frac{1}{z^2}\right)^n$

RESIDUES

4.4. SINGULAR POINTS

A point at which a function $f(z)$ ceases to be analytic is called a singular point or singularity of $f(z)$. For example $z = 2$ is a singular point of $f(z) = \frac{1}{z-2}$.

4.5. ISOLATED SINGULAR POINT AND NON-ISOLATED SINGULAR POINT

A singular point $z = a$ of a function $f(z)$ is called an isolated singular point if there exists a small circle with centre a which contains no other singular point of $f(z)$. Otherwise it is called non-isolated singular point.

For example; consider $f(z) = \frac{z}{z^2 - 1}$

The function $f(z)$ is analytic everywhere except at $z = -1, 1$ $\therefore$ $z = -1, 1$ are the isolated singularities of $f(z)$ as there are no other singularities of $f(z)$ in the neighbourhood of $z = -1$ and 1.

Again consider $$f(z) = \cot\left(\frac{\pi}{z}\right) = \frac{1}{\tan\frac{\pi}{z}}$$

It is not analytic at points where $\tan\frac{\pi}{z} = 0$ *i.e.,* $\frac{\pi}{z} = n\pi$ or $z = \frac{1}{n}$

Thus $z = 1, \frac{1}{2}, \frac{1}{3}, \ldots\ldots, 0$ are the singularities of $f(z)$, all of which are isolated except $z = 0$ $\because$ in the neighbourhood of $z = 0$, there are infinite number of other singularities $\left(z = \frac{1}{n} \text{ where } n \text{ is large}\right)$

$\therefore$ $z = 0$ is non-isolated singularity of the given function.

4.6. TYPES OF SINGULARITIES

Let $f(z)$ be analytic within a domain D except at $z = a$ (an isolated singularity). Then with $z = a$ as centre we can draw two circles C (with small radius) and C′ (with large radius) within D so that $f(z)$ is analytic within the annulus between C and C′. If z is any point in the annulus then we can expand $f(z)$ is a Laurent's series about $z = a$.

$$\therefore \quad f(z) = \sum_{n=0}^{\infty} a_n (z-a)^n + \sum_{n=1}^{\infty} a_{-n} (z-a)^{-n}$$

$$= \sum_{n=0}^{\infty} a_n (z-a)^n + \left[a_{-1}(z-a)^{-1} + a_{-2}(z-a)^{-2} + \cdots + a_{-m}(z-a)^{-m} + \cdots \infty\right] \quad \ldots(1)$$

It gives rise to three types of singularities

(*i*) **Removable Singularity**

If in $f(z)$ the series has no negative power terms *i.e.,* $a_{-1}, -a_{-2}, a_{-3}, \ldots \infty$ are all zero then

$f(z) = \sum_{n=0}^{\infty} a_n (z-a)^n$. In this case $z = a$ is called a removable singularity.

Note. This type of singularity can be made to disappear by re-defining $f(z)$ at $z = a$ in such a way that it becomes analytic at $z = a$.

e.g., $f(z) = \frac{\sin(z-a)}{z-a}$ has removable singularity at $z = a$ because

$$\frac{\sin(z-a)}{z-a} = \frac{1}{z-a}\left\{(z-a) - \frac{(z-a)^3}{3!} + \frac{(z-a)^5}{5!} - \ldots\ldots \infty\right\} = 1 - \frac{(z-a)^2}{3!} + \frac{(z-a)^4}{5!} - \ldots\ldots \infty$$

has no term containing negative powers of $(z - a)$. However this singularity can be removed and the function can be made analytic by redefining the function as follows:

$$f(z) = \frac{\sin(z-a)}{z-a}, \text{ when } z \neq a$$
$$= 1, \text{ when } z = a$$

(*ii*) **Essential Singularity.** If in (1) *i.e.*, in the expansion of $f(z)$ the series with negative powers does not terminate *i.e.*, has an infinite number of terms then $z = a$ is called an essential singularity of $f(z)$

e.g., $$f(z) = \sin \frac{1}{z-a} = \frac{1}{z-a} - \frac{1}{3!(z-a)^3} + \frac{1}{5!(z-a)^5} \ldots\ldots \infty$$

$$= (z-a)^{-1} - \frac{1}{3!}(z-a)^{-3} + \frac{1}{5!}(z-a)^{-5} \ldots\ldots \infty$$

has an infinite number of negative power of $z - a$

(*iii*) **Pole** (*K.U.K. 2013*)

If in (1) *i.e.*, in $f(z) = \sum_{n=1}^{\infty} a_n (z-a)^n + \sum_{n=1}^{\infty} a_{-n} (z-a)^{-n}$, the series with negative powers has a finite number of terms, say m, so that $a_{-m} \neq 0$ and all further coefficients *i.e.*, a_{-m-1}, a_{-m-2}, a_{-m-3}, ... are zero, then $z = a$ is called a pole of order m.

e.g., $f(z) = \dfrac{\sin(z-a)}{(z-a)^4}$ has a pole of order 3 at $z = a$.

$$\because \quad f(z) = \frac{1}{(z-a)^4} \cdot \left\{ (z-a) - \frac{1}{3!}(z-a)^3 + \frac{1}{5!}(z-a)^5 \ldots \infty \right\}$$

$$= \frac{1}{(z-a)^3} - \frac{1}{3!}\frac{1}{z-a} + \frac{1}{5!}(z-a) \ldots \infty$$

$$= (z-a)^{-3} - \frac{1}{3!}(z-a)^{-1} + \frac{1}{5!}(z-1) \ldots \infty$$

It has a finite number of (only two) negative power terms and all other are zero

$\therefore$ $z = a$ is a pole of $f(z)$ of order 3. ($a_{-3} = 1 \neq 0$ and $a_{-4} = a_{-5} = \ldots = 0$)

Note. Poles are always isolated singularities

4.7. HOW TO DETECT REMOVABLE SINGULAR POINTS

If $\underset{z \to a}{\text{Lt}}\, f(z)$ exists and is finite then $z = a$ is a **removable singular point.**

4.8. RESIDUE AT A POLE

If $z = a$ is an isolated singularity of $f(z)$ then $f(z)$ can be expressed expanded in Laurents series about $z = a$,

$$\therefore \quad f(z) = \sum_{n=0}^{\infty} a_n (z-a)^n + \sum_{n=1}^{\infty} a_{-n} (z-a)^{-n}$$

$$= \sum_{n=0}^{\infty} a_n (z-a)^n + a_{-1}(z-a)^{-1} + a_{-2}(z-a)^{-2} + a_{-3}(z-a)^{-3} + \cdots \infty$$

Then coefficient of $(z-a)^{-1}$ *i.e.*, a_{-1} is called the residue of $f(z)$ at $z = a$ and is written as

$$a_{-1} = \text{Res}\{f(z), a\}$$

Since $a_{-1} = \dfrac{1}{2\pi i} \oint_C f(z)\, dz$

$\therefore \quad \oint_C f(z)\,dz = 2\pi i \text{ Res } \{ f(z), a\}$

Note: Residue at infinity

Residue of $f(z)$ at $z = \infty$ is defined as $-\frac{1}{2\pi i}\int_C f(z)\,dz$ where the integration is taken round C in anti-clockwise direction.

4.9. CAUCHY'S RESIDUE THEOREM

(M.T.U. 2013; M.D.U. 2012)

If f(z) is analytic at all points inside and on a simple closed curve C, except at a finite number of isolated singular points within C, then

$\oint_C f(z)\,dz = 2\pi i$ *(sum of residues at singular points within C).*

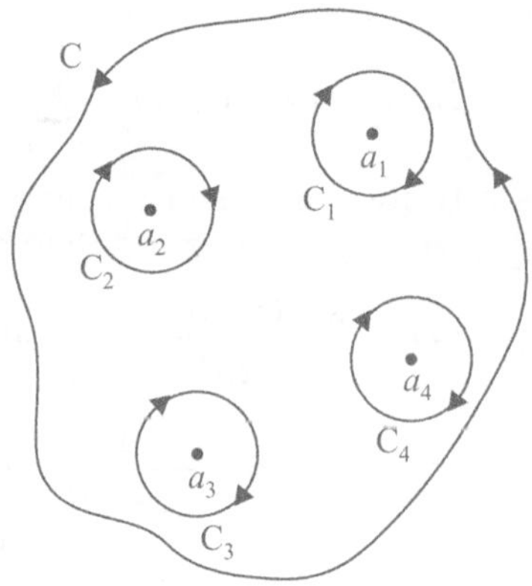

Proof. Around each of the isolated singular points $a_1, a_2, \ldots\ldots, a_n$ draw small non-intersecting circles $C_1, C_2, \ldots\ldots, C_n$ lying wholly inside C, with centres at $z = a_1, a_2, \ldots\ldots, a_n$ respectively.

Since $f(z)$ is analytic in the multiply connected region bounded by C, $C_1, C_2, \ldots\ldots, C_n$ we have by Cauchy's theorem for multiply connected region.

$$\oint_C f(z)dz = \oint_{C_1} f(z)dz + \oint_{C_2} f(z)dz + \cdots\cdots + \oint_{C_n} f(z)dz$$

$$= 2\pi i\,[\text{Res }\{(z), a_1\} + \text{Res }\{f(z), a_2\} + \cdots\cdots + \text{Res }\{f(z), a_n\}]$$

$$= 2\pi i\,(\text{sum of residues at } a_1, a_2, \cdots\cdots, a_n).$$

4.10. CALCULATION OF RESIDUES

(1) *If f(z) has a simple pole (i.e., pole of order 1) at z = a, then Res* $\{f(z), a\} = \lim_{z \to a} (z - a) f(z)$.

Since $z = a$ is a pole of order 1, the Laurent's series becomes

$$f(z) = a_0 + a_1(z-a) + a_2(z-a)^2 + \cdots\cdots + a_{-1}(z-a)^{-1}.$$

Multiplying both sides by $(z - a)$, we get

$$(z-a)\,f(z) = a_0(z-a) + a_1(z-a)^2 + a_2(z-a)^3 + \cdots\cdots + a_{-1}$$

$$\therefore \quad \lim_{z \to a} (z-a)\,f(z) = a_{-1} = \text{Res }\{f(z), a\}$$

(2) *If f(z) has a pole of order m at z = a, then*

$$\textbf{Res }\{\mathbf{f(z), a}\} = \frac{1}{(m-1)!}\lim_{z \to a}\frac{d^{m-1}}{dz^{m-1}}[(z-a)^m f(z)]$$

Since $z = a$ is a pole of order m, the Laurent's series becomes

$$f(z) = a_0 + a_1(z-a) + a_2(z-a)^2 + \cdots\cdots + a_{-1}(z-a)^{-1} + a_{-2}(z-a)^{-2} + \cdots\cdots + a_{-m}(z-a)^{-m}$$

Multiplying both sides by $(z-a)^m$, we get

$$(z-a)^m f(z) = a_0(z-a)^m + a_1(z-a)^{m+1} + a_2(z-a)^{m+2} + \cdots\cdots$$
$$+ a_{-1}(z-a)^{m-1} + a_{-2}(z-a)^{m-2} + \cdots\cdots + a_{-m}$$

Differentiating both sides $(m-1)$ times w.r.t. z and taking the limit as $z \to a$, we get

$$\underset{z \to a}{\text{Lt}} \frac{d^{m-1}}{dz^{m-1}} [(z-a)^m f(z)] = a_{-1}(m-1)\,!$$

or $$\frac{1}{(m-1)\,!} \underset{z \to a}{\text{Lt}} \frac{d^{m-1}}{dz^{m-1}} [(z-a)^m f(z)] = a_{-1} = \text{Res}\,\{f(z), a\}.$$

(3) *Another formula for Res {f(z), a}*

Let $$f(z) = \frac{\phi(z)}{\psi(z)}, \text{ where } \psi(z) = (z-a)\,F(z),\ F(a) \neq 0$$

Then $\text{Res}\,\{f(z), a\} = \lim\limits_{z \to a} (z-a) f(z)$

$$= \lim_{z \to a} \frac{(z-a)\,\phi(z)}{\psi(z)} = \lim_{z \to a} \frac{(z-a)\,\phi\{a+(z-a)\}}{\psi\{a+(z-a)\}}$$

[Expanding $\phi(z)$ and $\psi(z)$ in ascending powers of $(z-a)$]

$$= \lim_{z \to a} \frac{(z-a)\left[\phi(a)+(z-a)\,\phi'(a)+\cdots\right]}{\psi(a)+(z-a)\,\psi'(a)+\dfrac{(z-a)^2}{2!}\psi''(a)+\cdots}$$

$$= \lim_{z \to a} \frac{(z-a)\left[\phi(a)+(z-a)\,\phi'(a)+\cdots\right]}{(z-a)\left[\psi'(a)+\dfrac{z-a}{2!}\psi''(a)+\cdots\right]} \quad \text{since } \psi(a) = 0$$

$$= \lim_{z \to a} \frac{\phi(a)+(z-a)\,\phi'(a)+\cdots}{\psi'(a)+\dfrac{z-a}{2!}\psi''(a)+\cdots} = \frac{\phi(a)}{\psi'(a)} = \lim_{z \to a} \frac{\phi(z)}{\psi'(z)}.$$

(4) *Residue of f(z) at z = ∞ is given by* $\lim\limits_{z \to \infty} \{-zf(z)\}$.

Or

–[*co-efficient of* $\frac{1}{z}$ *in the expansion of f(z) for values of z in the neighbourhood of z = ∞*]

ILLUSTRATIVE EXAMPLES

Example 1. *Discuss the nature of singularities of the following functions:*

(i) $f(z) = \dfrac{z - \sin z}{z^3}$ at $z = 0$ (ii) $f(z) = e^{\frac{1}{z-a}}$ at $z = a$

(iii) $f(z) = \dfrac{1}{\sin z - \cos z}$ at $z = \dfrac{\pi}{4}$.

Sol. (*i*) $$f(z) = \frac{z - \sin z}{z^3} = \frac{1}{z^3}\left[z - \left(z - \frac{z^3}{3!} + \frac{z^5}{5!} - \frac{z^7}{7!} + \ldots\ldots \infty\right)\right]$$

$$= \frac{1}{3!} - \frac{z^2}{5!} + \frac{z^4}{7!} \ldots\ldots \infty$$

Since there are no negative power terms of z $\therefore$ $z = 0$ is a removable singularity.

Aliter: $$\underset{z\to 0}{\text{Lt}} \frac{z - \sin z}{z^3} \qquad \left[\frac{0}{0} \text{ form, apply L-Hospital's rule}\right]$$

$$= \underset{z\to 0}{\text{Lt}} \frac{1 - \cos z}{3z^2} \qquad \left[\frac{0}{0} \text{ form}\right]$$

$$= \underset{z\to 0}{\text{Lt}} \frac{\sin z}{6z} = \frac{1}{6} \qquad \left[\underset{z\to 0}{\text{Lt}} \frac{\sin z}{z} = 1\right]$$

Limit is finite $\therefore$ $z = 0$ is a removable singularity

(*ii*) $$f(z) = e^{\frac{1}{z-a}} = 1 + \frac{1}{z-a} + \frac{1}{2!}\frac{1}{(z-a)^2} + \frac{1}{3!}\frac{1}{(z-a)^3} + \ldots\ldots \infty$$

$$= 1 + (z-a)^{-1} + \frac{1}{2!}(z-a)^{-2} + \frac{1}{3!}(z-a)^{-3} + \ldots\ldots \infty$$

It has infinite number of negative power terms of $z - a$ $\therefore$ $z = a$ is an essential singularity.

(*iii*) $f(z) = \dfrac{1}{\sin z - \cos z}$ at $z = \dfrac{\pi}{4}$

Put $z = \dfrac{\pi}{4} + t$

$\therefore$ $$\phi(t) = \frac{1}{\sin\left(\frac{\pi}{4} + t\right) - \cos\left(\frac{\pi}{4} + t\right)} = \frac{1}{\sqrt{2}\sin t} = \frac{1}{\sqrt{2}\left[t - \frac{t^3}{3!} + \frac{t^5}{5!} - \ldots \infty\right]}$$

$$= \frac{1}{\sqrt{2}t}\left[1 - \left(\frac{t^2}{3!} - \frac{t^4}{5!} + \ldots \infty\right)\right]^{-1}$$

$$= \frac{1}{\sqrt{2}t}\left[1 + \left(\frac{t^2}{3!} - \frac{t^4}{5!} + \ldots \infty\right) + \left(\frac{t^2}{3!} - \frac{t^4}{5!} + \ldots \infty\right)^2 + \ldots\right]$$

$$= \frac{1}{\sqrt{2}}\left[\frac{1}{t} + \frac{1}{3!}t + \left(\frac{1}{(3!)^2} - \frac{1}{5!}\right)t^3 + \ldots\right]$$

Since $a_{-1} = \dfrac{1}{\sqrt{2}} \neq 0$ and $a_{-2} = a_{-3} = \ldots = 0$ $\therefore$ $t = 0$ is a simple pole of $f(t)$. Hence $z = \dfrac{\pi}{4}$ is a simple pole of $f(z)$.

Example 2. *What type of singularity have the following functions:*

(*i*) $\dfrac{e^{2z}}{(z-1)^4}$ (*ii*) $ze^{\frac{1}{z^2}}$ (*iii*) $\dfrac{1}{1 - e^z}$. (M.D.U. 2011)

Sol. (*i*) $\dfrac{e^{2z}}{(z-1)^4} = \dfrac{e^{2(t+1)}}{t^4}$, where $t = z - 1$

$$= \frac{e^2 . e^{2t}}{t^4}$$

$$= \frac{e^2}{t^4}\left[1 + \frac{2t}{1!} + \frac{(2t)^2}{2!} + \frac{(2t)^3}{3!} + \frac{(2t)^4}{4!} + \frac{(2t)^5}{5!} + \ldots\ldots\right]$$

$$= e^2\left[t^{-4} + 2t^{-3} + 2t^{-2} + \frac{4}{3}t^{-1} + \frac{2}{3} + \frac{4}{15}t + \ldots\ldots\right]$$

$$= e^2\left[(z-1)^{-4} + 2(z-1)^{-3} + 2(z-1)^{-2} + \frac{4}{3}(z-1)^{-1} + \frac{2}{3} + \frac{4}{15}(z-1) + \ldots\ldots\right]$$

It has a finite number of negative power terms.

$$a_{-4} = e^2 \neq 0 \quad \text{and} \quad a_{-5} = a_{-6} = \ldots\ldots = 0$$

$\Rightarrow \quad z = 1$ is a pole of order 4.

(*ii*) $$ze^{\frac{1}{z^2}} = z\left[1 + \frac{\frac{1}{z^2}}{1!} + \frac{\left(\frac{1}{z^2}\right)^2}{2!} + \frac{\left(\frac{1}{z^2}\right)^3}{3!} + \ldots\ldots\right] = z + z^{-1} + \frac{1}{2}z^{-3} + \frac{1}{6}z^{-5} + \ldots\ldots$$

It has an infinite number of negative powers of z

$\Rightarrow \quad z = 0$ is an essential singularity of $ze^{\frac{1}{z^2}}$.

(*iii*) $$\frac{1}{1-e^z} = \frac{1}{1 - \left(1 + \frac{z}{1!} + \frac{z^2}{2!} + \frac{z^3}{3!} + \ldots\ldots\right)}$$

$$= \frac{1}{-z\left(1 + \frac{z}{2} + \frac{z^2}{6} + \ldots\ldots\right)} = -\frac{1}{z}\left[1 + \left(\frac{z}{2} + \frac{z^2}{6} + \ldots\ldots\right)\right]^{-1}$$

$$= -\frac{1}{z}\left[1 - \left(\frac{z}{2} + \frac{z^2}{6} + \ldots\ldots\right) + \left(\frac{z}{2} + \frac{z^2}{6} + \ldots\ldots\right)^2 \ldots\ldots\right]$$

$$= -\frac{1}{z}\left[1 - \frac{z}{2} + \left(\frac{1}{4} - \frac{1}{6}\right)^{z^2} \ldots\ldots\right] = -z^{-1} + \frac{1}{2} - \frac{1}{12}z + \ldots\ldots$$

It has a finite number of negative power terms.

$$a_{-1} = -1 \neq 0 \quad \text{and} \quad a_{-2} = a_{-3} = \ldots\ldots = 0$$

Also $1 - e^z = 0 \quad \Rightarrow \quad e^z = 1 = e^{2n\pi i}$, n being an integer.

$$z = 2n\pi i \ (n = 0, \pm 1, \pm 2, \ldots\ldots)$$

$\therefore \quad \dfrac{1}{1-e^z}$ has a simple poles at $z = 0, \pm 2\pi i, \pm 4\pi i, \ldots\ldots$

Example 3. *Find the residue at $z = 0$ of*

(i) $f(z) = z \cos \dfrac{1}{z}$ *(ii)* $f(z) = \operatorname{cosec}^2 z$ *(iii)* $f(z) = \dfrac{1+e^z}{\sin z + z \cos z}$.

(M.D.U. May 2014)

Sol. (*i*)

$$f(z) = z \cos \frac{1}{z} = z . \left\{1 - \frac{1}{2!} . \frac{1}{z^2} + \frac{1}{4!} \frac{1}{z^4} \ldots\ldots \infty\right\}$$

$$= z - \frac{1}{2!} . \frac{1}{z} + \frac{1}{4!} . \frac{1}{z^3} \ldots\ldots \infty$$

$$= z - \frac{1}{2!} . z^{-1} + \frac{1}{4!} z^{-3} \ldots\ldots \infty$$

which is Laurent's expansion about $z = 0$

$$\therefore \qquad a_{-1} = \text{coeff. of } z^{-1} = -\frac{1}{2}$$

By definition of residue (Art 4.8)

Residue of $f(z)$ at $z = 0$ is $-\dfrac{1}{2}$.

(*ii*) $f(z) = \operatorname{cosec}^2 z = \dfrac{1}{(\sin z)^2} = (\sin z)^{-2}$

$$= \left[z - \frac{z^3}{3!} + \frac{z^5}{5!} \ldots \infty\right]^{-2} = \frac{1}{z^2}\left[1 - \left(\frac{z^2}{3!} - \frac{z^4}{5!} + \ldots \infty\right)\right]^{-2}$$

$$= \frac{1}{z^2}\left[1 + 2\left(\frac{z^2}{3!} - \frac{z^4}{5!} + \ldots \infty\right) + 3\left(\frac{z^2}{3!} - \frac{z^4}{5!} + \ldots \infty\right) \ldots\right]^2$$

$$= \frac{1}{z^2}\left[1 + \frac{1}{3} z^2 + \frac{1}{15} z^4 + \ldots \infty\right]$$

$$= \frac{1}{z^2} + \frac{1}{3} + \frac{1}{15} z^2 + \ldots \infty = z^{-2} + \frac{1}{3} + \frac{1}{15} z^2 + \ldots \infty$$

Here $\qquad a_{-1} = \text{Coeff. of } z^{-1} = 0$

$\therefore$ Res. of $f(z)$ at $z = 0$ is 0

(*iii*) $f(z) = \dfrac{1+e^z}{\sin z + z \cos z}$

$z = 0$ is a simple pole of $f(z)$

$$\therefore \qquad \text{Res}\,[f(z), 0)] = \operatorname*{Lt}_{z \to 0} z . \frac{1+e^z}{\sin z + z \cos z} = \operatorname*{Lt}_{z \to 0} \frac{1+e^z}{\dfrac{\sin z}{z} + \cos z} = \frac{1+1}{1+1} = \frac{2}{2} = 1.$$

Example 4. *Determine the poles of the function* $f(z) = \dfrac{z^2}{(z-1)^2 (z+2)}$ *and the residue at each pole.* (U.K.T.U. 2010; G.B.T.U. 2011)

Sol. The function $f(z)$ has a pole of order 2 at $z = 1$ and a simple pole at $z = -2$.
Residue of $f(z)$ at $z = 1$ is

$$\lim_{z \to 1} \frac{d}{dz} [(z-1)^2 f(z)] = \lim_{z \to 1} \frac{d}{dz}\left(\frac{z^2}{z+2}\right)$$

$$= \lim_{z \to 1} \frac{(z+2).2z - z^2 .1}{(z+2)^2} = \lim_{z \to 1} \frac{z^2+4z}{(z+2)^2} = \frac{5}{9}$$

Residue of $f(z)$ at $z = -2$, is

$$= \lim_{z \to -2} [(z+2) f(z)] = \lim_{z \to -2} \frac{z^2}{(z-1)^2} = \frac{4}{9}.$$

Example 5. *Find the sum of the residues of the function* $f(z) = \frac{\sin z}{z \cos z}$ *at its poles inside the circle* $| z | = 2$. (M.D.U. Dec. 2015)

Sol. The function $f(z)$ has simple poles at $z = 0, \pm\frac{\pi}{2}, \pm\frac{3\pi}{2}, \ldots\ldots$ of these, $z = 0, \pm\frac{\pi}{2}$ lie inside $| z | = 2$.

Residue of $f(z)$ at $z = 0$ is $\lim_{z \to 0} [z . f(z)] = \lim_{z \to 0} \frac{\sin z}{\cos z} = 0$

Residue of $f(z)$ at $z = \frac{\pi}{2}$ is

$$\lim_{z \to \frac{\pi}{2}} \left(z - \frac{\pi}{2}\right) f(z) = \lim_{z \to \frac{\pi}{2}} \frac{\left(z - \frac{\pi}{2}\right)\sin z}{z \cos z} \qquad \left|\ \text{Form } \frac{0}{0}\right.$$

$$= \lim_{z \to \frac{\pi}{2}} \frac{\left(z - \frac{\pi}{2}\right)\cos z + \sin z}{\cos z - z \sin z} \qquad \text{[By L-Hospital's Rule]}$$

$$= \frac{1}{-\frac{\pi}{2}} = -\frac{2}{\pi}$$

Similarly, residue of $f(z)$ at $z = -\frac{\pi}{2}$ is $\frac{2}{\pi}$.

$\therefore$ Sum of residues $= 0 - \frac{2}{\pi} + \frac{2}{\pi} = 0$.

Note. Using Art. 4.10.(3). Another formula for Res $\{f(z), a\}$

$$\text{Res}\left\{f(z), \frac{\pi}{2}\right\} = \lim_{z \to \frac{\pi}{2}} \frac{\sin z}{\frac{d}{dz}(z \cos z)} = \lim_{z \to \frac{\pi}{2}} \frac{\sin z}{\cos z - z \sin z} = \frac{1}{0 - \frac{\pi}{2} \times 1} = -\frac{2}{\pi}$$

Similarly, $\text{Res}\left\{f(z), -\frac{\pi}{2}\right\} = \frac{2}{\pi}$.

Example 6. *Evaluate the residues of* $\frac{z^2}{(z-1)(z-2)(z-3)}$ *at* $z = 1, 2, 3$ *and infinity and show that their sum is zero.* (M.D.U. May 2014)

Sol. Let $$f(z) = \frac{z^2}{(z-1)(z-2)(z-3)}$$

Residue of $f(z)$ at $z = 1$ is

$$\lim_{z \to 1} (z-1) f(z) = \lim_{z \to 1} (z-1) \cdot \frac{z^2}{(z-1)(z-2)(z-3)}$$

$$= \lim_{z \to 1} \cdot \frac{z^2}{(z-2)(z-3)} = \frac{1^2}{(1-2)(1-3)} = \frac{1}{2}$$

Residue of $f(z)$ at $z = 2$ is

$$\lim_{z \to 2} (z-2) f(z) = \lim_{z \to 2} (z-2) \cdot \frac{z^2}{(z-1)(z-2)(z-3)}$$

$$= \lim_{z \to 2} \cdot \frac{z^2}{(z-1)(z-3)} = \frac{2^2}{(2-1)(2-3)} = -4$$

Residue of $f(z)$ at $z = 3$ is

$$\lim_{z \to 3} (z-3) f(z) = \lim_{z \to 3} (z-3) \cdot \frac{z^2}{(z-1)(z-2)(z-3)}$$

$$= \lim_{z \to 3} \cdot \frac{z^2}{(z-1)(z-2)} = \frac{3^2}{(3-1)(3-2)} = \frac{9}{2}$$

Residue of $f(z)$ at $z = \infty$ is

$$\lim_{z \to \infty} -z f(z) = \lim_{z \to \infty} -z \cdot \frac{z^2}{(z-1)(z-2)(z-3)}$$

$$= \lim_{z \to \infty} \cdot \frac{-1}{\left(1 - \frac{1}{z}\right)\left(1 - \frac{2}{z}\right)\left(1 - \frac{3}{z}\right)} = -1$$

Sum of residues at $z = 1, 2, 3$ and infinity is $\frac{1}{2} + (-4) + \left(\frac{9}{2}\right) + (-1) = 5 - 5 = 0.$

Example 7. *Evaluate* $\oint_C \frac{e^z}{(z+1)^2} dz$, *where C is the circle* $|z - 1| = 3$.

Sol. Here $f(z) = \frac{e^z}{(z+1)^2}$ has only one singular point $z = -1$ which is a pole of order 2 and it lies inside the circle $|z - 1| = 3$.

Residue of $f(z)$ at $z = -1$ is $\lim_{z \to -1} \frac{d}{dz} [(z+1)^2 f(z)] = \lim_{z \to -1} \frac{d}{dz} (e^z) = \lim_{z \to -1} e^z = e^{-1}$

$\therefore$ By residue theorem, we have $\oint_C \frac{e^z}{(z+1)^2} dz = 2\pi i (e^{-1}) = \frac{2\pi i}{e}$.

Example 8. *Evaluate* $\oint_C \frac{2z-1}{z(z+1)(z-3)}\,dz$, *where C is the circle* $|z| = 2$.

Sol. Here $f(z) = \frac{2z-1}{z(z+1)(z-3)}$ has three simple poles at $z = 0, -1, 3$ of which only $z = 0, -1$ lie inside the circle $|z| = 2$.

Residue of $f(z)$ at $z = 0$ is $\lim_{z\to 0} z f(z) = \lim_{z\to 0} \frac{2z-1}{(z+1)(z-3)} = \frac{1}{3}$

Residue of $f(z)$ at $z = -1$ is $\lim_{z\to -1} (z+1) f(z) = \lim_{z\to -1} \frac{2z-1}{z(z-3)} = -\frac{3}{4}$

$\therefore$ By residue theorem,

$$\oint_C \frac{2z-1}{z(z+1)(z-3)}\,dz = 2\pi i \text{ (sum of residues)}$$

$$= 2\pi i\left(\frac{1}{3}-\frac{3}{4}\right) = -\frac{5\pi i}{6}.$$

Example 9. *Evaluate* $\oint_C \frac{z-1}{(z+1)^2(z-2)}\,dz$, *where C is given by* $|z-i| = 2$.

(DCRUST Murthal, May 2014)

Sol. Poles of $f(z) = \frac{z-1}{(z+1)^2(z-2)}$ are given by

$$(z+1)^2(z-2) = 0$$

$\Rightarrow$ $z = -1$ is a double pole and $z = 2$ is a simple pole.

The given curve $C : |z-i| = 2$ is a circle with centre at $z = i$, *i.e.*, (0, 1) and radius 2.

Only the pole $z = -1$ lies inside C.

Residue of $f(z)$ at $z = -1$ is

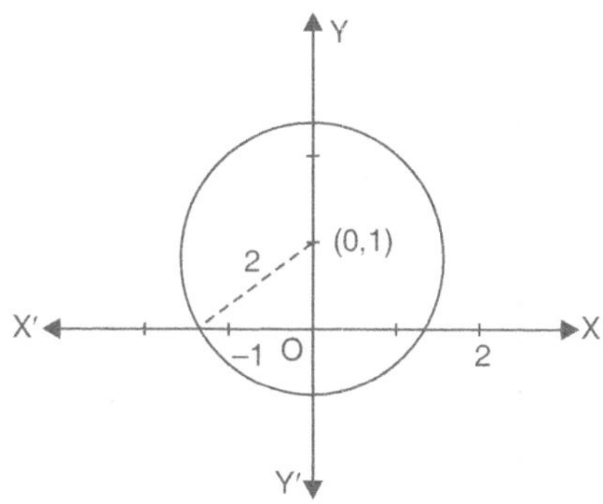

$$\frac{1}{(2-1)!}\lim_{z\to -1}\left[\frac{d}{dz}\left\{(z+1)^2 \cdot \frac{z-1}{(z+1)^2(z-2)}\right\}\right]$$

$$= \lim_{z\to -1}\left[\frac{d}{dz}\left(\frac{z-1}{z-2}\right)\right] = \lim_{z\to -1}\frac{-1}{(z-2)^2} = \frac{-1}{9}$$

$\therefore$ By Cauchy's residue theorem, we have

$$\oint_C f(z)\,dz = 2\pi i\left(-\frac{1}{9}\right) = -\frac{2\pi i}{9}.$$

Example 10. *Evaluate* $\oint_C \frac{z-3}{z^2+2z+5}\,dz$, *where C is the circle*

(*i*) $|z| = 1$ (*ii*) $|z+1-i| = 2$

(M.D.U. Dec., 2013; G.B.T.U. 2013)

(*iii*) $|z+1+i| = 2$.

Sol. The poles of $f(z) = \dfrac{z-3}{z^2+2z+5}$ are given by

$$z^2 + 2z + 5 = 0 \quad i.e., \quad z = \frac{-2 \pm \sqrt{4-20}}{2} = -1 \pm 2i$$

(*i*) Both the poles lie outside the circle $|z| = 1$.

$\therefore$ By Cauchy's theorem, we have $\oint_C \dfrac{z-3}{z^2+2z+5}\, dz = 0$

(*ii*) Only the pole $z = -1 + 2i$ lies inside the circle $|z+1-i| = 2$ *i.e.*, $|z-(-1+i)| = 2$ whose centre is at $-1+i$ and radius is 2.

Residue of $f(z)$ at $z = -1 + 2i$ is

$$\lim_{z \to -1+2i} (z+1-2i)\, f(x) = \lim_{z \to -1+2i} (z+1-2i) \frac{z-3}{z^2+2z+5}$$

$$= \lim_{z \to -1+2i} \frac{(z+1-2i)(z-3)}{(z+1+2i)(z+1-2i)}$$

$[\because \; z^2 + 2z + 5 = (z+1)^2 - 4i^2 = (z+1+2i)(z+1-2i)$
or roots of $z^2 + 2z + 5 = 0$ are $-1 \pm 2i$
$\therefore \; z^2 + 2z + 5 = (z-(-1+2i))(z-(-1-2i))$
$= (z+1-2i)(z+1-2i)]$

$$= \lim_{z \to -1+2i} \frac{z-3}{z+1+2i} = \frac{-1+2i-3}{-1+2i+1+2i} = \frac{2(i-2)}{4i} = \frac{i-2}{2i}.$$

$\therefore$ By residue theorem, $\oint_C \dfrac{z-3}{z^2+2z+5}\, dz = 2\pi i \left(\dfrac{i-2}{2i}\right) = \pi(i-2)$

(*iii*) Only the pole $z = -1 - 2i$ lies inside the circle $|z+1+i| = 2$ *i.e.*, $|z-(-1-i)| = 2$ whose centre is at $-1-i$ and radius 2.

Residue of $f(z)$ at $z = -1 - 2i$ is

$$\lim_{z \to -1-2i} (z+1+2i)\, f(z) = \lim_{z \to -1-2i} (z+1+2i) \frac{z-3}{z^2+2z+5}$$

$$= \lim_{z \to -1-2i} \frac{(z+1+2i)(z-3)}{(z+1+2i)(z+1-2i)}$$

$$= \lim_{z \to -1-2i} \frac{z-3}{z+1-2i} = \frac{-1-2i-3}{-1-2i+1-2i} = \frac{-2(i+2)}{-4i} = \frac{i+2}{2i}.$$

Example 11. *Evaluate* $\oint_C \dfrac{e^z}{\cos \pi z}$, *where C is the unit circle* $|z| = 1$.

(M.D.U. 2008)

Sol. Here $f(z) = \dfrac{e^z}{\cos \pi z}$ has simple poles at $z = \pm \dfrac{1}{2}, \pm \dfrac{3}{2}, \pm \dfrac{5}{2}, \ldots\ldots,$ of which only $z = \pm \dfrac{1}{2}$ lie inside the circle $| z | = 1$.

Residue of $f(z)$ at $z = \frac{1}{2}$ is

$$\lim_{z \to \frac{1}{2}} (z - \tfrac{1}{2}) f(z) = \lim_{z \to \frac{1}{2}} \frac{(z - \frac{1}{2}) e^z}{\cos \pi z} \qquad \Big|\ \text{Form } \frac{0}{0}$$

$$= \lim_{z \to \frac{1}{2}} \frac{(z - \frac{1}{2}) e^z + e^z}{-\pi \sin \pi z} \qquad |\ \text{By L' Hospital's Rule}$$

$$= \frac{e^{1/2}}{-\pi}.$$

Similarly, residue of $f(z)$ at $z = -\dfrac{1}{2}$ is $\dfrac{e^{-1/2}}{\pi}$.

$\therefore$ By residue theorem, $\displaystyle\oint_C \frac{e^z}{\cos \pi z}\, dz = 2\pi i$ (sum of residues)

$$= 2\pi i \left(-\frac{e^{1/2}}{\pi} + \frac{e^{-1/2}}{\pi} \right) = -4i \left(\frac{e^{1/2} - e^{-1/2}}{2} \right) = -4i \sinh \tfrac{1}{2}.$$

EXERCISE 4.2

Determine the poles of the following functions and the residue at each pole:

1. $\dfrac{2z + 1}{z^2 - z - 2}$
2. $\dfrac{z + 1}{z^2(z - 2)}$
3. $\dfrac{z^2}{(z - 1)(z - 2)^2}$
4. $\dfrac{1 - e^{2z}}{z^4}$
5. $\dfrac{e^z}{z^2 + \pi^2}$
6. $\dfrac{z}{\cos z}$
7. $\dfrac{z^2 - 2z}{(z + 1)^2 (z^2 + 4)}$.

Evaluate the following integrals:

8. $\displaystyle\oint_C \frac{z^2 + 2z - 2}{z - 4}\, dz$, where C is a closed curve containing the point $z = 4$ in its interior.
9. $\displaystyle\oint_C \frac{1 - 2z}{z(z - 1)(z - 2)}\, dz$, where C is the circle $| z | = 1.5$.
10. $\displaystyle\oint_C \frac{z}{(z - 1)(z - 2)^2}\, dz$, where C is the circle $| z - 2 | = \frac{1}{2}$. *(M.D.U. May 2015)*
11. $\displaystyle\oint_C \frac{12z - 7}{(z - 1)^2 (2z + 3)}\, dz$, where C is the circle $| z | = 2$. *(G.B.T.U. 2011)*
12. $\displaystyle\oint_C \frac{\sin \pi z^2 + \cos \pi z^2}{(z - 1)^2 (z - 2)}\, dz$, where C is the circle $| z | = 3$. *(P.T.U. 2010; M.D.U. Dec. 2013)*

13. $\oint_C \frac{1-\cos 2(z-3)}{(z-3)^3}\,dz$, where C: $|\, z-3 \,| = 1$.

14. $\oint_C \frac{z \sec z}{(1-z)^2}\,dz$, where C is the circle $|\, z \,| = 1.5$.

15. $\oint_C \frac{dz}{z^2+9}$, where C is

(*i*) $|\, z+3i \,| = 2$ (*ii*) $|\, z \,| = 5$

16. $\oint_C \frac{z^2}{(z-1)^2\,(z+2)}\,dz$, where C: $|\, z \,| = \frac{5}{2}$.

17. $\oint_C \frac{z^2-2z}{(z+1)^2\,(z^2+4)}\,dz$, where C: $|\, z \,| = 10$.

18. $\oint_C \frac{z^3}{(z-1)^4\,(z-2)\,(z-3)}\,dz$, where C: $|\, z \,| = \frac{5}{2}$.

19. $\oint_C \frac{dz}{\sinh z}$, where C is the circle $|\, z \,| = 4$.

20. $\oint_C \tan z\, dz$, where C is the circle $|\, z \,| = 2$.

Answers

1. $z = -1, 2;\ \frac{1}{3}, \frac{5}{3}$ **2.** $z = 0, 2;\ -\frac{3}{4}, \frac{3}{4}$ **3.** $z = 1, 2;\ 1, 0$ **4.** $z = 0;\ -\frac{4}{3}$

5. $z = \pm \pi i;\ \pm \frac{i}{2\pi}$ **6.** $z = (2n+1)\frac{\pi}{2};\ (-1)^{n+1}\,(2n+1)\,\frac{\pi}{2}$, where n is an integer.

7. $z = -1, \pm 2i;\ -\frac{14}{25}, \frac{7 \pm i}{25}$ **8.** $44\pi i$ **9.** $3\pi i$

10. $-2\pi i$ **11.** 0 **12.** $4\pi i(\pi + 1)$ **13.** $4\pi i$

14. $2\pi i \sec 1(1 + \tan 1)$ **15.** $-\frac{\pi}{3}, 0$ **16.** $2\pi i$

17. 0 **18.** $-\frac{27\,\pi i}{8}$ **19.** $-2\pi i$ **20.** $-4\pi i$.

4.11. APPLICATION OF RESIDUES TO EVALUATE REAL INTEGRALS

The residue theorem provides a simple and elegant method for evaluating many important definite integrals of real variables. Some of these are illustrated below:

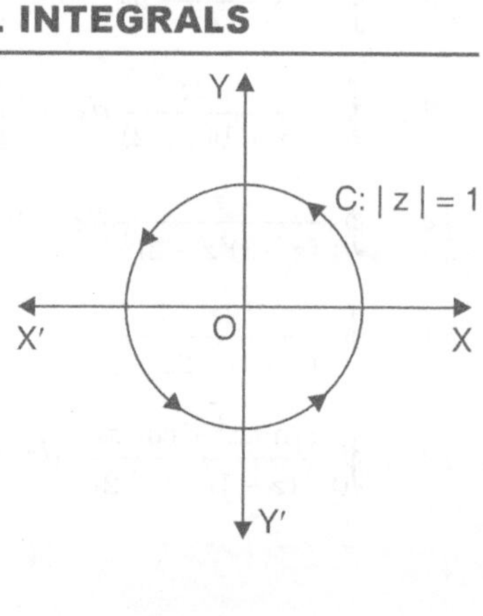

(1) **Integrals of the type $\int_0^{2\pi} F(\cos\theta, \sin\theta)\, d\theta$,**

where F(cos θ, sin θ) is a rational function of cos θ and sin θ.

Such integrals can be reduced to complex line integrals by the substitution $z = e^{i\theta}$, so that

$$dz = ie^{i\theta}\,d\theta, \text{ i.e., } d\theta = \frac{dz}{iz}.$$

Also,
$$\cos\theta = \frac{e^{i\theta}+e^{-i\theta}}{2} = \frac{1}{2}\left(z+\frac{1}{z}\right)$$

$$\sin\theta = \frac{e^{i\theta}-e^{-i\theta}}{2i} = \frac{1}{2i}\left(z-\frac{1}{z}\right).$$

As θ varies from 0 to 2π, z moves once round the unit circle in the anti-clockwise direction.

$$\therefore \quad \int_0^{2\pi} F(\cos\theta, \sin\theta)\,d\theta = \oint_C F\left(\frac{z+z^{-1}}{2}, \frac{z-z^{-1}}{2i}\right)\frac{dz}{iz}$$

where C is the unit circle $|z| = 1$.

The integral on the right can be evaluated by using the residue theorem.

ILLUSTRATIVE EXAMPLES

Example 1. *Evaluate* $\displaystyle\int_0^{2\pi} \frac{d\theta}{2+\cos\theta}$. (DCRUST Murthal May 2014; M.D.U. Dec. 2015)

Sol. Put $z = e^{i\theta}$ so that $\cos\theta = \frac{1}{2}\left(z+\frac{1}{z}\right)$ and $d\theta = \frac{dz}{iz}$

Then
$$\int_0^{2\pi} \frac{d\theta}{2+\cos\theta} = \oint_C \frac{1}{2+\frac{1}{2}\left(z+\frac{1}{z}\right)}\frac{dz}{iz} = \frac{2}{i}\oint_C \frac{1}{z^2+4z+1}\,dz$$

where C is the circle $|z| = 1$.

The poles of the integrand are the roots of $z^2 + 4z + 1 = 0$, which are $z = -2 \pm \sqrt{3}$.

Of the two poles, only $z = -2 + \sqrt{3}$ lies inside the circle C.

Residue at $z = -2 + \sqrt{3}$ is

$$= \lim_{z\to\alpha} (z-\alpha)\frac{1}{z^2+4z+1}, \quad \text{where } \alpha = -2+\sqrt{3} \qquad \left|\ \text{Form } \frac{0}{0}\right.$$

$$= \lim_{z\to\alpha} \frac{1}{2z+4} = \frac{1}{2\alpha+4} = \frac{1}{2\sqrt{3}}$$

$\therefore$ By residue theorem, $\displaystyle\oint_C \frac{dz}{z^2+4z+1} = 2\pi i\left(\frac{1}{2\sqrt{3}}\right) = \frac{\pi i}{\sqrt{3}}$

Hence
$$\int_0^{2\pi} \frac{d\theta}{2+\cos\theta} = \frac{2}{i}\oint_C \frac{dz}{z^2+4z+1} = \frac{2}{i}\left(\frac{\pi i}{\sqrt{3}}\right) = \frac{2\pi}{\sqrt{3}}.$$

Example 2. *Evaluate* $\displaystyle\int_0^{2\pi} \frac{d\theta}{1-2a\sin\theta+a^2}$, $0 < a < 1$. (M.D.U. 2011)

Sol. Put $z = e^{i\theta}$ so that $\sin\theta = \frac{1}{2i}\left(z - \frac{1}{z}\right)$ and $d\theta = \frac{dz}{iz}$.

Then, $$\int_0^{2\pi} \frac{d\theta}{1 - 2a\sin\theta + a^2} = \oint_C \frac{1}{1 - 2a \cdot \frac{1}{2i}\left(z - \frac{1}{z}\right) + a^2} \frac{dz}{iz}$$

$$= -\oint_C \frac{dz}{az^2 - i(a^2+1)z - a} = -\oint_C \frac{dz}{(az - i)(z - ia)}$$

where C is the circle $|z| = 1$.

The integrand has simple poles at $z = \frac{i}{a}$ and $z = ia$ of which only $z = ia$ lies inside the circle C. $[\because \quad 0 < a < 1]$

Residue at $z = ia$ is $\lim_{z \to ia} (z - ia) \frac{1}{(az - i)(z - ia)} = \lim_{z \to ia} \frac{1}{az - i} = \frac{1}{i(a^2 - 1)}$

$\therefore$ By residue theorem,

$$\oint_C \frac{dz}{(az - i)(z - ia)} = 2\pi i \cdot \frac{1}{i(a^2 - 1)} = \frac{2\pi}{a^2 - 1}$$

Hence $$\int_0^{2\pi} \frac{d\theta}{1 - 2a\sin\theta + a^2} = -\oint_C \frac{dz}{(az - i)(z - ia)} = \frac{2\pi}{1 - a^2}.$$

Example 3. *Prove that* $\int_0^{2\pi} \frac{\sin^2\theta}{a + b\cos\theta} d\theta = \frac{2\pi}{b^2}(a - \sqrt{a^2 - b^2})$, *where* $0 < b < a$.

Sol. Let $$I = \int_0^{2\pi} \frac{\sin^2\theta}{a + b\cos\theta} d\theta = \int_0^{2\pi} \frac{1 - \cos 2\theta}{2(a + b\cos\theta)} d\theta$$

$$= \text{Real part of } \int_0^{2\pi} \frac{1 - e^{2i\theta}}{2a + 2b\cos\theta} d\theta$$

Put $z = e^{i\theta}$ so that $\cos\theta = \frac{1}{2}\left(z + \frac{1}{z}\right)$ and $d\theta = \frac{dz}{iz}$

Then $$\int_0^{2\pi} \frac{1 - e^{2i\theta}}{2a + 2b\cos\theta} d\theta = \oint_C \frac{1 - z^2}{2a + b\left(z + \frac{1}{z}\right)} \frac{dz}{iz} = \oint_C \frac{1 - z^2}{i(bz^2 + 2az + b)} dz$$

where C is the circle $|z| = 1$.

The poles of the integrand are the roots of $bz^2 + 2az + b = 0$, *viz.*

$$z = \frac{-2a \pm \sqrt{4a^2 - 4b^2}}{2b} = \frac{-a \pm \sqrt{a^2 - b^2}}{b}$$

Let $$\alpha = \frac{-a + \sqrt{a^2 - b^2}}{b} \quad \text{and} \quad \beta = \frac{-a - \sqrt{a^2 - b^2}}{b}$$

Clearly, $|\beta| > 1$ so that $z = \alpha$ is the only simple pole inside C.

Also $\quad bz^2 + 2az + b = b(z - \alpha)(z - \beta)$

Residue at $z = \alpha$ is

$$\lim_{z \to \alpha} (z - \alpha) \cdot \frac{1 - z^2}{ib(z - \alpha)(z - \beta)} = \lim_{z \to \alpha} \frac{1 - z^2}{ib(z - \beta)} = \frac{1 - \alpha^2}{ib(\alpha - \beta)}$$

$$= \frac{\alpha\left(\frac{1}{\alpha} - \alpha\right)}{ib(\alpha - \beta)} = \frac{\alpha(\beta - \alpha)}{ib(\alpha - \beta)} \qquad [\because \ \alpha\beta = 1]$$

$$= -\frac{\alpha}{ib} = \frac{a - \sqrt{a^2 - b^2}}{ib^2}$$

$\therefore$ By residue theorem,

$$\oint_C \frac{1 - z^2}{i(bz^2 + 2az + b)} dz = 2\pi i \cdot \frac{a - \sqrt{a^2 - b^2}}{ib^2} = \frac{2\pi}{b^2}\left(a - \sqrt{a^2 - b^2}\right)$$

Hence I = Real part of $\oint_C \frac{1 - z^2}{i(bz^2 + 2az + b)} dz = \frac{2\pi}{b^2}\left(a - \sqrt{a^2 - b^2}\right)$.

Example 4. *Evaluate* $\int_0^{\pi} \frac{d\theta}{a + b \cos \theta}$, *where* $a > |b|$. (M.D.U. May 2013; P.T.U. 2010)

Sol. Since $\int_0^{2a} f(x)\, dx = 2 \int_0^{a} f(x)\, dx$, if $f(2a - x) = f(x)$

$$\therefore \quad \int_0^{2\pi} \frac{d\theta}{a + b \cos \theta} = 2 \int_0^{\pi} \frac{d\theta}{a + b \cos \theta} \qquad [\because \ \cos (2\pi - \theta) = \cos \theta]$$

or
$$\int_0^{\pi} \frac{d\theta}{a + b \cos \theta} = \frac{1}{2} \int_0^{2\pi} \frac{d\theta}{a + b \cos \theta} \qquad \dots(1)$$

Putting $z = e^{i\theta}$, so that $\cos \theta = \frac{1}{2}\left(z + \frac{1}{z}\right)$ and $d\theta = \frac{dz}{iz}$, we have

$$\int_0^{2\pi} \frac{d\theta}{a + b \cos \theta} = \oint_C \frac{1}{a + \frac{b}{2}\left(z + \frac{1}{z}\right)} \frac{dz}{iz} = \oint_C \frac{2dz}{i(bz^2 + 2az + b)}$$

where C is the circle $|z| = 1$.

Proceeding as in example 3, residue at $z = \alpha$ is

$$= \lim_{z \to \alpha} (z - \alpha) \cdot \frac{2}{ib(z - \alpha)(z - \beta)} = \lim_{z \to \alpha} \frac{2}{ib(z - \beta)} = \frac{2}{ib(\alpha - \beta)}$$

$$= \frac{2}{ib} \cdot \frac{b}{2\sqrt{a^2 - b^2}} = \frac{1}{i\sqrt{a^2 - b^2}}$$

$\therefore$ By residue theorem,

$$\oint_C \frac{2dz}{i(bz^2 + 2az + b)} = 2\pi i \cdot \frac{1}{i\sqrt{a^2 - b^2}} = \frac{2\pi}{\sqrt{a^2 - b^2}}$$

Hence from (1), $\int_0^{\pi} \frac{d\theta}{a + b \cos \theta} = \frac{\pi}{\sqrt{a^2 - b^2}}$.

(2) Integrals of the type $\int_{-\infty}^{\infty} \frac{f(x)}{F(x)} dx$, where f(x) and F(x) are polynomials in x such that $\frac{x f(x)}{F(x)} \to 0$ as $x \to \infty$ and F(x) has no zeros on the real axis.

Consider the integral $\oint_C \frac{f(z)}{F(z)} dz$ over the closed contour C consisting of the real axis from – R to R and the semi-circle C_1 of radius R in the upper half plane.

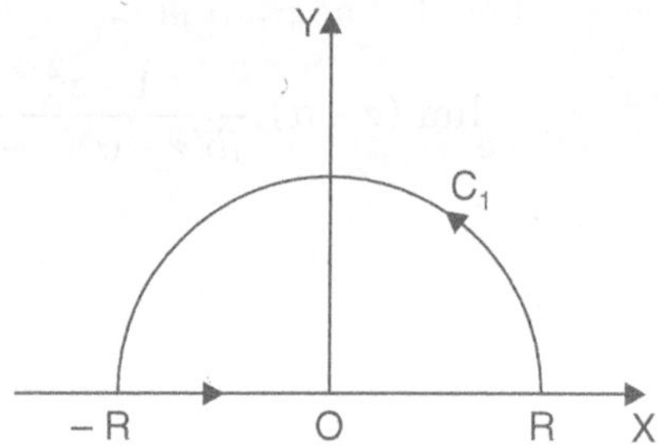

We take R large enough so that all the poles of $\frac{f(z)}{F(z)}$ in the upper half plane lie within C.

By residue theorem, we have

$$\oint_C \frac{f(z)}{F(z)} dz = 2\pi i \left[\text{sum of the residues of } \frac{f(z)}{F(z)} \text{ in the upper half plane}\right]$$

or $$\oint_{C_1} \frac{f(z)}{F(z)} dz + \int_{-R}^{R} \frac{f(x)}{F(x)} dx = 2\pi i \left[\text{sum of the residues of } \frac{f(z)}{F(z)} \text{ in the upper half plane}\right] \quad ...(1)$$

($\because$ on the real axis, $z = x$)

If we put $z = Re^{i\theta}$ in the first integral on the left side, then R is constant on C_1 and as z moves along C_1, θ varies from 0 to π.

$$\therefore \quad \int_{C_1} \frac{f(z)}{F(z)} dz = \int_0^{\pi} \frac{f(Re^{i\theta})}{F(Re^{i\theta})} Re^{i\theta}\, i\, d\theta$$

For large R, $\int_0^{\pi} \frac{f(Re^{i\theta})}{F(Re^{i\theta})} Re^{i\theta}\, i d\theta$ is of the order $\frac{Rf(R)}{F(R)}$

$$\therefore \quad \int_0^{\pi} \frac{f(Re^{i\theta})}{F(Re^{i\theta})} Re^{i\theta}\, i d\theta \to 0 \text{ when } R \to \infty$$

Hence from (1), we have

$$\int_{-\infty}^{\infty} \frac{f(x)}{F(x)} dx = 2\pi i \times \left[\text{sum of the residues of } \frac{f(z)}{F(z)} \text{ in the upper half plane}\right]$$

ILLUSTRATIVE EXAMPLES

Example 1. *Evaluate* $\int_{-\infty}^{\infty} \frac{x^2}{(x^2+a^2)(x^2+b^2)} dx$ *$(a > 0, b > 0)$.*

Sol. Consider the integral $\oint_C \frac{z^2}{(z^2+a^2)(z^2+b^2)} dz$ over the closed contour C consisting of the real axis from – R to R and the semi- circle C_1 of radius R in the upper half plane.

$$\therefore \quad \oint_C \frac{z^2}{(z^2+a^2)(z^2+b^2)} dz = \int_{C_1} \frac{z^2}{(z^2+a^2)(z^2+b^2)} dz + \int_{-R}^{R} \frac{z^2}{(z^2+a^2)(z^2+b^2)} dz \quad ...(1)$$

(*i*) To evaluate $\oint_C \frac{z^2}{(z^2+a^2)(z^2+b^2)}dz$

Poles of $\phi(z) = \frac{z^2}{(z^2+a^2)(z^2+b^2)}$ are given by $z^2 + a^2 = 0,\ z^2 + b^2 = 0$

i.e., $z = \pm ia,\ \pm ib$

of these, only $z = ia$ and $z = ib$ lie in the upper half plane.

Residue of $\phi(z)$ at $z = ia$

$$= \lim_{z \to ia} (z - ia)\frac{z^2}{(z+ia)(z-ia)(z^2+b^2)}$$

$$= \lim_{z \to ia} \frac{z^2}{(z+ia)(z^2+b^2)} = \frac{-a^2}{2ia(-a^2+b^2)} = \frac{a}{2i(a^2-b^2)}$$

Residue of $\phi(z)$ at $z = ib$

$$= \lim_{z \to ib} (z - ib)\frac{z^2}{(z^2+a^2)(z+ib)(z-ib)}$$

$$= \lim_{z \to ib} \frac{z^2}{(z^2+a^2)(z+ib)} = \frac{-b^2}{(-b^2+a^2)(2ib)} = \frac{-b}{2i(a^2-b^2)}$$

$\therefore$ By Cauchy Residue Theorem

$$\oint_C \frac{z^2}{(z^2+a^2)(z^2+b^2)}dz = 2\pi i\ [\text{Sum of residues of } \phi(z) \text{ within C}]$$

$$= 2\pi i\left[\frac{a}{2i(a^2-b^2)} - \frac{b}{2i(a^2-b^2)}\right] = \pi\left(\frac{a-b}{a^2-b^2}\right) = \frac{\pi}{a+b} \qquad \ldots(2)$$

(*ii*) Now $\int_{-R}^{R} \frac{z^2}{(z^2+a^2)(z^2+b^2)}dz = \int_{-\infty}^{\infty} \frac{x^2}{(x^2+a^2)(x^2+b^2)}dx \qquad \ldots(3)$

[$\because$ Along *x*-axis, $z = x$ and x varies from $-R$ to R where $R \to \infty$]

(*iii*) Also $\int_{C_1} \frac{z^2}{(z^2+a^2)(z^2+b^2)}dz$ Put $z = Re^{i\theta}$ (any point on C_1)

$$= \int_0^{\pi} \frac{R^2 e^{2i\theta}}{(R^2 e^{2i\theta}+a^2)(R^2 e^{2i\theta}+b^2)} \,.\, Re^{i\theta} \,.\, id\theta$$

$$= \int_0^{\pi} \frac{i\,R^3 e^{3i\theta}}{R^2\left(e^{2i\theta} + \frac{a^2}{R^2}\right) R^2\left(e^{2i\theta} + \frac{b^2}{R^2}\right)} d\theta$$

$$= \int_0^{\pi} \frac{i\,e^{3i\theta}}{R\left(e^{2i\theta} + \frac{a^2}{R^2}\right)\left(e^{2i\theta} + \frac{b^2}{R^2}\right)} d\theta \to 0 \text{ as } R \to \infty$$

$$\therefore \quad \int_{C_1} \frac{z^2}{(z^2 + a^2)(z^2 + b^2)} dz = 0 \qquad \ldots(4)$$

From (1), (2), (3) and (4)

$$\frac{\pi}{a+b} = 0 + \int_{-\infty}^{\infty} \frac{x^2}{(x^2 + a^2)(x^2 + b^2)} dx$$

Hence $\displaystyle\int_{-\infty}^{\infty} \frac{x^2}{(x^2 + a^2)(x^2 + b^2)} dx = \frac{\pi}{a+b}$

Example 2. *Evaluate* $\displaystyle\int_0^{\infty} \frac{dx}{x^4 + 1}$. (M.D.U. Dec. 2013)

Sol. $$\int_0^{\infty} \frac{dx}{x^4 + 1} = \frac{1}{2}\int_{-\infty}^{\infty} \frac{dx}{x^4 + 1} \qquad \left[\because \int_{-a}^{a} f(x)\,dx = 2\int_0^{a} f(x)dx \text{ if } f(x) \text{ is even}\right] \qquad \ldots(1)$$

Consider the integral $\displaystyle\oint_C \frac{dz}{z^4 + 1}$ over the closed contour C consisting of the real axis from –R to R and the semi-circle C_1 of radius R in the upper half plane.

$$\therefore \quad \oint_C \frac{dz}{z^4 + 1} = \int_{C_1} \frac{dz}{z^4 + 1} + \int_{-R}^{R} \frac{dz}{z^4 + 1} \qquad \ldots(2)$$

(*i*) To evaluate $\displaystyle\oint_C \frac{dz}{z^4 + 1}$.

Poles of $\phi(z) = \dfrac{1}{z^4 + 1}$ are obtained by solving $z^4 + 1 = 0$

Now $z^4 + 1 = 0$

$$\Rightarrow \qquad z = (-1)^{\frac{1}{4}} = (\cos\pi + i\sin\pi)^{\frac{1}{4}} = [\cos(2n\pi+\pi) + i\sin(2n\pi+\pi)]^{\frac{1}{4}}$$

$$= \cos\frac{(2n+1)\pi}{4} + i\sin\frac{(2n+1)\pi}{4} \quad \text{(By De Moivre's theorem)}$$

where $n = 0, 1, 2, 3$

When $n = 0$, $z = \cos\frac{\pi}{4} + i\sin\frac{\pi}{4} = \frac{1}{\sqrt{2}} + i\frac{1}{\sqrt{2}}$

When $n = 1$, $z = \cos\frac{3\pi}{4} + i\sin\frac{3\pi}{4} = -\frac{1}{\sqrt{2}} + i\frac{1}{\sqrt{2}}$

When $n = 2$, $z = \cos\frac{5\pi}{4} + i\sin\frac{5\pi}{4} = -\frac{1}{\sqrt{2}} - i\frac{1}{\sqrt{2}}$

When $n = 3$, $z = \cos\frac{7\pi}{4} + i\sin\frac{7\pi}{4} = \frac{1}{\sqrt{2}} - i\frac{1}{\sqrt{2}}$

Of these, only $z = \frac{1}{\sqrt{2}} + i\frac{1}{\sqrt{2}} = e^{\frac{i\pi}{4}}$ and $z = -\frac{1}{\sqrt{2}} + i\frac{1}{\sqrt{2}} = e^{\frac{3i\pi}{4}}$

lie in the upper half of z-plane.

Residue of $\phi(z)$ at $z = e^{\frac{i\pi}{4}}$ is

$$\lim_{z \to e^{\frac{i\pi}{4}}} \frac{z - e^{\frac{i\pi}{4}}}{z^4 + 1} \qquad \left| \text{Form } \frac{0}{0} \right.$$

$$= \lim_{z \to e^{i\frac{\pi}{4}}} \frac{1}{4z^3} \qquad | \text{ By L-Hospital's Rule}$$

$$= \frac{1}{4e^{\frac{3i\pi}{4}}} = \frac{1}{4}e^{-\frac{3i\pi}{4}} = \frac{1}{4}\left[\cos\frac{3\pi}{4} - i\sin\frac{3\pi}{4}\right]$$

$$= \frac{1}{4}\left(-\frac{1}{\sqrt{2}} - i\frac{1}{\sqrt{2}}\right) = -\frac{1+i}{4\sqrt{2}}$$

Residue of $\phi(z)$ at $z = e^{\frac{3i\pi}{4}}$ is

$$\lim_{z \to e^{\frac{3i\pi}{4}}} \frac{z - e^{\frac{3i\pi}{4}}}{z^4 + 1} \qquad \left| \text{Form } \frac{0}{0} \right.$$

$$= \lim_{z \to e^{\frac{3i\pi}{4}}} \frac{1}{4z^3} \quad | \text{ By L-Hospital's Rule}$$

$$= \frac{1}{4\, z^{9i\frac{\pi}{4}}} = \frac{1}{4\, e^{2\pi i + i\frac{\pi}{4}}} = \frac{1}{4\, e^{i\frac{\pi}{4}}} \qquad (\because e^{2\pi i} = 1)$$

$$= \frac{1}{4} e^{-i\frac{\pi}{4}} = \frac{1}{4}\left(\cos\frac{\pi}{4} - i\sin\frac{\pi}{4}\right) = \frac{1}{4}\left(\frac{1}{\sqrt{2}} - i\frac{1}{\sqrt{2}}\right) = \frac{1-i}{4\sqrt{2}}$$

$\therefore$ By Cauchy Residue Theorem

$$\oint_C \frac{dz}{z^4+1} = 2\pi i \text{ [Sum of residues of } \phi(z) \text{ within C]}$$

$$= 2\pi i\left(-\frac{1+i}{4\sqrt{2}} + \frac{1-i}{4\sqrt{2}}\right) = 2\pi i\left(\frac{-2i}{4\sqrt{2}}\right) = \frac{\pi}{\sqrt{2}} \qquad \ldots(3)$$

(*ii*) Now $\displaystyle\int_{-R}^{R} \frac{dz}{z^4+1} = \int_{-\infty}^{\infty} \frac{dx}{x^4+1}$ ...(4)

[$\because$ Along x-axis, $z = x$ and x varies from –R to R where $R \to \infty$]

(*iii*) Also $\displaystyle\int_{C_1} \frac{dz}{z^4+1}$ Put $z = Re^{i\theta}$ (any point on C_1)

$$= \int_0^{\pi} \frac{R e^{i\theta}.\, id\theta}{R^4 e^{4i\theta}+1} = \int_0^{\pi} \frac{ie^{i\theta} d\theta}{R^3\left(e^{4i\theta} + \frac{1}{R^4}\right)} \to 0 \text{ as } R \to \infty$$

$$\therefore \qquad \int_{C_1} \frac{dz}{z^4+1} = 0 \qquad \ldots(5)$$

From (2), (3), (4) and (5)

$$\frac{\pi}{\sqrt{2}} = 0 + \int_{-\infty}^{\infty} \frac{dx}{x^4+1}$$

$$\Rightarrow \qquad \frac{\pi}{\sqrt{2}} = 2\int_0^{\infty} \frac{dx}{x^4+1} \qquad \text{[Using (1)]}$$

$$\Rightarrow \qquad \int_0^{\infty} \frac{dx}{x^4+1} = \frac{\pi}{2\sqrt{2}}.$$

The above method can also be applied to some cases, where f (x) contains trigonometric functions (sin ax or cos ax) *i.e.*, integrals of the type $\int_{-\infty}^{\infty} \frac{\sin ax}{F(x)} dx$ or $\int_{-\infty}^{\infty} \frac{\cos ax}{F(x)} dx$, where a ≥ 0.

Example 3. *Evaluate* $\int_0^{\infty} \frac{\cos ax}{x^2+1} dx \ (a \geq 0)$.

Sol. $$\int_0^{\infty} \frac{\cos ax}{x^2+1} dx = \text{Real part of } \int_0^{\infty} \frac{e^{iax}}{x^2+1} dx$$

Consider $\oint_C \frac{e^{iaz}}{z^2+1} dz$, where C consists of

(*i*) the semicircle C_1 of radius R in the upper half plane

(*ii*) the real axis from – R to R

$$\therefore \quad \oint_C \frac{e^{iaz}}{z^2+1} dz = \oint_{C_1} \frac{e^{iaz}}{z^2+1} dz + \int_{-R}^{R} \frac{e^{iaz}}{z^2+1} dz \quad ...(1)$$

(*i*) To evaluate $\oint_C \frac{e^{iaz}}{z^2+1} dz$

Poles of the integral are given by $z^2 + 1 = 0 \Rightarrow z = \pm i$

Only $z = i$ lies within C

∴ Residue at $z = i$ (pole of order one) $= \underset{z \to i}{\text{Lt}} (z-i) \frac{e^{iaz}}{(z^2+1)} = \underset{z \to i}{\text{Lt}} \frac{e^{iaz}}{z+i} = \frac{e^{-a}}{2i}$

$$\therefore \quad \oint_C \frac{e^{iaz}}{z^2+1} dz = 2\pi i \cdot \frac{e^{-a}}{2i} = \pi e^{-a} \quad ...(2)$$

(*ii*) To evaluate $\oint_{C_1} \frac{e^{iaz}}{z^2+1} dz$ Put $z = Re^{i\theta}$

Now $$z \cdot \frac{e^{iaz}}{z^2+1} = Re^{i\theta} \frac{e^{iaR(\cos\theta + i\sin\theta)}}{R^2 e^{i2\theta}+1} = \frac{e^{i\theta} \cdot e^{iaR\cos\theta} \cdot e^{-aR\sin\theta}}{R \cdot \left[e^{i2\theta} + \frac{1}{R^2}\right]}$$

Since $$\left| e^{iaR\cos\theta} e^{i\theta} \right| = |\cos(aR\cos\theta) + i\sin(aR\cos\theta)| \cdot 1 \leq 1$$

$$\left| e^{-iaR\sin\theta} \right| \leq 1, \left| e^{2i\theta} + \frac{1}{R^2} \right| \neq 0$$

$$\therefore \quad z\frac{e^{iaz}}{z^2+1} \to 0 \text{ as } R \to \infty$$

$$\therefore \quad \int_{C_1} \frac{e^{iaz}}{z^2+1}\,dz \to 0 \quad \text{...(3)}$$

(*iii*)
$$\int_{-R}^{R} \frac{e^{iaz}}{z^2+1}\,dz = \int_{-\infty}^{\infty} \frac{e^{iax}}{x^2+1}\,dx \quad \text{...(4)}$$

[$\because$ Along x-axis, $z = x$ and x varies from $-$ R to R where R $\to \infty$]

$\therefore$ From (1), (2), (3), (4) $\pi e^{-a} = 0 + \int_{-\infty}^{\infty} \frac{e^{iax}}{x^2+1}\,dx$

$$\int_{-\infty}^{\infty} \frac{e^{iax}}{x^2+1}\,dx = \pi e^{-a}$$

$$\int_{-\infty}^{\infty} \frac{\cos ax}{x^2+1}\,dx = \text{Real part of } \int_{-\infty}^{\infty} \frac{e^{iax}}{x^2+1}\,dx = \pi e^{-a}$$

$$\therefore \quad \int_{0}^{\infty} \frac{\cos ax}{x^2+1}\,dx = \frac{1}{2}\int_{-\infty}^{\infty} \frac{\cos ax}{x^2+1}\,dx = \frac{\pi e^{-a}}{2}.$$

EXERCISE 4.3

Evaluate the following integrals by contour integration:

1. $\int_0^{2\pi} \frac{d\theta}{5 - 3\cos\theta}$ (*U.K.T.U. 2011*)

2. $\int_0^{2\pi} \frac{d\theta}{a + b\sin\theta}$, where $a > |\,b\,|$ (*G.B.T.U. 2012*)

3. $\int_0^{2\pi} \frac{\cos 3\theta}{5 - 4\cos\theta}\,d\theta$ (*M.D.U. Dec. 2010; K.U.K. 2013*)

4. $\int_0^{2\pi} \frac{\cos 2\theta}{5 + 4\cos\theta}\,d\theta$ (*M.T.U. 2012; G.B.T.U. 2010*)

5. $\int_0^{\pi} \frac{d\theta}{1 - 2r\cos\theta + r^2}$, $(0 < r < 1)$

6. $\int_0^{2\pi} \frac{\cos 2\theta\, d\theta}{1 - 2a\cos\theta + a^2}$, $(0 < a < 1)$

7. $\int_0^{2\pi} \frac{d\theta}{a + b\cos\theta}$, $(a > b > 0)$ (*U.K.T.U. 2010*)

8. $\int_0^{2\pi} \frac{d\theta}{1 + a\cos\theta}$, $(a^2 < 1)$

9. $\int_0^{\pi} \frac{d\theta}{17 - 8\cos\theta}$

10. $\int_0^{\pi} \frac{a\, d\theta}{a^2 + \sin^2\theta}$, $(a > 0)$

11. $\int_0^{2\pi} \frac{d\theta}{(5 - 3\cos\theta)^2}$

12. $\int_0^{2\pi} \frac{d\theta}{(a + b\cos\theta)^2}$, $(a > b)$

13. $\int_0^{2\pi} \frac{d\theta}{5 - 4\sin\theta}$

14. $\int_0^{2\pi} \frac{\cos\theta}{3 + \sin\theta} d\theta$

15. (a) $\int_0^{\pi} \frac{1 + 2\cos\theta}{5 + 4\cos\theta} d\theta$ (b) $\int_0^{2\pi} \frac{1 + \sin\theta}{3 + \cos\theta} d\theta$

16. $\int_0^{2\pi} \frac{\sin^2\theta}{5 - 4\cos\theta} d\theta$

17. $\int_{-\infty}^{\infty} \frac{x^2\,dx}{(x^2 + 1)(x^2 + 4)}$

18. $\int_{-\infty}^{\infty} \frac{x^2 - x + 2}{x^4 + 10x^2 + 9} dx$

19. $\int_{-\infty}^{\infty} \frac{dx}{x^4 + 1}$

20. $\int_0^{\infty} \frac{dx}{x^6 + 1}$

21. $\int_0^{\infty} \frac{x^2}{x^6 + 1} dx$ *(M.D.U. 2012, Dec. 2014)*

22. $\int_0^{\infty} \frac{dx}{(a^2 + x^2)^2}$ *(M.T.U. 2013)*

23. $\int_0^{\infty} \frac{x^2\,dx}{(x^2 + 9)(x^2 + 4)^2}$

24. $\int_{-\infty}^{\infty} \frac{dx}{(x^2 + 1)^3}$ *(M.D.U. 2010)*

25. $\int_{-\infty}^{\infty} \frac{x^2}{(1 + x^2)^3} dx$

Answers

1. $\frac{\pi}{2}$ 2. $\frac{2\pi}{\sqrt{a^2 - b^2}}$ 3. $\frac{\pi}{12}$ 4. $\frac{\pi}{6}$

5. $\frac{\pi}{1 - r^2}$ 6. $\frac{2\pi a^2}{1 - a^2}$ 7. $\frac{2\pi}{\sqrt{a^2 - b^2}}$ 8. $\frac{2\pi}{\sqrt{1 - a^2}}$

9. $\frac{\pi}{15}$ 10. $\frac{\pi}{\sqrt{1 + a^2}}$ 11. $\frac{5\pi}{32}$ 12. $\frac{2\pi a}{(a^2 - b^2)^{3/2}}$

13. $\frac{2\pi}{3}$ 14. 0 15. (a) 0 (b) $\frac{\pi}{\sqrt{2}}$

16. $\frac{\pi}{4}$ 17. $\frac{\pi}{3}$ 18. $\frac{5\pi}{12}$ 19. $\frac{\pi}{\sqrt{2}}$

20. $\frac{\pi}{3}$ 21. $\frac{\pi}{6}$ 22. $\frac{\pi}{4a^3}$ 23. $\frac{\pi}{200}$

24. $\frac{3\pi}{8}$ 25. $\frac{\pi}{8}$.

❑❑❑

CHAPTER 5

Probability Distributions

5.1. PROBABILITY

Here we define and explain certain terms which are used frequently.

(*a*) **Trial and event.** Let an experiment be repeated under essentially the same conditions and let it result in any one of the several possible outcomes. Then, the experiment is called a *trial* and the possible outcomes are known as *events* or *cases*.

For example: (*i*) Tossing of a coin is a trial and the turning up of head or tail is an event.

(*ii*) Throwing a die is a trial and getting 1 or 2 or 3 or 4 or 5 or 6 is an event.

(*b*) **Exhaustive events.** The total number of all possible outcomes in any trial is known as *exhaustive events* or *exhaustive cases*.

For example: (*i*) In tossing a coin, there are two exhaustive cases, head and tail.

(*ii*) In throwing a die, there are 6 exhaustive cases, for any one of the six faces may turn up.

(*iii*) In throwing two dice, the exhaustive cases are $6 \times 6 = 6^2$ for any of the 6 numbers from 1 to 6 on one die can be associated with any of the 6 numbers on the other die.

In general, in throwing n dice, the exhaustive cases are 6^n.

(*c*) **Favourable events or cases.** The cases which entail the happening of an event are said to be *favourable* to the event. It is the total number of possible outcomes in which the specified event happens.

For example: (*i*) In throwing a die, the number of cases favourable to the appearance of a multiple of 3 are two *viz.* 3 and 6 while the number of cases favourable to the appearance of an even number are three, *viz.*, 2, 4 and 6.

(*ii*) In a throw of two dice, the number of cases favourable to getting a sum 6 is 5, *viz.*, (1, 5); (5, 1); (2, 4); (4, 2); (3, 3).

(*d*) **Mutually exclusive events.** Events are said to be *mutually exclusive* or *incompatible* if the happening of any one of them precludes (*i.e.,* rules out) the happening of all others, *i.e.*, if no two or more than two of them can happen simultaneously in the same trial.

For example: (*i*) In tossing a coin, the events head and tail are mutually exclusive, since if the outcome is head, the possibility of getting tail in the same trial is ruled out.

(*ii*) In throwing a die, all the six faces numbered, 1, 2, 3, 4, 5, 6 are mutually exclusive since any outcome rules out the possibility of getting any other.

(*e*) **Equally likely events.** Events are said to be *equally likely* if there is no reason to expect any one in preference to any other.

For example: (*i*) When a card is drawn from a well shuffled pack, any card may appear in the draw so that the 52 different cases are equally likely.

(*ii*) In throwing a die, all the six faces are equally likely to come.

(*f*) **Independent and dependent events.** Two or more events are said to be *independent* if the happening or non-happening of any one does not depend (or is not affected) by the happening or non-happening of any other. Otherwise they are said to be *dependent*.

For example. If a card is drawn from a pack of well shuffled cards and replaced before drawing the second card, the result of the second draw is independent of the first draw. However, if the first card drawn is not replaced, then, the second draw is dependent on the first draw.

5.2. MATHEMATICAL (or Classical) DEFINITION OF PROBABILITY

If a trial results in n *exhaustive, mutually exclusive* and *equally likely cases* and m of them are favourable to the happening of an event E, then the probability of happening of E is given by

$$p \text{ or } P(E) = \frac{\text{Favourable number of cases}}{\text{Exhaustive number of cases}} = \frac{m}{n}.$$

Note 1. Since the number of cases favourable to happening of E is m and the exhaustive number of cases in n, therefore, the number of cases unfavourable to happening of E are $n - m$.

Note 2. The probability that the event E will not happen is given by

$$q \text{ or } P(E') = \frac{\text{Unfavourable number of cases}}{\text{Exhaustive number of cases}} = \frac{n-m}{n} = 1 - \frac{m}{n} = 1 - p$$

$$p + q = 1 \quad i.e., \quad P(E) + P(E') = 1$$

Obviously, p and q are non-negative and cannot exceed unity, *i.e.*, $0 \le p \le 1, 0 \le q \le 1$.

Note 3. If P(E) = 1, E is called a *certain event i.e.*, the chance of its happening is cent per cent. If P(E) = 0, then E is an *impossible event*.

Note 4. If n cases are favourable to E and m cases are favourable to E′ (*i.e.*, unfavourable to E), then exhaustive number of cases = $n + m$.

$$P(E) = \frac{n}{n+m} \quad \text{and} \quad P(E') = \frac{m}{n+m}$$

We say that "odds in *favour of E*" are $n : m$ and "odds *against* E" are $m : n$.

ILLUSTRATIVE EXAMPLES

Example 1. *A bag contains 7 white, 6 red and 5 black balls. Two balls are drawn at random. Find the probability that they will both be white.*

Sol. Total number of balls = 7 + 6 + 5 = 18.

Out of 18 balls, 2 can be drawn in $^{18}C_2$ ways.

$\therefore$ Exhaustive number of cases $= {}^{18}C_2 = \dfrac{18 \times 17}{2 \times 1} = 153$

Out of 7 white balls, 2 can be drawn in $^{7}C_2 = \dfrac{7 \times 6}{2 \times 1} = 21$ ways.

$\therefore$ Favourable number of cases $= 21$

Required probability $= \dfrac{21}{153} = \dfrac{7}{51}$.

Example 2. *Four cards are drawn from a pack of cards. Find the probability that (i) all are diamonds, (ii) there is one card of each suit, and (iii) there are two spades and two hearts.*

Sol. 4 cards can be drawn from a pack of 52 cards in ${}^{52}C_4$ ways.

$$\therefore \quad \text{Exhaustive number of cases} = {}^{52}C_4 = \frac{52 \times 51 \times 50 \times 49}{4 \times 3 \times 2 \times 1} = 270725.$$

(*i*) There are 13 diamonds in the pack and 4 can be drawn out of them in ${}^{13}C_4$ ways.

$$\therefore \quad \text{Favourable number of cases} = {}^{13}C_4 = \frac{13 \times 12 \times 11 \times 10}{4 \times 3 \times 2 \times 1} = 715.$$

$$\text{Required probability} = \frac{715}{270725} = \frac{143}{54145} = \frac{11}{4165}.$$

(*ii*) There are 4 suits, each containing 13 cards.

$$\therefore \quad \text{Favourable number of cases} = {}^{13}C_1 \times {}^{13}C_1 \times {}^{13}C_1 \times {}^{13}C_1 = 13 \times 13 \times 13 \times 13.$$

$$\text{Required probability} = \frac{13 \times 13 \times 13 \times 13}{270725} = \frac{2197}{20825}.$$

(*iii*) 2 spades out of 13 can be drawn in ${}^{13}C_2$ ways.

2 hearts out of 13 can be drawn in ${}^{13}C_2$ ways.

$$\therefore \quad \text{Favourable number of cases} = {}^{13}C_2 \times {}^{13}C_2 = 78 \times 78$$

$$\text{Required probability} = \frac{78 \times 78}{270725} = \frac{468}{20825}.$$

Example 3. *A bag contains 50 tickets numbered 1, 2, 3, ..., 50, of which five are drawn at random and arranged in ascending order of magnitude* ($x_1 < x_2 < x_3 < x_4 < x_5$). *What is the probability that* $x_3 = 30$ *?*

Sol. Exhaustive number of cases ${}^{50}C_5$.

If $x_3 = 30$, then the two tickets with numbers x_1 and x_2 must come out of 29 tickets numbered 1 to 29 and this can be done in ${}^{29}C_2$ ways. The other two tickets with numbers x_4 and x_5 must come out of the 20 tickets number 31 to 50 and this can be done in ${}^{20}C_2$ ways.

$$\therefore \quad \text{Favourable number of cases} = {}^{29}C_2 \times {}^{20}C_2.$$

$$\text{Required probability} = \frac{{}^{29}C_2 \times {}^{20}C_2}{{}^{50}C_5} = \frac{551}{15134}.$$

5.3. RANDOM EXPERIMENT

Occurrences which can be repeated a number of times, essentially under the same conditions, and whose result cannot be predicted before hand are known as **random experiments.**

For example, rolling of a die, tossing a coin, taking out balls from an urn.

Sample Space. Out of the several possible outcomes of a random experiment, one and only one can take place in a trial. The set of all these possible outcomes is called the **sample space** for the particular experiment and is denoted by S.

For example, if a coin is tossed, the possible outcomes are H (Head) and T (Tail).

Thus $\qquad S = \{H, T\}$.

Sample Point. The elements of S, the sample space, are called **sample points.**

For example, if a coin is tossed and H and T denote 'Head' and 'Tail' respectively, then S = {H, T}.

The two sample points are H and T.

Finite Sample Space. If the number of sample points in a sample space is finite, we call it a **finite sample space.** (In this chapter, we shall deal with finite sample spaces only).

Event. Every subset of S, the sample space, is called an event.

Since $S \subset S$, S itself is an event; called a **certain event.**

Also, $\phi \subset S$, the null set is also an event, called an **impossible event.**

If $e \in S$, then e is called an **elementary event.** Every elementary event contains only one sample point.

5.4. AXIOMS

(*i*) With each event E (*i.e.*, a sample point) is associated a real number between 0 and 1, called the probability of that event and is denoted by P(E). Thus $0 \le P(E) \le 1$.

(*ii*) The sum of the probabilities of all simple (elementary) events constituting the sample space is 1. Thus P(S) = 1.

(*iii*) The probability of a compound event (*i.e.*, an event made up of two or more sample events) is the sum of the probabilities of the simple events comprising the compound event.

Thus, if there are n *equally likely* possible outcomes of a random experiment, then the sample space S contains n sample points and the probability associated with each sample point is $\frac{1}{n}$. [By Axiom (*ii*)]

Now, if an event E consists of m sample points, then the probability of E is

$$P(E) = \frac{1}{n} + \frac{1}{n} + \cdots\cdots m \text{ times} = \frac{m}{n}$$

$$= \frac{\text{Number of sample points in E}}{\text{Number of sample points in S}}.$$

This closely agrees with the classical definition of probability.

5.5. PROBABILITY OF THE IMPOSSIBLE EVENT IS ZERO, *i.e.*, P (ϕ) = 0

Proof. Impossible event contains no sample point. As such, the sample space S and the impossible event ϕ are *mutually exclusive*.

$$\Rightarrow \quad S \cup \phi = S \quad \Rightarrow \quad P(S \cup \phi) = P(S)$$

$$\Rightarrow \quad P(S) + P(\phi) = P(S) \quad \Rightarrow \quad P(\phi) = 0.$$

5.6. PROBABILITY OF THE COMPLEMENTARY EVENT A′ OF A IS GIVEN BY P (A′) = 1 – P(A)

Proof. A′ and A are mutually exclusive events. Also $A \cup A' = S$

$$\therefore \quad P(A \cup A') = P(S)$$

$$\Rightarrow \quad P(A) + P(A') = 1 \text{ Hence } P(A') = 1 - P(A).$$

5.7. FOR ANY TWO EVENTS A AND B, P(A′ ∩ B) = P(B) – P(A ∩ B)

Proof. $A' \cap B = \{p : p \in B \text{ and } p \notin A\}$

Now $A' \cap B$ and $A \cap B$ are disjoint sets and

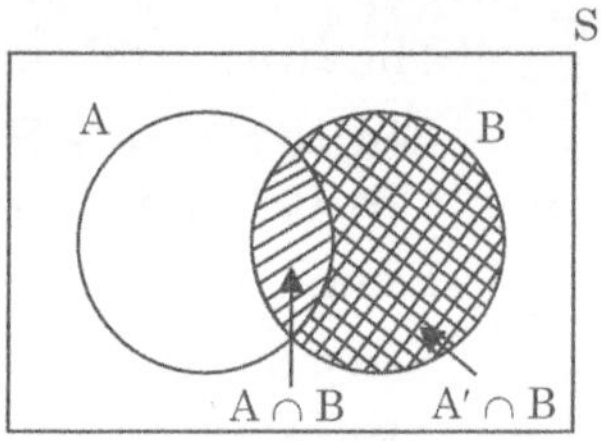

$$(A' \cap B) \cup (A \cap B) = B$$

$$\Rightarrow \quad P[(A' \cap B) \cup (A \cap B)] = P(B)$$

$$\Rightarrow \quad P(A' \cap B) + P(A \cap B) = P(B)$$

$$\Rightarrow \quad P(A' \cap B) = P(B) - P(A \cap B).$$

Note. Similarly, it can be proved that $P(A \cap B') = P(A) - P(A \cap B)$.

5.8. IF B ⊂ A, THEN (*i*) P(A ∩ B′) = P(A) – P(B) (*ii*) P(B) ≤ P(A)

Proof. When $B \subset A$, B and $A \cap B'$ are disjoint and their union is A.

$$\Rightarrow \quad B \cup (A \cap B') = A$$

$$\Rightarrow \quad P[B \cup (A \cap B')] = P(A)$$

$$\Rightarrow \quad P(B) + P(A \cap B') = P(A)$$

$$\Rightarrow \quad P(A \cap B') = P(A) - P(B) \quad \ldots(1)$$

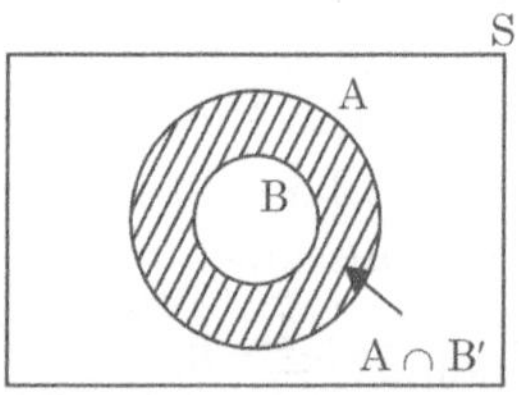

Now, if E is any event,

then $0 \le P(E) \le 1$, *i.e.*, $P(E) \ge 0$

$\therefore \quad P(A \cap B') \ge 0 \Rightarrow P(A) - P(B) \ge 0$ [Using 1]

$\Rightarrow \quad P(B) \le P(A).$

5.9. P(A ∩ B) ≤ P(A) and P(A ∩ B) ≤ P(B)

Proof. By 5.12 $B \subset A \Rightarrow P(B) \le P(A)$

Since $(A \cap B) \subset A$ and $(A \cap B) \subset B$

$\therefore \quad P(A \cap B) \le P(A)$ and $P(A \cap B) \le P(B)$.

5.10. ADDITION THEOREM OF PROBABILITY

Statement. *If A and B are any two events, then*

$$P(A \cup B) = P(A) + P(B) - P(A \cap B)$$

i.e., $$P(A \text{ or } B) = P(A) + P(B) - P(A \text{ and } B).$$

Proof. A and $A' \cap B$ are *disjoint sets* and their union is $A \cup B$.

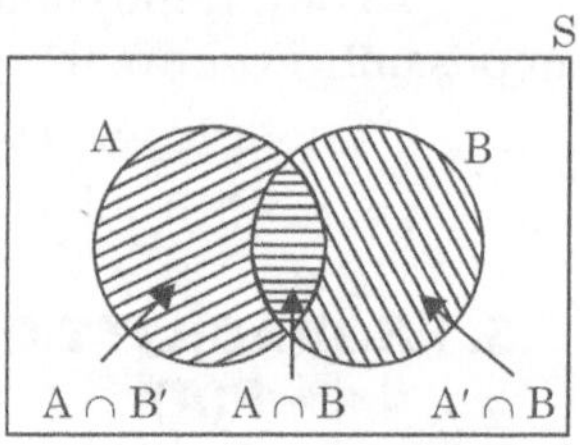

$$\Rightarrow \quad A \cup B = A \cup (A' \cap B)$$

$$\Rightarrow \quad P(A \cup B) = P[(A \cup (A' \cap B)] = P(A) + P(A' \cap B)$$

$$= P(A) + [P(A' \cap B) + P(A \cap B) - P(A \cap B)]$$

$$= P(A) + P[(A' \cap B) \cup (A \cap B)] - P(A \cap B)]$$

[$\because$ $A' \cap B$ and $A \cap B$ are disjoint]

$$= P(A) + P(B) - P(A \cap B)$$

[$\because$ $(A' \cap B) \cup (A \cap B) = B$]

$$P(A \cup B) = P(A) + P(B) - P(A \cap B).$$

Note 1. If A and B are two mutually disjoint events, then $A \cap B = \phi$, so that $P(A \cap B) = P(\phi) = 0$.

$$\therefore \qquad P(A \cup B) = P(A) + P(B).$$

Note 2. $P(A \cup B)$ is also written as $P(A + B)$. Thus, for mutually disjoint events A and B,

$$P(A + B) = P(A) + P(B).$$

$P(A \cap B)$ is also written as $P(AB)$.

5.11. IF A, B AND C ARE ANY THREE EVENTS, THEN

$$\mathbf{P(A \cup B \cup C) = P(A) + P(B) + P(C) - P(A \cap B) - P(B \cap C) - P(C \cap A) + P(A \cap B \cap C)}$$

Or

$$\mathbf{P(A + B + C) = P(A) + P(B) + P(C) - P(AB) - P(BC) - P(CA) + P(ABC).}$$

Proof. Using the above Art. 5.10 for two events, we have

$$\begin{aligned} P(A \cup B \cup C) &= P[(A \cup B) \cup C] \\ &= P(A \cup B) + P(C) - P[(A \cup B) \cap C] \\ &= [P(A) + P(B) - P(A \cap B)] + P(C) - P[(A \cap C) \cup (B \cap C)] \quad \text{[By distributive law]} \\ &= P(A) + P(B) + P(C) - P(A \cap B) - [P(A \cap C) \\ &\qquad + P(B \cap C) - P\{(A \cap C) \cap (B \cap C)\}] \quad \text{[By 5.10]} \\ &= P(A) + P(B) + P(C) - P(A \cap B) - P(A \cap C) - P(B \cap C) + P(A \cap B \cap C) \\ &\qquad [\because (A \cap C) \cap (B \cap C) = A \cap B \cap C] \\ &= P(A) + P(B) + P(C) - P(A \cap B) - P(B \cap C) - P(C \cap A) + P(A \cap B \cap C) \\ &\qquad [\because A \cap C = C \cap A] \end{aligned}$$

or $\quad P(A + B + C) = P(A) + P(B) + P(C) - P(AB) - P(BC) - P(CA) + P(ABC).$

5.12. IF $A_1, A_2, ..., A_n$ ARE n MUTUALLY EXCLUSIVE EVENTS, THEN THE PROBABILITY OF THE HAPPENING OF ONE OF THEM IS

$$\mathbf{P(A_1 \cup A_2 \cup \ldots\ldots \cup A_n) = P(A_1 + A_2 + \ldots + A_n) = P(A_1) + P(A_2) + \cdots + P(A_n)}$$

Proof. Let N be the total number of mutually exclusive, exhaustive and equally likely cases of which m_1 are favourable to A_1, m_2 are favourable to A_2 and so on.

$$\left.\begin{aligned} &\text{Probability of occurrence of event } A_1 = P(A_1) = \frac{m_1}{N} \\ &\text{Probability of occurrence of event } A_2 = P(A_2) = \frac{m_2}{N} \\ &\ldots\ldots\ldots\ldots\ldots\ldots\ldots\ldots\ldots\ldots\ldots\ldots \\ &\text{Probability of occurrence of event } A_n = P(A_n) = \frac{m_n}{N} \end{aligned}\right\} \qquad \ldots(1)$$

The events being mutually exclusive and equally likely, the number of cases favourable to the event

$$A_1 \text{ or } A_2 \text{ or } \ldots \text{ or } A_n \text{ is } m_1 + m_2 + \cdots + m_n.$$

$\therefore$ Probability of occurrence of one of the events $A_1, A_2, \cdots\cdots, A_n$ is $P(A_1 + A_2 + \cdots + A_n)$

$$\begin{aligned} &= \frac{m_1 + m_2 + \cdots + m_n}{N} = \frac{m_1}{N} + \frac{m_2}{N} + \cdots\cdots + \frac{m_n}{N} \\ &= P(A_1) + P(A_2) + \cdots\cdots + P(A_n) \qquad | \text{ Using (1)} \end{aligned}$$

Note. The student should not get confused with Theorems 5.10, 5.11 and 5.12. Theorems 5.10 and 5.11 are for ANY events (mutually exclusive or not) whereas Theorem 5.11 is for mutually exclusive events.

ILLUSTRATIVE EXAMPLES

Example 1. *In a given race, the odds in favour of four horses A, B, C, D are 1 : 3, 1 : 4, 1 : 5, 1 : 6 respectively. Assuming that a dead heat is impossible; find the chance that one of them wins the race.* (K.U.K. 2013)

Sol. Let p_1, p_2, p_3, p_4 be the probabilities of winning of the horses A, B, C, D respectively.

Since a *dead heat* (in which all the four horses cover *same distance in same time*) is not possible, the events are mutually exclusive.

Odds in favour of A are 1 : 3 $\therefore \quad p_1 = \frac{1}{1+3} = \frac{1}{4}$

Similarly, $\quad p_2 = \frac{1}{5}, p_3 = \frac{1}{6}, p_4 = \frac{1}{7}.$

If p is the chance that one of them wins, then

$$p = p_1 + p_2 + p_3 + p_4 = \frac{1}{4} + \frac{1}{5} + \frac{1}{6} + \frac{1}{7} = \frac{319}{420}.$$

Example 2. *A card is drawn from a well-shuffled pack of playing cards. What is the probability that it is either a spade or an ace ?*

Sol. Let A = the event of drawing a spade

and B = the event of drawing an ace

A and B are *not* mutually exclusive.

AB = the even of drawing the ace of spades

$$P(A) = \frac{13}{52}, P(B) = \frac{4}{52}, P(A \cap B) = \frac{1}{52}$$

$$\therefore \quad P(A \text{ or } B) = P(A) + P(B) - P(A \cap B) = \frac{13}{52} + \frac{4}{52} - \frac{1}{52} = \frac{16}{52} = \frac{4}{13}.$$

5.13. CONDITIONAL PROBABILITY

The probability of the happening of an event E_1 when another event E_2 is known to have already happened is called *Conditional Probability* and is denoted by $P(E_1/E_2)$.

Mutually Independent Events. An event E_1 is said to be independent of an event E_2 if

$$P(E_1/E_2) = P(E_1).$$

i.e., if the probability of happening of E_1 is independent of the happening of E_2.

5.14. MULTIPLICATIVE LAW OF PROBABILITY (Or Theorem of Compound Probability)

The probability of simultaneous occurrence of two events is equal to the probability of one of the events multiplied by the conditional probability of the other, *i.e.*, **for two events A and B,**

$$\mathbf{P(A \cap B) = P(A) \times P(B/A)}$$

where P(B/A) represents the conditional probability of occurrence of B when the event A has already happened.

Note. $P(A \cap B)$ is also written as $P(AB)$.

Thus $P(AB) = P(A) \times P(B/A)$.

Cor. 1. Interchanging A and B

$$P(BA) = P(B) \times P(A/B)$$

or $$P(AB) = P(B) \times P(A/B) \qquad [\because \; B \cap A = A \cap B]$$

Cor. 2. If A and B are independent events, then $P(B/A) = P(B)$

$\therefore$ $$P(AB) = P(A) \times P(B).$$

Generalisation. If $A_1, A_2, \ldots\ldots, A_n$ are n independent events, then

$$P(A_1 A_2 \ldots A_n) = P(A_1) \times P(A_2) \times \ldots\ldots \times P(A_n).$$

Cor. 3. If p is the chance that an event will happen in one trial then the chance that it will happen in a succession of r trials is

$$p.p \ldots\ldots p \,.\, (r \text{ times}) = p^r.$$

Cor. 4. If $p_1, p_2, \ldots\ldots, p_n$ are the probabilities that certain events happen, then the probabilities of their non-happening are $1-p_1, 1-p_2, \ldots\ldots, 1-p_n$ and, therefore, the probability of all of these failing is

$$(1-p_1)(1-p_2)\ldots\ldots(1-p_n).$$

Hence the chance in which at least one of these events much happen is

$$1-(1-p_1)(1-p_2) \ldots\ldots (1-p_n).$$

Cor. 5. $$P(A/B) = \frac{P(A \cap B)}{P(B)}, \quad P(B/A) = \frac{P(A \cap B)}{P(A)}$$

Cor. 6. Theorem of Total Probability

If events E_1, E_2, E_3 are mutually exclusive and exhaustive, then for any event A, we have

$$P(A) = P(E_1)\, P(A/E_1) + P(E_2)\, P(A/E_2) + P(E_3)\, P(A/E_3)$$

ILLUSTRATIVE EXAMPLES

Example 1. *A problem in mechanics is given to three students A, B, C whose chances of solving it are $\frac{1}{2}, \frac{1}{3}, \frac{1}{4}$ respectively. What is the probability that the problem will be solved?*

Sol. The probabilities of A, B, C solving the problem are $\frac{1}{2}, \frac{1}{3}, \frac{1}{4}$.

The probabilities of A, B, C not solving the problem are $1-\frac{1}{2}, 1-\frac{1}{3}, 1-\frac{1}{4}$ *i.e.,* $\frac{1}{2}, \frac{2}{3}, \frac{3}{4}$.

$\therefore$ The probability that the problem is not solved by any of them $= \frac{1}{2} \times \frac{2}{3} \times \frac{3}{4} = \frac{1}{4}$.

Hence the probability that the problem is solved by at least one of them $= 1 - \frac{1}{4} = \frac{3}{4}$.

Example 2. *The odds that a book will be favourably reviewed by three independent critics are 5 to 2, 4 to 3 and 3 to 4 respectively. What is the probability that, of the three reviews, a majority will be favourable ?*

Sol. Let the three critics be A, B, C. The probabilities p_1, p_2, p_3 of the book being favourably reviewed by A, B, C are $\frac{5}{7}, \frac{4}{7}, \frac{3}{7}$ respectively.

$\therefore$ The probabilities that the book is unfavourably reviewed by A, B, C are

$$1-\frac{5}{7}=\frac{2}{7},\ 1-\frac{4}{7}=\frac{3}{7},\ 1-\frac{3}{7}=\frac{4}{7}.$$

A majority will be favourable if the reviews of at least two are favourable.

(*i*) If A, B, C all review favourably, the probability is

$$\frac{5}{7}\times\frac{4}{7}\times\frac{3}{7}=\frac{60}{343} \qquad | \ p_1p_2p_3$$

(*ii*) If A, B review favourably and C reviews unfavourably, the probability is

$$\frac{5}{7}\times\frac{4}{7}\times\frac{4}{7}=\frac{80}{343} \qquad | \ p_1p_2(1-p_3)$$

(*iii*) If A, C review favourably and B reviews unfavourably, probability is

$$\frac{5}{7}\times\frac{3}{7}\times\frac{3}{7}=\frac{45}{343} \qquad | \ p_1(1-p_2)p_3$$

(*iv*) If B, C review favourably and A reviews unfavourably, the probability is

$$\frac{2}{7}\times\frac{4}{7}\times\frac{3}{7}=\frac{24}{343} \qquad | \ (1-p_1)p_2p_2$$

Hence the probability that a majority will be favourable is

$$\frac{60}{343}+\frac{80}{343}+\frac{45}{343}+\frac{24}{343}=\frac{209}{343}.$$

Example 3. *A can hit a target 4 times in 5 shots; B 3 times 4 shots; C twice in 3 shots. They fire a volley. What is the probability that at least two shots hit?*

Sol. Probability of A's hitting the target $=\frac{4}{5}$

Probability of B's hitting the target $=\frac{3}{4}$

Probability of C's hitting the target $=\frac{2}{3}$.

For at least two hits, we may have

(*i*) A, B, C all hit the target, the probability for which is

$$\frac{4}{5}\times\frac{3}{4}\times\frac{2}{3}=\frac{24}{60}.$$

(*ii*) A, B hit the target and C misses it, the probability for which is

$$\frac{4}{5}\times\frac{3}{4}\times\left(1-\frac{2}{3}\right)=\frac{4}{5}\times\frac{3}{4}\times\frac{1}{3}=\frac{12}{60}.$$

(*iii*) A, C hit the target and B misses it, the probability for which is

$$\frac{4}{5}\times\left(1-\frac{3}{4}\right)\times\frac{2}{3}=\frac{4}{5}\times\frac{1}{4}\times\frac{2}{3}=\frac{8}{60}.$$

(*iv*) B, C hit the target and A misses it, the probability for which is

$$\left(1-\frac{4}{5}\right)\times\frac{3}{4}\times\frac{2}{3}=\frac{1}{5}\times\frac{3}{4}\times\frac{2}{3}=\frac{6}{60}.$$

Since these are mutually exclusive events, required probability

$$=\frac{24}{60}+\frac{12}{60}+\frac{8}{60}+\frac{6}{60}=\frac{50}{60}=\frac{5}{6}.$$

Example 4. *A has 2 shares in a lottery in which there are 3 prizes and 5 blanks; B has 3 shares in a lottery in which there are 4 prizes and 6 blanks. Show that A's chance of success is to B's as 27 : 35.*

Sol. A can draw two tickets (out of 3 + 5 = 8) in $^8C_2 = 28$ ways.

A will get all blanks in $^5C_2 = 10$ ways. $\therefore$ A can win a prize in 28 – 10 = 18 ways

Hence A's chance of success $=\frac{18}{28}=\frac{9}{14}$

B can draw 3 tickets (out of 4 + 6 = 10) in $^{10}C_3 = 120$ ways; B will get all blanks in $^6C_3 = 20$ ways.

$\therefore$ B can win a prize in 120 – 20 = 100 ways.

Hence B's chance of success $=\frac{100}{120}=\frac{5}{6}.$

$\therefore$ A's chance : B's chance $=\frac{9}{14}:\frac{5}{6}=27:35.$

Example 5. *A and B throw alternately with a single die, A having the first throw. The person who first throws ace is to win. What are their respective chances of winning ?*

Sol. The chance of throwing an ace with a single die $=\frac{1}{6}$

The chance of not throwing an ace with a single die $=1-\frac{1}{6}=\frac{5}{6}$.

If A is to win, he should throw an ace in the first or third or fifth, ..., throws.

If B is to win, he should throw an ace in the second or fourth or sixth, ..., throws.

The chances that an ace is thrown in the first, second, third, ..., throws are

$$\frac{1}{6},\frac{5}{6}.\frac{1}{6},\frac{5}{6}.\frac{5}{6}.\frac{1}{6},\frac{5}{6}.\frac{5}{6}.\frac{5}{6}.\frac{1}{6}\ ......\quad\text{or}\quad\frac{1}{6},\frac{5}{6}.\frac{1}{6},\left(\frac{5}{6}\right)^2.\frac{1}{6},\left(\frac{5}{6}\right)^3.\frac{1}{6},\ ......$$

$$\therefore\ \text{A's chance}=\frac{1}{6}+\left(\frac{5}{6}\right)^2.\frac{1}{6}+\left(\frac{5}{6}\right)^4.\frac{1}{6}+......=\frac{\frac{1}{6}}{1-\left(\frac{5}{6}\right)^2}=\frac{6}{11}$$

Sum of an infinite G.P. $=\frac{a}{1-r}$

B's chance $=1-\frac{6}{11}=\frac{5}{11}.$

Example 6. *Cards are dealt one by one from a well-shuffled pack until an ace appears. Show that the probability that exactly n cards are dealt before the first ace appears is* $\frac{4(51-n)(50-n)(49-n)}{52.51.50.49}$.

Sol. Let A be the event of drawing n non-ace cards and B, the event of drawing an ace in the $(n+1)$th draw.

Consider the event A

n cards can be drawn out of 52 cards in ${}^{52}C_n$ ways.

$\Rightarrow$ Exhaustive cases $= {}^{52}C_n$

n non-ace cards can be drawn out of 52 cards in ${}^{48}C_n$ ways.

$\Rightarrow$ Favourable cases $= {}^{48}C_n$

$$\therefore \text{P(A)} = {}^{48}C_n/{}^{52}C_n = \frac{48!}{(48-n)!\,n!} \times \frac{(52-n)!\,(n)!}{52!}$$

$$= \frac{48!\cdot(52-n)(51-n)(50-n)(49-n)(48-n)!}{(48-n)!\cdot 52\cdot 51\cdot 50\cdot 49\cdot(48)!} = \frac{(52-n)(51-n)(50-n)(49-n)}{52\cdot 51\cdot 50\cdot 49}.$$

Consider the event B

n cards have already been drawn in the first n draws.

Exhaustive cases $= {}^{52-n}C_1 = 52-n$; Favourable cases $= {}^{4}C_1 = 4$

$$\therefore \qquad \text{P(B/A)} = \frac{4}{52-n}$$

Reqd. Probability = P(A) . P(B/A)

$$= \frac{(52-n)(51-n)(50-n)(49-n)}{52.51.50.49} \times \frac{4}{52-n} = \frac{4(51-n)(50-n)(49-n)}{52.51.50.49}.$$

Example 7. *An urn contains 10 white and 3 black balls, while another urn contains 3 white and 5 black balls. Two balls are drawn from the first urn and put into the second urn and then a ball is drawn from the latter. What is the probability that it is a white ball?*

Sol. The two balls drawn from the first urn may be

(*i*) both white (*ii*) both black (*iii*) one white and one black.

Let these events be denoted by A, B, C respectively.

$$\text{P(A)} = \frac{{}^{10}C_2}{{}^{13}C_2} = \frac{10\times 9}{13\times 12} = \frac{15}{26}; \qquad \text{P(B)} = \frac{{}^{3}C_2}{{}^{13}C_2} = \frac{3\times 2}{13\times 12} = \frac{1}{26}$$

$$\text{P(C)} = \frac{{}^{10}C_1\times{}^{3}C_1}{{}^{13}C_2} = \frac{10\times 3}{\frac{13\times 12}{2\times 1}} = \frac{10}{26}$$

When two balls are transferred from first urn to second urn, the second urn will contain.

(*i*) 5 white and 5 black balls (*ii*) 3 white and 7 black balls

(*iii*) 4 white and 6 black balls.

Let W denote the event of drawing a white ball from the second urn in the three cases (*i*), (*ii*) and (*iii*).

Now $$\text{P(W/A)} = \frac{5}{10}, \quad \text{P(W/B)} = \frac{3}{10}, \quad \text{P(W/C)} = \frac{4}{10}$$

$\therefore$ Reqd. probability $= P(A) \cdot P(W/A) + P(B) \cdot P(W/B) + P(C) \cdot P(W/C)$

$$= \frac{15}{26} \cdot \frac{5}{10} + \frac{1}{26} \cdot \frac{3}{10} + \frac{10}{26} \cdot \frac{4}{10} = \frac{75 + 3 + 40}{260} = \frac{118}{260} = \frac{59}{130}.$$

Example 8. *Three bags A, B, C contain 4 red, 3 black, 2 white; 3 red, 4 black, 4 white; and 5 red, 2 black, 6 white balls respectively. If a bag is selected at random and a ball is drawn from it, find the probability that the ball drawn is red.*

Sol. Let E_1, E_2, E_3 denote the events of choosing bags A, B, C respectively,

then $$P(E_1) = P(E_2) = P(E_3) = \frac{1}{3}$$

Let R denote the event of drawing a red ball, then

$$P(R/E_1) = \frac{4}{4 + 3 + 2} = \frac{4}{9}, \quad P(R/E_2) = \frac{3}{3 + 4 + 4} = \frac{3}{11},$$

$$P(R/E_3) = \frac{5}{5 + 2 + 6} = \frac{5}{13}$$

$$\therefore \quad P(R) = P(E_1)\, P(R/E_1) + P(E_2)\, P(R/E_2) + P(E_3)\, P(R/E_3)$$

$$= \frac{1}{3} \times \frac{4}{9} + \frac{1}{3} \times \frac{3}{11} + \frac{1}{3} \times \frac{5}{13} = \frac{4}{27} + \frac{1}{11} + \frac{5}{39}$$

$$= 0.148 + 0.090 + 0.128 = 0.366.$$

Example 9. *If six dice are rolled, then find the probability that all show different faces.*

Sol. When six dice are rolled, there are 6^6 possible outcomes. Let E denote the event that six dice show different faces. If the first die shows any of the six faces viz. 1, 2, 3, 4, 5, 6; the second die shows any of the remaining five faces, the third die shows any of the remaining four faces and so on.

$\therefore$ By the fundamental principle of counting, the number of cases favourable to the occurrence of E is $6 \times 5 \times 4 \times 3 \times 2 \times 1 = 6!$

Hence $P(E) = \dfrac{6!}{6^6}$.

Example 10. *If 10 men, among whom are A and B, stand in a row, what is the probability that there will be exactly 3 men between A and B?*

Sol. 10 men can be arranged in a row in 10! ways.

When there are 3 men between A and B, then they can occupy

(*i*) 1st and 5th positions (*ii*) 2nd and 6th positions
(*iii*) 3rd and 7th positions (*iv*) 4th and 8th positions
(*v*) 5th and 9th positions (*vi*) 6th and 10th positions

Also A and B can interchange their positions.

$\therefore$ Total number of favourable cases $= 6 \times 2 \times 8!$

Hence required probability $= \dfrac{6 \times 2 \times 8!}{10!} = \dfrac{6 \times 2 \times 8!}{10 \times 9 \times 8!} = \dfrac{2}{15}$.

EXERCISE 5.1

1. When a coin is tossed four times, find the probability of getting (*i*) exactly one head, (*ii*) at most three heads and (*iii*) at least two heads ?
2. What is the chance that a (*i*) non-leap year (*ii*) leap year should have fifty three Sundays ? (*M.D.U. Dec. 2014; K.U.K. 2013*)
3. A card is drawn from an ordinary pack and a gambler bets that it is a spade or an ace. What are the odds against his winning the bet ?
4. An integer is chosen at random from the first two hundred positive integers. What is the probability that the integer chosen is divisible by 6 or 8 ?
5. Six cards are drawn at random from a pack of 52 cards. What is the probability that 3 will be red and 3 black ?
6. (*a*) From a set of raffle tickets numbered 1 to 100, three are drawn at random. What is the probability that all are odd numbered ?

   (*b*) A box contains 9 tickets numbered 1 to 9 inclusive. If 3 tickets are drawn from the box, one at a time, find the probability they are alternatively either odd, even, odd or even, odd, even.
7. (*a*) If from a lottery of 30 tickets, marked, 1, 2, 3, ..., 30, four tickets be drawn, what is the chance that those marked 1 and 2 are among them ?

   (*b*) An urn contains 5 red and 10 black balls. Eight of them are placed in another urn. What is the chance that the latter then contains 2 red and 6 black balls ?
8. A party of n persons sit at a round table. Find the odds against two specified individuals sitting next to each other.
9. A five-figured number is formed by the digits 0, 1, 2, 3, 4 (without repetition). Find the probability that the number formed is divisible by 4.
10. Three newspapers A, B, C are published in a city and a survey of readers indicates the following:

    20% read A, 16% read B, 14% read C,

    8% read both A and B, 5% read both A and C,

    4% read both B and C, 2% read all the three.

    For a person chosen at random, find the probability that he reads none of the papers.
11. A problem is statistics is given to five students. Their chances of solving it are $\frac{1}{2}, \frac{1}{3}, \frac{1}{4}, \frac{1}{4}$ and $\frac{1}{5}$. What is the probability that the problem will be solved ?
12. A can hit a target 3 times in 5 shots, B 2 times in 5 shots and C 3 times in 4 shots. They fire a volley. What is the probability that (*i*) two shots hit, (*ii*) at least two shots hit the target ?
13. Three groups of children contain respectively 3 girls and 1 boy; 2 girls and 2 boys; 1 girl and 3 boys. One child is selected at random from each group. Show that the chance that the three selected consist of 1 girl and 2 boys is $\frac{13}{32}$.
14. Suppose that 5 men out of 100 and 25 women out of 1000 are good orators. Assuming that there are equal number of men and women, find the probability of choosing a good orator.
15. A bag contains 10 balls, two of which are red, three blue and five black. Three balls are drawn at random from the bag. What is the probability that

    (*i*) the three balls are of different colours, (*ii*) two balls are of the same colour,

    (*iii*) the balls are all of the same colour.
16. It is 8 : 5 against a person who is 40 years old living till he is 70 and 4 : 3 against a person now 50 living till he is 80. Find the probability that one at least of these persons will be alive 30 years hence.

17. An old carton in a pharmacy shop is found to contain 60 capsules and 180 tablets. Half of capsules and tablets are of expiry date. Find the probability that an item picked at random from the carton is of expiry date or a capsule.

18. A has 3 shares in a lottery where there are 3 prizes and 6 blanks. B has one share in another, where there is just one prize and two blanks. Show that A has a better chance of winning a prize than B in the ratio 16 : 7.

19. A, B and C, in order, toss a coin. The first one to throw a head wins. If A starts, find their respective chances of winning.

20. (*a*) A speaks truth in 75% cases and B in 80% cases. In what percentages of cases are they likely to contradict each other in stating the same fact ?

(*b*) A pair of dice is tossed twice. Find the probability of scoring 7 points (*i*) once, (*ii*) at least once (*iii*) twice.

21. (*a*) Two persons A and B toss an unbiased coin alternately on the understanding that the first who gets the head wins. If A starts the game, find their respective chances of winning.

(*b*) A and B throw alternately with a pair of ordinary dice. A wins if he throws 6 before B throws 7 and B wins if he throws 7 before A throws 6. If A begins, find their respective chances of winning. (Huyghen's Problem)

22. (*a*) Two cards are randomly drawn from a deck of 52 cards and thrown away. What is the probability of drawing an ace in a single draw from the remaining 50 cards ?

(*b*) A box A contains 2 white and 4 black balls. Another box B contains 5 white and 7 black balls. A ball is transferred from the box A to the box B, then a ball is drawn from the box B. Find the probability that it is white.

23. A committee consists of 9 students two of which are from 1st year, three from 2nd year and four from 3rd year. Three students are to be removed at random. What is the chance that

(*i*) the three students belong to different classes,

(*ii*) two belong to the same class and third to the different class, and

(*iii*) the three belong to the same class ?

24. Five men in a company of twenty are graduates. If 3 men and picked out of 20 at random, what is the probability that

(*i*) they are all graduates? (*ii*) at least one is graduate?

25. If A, B, C are events such that

$P(A) = 0.3, P(B) = 0.4, P(C) = 0.8$

$P(A \cap B) = 0.08, P(A \cap C) = 0.28, P(A \cap B \cap C) = 0.09$

If $P(A \cup B \cup C) \geq 0.75$, then show that $0.23 \leq P(B \cap C) \leq 0.48$.

26. (*a*) For two events A and B, let $P(A) = 0.4$, $P(B) = p$ and $P(A \cup B) = 0.6$

(*i*) Find p so that A and B are independent events.

(*ii*) For what value of p are A and B are mutually exclusive ?

(*b*) If A and B are two events such that $P(A) = \frac{1}{4}$, $P(B) = \frac{1}{3}$ and $P(A \cup B) = \frac{1}{2}$, show that A and B are independent events. *(M.D.U. 2012)*

(*c*) Let A and B be two events with their probabilities $P(A) = \frac{1}{2}$, $P(B) = \frac{1}{3}$ and $P(A \cap B) = \frac{1}{4}$. Find P(A/B), $P(A \cup B)$ and P(A′/B′). *(K.U.K. Jan. 2013)*

(*d*) Given that $P(A) = \frac{1}{4}$, $P(B) = \frac{1}{3}$ and $P(A \cup B) = \frac{1}{2}$, evaluate P(A/B), P(B/A), $P(A \cap B^C)$, $P(A/B^C)$. *(K.U.K. May 2013)*

(*e*) If $P(A) = \frac{6}{11}$, $P(B) = \frac{5}{11}$ and $P(A \cup B) = \frac{7}{11}$, find P(B/A). *(M.D.U. Dec. 2014)*

27. A husband and wife appear in an interview for two vacancies in the same post. The probability of husband's selection is $\frac{1}{7}$ and that of wife's selection is $\frac{1}{5}$. What is the probability that

(*i*) both of them will be selected, (*ii*) only one of them will be selected, and

(*iii*) none of them will be selected?

28. Two dice are tossed once. Find the probability of getting an even number on the first die or a total of 8.

29. There are two bags. The first bag contains 5 white and 3 black balls and the second bag contains 3 white and 5 black balls. Two balls are drawn at random from the first bag and without noting their colours, put into the second bag. Then two balls are drawn from the second bag. Find the probability that the balls drawn are white and black.

30. A purse contains 2 silver and 4 copper coins. A second purse contains 4 silver and 3 copper coins. If a coin is pulled out at random from one of the two purses, what is the probability that it is a silver coin?

31. A man wants to marry a girl having qualities: white complexion—the probability of getting such a girl is one in twenty; handsome dowry—the probability of getting this is one in fifty ; westernised manners and etiquettes—the probability here is one in hundred. Find out the probability of his getting married to such a girl when the possession of these three attributes is independent.

32. A class consists of 80 students, 25 of them are girls and 55 boys, 10 of them are rich and the remaining poor, 20 of them are fair complexioned. What is the probability of selecting a fair complexioned rich girl?

33. Of the students attending a lecture, 50% could not see what was written on the board and 40% could not hear what the lecturer was saying. Most unfortunate 30% fell into both of these categories. What is the probability that a student picked at random was able to see and hear satisfactorily?

34. The probabilities of A, B, C solving a problem are $\frac{1}{3}, \frac{2}{7}$ and $\frac{3}{8}$ respectively. If all the three try to solve the problem simultaneously, find the probability that exactly one of them will solve it.

35. A student takes his examination in four subjects $\alpha, \beta, \gamma, \delta$. He estimates his chance of passing in α as $\frac{4}{5}$, in β as $\frac{3}{4}$, in γ as $\frac{5}{6}$ and in δ as $\frac{2}{3}$. To qualify he must pass in α and at least two other subjects. What is the probability that the qualifies?

36. For any two events A and B, prove that

$$P(A \cap B) \le P(A) \le P(A \cup B) \le P(A) + P(B).$$

37. A student takes his examination in four subjects P, Q, R, S. He estimates his chances of passing in P as $\frac{4}{5}$, in Q as $\frac{3}{4}$, in R as $\frac{5}{6}$ and in S as $\frac{2}{3}$. To qualify, he must pass in P and at least two other subjects. What is the probability that he qualifies?

38. Ten cards numbered 1 to 10 are placed in a box, mixed up thoroughly and then one card is drawn randomly. If it is known that the number on the drawn card is more than 3, what is the probability that it is an even number?

39. A die is thrown three times. Events A and B are defined as below:

A: 4 appears on third throw

B: 6 and 5 appear respectively on first two throws

Find the probability of A given that B has already occurred.

40. A fair die is rolled. If $E = \{1, 3, 5\}$, $F = \{2, 3\}$ and $G = \{2, 3, 4, 5\}$, find $P((E \cup F)/G)$ and $P((E \cap F)/G)$.

41. A person has undertaken a construction job. The probabilities are 0.65 that there will be a strike, 0.80 that the construction job will be completed on time if there is no strike, and 0.32 that the construction job will be completed on time if there is a strike. Determine the probability that the construction job will be completed on time.

42. In a random arrangement of the letters of the word 'COMMERCE', find the probability that all the vowels come together.

43. If P(A) = 0.4, P(B) = 0.7 and P(at least one of A and B) = 0.8, find P(only one of A and B).

44. An investment consultant predicts that the odds against the price of a certain stock going up are 2 : 1 and the odds in favour of the price remaining the same are 1 : 3. What is the probability that the price of stock will go down?

45. A piece of electronic equipment has two essential parts A and B. In the past, part A failed 30% of the times, part B failed 20% of the times and both parts failed simultaneously 5% of the times. Assuming that both parts must operate to enable the equipment to function, what is the probability that the equipment will function?

46. If P(not A) = 0.4, $P(A \cup B) = 0.7$ and A and B are given to be independent events, find the value of P(B).

Answers

1. $\frac{1}{4}, \frac{15}{16}, \frac{11}{16}$ **2.** (*i*) $\frac{1}{7}$ (*ii*) $\frac{2}{7}$ **3.** 9: 4 **4.** $\frac{1}{4}$

5. $\frac{13000}{39151}$ **6.** (*a*) $\frac{4}{33}$ (*b*) $\frac{5}{18}$ **7.** (*a*) $\frac{2}{145}$, (*b*) $\frac{140}{429}$

8. $(n - 3)$: 2 **9.** $\frac{5}{16}$ **10.** $\frac{13}{20}$ **11.** $\frac{17}{20}$

12. 0.45, 0.63 **14.** 0.0375 **15.** (*i*) $\frac{1}{4}$ (*ii*) $\frac{79}{120}$ (*iii*) $\frac{11}{120}$

16. $\frac{59}{91}$ **17.** $\frac{5}{8}$ **19.** $\frac{4}{7}, \frac{2}{7}, \frac{1}{7}$

20. (*a*) 35% (*b*) (*i*) $\frac{5}{18}$ (*ii*) $\frac{11}{36}$ (*iii*) $\frac{1}{36}$ **21.** (*a*) $\frac{2}{3}, \frac{1}{3}$ (*b*) $\frac{30}{61}, \frac{31}{61}$

22. (*a*) $\frac{1}{13}$, (*b*) $\frac{16}{39}$ **23.** (*i*) $\frac{2}{7}$ (*ii*) $\frac{55}{84}$ (*iii*) $\frac{5}{84}$ **24.** (*i*) $\frac{1}{114}$ (*ii*) $\frac{137}{228}$

26. (*a*) (*i*) $\frac{1}{3}$ (*ii*) 0.2 (*c*) $\frac{3}{4}, \frac{7}{12}, \frac{5}{8}$ (*d*) $\frac{1}{4}, \frac{1}{3}, \frac{1}{6}, \frac{1}{4}$ (*e*) $\frac{2}{3}$

27. (*i*) $\frac{1}{35}$ (*ii*) $\frac{2}{7}$ (*iii*) $\frac{24}{35}$ **28.** $\frac{5}{9}$ **29.** $\frac{673}{1260}$

30. $\frac{19}{42}$ **31.** 0.000001 **32.** $\frac{5}{512}$ **33.** $\frac{2}{5}$

34. $\frac{25}{56}$ **35.** $\frac{61}{90}$ **37.** $\frac{61}{90}$ **38.** $\frac{4}{7}$

39. $\frac{1}{6}$ **40.** $\frac{3}{4}, \frac{1}{4}$ **41.** 0.488 **42.** $\frac{3}{28}$

43. 0.5 **44.** $\frac{5}{12}$ **45.** 0.55 **46.** 0.25

5.15. BAYE'S THEOREM

(M.D.U. 2010, May 2013)

If E_1, E_2, ..., E_n are mutually exclusive and exhaustive events with $P(E_i) \neq 0$, $(i = 1, 2, ..., n)$ of a random experiment then for any arbitrary event A of the sample space of the above experiment with $P(A) > 0$, we have

$$P(E_i/A) = \frac{P(E_i)P(A/E_i)}{\sum_{i=1}^{n} P(E_i)P(A/E_i)}.$$

Proof. Let S be the sample space of the random experiment.

The events E_1, E_2, ..., E_n being exhaustive

$$S = E_1 \cup E_2 \cup \cdots\cdots \cup E_n$$

$$\therefore \quad A = A \cap S \qquad [\because \ A \subset S]$$

$$= A \cap (E_1 \cup E_2 \cup \cdots\cdots \cup E_n)$$

$$= (A \cap E_1) \cup (A \cap E_2) \cup \cdots\cdots \cup (A \cap E_n) \qquad \text{[Distributive Law]}$$

$$\Rightarrow \quad P(A) = P(A \cap E_1) + P(A \cap E_2) + \cdots\cdots + P(A \cap E_n)$$

$$= P(E_1)P(A/E_1) + P(E_2)P(A/E_2) + \cdots\cdots + P(E_n)P(A/E_n)$$

$$= \sum_{i=1}^{n} P(E_i)P(A/E_i) \qquad ...(1)$$

Now $\quad P(A \cap E_i) = P(A)P(E_i/A)$

$$\Rightarrow \quad P(E_i/A) = \frac{P(A \cap E_i)}{P(A)} = \frac{P(E_i)P(A/E_i)}{\sum_{i=1}^{n} P(E_i)P(A/E_i)} \qquad \text{[Using (1)]}$$

Note. The significance of Baye's Theorem may be understood in the following manner:

$P(E_i)$ is the probability of occurrence of E_i. The experiment is performed and we are told that the event A has occurred. With this information, the probability $P(E_i)$ is changed to $P(E_i/A)$. Baye's Theorem enables us to evaluate $P(E_i/A)$ if all the $P(E_i)$ and the conditional probabilities $P(A/E_i)$ are known.

ILLUSTRATIVE EXAMPLES

Example 1. *A bag X contains 2 white and 3 red balls and a bag Y contains 4 white and 5 red balls. One ball is drawn at random from one of the bags and is found to be red. Find the probability that it was drawn from bag Y.*

Sol. Let $\quad E_1$: the ball is drawn from bag X; E_2: the ball is drawn from bag Y

and $\quad A$: the ball is red.

We have to find $P(E_2/A)$. By Baye's Theorem,

$$P(E_2/A) = \frac{P(E_2)P(A/E_2)}{P(E_1)P(A/E_1) + P(E_2)P(A/E_2)} \qquad ...(1)$$

Since the two bags are equally likely to be selected, $P(E_1) = P(E_2) = \frac{1}{2}$

Also $\quad P(A/E_1) = P(\text{a red ball is drawn from bag X}) = \frac{3}{5}$

$$P(A/E_2) = P(\text{a red ball is drawn from bag Y}) = \frac{5}{9}$$

$\therefore$ From (1), we have $P(E_2/A) = \dfrac{\frac{1}{2} \times \frac{5}{9}}{\frac{1}{2} \times \frac{3}{5} + \frac{1}{2} \times \frac{5}{9}} = \dfrac{25}{52}$.

Example 2. *In a bolt factory, machines A, B and C manufacture respectively 25%, 35% and 40% of the total. Of their output 5%, 4% and 2% per cent are defective bolts. A bolt is drawn at random from the product and is found to be defective. What is the probability that it was manufactured by machine B?* (M.D.U. Dec. 2013, May 2015; K.U.K. 2010, Jan. 2014)

Sol. Let E_1, E_2 and E_3 denote the events that a bolt selected at random is manufacture by the machines A, B and C respectively and let H denote the event of its being defective. Then

$$P(E_1) = 0.25, \; P(E_2) = 0.35, \; P(E_3) = 0.40$$

The probability of drawing a defective bolt manufactured by machine A is $P(H/E_1) = 0.05$

Similarly, $\quad P(H/E_2) = 0.04$ and $P(H/E_3) = 0.02$

By Baye's Theorem, we have

$$P(E_2/H) = \frac{P(E_2)P(H/E_2)}{P(E_1)P(H/E_1) + P(E_2)P(H/E_2) + P(E_3)P(H/E_3)}$$

$$= \frac{0.35 \times 0.04}{0.25 \times 0.05 + 0.35 \times 0.04 + 0.40 \times 0.02} = \frac{0.0140}{0.0345} = 0.41.$$

Example 3. *The contents of urns I, II and III are as follows:*

1 white, 2 black and 3 red balls,

2 white, 1 black and 1 red balls, and

4 white, 5 black and 3 red balls.

One urn is chosen at random and two balls drawn. They happen to be white and red. What is the probability that they come from urns I, II or III? (M.D.U. 2012)

Sol. Let $\quad E_1$: urn I is chosen; E_2: urn II is chosen; E_3: urn III is chosen

and $\quad$ A: the two balls are white and red.

We have to find $P(E_1/A)$, $P(E_2/A)$ and $P(E_3/A)$.

Now $\quad P(E_1) = P(E_2) = P(E_3) = \dfrac{1}{3}$

$$P(A/E_1) = P\text{ (a white and a red ball are drawn from urn I)} = \frac{{}^1C_1 \times {}^3C_1}{{}^6C_2} = \frac{1}{5}$$

$$P(A/E_2) = \frac{{}^2C_1 \times {}^1C_1}{{}^4C_2} = \frac{1}{3}; \; P(A/E_3) = \frac{{}^4C_1 \times {}^3C_1}{{}^{12}C_2} = \frac{2}{11}$$

By Baye's Theorem, we have

$$P(E_1/A) = \frac{P(E_1)P(A/E_1)}{P(E_1)P(A/E_1) + P(E_2)P(A/E_2) + P(E_3)P(A/E_3)} = \frac{\frac{1}{3} \times \frac{1}{5}}{\frac{1}{3} \times \frac{1}{5} + \frac{1}{3} \times \frac{1}{3} + \frac{1}{3} \times \frac{2}{11}} = \frac{33}{118}$$

Similarly $P(E_2/A) = \frac{55}{118}$, $P(E_3/A) = \frac{15}{59}$.

EXERCISE 5.2

1. (*a*) Two urns contain 4 white, 6 blue and 4 white, 5 blue balls respectively. One of the urns is selected at random and a ball is drawn from it. If the ball drawn is white, find the probability that it is drawn from the

(*i*) first urn (*ii*) second urn.

(*b*) Of the cigarette smoking population, 70% are men and 30% women, 10% of these men and 20% of these women smoke 'WILLS'. What is the probability that a person seen smoking a 'WILLS' will be a man?

2. (*a*) Three urns contain 6 red, 4 black; 4 red, 6 black and 5 red, 5 black balls respectively. One of the urns is selected at random and a ball is drawn from it. If the ball drawn is red, find the probability that it is drawn from the first urn. (*M.D.U. 2010*)

(*b*) There are three bags: first containing 1 white, 2 red, 3 green balls; second 2 white, 3 red, 1 green balls and third 3 white, 1 red, 2 green balls. Two balls are drawn from a bag chosen at random. They are found to be 1 red and 1 white. Find the probability that balls so drawn came from the second bag.

3. A factory has two machines A and B. Past record shows that machine A produced 60% of the items of output and machine B produced 40% of the items. Further, 2% of the items produced by machine A were defective and 1% produced by machine B were defective. If a defective item is drawn at random, what is the probability that it was produced by machine A?

4. An insurance company insured 2000 scooter drivers, 4000 car drivers and 6000 truck drivers. The probability of accident is 0.01, 0.03 and 0.15 respectively. One of the insured persons meets an accident. What is the probability that he is a scooter driver? (*K.U.K. Dec. 2015*)

5. A company has two plants to manufacture scooters. Plant I manufactures 70% of scooters and plant II manufactures 30%. At plant I, 80% of the scooters are rated standard quality and at plant II, 90% of the scooters are rated standard quality. A scooter is chosen at random and is found to be of standard quality. What is the chance that it has come from plant II?

6. In a bolt factory, there are four machines A, B, C, D manufacturing 20%, 15%, 25% and 40% of the total output respectively. Of their outputs 5%, 4%, 3% and 2%, in the same order, are defective bolts. A bolt is chosen at random from the factory's production and is found defective. What is the probability that the bolt was manufactured by machine A or machine D?

(*M.D.U. May 2013, Dec. 2014*)

7. A doctor is to visit a patient. From the past experience, it is known that the probabilities that he will come by train, bus, scooter or by other means of transport are respectively $\frac{3}{10}, \frac{1}{5}, \frac{1}{10}$ and $\frac{2}{5}$.

The probabilities that he will be late are $\frac{1}{4}, \frac{1}{3}$ and $\frac{1}{12}$, if he comes by train, bus and scooter respectively, but if he comes by other means of transport, then he will not be late. When he arrives, he is late. What is the probability that he comes by train?

8. A man is known to speak truth 3 out of 4 times. He throws a die and reports that it is a six. Find the probability that it is actually a six.

9. A survey was conducted to find the supplies of the consumer durables for the market. It was found that the three major companies A, B and C have market share of 35%, 25% and 40% respectively out of which 2%, 1% and 3% are not upto the satisfaction. A consumer buys a product and is dissatisfied with it. What is the probability that it might be from the company C?

(*M.D.U. Dec. 2015*)

10. By examining the chest X-ray, the probability that T.B. is detected when a person is actually suffering is 0.99. The probability that the doctor diagnoses incorrectly that a person has T.B. on the basis of X-ray is 0.001. In a certain city, 1 in 1000 persons suffer from T.B. A person is selected at random and is diagnosed to have T.B. What is the chance that he actually has T.B.?

11. Assume that the chance of a patient having a heart attack is 40%. It is also assumed that a meditation and yoga course reduce the risk of heart attack by 30% and prescription of certain drug reduces its chances by 25%. At a time a patient can choose any one of the two options with equal probabilities. It is given that after going through one of the two options the patient selected at random suffers a heart attack. Find the probability that the patient followed a course of meditation and yoga.

12. There are three coins. One is a two headed coin (having head on both faces), another is a biased coin that comes up heads 75% of the time and third is an unbiased coin. One of the three coins is chosen at random and tossed, it shows heads. What is the probability that it was the two headed coin?

13. In a certain college, 4% of the boys and 1% of girls are taller than 1.8 m. Further more 60% of the students are girls. If a student is selected at random and is found to be taller than 1.8 m, what is the probability that the student is a girl? (*K.U.K. Jan. 2013*)

14. In a certain recruitment test there are multiple choice questions. There are 4 possible answers to each question, of which one is correct. The probability that a student knows the answer to a question is 90%. If he gets the correct answer to a question, what is the probability that he was guessing ?

15. In a certain competitive examination, an examinee either guesses or copies or knows the answer to a multiple-choice question with four choices. The probability that he makes a guess is 0.35 and the probability that he copies the answer is 0.20. The probability that his answer is correct, given that he copied it, is 0.15. Find the probability that he guessed the answer to the question, given that he correctly answered it.

16. If a machine is set correctly it produces 10% of the defective items. If it is incorrectly set up then it produces 10% of good items. Chances for a setting to be correct and incorrect are in the ratio 7 : 3. After a setting is made, the first two items produced are found to be good items. What is the chance that the set up was correct?

17. Two-thirds of the students in a class are boys and the rest are girls. It is known that the probability of a girl getting a first class is 0.25 and that of a boy getting a first class is 0.28. A student is selected at random and is found to get the first class. What is the probability that the student is a boy?

18. Given three identical boxes I, II and III, each containing two coins. In box I, both coins are gold coins, in box II, both are silver coins and in box III, there is one gold and one silver coin. A person chooses a box at random and takes out a coin. If the coin is of gold, what is the probability that the other coin in the box is also of gold?

19. Suppose a girl throws a die. If she gets 5 or 6, she tosses a coin three times and notes the number of heads. If she gets 1, 2, 3 or 4, she tosses a coin once and notes whether a head or tail is obtained. If she obtained exactly one head, what is the probability that she threw 1, 2, 3 or 4 with the die?

20. A card from a pack of 52 cards is lost. From the remaining cards of the pack, two cards are drawn and are found to be both diamonds. Find the probability of the lost card being a diamond.

Answers

1. (*a*) (*i*) $\frac{9}{19}$ (*ii*) $\frac{10}{19}$ (*b*) $\frac{7}{13}$ 2. (*a*) $\frac{2}{5}$ (*b*) $\frac{6}{11}$

3. $\frac{3}{4}$ 4. $\frac{1}{52}$ 5. $\frac{27}{83}$

6. 0.3175, 0.254 7. $\frac{1}{2}$ 8. $\frac{3}{8}$

9. $\frac{24}{43}$ 10. $\frac{110}{221}$ 11. $\frac{14}{29}$ 12. $\frac{4}{9}$ 13. $\frac{9}{25}$

14. $\frac{1}{37}$ 15. $\frac{35}{227}$ 16. 0.995 17. $\frac{56}{81}$ 18. $\frac{2}{3}$

19. $\frac{8}{11}$ 20. $\frac{11}{50}$.

5.16. RANDOM VARIABLE

If the numerical values assumed by a variable are the result of some chance factors, so that a particular value cannot be exactly predicted in advance, the variable is then called a *random variable*. A random variable is also called '*chance variable*' or '*stochastic variable*'.

Random variables are denoted by capital letters, usually, from the last part of the alphabet, for instance, X, Y, Z etc.

Continuous and Discrete Random Variables

A *continuous random variable* is one which can assume any value within an interval, *i.e.*, all values of a continuous scale. For example (*i*) the weights (in kg) of a group of individuals, (*ii*) the heights of a group of individuals.

A *discrete random variable* is one which can assume only isolated values. For example,

(*i*) the number of heads in 4 tosses of a coin is a discrete random variable as it cannot assume values other than 0, 1, 2, 3, 4.

(*ii*) the number of aces in a draw of 2 cards from a well shuffled deck is a random variable as it can take the values 0, 1, 2 only.

5.17. (a) DISCRETE PROBABILITY DISTRIBUTION (*M.D.U. May 2015, Dec. 2015*)

Let a random variable X assume values $x_1, x_2, x_3, \ldots\ldots, x_n$ with probabilities $p_1, p_2, p_3, \ldots\ldots, p_n$ respectively, where $P(X = x_i) = p_i \geq 0$ for each x_i and $p_1 + p_2 + p_3 + \cdots\cdots + p_n = \sum_{i=1}^{n} p_i = 1$.

Then

$$\begin{array}{lll} X & : & x_1, \quad x_2, \quad x_3, \ldots\ldots, x_n \\ P(X) & : & p_1, \quad p_2, \quad p_3, \ldots\ldots, p_n \end{array}$$

is called the discrete probability distribution for X and it spells out how a total probability of 1 is distributed over several values of the random variable.

5.17. (b) CONTINUOUS PROBABILITY DISTRIBUTION

(*M.D.U. May 2015, Dec. 2015*)

Let X be a continuous random variable taking values in the interval $(-\infty, \infty)$. Let $f(x)$ be a function satisfying the following properties:

(*i*) $f(x)$ is integrable on $(-\infty, \infty)$

(*ii*) $f(x) \geq 0$ for all x in $(-\infty, \infty)$

(*iii*) $\int_{-\infty}^{\infty} f(x)\, dx = 1$

Then $f(x)$ is called the **probability distribution (or density) function** (p.d.f.) of the continuous random variable X.

The probability for a continuous random variable X to fall in the interval $[a, b]$ is

$$P(a \leq X \leq b) = \int_a^b f(x)\, dx$$

which is nothing but the area between the continuous curve $y = f(x)$, $x = a$, $x = b$ and x-axis. The curve $y = f(x)$ is called the **probability curve.**

Taking $b = a$, $P(X = a) = \int_a^a f(x)\, dx = 0$. Also,

$$P(a \leq X \leq b) = P(a \leq X < b) = P(a < X \leq b) = P(a < X < b)$$

If X is a continuous random variable with p.d.f. $f(x)$, then

$$F(x) = P(X \leq x) = \int_{-\infty}^{x} f(x)\, dx$$

is called **cumulative distribution function** or simply **distribution function** of the continuous random variable X. The distribution function has the following properties:

(*i*) $0 \leq F(x) \leq 1, \quad -\infty < x < \infty$

(*ii*) $F'(x) = f(x) \geq 0$

(*iii*) $P(a \leq X \leq b) = \int_a^b f(x)\, dx = \int_{-\infty}^{b} f(x)\, dx - \int_{-\infty}^{a} f(x)\, dx = P(X \leq b) - P(X \leq a) = F(b) - F(a)$

5.18. MEAN AND VARIANCE OF RANDOM VARIABLES

(*a*) Let

$$X: x_1, \quad x_2, \quad x_3, \ldots\ldots, x_n$$
$$P(X): p_1, \quad p_2, \quad p_3, \ldots\ldots, p_n$$

be a discrete probability distribution.

We denote the *mean* by μ and define $\mu = \dfrac{\Sigma p_i x_i}{\Sigma p_i} = \Sigma p_i x_i \qquad (\because \ \Sigma p_i = 1)$

Other names for the mean are *average or expected value E(X).*

We denote the *variance* by σ^2 and define $\sigma^2 = \Sigma p_i (x_i - \mu)^2$

If μ is not a whole number, then $\sigma^2 = \Sigma p_i x_i^2 - \mu^2$

Standard deviation $\sigma = +\sqrt{\text{Variance}}$.

The **r^th^ moment** about the mean is denoted by μ_r and defined as

$$\mu_r = \sum (x_i - \mu)^r f(x_i) \quad \text{where } f(x_i) = p_i$$

Putting, $r = 0, 1, 2, 3$ and 4, we get

$$\mu_0 = \sum f(x_i) = \sum p_i = 1$$

$$\mu_1 = \sum (x_i - \mu) f(x_i) = \sum (x_i - \mu) p_i$$

$$= \sum p_i x_i - \mu \sum p_i = \mu - \mu = 0$$

$$\mu_2 = \sum (x_i - \mu)^2 f(x_i) = \sigma^2$$

$$\mu_3 = \sum (x_i - \mu)^3 f(x_i)$$

$$\mu_4 = \sum (x_i - \mu)^4 f(x_i)$$

In practice the first four moments suffice.

Mean deviation from mean $= \sum |x_i - \mu| f(x_i)$

(*b*) If X is a continuous random variable with probability density function $f(x)$, then

$$E(X) = \int_{-\infty}^{\infty} x f(x)\, dx$$

$$E(X^r) = \int_{-\infty}^{\infty} x^r f(x)\, dx$$

$$\text{Var}(X) = \int_{-\infty}^{\infty} (x - \mu)^2 f(x)\, dx = E(X^2) - [E(X)]^2$$

Mean deviation from mean $= \int_{-\infty}^{\infty} |x - \mu| f(x)\, dx$

The rth **moment** about the mean is defined as

$$\mu_r = \int_{-\infty}^{\infty} (x - \mu)^r f(x)\, dx$$

Moments about the mean are called **central moments.**

(*c*) **Moment Generating Function**

Consider the function $M_a(t) = \sum p_i e^{t(x_i - a)}$ ...(1)

This function is a function of the parameter t and gives the mean of the probability distribution of $e^{t(x_i - a)}$. Expanding the exponential in equation (1), we get

$$M_a(t) = \sum p_i \left[1 + t(x_i - a) + \frac{t^2}{2!}(x_i - a)^2 + \dots + \frac{t^r}{r!}(x_i - a)^r + \dots\right]$$

$$= \sum p_i + t\sum p_i (x_i - a) + \frac{t^2}{2!}\sum p_i (x_i - a)^2 + \dots + \frac{t^r}{r!}\sum p_i (x_i - a)^r + \dots$$

$$= 1 + t\mu_1' + \frac{t^2}{2!}\mu'_2 + \dots + \frac{t^r}{r!}\mu'_r + \dots \qquad \text{...(2)}$$

where μ'_r is the moment of order r about a. Thus $M_a(t)$ generates moments and is, therefore, called the moment generating function (m.g.f.) of the discrete probability distribution of the variate X about the value $x = a$. We observe that μ'_r is equal to the co-efficient of $\frac{t^r}{r!}$ in the expansion of m.g.f. $M_a(t)$.

Alternately, μ'_r can be obtained by differentiating (2) r times with respect to t and then putting $t = 0$.

Thus, $$\mu'_r = \left[\frac{d^r}{dt^r} M_a(t)\right]_{t=0}$$

Rewriting equation (1) as

$$M_a(t) = \sum p_i e^{t(x_i - a)} = e^{-at} \sum p_i e^{tx_i} = e^{-at} M_0(t)$$

Hence the m.g.f. about the value a is e^{-at} times the m.g.f. about the origin.

If X is a continuous random variable with probability density function $f(x)$, then

$$M_a(t) = \int_{-\infty}^{\infty} e^{t(x-a)} f(x)\, dx$$

Note 1. If X is a random variable and a, b are constants, then

$$E(a) = a$$
$$E(aX + b) = aE(X) + b$$
$$\text{Var.}(aX + b) = a^2\, \text{Var.}(X)$$

Note 2. Relation between central moments and moments about any arbitrary origin a

$$\mu'_1 = \mu - a$$
$$\mu_2 = \mu'_2 - (\mu'_1)^2$$
$$\mu_3 = \mu'_3 - 3\mu'_2\mu'_1 + 2(\mu'_1)^3$$
$$\mu_4 = \mu'_4 - 4\mu'_3\mu'_1 + 6\mu'_2(\mu'_1)^2 - 3(\mu'_1)^4$$

ILLUSTRATIVE EXAMPLES

Example 1. *Five defective bulbs are accidentally mixed with twenty good ones. It is not possible to just look at a bulb and tell whether or not it is defective. Find the probability distribution of the number of defective bulbs, if four bulbs are drawn at random from this lot.*

Sol. Let X denote the number of defective bulbs in 4. Clearly X can take the values 0, 1, 2, 3 or 4.

Number of defective bulbs = 5
Number of good bulbs = 20
Total number of bulbs = 25

P(X = 0) = P (no defective) = P (all 4 good ones)

$$= \frac{{}^{20}C_4}{{}^{25}C_4} = \frac{20 \times 19 \times 18 \times 17}{25 \times 24 \times 23 \times 22} = \frac{969}{2530}$$

$$P(X = 1) = P(\text{one defective and 3 good ones}) = \frac{{}^5C_1 \times {}^{20}C_3}{{}^{25}C_4} = \frac{1140}{2530}$$

$$P(X = 2) = P \text{ (2 defectives and 2 good ones)} = \frac{{}^5C_2 \times {}^{20}C_2}{{}^{25}C_4} = \frac{380}{2530}$$

$$P(X = 3) = P\text{(3 defectives and 1 good one)} = \frac{{}^5C_3 \times {}^{20}C_1}{{}^{25}C_4} = \frac{40}{2530}$$

$$P(X = 4) = P \text{ (all 4 defective)} = \frac{{}^5C_4}{{}^{25}C_4} = \frac{1}{2530}$$

∴ The probability distribution of the random variable X is

X :	0	1	2	3	4
P(X) :	$\frac{969}{2530}$	$\frac{1140}{2530}$	$\frac{380}{2530}$	$\frac{40}{2530}$	$\frac{1}{2530}$.

Example 2. *A die is tossed thrice. A success is 'getting 1 or 6' on a toss. Find the mean and the variance of the number of successes.*

Sol. Let X denote the number of success. Clearly X can take the values 0, 1, 2 or 3.

Probability of success $= \frac{2}{6} = \frac{1}{3}$; Probability of failure $= 1 - \frac{1}{3} = \frac{2}{3}$

$$P(X = 0) = P \text{ (no success)} = P \text{ (all 3 failures)} = \frac{2}{3} \times \frac{2}{3} \times \frac{2}{3} = \frac{8}{27}$$

$$P(X = 1) = P \text{ (one success and 2 failures)} = {}^3C_1 \times \frac{1}{3} \times \frac{2}{3} \times \frac{2}{3} = \frac{12}{27}$$

$$P(X = 2) = P \text{ (two successes and one failure)} = {}^3C_2 \times \frac{1}{3} \times \frac{1}{3} \times \frac{2}{3} = \frac{6}{27}$$

$$P(X = 3) = P \text{ (all 3 successes)} = \frac{1}{3} \times \frac{1}{3} \times \frac{1}{3} = \frac{1}{27}$$

∴ The probability distribution of the random variable X is

X :	0	1	2	3
P(X) :	$\frac{8}{27}$	$\frac{12}{27}$	$\frac{6}{27}$	$\frac{1}{27}$

To find the mean and variance

x_i	p_i	$p_i x_i$	$p_i x_i^2$
0	$\frac{8}{27}$	0	0
1	$\frac{12}{27}$	$\frac{12}{27}$	$\frac{12}{27}$
2	$\frac{6}{27}$	$\frac{12}{27}$	$\frac{24}{27}$
3	$\frac{1}{27}$	$\frac{3}{27}$	$\frac{9}{27}$
		1	$\frac{5}{3}$

Mean $\mu = \Sigma p_i x_i = 1$

Variance $\sigma^2 = \Sigma p_i x_i^2 - \mu^2 = \frac{5}{3} - 1 = \frac{2}{3}$.

Example 3. *A random variable X has the following probability function:*

Values of X, x :	0	1	2	3	4	5	6	7
$p(x)$:	0	k	$2k$	$2k$	$3k$	k^2	$2k^2$	$7k^2 + k$

(i) Find k, *(ii) Evaluate* $P(X < 6)$, $P(X \geq 6)$, $P(3 < X \leq 6)$

(iii) Find the minimum value of x so that $P(X \leq x) > \frac{1}{2}$.

Sol. (*i*) Since $\sum_{x=0}^{7} p(x) = 1$, we have

$$0 + k + 2k + 2k + 3k + k^2 + 2k^2 + 7k^2 + k = 1$$

$$\Rightarrow \quad 10k^2 + 9k - 1 = 0 \quad \Rightarrow \quad (10k - 1)(k + 1) = 0$$

$$\Rightarrow \quad k = \frac{1}{10} \qquad [\because \ p(x) \geq 0]$$

(*ii*) $P(X < 6) = P(X = 0) + P(X = 1) + \cdots\cdots + P(X = 5)$

$$= 0 + k + 2k + 2k + 3k + k^2 = 8k + k^2 = \frac{8}{10} + \frac{1}{100} = \frac{81}{100}$$

$P(X \geq 6) = P(X = 6) + P(X = 7)$

$$= 2k^2 + 7k^2 + k = \frac{9}{100} + \frac{1}{10} = \frac{19}{100}$$

$P(3 < X \leq 6) = P(X = 4) + P(X = 5) + P(X = 6)$

$$= 3k + k^2 + 2k^2 = \frac{3}{10} + \frac{3}{100} = \frac{33}{100}$$

(*iii*) $P(X \leq 1) = k = \frac{1}{10} < \frac{1}{2}$; $\qquad P(X \leq 2) = k + 2k = \frac{3}{10} < \frac{1}{2}$

$P(X \leq 3) = k + 2k + 2k = \frac{5}{10} = \frac{1}{2}$; $\qquad P(X \leq 4) = k + 2k + 2k + 3k = \frac{8}{10} > \frac{1}{2}$

$\therefore$ The minimum value of x so that $P(X \leq x) > \frac{1}{2}$ is 4.

Example 4. *A random variable X can take all non-negative integral values and the probability that X takes the value r is proportional to* a^r $(0 < a < 1)$. *Find* $P(X = 0)$.

Sol. Given $P(X = r) \propto a^r \Rightarrow P(X = r) = ka^r$

where k is a constant and $r = 0, 1, 2, \ldots\ldots$

Since, $\sum_{r=0}^{\infty} P(X = r) = 1$, we have

$$P(X = 0) + P(X = 1) + P(X = 2) + \cdots = 1$$

$$\Rightarrow \quad ka^0 + ka^1 + ka^2 + \cdots = 1$$

$\Rightarrow \qquad k(1 + a + a^2 + \cdots) = 1$

$$\Rightarrow \qquad k\left(\frac{1}{1-a}\right) = 1 \;\Rightarrow\; k = 1 - a$$

$\therefore \qquad P(X = r) = (1 - a)a^r$

Hence, $\qquad P(X = 0) = (1 - a)a^0 = 1 - a.$

Example 5. *In a lottery, m tickets are drawn at a time out of n tickets numbered from 1 to n. Find the expected value of the sum of the numbers on the tickets drawn.*

(M.D.U. 2011)

Sol. Let X denote the number on a ticket then X assumes vlaues 1, 2, 3, ..., n.

The probability of drawing a ticket out of n tickets is $\frac{1}{n}$.

$\therefore$ Values of X,

$$x_i \;:\; 1 \quad 2 \quad 3 \;......\; n$$

$$P(X = x_i) \;:\; \frac{1}{n} \quad \frac{1}{n} \quad \frac{1}{n} \;......\; \frac{1}{n}$$

$$\Rightarrow \qquad E(x_i) = \sum p_i\, x_i$$

$$= 1 \cdot \frac{1}{n} + 2 \cdot \frac{1}{n} + 3 \cdot \frac{1}{n} + ... + n \cdot \frac{1}{n}$$

$$= \frac{1}{n}\,(1 + 2 + 3 + ... + n)$$

$$= \frac{1}{n} \cdot \frac{n(n+1)}{2} = \frac{n+1}{2}$$

If the numbers on the m tickets drawn are $n_1, n_2, ..., n_m$ where $n_i \in \{1, 2, 3, ..., n\}$, $i = 1, 2, ... m$, then the expected value of the sum of numbers on m tickets

$$= E(n_1 + n_2 + ... + n_m)$$

$$= E(n_1) + E(n_2) + ... + E(n_m)$$

$$= m\, E(x_i) = \frac{m(n+1)}{2}.$$

Example 6. *Is the function defined as follows a density function?*

$$f(x) = \begin{cases} e^{-x}, & x \geq 0 \\ 0, & otherwise \end{cases}$$

If so, find P $(1 \leq X \leq 2)$.

Sol. $f(x) \geq 0$ for every x in $(-\infty, \infty)$ and

$$\int_{-\infty}^{\infty} f(x)\,dx = \int_{-\infty}^{0} f(x)\,dx + \int_{0}^{\infty} f(x)\,dx$$

$$= \int_{-\infty}^{0} 0\,dx + \int_{0}^{\infty} e^{-x}\,dx = 0 + \left[\frac{e^{-x}}{-1}\right]_0^{\infty} = -(0 - 1) = 1$$

Since $f(x)$ satisfies the requirements for a density function, therefore, $f(x)$ is a density function.

$$P(1 \le X \le 2) = \int_1^2 e^{-x}\, dx = \left[\frac{e^{-x}}{-1}\right]_1^2 = e^{-1} - e^{-2}$$

$$= 0.368 - 0.135 = 0.233.$$

Example 7. *X is a continuous random variable with probability density function given by*

$$f(x) = \begin{cases} kx, & \text{if } 0 \le x < 2 \\ 2k, & \text{if } 2 \le x < 4 \\ -kx + 6k, & \text{if } 4 \le x < 6 \end{cases}$$

Find k and mean value of X.

Sol. Since total probability = 1, we have

$$\int_0^6 f(x)\, dx = 1$$

$$\Rightarrow \quad \int_0^2 f(x)\, dx + \int_2^4 f(x)\, dx + \int_4^6 f(x)\, dx = 1$$

$$\Rightarrow \quad \int_0^2 kx\, dx + \int_2^4 2k\, dx + \int_4^6 (-kx + 6k)\, dx = 1$$

$$\Rightarrow \quad k\left[\frac{x^2}{2}\right]_0^2 + 2k\,[x]_2^4 + \left[-k\frac{x^2}{2} + 6kx\right]_4^6 = 1$$

$$\Rightarrow \quad k(2 - 0) + 2k(4 - 2) + (-18k + 36k) - (-8k + 24k) = 1$$

$$\Rightarrow \quad 2k + 4k + 18k - 16k = 1$$

$$\Rightarrow \quad 8k = 1 \quad \therefore \quad k = \frac{1}{8}$$

$$\text{Mean value of X} = E(X) = \int_0^6 x\, f(x)\, dx$$

$$= \int_0^2 x\,.\,kx\, dx + \int_2^4 x\,.\,2k\, dx + \int_4^6 x(-kx + 6k)\, dx$$

$$= k\left[\frac{x^3}{3}\right]_0^2 + 2k\left[\frac{x^2}{2}\right]_2^4 + \left[-k\frac{x^3}{3} + 6k\,.\,\frac{x^2}{2}\right]_4^6$$

$$= k\left(\frac{8}{3}\right) + k(12) - \frac{k}{3}(152) + 3k(20)$$

$$= 24k = 24\left(\frac{1}{8}\right) = 3.$$

Example 8. *The density function of a random variable X is given by $f(x) = kx\,(2 - x)$, $0 \le x \le 2$. Find (i) k (ii) mean (iii) variance (iv) mean deviation about the mean.*

(K.U.K. May 2013)

Sol. (*i*) Since total probability = 1, we have

$$\int_0^2 f(x)\,dx = 1$$

$$\Rightarrow \quad \int_0^2 kx\,(2-x)\,dx = 1 \;\Rightarrow\; k\left[x^2 - \frac{x^3}{3}\right]_0^2 = 1$$

$$\Rightarrow \quad k\left(4 - \frac{8}{3}\right) = 1 \;\Rightarrow\; k\,.\,\frac{4}{3} = 1 \quad \therefore \quad k = \frac{3}{4}$$

Hence $\qquad f(x) = \frac{3}{4}\,x\,(2-x), \quad 0 \le x \le 2$

(*ii*) $\qquad \text{Mean} = E(X) = \int_0^2 x\,f(x)\,dx = \frac{3}{4}\int_0^2 x^2\,(2-x)\,dx$

$$= \frac{3}{4}\left[\frac{2x^3}{3} - \frac{x^4}{4}\right]_0^2 = \frac{3}{4}\left(\frac{16}{3} - 4\right) = \frac{3}{4}\,.\,\frac{4}{3} = 1$$

(*iii*) $\qquad \text{Var}(X) = E(X^2) - [E(X)]^2$

$$= \int_0^2 x^2\,f(x)\,dx - (1)^2 = \frac{3}{4}\int_0^2 x^3\,(2-x)\,dx - 1$$

$$= \frac{3}{4}\left[2\,.\,\frac{x^4}{4} - \frac{x^5}{5}\right]_0^2 - 1 = \frac{3}{4}\left(8 - \frac{32}{5}\right) - 1$$

$$= \frac{3}{4}\,.\,\frac{8}{5} - 1 = \frac{6}{5} - 1 = \frac{1}{5}$$

(*iv*) Mean deviation about the mean

$$= \int_0^2 |\,x - \bar{x}\,|\,f(x)\,dx = \int_0^2 |\,x - 1\,|\,f(x)\,dx$$

$$= \int_0^1 |\,x-1\,|\,f(x)\,dx + \int_1^2 |\,x-1\,|\,f(x)\,dx$$

$$= \int_0^1 (1-x)\,.\,\frac{3}{4}\,x\,(2-x)\,dx + \int_1^2 (x-1)\,.\,\frac{3}{4}\,x\,(2-x)\,dx$$

$$= \frac{3}{4}\int_0^1 (2x - 3x^2 + x^3)\,dx + \frac{3}{4}\int_1^2 (-2x + 3x^2 - x^3)\,dx$$

$$= \frac{3}{4}\left[x^2 - x^3 + \frac{x^4}{4}\right]_0^1 + \frac{3}{4}\left[-x^2 + x^3 - \frac{x^4}{4}\right]_1^2$$

$$= \frac{3}{4}\left(\frac{1}{4}\right) + \frac{3}{4}\left(\frac{1}{4}\right) = \frac{3}{8}$$

Example 9. *Find the moment generating function of the exponential distribution* $f(x) = \frac{1}{c}e^{-x/c}$, $0 \le x < \infty$, $c > 0$. *Hence find its mean and standard deviation.* (K.U.K. 2010)

Sol. The moment generating function about the origin is

$$M_0(t) = \int_0^{\infty} e^{tx} f(x)\, dx = \int_0^{\infty} e^{tx} \cdot \frac{1}{c} e^{-x/c}\, dx$$

$$= \frac{1}{c}\int_0^{\infty} e^{\left(t - \frac{1}{c}\right)x} dx = \frac{1}{c}\left[\frac{e^{\left(t-\frac{1}{c}\right)x}}{t - \frac{1}{c}}\right]_0^{\infty}$$

$$= \frac{1}{c} \cdot \frac{c}{ct-1}(e^{-\infty} - 1) = \frac{1}{1-ct} = (1-ct)^{-1} \qquad \left[\because\ |t| < \frac{1}{c}\right]$$

$$= 1 + ct + c^2t^2 + c^3t^3 + \ldots$$

$$\mu'_1 = \left[\frac{d}{dt} M_0(t)\right]_{t=0} = [c + 2c^2t + 3c^3t^2 + \ldots]_{t=0} = c$$

$\therefore$ $\quad$ Mean $= 0 + \mu'_1 = c$ $\quad$ $(\because\ \mu'_1 = \mu - a$, here, $a = 0)$

$$\mu'_2 = \left[\frac{d^2}{dt^2} M_0(t)\right]_{t=0} = [2c^2 + 6c^3t + \ldots]_{t=0} = 2c^2$$

$\therefore$ Variance $\sigma^2 = \mu_2 = \mu'_2 - (\mu'_1)^2 = 2c^2 - c^2 = c^2$

$\Rightarrow$ Standard deviation $= c$

EXERCISE 5.3

1. Two bad eggs are mixed accidently with 10 good ones. Find the probability distribution of the number of bad eggs in 3, drawn at random, without replacement, from this lot.
2. A die is tossed twice. Getting a number greater than 4 is considered a success. Find the variance of the probability distribution of the number of successes.
3. Two cards are drawn simultaneously from a well-shuffled deck of 52 cards. Compute the variance for the number of aces.
4. An urn contains 4 white and 3 red balls. Three balls are drawn, with replacement, from this urn. Find μ, σ^2 and σ for the number of red balls drawn.
5. Compute the variance of the probability distribution of the number of doublets in four throws of a pair of dice.
6. Obtain the distribution function and mean of the total number of heads occurring in three tosses of an unbiased coin.
7. Let X be a number on a die when thrown. Find the mean and variance of X. (*Cochin 2013*)
8. Suppose that X is a random variable for which E(X) = 10 and Var(X) = 25. Find the positive values of a and b such that $Y = aX - b$ has expectation 0 and variance 1.

9. A variate X has the probability distribution

x :	-3	6	9
$P(X = x)$:	1/6	1/2	1/3

Find $E(X)$ and $E(X^2)$. Hence evaluate $E(2X + 1)^2$.

10. A random variable X has the following probability distribution:

X :	-1	0	1	2
$P(X = x)$:	$\frac{1}{3}$	$\frac{1}{6}$	$\frac{1}{6}$	$\frac{1}{3}$

Compute:

(*i*) $E(X)$ (*ii*) $E(X + 2)$ (*iii*) $E(4X)$

(*iv*) Var (X) (*v*) Var (X + 4) (*vi*) Var (2X).

11. A random variable X has the following probability distribution:

x:	-2	-1	0	1	2	3
$p(x)$:	0.1	k	0.2	$2k$	0.3	k

Find the value of k and calculate mean and variance.

(M.D.U. Dec. 2013; K.U.K. Jan. 2014; Cochin 2011)

12. From an urn containing 3 red and 2 white balls, a man is to draw 2 balls at random without replacement, being promised ₹ 20 for each red ball he draws and ₹ 10 for each white one. Find his expectation.

13. A random variable X has the following probability distribution :

Values of X, x :	0	1	2	3	4	5	6	7	8
$p(x)$:	a	$3a$	$5a$	$7a$	$9a$	$11a$	$13a$	$15a$	$17a$

(*i*) Determine the value of a. (*ii*) Find $P(X < 3)$, $P(X \geq 3)$, $P(2 \leq X < 5)$

(*iii*) What is the smallest value of x for which $P(X \leq x) > 0.5$?

14. Find the standard deviation for the following discrete distribution:

x :	8	12	16	20	24
$p(x)$:	$\frac{1}{8}$	$\frac{1}{6}$	$\frac{3}{8}$	$\frac{1}{4}$	$\frac{1}{12}$

15. Is the function given below a density function?

$$f(x) = \begin{cases} 0 & \text{for} \quad x < 2 \\ \frac{1}{18}(3 + 2x) & \text{for} \quad 2 \leq x \leq 4 \\ 0 & \text{for} \quad x > 4 \end{cases}$$

Also find $P(2 \leq X \leq 3)$.

16. (*a*) Find the mean and variance of the following density function :

(*i*) $$f(x) = \begin{cases} x & \text{for} \quad 0 < x < 1 \\ 2 - x & \text{for} \quad 1 < x < 2 \\ 0 & \text{otherwise} \end{cases}$$

(*ii*) $f(x) = k(1 - x^2)$, $0 < x < 1$ *(Cochin 2009)*

(*b*) The frequency distribution of a measurable characteristic varying between 0 and 2 is as under:

$$f(x) = \begin{cases} x^3, & 0 \leq x \leq 1 \\ (2 - x)^3, & 1 \leq x \leq 2 \end{cases}$$

Calculate the standard deviation and also the mean deviation about the mean.

(K.U.K. Dec. 2015)

17. Given that $f(x) = kx^4e^{-x}$, $0 < x < \infty$ is a pdf, determine the value of k. *(Cochin 2010)*

[**Hint.** $\int_0^\infty e^{-x} \cdot x^4\, dx = \Gamma(5) = 4!$]

18. The probability density $p(x)$ of a continuous random variable is given by

$$p(x) = y_0\, e^{-|x|}, \; -\infty < x < \infty.$$

Prove that $y_0 = \frac{1}{2}$. Find the mean and variance of the distribution.

19. If $f(x) = \begin{cases} \frac{1}{2}(x+1), & -1 < x < 1 \\ 0 &, \text{ elsewhere} \end{cases}$

represents the density of a random variable X, find E(X) and Var. (X).

20. Find the expectation and variance of the random variable X, whose pdf is given by

$$f(x) = \begin{cases} 2e^{-2x}, & \text{if } x > 0 \\ 0, & \text{otherwise} \end{cases}.$$

21. Let X be a random variable defined by the density function

$$f(x) = \begin{cases} 3x^2, & 0 \le x \le 1 \\ 0, & \text{otherwise} \end{cases}$$

Find E(X), E $(3X - 2)$, E(X^2). *(M.D.U. Dec. 2014)*

Answers

1.

X:	0	1	2
P(X):	$\frac{12}{22}$	$\frac{9}{22}$	$\frac{1}{22}$

2. $\frac{4}{9}$ **3.** $\frac{400}{2873}$ **4.** $\frac{9}{7}, \frac{36}{49}, \frac{6}{7}$ **5.** $\frac{5}{9}$

6. $\frac{3}{2}$ **7.** $\frac{7}{2}, \frac{175}{18}$ **8.** $a = \frac{1}{5}$, $b = 2$ **9.** $\frac{11}{2}, \frac{93}{2}, 209$

10. (*i*) $\frac{1}{2}$ (*ii*) $\frac{5}{2}$ (*iii*) 2 (*iv*) $\frac{19}{12}$

(*v*) $\frac{19}{12}$ (*vi*) $\frac{19}{3}$ **11.** $k = 0.1$, $\mu = 0.8$, $\sigma^2 = 2.232$

12. ₹ 32 **13.** (*i*) $a = \frac{1}{81}$ (*ii*) $\frac{1}{9}, \frac{8}{9}, \frac{7}{27}$ (*iii*) 5

14. $2\sqrt{5}$ **15.** Yes, $\frac{4}{9}$

16. (*a*) (*i*) $1, \frac{1}{6}$ (*ii*) $\frac{3}{8}, \frac{19}{320}$ (*b*) $\frac{1}{\sqrt{15}}, \frac{1}{5}$ **17.** $\frac{1}{24}$

18. 0, 2 **19.** $\frac{1}{3}, \frac{2}{9}$ **20.** $\frac{1}{2}, \frac{1}{4}$ **21.** $\frac{3}{4}, \frac{1}{4}, \frac{3}{5}$.

5.19. THEORETICAL DISTRIBUTIONS

Frequency distributions can be classified under two heads:

(*i*) Observed Frequency Distributions.

(*ii*) Theoretical or Expected Frequency Distributions.

Observed frequency distributions are based on actual observation and experimentation. If certain hypothesis is assumed, it is sometimes possible to derive mathematically what the frequency distribution of certain universe should be. Such distributions are called **Theoretical Distributions**.

There are many types of theoretical frequency distributions but we shall consider only three which are of great importance:

(*i*) Binomial Distribution (or Bernoulli's Distribution);

(*ii*) Poisson's Distribution;

(*iii*) Normal Distribution.

BINOMIAL (OR BERNOULLI'S) DISTRIBUTION

5.20. BINOMIAL PROBABILITY DISTRIBUTION

Let there be n independent trials in an experiment. Let a random variable X denote the number of successes in these n trials. Let p be the probability of a success and q that of a failure in a single trial so that $p + q = 1$. Let the trials be independent and p be constant for every trial.

Let us find the probability of r successes in n trials.

r successes can be obtained in n trials in ${}^nC_r$ ways.

$$\therefore \quad P(X = r) = {}^nC_r P(\underbrace{S\,S\,S \ldots\ldots S}_{r \text{ times}} \quad \underbrace{F\,F\,F \ldots\ldots F}_{(n-r) \text{ times}})$$

$$= {}^nC_r \underbrace{P(S)\,P(S) \ldots\ldots P(S)}_{r \text{ factors}} \quad \underbrace{P(F)\,P(F) \ldots\ldots P(F)}_{(n-r) \text{ factors}}$$

$$= {}^nC_r \underbrace{p\,p\,p \ldots\ldots p}_{r \text{ factors}} \quad \underbrace{q\,q\,q \ldots\ldots q}_{(n-r) \text{ factors}}$$

$$= {}^nC_r p^r q^{n-r} \qquad \ldots(1)$$

Hence, $\mathbf{P(X = r) = {}^nC_r\, q^{n-r}\, p^r}$, where $\mathbf{p + q = 1}$ **and** $\mathbf{r = 0, 1, 2, \ldots, n}$.

The distribution (1) is called the *binomial probability distribution* and X is called the *binomial variate*.

Note 1. $P(X = r)$ is usually written as $P(r)$.

Note 2. The successive probabilities $P(r)$ in (1) for $r = 0, 1, 2, \ldots\ldots, n$ are

$${}^nC_0 q^n,\ {}^nC_1 q^{n-1} p,\ {}^nC_2 q^{n-2} p^2,\ \ldots\ldots,\ {}^nC_n p^n$$

which are the successive terms of the binomial expansion of $(q + p)^n$. That is why this distribution is called "binomial" distribution.

Note 3. n and p occurring in the binomial distribution are called the *parameters* of the distribution.

Note 4. In a binomial distribution:

(*i*) n, the number of trials is finite.

(*ii*) each trial has only two possible outcomes usually called success and failure.

(*iii*) all the trials are independent.

(*iv*) p (and hence q) is constant for all the trials.

Note 5. A binomial distribution with n trials and probability of success in each trial as p, is denoted by $B(n, p)$.

5.21. RECURRENCE OR RECURSION FORMULA FOR THE BINOMIAL DISTRIBUTION

In a binomial distribution,

$$P(r) = {}^nC_r\, q^{n-r} p^r = \frac{n!}{(n-r)!\,r!}\, q^{n-r} p^r$$

$$P(r+1) = {}^nC_{r+1}\, q^{n-r-1} p^{r+1} = \frac{n!}{(n-r-1)!\,(r+1)!}\, q^{n-r-1} p^{r+1}$$

$$\therefore \quad \frac{P(r+1)}{P(r)} = \frac{(n-r)!}{(n-r-1)!} \times \frac{r!}{(r+1)!} \times \frac{p}{q} = \frac{(n-r)\times(n-r-1)!}{(n-r-1)!} \times \frac{r!}{(r+1)\times r!} \times \frac{p}{q} = \frac{n-r}{r+1}\cdot\frac{p}{q}$$

$$\Rightarrow \quad P(r+1) = \frac{n-r}{r+1}\cdot\frac{p}{q}\, P(r)$$

which is the required recurrence formula. Applying this formula successively, we can find P(1), P(2), P(3), ..., if P(0) is known.

5.22. MEAN AND VARIANCE OF THE BINOMIAL DISTRIBUTION

(Cochin 2010, 2011)

For the binomial distribution, $P(r) = {}^nC_r\, q^{n-r} p^r$

$$\text{Mean } (\mu) = \sum_{r=0}^{n} rP(r) = \sum_{r=0}^{n} r\,.\,{}^nC_r q^{n-r} p^r$$

$$= 0 + 1\,.\,{}^nC_1\, q^{n-1} p + 2\,.\,{}^nC_2\, q^{n-2} p^2 + 3\,.\,{}^nC_3\, q^{n-3} p^3 + \cdots\cdots + n\,.\,{}^nC_n\, p^n$$

$$= nq^{n-1} p + 2\,.\,\frac{n(n-1)}{2\,.\,1}\, q^{n-2}p^2 + 3\,.\,\frac{n(n-1)(n-2)}{3\,.\,2\,.\,1}\, q^{n-3} p^3 + \cdots\cdots + n.p^n$$

$$= nq^{n-1}p + n(n-1)\, q^{n-2}p^2 + \frac{n(n-1)(n-2)}{2\,.\,1}\, q^{n-3}p^3 + \cdots\cdots + np^n$$

$$= np\left[q^{n-1} + (n-1)q^{n-2}\, p + \frac{(n-1)(n-2)}{2.1}\, q^{n-3}p^2 + \cdots\cdots + p^{n-1}\right]$$

$$= np[{}^{n-1}C_0 q^{n-1} + {}^{n-1}C_1 q^{n-2}p + {}^{n-1}C_2 q^{n-3}p^2 + \cdots + {}^{n-1}C_{n-1} p^{n-1}]$$

$$= np(q+p)^{n-1} = np \qquad (\because\ p + q = 1)$$

Hence the mean of the binomial distribution is np.

$$\text{Variance } \sigma^2 = \sum_{r=0}^{n} r^2P(r) - \mu^2 = \sum_{r=0}^{n} [r + r(r-1)]\, P(r) - \mu^2$$

$$= \sum_{r=0}^{n} rP(r) + \sum_{r=0}^{n} r(r-1)\, P(r) - \mu^2 = \mu + \sum_{r=2}^{n} r(r-1)\, {}^nC_r q^{n-r}p^r - \mu^2$$

(Since the contribution due to $r = 0$ and $r = 1$ is zero)

$$= \mu + [2\cdot 1\cdot {}^nC_2 q^{n-2}p^2 + 3.2\cdot {}^nC_3 q^{n-3}p^3 + \ldots + n(n-1)\, {}^nC_n p^n] - \mu^2$$

$$= \mu + \left[2\,.\,1.\frac{n(n-1)}{2\cdot 1}\, q^{n-2}p^2 + 3\cdot 2\cdot\frac{n(n-1)(n-2)}{3\cdot 2\cdot 1}\, q^{n-3}p^3 + \cdots + n(n-1)p^n\right] - \mu^2$$

$$= \mu + [n(n-1)q^{n-2}p^2 + n(n-1)(n-2)q^{n-3}p^3 + \cdots + n(n-1)p^n] - \mu^2$$
$$= \mu + n(n-1)\, p^2[q^{n-2} + (n-2)q^{n-3}p + \cdots + p^{n-2}] - \mu^2$$
$$= \mu + n\,(n-1)p^2\,[{}^{n-2}C_0q^{n-2} + {}^{n-2}C_1q^{n-3}p + \cdots + {}^{n-2}C_{n-2}p^{n-2}] - \mu^2$$
$$= \mu + n(n-1)p^2(q+p)^{n-2} - \mu^2 = \mu + n(n-1)p^2 - \mu^2 \qquad [\because\ q+p=1]$$
$$= np + n(n-1)p^2 - n^2p^2 \qquad [\because\ \mu = np]$$
$$= np[1 + (n-1)p - np] = np[1-p] = npq.$$

Hence the variance of the binomial distribution is npq.

Standard deviation of the binomial distribution is $\sqrt{npq}$.

Similarly, we can prove that

$$\beta_1 = \frac{{\mu_3}^2}{{\mu_2}^3} = \frac{(q-p)^2}{npq} = \frac{(1-2p)^2}{npq}; \qquad \beta_2 = \frac{\mu_4}{{\mu_2}^2} = 3 + \frac{1-6pq}{npq}$$

Hence $\quad \gamma_1 = \sqrt{\beta_1} = \dfrac{q-p}{\sqrt{npq}} = \dfrac{1-2p}{\sqrt{npq}}; \qquad \gamma_2 = \beta_2 - 3 = \dfrac{1-6pq}{\sqrt{npq}}.$

Note. $\gamma_1 = \dfrac{q-p}{\sqrt{npq}} = \dfrac{1-2p}{\sqrt{npq}}$ gives a **measure of skewness** of the binomial distribution. If $p < \frac{1}{2}$, skewness is positive, if $p > \frac{1}{2}$, skewness is negative and if $p = \frac{1}{2}$, it is zero.

$\beta_2 = 3 + \dfrac{1-6pq}{npq}$ gives a **measure of the kurtosis** of the binomial distribution.

ILLUSTRATIVE EXAMPLES

Example 1. *During war, 1 ship out of 9 was sunk on an average in making a certain voyage. What was the probability that exactly 3 out of a convoy of 6 ships would arrive safely?*

Sol. p, the probability of a ship arriving safely $= 1 - \dfrac{1}{9} = \dfrac{8}{9}; \quad q = \dfrac{1}{9}, n = 6.$

Binomial distribution is $\left(\dfrac{1}{9} + \dfrac{8}{9}\right)^6$

The probability that exactly 3 ships arrive safely $= {}^6C_3\left(\dfrac{1}{9}\right)^3\left(\dfrac{8}{9}\right)^3 = \dfrac{10240}{9^6}.$

Example 2. *Assume that on the average one telephone number out of fifteen called between 2 P.M. and 3 P.M. on week-days is busy. What is the probability that if 6 randomly selected telephone numbers are called (i) not more than three, (ii) at least three of them will be busy?*

Sol. p, the probability of a telephone number being busy between 2 P.M. and 3 P.M. on week-days $= \dfrac{1}{15}$

$$q = 1 - \frac{1}{15} = \frac{14}{15}, \quad n = 6; \text{ Binomial distribution is } \left(\frac{14}{15} + \frac{1}{15}\right)^6$$

The probability that not more than three will be busy

$$= P(0) + P(1) + P(2) + P(3)$$

$$= {}^6C_0\left(\frac{14}{15}\right)^6 + {}^6C_1\left(\frac{14}{15}\right)^5\left(\frac{1}{15}\right) + {}^6C_2\left(\frac{14}{15}\right)^4\left(\frac{1}{15}\right)^2 + {}^6C_3\left(\frac{14}{15}\right)^3\left(\frac{1}{15}\right)^3$$

$$= \frac{(14)^3}{(15)^6}[2744 + 1176 + 210 + 20] = \frac{2744 \times 4150}{(15)^6} = 0.9997$$

The probability that at least three of them will be busy

$$= P(3) + P(4) + P(5) + P(6)$$

$$= {}^6C_3\left(\frac{14}{15}\right)^3\left(\frac{1}{15}\right)^3 + {}^6C_4\left(\frac{14}{15}\right)^2\left(\frac{1}{15}\right)^4 + {}^6C_5\left(\frac{14}{15}\right)\left(\frac{1}{15}\right)^5 + {}^6C_6\left(\frac{1}{15}\right)^6 = 0.005.$$

Example 3. *Six dice are thrown 729 times. How many times do you expect at least three dice to show a five or six?*

Sol. p = the chance of getting 5 or 6 with one die $= \frac{2}{6} = \frac{1}{3}$

$$q = 1 - \frac{1}{3} = \frac{2}{3},\ n = 6,\ N = 729$$

since dice are in sets of 6 and there are 729 sets.

The binomial distribution is $N(q + p)^n = 729\left(\frac{2}{3} + \frac{1}{3}\right)^6$

The expected number of times *at least three* dice showing five or six

$$= 729\left[{}^6C_3\left(\frac{2}{3}\right)^3\left(\frac{1}{3}\right)^3 + {}^6C_4\left(\frac{2}{3}\right)^2\left(\frac{1}{3}\right)^4 + {}^6C_5\left(\frac{2}{3}\right)\left(\frac{1}{3}\right)^5 + {}^6C_6\left(\frac{1}{3}\right)^6\right]$$

$$= \frac{729}{3^6}[160 + 60 + 12 + 1] = 233.$$

Example 4. *Out of 800 families with 4 children each, how many families would be expected to have (i) 2 boys and 2 girls (ii) at least one boy (iii) no girl (iv) at most two girls? Assume equal probabilities for boys and girls.* (DCRUST Murthal May 2014; M.D.U. 2010)

Sol. Since probabilities for boys and girls are equal

p = probability of having a boy $= \frac{1}{2}$; q = probability of having a girl $= \frac{1}{2}$

$n = 4$, $N = 800$ $\therefore$ The binomial distribution is $800\left(\frac{1}{2} + \frac{1}{2}\right)^4$.

(*i*) The expected number of families having 2 *boys* and 2 girls

$$= 800\ {}^4C_2\left(\frac{1}{2}\right)^2\left(\frac{1}{2}\right)^2 = 800 \times 6 \times \frac{1}{16} = 300.$$

(*ii*) The expected number of families having *at least one boy*

$$= 800\left[{}^4C_1\left(\frac{1}{2}\right)^3\left(\frac{1}{2}\right) + {}^4C_2\left(\frac{1}{2}\right)^2\left(\frac{1}{2}\right)^2 + {}^4C_3\left(\frac{1}{2}\right)\left(\frac{1}{2}\right)^3 + {}^4C_4\left(\frac{1}{2}\right)^4\right]$$

$$= 800 \times \frac{1}{16}\ [4 + 6 + 4 + 1] = 750.$$

(*iii*) The expected number of families having no girl, *i.e.*, having 4 boys

$$= 800\ .\ {}^4C_4 \left(\frac{1}{2}\right)^4 = 50.$$

(*iv*) The expected number of families having *at most two girls i.e.*, having *at least 2 boys*

$$= 800 \left[{}^4C_2 \left(\frac{1}{2}\right)^2 \left(\frac{1}{2}\right)^2 + {}^4C_3 \left(\frac{1}{2}\right)\left(\frac{1}{2}\right)^3 + {}^4C_4 \left(\frac{1}{2}\right)^4\right] = 800 \times \frac{1}{16}\ [6 + 4 + 1] = 550.$$

Example 5. *Suppose X has a binomial distribution B* $\left(6, \frac{1}{2}\right)$. *Show that X = 3 is the most likely outcome.*

Sol. Here $n = 6,\ \ p = \frac{1}{2}\ \ \therefore\ \ q = 1 - p = \frac{1}{2}$

$$\text{P}(0) = {}^6C_0\, q^6 = \left(\frac{1}{2}\right)^6 = \frac{1}{64}; \qquad \text{P}(1) = {}^6C_1\, q^5 p = 6\left(\frac{1}{2}\right)^6 = \frac{6}{64}$$

$$\text{P}(2) = {}^6C_2\, q^4 p^2 = 15\left(\frac{1}{2}\right)^6 = \frac{15}{64}; \qquad \text{P}(3) = {}^6C_3\, q^3 p^3 = 20\left(\frac{1}{2}\right)^6 = \frac{20}{64}$$

$$\text{P}(4) = {}^6C_4\, q^2 p^4 = 15\left(\frac{1}{2}\right)^6 = \frac{15}{64}; \qquad \text{P}(5) = {}^6C_5\, q\, p^5 = 6\left(\frac{1}{2}\right)^6 = \frac{6}{64}$$

$$\text{P}(6) = {}^6C_6\, p^6 = \left(\frac{1}{2}\right)^6 = \frac{1}{64}$$

Since P(3) is the maximum among all P(r), r = 0, 1, 2, 3, 4, 5, 6, therefore, X = 3 is the most likely outcome.

Example 6. *Find the mean of the binomial distribution B* $\left(4, \frac{1}{3}\right)$.

Sol. Let X be the random variable whose probability distribution is B $\left(4, \frac{1}{3}\right)$. Then X = 0, 1, 2, 3, 4.

Here $n = 4,\ p = \frac{1}{3}\ \ \therefore\ \ q = 1 - \frac{1}{3} = \frac{2}{3}$

$$\text{P}(0) = {}^4C_0 \left(\frac{2}{3}\right)^4 = \frac{16}{81}; \qquad \text{P}(1) = {}^4C_1 \left(\frac{2}{3}\right)^3 \left(\frac{1}{3}\right) = \frac{32}{81};$$

$$\text{P}(2) = {}^4C_2 \left(\frac{2}{3}\right)^2 \left(\frac{1}{3}\right)^2 = \frac{24}{81};$$

$$\text{P}(3) = {}^4C_3 \left(\frac{2}{3}\right)\left(\frac{1}{3}\right)^3 = \frac{8}{81}; \qquad \text{P}(4) = {}^4C_4 \left(\frac{1}{3}\right)^4 = \frac{1}{81}$$

$$\text{Mean } (\mu) = \sum_{i=0}^{4} p_i x_i = \frac{16}{81} \times 0 + \frac{32}{81} \times 1 + \frac{24}{81} \times 2 + \frac{8}{81} \times 3 + \frac{1}{81} \times 4$$

$$= \frac{32 + 48 + 24 + 4}{81} = \frac{108}{81} = \frac{4}{3}.$$

Example 7. *The probability of a shooter hitting a target is* $\frac{3}{4}$. *How many minimum number of times must he/she fire so that the probability of hitting the target at least once is more than 0.99?*

Sol. Let the shooter fire n times. Here n fires are n Bernoulli trials with

$$p = \frac{3}{4} \quad \text{and} \quad q = 1 - \frac{3}{4} = \frac{1}{4}.$$

Now $\qquad P(X \geq 1) > 0.99 \qquad$ (given)

$\Rightarrow \qquad 1 - P(X = 0) > 0.99$

$\Rightarrow \qquad P(X = 0) < 0.01$

$\Rightarrow \qquad {}^nC_0 \left(\frac{1}{4}\right)^n < \frac{1}{100}$

$\Rightarrow \qquad 4^n > 100 \qquad$ ...(1)

The minimum value of n satisfying (1) is 4.

Hence the shooter must fire 4 times.

Example 8. *Fit a binomial distribution to the following data:*

x:	*0*	*1*	*2*	*3*	*4*
f:	*30*	*62*	*46*	*10*	*2*

(M.D.U. 2009; K.U.K. Dec. 2015)

Sol. The table is as follows:

x	f	fx
0	30	0
1	62	62
2	46	92
3	10	30
4	2	8
	$\Sigma f = 150$	$\Sigma fx = 192$

$$\text{Mean of observations} = \frac{\Sigma fx}{\Sigma f} = \frac{192}{150} = 1.28$$

$\Rightarrow \qquad np = 1.28$

$\Rightarrow \qquad 4p = 1.28 \qquad (\because \; n = 4$

$\Rightarrow \qquad p = 0.32 \qquad \therefore \; q = 1 - p = 1 - 0.32 = 0.68)$

Also $N = 150$ $[\because \quad N = \Sigma f]$

Hence, the binomial distribution is

$$N(q + p)^n = 150\,(0.68 + 0.32)^4.$$

EXERCISE 5.4

1. Ten coins are tossed simultaneously. Find the probability of getting at least seven heads.

2. The probability of any ship of a company being destroyed on a certain voyage is 0.02. The company owns 6 ships for the voyage. What is the probability of:

(*i*) losing one ship (*ii*) losing at most two ships (*iii*) losing none?

3. (*a*) The probability that a man aged 60 will live to be 70 is 0.65. What is the probability that out of ten men now 60, at least 7 would live to be 70?

(*b*) The probability that a man aged 50 years will die within a year is 0.01125. What is the probability that out of 12 such men, at least 11 will reach their fifty first birthday? (*M.D.U. 2011*)

4. (*a*) The incidence of occupational disease in an industry is such that the workers have a 20% chance of suffering from it. What is the probability that out of six workers chosen at random, four or more will suffer from the disease?

(*b*) A binomial variate X satisfies the relation $9P(X = 4) = P(X = 2)$ when $n = 6$. Find the value of the parameter p and $P(X = 1)$.

5. (*a*) The probability that a pen manufactured by a company will be defective is $\frac{1}{10}$. If 12 such pens are manufactured, find the probability that

(*i*) exactly two will be defective (*ii*) at least two will be defective

(*iii*) none will be defective.

(*b*) If 10 per cent of the rivets produced by a machine are defective, find the probability that out of 5 rivets chosen at random (*i*) none will be defective, (*ii*) one will be defective and (*iii*) at least two will be defective. (*K.U.K. Jan. 2013*)

6. If the chance that one of the ten telephone lines is busy at an instant is 0.2.

(*i*) What is the chance that 5 of the lines are busy?

(*ii*) What is the probability that all the lines are busy?

7. If on an average 1 vessel in every 10 is wrecked, find the probability that out of 5 vessels expected to arrive, at least 4 will arrive safely.

8. A policeman fires 6 bullets on a dacoit. The probability that the dacoit will be killed by a bullet is 0.6. What is the probability that the dacoit is still alive?

9. A bag contains 5 white, 7 red and 8 black balls. If four balls are drawn, one by one, with replacement, what is the probability that

(*i*) none is white (*ii*) all are white

(*iii*) at least one is white (*iv*) only two are white?

10. In a hurdle race, a player has to cross 10 hurdles. The probability that he will clear each hurdle is $\frac{5}{6}$. What is the probability that he will knock down fewer than 2 hurdles? (*Calicut 2012*)

11. Fit a Binomial distribution for the following data and compare the theoretical frequencies with the actual ones:

x:	0	1	2	3	4	5
f:	2	14	20	34	22	8

12. If the sum of mean and variance of a binomial distribution is 4.8 for five trials, find the distribution.

13. If the mean of a binomial distribution is 3 and the variance is $\frac{3}{2}$, find the probability of obtaining at least 4 success.

14. In 800 families with 5 children each, how many families would be expected to have (*i*) 3 boys and 2 girls, (*ii*) 2 boys and 3 girls, (*iii*) no girl (*iv*) at the most two girls. (Assume probabilities for boys and girls to be equal.)

15. (*a*) In 100 sets of ten tosses of an unbiased coin, in how many cases do you expect to get

(*i*) 7 heads and 3 tails (*ii*) at least 7 heads?

(*b*) In 256 sets of 12 tosses of a coin, in how many cases one can expect 8 heads and 4 tails?

(M.D.U. 2011)

16. (*a*) The following data are the number of seeds germinating out of 10 on damp filter for 80 sets of seeds. Fit a Binomial distribution to this data:

x:	0	1	2	3	4	5	6	7	8	9	10	Total
f:	6	20	28	12	8	6	0	0	0	0	0	80

[**Hint.** Here $n = 10$, $N = 80$, Mean $= \frac{\Sigma fx}{\Sigma f} = 2.175$ $\therefore$ $np - 2.175$ etc.]

(*b*) Fit a Binomial distribution to the following frequency distribution:

x:	0	1	2	3	4	5	6
f:	13	25	52	58	32	16	4

(M.D.U. 2013)

17. A bag contains 10 balls each marked with one of the digits 0 to 9. If four balls are drawn successively with replacement from the bag, what is the probability that none is marked with the digit 0?

18. A box contains 100 tickets each bearing one of the numbers from 1 to 100. If 5 tickets are drawn successively with replacement from the box, find the probability that all the tickets bear numbers divisible by 10.

19. The probability that a bomb dropped from a plane will strike the target is $\frac{1}{5}$. If six bombs are dropped, find the probability that (*i*) exactly two will strike the target, (*ii*) at least two will strike the target. *(K.U.K. May 2013)*

20. In sampling a large number of parts manufactured by a machine, the mean number of defectives in a sample of 20 is 2. Out of 1000 such samples, how many would be expected to contain at least 3 defective parts. *(K.U.K. 2010; M.D.U. 2010)*

21. In a lot of 500 solenoids 25 are defective. Find the probability of 0, 1, 2, 3, defective soleniods in a random sample of 20 solenoids.

22. Find the probability that in five tosses of a fair die the number '3' appears atmost once.

(Cochin 2011)

23. If on the average rain falls on 10 days in every 30 days, obtain the probability that

(*i*) rain will fall on at least 3 days of a given week

(*ii*) first three days of a given week will be fine and the remaining 4 days wet.

[**Hint.** (*ii*) Reqd. probability $= q^3p^4$]

24. How many tosses of a coin are needed so that the probability of getting at least one head is 0.875?

25. In a binomial distribution with 6 independent trials, the probabilities of 3 and 4 successes are found to be 0.2457 and 0.0819 respectively. Find the parameters 'p' and 'q' of the binomial distribution.

26. A die is thrown again and again until three sixes are obtained. Find the probability of obtaining the third six in the sixth throw of the die.

Answers

1. $\frac{11}{64}$ **2.** (*i*) 0.1085, (*ii*) 0.9997, (*iii*) 0.8858

3. (*a*) 0.514 (*b*) 0.9923

4. (*a*) 0.017 (*b*) $\frac{1}{4}$; 0.3559

5. (*a*) (*i*) 0.2301 (*ii*) 0.3310 (*iii*) 0.2824 (*b*) 0.590, 0.328, 0.082

6. (*i*) 0.02642 (*ii*) 1.024×10^{-7} **7.** 0.91854 **8.** 0.004096

9. (*i*) $\frac{81}{256}$ (*ii*) $\frac{1}{256}$ (*iii*) $\frac{175}{256}$ (*iv*) $\frac{27}{128}$ **10.** $\frac{5}{2}\left(\frac{5}{6}\right)^9$

11. $100\ (0.432 + 0.568)^5$ **12.** $\left(\frac{1}{4}+\frac{5}{6}\right)^5$ **13.** $\frac{11}{32}$

14. (*i*) 250, (*ii*) 250, (*iii*) 25, (*iv*) 400 **15.** (*a*) (*i*) 12 nearly (*ii*) 17 nearly (*b*) 31

16. (*a*) $80\ (0.7825 + 0.2175)^{10}$ (*b*) $200\ (0.554 + 0.446)^6$

17. $\left(\frac{9}{10}\right)^4$ **18.** 0.00001 **19.** (*i*) 0.246 (*ii*) 0.345 **20.** 323

21. 0.3585, 0.3774, 0.1887, 0.0596 **22.** $2\left(\frac{5}{6}\right)^5$

23. (*i*) 0.4294 (*ii*) 0.0037 **24.** 3 **25.** $p = \frac{4}{13}, q = \frac{9}{13}$

26. $\frac{625}{23328}$.

POISSON DISTRIBUTION

5.23. POISSON DISTRIBUTION AS A LIMITING CASE OF BINOMIAL DISTRIBUTION

(M.D.U. 2012; Calicut 2010, 2011; Cochin 2013)

If the parameters n and p of a binomial distribution are known, we can find the distribution. But in situations where n is very large and p is very small, application of binomial distribution is very labourious. However, if we assume that as $n \to \infty$ and $p \to 0$ such that np always remains finite, say λ, we get the Poisson approximation to the binomial distribution.

Now, for a Binomial distribution

$$P(X = r) = {}^nC_r q^{n-r} p^r$$

$$= \frac{n(n-1)(n-2)\dots(n-r+1)}{r!} \times (1-p)^{n-r} \times p^r$$

$$= \frac{n(n-1)(n-2)\dots(n-r+1)}{r!} \times \left(1-\frac{\lambda}{n}\right)^{n-r} \times \left(\frac{\lambda}{n}\right)^r \quad \text{since } np = \lambda \quad \therefore \quad p = \frac{\lambda}{n}$$

$$= \frac{\lambda^r}{r!} \times \frac{n(n-1)(n-2)\dots(n-r+1)}{n^r} \times \frac{\left(1-\frac{\lambda}{n}\right)^n}{\left(1-\frac{\lambda}{n}\right)^r}$$

$$= \frac{\lambda^r}{r!}\left(\frac{n}{n}\right)\left(\frac{n-1}{n}\right)\left(\frac{n-2}{n}\right)\dots\dots\left(\frac{n-r+1}{n}\right) \times \frac{\left(1-\frac{\lambda}{n}\right)^n}{\left(1-\frac{\lambda}{n}\right)^r}$$

$$= \frac{\lambda^r}{r!}\left(1-\frac{1}{n}\right)\left(1-\frac{2}{n}\right)\dots\dots\left(1-\frac{r-1}{n}\right) \times \frac{\left[\left(1-\frac{\lambda}{n}\right)^{-\frac{n}{\lambda}}\right]^{-\lambda}}{\left(1-\frac{\lambda}{n}\right)^r}$$

As $n \to \infty$, each of the $(r-1)$ factors

$$\left(1-\frac{1}{n}\right), \left(1-\frac{2}{n}\right), \dots\dots, \left(1-\frac{r-1}{n}\right) \quad \text{tends to 1. Also } \left(1-\frac{\lambda}{n}\right)^r \text{ tends to 1.}$$

Since $\underset{x \to \pm\infty}{\text{Lt}} \left(1+\frac{1}{x}\right)^x = e$, the Naperian base. $\therefore \quad \left[\left(1-\frac{\lambda}{n}\right)^{-\frac{n}{\lambda}}\right]^{-\lambda} \longrightarrow e^{-\lambda}$ as $n \to \infty$

Hence in the limiting case when $n \to \infty$, we have

$$P(X = r) = \lambda^r \frac{e^{-\lambda}}{r!} \quad (r = 0, 1, 2, 3, \dots) \qquad \dots(A)$$

where λ is a finite number $= np$.

(A) represents a probability distribution which is called the Poisson probability distribution.

Note 1. λ is called the parameter of the distribution.

Note 2. $e^x = 1 + \frac{x}{1!} + \frac{x^2}{2!} + \dots\dots + \frac{x^n}{n!} + \dots\dots$ to ∞.

Note 3. The sum of the probabilities P(r) for $r = 0, 1, 2, 3, \ldots\ldots$ is 1, since

$$P(0) + P(1) + P(2) + P(3) + \cdots\cdots = e^{-\lambda} + \frac{\lambda e^{-\lambda}}{1!} + \frac{\lambda^2 e^{-\lambda}}{2!} + \frac{\lambda^3 e^{-\lambda}}{3!} + \cdots\cdots$$

$$= e^{-\lambda}\left(1 + \frac{\lambda}{1!} + \frac{\lambda^2}{2!} + \frac{\lambda^3}{3!} + \cdots\cdots\right) = e^{-\lambda} \cdot e^{\lambda} = 1.$$

5.24. RECURRENCE FORMULA FOR THE POISSON DISTRIBUTION

For Poisson distribution, $P(r) = \dfrac{\lambda^r e^{-\lambda}}{r!}$ and $P(r+1) = \dfrac{\lambda^{r+1} e^{-\lambda}}{(r+1)!}$

$$\therefore \quad \frac{P(r+1)}{P(r)} = \frac{\lambda r!}{(r+1)!} = \frac{\lambda}{r+1} \quad \text{or} \quad P(r+1) = \frac{\lambda}{r+1} P(r),\ r = 0, 1, 2, 3, \ldots\ldots$$

This is called the recurrence formula for the Poisson distribution.

5.25. MEAN AND VARIANCE OF THE POISSON DISTRIBUTION *(M.D.U. Dec.2015)*

For the Poisson distribution, $P(r) = \dfrac{\lambda^r e^{-\lambda}}{r!}$

$$\text{Mean } \mu = \sum_{r=0}^{\infty} rP(r) = \sum_{r=0}^{\infty} r \cdot \frac{\lambda^r e^{-\lambda}}{r!}$$

$$= e^{-\lambda} \sum_{r=1}^{\infty} \frac{\lambda^r}{(r-1)!} = e^{-\lambda}\left(\lambda + \frac{\lambda^2}{1!} + \frac{\lambda^3}{2!} + \cdots\cdots\right)$$

$$= \lambda e^{-\lambda}\left(1 + \frac{\lambda}{1!} + \frac{\lambda^2}{2!} + \cdots\cdots\right) = \lambda e^{-\lambda} \cdot e^{\lambda} = \lambda$$

Thus, the mean of the Poisson distribution is equal to the parameter λ.

$$\text{Variance } \sigma^2 = \sum_{r=0}^{\infty} r^2 P(r) - \mu^2 = \sum_{r=0}^{\infty} r^2 \cdot \frac{\lambda^r e^{-\lambda}}{r!} - \lambda^2 = e^{-\lambda} \sum_{r=1}^{\infty} \frac{r^2 \lambda^r}{r!} - \lambda^2$$

$$= e^{-\lambda}\left[\frac{1^2 \cdot \lambda}{1!} + \frac{2^2 \cdot \lambda}{2!} + \frac{3^2 \lambda^3}{3!} + \frac{4^2 \lambda^4}{4!} + \cdots\cdots\right] - \lambda^2$$

$$= \lambda e^{-\lambda}\left[1 + \frac{2\lambda^2}{1!} + \frac{3\lambda^2}{2!} + \frac{4\lambda^3}{3!} + \cdots\cdots\right] - \lambda^2$$

$$= \lambda e^{-\lambda}\left[1 + \frac{(1+1)\lambda}{1!} + \frac{(1+2)\lambda^2}{2!} + \frac{(1+3)\lambda^3}{3!} + \cdots\cdots\right] - \lambda^2$$

$$= \lambda e^{-\lambda}\left[\left(1+\frac{\lambda}{1!}+\frac{\lambda^2}{2!}+\frac{\lambda^3}{3!}+\ldots\ldots\right)+\left(\frac{\lambda}{1!}+\frac{2\lambda^2}{2!}+\frac{3\lambda^3}{3!}+\ldots\ldots\right)\right]-\lambda^2$$

$$= \lambda e^{-\lambda}\left[e^{\lambda}+\lambda\left(1+\frac{\lambda}{1!}+\frac{\lambda^2}{2!}+\ldots\ldots\right)\right]-\lambda^2$$

$$= \lambda e^{-\lambda}\,[e^{\lambda}+\lambda e^{\lambda}]-\lambda^2 = \lambda e^{-\lambda}\,.\,e^{\lambda}\,(1+\lambda)-\lambda^2 = \lambda(1+\lambda)-\lambda^2 = \lambda.$$

Hence, the variance of the Poisson distribution is also λ.

Thus, the mean and the variance of the Poisson distribution are each equal to the parameter λ.

Note. The mean and the variance of the Poisson distribution can also be derived from those of the binomial distribution in the limiting case when $n \to \infty$, $p \to 0$ and $np = \lambda$.

Mean of binomial distribution is np.

$\therefore$ Mean of Poisson distribution $= \underset{n\to\infty}{\text{Lt}}\ np = \underset{n\to\infty}{\text{Lt}}\ \lambda = \lambda$

Variance of binomial distribution is $npq = np\,(1-p)$

$\therefore$ Variance of Poisson distribution $= \underset{n\to\infty}{\text{Lt}}\ np\,(1-p) = \underset{n\to\infty}{\text{Lt}}\ \lambda\left(1-\frac{\lambda}{n}\right) = \lambda.$

ILLUSTRATIVE EXAMPLES

Example 1. *If the variance of the Poisson distribution is 2, find the probabilities for r = 1, 2, 3, 4 from the recurrence relation of the Poisson distribution.*

Sol. λ, the parameter of Poisson distribution = Variance = 2

Recurrence relation for the Poisson distribution is

$$P(r+1) = \frac{\lambda}{r+1}\,P(r) = \frac{2}{r+1}\,P(r) \qquad \ldots(1)$$

Now $\quad P(r) = \dfrac{\lambda^r\, e^{-\lambda}}{r!} \;\Rightarrow\; P(0) = \dfrac{e^{-2}}{0!} = e^{-2} = 0.1353$

Putting $r = 0, 1, 2, 3$ in (1), we get

$$P(1) = 2P(0) = 2 \times 0.1353 = 0.2706;\;\; P(2) = \frac{2}{2}\,P(1) = 0.2706$$

$$P(3) = \frac{2}{3}\,P(2) = \frac{2}{3} \times 0.2706 = 0.1804;\; P(4) = \frac{2}{4}\,P(3) = \frac{1}{2} \times 0.1804 = 0.0902.$$

Example 2. *Assume that the probability of an individual coalminer being killed in a mine accident during a year is* $\dfrac{1}{2400}$. *Use Poisson's distribution to calculate the probability that in a mine employing 200 miners there will be at least one fatal accident in a year.*

Sol. Here $\quad p = \dfrac{1}{2400},\ n = 200; \quad \therefore\ \lambda = np = \dfrac{200}{2400} = \dfrac{1}{12} = 0.083$

$$\therefore \qquad P(r) = \frac{\lambda^r e^{-\lambda}}{r!} = \frac{(0.083)^r e^{-0.083}}{r!}$$

P(at least one fatal accident) = 1 – p (no fatal accident)

$$= 1 - P(0) = 1 - \frac{(0.083)^0 e^{-0.83}}{0!} = 1 - 0.92 = 0.08.$$

Example 3. *Data was collected over a period of 10 years, showing number of deaths from horse kicks in each of the 200 army corps. The distribution of deaths was as follows:*

No. of deaths:	*0*	*1*	*2*	*3*	*4*	*Total*
Frequency:	*109*	*65*	*22*	*3*	*1*	*200*

Fit a Poisson distribution to the data and calculate the theoretical frequencies:

Sol. Mean of given distribution $= \frac{\Sigma fx}{\Sigma f} = \frac{65 + 44 + 9 + 4}{200} = \frac{122}{200} = 0.61$

This is the parameter (λ) of the Poisson distribution.

$\therefore$ Required Poisson distribution is $N \cdot \frac{\lambda^r e^{-\lambda}}{r!}$ where $N = \Sigma f = 200$

$$= 200e^{-0.61} \cdot \frac{(0.61)^r}{r!} = 200 \times 0.5435 \frac{(0.61)^r}{r!} = 108.7 \times \frac{(0.61)^r}{r!}.$$

r	$P(r)$	*Theoretical frequency*
0	108.7	109
1	$108.7 \times 0.61 = 66.3$	66
2	$108.7 \times \frac{(0.61)^2}{2!} = 20.2$	20
3	$108.7 \times \frac{(0.61)^3}{3!} = 4.1$	4
4	$108.7 \times \frac{(0.61)^4}{4!} = 0.7$	1
		Total = 200

Example 4. *A car-hire firm has two cars, which it hires out day by day. The number of demands for a car on each day is distributed as a Poisson distribution with mean 1.5. Calculate the proportion of days on which neither car is used and the proportion of days on which some demand is refused.* ($e^{-1.5} = 0.2231$) (M.D.U. 2010)

Sol. Since the number of demands for a car is distributed as a Poisson distribution with mean $\lambda = 1.5$.

∴ Proportion of days on which neither car is used

= Probability of there being no demand for the car

$$= \frac{\lambda^0 e^{-\lambda}}{0!} = e^{-1.5} = 0.2231$$

Proportion of days on which some demand is refused

= probability for the number of demands to be more than two

$$= 1 - P(x \le 2) = 1 - \left(e^{-\lambda} + \frac{\lambda e^{-\lambda}}{1!} + \frac{\lambda^2 e^{-\lambda}}{2!}\right)$$

$$= 1 - e^{-1.5}\left(1 + 1.5 + \frac{(1.5)^2}{2}\right) = 1 - 0.2231\ (1 + 1.5 + 1.125)$$

$$= 1 - 0.2231 \times 3.625 = 1 - 0.8087375 = 0.1912625.$$

Example 5. *Six coins are tossed 6400 times. Using the Poisson distribution, determine the approximate probability of getting six heads x times.*

Sol. Probability of getting one head with one coin $= \frac{1}{2}$.

∴ The probability of getting six heads with six coins $= \left(\frac{1}{2}\right)^6 = \frac{1}{64}$

∴ Average number of six heads with six coins in 6400 throws $= np = 6400 \times \frac{1}{64} = 100$

∴ The mean of the Poisson distribution = 100.

Approximate probability of getting six heads x times when the distribution is Poisson

$$= \frac{\lambda^x e^{-\lambda}}{x!} = \frac{(100)^x \cdot e^{-100}}{x!}; x = 0, 1, 2, ...\ .$$

EXERCISE 5.5

1. If the probability that an individual suffers a bad reaction from a certain injection is 0.001, determine the probability that out of 2000 individuals
   (*i*) exactly 3
   (*ii*) more than 2 individuals
   (*iii*) none
   (*iv*) more than one individual will suffer a bad reaction. *(K.U.K. Jan. 2014; M.D.U. May 2014)*
2. If X is a Poisson variate such that P(X = 2) = 9P(X = 4) + 90P(X = 6), find the standard deviation.
3. If a random variable has a Poisson distribution such that P(1) = P(2), find
   (*i*) mean of the distribution *(M.D.U. 2011)* (*ii*) P(4).
4. Suppose that X has a Poisson distribution. If $P(X = 2) = \frac{2}{3} P(X = 1)$ find, (*i*) P(X = 0) (*ii*) P(X = 3). *(Cochin 2010)*
5. A certain screw making machine produces on average 2 defective screws out of 100, and packs them in boxes of 500. Find the probability that a box contains 15 defective screws.

6. A manufacturer knows that the condensers he makes contain on an average 2% defective. He packs them in boxes of 100. What is the probability that a box selected at random will contain 3 or more defective condensers?

7. Fit a Poisson distribution to the following:

x:	0	1	2	3	4
f:	192	100	24	3	1

8. (*a*) Fit a Poisson distribution to the following and calculate theoretical frequencies:

x:	0	1	2	3	4
f:	122	60	15	2	1

(*b*) Fit a Poisson distribution to the following data and calculate the theoretical frequencies:

x:	0	1	2	3	4
f:	46	38	22	9	1

$(e^{-0.97} = 0.379)$

(*K.U.K. May 2013; M.D.U. 2012*)

9. Fit a Poisson distribution to the following data given the number of yeast cells per square for 400 squares:

No. of cells per sq.:	0	1	2	3	4	5	6	7	8	9	10
No. of squares:	103	143	98	42	8	4	2	0	0	0	0

10. Show that in a Poisson distribution with unit mean, mean deviation about mean is $\left(\frac{2}{e}\right)$ times the standard deviation.

11. (*a*) In a certain factory turning razor blades, there is a small chance of 0.002 for any blade to be defective. The blades are supplied in packets of 10. Use Poisson distribution to calculate the approximate number of packets containing no defective, one defective and two defective blades respectively in a consignment of 10000 packets.

(*b*) In a certain factory turning out optical lenses, there is a small chance of $\frac{1}{500}$ for any one lense to be defective. The lenses are supplied in packets of 10. Use Poisson distribution to calculate the approximate number of packets containing no defective, one defective, two defective, three defective lenses respectively in a consignment of 20,000 packets.

12. The probability that a man aged 35 years will die before reaching the age of 40 years may be taken as 0.018. Out of a group of 400 men, now aged 35 years, what is the probability that 2 men will die within the next 5 years? (*Calicut 2011*)

13. Suppose a book of 585 pages contains 43 typographical errors. If these errors are randomly distributed throughout the book, what is the probability that 10 pages, selected at random, will be free from errors?

14. The side effects of a certain drug cause discomfort to only a few patients. The probability that any individual will suffer from these side effects is 0.002. If the drug is given to 3000 patients, what is the probability that (*i*) exactly 3 (*ii*) 5 or more than 5 will suffer side effects?

15. The probability that a man aged 50 years will die within a year is 0.01125. What is the probability that out of 12 such men, at least 11 will reach their fifty first birthday?

(Given $e^{-0.135} = 0.87371$)

Answers

1. (*i*) 0.1804 (*ii*) 0.3235 (*iii*) 0.1353 (*iv*) 0.5941

2. 1 **3.** (*i*) 2 (*ii*) $\frac{2}{3e^2}$ **4.** (*i*) e^{-4} (*ii*) $4e^{-4}$ **5.** $\frac{(10)^{15}\, e^{-10}}{(15)!} = 0.035$

6. 0.3235 **7.** $320 \times \frac{e^{0.503}\,(9.503)^r}{r!}$

8. (*a*) $P(r) = 121.36 \times \frac{(0.5)^{r!}}{r!}$, where $r = 0, 1, 2, 3, 4$

Theoretical frequencies are 121, 61, 15, 3, 0 respectively

(*b*) $P(r) = 43.964 \times \frac{(0.97)^r}{r!}$, where $r = 0, 1, 2, 3, 4$

Theoretical frequencies are 44, 43, 14, 3, 1 respectively

9. Theoretical frequencies are 109, 142, 92, 40, 13, 3, 1, 0, 0, 0, 0

11. (*a*) 9802, 196, 2 (*b*) 19604, 392, 4, 0 **12.** 0.01936 **13.** 0.4795

14. (*i*) 0.1784 (*ii*) 0.7150 **15.** 0.99166.

NORMAL DISTRIBUTION

5.26. NORMAL DISTRIBUTION

(*K.U.K. 2009, Dec. 2015*)

The normal distribution is a continuous distribution. It can be derived from the Binomial distribution in the limiting case when n, the number of trials is very large and p, the probability of a success, is close to $\frac{1}{2}$. The general equation of the normal distribution is given by

$$f(x) = \frac{1}{\sigma\sqrt{2\pi}}\, e^{-\frac{1}{2}\left(\frac{x-\mu}{\sigma}\right)^2}$$

where the variable x can assume all values from $-\infty$ to $+\infty$. μ and σ, called the parameters of the distribution, are respectively the mean and the standard deviation of the distribution and $-\infty < \mu < \infty$, $\sigma > 0$. x is called the normal variate and $f(x)$ is called probability density function of the normal distribution.

If a variable x has the normal distribution with mean μ and standard deviation σ, we briefly write x: $N(\mu, \sigma^2)$.

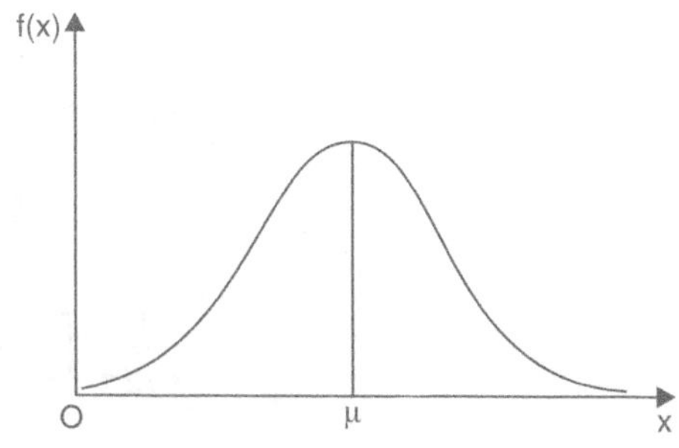

The graph of the normal distribution is called the *normal curve*. It is bell-shapped and symmetrical about the mean μ. The two tails of the curve extend to $+\infty$ and $-\infty$ towards the positive and negative directions of the x-axis respectively and gradually approach the x-axis without ever metting it. The curve is unimodal and the mode of the normal distribution coincides with its mean μ. The line $x = \mu$ divides the area under the normal curve above x-axis into two equal parts. Thus, the median of the distribution also coincides with its mean and mode. The area under the normal curve between any two given ordinates $x = x_1$ and $x = x_2$ represents the probability of values falling into the given interval. The total area under the normal curve above the x-axis is 1.

5.27. BASIC PROPERTIES OF THE NORMAL DISTRIBUTION

The probability density function of the normal distribution is given by

$$f(x) = \frac{1}{\sigma\sqrt{2\pi}} e^{-\frac{1}{2}\left(\frac{x-\mu}{\sigma}\right)^2}$$

1. *The total area under normal probability curve is unity.*

Normal probability curve is given by

$$f(x) = \frac{1}{\sigma\sqrt{2\pi}} e^{-\frac{(x-\mu)^2}{2\sigma^2}}$$

Area under this curve

$$= \int_{-\infty}^{\infty} f(x)\, dx = \frac{1}{\sigma\sqrt{2\pi}} \int_{-\infty}^{\infty} e^{-\frac{(x-\mu)^2}{2\sigma^2}} dx$$

Put $\frac{x-\mu}{\sigma\sqrt{2}} = t, \quad dx = \sigma\sqrt{2}\, dt$

Required area $= \frac{1}{\sigma\sqrt{2\pi}} \int_{-\infty}^{\infty} e^{-t^2} \sigma\sqrt{2}\, dt = \frac{1}{\sqrt{\pi}} \int_{-\infty}^{\infty} e^{-t^2} dt = \frac{1}{\sqrt{\pi}} \sqrt{\pi} = 1.$

2. *The mean of the normal distribution.*

The general form of the normal curve is

$$f(x) = \frac{1}{\sigma\sqrt{2\pi}} e^{-\frac{(x-\mu)^2}{2\sigma^2}}$$

Mean $= \int_{-\infty}^{\infty} x f(x)\, dx$ [See Art. 5.18(*b*)]

$$= \frac{1}{\sigma\sqrt{2\pi}} \int_{-\infty}^{\infty} x e^{-\frac{(x-\mu)^2}{2\sigma^2}} dx$$

Putting $\frac{x-\mu}{\sigma\sqrt{2}} = t, \; dx = \sigma\sqrt{2}\, dt$

∴ Mean $= \frac{1}{\sigma\sqrt{2\pi}} \int_{-\infty}^{\infty} (\mu + \sigma\sqrt{2}t) e^{-t^2} \sigma\sqrt{2}\, dt$

$$= \frac{1}{\sqrt{\pi}} \int_{-\infty}^{\infty} (\mu + \sigma\sqrt{2}t) e^{-t^2} dt = \frac{\mu}{\sqrt{\pi}} \int_{-\infty}^{\infty} e^{-t^2} dt + \frac{\sigma\sqrt{2}}{\sqrt{\pi}} \int_{-\infty}^{\infty} te^{-t^2} dt$$

$$= \frac{\mu}{\sqrt{\pi}} \sqrt{\pi} + \frac{\sigma\sqrt{2}}{\sqrt{\pi}} (0) = \mu \qquad | \because \; te^{-t^2} \text{ is an odd function of } t$$

3. *For a normal curve, the ordinate at the mean is the maximum ordinate.*

The equation of the normal curve is

$$y = \frac{1}{\sigma\sqrt{2\pi}} e^{-\frac{(x-\mu)^2}{2\sigma^2}}, \quad \text{Mean} = \mu$$

$$\frac{dy}{dx} = \frac{1}{\sigma\sqrt{2\pi}} e^{-\frac{(x-\mu)^2}{2\sigma^2}} \cdot \frac{-2(x-\mu)}{2\sigma^2} = -\frac{(x-\mu)}{\sigma^3\sqrt{2\pi}} e^{-\frac{(x-\mu)^2}{2\sigma^2}}$$

$$\frac{d^2y}{dx^2} = -\frac{1}{\sigma^3\sqrt{2\pi}} \left[1.e^{-\frac{(x-\mu)^2}{2\sigma^2}} + (x-\mu).e^{-\frac{(x-\mu)^2}{2\sigma^2}} \cdot \frac{-2(x-\mu)}{2\sigma^2} \right]$$

$$= -\frac{1}{\sigma^3\sqrt{2\pi}} . e^{-\frac{(x-\mu)^2}{2\sigma^2}} \left[1 - \frac{(x-\mu)^2}{\sigma^2} \right]$$

Now $\quad \dfrac{dy}{dx} = 0 \quad \text{when } x = \mu$

and $\quad \left[\dfrac{d^2y}{dx^2}\right]_{x=\mu} = -\dfrac{1}{\sigma^3\sqrt{2\pi}} < 0.$

Hence y, the ordinate is maximum when $x = \mu$ *i.e.*, the ordinate at the mean is the maximum ordinate.

4. *The mode of the normal distribution.*

The equation of the normal curve is

$$y = \frac{1}{\sigma\sqrt{2\pi}} e^{-\frac{(x-\mu)^2}{2\sigma^2}}$$

Mode is the value of x corresponding to $y = y_0$, where y_0 is the maximum frequency.

Proceeding as in Example 3, y is maximum when $x = \mu$.

Hence **the mode = the mean =** μ.

5. *The median of the normal distribution.*

If M is the median of the normal distribution, we have

$$\int_{-\infty}^{M} \frac{1}{\sigma\sqrt{2\pi}} e^{-\frac{(x-\mu)^2}{2\sigma^2}} dx = \frac{1}{2}$$

$$\Rightarrow \quad \frac{1}{\sigma\sqrt{2\pi}}\int_{-\infty}^{M} e^{-\frac{(x-\mu)^2}{2\sigma^2}}\,dx = \frac{1}{2}$$

$$\Rightarrow \quad \frac{1}{\sigma\sqrt{2\pi}}\int_{-\infty}^{\mu} e^{-\frac{(x-\mu)^2}{2\sigma^2}}\,dx + \frac{1}{\sigma\sqrt{2\pi}}\int_{\mu}^{M} e^{-\frac{(x-\mu)^2}{2\sigma^2}}\,dx = \tfrac{1}{2} \qquad \ldots(1)$$

Now $\quad \dfrac{1}{\sigma\sqrt{2\pi}}\displaystyle\int_{-\infty}^{\mu} e^{-\frac{(x-\mu)^2}{2\sigma^2}}\,dx$

$$= \frac{1}{\sigma\sqrt{2\pi}}\int_{\infty}^{0} e^{-t^2}(-\sigma\sqrt{2})\,dt, \quad \text{where} \quad -\frac{x-\mu}{\sigma\sqrt{2}} = t$$

$$= \frac{1}{\sqrt{\pi}}\int_{0}^{\infty} e^{-t^2}\,dt = \frac{1}{\sqrt{\pi}}\cdot\tfrac{1}{2}\sqrt{\pi} = \tfrac{1}{2}$$

$\therefore$ From (1)

$$\frac{1}{2} + \frac{1}{\sigma\sqrt{2\pi}}\int_{\mu}^{M} e^{-\frac{(x-\mu)^2}{2\sigma^2}}\,dx = \tfrac{1}{2}$$

$$\Rightarrow \quad \int_{m}^{M} e^{-\frac{(x-\mu)^2}{2\sigma^2}}\,dx = 0 \quad \Rightarrow \quad M = \mu$$

Hence, for the normal distribution, mean, median and mode coincide.

6. *The variance and standard deviation of a normal distribution.*

The equation of the normal curve is

$$f(x) = \frac{1}{\sigma\sqrt{2\pi}} e^{-\frac{(x-\mu)^2}{2\sigma^2}}, \text{ Mean} = \mu$$

$$\text{Variance} = \int_{-\infty}^{\infty} (x-\mu)^2 f(x)dx = \frac{1}{\sigma\sqrt{2\pi}}\int_{-\infty}^{\infty} (x-\mu)^2 . e^{-\frac{(x-\mu)^2}{2\sigma^2}}\,dx$$

Putting $\quad \dfrac{x-\mu}{\sigma\sqrt{2}} = t, \quad dx = \sigma\sqrt{2}\,dt$

$$\text{Variance} = \frac{1}{\sigma\sqrt{2\pi}}\int_{-\infty}^{\infty} 2\sigma^2 t^2 . e^{-t^2} . \sigma\sqrt{2}\,dt$$

$$= \frac{2\sigma^2}{\sqrt{\pi}}\int_{-\infty}^{\infty} t^2 e^{-t^2}\,dt = \frac{4a^2}{\sqrt{\pi}}\int_{0}^{\infty} t^2 e^{-t^2}\,dt$$

Putting $t^2 = z$, $2t\,dt = dz$ or $dt = \dfrac{dz}{2\sqrt{z}}$

$$\therefore \quad \text{Variance} = \frac{4\sigma^2}{\sqrt{\pi}}\int_{0}^{\infty} ze^{-z} . \frac{dz}{2\sqrt{z}} = \frac{2\sigma^2}{\sqrt{\pi}}\int_{0}^{\infty} z^{1/2} e^{-z}\,dz$$

$$= \frac{2\sigma^2}{\sqrt{\pi}}\int_{0}^{\infty} z^{3/2-1} e^{-2}\,dz = \frac{2\sigma^2}{\sqrt{\pi}}\Gamma(\tfrac{3}{2}) = \frac{2\sigma^2}{\sqrt{\pi}}.\tfrac{1}{2}\sqrt{\pi} = \sigma^2.$$

Standard deviation $= \sqrt{\textbf{Variance}} = \sigma$.

7. *The points of inflexion of the normal curve.*

Let the equation of the normal curve be

$$y = \frac{1}{\sigma\sqrt{2\pi}} e^{-\frac{(x-\mu)^2}{2\sigma^2}}$$

Taking logarithms

$$\log y = \log \frac{1}{\sigma\sqrt{2\pi}} - \frac{(x-\mu)^2}{\sigma^2}$$

Differentiating w.r.t. x

$$\frac{1}{y} \cdot \frac{dy}{dx} = -\frac{x-\mu}{\sigma^2}$$

Differentiating again

$$\frac{1}{y} \cdot \frac{d^2y}{dx^2} - \frac{1}{y^2} \cdot \left(\frac{dy}{dx}\right)^2 = -\frac{1}{\sigma^2}$$

$$\frac{d^2y}{dx^2} = \frac{1}{y} \cdot \left(\frac{dy}{dx}\right)^2 - \frac{y}{\sigma^2} = \frac{1}{y} \cdot \frac{y^2(x-\mu)^2}{\sigma^4} - \frac{y}{\sigma^2}$$

$$= \frac{y}{\sigma^4} [(x-\mu)^2 - \sigma^2]$$

At a point of inflexion.

$$\frac{d^2y}{dx^2} = 0 \qquad \left(\text{and } \frac{d^3y}{dx^3} \neq 0 \text{ which can be shown}\right)$$

$\therefore \qquad (x-\mu)^2 = \sigma^2$

or $\qquad x - \mu = \pm\, \sigma \quad$ or $\quad x = \mu \pm \sigma.$

Thus, the curve has two points of inflexion, one at $\mu - \sigma$ and the other at $\mu + \sigma$, *i.e.*, at a distance from the mean, equal to the standard deviation.

8. *The mean deviation from the mean of the normal distribution is about $\frac{4}{5}$ of its standard deviation.* (M.D.U. 2011)

Let the equation of the normal curve be

$$y = \frac{1}{\sigma\sqrt{2\pi}} e^{-\frac{(x-\mu)^2}{2\sigma^2}}$$

Standard deviation $= \sigma$.

Mean deviation from the mean

$$= \int_{-\infty}^{\infty} y\,|x-\mu|\,dx = \frac{1}{\sigma\sqrt{2\pi}}\int_{-\infty}^{\infty} |x-\mu|.e^{-\frac{(x-\mu)^2}{2\sigma^2}}\,dx$$

$$= \frac{\sigma^2}{\sigma\sqrt{2\pi}}\int_{-\infty}^{\infty} |z|\,e^{-\frac{1}{2}z^2}\,dz, \text{ where } z = \frac{x-\mu}{\sigma}$$

$$= \frac{\sigma}{\sqrt{2\pi}}\left[\int_{-\infty}^{0} -z.e^{-\frac{1}{2}z^2}\,dz + \int_{0}^{\infty} ze^{-\frac{1}{2}z^2}\,dz\right] \qquad \because\ |z| = \begin{cases} -z & \text{if } z<0 \\ z & \text{if } z \ge 0 \end{cases}$$

$$= \frac{\sigma}{\sqrt{2\pi}}\left[-\int_{\infty}^{0} te^{-\frac{1}{2}t^2}\,dt + \int_{0}^{\infty} z\,e^{-\frac{1}{2}z^2}\,dz\right]$$

where $t = -z$

$$= \frac{\sigma}{\sqrt{2\pi}}\left[\int_{0}^{\infty} ze^{-\frac{1}{2}z^2}\,dz + \int_{0}^{\infty} z\,e^{-\frac{1}{2}z^2}\,dz\right]$$

$$\left(\because\ \int_a^b f(x)\,dx = -\int_b^a f(x)\,dx \text{ and } \int_a^b f(x)\,dx = \int_a^b f(z)\,dz\right)$$

$$= \sigma\sqrt{\frac{2}{\pi}}\int_{0}^{\infty} ze^{-\frac{1}{2}z^2}\,dz = \sigma\sqrt{\frac{2}{\pi}}\int_{0}^{\infty} \tfrac{1}{2}e^{-\frac{1}{2}t}\,dt, \text{ where } t = z^2$$

$$= \sigma\sqrt{\frac{2}{\pi}}\left[-e^{-\frac{1}{2}t}\right]_0^{\infty} = -\sigma\sqrt{\frac{2}{\pi}}\,[0-1] = \sqrt{\frac{2}{\pi}}\,.\,\sigma = 0.7979\sigma = \tfrac{4}{5}\sigma \text{ (approx.)}$$

$$= \frac{4}{5} \times \text{standard deviation (approx.)}.$$

9. *Standard deviation for a normal distribution is approximately 25% more than the mean deviation.* (K.U.K. Dec. 2015)

$$\text{Standard deviation} - \text{Mean deviation} = \sigma - \frac{4}{5}\sigma \text{ (approx.)}$$

$$= \frac{1}{5}\sigma = \frac{1}{4}\left(\frac{4}{5}\sigma\right) = \frac{1}{4} \text{ (Mean deviation)}$$

= 25% of Mean Deviation.

5.28. STANDARD FORM OF THE NORMAL DISTRIBUTION

If X is a normal random variable with mean μ and standard deviation σ, then the random variable $Z = \frac{X-\mu}{\sigma}$ has the normal distribution with mean 0 and standard deviation 1. The random variable Z is called the *standardized* (or *standard*) normal random variable.

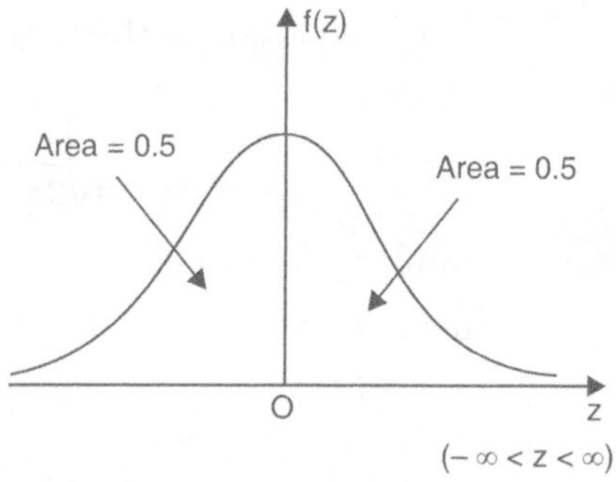

The probability density function for the normal distribution in standard form is given by

$$f(z) = \frac{1}{\sqrt{2\pi}}e^{-\frac{1}{2}z^2}$$

It is free from any parameter. This helps us to compute areas under the normal probability curve by making use of standard tables.

Note 1. If $f(z)$ is the probability density function for the normal distribution, then

$$P(z_1 \le Z \le z_2) = \int_{z_1}^{z_2} f(z)\, dz = F(z_2) - F(z_1), \quad \text{where} \quad F(z) = \int_{-\infty}^{z} f(z)\, dz = P(Z \le z)$$

The function F(z) defined above is called the *distribution function* for the normal distribution.

Note 2. The probabilities $P(z_1 \le Z \le z_2)$, $P(z_1 < Z \le z_2)$, $P(z_1 \le Z < z_2)$ and $P(z_1 < Z < z_2)$ are all regarded to be the same.

Note 3. $F(-z_1) = 1 - F(z_1)$.

ILLUSTRATIVE EXAMPLES

Example 1. *A sample of 100 dry battery cells tested to find the length of life produced the following results:*

$$\bar{x} = 12 \text{ hours}, \sigma = 3 \text{ hours}.$$

Assuming the data to be normally distributed, what percentage of battery cells are expected to have life

(i) more than 15 hours *(ii) less than 6 hours*

(iii) between 10 and 14 hours? (Cochin 2011)

Sol. Here x denotes the length of life of dry battery cells.

Also $z = \dfrac{x - \bar{x}}{\sigma} = \dfrac{x - 12}{3}$.

(i) When $x = 15$, $z = 1$

$\therefore \quad P(x > 15) = P(z > 1)$

$= P(0 < z < \infty) - P(0 < z < 1)$

$= 0.5 - 0.3413 = 0.1587 = 15.87\%.$

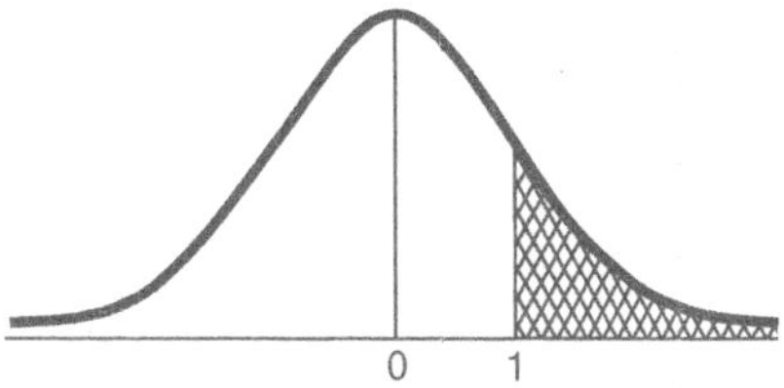

(ii) When $x = 6$, $z = -2$

$\therefore \quad P(x < 6) = P(z < -2)$

$= P(z > 2) = P(0 < z < \infty) - P(0 < z < 2)$

$= 0.5 - 0.4772 = 0.0228 = 2.28\%.$

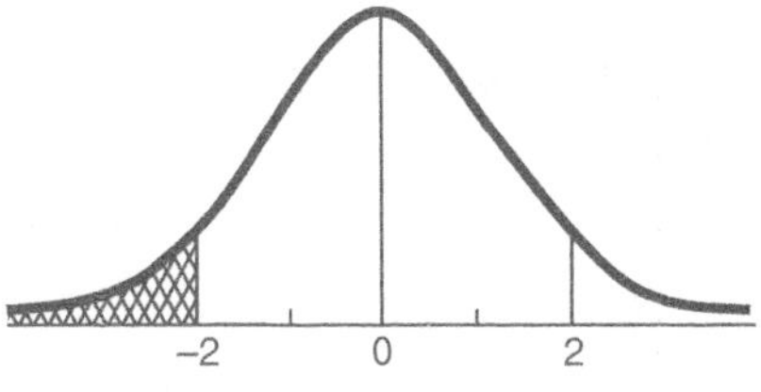

(iii) When $x = 10$, $z = -\dfrac{2}{3} = -0.67$

When $x = 14$, $z = \dfrac{2}{3} = 0.67$

$P(10 < x < 14)$

$= P(-0.67 < z < 0.67)$

$= 2P(0 < z < 0.67) = 2 \times 0.2485$

$= 0.4970 = 49.70\%.$

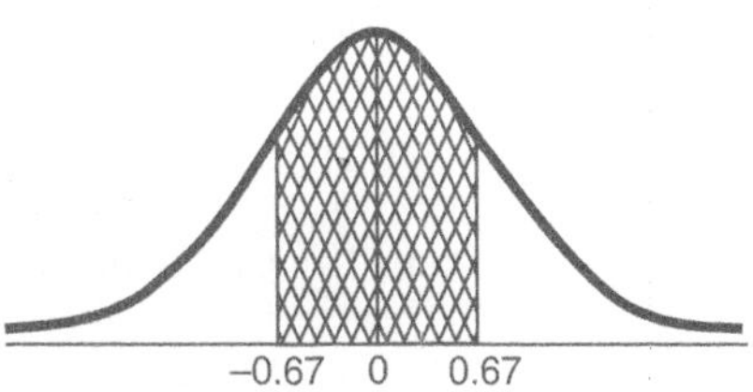

Example 2. *In a normal distribution, 31% of the items are under 45 and 8% are over 64. Find the mean and standard deviation of the distribution.*

(Cochin 2010; Calicut 2011; K.U.K., 2010, Jan. 2014; M.D.U. 2012, May 2015)

Sol. Let $\bar{x}$ and σ be the mean and S.D. respectively.

31% of the items are under 45.

$\Rightarrow$ Area to the left of the ordinate $x = 45$ is 0.31

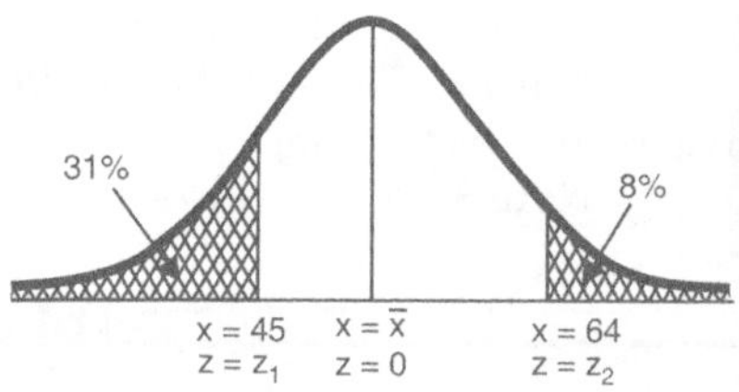

When $x = 45$, let $z = z_1$

$$P(z_1 < z < 0) = .5 - .31 = .19$$

From the tables, the value of z corresponding to this area is 0.5

$$\therefore \qquad z_1 = -0.5 \qquad [z_1 < 0]$$

When $x = 64$, let $z = z_2$

$$P(0 < z < z_2) = 0.5 - 0.08 = 0.42$$

From the tables, the value of z corresponding to this area is 1.4.

$$z_2 = 1.4$$

Since $$z = \frac{x - \bar{x}}{\sigma}$$

$$-0.5 = \frac{45 - \bar{x}}{\sigma} \quad \text{and} \quad 1.4 = \frac{64 - \bar{x}}{\sigma}$$

$$\Rightarrow \qquad 45 - \bar{x} = -0.5\sigma \qquad \ldots(1)$$

and $$64 - \bar{x} = 1.4\sigma \qquad \ldots(2)$$

Subtracting $\quad -19 = -1.9\sigma \quad \therefore \quad \sigma = 10$

From (1), $\quad 45 - \bar{x} = -0.5 \times 10 = -5 \quad \therefore \quad \bar{x} = 50.$

Example 3. *Obtain the equation of the normal curve that may be fitted to the following data:*

Class:	*60–65*	*65–70*	*70–75*	*75–80*	*80–85*	*85–90*	*90–95*	*95–100*
Frequency:	*3*	*21*	*150*	*335*	*326*	*135*	*26*	*4*

Sol. The equation of the normal curve is

$$y = \frac{N}{\sigma\sqrt{2\pi}} e^{-\frac{(x-\mu)^2}{2\sigma^2}}$$

Now, from the given data, we shall find $N = \Sigma f$; μ, the mean and the S.D. σ.

Class	*Mid-values (x)*	*f*	$u = \dfrac{x-77.5}{5}$	*fu*	*fu*^{2}
60–65	62.5	3	– 3	– 9	27
65–70	67.5	21	– 2	– 42	84
70–75	72.5	150	– 1	– 150	150
75–80	77.5	335	0	0	0
80–85	82.5	326	1	326	326
85–90	87.5	135	2	270	540
90–95	92.5	26	3	78	234
95–100	97.5	4	4	16	64
		N = 1000		489	1425

$$\text{Mean } \mu = A + \frac{h}{N}\sum fu = 77.5 + \frac{5}{1000} \times 489$$

$$= 77.5 + 2.445 = 79.945$$

$$\text{S.D. } \sigma = h\sqrt{\frac{1}{N}\sum fu^2 - \left(\frac{\sum fu}{N}\right)^2} = 5\sqrt{\frac{1425}{1000} - \left(\frac{489}{1000}\right)^2}$$

$$= \frac{5}{1000}\sqrt{1425000 - 239121} = \frac{5}{1000}\sqrt{1185879}$$

$$= \frac{5}{1000} \times 1088.98 = 5.445$$

Hence the equation of the normal curve fitted to the given data is

$$y = \frac{1000}{5.445\sqrt{2\pi}}\, e^{-\frac{1}{2}\left(\frac{x-79.945}{5.445}\right)^2}$$

EXERCISE 5.6

1. If z is the standard normal variate, then find the following probabilities:

   (*i*) P $(1 \le z \le 2)$ (*ii*) P $(-2.3 \le z \le -1.5)$

   (*iii*) P $(-0.42 \le z \le 2)$ (*iv*) P $(z \le -0.56)$

   (*v*) P $(|z| \le 1)$ (*vi*) P $(|z| \ge 1)$

2. (*a*) Let X be a random variable having a normal distribution with mean 30 and standard deviation 5. Find the probability that

   (*i*) $20 \le x \le 40$ (*ii*) $|x - 30| > 5$

   (*iii*) $|x - 24| < 8$ (*iv*) $x \ge 45$ (*v*) $x \le 45$

   (*M.D.U. 2010, May 2013*)

(*b*) For a normally distributed variate with mean 1 and S.D. 3, find the probabilities that:

(*i*) $3.43 \le x \le 6.19$ (*ii*) $-1.42 \le x \le 6.18$ (*K.U.K. Jan. 2013*)

3. The mean height of 500 male students in a certain college is 151 cm and the standard deviation is 15 cm. Assuming the heights are normally distributed, find how many students have heights between 120 and 155 cm?

4. An aptitude test for selecting officers in a bank is conducted on 1000 candidates. The average score is 42 and the standard deviation of score is 24. Assuming normal distribution for the scores, find

(*i*) the number of candidates whose scores exceed 60

(*ii*) the number of candidates whose scores lie between 30 and 60.

5. If the mean height of an Indian police inspector be 170 cm with variance 25 cm^2, how many inspectors out of 1000 would you expect

(*i*) between 170 cm and 180 cm (*ii*) less than 160 cm?

6. The marks obtained by a large group of students in a final examination in statistics have a mean of 58 and a standard deviation of 8.5. Assuming that these marks are approximately normally distributed, what percentage of the students can be expected to have obtained marks from 60 to 69?

7. The income of a group of 10,000 persons was found to be normally distributed with mean ₹ 750 p.m. and standard deviation ₹ 50. Show that, of this group, about 95% had income exceeding ₹ 668 and only 5% had income exceeding ₹ 832. What was the lowest income among the richest 100?

8. (*a*) In a normal distribution, 7% of the items are under 35 and 89% are under 63. What are the mean and standard deviation of the distribution?

(*b*) In a normal distribution 17% of the items are below 30 and 17% of the items are above 60. Find the mean and standard deviation.

(*c*) In a normal distribution 10% of the items are below 55 and 20% are above 59. Find the mean and variance of the normal distribution.

9. Let X denote the number of scores on a test. If X is normally distributed with mean 100 and standard deviation 15, find the probability that X does not exceed 130.

10. It is known from the past experience that the number of telephone calls made daily in a certain community between 3 p.m. and 4 p.m. have a mean of 352 and a standard deviation of 31. What percentage of the time will there be more than 400 telephone calls made in this community between 3 p.m. and 4 p.m.?

11. Students of a class were given a mechanical aptitude test. Their marks were found to be normally distributed with mean 60 and standard deviation 5. What per cent of students scored

(*i*) more than 60 marks? (*ii*) less than 56 marks?

(*iii*) between 45 and 65 marks?

12. In an examination taken by 500 candidates, the average and the standard deviation of marks obtained (normally distributed) are 40% and 10%. Find approximately

(*i*) how many will pass, if 50% is fixed as a minimum?

(*ii*) what should be the minimum if 350 candidates are to pass?

(*iii*) how many have scored marks above 60%?

13. In a distribution, exactly normal, 9.85% of the items are under 40 and 89.97% are under 60. What are the mean and standard deviation of the distribution?

14. The income distribution of workers in a certain factory was found to be normal with mean of ₹ 500 and standard deviation of ₹ 50. There were 228 workers getting above ₹ 600. How many workers were there in all?

15. The mean inside diameter of a sample of 500 washers produced by a machine is 5.02 mm and the standard deviation is 0.05 mm. The purpose for which these washers are intended allows a maximum tolerance in the diameter of 4.96 to 5.08 mm, otherwise the washers are considered defective. Determine the percentage of defective washers produced by the machine, assuming the diameters are normally distributed.

16. Assuming that the diameters of 1000 brass plugs taken consecutively from a machine, form a normal distribution with mean 0.7515 cm and standard deviation 0.0020 cm. How many of the plugs are likely to be rejected if the approved diameter is 0.752 ± 0.004 cm?

(DCRUST Murthal, May 2014)

17. In a test on 2000 electric bulbs, it was found that the life of a particular make was normally distributed with an average of 2040 hours and S.D. of 60 hours. Estimate the number of bulbs likely to burn for

(*i*) more than 2150 hours

(*ii*) less than 1950 hours

(*iii*) more than 1920 hours but less than 2160 hours

(*iv*) not less than 2000 hours. *(Calicut 2011)*

18. Obtain the equation of the normal curve that may be fitted to the following data:

Variable (x)	0	1	2	3	4	5
Frequency (f)	10	14	19	8	5	4

19. Fit a normal curve to the following distribution:

x:	2	4	6	8	10
f:	1	4	6	4	1

(M.D.U. 2008, 2011, May 2013, May 2014)

20. The average test marks in a particular test is 79. The standard deviation is 5. If the marks are normally distributed, how many students in a class of 200 did not receive marks between 75 and 82?

Answers

1. (*i*) 0.1359 (*ii*) 0.0561 (*iii*) 0.64
(*iv*) 0.2877 (*v*) 0.6826 (*vi*) 0.3174

2. (*a*) (*i*) 0.9544 (*ii*) 0.3174 (*iii*) 0.6528
(*iv*) 0.0013 (*v*) 0.9987 (*b*) (*i*) 0.1672
(*ii*) 0.7492

3. 294

4. (*i*) 227 (*ii*) 465

5. (*i*) 477 (*ii*) 23

6. 30.67%

7. ₹ 866.50 **8.** (*a*) $\mu = 50.29, \sigma = 10.33$ (*b*) $\mu = 45, \sigma = 15.8$

(*c*) $\mu = 57.41, \sigma^2 = 3.469$ **9.** 0.9772 **10.** 6.06%

11. (*i*) 50% (*ii*) 21.2% (*iii*) 84%

12. (*i*) 79 (*ii*) 35% (*iii*) 11

13. $\mu = 50.04, \sigma = 7.78$ **14.** 10,000 **15.** 23 **16.** 52

17. (*i*) 72 (*ii*) 134 (*iii*) 1909

(*iv*) 14.97

18. $y = \dfrac{60}{1.45\sqrt{2\pi}} e^{-\frac{1}{2}\left(\frac{x-1.93}{2.09}\right)^2}$ **19.** $y = \dfrac{1}{2\sqrt{2\pi}} e^{-\frac{(x-6)^2}{8}}$ **20.** 97.

❑ ❑ ❑

SECTION-D

CHAPTER 6

Hypothesis Testing

SAMPLING AND TESTS OF SIGNIFICANCE

6.1. POPULATION OR UNIVERSE

An aggregate of objects (animate or inanimate) under study is called **population or universe**. It is thus a collection of individuals or of their attributes (qualities) or of results of operations which can be numerically specified.

A universe containing a finite number of individuals or members is called a **finite universe**. For example, the universe of the weights of students in a particular class or the universe of smokers in Rohtak district.

A universe with infinite number of members is known as an **infinite universe**. For example, the universe of pressures at various points in the atmosphere.

In some cases, we may be even ignorant whether or not a particular universe is infinite, *e.g.*, the universe of stars.

The universe of concrete objects is an **existent universe**. The collection of all possible ways in which a specified event can happen is called a **hypothetical universe**. The universe of heads and tails obtained by tossing a coin an infinite number of times (provided that it does not wear out) is a hypothetical one.

6.2. SAMPLING

The statistician is often confronted with the problem of discussing universe of which he cannot examine every member *i.e.*, of which complete enumeration is impracticable. For example, if we want to have an idea of the average per capita income of the people of India, enumeration of every earning individual in the country is a very difficult task. Naturally, the question arises: What can be said about a universe of which we can examine only a limited number of members? This question is the origin of the Theory of Sampling.

A finite sub-set of a universe is called a **sample**. A sample is thus a small portion of the universe. The number of individuals in a sample is called the **sample size**. The process of selecting a sample from a universe is called **sampling**.

The theory of sampling is a study of relationship existing between a population and samples drawn from the population. The fundamental object of sampling is to get as much information as possible of the whole universe by examining only a part of it. An attempt is thus made through sampling to give the maximum information about the parent universe with the minimum effort.

Sampling is quite often used in our day-to-day practical life. For example, in a shop we asses the quality of sugar, rice or any other commodity by taking only a handful of it from the bag and then decide whether to purchase it or not. A housewife normally tests the cooked products to find if they are properly cooked and contain the proper quantity of salt or sugar, by taking a spoonful of it.

6.3. PARAMETERS OF STATISTICS

The statistical constants of the population such as mean, the variance etc. are known as the parameters. The statistical concepts of the sample computed from the members of the sample to estimate the parameters of the population from which the sample has been drawn are known as *statistic*.

Population mean and variance are denoted by μ and σ^2, while those of the samples are given by $\overline{x}$, s^2.

6.4. STANDARD ERROR (S.E.)

The standard deviation of the sampling distribution of a statistic is known as the standard error (S.E.).

It plays an important role in the theory of large samples and it forms a basis of the testing of hypothesis. If t is any statistic, for large sample,

$$z = \frac{t - \mathrm{E}(t)}{\mathrm{S.E}(t)} \text{ is normally distributed with mean 0 and variance unity.}$$

For large sample, the standard errors of some of the well known statistic are listed below:

n—sample size; σ^2—population variance; s^2—sample variance; P—population proportion ; $Q = 1 - P$; n_1, n_2—are sizes of two independent random samples.

S. No.	*Statistic*	*Standard error*
1.	$\overline{x}$	$\sigma/\sqrt{n}$
2.	s	$\sqrt{\sigma^2/2n}$
3.	Difference of two sample means $\overline{x}_1 - \overline{x}_2$	$\sqrt{\frac{{\sigma_1}^2}{n_1} + \frac{{\sigma_2}^2}{n_2}}$
4.	Difference of two sample standard deviations $s_1 - s_2$	$\sqrt{\frac{{\sigma_1}^2}{2n_1} + \frac{{\sigma_2}^2}{2n_2}}$
5.	Difference of two sample proportions $p_1 - p_2$	$\sqrt{\frac{P_1Q_1}{n_1} + \frac{P_2Q_2}{n_2}}$
6.	Observed sample proportion p	$\sqrt{PQ/n}$

6.5. TEST OF SIGNIFICANCE *(P.T.U. 2008)*

An important aspect of the sampling theory is to study the test of significance, which will enable us to decide, on the basis of the results of the sample, whether (*i*) *the deviation between the observed sample statistic and the hypothetical parameter value or*

(*ii*) *the deviation between two sample statistics*

is significant or might be attributed due to chance or the fluctuations of the sampling.

For applying the tests of significance, we first set up a hypothesis which is a definite statement about the population parameter. Usually, such a hypothesis is a hypothesis of no difference and it is called the **Null hypothesis** denoted by H_0.

Any hypothesis which is complementary to the null hypothesis ($\mathbf{H_0}$) is called an **Alternative hypothesis** denoted by $\mathbf{H_1}$.

For example if we want to test the null hypothesis that the population has a specified mean μ_0, then we have

$$\mathbf{H_0}: \mu = \mu_0$$

Alternative hypothesis will be

(*i*) $H_1: \mu \neq \mu_0$ ($\mu > \mu_0$ *or* $\mu < \mu_0$) (two tailed alternative hypothesis).

(*ii*) $H_1: \mu > \mu_0$ (right tailed alternative hypothesis (*or*) single tailed).

(*iii*) $H_1: \mu < \mu_0$ (left tailed alternative hypothesis (*or*) single tailed).

Hence alternative hypothesis helps to know whether the test is two tailed test or one tailed test.

6.6. CRITICAL REGION

A region corresponding to a statistic t, in the sample space S which amounts to rejection of the null hypothesis H_0 is called as **critical region** or **region of rejection**. The region of the sample space S which amounts to the acceptance of H_0 is called *acceptance region*.

6.7. LEVEL OF SIGNIFICANCE

The probability of the value of the variate falling in the critical region is known as level of significance.

The probability α that a random value of the statistic t belongs to the critical region is known as the level of significance.

$$P(t \in \omega \mid H_0) = \alpha$$

i.e., the level of significance is the size of the type I error or the maximum producer's risk.

In testing of hypothesis, the levels of significance are usually taken as 5% and 1%.

6.8. ERRORS IN SAMPLING

The main aim of the sampling theory is to draw a valid conclusion about the population parameters on the basis of the sample results. In doing this we may commit the following two types of errors:

Type I. Error. When H_0 is true, we may reject it.

P(Reject H_0 when it is true) = P(Reject H_0/H_0) = α

α is called the size of the type I error also referred to as **producer's risk**.

Type I error amounts to rejecting a lot when it is good.

Type II. Error. When H_0 is wrong we may accept it

P(Accept H_0 when it is wrong) = P(Accept H_0/H_1) = β.

β is called the size of the type II error, also referred to as **consumer's risk.**

Type II error amounts to accepting a lot when it is bad.

Critical values or significant values

The values of the test statistic which separates the critical region and acceptance region is called the **critical values** or **significant value.**

This value is dependent on (*i*) the level of significance used and (*ii*) the alternative hypothesis, whether it is one tailed or two tailed.

For larger samples corresponding to the statistic t, the variable $z = \dfrac{t - \text{E}(t)}{\text{S.E.}(t)}$ is normally distributed with mean 0 and variance 1. The value of z given above under the null hypothesis is known as **test statistic.**

The critical value of z_α of the test statistic at level of significance α for a two tailed test is given by

$$p(|\, z\, | > z_\alpha) = \alpha \qquad \ldots(1)$$

i.e., z_α is the value of z so that the total area of the critical region on both tails is α. Since the normal curve is symmetrical, from equation (1), we get

$$p(z > z_\alpha) + p(z < -z_\alpha) = \alpha;\ i.e.,\ 2p(z > z_\alpha) = \alpha;\ p(z > z_\alpha) = \alpha/2$$

i.e., the area of each tail is $\alpha/2$.

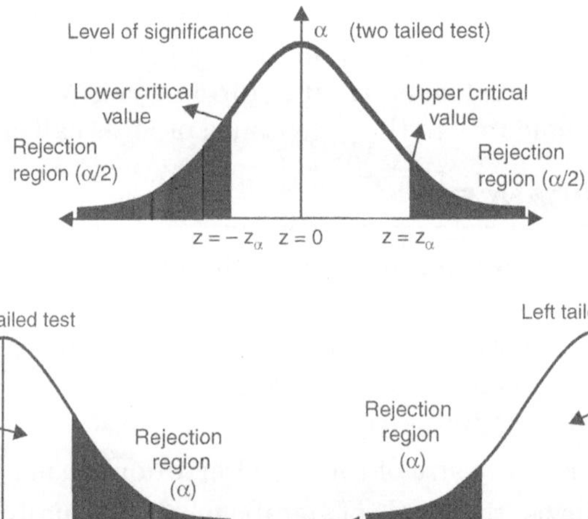

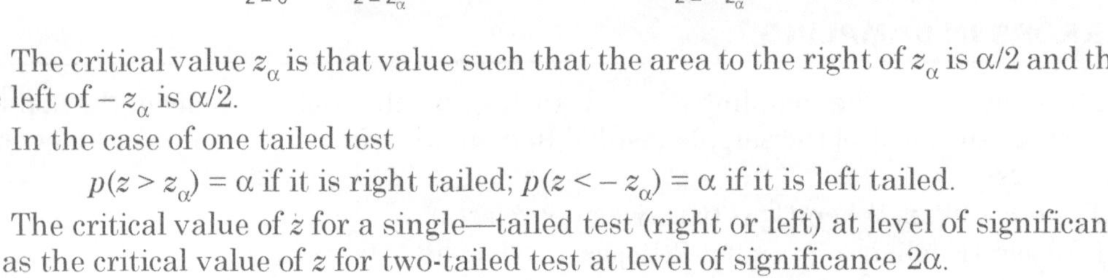

The critical value z_α is that value such that the area to the right of z_α is $\alpha/2$ and the area to the left of $-z_\alpha$ is $\alpha/2$.

In the case of one tailed test

$$p(z > z_\alpha) = \alpha \text{ if it is right tailed; } p(z < -z_\alpha) = \alpha \text{ if it is left tailed.}$$

The critical value of z for a single—tailed test (right or left) at level of significance α is same as the critical value of z for two-tailed test at level of significance 2α.

Using the equation, also using the normal tables, the critical value of z at different level of significance (α) for both single tailed and two tailed test are calculated and listed below. The equations are

$$p(|\, z\, | > z_\alpha) = \alpha;\ p(z > z_\alpha) = \alpha;\ p(z < -z_\alpha) = \alpha$$

	Level of significance		
	1% (0.01)	5% (0.05)	10% (0.1)
Two tailed test	$\mid z_\alpha \mid = 2.58$	$\mid z_\alpha \mid = 1.966$	$\mid z_\alpha \mid = 0.645$
Right tailed	$z_\alpha = 2.33$	$z_\alpha = 1.645$	$z_\alpha = 1.28$
Left tailed	$z_\alpha = -2.33$	$z_\alpha = -1.645$	$z_\alpha = -1.28$

Note. The following steps may be adopted in testing of statistical hypothesis:

Step 1: Null hypothesis. Set up H_0 in clear terms.

Step 2: Alternative hypothesis. Set up H_1, so that we could decide whether we should use one-tailed test or two tailed test.

Step 3: Level of significance. Select the appropriate level of significance in advance depending on the reliability of the estimates.

Step 4: Test statistic. Compute the test statistic $z = \dfrac{t - E(t)}{S.E.(t)}$ under the null hypothesis.

Step 5: Conclusion. Compare the computed value of z with the critical value z_α at level of significance (α).

If $\mid z \mid > z_\alpha$, we reject H_0 and conclude that there is significant difference. If $\mid z \mid < z_\alpha$, we accept H_0 and conclude that there is no significant difference.

TEST OF SIGNIFICANCE FOR LARGE SAMPLES

If the sample size $n > 30$, the sample is taken as large sample. For such sample we apply normal test, as Binomial, Poisson, chisquare etc. are closely approximated by normal distributions assuming the population as normal.

Under large sample test, the following are the important tests to test the significance.

1. *Testing of significance for single proportion.*
2. *Testing of significance for difference of proportions.*
3. *Testing of significance for single mean.*
4. *Testing of significance for difference of means.*
5. *Testing of significance for difference of standard deviations.*

6.9. TESTING OF SIGNIFICANCE FOR SINGLE PROPORTION

This test is used to find the significant difference between proportion of the sample and the population.

Let X be the number of successes in n independent trials with constant probability P of success for each trial.

$$E(X) = nP;\ V(X) = nPQ;\ Q = 1 - P = \text{Probability of failure.}$$

Let $p = X/n$ called the observed proportion of success.

$$E(p) = E(X/n) = \frac{1}{n} E(X) = \frac{nP}{n} = P;$$

$$V(p) = V(X/n) = \frac{1}{n^2} V(X) = \frac{nPQ}{n^2} = PQ/n$$

$$\text{S.E.}(p) = \sqrt{\frac{\text{PQ}}{n}}\,;\; z = \frac{p - \text{E}(p)}{\text{S.E.}(p)} = \frac{p - \text{P}}{\sqrt{\text{PQ}/n}} \sim \text{N}(0, 1)$$

This z is called test statistic which is used to test the significant difference of sample and population proportion.

Note 1. The probable limits for the observed proportion of successes are $\text{P} \mp z_\alpha \sqrt{\frac{\text{PQ}}{n}}$ where z_α is the significant value of z at level of significance α.

Note 2. If P is not known, the limits for the proportion in the population are $p \mp z_\alpha \sqrt{\frac{pq}{n}}$ at level of significance α and $q = 1 - p$.

These limits are called confidence limits or fiducial limits at level of significance α.

Note 3. If α is not given, then

(*i*) Confidence limits for observed proportion p are $\text{P} \mp 3\sqrt{\frac{\text{PQ}}{n}}$, and

(*ii*) Confidence limits for the population proportion P are $p \mp 3\sqrt{\frac{pq}{n}}$.

ILLUSTRATIVE EXAMPLES

Example 1. *A coin was tossed 400 times and the head turned up 216 times. Test the hypothesis that the coin is unbiased.*

Sol. H_0: The coin is unbiased *i.e.*, P = 0.5.

H_1: The coin is not unbiased. (biased); $\text{P} \neq 0.5$

Hence we use two tailed test.

Here $n = 400$; X = No. of successes = 216

$$\therefore \quad p = \text{proportion of success in the sample} = \frac{\text{X}}{n} = \frac{216}{400} = 0.54$$

P = population proportion = 0.5; Q = 1 – P = 1 – 0.5 = 0.5

Under H_0, test statistic $z = \dfrac{p - \text{P}}{\sqrt{\text{PQ}/n}}$

$$|z| = \left| \frac{0.54 - 0.5}{\sqrt{\frac{0.5 \times 0.5}{400}}} \right| = 1.6$$

Conclusion. Since $|z| = 1.6 < 1.96$

i.e., $|z| < z_\alpha$, where z_α is the significant value of z at 5% level of significance. Hence we accept H_0 and conclude that the coin is unbiased.

Example 2. *A cubical die is thrown 9000 times and a throw of 3 or 4 is observed 3240 times. Show that the die cannot be regarded as an unbiased one and find the extreme limits between which the probability of a throw of 3 or 4 lies.* (M.D.U. 2009)

Sol. Here $n = 9000$

P = probability of success (*i.e.*, getting 3 or 4 on die)

$P = 2/6 = 1/3, \quad Q = 1 - 1/3 = 2/3$

$$p = \frac{X}{n} = \frac{3240}{9000} = 0.36$$

H_0: The die is unbiased, *i.e.*, $P = 1/3$; H_1: $P \neq 1/3$

Hence we use two tailed test.

Under H_0, the test statistic $$z = \frac{p - P}{\sqrt{\frac{PQ}{n}}} = \frac{0.36 - 0.33}{\sqrt{\frac{1}{3} \times \frac{2}{3} \times \frac{1}{9000}}} = 6.0376$$

$$|z| = 6.0376 > 1.966$$

Conclusion.

As $|z| > z_\alpha$, z_α is the tabulated value of z at 5% level of significance.

$\therefore$ H_0 is rejected and we conclude that the die is biased.

To find 95% confidence limits of the proportion.

It is given by $$P \mp z_\alpha \sqrt{\frac{PQ}{n}}$$

i.e., $$0.33 \mp 1.96 \sqrt{\frac{0.33 \times 0.67}{9000}} = 0.33 \mp 0.0097 = 0.3203 \text{ and } 0.3397.$$

Example 3. *A manufacturer claims that only 4% of his products supplied by him are defective. A random sample of 600 products contained 36 defectives. Test the claim of the manufacturer.*

Sol. (*i*) p = observed proportion of success.

i.e., p = proportion of defective in the sample $= \frac{36}{600} = 0.06$

P = proportion of defectives in the population = 0.04

H_0: $P = 0.04$ is true, *i.e.*, the claim of the manufacturer is accepted.

H_1: (*i*) $P \neq 0.04$ (two tailed test)

(*ii*) If we want to reject only if $P > 0.04$ then (right tailed test).

Under H_0, $$z = \frac{p - P}{\sqrt{PQ/n}} = \frac{0.06 - 0.04}{\sqrt{\frac{0.04 \times 0.96}{600}}} = 2.5.$$

Conclusion. Since $|z| = 2.5 > 1.96$, we reject the hypothesis H_0 at 5% level of significance (two tailed).

If H_1 is taken as $p > 0.04$ we apply right tailed test.

$|z| = 2.5 > 1.645 \ (= z_\alpha)$ we reject the null hypothesis here also.

In both cases, manufacturer's claim is not acceptable.

Example 4. *A machine is producing bolts of which a certain fraction is defective. A random sample of 400 is taken from a large batch and is found to contain 30 defective bolts. Does this indicate that the proportion of defectives is larger than that claimed by the manufacturer where the manufacturer claims that only 5% of his product are defective. Find 95% confidence limits of the proportion of defective bolts in batch.*

Sol. Null hypothesis. H_0: The manufacturer's claim is accepted *i.e.*, $P = \frac{5}{100} = 0.05$

$$Q = 1 - P = 1 - 0.05 = 0.95$$

Alternative hypothesis. $H_1 : P > 0.05$ (Right tailed test).

$$p = \text{observed proportion of sample} = \frac{30}{400} = 0.075$$

Under H_0, the test statistic $z = \dfrac{p - P}{\sqrt{PQ/n}}$ $\therefore$ $z = \dfrac{0.075 - 0.05}{\sqrt{\dfrac{0.05 \times 0.95}{400}}} = 2.2941.$

Conclusion. The tabulated value of z at 5% level of significance for right tailed test is $z_\alpha = 1.645$. Since $|z| = 2.2941 > 1.645$,

H_0 is rejected at 5% level of significance. *i.e.*, the proportion of defective bolts is larger than the manufacturer's claim.

To find 95% confidence limits of the proportion.

It is given by $P \mp z_\alpha \sqrt{PQ/n}$

i.e., $$0.05 \mp 1.96 \sqrt{\frac{0.05 \times 0.95}{400}} = 0.05 \mp 0.02135 = 0.02865 \text{ and } 0.07136.$$

Hence 95% confidence limits for the proportion of defective bolt are (0.07136, 0.02865).

Example 5. *A bag contains defective article, the exact number of which is not known. A sample of 100 from the bag gives 10 defective articles. Find the limits for the proportion of defective articles in the bag.*

Sol. Here p = proportion of defective articles $= \frac{10}{100} = 0.1$; $q = 1 - p = 1 - 0.1 = 0.9$

Here the confidence limits are not given.

Also P, the proportion of population is not given. Therefore the confidence limits are

$$p \pm 3\sqrt{pq/n} = 0.1 \pm 3\sqrt{\frac{0.1 \times 0.9}{100}} = 0.1 \mp 0.09 = 0.01 \text{ and } 0.19$$

Hence confidence limits for defective article in the bag are 0.01 and 0.19.

EXERCISE 6.1

1. A coin is tossed 900 times and heads appeared 490 times. Can you say that the coin is unbiased?
2. In a random sample of 400 persons from a large population, 280 are males. At 1% level of significance, discuss the hypothesis that females and males are in the ratio 3 : 5.
3. A sample of 900 days was taken for meteorological records of a certain district and 100 of them were found to be foggy. What are the possible limits to the percentage of foggy days in the district?

4. A sample of size of 600 persons selected at random from a large city shows that the percentage of males in the sample is 53. It is believed that the ratio of males to the total population in the city is 0.5. Test whether the belief is confirmed by the observation.
5. In a city a sample of 1000 people were taken and out of them 540 are vegetarian and the rest are non-vegetarian. Can we say that the both habits of eating (Vegetation or non vegetation) are equally popular in the city at (*i*) 1% level of significance (*ii*) 5% level of significance?
6. 325 men out of 600 men chosen from a big city were found to be smokers. Does this information support the conclusion that the majority of men in the city are smokers.
7. A random sample of 500 bolts was taken from a large consignment and 65 were found to be defective. Find the percentage of defective bolts in the consignment.
8. In a hospital 475 female and 525 male babies were born in a week. Do these figure confirm the hypothesis that males and females are born in equal number? *(M.D.U. Dec. 2014)*
9. 400 apples are taken at random from a large basket and 40 are found to be bad. Estimate the proportion of bad apples in the basket and assign limits within which the percentage most probably lies.
10. A manufacturer claims that only 10% of the articles produced are below the standard quality. Out of a random inspection of 300 articles, 37 are found to be of poor quality. Test the manufacturer's claim at 5% level of significance.
11. In a sample of 100 residents of a colony 60 are found to be wheat eaters and 40 are rice eaters. Can we assume that both food articles are equally popular? *(DCRUST Murthal, May 2014)*

Answers

1. H_0 rejected **2.** H_0 rejected; not in the ratio 3 : 5
3. 8% and 14.2% **4.** H_0 accepted at 5% level
5. H_0 rejected at 5% level, accepted at 1% level
6. H_0 rejected at 5% level **7.** Between 8.7% and 17.5%
8. H_0 accepted at 5% level **9.** 5.5% and 14.5% **10.** H_0 accepted at 5% level
11. At 5% level, No.; at 1% level, yes.

6.10. TEST OF DIFFERENCE BETWEEN PROPORTIONS

Consider two samples X_1 and X_2 of sizes n_1 and n_2 respectively taken from two different populations. To test the significance of the difference between the sample proportion p_1 and p_2, the test statistic under the null hypothesis H_0, that there is no significant difference between the two sample proportions, we have

$$z = \frac{p_1 - p_2}{\sqrt{PQ\left(\frac{1}{n_1} + \frac{1}{n_2}\right)}}, \qquad \text{where } P = \frac{n_1 p_1 + n_2 p_2}{n_1 + n_2} \text{ and } Q = 1 - P.$$

ILLUSTRATIVE EXAMPLES

Example 1. *Before an increase in excise duty on tea, 800 people out of a sample of 1000 persons were found to be tea drinkers. After an increase in the duty, 800 persons were known to be tea drinkers in a sample of 1200 people. Do you think that there has been a significant decrease in the consumption of tea after the increase in the excise duty?*

Sol. Here $n_1 = 1000,\ n_2 = 1200$

$$p_1 = \frac{X_1}{n_1} = \frac{800}{1000} = \frac{4}{5}\,;\ p_2 = \frac{X_2}{n_2} = \frac{800}{1200} = \frac{2}{3}$$

$$P = \frac{n_1p_1 + n_2p_2}{n_1 + n_2} = \frac{X_1 + X_2}{n_1 + n_2} = \frac{800 + 800}{1000 + 1200} = \frac{8}{11}; \ Q = \frac{3}{11}$$

Null hypothesis H_0: $p_1 = p_2$ *i.e.*, there is no significant difference in the consumption of tea before and after increase of excise duty.

$$H_1: p_1 > p_2$$

Hence we use right tailed test.

The test statistic $$z = \frac{p_1 - p_2}{\sqrt{PQ\left(\frac{1}{n_1} + \frac{1}{n_2}\right)}} = \frac{0.8 - 0.6666}{\sqrt{\frac{8}{11} \times \frac{3}{11}\left(\frac{1}{1000} + \frac{1}{1200}\right)}} = 6.842.$$

Conclusion. Since the calculated value of $|z| > 1.645$ and also $|z| > 2.33$, both the significant value of z at 5% and 1% level of significance, hence H_0 is rejected *i.e.*, there is a significant decrease in the consumption of tea due to increase in excise duty.

Example 2. *A machine produced 16 defective articles in a batch of 500. After overhauling, it produced 3 defectives in a batch of 100. Has the machine improved?*

Sol. $$p_1 = \frac{16}{500} = 0.032; \quad n_1 = 500$$

$$p_2 = \frac{3}{100} = 0.03; \quad n_2 = 100$$

Null hypothesis H_0: The machine has not improved due to overhauling *i.e.*, $p_1 = p_2$.

$$H_1: p_1 > p_2$$

Hence we use right tailed test.

$$\therefore \quad P = \frac{n_1p_1 + n_2p_2}{n_1 + n_2} = \frac{19}{600} \cong 0.032, \ Q = 1 - P = 0.968.$$

Under H_0, the test statistic

$$z = \frac{p_1 - p_2}{\sqrt{PQ\left(\frac{1}{n_1} + \frac{1}{n_2}\right)}} = \frac{0.032 - 0.03}{\sqrt{(0.032)(0.968)\left(\frac{1}{500} + \frac{1}{100}\right)}} = 0.104.$$

Conclusion. The calculated value of $|z| < 1.645$, the significant value of z at 5% level of significance. H_0 is accepted. *i.e.*, the machine has not improved due to overhauling.

Example 3. *In two large populations there are 30% and 25% respectively of fair haired people. Is this difference likely to be hidden in samples of 1200 and 900 respectively from the two populations.* (M.D.U. 2010)

Sol. P_1 = proportion of fair haired people in the first population = 30% = 0.3;
P_2 = 25% = 0.25; $Q_1 = 0.7$, $Q_2 = 0.75$

H_0: Sample proportions are equal *i.e.*, the difference in population proportions is likely to be hidden in sampling.

$H_1: p_1 \neq p_2$ We use two tailed test.

$$z = \frac{P_1 - P_2}{\sqrt{\left(\frac{P_1Q_1}{n_1} + \frac{P_2Q_2}{n_2}\right)}} = \frac{0.3 - 0.25}{\sqrt{\frac{0.3 \times 0.7}{1200} + \frac{0.25 \times 0.75}{900}}} = 2.553769.$$

Conclusion. Since $|z| > 1.96$, the significant value of z at 5% level of significance, H_0 is rejected. However $|z| < 2.58$, the significant value of z at 1% level of significance. H_0 is accepted. At 5% level these samples will reveal the difference in the population proportions.

Example 4. *500 articles from a factory are examined and found to be 2% defective. 800 similar articles from a second factory are found to have only 1.5% defective. Can it reasonably be concluded that the product of the first factory are inferior to those of second?*

Sol. $n_1 = 500, n_2 = 800$

p_1 = proportion of defective from first factory = 2% = 0.02

p_2 = proportion of defective from second factory = 1.5% = 0.015

H_0: There is no significant difference between the two products *i.e.*, the products do not differ in quality.

H_1: $p_1 < p_2$ (one tailed test)

Under H_0,
$$z = \frac{p_1 - p_2}{\sqrt{PQ\left(\frac{1}{n_1} + \frac{1}{n_2}\right)}}$$

$$P = \frac{n_1 p_1 + n_2 p_2}{n_1 + n_2} = \frac{0.02(500) + (0.015)(800)}{500 + 800} = 0.01692;\ Q = 1 - P = 0.9830$$

$$z = \frac{0.02 - 0.015}{\sqrt{0.01692 \times 0.983\left(\frac{1}{500} + \frac{1}{800}\right)}} = 0.68$$

Conclusion. As $|z| < 1.645$, the significant value of z at 5% level of significance. H_0 is accepted. *i.e.*, the products do not differ in quality.

EXERCISE 6.2

1. Random sample of 400 men and 600 women were asked whether they would like to have a school near their residence. 200 men and 325 women were in favour of proposal. Test the hypothesis that the proportion of men and women in favour of the proposal are same at 5% level of significance.
2. In a town A, there were 956 births of which 52.5% was males while in towns A and B combined, this proportion in total of 1406 births was 0.496. Is there any significant difference in the proportion of male births in the two towns?
3. In a referendum submitted to the students body at a university, 850 men and 560 women voted. 500 men and 320 women voted yes. Does this indicate a significant difference of opinion between men and women on this matter at 1% level?
4. A manufacturing firm claims that its brand A product outsells its brand B product by 8%. If it is found that 42 out of a sample of 200 person prefer brand A and 18 out of another sample of 100 person prefer brand B. Test whether the 8% difference is a valid claim.
5. During a countrywide investigation, the incidence of T.B. was found to be 1%. In a college 400 strong 5 were affected, whereas in another, 1200 strong, 10 were affected. Does this indicate any significant difference?
6. A machine puts out 10 imperfect articles in a sample of 200. After the machine is overhauled, it puts out 4 imperfect articles in a batch of 100. Has the machine been improved due to overhauling?

Answers

1. H_0: accepted **2.** H_0: rejected **3.** H_0: accepted **4.** H_0: accepted
5. No **6.** No:

6.11. TEST OF SIGNIFICANCE FOR SINGLE MEAN

To test whether the difference between sample mean and population mean is significant or not.

Let $x_1, x_2, ..., x_n$ be a random sample of size n from a large population $X_1, X_2, ..., X_N$ of size N with mean μ and variance σ^2. Then the standard error of mean of a random sample of size n from a population with variance σ^2 is $\sigma/\sqrt{n}$.

To test whether the given sample of size n has been drawn from a population with mean μ *i.e.,* to test whether the difference between the sample mean and population mean is significant or not, we set up the null hypothesis that there is no difference between the sample mean and population mean.

The test statistic is $z = \dfrac{\bar{x} - \mu}{\sigma/\sqrt{n}}$, where σ is the standard deviation of the population.

If σ is not known, we use the test statistic $z = \dfrac{\bar{x} - \mu}{s/\sqrt{n}}$, where s is the standard deviation of the sample.

Note. If the level of significance is α and z_α is the critical value, then

$$| z | < z_\alpha$$

$$\Rightarrow \quad \left| \frac{\bar{x} - \mu}{\sigma/n} \right| < z_\alpha \;\Rightarrow\; -z_\alpha < \frac{\bar{x} - \mu}{\sigma/n} < z_\alpha$$

$$\Rightarrow \quad -z_\alpha \cdot \frac{\sigma}{n} < \bar{x} - \mu < z_\alpha \cdot \frac{\sigma}{n}$$

$$\Rightarrow \quad -\bar{x} - z_\alpha \cdot \frac{\sigma}{n} < -\mu < -\bar{x} + z_\alpha \cdot \frac{\sigma}{n}$$

$$\Rightarrow \quad \bar{x} - z_\alpha \cdot \frac{\sigma}{n} < \mu < \bar{x} + z_\alpha \cdot \frac{\sigma}{n}$$

At 5% of level of significance, 95% confidence limits are $\bar{x} - 1.96 \dfrac{\sigma}{\sqrt{n}} < \mu < \bar{x} + 1.96 \dfrac{\sigma}{\sqrt{n}}$

At 1% level of significance, 99% confidence limits are $\bar{x} - 2.58 \dfrac{\sigma}{\sqrt{n}} < \mu < \bar{x} + 2.58 \dfrac{\sigma}{\sqrt{n}}$.

These limits are called confidence limits or fiducial limits.

ILLUSTRATIVE EXAMPLES

Example 1. *A normal population has a mean of 6.8 and standard deviation of 1.5. A sample of 400 members gave a mean of 6.75. Is the difference significant?*

Sol. H_0: There is no significant difference between $\bar{x}$ and μ.

H_1: There is significant difference between $\bar{x}$ and μ.

Given $\mu = 6.8$, $\sigma = 1.5$, $\bar{x} = 6.75$ and $n = 400$

$$| z | = \left| \frac{\bar{x} - \mu}{\sigma/\sqrt{n}} \right| = \left| \frac{6.75 - 6.8}{1.5/\sqrt{400}} \right| = | -0.67 | = 0.67$$

Conclusion. As the calculated value of $| z | < z_\alpha = 1.96$ at 5% level of significance, H_0 is accepted *i.e.*, there is no significant difference between $\bar{x}$ and μ.

Example 2. *A random sample of 900 members has a mean 3.4 cms. Can it be reasonably regarded as a sample from a large population of mean 3.2 cms and S.D. 2.3 cms?*

Sol. Here $n = 900$, $\bar{x} = 3.4$, $\mu = 3.2$, $\sigma = 2.3$

H_0: Assume that the sample is drawn from a large population with mean 3.2 and S.D. = 2.3

H_1: $\mu \neq 3.25$ (Apply two tailed test)

Under H_0; $z = \dfrac{\bar{x} - \mu}{\sigma/\sqrt{n}} = \dfrac{3.4 - 3.2}{2.3/\sqrt{900}} = 2.6086.$

Conclusion. As the calculated value of $| z | = 2.6086 > 1.96$ the significant value of z at 5% level of significance, H_0 is rejected *i.e.*, the sample is not drawn from the population with mean 3.2 and S.D. = 2.3.

Example 3. *The mean weight obtained from a random sample of size 100 is 64 gms. The S.D. of the weight distribution of the population is 3 gms. Test the statement that the mean weight of the population is 67 gms at 5% level of significance. Also set up 99% confidence limits of the mean weight of the population.* (Calicut 2011)

Sol. Here $n = 100$, $\mu = 67$, $\bar{x} = 64$, $\sigma = 3$

H_0: There is no significant difference between sample and population mean.

i.e., the sample is drawn from the population with $\mu = 67$

H_1: $\mu \neq 67$ (Two tailed test)

Under H_0, $z = \dfrac{\bar{x} - \mu}{\sigma/\sqrt{n}} = \dfrac{64 - 67}{3/\sqrt{100}} = -10 \quad \therefore \quad | z | = 10.$

Conclusion. Since the calculated value of $| z | > 1.96$, the significant value of z at 5% level of significance, H_0 is rejected *i.e.*, the sample is not drawn from the population with mean 67.

To find 99% confidence limits. It is given by $\bar{x} \mp 2.58\, \sigma/\sqrt{n} = 64 \mp 2.58 \times 3/\sqrt{100}$ $= 63.226$ and 64.774.

Example 4. *The average marks in Mathematics of a sample of 100 students was 51 with a S.D. of 6 marks. Could this have been a random sample from a population with average marks 50?* (M.D.U. 2011)

Sol. Here $n = 100$, $\bar{x} = 51$, $s = 6$, $\mu = 50$; σ is unknown.

H_0: The sample is drawn from a population with mean 50, $\mu = 50$

H_1: $\mu \neq 50$

Under H_0, $z = \dfrac{\bar{x} - \mu}{s/\sqrt{n}} = \dfrac{51 - 50}{6/\sqrt{100}} = \dfrac{10}{6} = 1.6666$

Conclusion. Since $| z | = 1.666 < 1.96$, z_α the significant value of z at 5% level of significance, H_0, is accepted *i.e.*, the sample is drawn from the population with mean 50.

EXERCISE 6.3

1. A sample of 1000 students from a university was taken and their average weight was found to be 112 pounds with a S.D. of 20 pounds. Could the mean weight of students in the population be 120 pounds?
2. A sample of 400 male students is found to have a mean height of 160 cms. Can it be reasonably regarded as a sample from a large population with mean height 162.5 cms and standard deviation 4.5 cms? *(M.D.U. Dec. 2013)*
3. A random sample of 200 measurements from a large population gave a mean value of 50 and a S.D. of 9. Determine 95% confidence interval for the mean of population.
4. The guaranteed average life of certain type of bulbs is 1000 hours with a S.D. of 125 hours. It is decided to sample the output so as to ensure that 90% of the bulbs do not fall short of the guaranteed average by more than 2.5%. What must be the minimum size of the sample.
5. The heights of college students in a city are normally distributed with S.D. 6 cms. A sample of 1000 students has mean height 158 cms. Test the hypothesis that the mean height of college students in the city is 160 cms. *(M.D.U. May 2014)*
6. The mean life of 100 fluorescent light tubes produced by a company is computed to be 1570 hours with S.D. 120 hours. The company claims that the average life of the tubes produced by the company is 1600 hours. Using level of significance 0.05, is the claim acceptable?

Answers

1. H_0 : rejected **2.** H_0 : rejected **3.** 48.8 and 51.2 **4.** $n = 4$

5. H_0 rejected both at 1% and 5% level of significance **6.** H_0 : rejected.

6.12. TEST OF SIGNIFICANCE FOR DIFFERENCE OF MEANS OF TWO LARGE SAMPLES

Let $\bar{x}_1$ be the mean of a sample of size n_1 from a population with mean μ_1, and variance σ_1^2. Let $\bar{x}_2$ be the mean of an independent sample of size n_2 from another population with mean μ_2 and variance σ_2^2. The test statistic is given by $z = \dfrac{\bar{x}_1 - \bar{x}_2}{\sqrt{\dfrac{\sigma_1^2}{n_1} + \dfrac{\sigma_2^2}{n_2}}}$.

Under the null hypothesis $H_0 : \mu_1 = \mu_2$, if the samples are drawn from the same population where $\sigma_1 = \sigma_2 = \sigma$ the test statistic is given by $z = \dfrac{\bar{x}_1 - \bar{x}_2}{\sigma\sqrt{\dfrac{1}{n_1} + \dfrac{1}{n_2}}}$.

Note 1. If σ_1, σ_2 are not known and $\sigma_1 \neq \sigma_2$ the test statistic in this case is $z = \dfrac{\bar{x}_1 - \bar{x}_2}{\sqrt{\dfrac{s_1^2}{n_1} + \dfrac{s_2^2}{n_2}}}$.

Note 2. If σ is not known and $\sigma_1 = \sigma_2$, we use $\sigma^2 = \dfrac{n_1 s_1^2 + n_2 s_2^2}{n_1 + n_2}$ to calculate σ;

Hence, $z = \dfrac{\bar{x}_1 - \bar{x}_2}{\sqrt{\dfrac{n_1 s_1^2 + n_2 s_2^2}{n_1 + n_2}\left(\dfrac{1}{n_1} + \dfrac{1}{n_2}\right)}} = \dfrac{\bar{x}_1 - \bar{x}_2}{\sqrt{\dfrac{n_1 s_1^2 + n_2 s_2^2}{n_1\, n_2}}}$.

ILLUSTRATIVE EXAMPLES

Example 1. *The average income of persons was ₹ 210 with a S.D. of ₹ 10 in sample of 100 people of a city. For another sample of 150 persons, the average income was ₹ 220 with S.D. of ₹ 12. The S.D. of incomes of the people of the city was ₹ 11. Test whether there is any significant difference between the average incomes of the localities.*

Sol. Here $n_1 = 100, n_2 = 150, \bar{x}_1 = 210, \bar{x}_2 = 220, s_1 = 10, s_2 = 12.$

Null hypothesis. The difference is not significant. *i.e.* there is no difference between the incomes of the localities.

$$H_0: \bar{x}_1 = \bar{x}_2, \; H_1: \bar{x}_1 \neq \bar{x}_2$$

Under H_0, $$z = \frac{\bar{x}_1 - \bar{x}_2}{\sqrt{\frac{s_1^2}{n_1} + \frac{s_2^2}{n_2}}} = \frac{210 - 220}{\sqrt{\frac{10^2}{100} + \frac{12^2}{150}}} = -7.1428 \quad \therefore \quad |z| = 7.1428.$$

Conclusion. As the calculated value of $|z| > 1.96$, the significant value of z at 5% level of significance, H_0 is rejected *i.e.*, there is significant difference between the average incomes of the localities.

Example 2. *Intelligence tests were given to two groups of boys and girls.*

	Mean	*S.D.*	*size*
Girls	*75*	*8*	*60*
Boys	*73*	*10*	*100*

Examine if the difference between mean scores is significant. (M.D.U. 2011)

Sol. Null hypothesis H_0: There is no significant difference between mean scores *i.e.*, $\bar{x}_1 = \bar{x}_2$

$$H_1: \bar{x}_1 \neq \bar{x}_2$$

Under the null hypothesis $$z = \frac{\bar{x}_1 - \bar{x}_2}{\sqrt{\frac{s_1^2}{n_1} + \frac{s_2^2}{n_2}}} = \frac{75 - 73}{\sqrt{\frac{8^2}{60} + \frac{10^2}{100}}} = 1.3912.$$

Conclusion. As the calculated value of $|z| < 1.96$, the significant value of z at 5% level of significance, H_0 is accepted. *i.e.*, there is no significant difference between mean scores.

Example 3. *For sample I, $n_1 = 1000$, $\Sigma x = 49,000$, $\Sigma(x - \bar{x})^2 = 7,84,000$.*

For sample II, $n_2 = 1,500$, $\Sigma x = 70,500$, $\Sigma(x - \bar{x})^2 = 24,00,000$. Discuss the significance of the difference of the sample means.

Sol. Null hypothesis H_0: There is no significant difference between the sample means.

$$H_0: \bar{x}_1 = \bar{x}_2; \; H_1: \bar{x}_1 \neq \bar{x}_2$$

To calculate sample variance

$$s_1^2 = \frac{1}{n_1} \Sigma(x_1 - \bar{x}_1)^2 = \frac{784000}{1000} = 784$$

$$s_2^2 = \frac{1}{n_2} \Sigma(x_2 - \bar{x}_2)^2 = \frac{1}{1500} (2400000) = 1600$$

$$\bar{x}_1 = \frac{\Sigma x_1}{n_1} = \frac{49000}{1000} = 49;\ \bar{x}_2 = \frac{\Sigma x_2}{n_2} = \frac{70500}{1500} = 47$$

Under the null hypothesis, the test statistic

$$z = \frac{\bar{x}_1 - \bar{x}_2}{\sqrt{\frac{s_1^2}{n_1} + \frac{s_2^2}{n_2}}} = \frac{49 - 47}{\sqrt{\frac{784}{1000} + \frac{1600}{1500}}} = 1.470.$$

Conclusion. As the calculated value of $|z| = 1.47 < 1.96$, the significant value of z at 5% level of significance, H_0 is accepted. *i.e.*, there is no significant difference between the sample means.

Example 4. *From the data given below, compute the standard error of the difference of the two sample means and find out if the two means significantly differ at 5% level of significance.*

	No. of items	*Mean*	*S.D.*
Group I	*50*	*181.5*	*3.0*
Group II	*75*	*179*	*3.6*

Sol. Null hypothesis H_0: There is no significant difference between the sample means.

$$\bar{x}_1 = \bar{x}_2;\ H_1: \bar{x}_1 \neq \bar{x}_2$$

$$\text{Under } H_0,\ z = \frac{\bar{x}_1 - \bar{x}_2}{\sqrt{\frac{s_1^2}{n_1} + \frac{s_2^2}{n_2}}} = \frac{181.5 - 179.0}{\sqrt{\frac{9}{50} + \frac{(3.6)^2}{75}}} = 4.2089.$$

Conclusion. As $|z| > 1.96$, the tabulated value of z at 5% level of significance, H_0 is rejected *i.e.*, there is significant difference between the samples.

Example 5. *A random sample of 200 villages from Coimbatore district gives the mean population per village at 485 with a S.D. of 50. Another random sample of the same size from the same district gives the mean population per village at 510 with a S.D. of 40. Is the difference between the mean values given by the two samples statistically significant? Justify your answer.*

Sol. Here $n_1 = 200$, $n_2 = 250$, $\bar{x}_1 = 485$, $\bar{x}_2 = 510$, $s_1 = 50$, $s_2 = 40$

Null hypothesis H_0: There is no significant difference between the mean values (*i.e.*,) $\bar{x}_1 = \bar{x}_2$; $H_1 : \bar{x}_1 \neq \bar{x}_2$

$$\text{Under } H_0, \text{ the test statistic is given by } z = \frac{\bar{x}_1 - \bar{x}_2}{\sqrt{\frac{s_1^2}{n_1} + \frac{s_2^2}{n_2}}} = \frac{485 - 510}{\sqrt{\frac{50^2}{200} + \frac{40^2}{200}}} = -5.52$$

$\therefore \quad |z| = 5.52.$

Conclusion. As the calculated value of $|z| > 1.96$, the significant value of z at 5% level of significance, H_0 is rejected. *i.e.*, there is significant difference between the mean values of two sample.

EXERCISE 6.4

1. Intelligence tests on two groups of boys and girls gave the following results. Examine if the difference between mean scores is significant.

	Mean	*S.D.*	*Size*
Girls	*70*	*10*	*70*
Boys	*75*	*11*	*100*

2. (*a*) Two random samples of sizes 1000 and 2000 farms gave an average yield of 2000 kg and 2050 kg respectively. The variance of wheat farms in the country may be taken as 100 kg. Examine whether the two samples differ significantly in yield?

   (*b*) The means of simple samples of sizes 1000 and 2000 are 67.5 and 68.0 cm respectively. Can the samples be regarded as drawn from the same population of S.D. 2.5 cm?

   (*M.D.U. 2011*)

3. A simple sample of heights of 6400 Englishmen has a mean of 170 cm and a S.D. of 6.4 cm, while a simple sample of heights of 1600 Americans has a mean of 172 cm and a S.D. of 6.3 cm. Does the data indicate that Americans are, on the average, taller than Englishmen? (

4. In a survey of buying habits, 400 women shoppers are chosen at random in supermarket A. Their average weekly food expenditure is ₹ 250 with a S.D. of ₹ 40. For 500 women shoppers chosen at supermarket B, the average weekly food expenditure is ₹ 220 with a S.D. of ₹ 45 test at 1% level of significance whether the average food expenditures of the two groups are equal.

5. The number of accident per day were studied for 144 days in town A and for 100 days in town B and the following information was obtained.

	Mean number of accident	*S.D.*
Town A	4.5	1.2
Town B	5.4	1.5

   Is the difference between the mean accidents of the two towns statistically significant?

6. An examination was given to 50 students of a college A and to 60 students of college B. For A, the mean grade was 75 with S.D. of 9 and for B, the mean grade was 79 with S.D. of 7. Is there any significant difference between the performance of the students of college A and those of college B?

7. The means of two large samples of 1000 and 2000 members are 168.75 cm and 170 cm respectively. Can the samples be regarded as drawn from the same population of standard deviation 6.25 cm?

Answers

1. Significant **2.** (*a*) Highly significant (*b*) Highly significant
3. Highly significant **4.** Highly significant **5.** Highly significant
6. Not significant **7.** No.

6.13. TEST OF SIGNIFICANCE FOR THE DIFFERENCE OF STANDARD DEVIATIONS

If s_1 and s_2 are the standard deviations of two independent samples, then under the null hypothesis H_0: $\sigma_1 = \sigma_2$, *i.e.*, the sample standard deviations don't differ significantly, the test statistic is

$$z = \frac{s_1 - s_2}{\sqrt{\frac{\sigma_1^2}{2n_1} + \frac{\sigma_2^2}{2n_2}}}, \text{ where } \sigma_1 \text{ and } \sigma_2 \text{ are population standard deviations.}$$

When population standard deviations are not known, then $z = \dfrac{s_1 - s_2}{\sqrt{\dfrac{s_1^2}{2n_1} + \dfrac{s_2^2}{2n_2}}}$.

ILLUSTRATIVE EXAMPLES

Example 1. *Random samples drawn from two countries gave the following data relating to the heights of adult males.*

	Country A	*Country B*
Mean height (in inches)	*67.42*	*67.25*
Standard deviation	*2.58*	*2.50*
Number in samples	*1000*	*1200*

(*i*) *Is the difference between the means significant?*

(*ii*) *Is the difference between the standard deviations significant?*

Sol. Given: $n_1 = 1000$, $n_2 = 1200$, $\bar{x}_1 = 67.42$; $\bar{x}_2 = 67.25$, $s_1 = 2.58$, $s_2 = 2.50$

Since the samples size are large we can take $\sigma_1 = s_1 = 2.58$; $\sigma_2 = s_2 = 2.50$.

(*i*) **Null Hypothesis:** $H_0 = \mu_1 = \mu_2$ *i.e.*, sample means do not differ significantly.

Alternative hypothesis: $H_1 : \mu_1 \neq \sigma_2$ (two tailed test)

$$z = \frac{\bar{x}_1 - \bar{x}_2}{\sqrt{\frac{s_1^2}{n_1} + \frac{s_2^2}{n_2}}} = \frac{67.42 - 67.25}{\sqrt{\frac{(2.58)^2}{1000} + \frac{(2.50)^2}{1200}}} = 1.56$$

since | z | < 1.96 we accept the null hypothesis at 5% level of significance.

(*ii*) We set up the null hypothesis.

H_0: $\sigma_1 = \sigma_2$ *i.e.*, the sample S.D.'s do not differ significantly.

Alternative Hypothesis: $H_1 = \sigma_1 \neq \sigma_2$ (two tailed test)

∴ The test statistic is given by

$$z = \frac{s_1 - s_2}{\sqrt{\frac{\sigma_1^2}{2n_1} + \frac{\sigma_2^2}{2n_2}}} = \frac{s_1 - s_2}{\sqrt{\frac{s_1^2}{2n_1} + \frac{s_2^2}{2n_2}}} \quad (\because \ \sigma_1 = s_1, \sigma_2 = s_2 \text{ for large samples})$$

$$= \frac{2.58 - 2.50}{\sqrt{\frac{(2.58)^2}{2 \times 1000} \times \frac{(2.50)^2}{2 \times 1200}}} = \frac{0.08}{\sqrt{\frac{6.6564}{2000} + \frac{6.25}{2400}}} = 1.0387$$

Since | z | < 1.96 we accept the null hypothesis at 5% level of significance.

Example 2. *Intelligence test of two groups of boys and girls gives the following results:*

Girls	mean = 84	S.D. = 10	N = 121
Boys	mean = 81	S.D. = 12	N = 81

(*a*) *Is the difference in mean scores significant?*

(*b*) *Is the difference between the standard deviations significant?*

Sol. Given: $n_1 = 121$, $n_2 = 81$, $\bar{x}_1 = 84$, $\bar{x}_2 = 81$, $s_1 = 10$, $s_2 = 12$

(*a*) **Null hypothesis:** $H_0 = \mu_1 = \mu_2$ *i.e.*, sample means do not differ significantly.

Alternative hypothesis: $H_1 = \mu_1 \neq \mu_2$ (two tailed)

The test statistic is $z = \dfrac{\bar{x}_1 - \bar{x}_2}{\sqrt{\dfrac{s_1^2}{n_1} + \dfrac{s_2^2}{n_2}}} = \dfrac{84 - 81}{\sqrt{\dfrac{(10)^2}{121} + \dfrac{(12)^2}{81}}} = 0.1859$

Since $|z| < 1.96$ we accept the null hypothesis at 5% level of significance.

(*b*) We set up the null hypothesis $H_0 = \sigma_1 = \sigma_2$ *i.e.*, the sample S.D.'s do not differ significantly. Alternative hypothesis: $H_1 = \sigma_1 \neq \sigma_2$ (two tailed)

The test statistic is $z = \dfrac{s_1 - s_2}{\sqrt{\dfrac{\sigma_1^2}{2n_1} + \dfrac{\sigma_2^2}{2n_2}}} = \dfrac{s_1 - s_2}{\sqrt{\dfrac{s_1^2}{2n_1} + \dfrac{s_2^2}{2n_2}}}$

($\because$ $\sigma_1 = s_1$, $\sigma_2 = s_2$ for large samples)

$$= \frac{10 - 12}{\sqrt{\dfrac{100}{2 \times 121} + \dfrac{144}{2 \times 81}}} = -1.7526 \quad \therefore \quad |z| = 1.7526$$

since $|z| = 1.75 < 1.96$ we accept the null hypothesis at 5% level of significance.

EXERCISE 6.5

1. The mean yield of two sets of plots and their variability are as given examine.
   (*i*) Whether the difference in the mean yield of the two sets of plots is significant.
   (*ii*) Whether the difference in the variability in yields is significant.

	Set of 40 plots	*Set of 60 plots*
Mean yield per plot	1258 lb	1243 lb
S.D. per plot	34	28

2. The yield of wheat in a random sample of 1000 farms in a certain area has a S.D. of 192 kg. Another random sample of 1000 farms gives a S.D. of 224 kg. Are the S.Ds significantly different?

Answers

1. $z = 2.321$ Difference significant at 5% level; $z = 1.31$ Difference not significant at 5% level
2. $z = 4.851$ The S.Ds are significantly different.

6.14. TEST OF SIGNIFICANCE OF SMALL SAMPLES

When the size of the sample is less than 30, then the sample is called small sample. For such sample it will not be possible for us to assume that the random sampling distribution of a statistic is approximately normal and the values given by the sample data are sufficiently close to the population values and can be used in their place for the calculation of the standard error of the estimate.

t-TEST

6.15. STUDENT'S-t-DISTRIBUTION

This t-distribution is used when sample size is ≤ 30 and the population standard deviation is unknown.

t-statistic is defined as $t = \dfrac{\bar{x} - \mu}{S/\sqrt{n}}$, d.f. (degrees of freedom) $= n - 1$ where $S^2 = \dfrac{\Sigma(x - \bar{x})^2}{n - 1}$.

is an unbiased estimate of the population variance σ^2.

Thus, $\bar{x}$ = sample mean $= \dfrac{1}{n} \sum x$, μ = population mean

$$S = \text{standard divation of population} = \sqrt{\frac{\Sigma (x - \bar{x})^2}{n - 1}}$$

n = sample size.

If the standard deviation of the sample 's' is given, then t-statistic is defined as

$$t = \frac{\bar{x} - \mu}{s/\sqrt{n - 1}}.$$

Note. Since $s^2 = \dfrac{1}{n} \sum (x - \bar{x})^2$ and $S^2 = \dfrac{1}{n - 1} \sum (x - \bar{x})^2$

$\therefore \quad ns^2 = (n - 1)S^2.$

The t-table

The t-table given at the end is the probability integral of t-distribution. The t-distribution has different values for each degrees of freedom and when the degrees of freedom are infinitely large, the t-distribution is equivalent to normal distribution and the probabilities shown in the normal distribution tables are applicable.

Application of t-distribution

Some of the applications of t-distribution are given below:

1. *To test if the sample mean* ($\bar{x}$) *differs significantly from the hypothetical value* μ *of the population mean.*
2. *To test the significance between two sample means.*
3. *To test the significance of observed partial and multiple correlation coefficients.*

Critical value of t

The critical value or significant value of t at level of significance α degrees of freedom γ for two tailed test is given by

$$P[\,|\,t\,| > t_\gamma(\alpha)] = \alpha$$
$$P[\,|\,t\,| \le t_\gamma(\alpha)] = 1 - \alpha$$

The significant value of t at level of significance α for a single tailed test can be got from those of two tailed test by referring to the values at 2α.

6.16. TEST I: t-TEST OF SIGNIFICANCE OF THE MEAN OF A RANDOM SAMPLE

To test whether the mean of a sample drawn from a normal population deviates significantly from a stated value when variance of the population is unknown.

H_0: There is no significant difference between the sample mean $\bar{x}$ and the population mean μ *i.e.*, we use the statistic

$$t = \frac{\bar{x} - \mu}{S/\sqrt{n}}, \qquad \text{where } \bar{x} \text{ is mean of the sample.}$$

$$S^2 = \frac{1}{n-1}\sum_{i=1}^{n}(x_i - \bar{x})^2 \text{ with degrees of freedom } (n-1).$$

At given level of significance α and degrees of freedom $(n - 1)$, we refer to t-table for t_α (two tailed or one tailed).

If calculated t value is such that $|\,t\,| < t_\alpha$ the null hypothesis is accepted. If $|\,t\,| > t_\alpha$, H_0 is rejected.

Fiducial limits of population mean

If t_α is the table of t at level of significance α at $(n - 1)$ degrees of freedom.

$$\left|\frac{\bar{x} - \mu}{S/\sqrt{n}}\right| < t_\alpha \text{ for acceptance of } H_0.$$

$$\bar{x} - t_\alpha\, S/\sqrt{n} < \mu < \bar{x} + t_\alpha\, S/\sqrt{n}$$

95% confidence limits (level of significance 5%) are $\bar{x} \mp t_{0.05}\, S/\sqrt{n}$.

99% confidence limits (level of significance 1%) are $\bar{x} \mp t_{0.01}\, S/\sqrt{n}$.

ILLUSTRATIVE EXAMPLES

Example 1. *A random sample of size 16 has 53 as mean. The sum of squares of the derivation from mean is 135. Can this sample be regarded as taken from the population having 56 as mean? Obtain 95% and 99% confidence limits of the mean of the population.*

Sol. H_0: There is no significant difference between the sample mean and hypothetical population mean.

$$H_0: \mu = 56;\ H_1: \mu \ne 56 \qquad \text{(Two tailed test)}$$

$$t = \frac{\bar{x} - \mu}{S/\sqrt{n}} \sim t(n - 1 \text{ d.f.})$$

Given: $\bar{x} = 53$, $\mu = 56$, $n = 16$, $(x - \bar{x})^2 = 135$

$$S = \sqrt{\frac{(x-\bar{x})^2}{n-1}} = \sqrt{\frac{135}{15}} = 3; \; t = \frac{53-56}{3/\sqrt{16}} = \frac{-3 \times 4}{3} = -4$$

$|t| = 4$. d.f. $= 16 - 1 = 15$.

Conclusion. $t_{0.05} = 2.13$.

Since $|t| = 4 > t_{0.05} = 2.13$ *i.e.*, the calculated value of t is more than the table value. The hypothesis is rejected. Hence the sample mean has not come from a population having 56 as mean.

95% confidence limits of the population mean

$$\bar{x} \mp \frac{S}{\sqrt{n}} t_{0.05} = 53 \mp \frac{3}{\sqrt{16}} (2.13) = 51.4025, 54.5975$$

99% confidence limits of the population mean

$$\bar{x} \mp \frac{S}{\sqrt{n}} t_{0.01} = 53 \mp \frac{3}{\sqrt{16}} (2.95) = 50.7875, 55.2125$$

Example 2. *The life time of electric bulbs for a random sample of 10 from a large consignment gave the following data:*

Item	*1*	*2*	*3*	*4*	*5*	*6*	*7*	*8*	*9*	*10*
Life in '000 hrs.	*4.2*	*4.6*	*3.9*	*4.1*	*5.2*	*3.8*	*3.9*	*4.3*	*4.4*	*5.6*

Can we accept the hypothesis that the average life time of bulb is 4000 hrs?

Sol. H_0: There is no significant difference in the sample mean and population mean. *i.e.*, $\mu = 4000$ hrs.

$H_1 : \mu \neq 4000$ hrs (Two tailed test)

Applying the t-test: $t = \dfrac{\bar{x} - \mu}{S/\sqrt{n}} \sim t(10 - 1 \text{ d.f.})$

x	4.2	4.6	3.9	4.1	5.2	3.8	3.9	4.3	4.4	5.6
$x - \bar{x}$	– 0.2	0.2	– 0.5	– 0.3	0.8	– 0.6	– 0.5	– 0.1	0	1.2
$(x - \bar{x})^2$	0.04	0.04	0.25	0.09	0.64	0.36	0.25	0.01	0	1.44

$$\bar{x} = \frac{\Sigma x}{n} = \frac{44}{10} = 4.4 \qquad \Sigma(x - \bar{x})^2 = 3.12$$

$$S = \sqrt{\frac{\Sigma(x-\bar{x})^2}{n-1}} = \sqrt{\frac{3.12}{9}} = 0.589; \; t = \frac{4.4 - 4}{\left(\dfrac{0.589}{\sqrt{10}}\right)} = 2.123$$

For $\gamma = 9$, $t_{0.05} = 2.26$.

Conclusion. Since the calculated value of t is less than table $t_{0.05}$. $\therefore$ The hypothesis $\mu = 4000$ hrs is accepted.

i.e., The average life time of bulbs could be 4000 hrs.

Example 3. *A sample of 20 items has mean 42 units and S.D. 5 units. Test the hypothesis that it is a random sample from a normal population with mean 45 units.*

(M.D.U. 2012)

Sol. H_0: There is no significant difference between the sample mean and the population mean.

i.e., $\mu = 45$ units

H_1: $\mu \neq 45$ (Two tailed test)

Given: $n = 20$, $\bar{x} = 42$, $s = 5$; $\gamma = 19$ d.f.

$$S^2 = \frac{n}{n-1} s^2 = \left[\frac{20}{20-1}\right](5)^2 = 26.31 \quad \therefore \quad S = 5.129$$

Applying t-test $$t = \frac{\bar{x} - \mu}{S/\sqrt{n}} = \frac{42-45}{5.129/\sqrt{20}} = -2.615; \ |t| = 2.615$$

The tabulated value of t at 5% level for 19 d.f. is $t_{0.05} = 2.09$.

Conclusion. Since $|t| > t_{0.05}$, the hypothesis H_0 is rejected. *i.e.,* there is significant difference between the sample mean and population mean.

i.e., The sample could not have come from this population.

Example 4. *The 9 items of a sample have the following values 45, 47, 50, 52, 48, 47, 49, 53, 51. Does the mean of these values differ significantly from the assumed mean 47.5.*

(M.D.U. 2011)

Sol. H_0: $\mu = 47.5$

i.e., there is no significant difference between the sample and population mean.

H_1: $\mu \neq 47.5$ (two tailed test); Given: $n = 9$, $\mu = 47.5$

x	45	47	50	52	48	47	49	53	51
$x - \bar{x}$	– 4.1	– 2.1	0.9	2.9	– 1.1	– 2.1	– 0.1	3.9	1.9
$(x - \bar{x})^2$	16.81	4.41	0.81	8.41	1.21	4.41	0.01	15.21	3.61

$$\bar{x} = \frac{\Sigma x}{n} = \frac{442}{9} = 49.11; \ \Sigma(x - \bar{x})^2 = 54.89; \ S^2 = \frac{\Sigma(x-\bar{x})^2}{(n-1)} = 6.86 \quad \therefore \quad S = 2.619$$

Applying t-test, $$t = \frac{\bar{x} - \mu}{S/\sqrt{n}} = \frac{49.1 - 47.5}{2.619/\sqrt{9}} = \frac{(1.6)3}{2.619} = 1.8327$$

$$t_{0.05} = 2.31 \text{ for } \gamma = 8.$$

Conclusion. Since $|t| < t_{0.05}$, the null hypothesis H_0 is accepted *i.e.,* there is no significant difference between their mean.

Example 5. *The following results are obtained from a sample of 10 boxes is biscuits.*

Mean weight content = 490 gm

S.D. of the weight = 9 gm could the sample come from a population having a mean of 500 gm?

Sol. Given: $n = 10,\ \bar{x} = 490;\ s = 9$ gm, $\mu = 500$

$$S = \sqrt{\frac{n}{n-1}}\, s = \sqrt{\frac{10}{9}} \times 9 = 9.486$$

H_0: The difference is not significant *i.e.*, $\mu = 500$; H_1: $\mu \neq 500$

Applying t-test, $$t = \frac{\bar{x} - \mu}{S/\sqrt{n}} = \frac{490 - 500}{9.486/\sqrt{10}} = -3.33$$

$$t_{0.05} = 2.26 \text{ for } \gamma = 9.$$

Conclusion. Since $|t| = 3.33 > t_{0.05}$ the hypothesis H_0 is rejected *i.e.*, $\mu \neq 500$

$\therefore$ The sample could not have come from the population having mean 500 gm.

EXERCISE 6.6

1. Find the student's t for the following variable values in a sample of eight:
$-4, -2, -2, 0, 2, 2, 3, 3$ taking the mean of the universe to be zero. *(M.D.U. 2010)*

2. Ten individuals are choosen at random from a normal population of students and their marks found to be 63, 63, 66, 67, 68, 69, 70, 70, 71, 71. In the light of these data discuss the suggestion that mean mark of the population of students is 66.

3. The following values gives the lengths of 12 samples of egyptian cotton taken from a consignment: 48, 46, 49, 46, 52, 45, 43, 47, 47, 46, 45, 50. Test if the mean length of the consignment can be taken as 46.

4. A sample of 18 items has a mean 24 units and standard deviation 3 units. Test the hypothesis that it is a random sample from a normal population with mean 27 units.

5. A random sample of 10 boys had the I.Q's 70, 120, 110, 101, 88, 83, 95, 98, 107 and 100. Do these data support the assumption of a population mean I.Q of 106?

6. A filling machine is expected to fill 5 kg of powder into bags. A sample of 10 bags gave the following weights. 4.7, 4.9, 5.0, 5.1, 5.4, 5.2, 4.6, 5.1, 4.6 and 4.7. Test whether the machine is working properly. *(M.D.U. 2010)*

7. A machinist is making engine parts with axle-diameter of 0.7 unit. A random sample of 10 parts shows mean diameter 0.742 unit with a standard deviation of 0.04 unit. On the basis of this sample, what would you say that the work in inferior?

8. A sample of 6 persons in an office revealed an average daily smoking of 10, 12, 8, 9, 16, 5 cigarettes. The average level of smoking in the whole office has to be estimated at 90% level of confidence. $t = 2.015$ for 5 degree of freedom. *(M.D.U. May 2014)*

Answers

1. $t = 0.2659$ **2.** accepted **3.** accepted

4. rejected **5.** accepted **6.** accepted

7. rejected **8.** 8.74.

6.17. TEST II: t-TEST FOR DIFFERENCE OF MEANS OF TWO SMALL SAMPLES (From a Normal Population)

This test is used to test whether the two samples $x_1, x_2, ..., x_{n_1}, y_1, y_2, ..., y_{n_2}$ of sizes n_1, n_2 have been drawn from two normal populations with mean μ_1 and μ_2 respectively under the assumption that the population variance are equal. $(\sigma_1 = \sigma_2 = \sigma)$.

H_0: The samples have been drawn from the normal population with means μ_1 and μ_2 *i.e.*, $H_0: \mu_1 \neq \mu_2$.

Let $\overline{X}, \overline{Y}$ be the means of the two samples.

Under this H_0 the test statistic t is given by $t = \dfrac{(\overline{X} - \overline{Y})}{S\sqrt{\dfrac{1}{n_1} + \dfrac{1}{n_2}}} \sim t(n_1 + n_2 - 2 \text{ d.f.})$

Note 1. If the two sample standard deviations s_1, s_2 are given then we have $S^2 = \dfrac{n_1 s_1^2 + n_2 s_2^2}{n_1 + n_2 - 2}$.

Note 2. If $n_1 = n_2 = n$, $t = \dfrac{\overline{X} - \overline{Y}}{\sqrt{\dfrac{s_1^2 + s_2^2}{n-1}}}$ can be used as a test statistic.

Note 3. If s_1, s_2 are not given, then $S^2 = \dfrac{\Sigma (x_1 - \overline{x}_1)^2 + \Sigma (x_2 - \overline{x}_2)^2}{n_1 + n_2 - 2}$

ILLUSTRATIVE EXAMPLES

Example 1. *Two samples of sodium vapour bulbs were tested for length of life and the following results were got:*

	Size	*Sample mean*	*Sample S.D.*
Type I	*8*	*1234 hrs*	*36 hrs*
Type II	*7*	*1036 hrs*	*40 hrs*

Is the difference in the means significant to generalise that type I is superior to type II regarding length of life.

Sol. $H_0: \mu_1 = \mu_2$ *i.e.*, two types of bulbs have same lifetime.

$H_1: \mu_1 > \mu_2$ *i.e.*, type I is superior to Type II

$$S^2 = \frac{n_1 s_1^2 + n_2 s_2^2}{n_1 + n_2 - 2} = \frac{8 \times (36)^2 + 7(40)^2}{8 + 7 - 2} = 1659.076 \quad \therefore \quad s = 40.7317$$

The t-statistic
$$t = \frac{\overline{X}_1 - \overline{X}_2}{S\sqrt{\dfrac{1}{n_1} + \dfrac{1}{n_2}}} = \frac{1234 - 1036}{40.7317\sqrt{\dfrac{1}{8} + \dfrac{1}{7}}} = 18.1480 \sim t(n_1 + n_2 - 2 \text{ d.f.})$$

$t_{0.05}$ at d.f. 13 is 1.77 (One tailed test)

Conclusion. Since calculated $|t| > t_{0.05}$ H_0 is rejected *i.e.*, H_1 is accepted.

$\therefore$ Type I is definitely superior to type II.

Example 2. *Samples of sizes 10 and 14 were taken from two normal populations with S.D. 3.5 and 5.2. The sample means were found to be 20.3 and 18.6. Test whether the means of the two populations are the same at 5% level.* (M.D.U. May 2015)

Sol. H_0: $\mu_1 = \mu_2$ *i.e.*, the means of the two populations are the same.

H_1: $\mu_1 \neq \mu_2$.

Given $\overline{X}_1 = 20.3$, $\overline{X}_2 = 18.6$; $n_1 = 10$, $n_2 = 14$, $s_1 = 3.5$, $s_2 = 5.2$

$$S^2 = \frac{n_1 s_1^2 + n_2 s_2^2}{n_1 + n_2 - 2} = \frac{10(3.5)^2 + 14(5.2)^2}{10 + 14 - 2} = 22.775 \quad \therefore \quad s = 4.772$$

$$t = \frac{\overline{X}_1 - \overline{X}_2}{S\sqrt{\frac{1}{n_1} + \frac{1}{n_2}}} = \frac{20.3 - 18.6}{\left(\sqrt{\frac{1}{10} + \frac{1}{14}}\right) 4.772} = 0.8604$$

The value of t at 5% level for 22 d.f. is $t_{0.05} = 2.0739$.

Conclusion. Since $|t| = 0.8604 < t_{0.05}$ the hypothesis is accepted. *i.e.*, there is no significant difference between their means.

Example 3. *The height of 6 randomly chosen sailors in inches are 63, 65, 68, 69, 71 and 72. Those of 9 randomly chosen soldiers are 61, 62, 65, 66, 69, 70, 71, 72 and 73. Test whether the sailors are on the average taller than soldiers.*

Sol. Let X_1 and X_2 be the two samples denoting the heights of sailors and soldiers.

Given the sample size $n_1 = 6$, $n_2 = 9$, H_0: $\mu_1 = \mu_2$.

i.e., the mean of both the population are the same.

H_1: $\mu_1 > \mu_2$ (one tailed test)

Calculation of two sample mean:

X_1	63	65	68	69	71	72
$X_1 - \overline{X}_1$	− 5	− 3	0	1	3	4
$(X_1 - \overline{X}_1)^2$	25	9	0	1	9	16

$$\overline{X}_1 = \frac{\Sigma X_1}{n_1} = 68; \quad \Sigma(X_1 - \overline{X}_1)^2 = 60$$

X_2	61	62	65	66	69	70	71	72	73
$X_2 - \overline{X}_2$	− 6.66	− 5.66	− 2.66	1.66	1.34	2.34	3.34	4.34	5.34
$(X_2 - \overline{X}_2)^2$	44.36	32.035	7.0756	2.7556	1.7956	5.4756	11.1556	18.8356	28.5156

$$\overline{X}_2 = \frac{\Sigma X_2}{n_2} = 67.66; \quad \Sigma(X_2 - \overline{X}_2)^2 = 152.0002$$

$$S^2 = \frac{1}{n_1 + n_2 - 2} [\Sigma(X_1 - \overline{X}_1)^2 + \Sigma(X_2 - \overline{X}_2)^2]$$

$$= \frac{1}{6+9-2}[60 + 152.0002] = 16.3077 \quad \Rightarrow \quad S = 4.038$$

Under H_0, $$t = \frac{\overline{X}_1 - \overline{X}_2}{S\sqrt{\frac{1}{n_1} + \frac{1}{n_2}}} = \frac{68 - 67.666}{4.0382\sqrt{\frac{1}{6} + \frac{1}{9}}} = 0.3031 \sim t(n_1 + n_2 - 2 \text{ d.f.})$$

The value of t at 10% level of significance ($\because$ the test is one tailed) for 13 d.f. is 1.77.

Conclusion. Since $|t| = 0.3031 < t_{0.05} = 1.77$ the hypothesis H_0 is accepted.

i.e., there is no significance difference between their average.

i.e., the sailors are not on the average taller than the soldiers.

Example 4. *A certain stimulus administered to each of 12 patients resulted in the following increases of blood pressure: 5, 2, 8, – 1, 3, 0, – 2, 1, 5, 0, 4, 6. Can it be concluded that the stimulus will in general be accompanied by an increase in blood pressure?*

(M.D.U. 2011)

Sol. To test whether the mean increase in blood pressure of all patients to whom the stimulus is administered will be positive. We have to assume that this population is normal with mean μ and S.D. σ which are unknown.

$$H_0: \mu = 0; \ H_1: \mu_1 > 0$$

The test statistic under H_0

$$t = \frac{\bar{d}}{s/\sqrt{n-1}} \sim t(n - 1 \text{ degrees of freedom})$$

$$\bar{d} = \frac{5 + 2 + 8 + (-1) + 3 + 0 + 6 + (-2) + 1 + 5 + 0 + 4}{12} = 2.583$$

$$s^2 = \frac{\Sigma d^2}{n} - \bar{d}^2 = \frac{1}{12}[5^2 + 2^2 + 8^2 + (-1)^2 + 3^2 + 0^2 + 6^2 + (-2)^2 + 1^2 + 5^2 + 0^2 + 4^2] - (2.583)^2$$

$$= 8.744 \quad \therefore \quad s = 2.9571$$

$$t = \frac{\bar{d}}{s/\sqrt{n-1}} = \frac{2.583}{2.9571/\sqrt{12-1}} = \frac{2.583\sqrt{11}}{2.9571} = 2.897 \sim t(n - 1 \text{ d.f.})$$

Conclusion. The tabulated value of $t_{0.05}$ at 11 d.f. is 2.2.

$\because \ |t| > t_{0.05}$, H_0 is rejected.

i.e., the stimulus does not increase the blood pressure. The stimulus in general will be accompanied by increase in blood pressure.

Example 5. *Memory capacity of 9 students was tested before and after a course of meditation for a month. State whether the course was effective or not from the data below (in same units)*

Before	*10*	*15*	*9*	*3*	*7*	*12*	*16*	*17*	*4*
After	*12*	*17*	*8*	*5*	*6*	*11*	*18*	*20*	*3*

Sol. Since the data are correlated and concerned with the same set of students we use paired t-test.

H_0: Training was not effective $\mu_1 = \mu_2$

H_1: $\mu_1 \neq \mu_2$ (Two tailed test).

Before training (X)	*After training (Y)*	$d = X - Y$	d^2
10	12	− 2	4
15	17	− 2	4
9	8	1	1
3	5	− 2	4
7	6	1	1
12	11	1	1
16	18	− 2	4
17	20	− 3	9
4	3	1	1
		$\Sigma d = -7$	$\Sigma d^2 = 29$

$$\bar{d} = \frac{\Sigma d}{n} = \frac{-7}{9} = -0.7778;\ s^2 = \frac{\Sigma d^2}{n} - (\bar{d})^2 = \frac{29}{9} - (-0.7778)^2 = 2.617$$

$$t = \frac{\bar{d}}{s/\sqrt{n-1}} = \frac{-0.7778}{\sqrt{2.6172}/\sqrt{8}} = \frac{-0.7778 \times \sqrt{8}}{1.6177} = -1.359$$

The tabulated value of $t_{0.05}$ at 8 d.f. is 2.31.

Conclusion. Since $|t| = 1.359 < t_{0.05}$, H_0 is accepted *i.e.*, training was not effective in improving performance.

Example 6. *The following figures refer to observations in live independent samples.*

Sample I	*25*	*30*	*28*	*34*	*24*	*20*	*13*	*32*	*22*	*38*
Sample II	*40*	*34*	*22*	*20*	*31*	*40*	*30*	*23*	*36*	*17*

Analyse whether the samples have been drawn from the populations of equal means.

Sol. H_0: The two samples have been drawn from the population of equal means. *i.e.*, there is no significant difference between their means

i.e., $\mu_1 = \mu_2$

H_1: $\mu_1 \neq \mu_2$ (Two tailed test)

Given n_1 = Sample I size = 10; n_2 = Sample II size = 10

To calculate the two sample mean and sum of squares of deviation from mean. Let X_1 be the sample I and X_2 be the sample II.

X_1	25	30	28	34	24	20	13	32	22	38
$X_1 - \bar{X}_1$	– 1.6	3.4	1.4	7.4	– 2.6	– 6.6	– 13.6	5.4	4.6	11.4
$(X_1 - \bar{X}_1)^2$	2.56	11.56	1.96	54.76	6.76	43.56	184.96	29.16	21.16	129.96
X_2	40	34	22	20	31	40	30	23	36	17
$X_2 - \bar{X}_2$	10.7	4.7	– 7.3	– 9.3	1.7	10.7	0.7	– 6.3	6.7	– 12.3
$(X_2 - \bar{X}_2)^2$	114.49	22.09	53.29	86.49	2.89	114.49	0.49	39.67	44.89	151.29

$$\bar{X}_1 = \frac{\Sigma X_1}{n_1} = 26.6 \qquad \bar{X}_2 = \frac{\Sigma X_2}{n_2} = \frac{293}{10} = 29.3$$

$$\Sigma(X_1 - \bar{X}_1)^2 = 486.4 \qquad \Sigma(X_2 - \bar{X}_2)^2 = 630.08$$

$$s^2 = \frac{1}{n_1 + n_2 - 2}[\Sigma(X_1 - \bar{X}_1)^2 + \Sigma(X_2 - \bar{X}_2)^2]$$

$$= \frac{1}{10 + 10 - 2}[486.4 + 630.08] = 62.026 \qquad \therefore \quad s = 7.875$$

Under H_0 the test statistic is given by

$$t = \frac{\bar{X}_1 - \bar{X}_2}{s\sqrt{\frac{1}{n_1} + \frac{1}{n_2}}} = \frac{26.6 - 29.3}{7.875\sqrt{\frac{1}{10} + \frac{1}{10}}} = -0.7666 \sim t(n_1 + n_2 - 2 \text{ d.f.})$$

$|t| = 0.7666.$

Conclusion. The tabulated value of t at 5% level of significance for 18 d.f. is 2.1. Since the calculated value $|t| = 0.7666 < t_{0.05}$, H_0 is accepted.

i.e., there is no significant difference between their means.

i.e., the two samples have been drawn from the populations of equal means.

EXERCISE 6.7

1. The mean life of 10 electric motors was found to be 1450 hrs with S.D. of 423 hrs. A second sample of 17 motors chosen from a different batch showed a mean life of 1280 hrs with a S.D. of 398 hrs. Is there a significant difference between means of the two samples?
2. The marks obtained by a group of 9 regular course students and another group of 11 part time course students in a test are given below

Regular	56	62	63	54	60	51	67	69	58		
Part-time	62	70	71	62	60	56	75	64	72	68	66

Examine whether the marks obtained by regular students and part-time students differ significantly at 5% and 1% level of significance.

3. A group of 10 rats fed on diet A and another group of 8 rats fed on a different diet B recorded the following increase in weight (gm).

Diet A	5	6	8	1	12	4	3	9	6	10
Diet B	2	3	6	8	10	1	2	8		

Does it show the superiority of diet A over the diet B? *(M.D.U. 2011)*

4. Two independent samples of sizes 7 and 9 have the following values:

Sample A	10	12	10	13	14	11	10		
Sample B	10	13	15	12	10	14	11	12	11

Test whether the difference between the mean is significant.

5. To compare the prices of a certain product in two cities, 10 shops when related at random in each town. The price was noted below:

City 1	61	63	56	63	56	63	59	56	44	61
City 2	55	54	47	59	51	61	57	54	64	58

Test whether the average prices can be said to be the same in two cities.

6. The average number of articles produced by two machines per day are 200 and 250 with standard deviation 20 and 25 respectively on the basis of records of 25 days production. Can you regard both the machines equally efficient at 5% level of significance?

7. Two salesman represent a firm in a certain company. One of them claims that he makes large sales than the other. A sample survey was made and the following results were obtained:

No. of sales	1st Salesman (18)	2nd Salesman (20)
Average sales	₹ 210	₹ 175
S.D.	₹ 25	₹ 20

Find if the average sales differ significantly.

8. Two types of drugs A and B were used on 5 and 7 patients respectively for reducing their weights. The decrease in the weight after using drugs for six months are as follows:

Drug A	10	12	13	11	14		
Drug B	8	9	12	14	15	10	9

Is there a significant difference in the efficiency of the two drugs? If not, which drug you should buy?

9. Nine patients, to whom a certain drug was administered, registered the following rise in blood pressure:

$3, 7, 4 - 1, -3, 6, -4, 1, 5$

Test the hypothesis that the drug does not raise the blood pressure.

10. The change in sleeping hours of 7 patients after taking a medicine are as follows:

$0.7, 0.1, -0.3, 1.2, 1.0, 0.3, -0.4$ hrs.

Do these data give evidence that the medicine produces additional hours of sleep?

Answers

1. accepted **2.** rejected at 5% level, accepted at 1% **3.** accepted
4. accepted **5.** accepted **6.** rejected **7.** rejected
8. accepted, buy any **9.** accepted **10.** rejected.

6.18. CHI-SQUARE (χ^2) TEST

When a coin is tossed 200 times, the theoretical considerations lead us to expect 100 heads and 100 tails. But in practice, these results are rarely achieved. The quantity χ^2 (a Greek letter, pronounced as chi-square) describes the magnitude of discrepancy between theory and observation. If $\chi = 0$, the observed and expected frequencies completely coincide. The greater the discrepancy between the observed and expected frequencies, the greater is the value of χ^2. Thus χ^2 **affords a measure of the correspondence between theory and observation.**

If O_i $(i = 1, 2, ..., n)$ is a set of observed (experimental) frequencies and E_i $(i = 1, 2, ..., n)$ is the corresponding set of expected (theoretical or hypothetical) frequencies, then, χ^2 **is defined as**

$$\chi^2 = \sum_{i=1}^{n} \left[\frac{(O_i - E_i)^2}{E_i} \right]$$

where $\Sigma O_i = \Sigma E_i = N$ (total frequency) and degrees of freedom (d.f.) $= (n - 1)$.

Note. (*i*) If $\chi^2 = 0$, the observed and theoretical frequencies agree exactly.

(*ii*) If $\chi^2 > 0$ they do not agree exactly.

6.19. DEGREE OF FREEDOM

While comparing the calculated value of χ^2 with the table value, we have to determine the degrees of freedom.

If we have to choose any four numbers whose sum is 50, we can exercise our independent choice for any three numbers only, the fourth being 50 minus the total of the three numbers selected. Thus, though we were to choose any four numbers, our choice was reduced to three because of one condition imposed. There was only one restraint on our freedom and our degrees of freedom were $4 - 1 = 3$. If two restrictions are imposed, our freedom to choose will be further curtailed and degrees of freedom will be $4 - 2 = 2$.

In general, the number of degrees of freedom is the total number of observations less the number of independent constraints imposed on the observations. Degrees of freedom (d.f.) are usually denoted by ν (the letter 'nu' of the Greek alphabet).

Thus, $\nu = n - k$, where k is the number of independent constraints in a set of data of n observations.

Note 1. For a $p \times q$ contingency table (p columns and q rows).

$$\nu = (p - 1)(q - 1)$$

Note 2. In the case of a contingency table, the expected frequency of any class

$$= \frac{\text{Total of row in which it occurs} \times \text{Total of col. in which it occurs}}{\text{Total number of observations}}$$

Note 3. χ^2 test is one of the simplest and the most general test known. It is applicable to a very large number of problems in practice which can be summed up under the following heads:

(*i*) as a test of goodness of fit. (*ii*) as a test of independence of attributes.

(*iii*) as a test of homogeneity of independent estimates of the population variance.

(*iv*) as a test if the hypothetical value of the population variance σ^2.

(*v*) as a list to the homogeneity of independent estimates of the population correlation coefficient.

6.20. CONDITIONS FOR APPLYING χ^2 TEST

Following are the conditions which should be satisfied before χ^2 test can be applied.

(*a*) N, the total number of frequencies should be large. It is difficult to say what constitutes largeness, but as an arbitrary figure, we may say that **N should be at least 50,** however, few the cells.

(*b*) No theoretical cell-frequency should be small. Here again, it is difficult to say what constitutes smallness, but 5 should be regarded as the very minimum and **10 is better**. If small theoretical frequencies occur (*i.e.*, < 10), the difficulty is overcome by grouping two or more classes together before calculating (O – E). **It is important to remember that the number of degrees of freedom is determined with the number of classes after regrouping.**

(*c*) The constraints on the cell frequencies, if any, should be linear.

Note. If any one of the theoretical frequency is less than 5, then we apply a correction given by F Yates, which is usually known as 'Yates correction for continuity', we add 0.5 to the cell frequency which is less than 5 and adjust the remaining cell frequency suitably so that the marginal total is not changed.

6.21. THE χ^2 DISTRIBUTION

For large sample sizes, the sampling distribution of χ^2 can be closely approximated by a continuous curve known as the chi-square distribution. The probability function of χ^2 distribution is given by

$$f(\chi)^2 = c(\chi^2)^{(v/2-1)} \; e^{-x^2/2}$$

where $e = 2.71828$, v = number of degrees of freedom; c = a constant depending only on v.

Symbolically, the degrees of freedom are denoted by the symbol v or by d.f. and are obtained by the rule $v = n - k$, where k refers to the number of independent constraints.

In general, when we fit a binomial distribution the number of degrees of freedom is one less than the number of classes; when we fit a Poisson distribution the degrees of freedom are 2 less than the number of classes, because we use the total frequency and the arithmetic mean to get the parameter of the Poisson distribution. When we fit a normal curve the number of degrees of freedom are 3 less than the number of classes, because in this fitting we use the total frequency, mean and standard deviation.

If the data is given in a series of "n" number then degrees of freedom = $n - 1$.

In the case of Binomial distribution d.f. = $n - 1$

In the case of Poisson distribution d.f. = $n - 2$

In the case of Normal distribution d.f. = $n - 3$.

6.22. χ^2 TEST AS A TEST OF GOODNESS OF FIT

χ^2 test enables us to ascertain how well the theoretical distributions such as Binomial, Poisson or Normal etc. fit empirical distributions, *i.e.*, distributions obtained from sample data. If the **calculated value of χ^2 is less than the table value** at a specified level (generally 5%) of significance, the **fit is considered to be good** *i.e.*, the divergence between actual and expected frequencies is attributed to fluctuations of simple sampling. If the calculated value of χ^2 is greater than the table value, the fit is considered to be poor.

ILLUSTRATIVE EXAMPLES

Example 1. *The following table gives the number of accidents that took place in an industry during various days of the week. Test if accidents are uniformly distributed over the week.*

Day	*Mon*	*Tue*	*Wed*	*Thu*	*Fri*	*Sat*
No. of accidents	*14*	*18*	*12*	*11*	*15*	*14*

Sol. H_0: Null hypothesis. The accidents are uniformly distributed over the week.

Under this H_0, the expected frequencies of the accidents on each of these days $= \frac{84}{6} = 14$

Observed frequency O_i	14	18	12	11	15	14
Expected frequency E_i	14	14	14	14	14	14
$(O_i - E_i)^2$	0	16	4	9	1	0

$$\chi^2 = \frac{\Sigma(O_i - E_i)^2}{E_i} = \frac{30}{14} = 2.1428.$$

Conclusion. Table value of χ^2 at 5% level for (6 – 1 = 5 d.f.) is 11.07.

Since the calculated value of χ^2 is less than the tabulated value H_0 is accepted *i.e.*, the accidents are uniformly distributed over the week.

Example 2. *A die is thrown 276 times and the results of these throws are given below:*

No. appeared on the die	*1*	*2*	*3*	*4*	*5*	*6*
Frequency	*40*	*32*	*29*	*59*	*57*	*59*

Test whether the die is biased or not.

Sol. Null hypothesis H_0: Die is unbiased.

Under this H_0, the expected frequencies for each digit is $\frac{276}{6} = 46$

To find the value of χ^2

O_i	40	32	29	59	57	59
E_i	46	46	46	46	46	46
$(O_i - E_i)^2$	36	196	289	169	121	169

$$\chi^2 = \frac{\Sigma(O_i - E_i)^2}{E_i} = \frac{980}{46} = 21.30.$$

Conclusion. Tabulated value of χ^2 at 5% level of significance for $(6 - 1 = 5)$ d.f. is 11.07. Since the calculated value of $\chi^2 = 21.30 > 11.07$ the tabulated value, H_0 is rejected.

i.e., Die is not unbiased or Die is biased.

Example 3. *The following table shows the distribution of digits in numbers chosen at random from a telephone directory.*

Digits	*0*	*1*	*2*	*3*	*4*	*5*	*6*	*7*	*8*	*9*
Frequency	*1026*	*1107*	*997*	*966*	*1075*	*933*	*1107*	*972*	*964*	*853*

Test whether the digits may be taken to occur equally frequently in the directory.

Sol. Null hypothesis H_0: The digits taken in the directory occur equally frequency. *i.e.*, there is no significant difference between the observed and expected frequency.

Under H_0, the expected frequency is given by $= \frac{10,000}{10} = 1000$

To find the value of χ^2

O_i	1026	1107	997	996	1075	1107	933	972	964	853
E_i	1000	1000	1000	1000	1000	1000	1000	1000	1000	1000
$(O_i - E_i)^2$	676	11449	9	1156	5625	11449	4489	784	1296	21609

$$\chi^2 = \frac{\Sigma(O_i - E_i)^2}{E_i} = \frac{58542}{1000} = 58.542.$$

Conclusion. The tabulated value of χ^2 at 5% level of significance for 9 d.f. is 16.919. Since the calculated value of χ^2 is greater than the tabulated value, H_0 is rejected.
i.e., there is significant difference between the observed and theoretical frequency.
i.e., the digits taken in the directory do not occur equally frequently.

Example 4. *Records taken of the number of male and female births in 800 families having four children are as follows:*

No. of male births	*0*	*1*	*2*	*3*	*4*
No. of female births	*4*	*3*	*2*	*1*	*0*
No. of families	*32*	*178*	*290*	*236*	*94*

Test whether the data are consistent with the hypothesis that the binomial law holds and the chance of male birth is equal to that of female birth, namely $p = q = 1/2$.

Sol. H_0: The data are consistent with the hypothesis of equal probability for male and female births. *i.e.*, $p = q = 1/2$.

We use Binomial distribution to calculate theoretical frequency given by:
$N(r) = N \times P(X = r)$
where N is the total frequency. $N(r)$ is the number of families with r male children:
$P(X = r) = {}^nC_r p^r q^{n-r}$
where p and q are probability of male and female birth, n is the number of children.

$$N(0) = \text{No. of families with 0 male children} = 800 \times {}^4C_0 \left(\frac{1}{2}\right)^4 = 800 \times 1 \times \frac{1}{2^4} = 50$$

$$N(1) = 800 \times {}^4C_1 \left(\frac{1}{2}\right)^1 \left(\frac{1}{2}\right)^3 = 200; \; N(2) = 800 \times {}^4C_2 \left(\frac{1}{2}\right)^2 \left(\frac{1}{2}\right)^2 = 300$$

$$N(3) = 800 \times {}^4C_3 \left(\frac{1}{2}\right)^1 \left(\frac{1}{2}\right)^3 = 200; \; N(4) = 800 \times {}^4C_4 \left(\frac{1}{2}\right)^0 \left(\frac{1}{2}\right)^4 = 50$$

Observed frequency O_i	32	178	290	236	94
Exp-frequency E_i	50	200	300	200	50
$(O_i - E_i)^2$	324	484	100	1296	1936
$\frac{(O_i - E_i)^2}{E_i}$	6.48	2.42	0.333	6.48	38.72

$$\chi^2 = \frac{\Sigma(O_i - E_i)^2}{E_i} = 54.433.$$

Conclusion. Table value of χ^2 at 5% level of significance for 5 – 1 = 4 d.f. is 9.49.

Since the calculated value of χ^2 is greater than the tabulated value, H_0 is rejected.

i.e., the data are not consistent with the hypothesis that the Binomial law holds and that the chance of a male birth is not equal to that of a female birth.

Note. Since the fitting is binomial, the degrees of freedom $\nu = n - 1$ *i.e.*, $\nu = 5 - 1 = 4$.

Example 5. *Verify whether Poisson distribution can be assumed from the data given below:*

No. of defects	*0*	*1*	*2*	*3*	*4*	*5*
Frequency	*6*	*13*	*13*	*8*	*4*	*3*

Sol. H_0: Poisson fit is a good fit to the data.

$$\text{Mean of the given distribution} = \frac{\Sigma f_i x_i}{\Sigma f_i} = \frac{94}{47} = 2$$

To fit a Poisson distribution we require m. Parameter $m = \overline{x} = 2$.

By Poisson distribution the frequency of r success is

$$N(r) = N \times e^{-m} \cdot \frac{m^r}{r!}, \text{ N is the total frequency.}$$

$$N(0) = 47 \times e^{-2} . \frac{(2)^0}{0!} = 6.36 \approx 6; N(1) = 47 \times e^{-2} . \frac{(2)^1}{1!} = 12.72 \approx 13$$

$$N(2) = 47 \times e^{-2} . \frac{(2)^2}{2!} = 12.72 \approx 13; N(3) = 47 \times e^{-2} . \frac{(2)^3}{3!} = 8.48 \approx 9$$

$$N(4) = 47 \times e^{-2} . \frac{(2)^4}{4!} = 4.24 \approx 4; N(5) = 47 \times e^{-2} . \frac{(2)^5}{5!} = 1.696 \approx 2.$$

X	0	1	2	3	4	5
O_i	6	13	13	8	4	3
E_i	6.36	12.72	12.72	8.48	4.24	1.696
$\frac{(O_i - E_i)^2}{E_i}$	0.2037	0.00616	0.00616	0.02716	0.0135	1.0026

$$\chi^2 = \frac{\Sigma(O_i - E_i)^2}{E_i} = 1.2864.$$

Conclusion. The calculated value of χ^2 is 1.2864. Tabulated value of χ^2 at 5% level of significance for $\gamma = 6 - 2 = 4$ d.f. is 9.49. Since the calculated value of χ^2 is less than that of tabulated value. H_0 is accepted *i.e.*, Poisson distribution provides a good fit to the data.

Example 6. *The theory predicts the proportion of beans in the four groups, G_1, G_2, G_3, G_4 should be in the ratio 9 : 3 : 3 : 1. In an experiment with 1600 beans the numbers in the four groups were 882, 313, 287 and 118. Does the experimental result support the theory?*

Sol. H_0. The experimental result support the theory. *i.e.*, there is no significant difference between the observed and theoretical frequency under H_0, the theoretical frequency can be calculated as follows:

$$E(G_1) = \frac{1600 \times 9}{16} = 900; E(G_2) = \frac{1600 \times 3}{16} = 300;$$

$$E(G_3) = \frac{1600 \times 3}{16} = 300; E(G_4) = \frac{1600 \times 1}{16} = 100$$

To calculate the value of χ^2.

Observed frequency O_i	882	313	287	118
Exp. frequency E_i	900	300	300	100
$\frac{(O_i - E_i)^2}{E_i}$	0.36	0.5633	0.5633	3.24

$$\chi^2 = \frac{\Sigma(O_i - E_i)^2}{E_i} = 4.7266.$$

Conclusion. Table value of χ^2 at 5% level of significance for 3 d.f. is 7.815. Since the calculated value of χ^2 is less than that of the tabulated value. Hence H_0 is accepted.
i.e., the experimental result support the theory.

EXERCISE 6.8

1. The following table gives the frequency of occupance of the digits 0, 1, ..., 9 in the last place in four logarithm of numbers 10-99. Examine if there is any peculiarity.

Digits	0	1	2	3	4	5	6	7	8	9
Frequency	6	16	15	10	12	12	3	2	9	5

2. The sales in a supermarket during a week are given below. Test the hypothesis that the sales do not depend on the day of the week, using a significant level of 0.05.

Days	Mon	Tues	Wed	Thus	Fri	Sat
Sales (in 1000 ₹)	65	54	60	56	71	84

3. A survey of 320 families with 5 children each revealed the following information:

No. of boys	5	4	3	2	1	0
No. of girls	0	1	2	3	4	5
No. of families	14	56	110	88	40	12

Is this result consistent with the hypothesis that male and female birth are equally probable?

(M.D.U. May 2013)

4. (*a*) 4 coins were tossed at a time and this operation is repeated 160 times. It is found that 4 heads occur 6 times, 3 heads occur 43 times, 2 heads occur 69 times, one head occur 34 times. Discuss whether the coin may be regarded as unbiased?

(*b*) A set of 5 coins is tossed 3200 times and the number of heads appearing each time is noted. The results are given below:

No. of heads	0	1	2	3	4	5
Frequency	80	570	1100	900	500	50

Test the hypothesis that coins are unbiased. *(M.D.U. Dec. 2014)*

5. Fit a Poisson distribution to the following data and test the goodness of fit.

x	0	1	2	3	4
f	109	65	22	3	1

(M.D.U. Dec. 2015)

6. Fit a Poisson distribution to the following data and test the goodness of fit.

x	0	1	2	3	4
f	46	38	22	9	1

(M.D.U. 2010)

7. (*a*) Fit a binomial distribution to the following data and test the goodness of fit.

x	0	1	2	3	4	5	6
f	13	25	52	58	32	16	4

(M.D.U. May 2015)

(*b*) A set of five similar coins is tossed 320 times and the result is

No. of heads	0	1	2	3	4	5
Frequency	6	27	72	112	71	32

Test the hypothesis that the data follow a Binomial distribution.

(*M.D.U. 2010; DCRUST Murthal, May 2014*)

8. 200 digits are chosen at random from a set of tables. The frequencies of the digits are as follows:

Digit	0	1	2	3	4	5	6	7	8	9
Frequency	18	19	23	21	16	25	22	20	21	15

Use chi-square test to assess the correctness of the hypothesis that the digits were distributed in equal numbers in the tables from where they are chosen.

9. In the accounting department of bank, 100 accounts are selected at random and estimated for errors. The following results were obtained:

No. of errors	0	1	2	3	4	5	6
No. of accounts	35	40	19	2	0	2	2

Does this information verify that the errors are distributed according to the Poisson probability law?

10. A sample analysis of examination results of 500 students, it was found that 280 students have failed, 170 have secured a third class, 90 have secured a second class and the rest, a first class. Do these figures support the general belief that above categories are in the ratio 4 : 3 : 2 : 1 respectively?

Answers

1. no **2.** accepted **3.** accepted

4. (*a*) unbiased (*b*) rejected

5. Poisson law fits the data **6.** Poisson law fits the data

7. (*a*) Binomial law does not fit the data (*b*) Rejected **8.** accepted **9.** yes

10. yes.

6.23. χ^2-TEST AS A TEST OF INDEPENDENCE

With the help of χ^2 test, we can find whether or not two attributes are associated. We take the null hypothesis that there is no association between the attributes under study, *i.e.*, **we assume that the two attributes are independent. If the calculated value of χ^2 is less than the table value** at a specified level (generally 5%) of significance, the hypothesis holds good, *i.e.*, **the attributes are independent** and do not bear any association. On the other hand, if the calculated value of χ^2 is greater than the table value at a specified level of significance, we say that the results of the experiment do not support the hypothesis. In other words, the attributes are associated. Thus a very useful application of χ^2 test is to investigate the relationship between trials or attributes which can be classified into two or more categories.

The sample data set out into two-way table, called **contingency table.**

Let us consider two attributes A and B divided into r classes $A_1, A_2, A_3, \ldots\ldots, A_r$ and B divided into s classes $B_1, B_2, B_3, \ldots\ldots, B_s$. If (A_i), (B_j) represents the number of persons possessing the attributes A_i, B_j respectively, ($i = 1, 2, ..., r, j = 1, 2, ..., s$) and ($A_i B_j$) represent the number of persons possessing attributes A_i and B_j. Also we have $\sum_{i=1}^{r} A_i = \sum_{j=1}^{s} B_j = N$ where N is the total frequency. The contingency table for $r \times s$ is given below:

B \ A	A_1	A_2	A_3	...A_r	Total
B_1	(A_1B_1)	(A_2B_1)	(A_3B_1)	...... (A_rB_1)	B_1
B_2	(A_1B_2)	(A_2B_2)	(A_3B_2)	...... (A_rB_2)	B_2
B_3	(A_1B_3)	(A_2B_3)	(A_3B_3)	...... (A_rB_3)	B_3
......	......	......	......	......	......
......	......	......	......	......	......
B_s	(A_1B_s)	(A_2B_s)	(A_3B_s)	...... (A_rB_s)	(B_s)
Total	(A_1)	(A_2)	(A_3)	...... (A_r)	N

H_0: Both the attributes are independent *i.e.,* A and B are independent under the null hypothesis, we calculate the expected frequency as follows:

$P(A_i)$ = Probability that a person possesses the attribute $A_i = \dfrac{(A_i)}{N}$, $i = 1, 2, ..., r$

$P(B_j)$ = Probability that a person possesses the attribute $B_j = \dfrac{(B_j)}{N}$

$P(A_iB_j)$ = Probability that a person possesses both attributes A_i and $B_j = \dfrac{(A_iB_j)}{N}$

If $(A_iB_j)_0$ is the expected number of persons possessing both the attributes A_i and B_j

$$(A_iB_j)_0 = NP(A_iB_j) = NP(A_i)(B_j)$$

$$= N\frac{(A_i)}{N}\frac{(B_j)}{N} = \frac{(A_iB_j)}{N} \qquad (\because \text{ A and B are independent})$$

Hence
$$\chi^2 = \sum_{i=1}^{r}\sum_{j=1}^{s}\left[\frac{[(A_iB_j) - (A_iB_j)_0]^2}{(A_iB_j)_0}\right]$$

which is distributed as a χ^2 variate with $(r - 1)(s - 1)$ degrees of freedom.

Note 1. For a 2×2 contingency table where the frequencies are $\dfrac{a}{b}\Big|\dfrac{b}{d}$ χ^2 can be calculated from independent frequencies as $\chi^2 = \dfrac{(a+b+c+d)(ad-bc)^2}{(a+b)(c+d)(b+d)(a+c)}$.

Note 2. If the contingency table is not 2×2, then the formula for calculating χ^2 as given in note 1, can't be used. Hence we have another formula for calculating the expected frequency $(A_iB_j)_0 = \dfrac{(A_i)(B_j)}{N}$

i.e., expected frequency in each cell is $= \dfrac{\textbf{Product of column total and row total}}{\textbf{whole total}}$.

Note 3. If $\frac{a}{c}\left|\frac{b}{d}\right.$ is the 2×2 contingency table with two attributes, $Q = \frac{ad - bc}{ad + bc}$ is called the coefficient of association.

If the attributes are independent then $\frac{a}{b} = \frac{c}{d}$.

Note 4. Yates's Correction. In a 2×2 table, if the frequencies of a cell is small, we make Yate's correction to make χ^2 continuous.

Decrease by $\frac{1}{2}$ those cell frequencies which are greater than expected frequencies, and increase by $\frac{1}{2}$ those which are less than expectation. This will not affect the marginal columns. This correction is known as Yate's correction to continuity.

After Yate's correction
$$\chi^2 = \frac{N\left(bc - ad - \frac{1}{2}N\right)^2}{(a+c)(b+d)(c+d)(a+b)} \text{ when } ad - bc < 0$$

$$\chi^2 = \frac{N\left(ad - bc - \frac{1}{2}N\right)^2}{(a+c)(b+d)(c+d)(a+b)} \text{ when } ad - bc > 0.$$

ILLUSTRATIVE EXAMPLES

Example 1. *What are the expected frequencies of 2 × 2 contingency tables given below:*

(*i*)

a	*b*
c	*d*

(*ii*)

2	10
6	6

Sol.

(*i*) Observed frequencies

a	*b*	*a* + *b*
c	*d*	*c* + *d*
a + *c*	*b* + *d*	*a* + *b* + *c* + *d* = N

→ Expected frequencies

$\frac{(a+c)(a+b)}{a+b+c+d}$	$\frac{(b+d)(a+b)}{a+b+c+d}$
$\frac{(a+c)(c+d)}{a+b+c+d}$	$\frac{(b+d)(c+d)}{a+b+c+d}$

(*ii*) Observed frequencies

2	10	12
6	6	12
8	16	24

→ Expected frequencies

$\frac{8 \times 12}{24} = 4$	$\frac{16 \times 12}{24} = 8$
$\frac{8 \times 12}{24} = 4$	$\frac{16 \times 12}{24} = 8$

Example 2. *From the following table regarding the colour of eyes of fathers and sons test if the colour of son's eye is associated with that of the father.*

		Eye colour of son	
		Light	Not light
Eye colour of father	Light	471	51
	Not light	148	230

Sol. Null hypothesis H_0: The colour of son's eye is not associated with that of the father. *i.e.,* they are independent.

Under H_0, we calculate the expected frequency in each cell as

$$= \frac{\text{Product of column total and row total}}{\text{whole total}}$$

Expected frequencies are

Eye colour of father \ Eye colour of son	Light	Not light	Total
Light	$\frac{619 \times 522}{900} = 359.02$	$\frac{289 \times 522}{900} = 167.62$	522
Not light	$\frac{619 \times 378}{900} = 259.98$	$\frac{289 \times 378}{900} = 121.38$	378
Total	619	289	900

$$\chi^2 = \frac{(471-359.02)^2}{359.02} + \frac{(51-167.62)^2}{167.62} + \frac{(148-259.98)^2}{259.98} + \frac{(230-121.38)^2}{121.38} = 261.498.$$

Conclusion. Tabulated value of χ^2 at 5% level for 1 d.f. is 3.841.

Since the calculated value of $\chi^2 >$ tabulated value of χ^2, H_0 is rejected. They are dependent *i.e.,* the colour of son's eye is associated with that of the father.

Example 3. *The following table gives the number of good and bad parts produced by each of the three shifts in a factors*

	Goods parts	Bad parts	Total
Day shift	960	40	1000
Evening shift	940	50	990
Night shift	950	45	995
Total	2850	135	2985

Test whether or not the production of bad parts is independent of the shift on which they were produced.

Sol. Null hypothesis H_0. The production of bad parts is independent of the shift on which they were produced.

i.e., the two attributes, production and shifts are independent.

Under H_0, $$\chi^2 = \sum_{i=1}^{2} \sum_{j=1}^{3} \left[\frac{[(A_iB_j)_0 - (A_iB_j)]^2}{(A_i\, B_j)_0} \right].$$

Calculation of expected frequencies

Let A and B be the two attributes namely production and shifts. A is divided into two classes A_1, A_2. B is divided into three classes B_1, B_2, B_3.

$$(A_1B_1)_0 = \frac{(A_1)(B_1)}{N} = \frac{(2850)\times(1000)}{2985} = 954.77;$$

$$(A_1B_2)_0 = \frac{(A_1)(B_2)}{N} = \frac{(2850)\times(990)}{2985} = 945.226$$

$$(A_1B_3)_0 = \frac{(A_1)(B_3)}{N} = \frac{(2850)\times(995)}{2985} = 950;$$

$$(A_2B_1)_0 = \frac{(A_2)(B_1)}{N} = \frac{(135)\times(1000)}{2985} = 45.27$$

$$(A_2B_2)_0 = \frac{(A_2)(B_2)}{N} = \frac{(135)\times(990)}{2985} = 44.773;$$

$$(A_2B_3)_0 = \frac{(A_2)(B_3)}{N} = \frac{(135)\times(995)}{2985} = 45$$

To calculate the value of χ^2

Class	O_i	E_i	$(O_i - E_i)^2$	$(O_i - E_i)^2/E_i$
(A_1B_1)	960	954.77	27.3529	0.02864
(A_1B_2)	940	945.226	27.3110	0.02889
(A_1B_3)	950	950	0	0
(A_2B_1)	40	45.27	27.7729	0.61349
(A_2B_2)	50	44.773	27.3215	0.61022
(A_2B_3)	45	45	0	0
				1.28126

Conclusion. The tabulated value of χ^2 at 5% level of significance for 2 degrees of freedom $(r-1)(s-1)$ is 5.991. Since the calculated value of χ^2 is less than the tabulated value, we accept H_0. *i.e.,* the production of bad parts is independent of the shift on which they were produced.

Example 4. *From the following data, find whether hair colour and sex are associated.*

Sex \ *Colour*	*Fair*	*Red*	*Medium*	*Dark*	*Black*	*Total*
Boys	*592*	*849*	*504*	*119*	*36*	*2100*
Girls	*544*	*677*	*451*	*97*	*14*	*1783*
Total	*1136*	*1526*	*955*	*216*	*50*	*3883*

Sol. Null hypothesis. H_0: The two attributes hair colour and sex are not associated. *i.e.*, They are independent.

Let A and B be the attributes hair colour and sex respectively. A is divides into 5 classes ($r = 5$). B is divided into 2 classes ($s = 2$).

$\therefore$ Degrees of freedom $= (r - 1)(s - 1) = (5 - 1)(2 - 1) = 4$

Under H_0, we calculate $\chi^2 = \sum_{i=1}^{5} \sum_{j=1}^{2} \frac{[(A_iB_j)_0 - (A_iB_j)]^2}{(A_iB_j)_0}$

To calculate the expected frequency $(A_i B_j)_0$ as follows:

$$(A_1B_1)_0 = \frac{(A_1)(B_1)}{N} = \frac{1136 \times 2100}{3883} = 614.37;$$

$$(A_1B_2)_0 = \frac{(A_1)(B_2)}{N} = \frac{1136 \times 1783}{3883} = 521.629$$

$$(A_2B_1)_0 = \frac{(A_2)(B_1)}{N} = \frac{1526 \times 2100}{3883} = 852.289;$$

$$(A_2B_2)_0 = \frac{(A_2)(B_2)}{N} = \frac{1526 \times 1783}{3883} = 700.71$$

$$(A_3B_1)_0 = \frac{(A_3)(B_1)}{N} = \frac{955 \times 2100}{3883} = 516.482;$$

$$(A_3B_2)_0 = \frac{(A_3)(B_2)}{N} = \frac{955 \times 1783}{3883} = 483.517$$

$$(A_4B_1)_0 = \frac{(A_4)(B_1)}{N} = \frac{216 \times 2100}{3883} = 116.816;$$

$$(A_4B_2)_0 = \frac{(A_4)(B_2)}{N} = \frac{216 \times 1783}{3883} = 99.183$$

$$(A_5B_1)_0 = \frac{(A_5)(B_1)}{N} = \frac{50 \times 2100}{3883} = 27.04;$$

$$(A_5B_2)_0 = \frac{(A_5)(B_2)}{N} = \frac{50 \times 1783}{3883} = 22.959$$

Calculation of χ^2

Class	O_i	E_i	$(O_i - E_i)^2$	$\frac{(O_i - E_i)^2}{E_i}$
A_1B_1	592	614.37	500.416	0.8145
A_1B_2	544	521.629	500.462	0.959
A_2B_1	849	852.289	10.8175	0.0127
A_2B_2	677	700.71	562.1641	0.8023
A_3B_1	504	516.482	155.800	0.3016
A_3B_2	451	438.517	155.825	0.3553
A_4B_1	119	116.816	4.7698	0.0408
A_4B_2	97	99.183	4.7654	0.0480
A_5B_1	36	27.04	80.2816	2.9689
A_5B_2	14	22.959	80.2636	3.495
				9.79975

$$\chi^2 = 9.799.$$

Conclusion. Table of χ^2 at 5% level of significance for 4 d.f. is 9.488.

Since the calculated value of χ^2 < tabulated value H_0 is rejected. *i.e.*, the two attributes are not independent. *i.e.*, the hair colour and sex are associated.

Example 5. *Can vaccination be regarded as preventive measure of small pox as evidenced by the following data of 1482 persons exposed to small pox in a locality. 368 in all were attacked of these 1482 persons and 343 were vaccinated and of these only 35 were attacked.*

Sol. For the given data we form the contingency table. Let the two attributes be vaccination and exposed to small pox. Each attributes is divided into two classes.

Vaccination A / *Disease small pox B*	*Vaccinated*	*Not*	*Total*
Attacked	35	333	368
Not	308	806	1114
Total	343	1139	1482

Null hypothesis H_0. The two attributes are independent *i.e.*, vaccination can't be regarded as preventive measure of small pox.

Degrees of freedom $v = (r - 1)(s - 1) = (2 - 1)(2 - 1) = 1$

Under H_0,
$$\chi^2 = \sum_{i=1}^{2} \sum_{j=1}^{2} \frac{[(A_iB_j)_0 - (A_iB_j)]^2}{(A_iB_j)_0}$$

Calculation of expected frequency

$$(A_1B_1)_0 = \frac{(A_1)(B_1)}{N} = \frac{343 \times 368}{1482} = 85.1713;$$

$$(A_1B_2)_0 = \frac{(A_1)(B_2)}{N} = \frac{343 \times 1114}{1482} = 257.828$$

$$(A_2B_1)_0 = \frac{(A_2)(B_1)}{N} = \frac{1139 \times 368}{1482} = 282.828;$$

$$(A_2B_2)_0 = \frac{(A_2)(B_2)}{N} = \frac{1139 \times 1114}{1482} = 856.171$$

Calculation of χ^2

Class	O_i	E_i	$(O_i - E_i)^2$	$\frac{(O_i - E_i)^2}{E_i}$
(A_1B_1)	35	85.1713	2517.159	29.554
(A_1B_2)	308	257.828	2517.229	8.1728
(A_2B_1)	333	282.828	2517.2295	7.5592
(A_2B_2)	806	856.171	2517.1292	2.9399
				48.2261

Calculated value of $\chi^2 = 48.2261$.

Conclusion. Tabulated value of χ^2 at 5% level of significance for 1 d.f. is 3.841.

Since the calculated value of χ^2 > tabulated value H_0 is rejected.

i.e., the two attributes are not independent. *i.e.,* the vaccination can be regarded as preventive measure of small pox.

EXERCISE 6.9

1. In a locality 100 persons were randomly selected and asked about their educational achievements. The results are given below:

		Education		
		Middle	*High school*	*College*
Sex	Male	10	15	25
	Female	25	10	15

Based on this information can you say the education depends on sex.

2. The following data is collected on two characters:

	Smokers	*Non-smokers*
Literate	83	57
Illiterate	45	68

Based on this information can you say that there is no relation between habit of smoking and literacy.

3. 500 students at school were graded according to their intelligences and economic conditions of their homes. Examine whether there is any association between economic condition and intelligence, from the following data:

Economic conditions	*Intelligence*	
	Good	*Bad*
Rich	85	75
Poor	165	175

4. In an experiment on the immunisation of goats from anthrex, the following results were obtained. Derive your inferences on the efficiency of the vaccine.

	Died	*Survived*
Inoculated with vaccine	2	10
Not inoculated	6	6

Answers

1. Yes **2.** No **3.** No **4.** Not effective.

❑❑❑

CHAPTER 7

Linear Programming

7.1. INTRODUCTION

Linear programming (briefly written as LP) came into existence during World War II (1939-45) when the British and American military management called upon a group of scientists to study and plan the war activities so that maximum damages could be inflicted on the enemy camps at minimum cost and loss with the limited resources available with them. Because of the success in military operations, the subject was found extremely useful for allocation of scarce resources for optimal results. Business and industry, agriculture and military sectors have, however, made the most significant use of the technique. But now it is being extensively used in all functional areas of management, airlines, oil refining, education, pollution control, transportation planning, health care system etc. The utility of the technique is enhanced by the availability of highly efficient computer codes. A lot of research work is being carried out all over the world. Kantorovich and Koopmans were awarded the noble prize in the year 1975 in economics for their pioneering work in linear programming. In India, it came into existence in 1949, with the opening of an operations research unit at the Regional Research Laboratory at Hyderabad.

7.2. DEFINITIONS AND BASIC CONCEPTS

The word '*programming*' means planning and refers to a process of determining a particular program or strategy or course of action among various alternatives to achieve the desired objective. The word **'linear'** means that all relationships involved in a particular programme are linear.

Linear Programming is the technique of optimizing (*i.e.*, maximizing or minimizing) a linear function of several variables subject to a number of constraints stated in the form of linear inequations/equations.

Linear Programming Problem. Any problem in which we apply linear programming is called a linear programming problem (briefly written as LPP).

The mathematical model of a general linear programming problem with n variables and m constraints can be stated as

Optimize (maximize or minimize)

$$Z = c_1x_1 + c_2x_2 + \cdots\cdots + c_nx_n$$

subject to

$$\left.\begin{array}{l} a_{11}x_1 + a_{12}x_2 + \dots\dots + a_{1n}x_n \ (\le, =, \ge)\ b_1 \\ a_{21}x_1 + a_{22}x_2 + \dots\dots + a_{2n}x_n \ (\le, =, \ge)\ b_2 \\ \dots\dots\dots\dots\dots\dots\dots\dots\dots \\ \dots\dots\dots\dots\dots\dots\dots\dots\dots \\ a_{m1}x_1 + a_{m2}x_2 + \dots\dots + a_{mn}x_n \ (\le, =, \ge)\ b_m \end{array}\right\} \quad \dots(1)$$

and $\quad x_1, x_2, \dots\dots, x_n \ge 0 \quad \dots(2)$

where

(*i*) the linear function Z which is to be maximized or minimized is called the **objective function** of the LPP.

(*ii*) $x_1, x_2, \dots\dots, x_n$ are called **decision** or **structural variables.**

(*iii*) The equations/inequations (1) are called the **constraints** of LPP. Due to limited resources, there is always a stage beyond which we cannot pursue our objective. Any linear equation or inequation in one or more decision variables is called a linear constraint.

(*iv*) The expression (≤, =, ≥) means that each constraint may take any one of the three signs.

(*v*) The set of inequations (2) is known as the set of **non-negative restrictions** of the general LPP.

(*vi*) The constant c_j ($j = 1, 2, \dots\dots, n$) represents the contribution to the objective function of the jth variable.

(*vii*) b_i ($i = 1, 2, \dots\dots, m$) is the constant representing the requirement or availability of the ith constraint.

(*viii*) a_{ij} = ($i = 1, 2, \dots\dots, m; j = 1, 2, \dots\dots, n$) is referred to as the technological co-efficient.

Now, we give below some definitions and concepts which pertain to general LPP.

Solution. A set of values of the decision variables which satisfy all the constraints of an LPP is called a **solution** to that problem.

Feasible Solution. Any solution of an LPP which also satisfies the non-negativity restrictions of the problem is called a **feasible solution** to that problem. (*K.U.K. Jan. 2013*)

Feasible Region. The set of all feasible solutions of an LPP is called its **feasible region.**

Optimal (or Optimum) Solution. Any feasible solution which optimizes the objective function of an LPP is called **an optimal (or optimum) solution** of the LPP. We have used the word 'an' with optimal solution. The reason being that an LPP may have many optimal solutions. (*K.U.K. Jan. 2013*)

The value of the objective function Z at an optimal solution is called an **optimal value.**

Optimization Technique. The process of obtaining an optimal value is called **optimization technique.**

Convex Region. A region is said to be convex if the line segment joining any two arbitrary points of the region lies entirely within the region. The feasible region of an LPP is always a polyhedral convex region, *i.e.*, a convex region whose boundary consists of line segments.

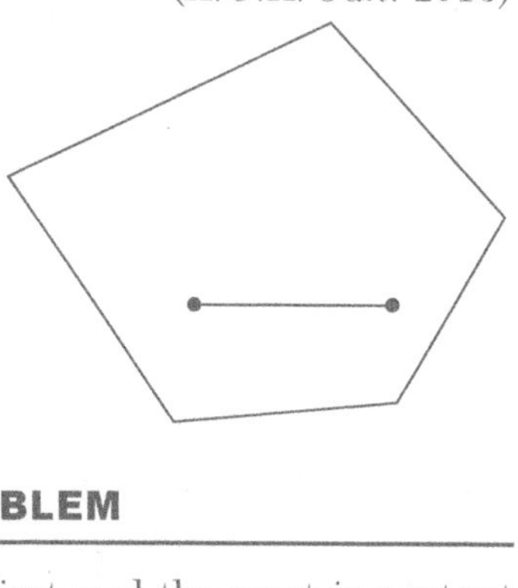

7.3. FORMULATION OF A LINEAR PROGRAMMING PROBLEM

Formulation of an LPP as a mathematical model is the first and the most important step in the solution of the LPP. It is an art in itself and needs sufficient practice. However, in general, the following steps are involved:

1. Identify the decision variables and denote them by $x_1, x_2, x_3, \ldots\ldots$.
2. Identify the objective function and express it as a linear function of the decision variables. In its general form, it is represented as:
   Optimize (Maximize or Minimize) $Z = c_1x_1 + c_2x_2 + \cdots\cdots + c_nx_n$
   (The optimal value of Z is obtained by the graphical method or simplex method. The graphical method is more suitable when there are two variables.)
3. Identify the set of constraints and express them as linear equations/inequations in terms of decision variables.
4. Add the non-negativity restrictions on the decision variables, as in the physical problems, negative values of decision variables have no valid interpretation.
   Thus $x_1 \geq 0,\quad x_2 \geq 0, \ldots\ldots, x_n \geq 0.$

ILLUSTRATIVE EXAMPLES

Example 1. *A dealer wishes to purchase a number of fans and sewing machines. He has only ₹ 9750 to invest and has space for at most 30 items. A fan costs him ₹ 480 and a sewing machine ₹ 360. His expectation is that he can sell a fan at a profit of ₹ 35 and a sewing machine at a profit of ₹ 24. Assume that he can sell all the items that he buys. Formulate this problem as an LPP, so that he can maximize his profit.*

Sol. Let x and y denote the number of fans and sewing machines respectively. (x and y are the decision variables)

Cost of x fans = ₹ $480x$

Cost of y sewing machines = ₹ $360y$

⇒ The total cost of x fans and y sewing machines = ₹ $(480x + 360y)$

Since the dealer has only ₹ 9750 to invest, the total cost cannot exceed ₹ 9750.

$$\therefore \quad 480x + 360y \leq 9750$$

Dividing by 30, we have $\mathbf{16x + 12y \leq 325}$

Since the dealer has space for at most 30 items

$$\therefore \quad \mathbf{x + y \leq 30}$$

Since the number of fans and the number of sewing machines cannot be negative

$$\therefore \quad \mathbf{x \geq 0, \quad y \geq 0} \qquad \text{(non-negativity constraints)}$$

Profit on x fans = ₹ $35x$

Profit on y sewing machines = ₹ $24y$

⇒ The total profit on x fans and y sewing machines

$$= ₹\ (35x + 24y)$$

The dealer wishes to maximize his profit

$$Z = 35x + 24y \quad \text{(objective function)}$$

∴ The mathematical formulation of the LPP is

Maximize $\quad Z = 35x + 24y$

subject to the constraints

$$16x + 12y \leq 325$$
$$x + y \leq 30$$
$$x \geq 0, \quad y \geq 0.$$

Example 2. *A manufacturer of patent medicines is preparing a production plan on medicines A and B. There are sufficient raw materials available to make 20000 bottles of A and 40000 bottles of B, but there are only 45000 bottles into which either of the medicines can be put. Further, it takes 3 hours to prepare enough material to fill 1000 bottles of A, it takes 1 hour to prepare enough material to fill 1000 bottles of B and there are 66 hours available for this operation. The profit is ₹ 8 per bottle for A and ₹ 7 per bottle for B. The manufacturer wants to schedule his production in order to maximize his profit. Formulate this problem as an LP model.*

Sol. Let x and y denote the number of bottles of type A and type B medicines respectively.

Total profit (in ₹) is $Z = 8x + 7y$

Raw material constraints are

$$x \leq 20000, \quad y \leq 40000$$

Since only 45000 bottles are available

$$\therefore \qquad x + y \leq 45000$$

It takes 3 hours to prepare enough material to fill 1000 bottles of type A.

$\therefore$ The number of hours required to prepare enough material to fill x bottles of type A $= \dfrac{3x}{1000}$.

Similarly, the number of hours required to prepare enough material to fill y bottles of type B $= \dfrac{y}{1000}$.

Since total number of hours available for this operation is 66

$$\therefore \qquad \frac{3x}{1000} + \frac{y}{1000} \leq 66 \quad \text{or} \quad 3x + y \leq 66000$$

Obviously $\qquad x \geq 0, \quad y \geq 0$

The LP model of the problem is

Maximize $\qquad Z = 8x + 7y$

subject to the constraints

$$x \leq 20000$$
$$y \leq 40000$$
$$x + y \leq 45000$$
$$3x + y \leq 66000$$
$$x \geq 0, \quad y \geq 0.$$

Example 3. *An electronic company produces three types of parts for automatic washing machines. It purchases casting of the parts from a local foundry and then finishes the part on drilling, shaping and polishing machines.*

The selling prices of parts A, B and C respectively are ₹ 8, ₹ 10 and ₹ 14. All parts made can be sold. Castings for parts A, B and C respectively cost ₹ 5, ₹ 6 and ₹ 10. The company possesses only one of each type of machine. Costs per hour to run each of the three machines are ₹ 20 for drilling, ₹ 30 for shaping and ₹ 30 for polishing. The capacities (parts per hour) for each part on each machine are shown in the following table:

Machine	*Capacity per hour*		
	Part A	*Part B*	*Part C*
Drilling	*25*	*40*	*25*
Shaping	*25*	*20*	*20*
Polishing	*40*	*30*	*40*

The management of the shop wants to know how many parts of each type it should produce per hour in order to maximize profit for an hour's run. Formulate this problem as an LP model so as to maximize total profit to the company.

Sol. Let x_1, x_2 and x_3 denote respectively the numbers of type A, B and C parts to be produced per hour.

Consider one type A part

Selling price $= ₹\ 8$

Cost of casting $= ₹\ 5$

Since 25 type A parts per hour can be run on the drilling machine at a cost of ₹ 20

$\therefore$ Drilling cost per type A part $= ₹\ \dfrac{20}{25} = ₹\ 0.80$

Similarly shaping cost $= ₹\ \dfrac{30}{25} = ₹\ 1.20$

Polishing cost $= ₹\ \dfrac{30}{40} = ₹\ 0.75$

$\therefore$ **Profit per type A part** $= ₹\ [8 - (5 + 0.80 + 1.20 + 0.75)] = ₹\ 0.25$

By similar reasoning

Profit per type B part $= ₹\left[10 - \left(6 + \dfrac{20}{40} + \dfrac{30}{20} + \dfrac{30}{30}\right)\right] = ₹\ 1.00$

Profit per type C part $= ₹\left[14 - \left(10 + \dfrac{20}{25} + \dfrac{30}{20} + \dfrac{30}{40}\right)\right] = ₹\ 0.95$

$\therefore$ Total profit is given by $Z = 0.25x_1 + 1.00x_2 + 0.95x_3$.

On the drilling machine, one type A part consumes $\dfrac{1}{25}$ th of the available hour, one type B part consumes $\dfrac{1}{40}$ th and one type C part consumes $\dfrac{1}{25}$ th of the available hour.

$\therefore$ **The drilling machine constraint** is

$$\frac{x_1}{25} + \frac{x_2}{40} + \frac{x_3}{25} \le 1$$

Similarly, the shaping machine constraint is

$$\frac{x_1}{25} + \frac{x_2}{20} + \frac{x_3}{20} \le 1$$

and the polishing machine constraint is

$$\frac{x_1}{40} + \frac{x_2}{30} + \frac{x_3}{40} \le 1$$

Also the non-negativity constraint is

$$x_1, x_2, x_3 \ge 0$$

∴ The LP model of the given problem is:

Maximize $Z = 0.25x_1 + 1.00x_2 + 0.95x_3$

subject to the constraints

$$\frac{x_1}{25} + \frac{x_2}{40} + \frac{x_3}{25} \le 1$$

$$\frac{x_1}{25} + \frac{x_2}{20} + \frac{x_3}{20} \le 1$$

$$\frac{x_1}{40} + \frac{x_2}{30} + \frac{x_3}{40} \le 1$$

$$x_1, x_2, x_3 \ge 0.$$

Example 4. *Consider the following problem faced by a production planner in a soft drink plant. He has two bottling machines A and B. A is designed for 8-ounce bottles and B for 16-ounce bottles. However, each can also be used for both types of bottles with some loss of efficiency. The manufacturing data is as follows:*

Machine	*8-ounce Bottles*	*16-ounce Bottles*
A	*100/minute*	*40/minute*
B	*60/minute*	*75/minute*

The machines can be run for 8 hours per day, 5 days per week. Profit on an 8-ounce bottle is 15 paise and on a 16-ounce bottle 25 paise. Weekly production of the drink cannot exceed 3,00,000 bottles and the market can absorb 25,000, 8-ounce bottles and 7,000, 16-ounce bottles per week. The planner wishes to maximize his profit, subject of course, to all the production and marketing restrictions. Formulate this problem as an LP model to maximize total profit.

Sol. Let x_1 and x_2 denote the number of 8-ounce and 16-ounce bottles respectively to be produced weekly.

Total profit (in ₹) is given by $Z = 0.15x_1 + 0.25x_2$

Machine time constraints

Total available time in a week on **machine A** = 8 × 5 × 60 minutes = 2400 minutes

Time for 100, 8-ounce bottles is 1 minute

∴ Time for x_1 8-ounce bottles is $\frac{x_1}{100}$ minutes

Similarly, time for x_2 16-ounce bottles is $\frac{x_2}{40}$

⇒ Machine time constraint on machine A is

$$\frac{x_1}{100} + \frac{x_2}{40} \le 2400$$

Similarly, machine time constraint on machine B is

$$\frac{x_1}{60} + \frac{x_2}{75} \le 2400$$

Production constraint. Since weekly production cannot exceed 3,00,000

$\therefore \qquad x_1 + x_2 \le 3,00,000$

Marketing constraints. The market can absorb 25,000, 8-ounce bottles and 7,000, 16-ounce bottles

$\therefore \qquad x_1 \le 25,000 \quad \text{and} \quad x_2 \le 7,000.$

Non-negativity constraints. The number of bottles cannot be negative

$\therefore \qquad x_1 \ge 0, \quad x_2 \ge 0$

Hence the LP model of the given problem is:

Maximize $\qquad Z = 0.15x_1 + 0.25x_2$

subject to the constraints

$$\frac{x_1}{100} + \frac{x_2}{40} \le 2400$$

$$\frac{x_1}{60} + \frac{x_2}{75} \le 2400$$

$$x_1 + x_2 \le 3,00,000$$

$$x_1 \le 25000, \quad x_2 \le 7000$$

$$x_1 \ge 0, \quad x_2 \ge 0.$$

Example 5. *A firm making castings uses electric furnace to melt iron with the following specifications:*

	Minimum	*Maximum*
Carbon	*3.20%*	*3.40%*
Silicon	*2.25%*	*2.35%*

Specifications and costs of various raw materials used for this purpose are given below:

Material	*Carbon %*	*Silicon %*	*Cost (₹)*
Steel scrap	*0.4*	*0.15*	*850/tonne*
Cast iron scrap	*3.80*	*2.40*	*900/tonne*
Remelt from foundry	*3.50*	*2.30*	*500/tonne*

If the total charge of iron metal required is 4 tonnes, find the weight in kg of each raw material that must be used in the optimal mix at minimum cost. Formulate this problem as an LP model.

Sol. Let x_1, x_2, x_3 be the weights (in kg) of the raw materials:

Steel scrap, cast iron scrap and remelt from foundry respectively.

Cost of 1 tonne *i.e.,* 1000 kg steel scrap is ₹ 850

$\Rightarrow$ cost of x_1 kg steel scrap $= ₹ \dfrac{850}{1000} x_1 = ₹\ 0.85x_1$

Similarly, cost of x_2 kg of cast iron scrap $= ₹ \dfrac{900}{1000} x_2 = ₹\ 0.9x_2$

and cost of x_3 kg of remelt from foundry $= ₹ \dfrac{500}{1000} x_3 = ₹\ 0.5x_3$

∴ Total cost of raw material is given by

$$Z = 0.85x_1 + 0.9x_2 + 0.5x_3$$

and the objective is to minimize it.

Total iron metal required is 4 tonnes *i.e.*, 4000 kg

$\Rightarrow \quad x_1 + x_2 + x_3 = 4000$

The iron melt is to have a minimum of 3.2% carbon

$\Rightarrow \quad 0.4x_1 + 3.8x_2 + 3.5x_3 \geq 3.2 \times 4000 \quad i.e., \quad 12800$

The iron melt is to have a maximum of 3.4% carbon

$\Rightarrow \quad 0.4x_1 + 3.8x_2 + 3.5x_3 \leq 3.4 \times 4000 \quad i.e., \quad 13600$

The iron melt is to have a minimum of 2.25% silicon

$\Rightarrow \quad 0.15x_1 + 2.4x_2 + 2.3x_3 \geq 2.25 \times 4000 \quad i.e., \quad 9000$

The iron melt is to have a maximum of 2.35% silicon

$\Rightarrow \quad 0.15x_1 + 2.4x_2 + 2.3x_3 \leq 2.35 \times 4000 \quad i.e., \quad 9400$

Since the amounts of raw material cannot be negative

$\therefore \quad x_1 \geq 0, \quad x_2 \geq 0, \quad x_3 \geq 0$

Hence the LP model of the given problem is:

Minimize $\quad Z = 0.85x_1 + 0.9x_2 + 0.5x_3$

subject to the constraints

$$x_1 + x_2 + x_3 = 4000$$
$$0.4x_1 + 3.8x_2 + 3.5x_3 \geq 12800$$
$$0.4x_1 + 3.8x_2 + 3.5x_3 \leq 13600$$
$$0.15x_1 + 2.4x_2 + 2.3x_3 \geq 9000$$
$$0.15x_1 + 2.4x_2 + 2.3x_3 \leq 9400$$
$$x_1, x_2, x_3 \geq 0.$$

Example 6. *There is a factory located at each of the two places P and Q. From these locations, a certain commodity is delivered to each of the three depots situated at A, B and C. The weekly requirements of the depots are, respectively, 5, 5 and 4 units of the commodity while the production capacity of the factories at P and Q are, respectively, 8 and 6 units. The cost of transportation per unit is given below:*

From \ *To*	*Cost (₹/unit)*		
	A	*B*	*C*
P	*16*	*10*	*15*
Q	*10*	*12*	*10*

How many units should be transported from each factory to each depot in order that the transportation cost is minimum? Formulate the above linear programming problem mathematically.

Sol. Let x units and y units of the commodity be transported from the factory P to the depots A and B respectively. Since the factory P has the capacity of producing 8 units of commodity, therefore, $(8 - x - y)$ units will be transported from the factory P to the depot C.

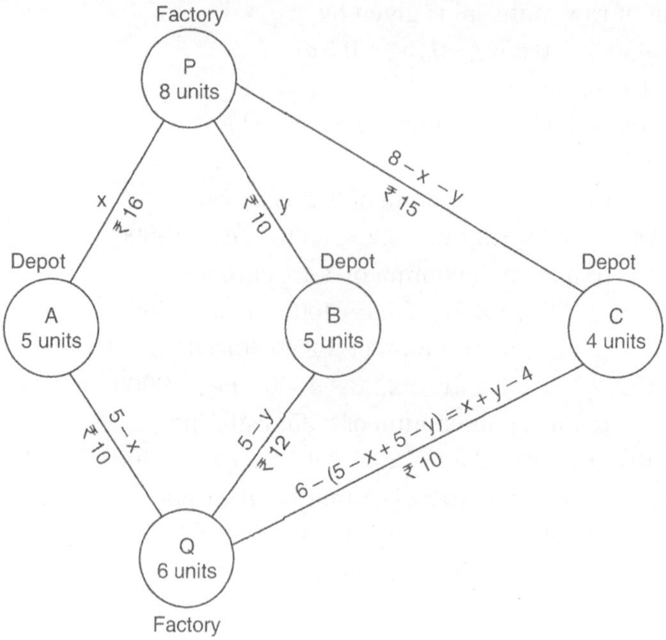

Clearly, $x \geq 0, \quad y \geq 0 \quad \text{and} \quad 8 - x - y \geq 0 \quad i.e., \quad x + y \leq 8.$

Now, the weekly requirement of the depot A is 5 units of the commodity and x units are already transported from the factory P, therefore, the remaining $(5 - x)$ units are to be transported from the factory Q. Similarly $(5 - y)$ and $6 - (5 - x + 5 - y) = x + y - 4$ units are to be transported from the factory Q to the depots B and C respectively.

Clearly $5 - x \geq 0, \quad 5 - y \geq 0 \quad \text{and} \quad x + y - 4 \geq 0$

i.e., $x \leq 5, \quad y \leq 5 \quad \text{and} \quad x + y \geq 4.$

The total transportation cost is given by

$$Z_1 = 16x + 10y + 15(8 - x - y) + 10(5 - x) + 12(5 - y) + 10(x + y - 4)$$

or $$Z_1 = x - 7y + 190.$$

The objective is to minimize $Z = x - 7y$ since 190 is a constant and does not affect the optimal solution.

Hence the above LPP can be stated mathematically as follows:

Minimize $Z = x - 7y$

subject to the constraints

$$x + y \leq 8$$
$$x \leq 5$$
$$y \leq 5$$
$$x + y \geq 4$$
$$x \geq 0, \quad y \geq 0.$$

EXERCISE 7.1

1. A company produces two types of models M_1 and M_2. Each M_1 model requires 4 hours of grinding and 2 hours of polishing whereas each M_2 model requires 2 hours of grinding and 5 hours of polishing. The company has 2 grinders and 3 polishers. Each grinder works for 40 hours a week and each polisher works for 60 hours a week. Profit on an M_1 model is ₹ 3 and on an M_2 model is ₹ 4. Whatever is produced in a week is sold in the market. How should the company allocate its production capacity to the two types of models so that it may make the maximum profit in a week? Formulate the problem as an LPP.
2. A housewife wishes to mix two types of foods X and Y in such a way that the vitamin contents of the mixture contain at least 8 units of vitamin A and 11 units of vitamin B. Food X costs ₹ 60 per kg and food Y costs ₹ 80 per kg. Food X contains 3 units per kg of vitamin A and 5 units per kg of vitamin B while food Y contains 4 units per kg of vitamin A and 2 units per kg of vitamin B. Formulate the above problem as an LPP to minimize the cost of mixture.
3. A manufacturer has 3 machines installed in his factory. Machines I and II are capable of being operated for at most 12 hours whereas machine III must be operated at least 5 hours a day. He produces only two items A and B each requiring the use of the three machines. The number of hours required for producing 1 unit each of the items on the three machines is given in the following table:

Item	*Number of hours required by the machine*		
	I	*II*	*III*
A	1	2	1
B	2	1	$\frac{5}{4}$

He makes a profit of ₹ 6 on item A and ₹ 4 on item B. Assuming that he can sell all he produces, how many of each item should he produce so as to maximize his profit. Formulate this LPP mathematically.

4. An aeroplane can carry a maximum of 200 passengers. A profit of ₹ 400 is made on each first class ticket and a profit of ₹ 300 is made on each economy class ticket. The airline reserves at least 20 seats for first class. However, at least 4 times as many passengers prefer to travel by economy class than by the first class. How many tickets of each class must be sold in order to maximize profit for the airline? Formulate the problem as an LP model.
5. Sudha wants to invest ₹ 12000 in Saving Certificates and in National Saving Bonds. According to rules, she has to invest at least ₹ 1000 in Saving Certificates and at least ₹ 2000 in National Saving Bonds. If the rate of interest on Saving Certificates is 8% p.a. and the rate of interest on the National Saving Bonds is 10% p.a., how should she invest her money to earn maximum yearly income. Formulate the problem as an LP model.
6. A firm manufactures 3 products A, B and C. The profits are ₹ 3, ₹ 2 and ₹ 4 respectively. The firm has two machines M_1 and M_2 and below is the required processing time in minutes for each machine on each product.

Machine	*Product*		
	A	*B*	*C*
M_1	4	3	5
M_2	2	2	4

Machines M_1 and M_2 have 2000 and 2500 machine-minutes respectively. The firm must manufacture 100 A's, 200 B's and 50 C's but not more than 150 A's. Set up an LPP to maximize profit.

7. A firm manufactures headache pills in two sizes A and B. Size A contains 2 grains of asprin, 5 grains of bicarbonate and 1 grain of codeine. Size B contains 1 grain of asprin, 8 grains of bicarbonate and 6 grains of codeine. It is found by users that it requires at least 12 grains of asprin, 74 grains of bicarbonate and 24 grains of codeine for providing immediate effect. It is required to determine the least number of pills a patient should take to get immediate relief. Formulate the above LPP mathematically.

8. The manager of an oil refinery must decide on the optimal mix of two possible blending processes of which the inputs and outputs per production run are as follows:

Processes	*Inputs (units)*		*Outputs (units)*	
	Crude 1	*Crude 2*	*Petrol (superior)*	*Petrol (ordinary)*
A	10	6	10	16
B	12	15	12	12

The availability of the two varieties of crude is limited to the extent of 400 units and 450 units respectively per day. The market demand indicates that at least 200 units and 240 units of the superior and ordinary quality petrol is required every day. The profitability analysis indicate that process A contributes ₹ 480 per run while the process B contributes ₹ 400 per run. The manager is interested in determining an optimal product-mix for maximizing the company's profits. Formulate the problem as an LP model.

9. A tape recorder company manufactures models A, B and C which have profit contributions per unit of ₹ 15, ₹ 40 and ₹ 60 respectively. The weekly minimum production requirements are 25 units, 130 units and 55 units for models A, B and C respectively. Each type of recorder requires a certain amount of time for the manufacturing of component parts, for assembling and for packing. Specifically, a dozen units of model A require 4 hours for manufacturing, 3 hours for assembling and 1 hour for packing. The corresponding figures for a dozen units of model B are 2.5, 4 and 2 and for a dozen units of model C are 6, 9 and 4. During the forthcoming week, the company has available 130 hours of manufacturing, 170 hours of assembling and 52 hours of packing time. Formulate this problem as an LPP model so as to maximize total profit to the company.

10. ABC Foods Company is developing a low-calorie high-portein diet supplement called Hi-Pro. The specifications for Hi-Pro have been established by a panel of medical experts. These specifications along with the calorie, protein and vitamin content of three basic foods, are given in the following table:

Nutritional Elements	*Units of Nutritional Elements (Per 100 gm Serving of Basic Foods)*			*Basic Foods Hi-Pro Specifications*
	1	*2*	*3*	
Calories	350	250	200	300
Proteins	250	300	150	200
Vitamin A	100	150	75	100
Vitamin C	75	125	150	100
Cost per serving (₹)	1.50	2.00	1.20	

What quantities of foods 1, 2 and 3 should be used? Formulate this problem as an LP model to minimize cost of serving.

11. A firm manufactures two items A and B. It purchases castings which are then machined, bored and polished. Castings for items A and B cost ₹ 3 and ₹ 4 each and are sold at ₹ 6 and ₹ 7 respectively. Running costs of these machines are ₹ 20, ₹ 14 and ₹ 17.50 per hour respectively.

Formulate the problem so that the product mix maximizes the profit. The capacities of the machines are

	Item–A	Item–B
Machining	25 per hr.	40 per hr.
Boring	28 per hr.	35 per hr.
Polishing	35 per hr.	25 per hr.

(*K.U.K. 2010, 2013*)

12. A brick manufacturer has two depots, A and B, with stocks of 30000 and 20000 bricks respectively. He receives orders from three building's P, Q and R for 15000, 20000 and 15000 bricks respectively. The cost of transporting 1000 bricks to the builders from the depots (in ₹) are given below:

To / *From*	*Transportation cost per 1000 bricks (in ₹)*		
	P	*Q*	*R*
A	40	20	20
B	20	60	40

How should the manufacturer fulfil the orders so as to keep the cost of transportation minimum? Formulate the above LPP mathematically.

Answers

1. Maximize $Z = 3x + 4y$

subject to the constraints
$$2x + y \le 40$$
$$2x + 5y \le 180$$
$$x \ge 0, \quad y \ge 0.$$

2. Minimize $Z = 60x + 80y$

subject to the constraints
$$3x + 4y \ge 8$$
$$5x + 2y \ge 11$$
$$x \ge 0, \quad y \ge 0.$$

3. Maximize $Z = 6x + 4y$

subject to the constraints
$$x + 2y \le 12$$
$$2x + y \le 12$$
$$4x + 5y \ge 20$$
$$x \ge 0, \quad y \ge 0.$$

4. Maximize $Z = 400x + 300y$

subject to the constraints
$$x + y \le 200$$
$$x \ge 20, \quad y \ge 4x$$
$$x \ge 0, \quad y \ge 0.$$

5. Maximize $Z = \frac{8}{100}x + \frac{10}{100}y$

subject to the constraints
$$x + y \le 12000$$
$$x \ge 1000, \quad y \ge 2000$$
$$x \ge 0, \quad y \ge 0.$$

6. Maximize $Z = 3x_1 + 2x_2 + 4x_3$

subject to the constraints
$$4x_1 + 3x_2 + 5x_3 \le 2000$$
$$2x_1 + 2x_2 + 4x_3 \le 2500$$
$$100 \le x_1 \le 150$$
$$x_2 \ge 200, \quad x_3 \ge 50.$$

7. Minimize $Z = x + y$
subject to the constraints

$$2x + y \geq 12$$
$$5x + 8y \geq 74$$
$$x + 6y \geq 24$$
$$x \geq 0, y \geq 0.$$

8. Maximize $Z = 480\,x + 400\,y$
subject to the constraints

$$10x + 12y \leq 400$$
$$6x + 15y \leq 450$$
$$10x + 12y \geq 200$$
$$16x + 12y \geq 240$$
$$x \geq 0, \; y \geq 0.$$

9. Maximize $Z = 15x_1 + 40x_2 + 60x_3$
subject to the constraints

$$x_1 \geq 25$$
$$x_2 \geq 130$$
$$x_3 \geq 55$$
$$4x_1 + 2.5x_2 + 6x_3 \leq 1560$$
$$3x_1 + 4x_2 + 9x_3 \leq 2040$$
$$x_1 + 2x_2 + 4x_3 \leq 624.$$

10. Minimize $Z = 1.5x_1 + 2x_2 + 1.2x_3$
subject to the constraints

$$350x_1 + 250x_2 + 200x_3 \geq 300$$
$$250x_1 + 300x_2 + 150x_3 \geq 200$$
$$100x_1 + 150x_2 + 75x_3 \geq 100$$
$$75x_1 + 125x_2 + 150x_3 \geq 100$$
$$x_1, x_2, x_3 \geq 0.$$

11. Maximize $Z = 1.2\,x + 1.4y$
subject to the constraints

$$40x + 25y \leq 1000$$
$$35x + 28y \leq 980$$
$$25x + 35y \leq 875$$
$$x, y \geq 0.$$

12. Minimize $Z = 40x - 20y$
subject to the constraints

$$x + y \geq 15$$
$$x \leq 15$$
$$y \leq 20$$
$$x + y \leq 30$$
$$x \geq 0, y \geq 0$$

where 1 unit of bricks = 1000 bricks
Total cost of transportation is $Z_1 = Z + 1500.$

7.4. GRAPH OF A LINEAR INEQUALITY

The constraints in the mathematical model of an LPP are in the form of linear inequalities. Let us see how to graph linear inequalities involving two variables.

A linear inequality in two variables x and y is of the form $ax + by < c$ or $ax + by \le c$ or $ax + by > c$ or $ax + by \ge c$, where at least one of a and b is non-zero. The graph of a linear inequality in two variables x and y is the set of all ordered pairs (x, y) for which the inequality holds. The following three steps are involved in graphing a linear inequality.

1. Graph the corresponding linear equation $ax + by = c$ which is a straight line. Find any two distinct points on the line and draw a straight line through them. (For convenience, choose one point on x-axis by putting $y = 0$ and the other on y-axis by putting $x = 0$.) The line divides the x-y plane into two half planes.

2. To decide which of the two half-planes satisfies the linear inequality, choose a point P not on the line. (If the line does not pass through the origin, then origin is the best choice). Check whether P satisfies the inequality.

3. If the coordinates of P satisfy the inequality, then every point in the half plane containing P satisfies the inequality. If the coordinates of P donot satisfy the inequality, then every point in the other half plane (not containing P) satisfies the inequality.

Remark. $x = 0$ is the equation of y-axis. $x = a$, $a > 0$ is a straight line parallel to y-axis and at a distance 'a' to its right. $x > a$ is the half plane to the right of the line $x = a$ and $x < a$ is the half plane to the left of the line $x = a$.

$x = -a$, $a > 0$ is a straight line parallel to y-axis and at a distance 'a' to its left. $x > -a$ is the half plane to the right of the line $x = -a$ and $x < -a$ is the half plane to the left of the line $x = -a$.

Similarly, $y = 0$ is the equation of x-axis. $y = a$ and $y = -a$, $a > 0$, are lines parallel to x-axis at a distance 'a' above and below the x-axis respectively. $y > a$ is the half plane above the line $y = a$ and $y < a$ is the half plane below the line $y = a$. Similar remarks apply to $y > -a$ and $y < -a$.

ILLUSTRATIVE EXAMPLES

Example 1. *Graph the linear inequality*:

$$3x + 4y \le 12.$$

Sol. The given inequation is $3x + 4y \le 12$ ...(1)

Replacing $\le$ by $=$, the corresponding equation is $3x + 4y = 12$ ...(2)

Putting $y = 0$ in (2), $3x = 12$ or $x = 4$

Putting $x = 0$ in (2), $4y = 12$ or $y = 3$

$\therefore$ The graph of line (2) passes through the points A(4, 0) and B(0, 3). The line through these two points in the graph of equation (2). This line divides the plane into two half-planes.

Putting $x = 0$ and $y = 0$ in (1), we get $0 \le 12$ which is true. Therefore the origin O(0, 0) lies in the feasible region. Hence, the shaded region below the line AB and the points on the line AB (as shown in the adjoining figure) constitute the graph of the inequality (1).

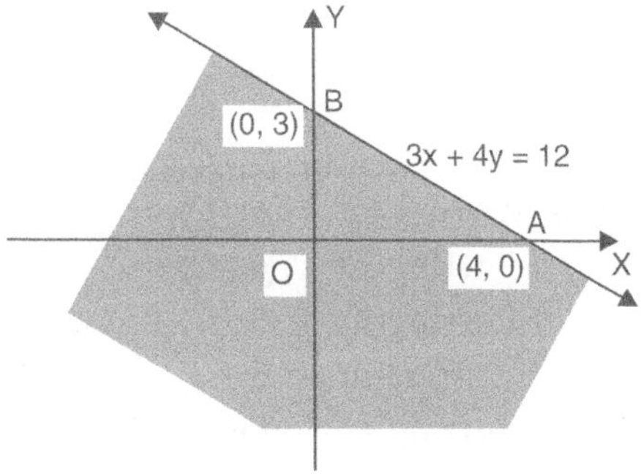

Example 2. *Solve graphically the following system of linear inequations*:

$$x \ge 4, y \ge 2.$$

Sol. The given system of inequations is

$x \ge 4$ ...(1), $y \ge 2$ ...(2)

The equation corresponding to inequation (1) is $x = 4$. This is a line parallel to y-axis and at a distance 4 units to the **right** of y-axis.

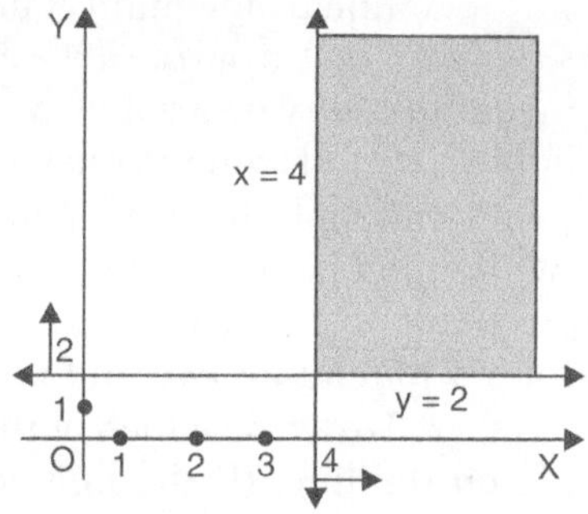

Putting $x = 0$, $y = 0$ in (1), $0 \geq 4$ which is not true.

$\therefore$ Graph of inequation (1) does not contain origin *i.e.*, $x \geq 4$ represents the region away from the origin.

$\therefore$ The inequation (1) represents the half-plane to the right of this line, including this line.

The equation corresponding to inequation (2) is $y = 2$. This is a line parallel to x-axis and at a distance 2 units above it.

Putting $x = 0$ and $y = 0$ in (2), $0 \geq 2$ which is not true.

$\therefore$ Graph of inequation (2) does not contain the origin.

$\therefore$ The inequation (2) represents the half-plane **above** this line, including this line.

Hence the common region is the shaded region. Any point in this shaded region represents a solution of the given system of inequations.

Example 3. *Solve graphically $x + 3y \leq 12$, $3x + y \leq 12$, $x \geq 0$, $y \geq 0$.*

Sol. The given system of inequations is

$$x + 3y \leq 12 \quad \ldots(1)$$

$$3x + y \leq 12 \quad \ldots(2)$$

and

$$x \geq 0,\ y \geq 0 \quad \ldots(3)$$

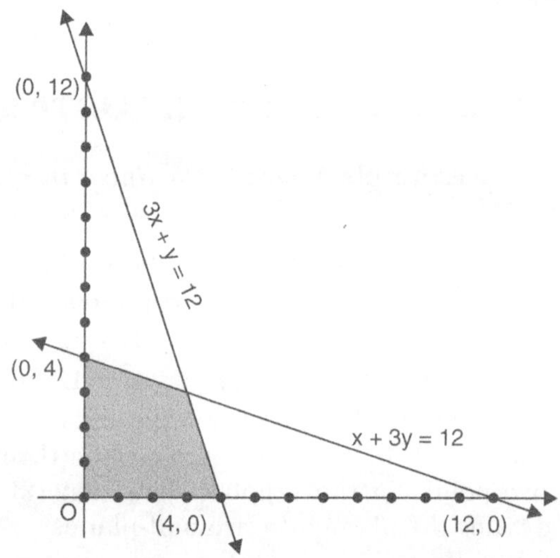

The equation corresponding to inequation (1) is $x + 3y = 12$

Putting $x = 0$, $3y = 12$ or $y = 4$

Putting $y = 0$, $x = 12$

$\therefore$ Graph of $x + 3y = 12$ is the straight line through the points (0, 4) and (12, 0).

Putting $x = 0$ and $y = 0$ in (1); $0 \leq 12$ which is true.

$\therefore$ Graph of inequation (1) contains the origin.

$\therefore$ The inequation (1) represents the half plane on the origin side of this line, including the line.

The equation corresponding to inequation (2) is $3x + y = 12$

Putting $x = 0$, $y = 12$

Putting $y = 0$, $3x = 12$ or $x = 4$

$\therefore$ Graph of $3x + y = 12$ is the straight line through the points (0, 12) and (4, 0).

Putting $x = 0$ and $y = 0$ in (2), $0 \leq 12$ which is also true.

$\therefore$ Graph of inequation (2) also contains the origin.

$\therefore$ The inequation (2) represents the half-plane on the origin side of this line, including the line.

Inequation (3) is $x \geq 0$, $y \geq 0$ $\Rightarrow$ feasible region is in first quadrant only.

$\therefore$ The common feasible region (*i.e.*, intersection of all feasible regions) in first quadrant is shown as shaded in the figure. Every point in this shaded region represents a solution of the given system of linear inequations.

Example 4. *Solve graphically*: $x + y \geq 3,\ 7x + 6y \leq 42,\ x \leq 5,\ y \leq 4,\ x \geq 0,\ y \geq 0.$

Sol. The graph of $x + y = 3$ is the line AB through the points (3, 0) and (0, 3). Since (0, 0) does not satisfy $x + y \geq 3$, the region represented by this inequation is the half-plane above the line AB, including this line.

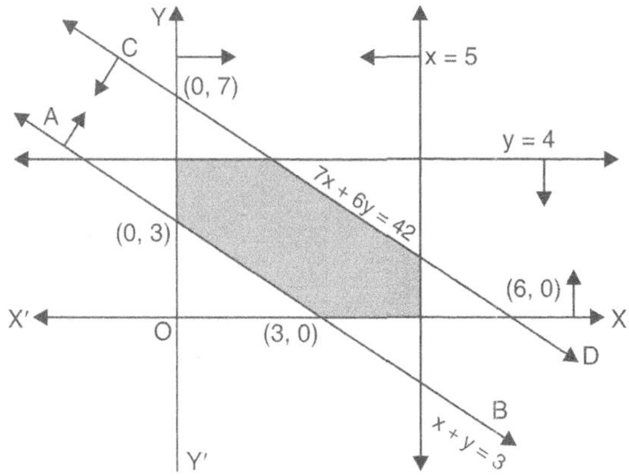

The graph of $7x + 6y = 42$ is the line CD through the points (6, 0) and (0, 7) (By putting $x = 0$ and $y = 0$ respectively in $7x + 6y = 42$). Since (0, 0) satisfies $7x + 6y < 42$, the region represented by this inequation is the half-plane on the origin side of line CD, including this line.

$x \leq 5$ represents the half-plane on the left of the line $x = 5$, including this line.

$y \leq 4$ represents the half-plane below the line $y = 4$, including this line.

$x \geq 0$ and $y \geq 0 \Rightarrow$ Feasible region is in first quadrant only.

The common feasible region (*i.e.*, intersection of all feasible regions) is shown shaded in the figure. Every point in this shaded region represents a solution of the given system of linear inequations.

Example 5. *Solve graphically the inequations* $3x + 2y \leq 24,\ x + 2y \leq 16,\ x + y \leq 10,\ x \geq 0,\ y \geq 0.$

Sol. The given system of inequations is

$$3x + 2y \leq 24 \quad \ldots(1)$$

$$x + 2y \leq 16 \quad \ldots(2)$$

$$x + y \leq 10 \quad \ldots(3)$$

and $$x \geq 0,\ y \geq 0 \quad \ldots(4)$$

The equation corresponding to inequation (1) is $3x + 2y = 24$.

Putting $x = 0$, $2y = 24$ or $y = 12$

Putting $y = 0$, $3x = 24$ or $x = 8$.

$\therefore$ Graph of $3x + 2y = 24$ is the straight line through the points (0, 12) and (8, 0).

Putting $x = 0$ and $y = 0$ in (1), $0 \leq 24$ which is true.

$\therefore$ Graph of inequation (1) contains the origin.

$\therefore$ The inequation (1) represents the half-plane on the origin side of this line, including the line.

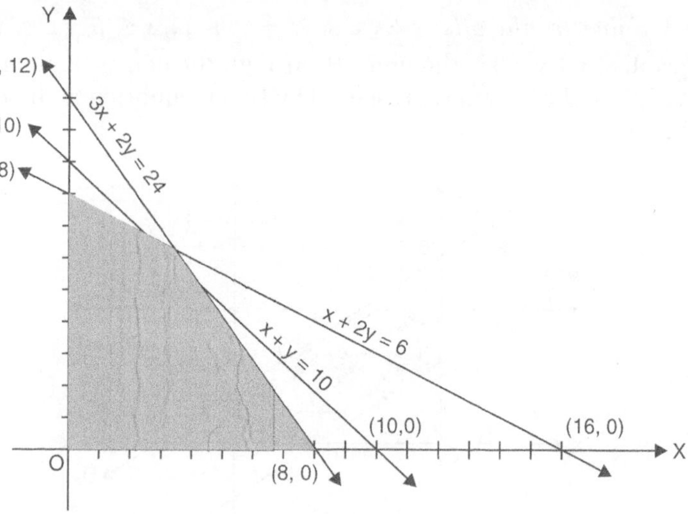

The equation corresponding to inequation (2) is

$$x + 2y = 16$$

Putting $\quad x = 0,\ 2y = 16 \quad \text{or} \quad y = 8$

Putting $\quad y = 0,\ x = 16$

$\therefore$ Graph of $x + 2y = 16$ is the straight line through the points (0, 8) and (16, 0).

Putting $x = 0$ and $y = 0$ in (2), $0 \leq 16$ which is true.

$\therefore$ Graph of inequation (2) also contains the origin.

$\therefore$ The inequation (2) represents the half plane on the origin side of this line, including the line.

The equation corresponding to inequation (3) is

$$x + y = 10$$

Putting $\quad x = 0,\ y = 10$

Putting $\quad y = 0,\ x = 10$

$\therefore$ Graph of $x + y = 10$ is the straight line through the points (0, 10) and (10, 0).

Putting $x = 0$ and $y = 0$ in (3), $0 \leq 10$ which is also true.

$\therefore$ Graph of inequation (3) also contains the origin.

$\therefore$ The inequation (3) represents the half-plane on the origin side of this line, including the line.

Inequation (4) is $x \geq 0,\ y \geq 0 \;\Rightarrow\;$ Feasible region is in first quadrant only.

$\therefore$ The common feasible region (*i.e.*, intersection of all feasible regions) in first quadrant is shown as shaded in the figure. Every point in this shaded region represents a solution of the given system of linear inequations.

EXERCISE 7.2

Graph the following systems of inequations and shade the feasible region:

1. $x \geq 1$, $x \leq 4$, $y \geq 1$, $y \leq 3$ (or $1 \leq x \leq 4$, $1 \leq y \leq 3$).
2. $0 \leq y \leq 3$, $y \leq x$, $2x + y \leq 9$.
3. $3x + 2y \leq 18$, $x + 2y \leq 10$, $x \geq 0$, $y \geq 0$.
4. $x + y \leq 9$, $y \geq x$, $x \geq 1$.
5. $3x + 4y \leq 60$, $x + 3y \leq 30$, $x \geq 0$, $y \geq 0$
6. $2x + y \leq 24$, $x + y \leq 11$, $2x + 5y \leq 40$, $x \geq 0$, $y \geq 0$.

Answers

1.

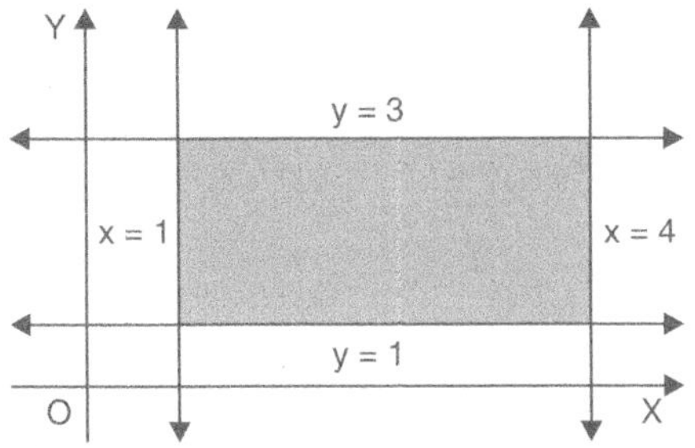

2.

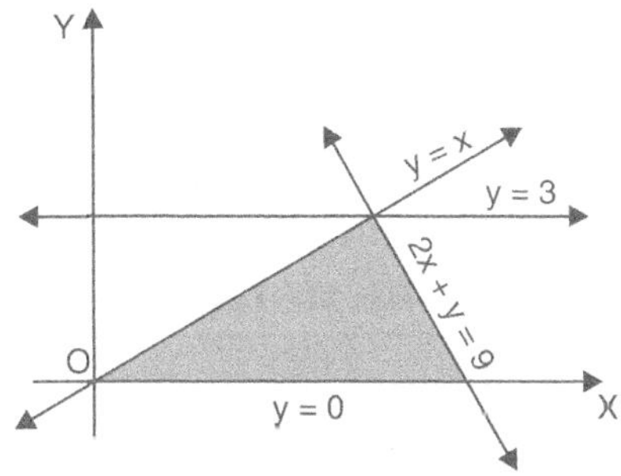

3.

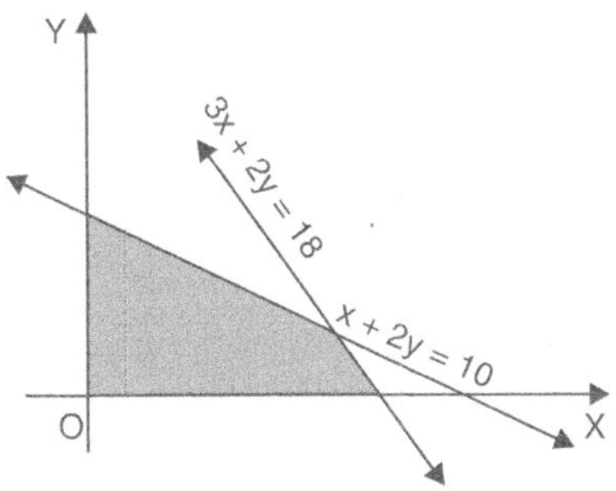

4.

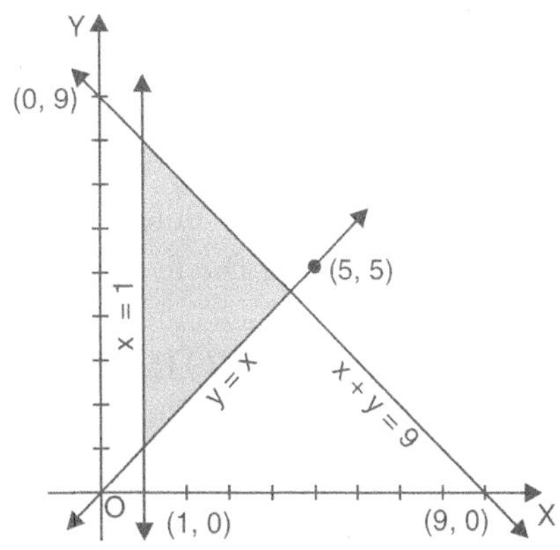

5.

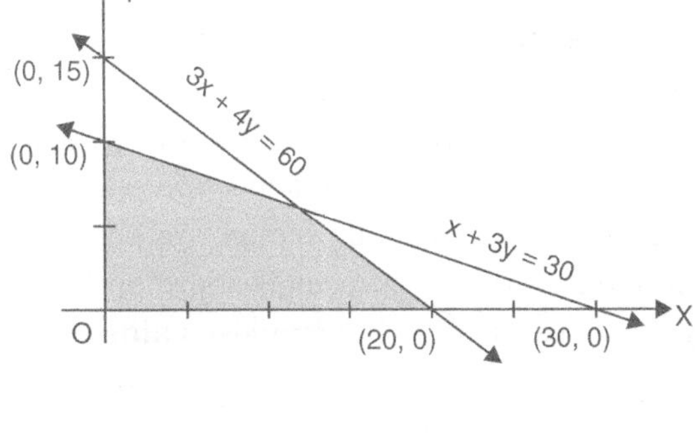

6.

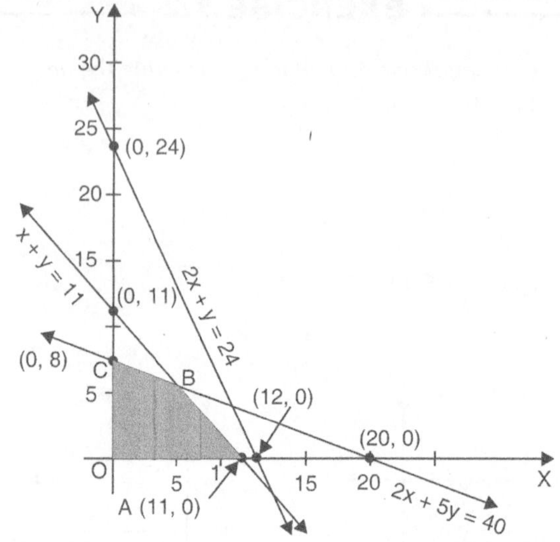

7.5. THE GRAPHICAL METHOD OF SOLVING AN LPP

For linear programming problems having only two variables, the set of all feasible solutions can be displayed graphically by determining the feasible region. The points lying within the feasible region satisfy all the constraints. The graphical approach gives an insight into the basic concepts and provides valuable understanding for solving LP problems involving more than two variables algebraically. Problems involving more than two variables cannot be solved graphically.

There are two techniques of solving an LPP by graphical method

(*i*) Corner point method.

(*ii*) Iso-profit or Iso-cost method.

We shall discuss only the first technique.

CORNER POINT METHOD

This method is based on a theorem called **extreme point theorem.**

Extreme Point Theorem. The optimal solution to a linear programming problem, if it exists, occurs at an extreme point (corner) of the feasible region.

The collection of all feasible solutions to an LPP constitutes a convex set whose extreme points correspond to the basic feasible solutions.

Working procedure to solve an LPP graphically:

1. Formulate the given problem as an LPP.
2. Plot the constraints and shade the common region that satisfies all the constraints simultaneously. The shaded area is called the feasible region.
3. Determine the coordinates of each corner of the feasible region.
4. Find the value of the objective function $Z = ax + by$ at each corner point.

   Let M and m denote respectively the largest and the smallest values of Z at the corner points.

5. When the feasible region is **bounded,** M and m are the maximum and the minimum values of Z.

6. When the feasible region is **unbounded**, we have:

(*a*) M is the maximum value of Z, if the open half plane determined by Z > M *i.e.*, $ax + by > M$ has no point in common with the feasible region. Otherwise Z has no maximum value.

(*b*) m is the minimum value of Z, if the open half plane determined by Z < m *i.e.*, $ax + by < m$ has no point in common with the feasible region. Otherwise Z has no minimum value.

ILLUSTRATIVE EXAMPLES

Example 1. *Solve the following linear programming problem graphically:*

Maximize $Z = 2x + 3y$

subject to the constraints

$$x + 2y \le 10$$
$$2x + y \le 14$$
$$x, y \ge 0.$$

Sol. Mathematical formulation of LPP is already given. The non negativity constraints $x \ge 0$, $y \ge 0$ imply that the feasible region is in first quadrant only. Plot each constraint by first treating it as a linear equation and then using the inequality condition of each constraint, mark the feasible region as shown in the figure. The feasible region is shown by the shaded area OABC in the figure.

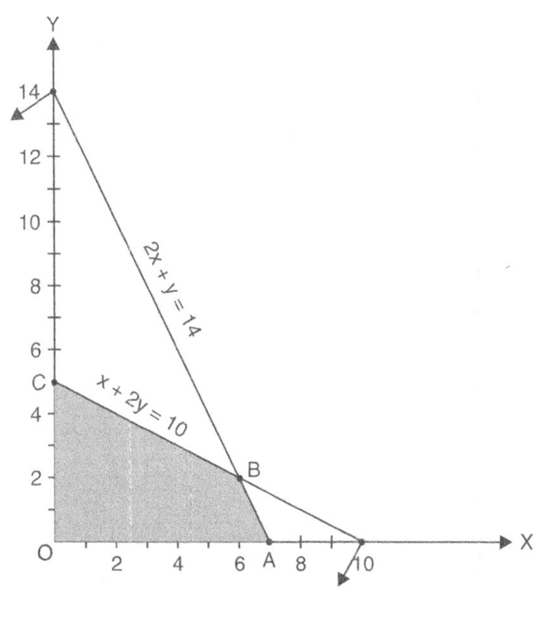

Since the optimal value of the objective function occurs at one of the corners of the feasible region, we determine their coordinates.

Clearly O = (0, 0), A = (7, 0),
B = (6, 2), C = (0, 5)

Now we find the value of the objective function Z = $2x + 3y$ at each corner point.

Corner Point	*Coordinates* (x, y)	*Value of Objective Function* $Z = 2x + 3y$
O	(0, 0)	0 + 0 = 0
A	(7, 0)	2(7) + 3(0) = 14
B	(6, 2)	2(6) + 3(2) = 18
C	(0, 5)	2(0) + 3(5) = 15

The maximum value of Z is 18 at B(6, 2). Hence the optimal solution to the given LP problem is

$$x = 6, \quad y = 2 \quad \text{and Max. } Z = 18$$

Remark. The coordinates of B are obtained by solving $x + 2y = 10$ and $2x + y = 14$.

Example 2. *Solve the following LP problem by the graphical method:*

Minimize $Z = 20x_1 + 30x_2$

subject to the constraints
$$x_1 + 2x_2 \le 40$$
$$3x_1 + x_2 \ge 30$$
$$4x_1 + 3x_2 \ge 60$$
$$x_1, x_2 \ge 0.$$
(K.U.K Jan. 2013, Jan. 2014)

Sol. Mathematical formulation of LPP is already given. The non-negativity constraints $x_1, x_2 \ge 0$ imply that the feasible region is in first quadrant only. Plot each constraint by first treating it as a linear equation and then using the inequality condition of each constraint, mark the feasible region as shown in the figure. The feasible region is shown by the shaded area ABCD in the figure.

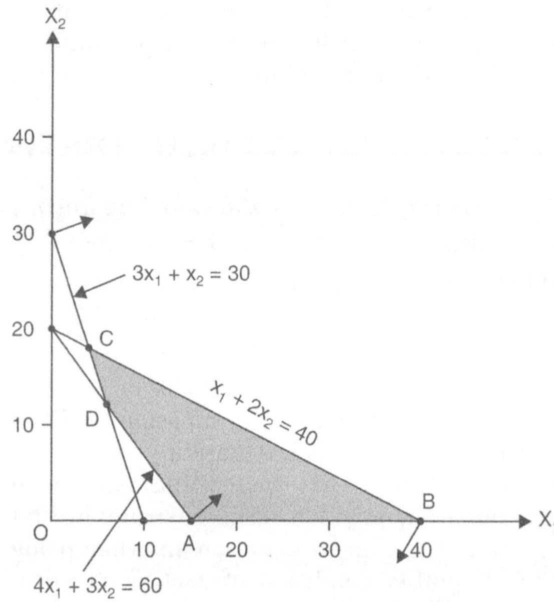

Since the optimal value of the objective function occurs at one of the corners of the feasible region, we determine their coordinates.

Clearly A = (15, 0), B = (40, 0)
C = (5, 18), D = (6, 12)

Now we find the value of the objective function Z = $20x_1 + 10x_2$ at each corner point.

Corner Point	*Coordinates* (x_1, x_2)	*Value of Objective Function* $Z = 20x_1 + 30x_2$
A	(15, 0)	20(15) + 30(0) = 300
B	(40, 0)	20(40) + 30(0) = 800
C	(5, 18)	20(5) + 30(18) = 640
D	(6, 12)	20(6) + 30(12) = 480

The minimum value of Z is 300 at A(15, 0). Hence the optimal solution to the given LP problem is

$$x_1 = 15, \quad x_2 = 0 \quad \text{and} \quad \min. Z = 300.$$

Example 3. *A manufacturer has 3 machines installed in his factory. Machines I and II are capable of being operated for at the most 12 hours, whereas machine III must be operated at least 5 hours a day. He produces only two items, each requiring the use of the three machines.*

The number of hours required for producing 1 unit of each of the items A and B on the three machines are given in the following table:

Items	*Number of hours required on the machines*		
	I	*II*	*III*
A	*1*	*2*	*1*
B	*2*	*1*	$\frac{5}{4}$

He makes a profit of ₹ 60 on item A and ₹ 40 on item B. Assuming that he can sell all that he produces, how many of each item should be produce so as to maximize his profit. Solve the LP problem graphically.

Sol. Suppose the manufacturer produces x and y units of items A and B respectively. His objective is to maximize the total profit = ₹ $(60x + 40y)$

$\therefore$ Objective function is given by $Z = 60x + 40y$

Machine hour constraints are:

Machine I $\quad x + 2y \leq 12$

Machine II $\quad 2x + y \leq 12$

Machine III $\quad x + \frac{5}{4}y \geq 5$ or $4x + 5y \geq 20$

Non-negativity constraints are $\quad x, y \geq 0$

(since it makes no sense to assign negative values to x and y).

Thus, the mathematical formulation of the LP problem is

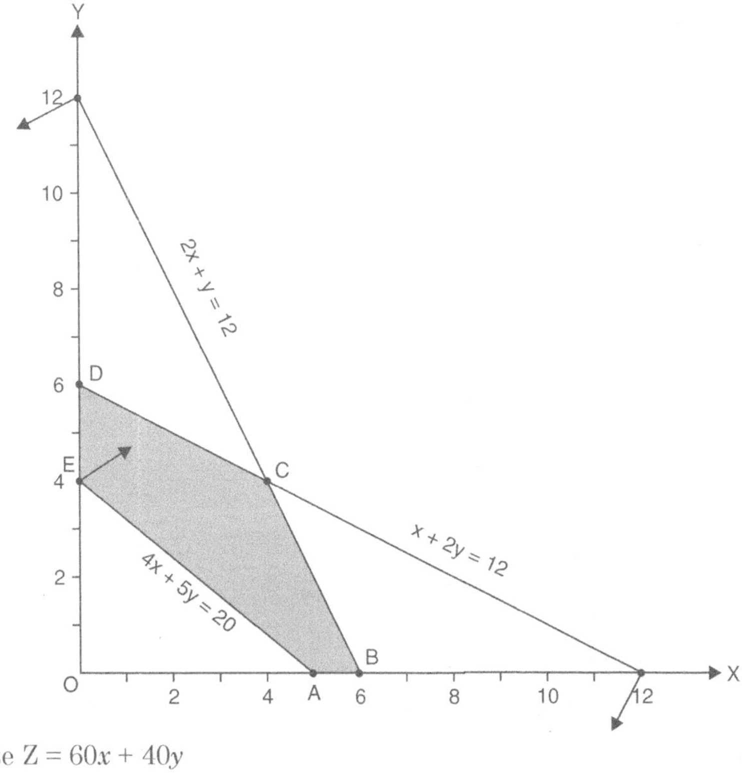

Maximize $Z = 60x + 40y$

subject to the constraints

$$x + 2y \leq 12$$
$$2x + y \leq 12$$
$$4x + 5y \geq 20$$
$$x, y \geq 0$$

The feasible region is shown by the shaded area ABCDE in the figure.

Since the optimal value of the objective function occurs at one of the corners of the feasible region, we determine their coordinates

Here $A = (5, 0)$, $B = (6, 0)$, $C = (4, 4)$, $D = (0, 6)$, $E = (0, 4)$

Now we find the value of the objective function $Z = 60x + 40y$ at each corner point.

Corner Point	*Coordinates* (x, y)	*Value of Objective Function* $Z = 60x + 40y$
A	(5, 0)	60(5) + 40(0) = 300
B	(6, 0)	60(6) + 40(0) = 360
C	(4, 4)	60(4) + 40(4) = $\boxed{400}$
D	(0, 6)	60(0) + 40(6) = 240
E	(0, 4)	60(0) + 40(4) = 160

The maximum value of Z is 400 at C(4, 4).

∴ The manufacturer should produce $x = 4$ units of item A and $y = 4$ units of item B to get the maximum profit of ₹ 400.

Example 4. *A company makes two kinds of leather belts, A and B. Belt A is a high quality belt and B is of lower quality. The respective profits are ₹ 4 and ₹ 3 per belt. Each belt of type A requires twice as much time as a belt of type B, and, if all belts were of type B, the company could make 1000 per day. The supply of leather is sufficient for only 800 belts per day (both A and B combined). Belt A requires a fancy buckle and only 400 per day are available. There are only 700 buckles a day available for belt B. What should be the daily production of each type of belt?*

Sol. Let x and y denote respectively the number of belts of type A and type B produced per day.

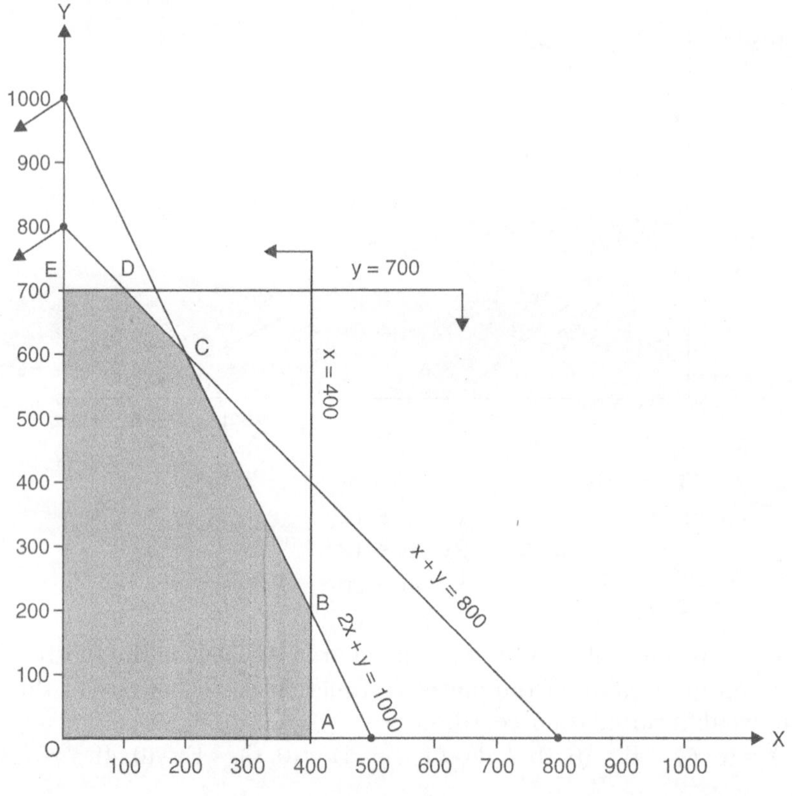

The objective is to maximize the profit Z (in rupees) given by

$$Z = 4x + 3y$$

Since the rate of producing type B belts is 1000 per day, the total time taken to produce y belts of type B is $\frac{y}{1000}$.

Also, each belt of type A requires twice as much time as a belt of type B, the rate of producing type A belts is 500 per day and the total time taken to produce x belts of type A is $\frac{x}{500}$.

$\therefore$ The time constraint is $\frac{x}{500} + \frac{y}{1000} \le 1$

or $$2x + y \le 1000$$

The constraint imposed by supply of leather is

$$x + y \le 800$$

The constraint imposed by supply of fancy buckles is

$$x \le 400$$

The constraint imposed by supply of buckles for type B belts is

$$y \le 700$$

Since the number of belts cannot be negative, we have non-negativity constraints

$$x \ge 0, \quad y \ge 0$$

$\therefore$ The mathematical formulation of the problem is:

Maximize $Z = 4x + 3y$

subject to the constraints

$$2x + y \le 1000$$
$$x + y \le 800$$
$$x \le 400$$
$$y \le 700$$
$$x, y \ge 0$$

The feasible region is shown by the shaded area OABCDE in the figure.

Since the optimal value of the objective function occurs at one of the corners of the feasible region, we determine their coordinates.

Here O = (0, 0), A = (400, 0), B = (400, 200), C = (200, 600), D = (100, 700), E = (0, 700)

Now we find the value of the objective function Z = $4x + 3y$ at each corner point.

Corner Point	*Coordinates* (x, y)	*Value of Objective Function* $Z = 4x + 3y$
O	(0, 0)	4(0) + 3(0) = 0
A	(400, 0)	4(400) + 3(0) = 1600
B	(400, 200)	4(400) + 3(200) = 2200
C	(200, 600)	4(200) + 3(600) = **2600**
D	(100, 700)	4(100) + 3(700) = 2500
E	(0, 700)	4(0) + 3(700) = 2100

The maximum value of Z is 2600 at C(200, 600).

∴ The company should produce 200 belts of type A and 600 belts of type B in order to get the maximum profit of ₹ 2600 per day.

Example 5. *Solve the following linear programming problem graphically:*

Minimize $Z = 3x + 5y$

subject to the constraints $x + y = 6$

$x \leq 4$

$y \leq 5$

$x \geq 0, \quad y \geq 0.$

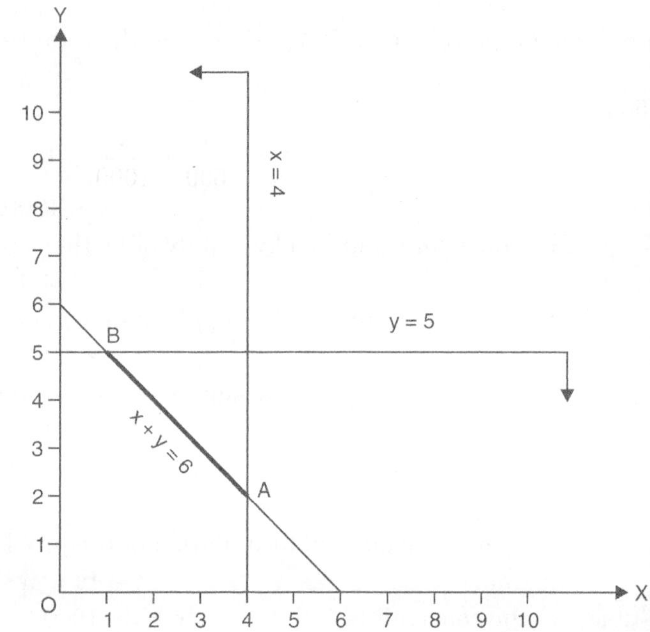

Sol. Here the feasible region of the LPP is the line segment AB with A = (4, 2) and B = (1, 5). These are the corner points of the feasible region.

At A(4, 2), $Z = 3(4) + 5(2) = 22$

At B(1, 5), $Z = 3(1) + 5(5) = 28$

The minimum value of Z is 22 which occurs at A(4, 2).

Hence, optimal solution is $x = 4$, $y = 2$ and optimal value = 22.

Example 6. *A person consumes two types of food, A and B, everyday to obtain 8 units of protein, 12 units of carbohydrates and 9 units of fat which is his daily minimum requirement. 1 kg of food A contains 2, 6, 1 units of protein, carbohydrates and fat, respectively. 1 kg of food B contains 1, 1 and 3 units of protein, carbohydrates and fat, respectively. Food A costs ₹ 8 per kg while food B costs ₹ 5 per kg. Form an LPP to find how many kgs of each food should he buy daily to minimize his cost of food and still meet minimal nutritional requirements and solve it.*

Sol. Suppose the person buys x kg of food A and y kg of food B daily. Total cost of food (in rupees) is given by

$$Z = 8x + 5y$$

The objective is to minimize Z.

x kg of food A contains $2x$ units of protein.

y kg of food B contains y units of protein.

Since minimum requirement of protein is 8 units, therefore, protein constraint is

$$2x + y \geq 8$$

Similarly $6x + y \geq 12$ (carbohydrate constraint)

$x + 3y \geq 9$ (fat constraint)

Also $x, y \geq 0$

∴ The mathematical formulation of the problem is:

Minimize $Z = 8x + 5y$

subject to the constraints $2x + y \geq 8$

$6x + y \geq 12$

$$x + 3y \geq 9$$
$$x, y \geq 0$$

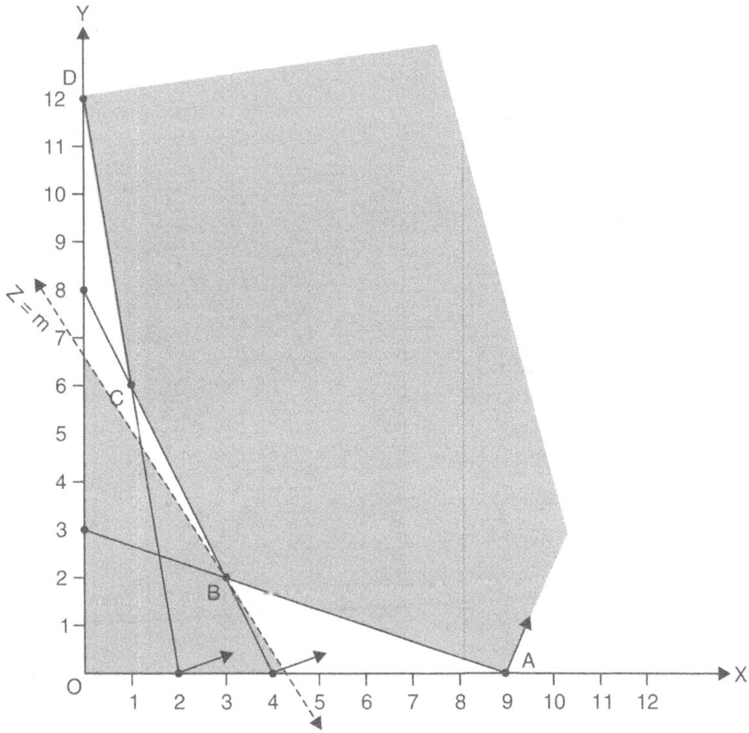

The feasible region is shown shaded in the figure. It is unbounded. The extreme points of the feasible region are:

$$A = (9, 0), \quad B = (3, 2), \quad C = (1, 6), \quad D = (0, 12)$$

Now, we find the value of the objective function $Z = 8x + 5y$ at each corner point.

Corner Point	*Coordinates* (x, y)	*Value of Objective Function* $Z = 8x + 5y$
A	(9, 0)	$8(9) + 5(0) = 72$
B	(3, 2)	$8(3) + 5(2) = \boxed{34} = m$
C	(1, 6)	$8(1) + 5(6) = 38$
D	(0, 12)	$8(0) + 5(12) = 60$

The minimum value of Z is 34 at B(3, 2).

But this value is doubtful. It is yet to be confirmed. Now we draw the graph of $Z < m$ *i.e.*, $8x + 5y < 34$. Since the half plane determined by $Z < m$ has no point in common with the feasible region, $m = 34$ is the minimum value of Z.

$\therefore$ The person should buy 3 kg of food A and 2 kg of food B at the minimum cost of ₹ 34.

7.6. SOME EXCEPTIONAL CASES

So far we have been solving LP problems which may be called 'well behaved' in the sense each problem has a unique optimal solution. However, there are certain exceptional cases which must be taken into consideration.

I. Infeasible solution. Wrong formulation of an LPP results into inconsistent constraints and no value of the variables satisfies all the constraints simultaneously. In such cases, the linear programming problem is said to have no feasible solution or infeasible solution. Infeasibility depends only on the constraints and has nothing to do with the objective function.

Example. *Maximize* $Z = 5x + 12y$ *subject to the constraints* $2x + 3y \leq 18$

$$x + y \geq 10$$
$$x, y \geq 0.$$

Sol. From the adjoining figure, it is clear that there is no point in the first quadrant satisfying all the constraints. Hence, there is no feasible solution to the problem because of conflicting constraints.

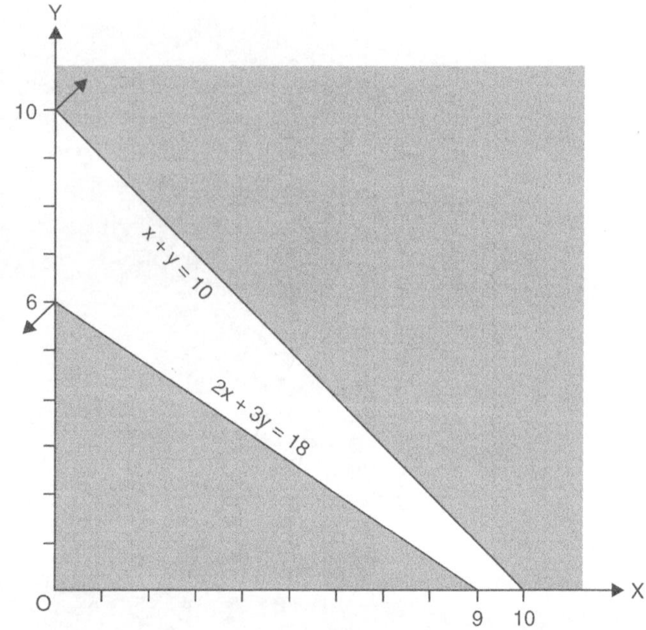

II. Unbounded Solutions. Some linear programming problems have unbounded feasible region so that the variables can take any value in the unbounded region without violating any constraint. If we wish to maximize the objective function Z, then for any value of Z we can find a feasible solution with a greater value of Z. Such problems are said to have unbounded solutions.

Example. *Maximize* $Z = 2x + 5y$ *subject to the constraints* $x + y \geq 3$, $x - y \leq 1$, $x, y \geq 0$.

Sol. The feasible region, shown shaded in the figure, is unbounded. The corner points of the feasible region are A(0, 3) and B(2, 1).

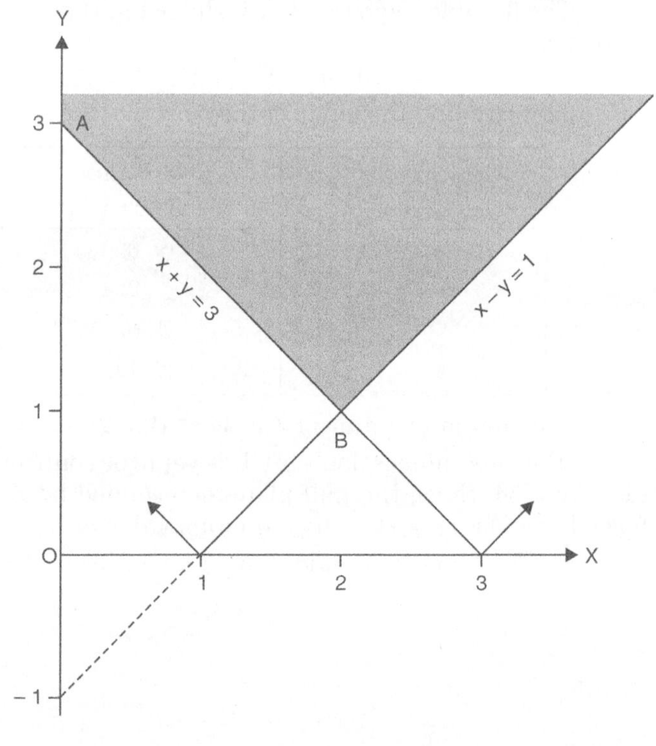

$$Z(A) = 2(0) + 5(3) = 15$$
$$Z(B) = 2(2) + 5(1) = 9$$

Since the given LP problem is of maximization. There are infinite number of points in the feasible region where the value of the objective function is more than this value Z(A) = 15. Both the variables x and y can be made arbitrarily large and there is no limit to the value of Z . Hence the problem has unbounded solutions.

III. Multiple Optimal Solutions. When there exist more than one points in the feasible region such that the objective function Z has the same optimal value, say k, then each such point corresponds to an optimal solution. The LPP has multiple solutions. Each of such optimal solutions is called multiple optimal solution or an alternative optimal solution.

The following two conditions must be satisfied for multiple optimal solutions to exist:

(*i*) the objective function must be parallel to a constraint which forms the boundary of the feasible region.

(*ii*) the constraint must form a boundary of the feasible region in the direction of the optimal movement of the objective function.

Example. *Solve the following problem graphically.*

Maximize $Z = 2.5x + y$

subject to the constraints $3x + 5y \le 15$

$5x + 2y \le 10$

$x, y \ge 0.$

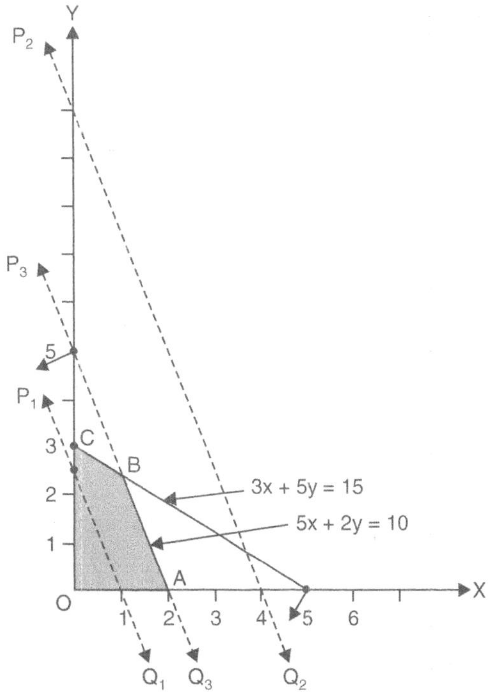

Sol. The feasible region is shown by the shaded area OABC in the figure.

We give a constant value 2.5 to Z and draw the line $2.5x + y = 2.5$, shown by P_1Q_1. Give another value 10 to Z and draw the line $2.5x + y = 10$, shown by P_2Q_2. Moving P_1Q_1 parallel to itself towards Z increasing *i.e.*, towards P_2Q_2 as far are possible, until the farthest point B within the feasible region is touched by the line, shown by P_3Q_3. Clearly P_3Q_3 contains the line segment AB which corresponds to the constraint $5x + 2y \le 10$ and forms the boundary of the feasible region. Solving $3x + 5y = 15$ and $5x + 2y = 10$, we have $B = \left(\frac{20}{19}, \frac{45}{19}\right)$ and

$Z(B) = \frac{5}{2}\left(\frac{20}{19}\right) + \frac{45}{19} = 5.$ Also $A = (2, 0)$ and $Z(A) = 2.5 \times 2 + 0 = 5$. All points on the line segment AB give the same optimal value Z = 5. The optimal value is unique but there are infinite number of optimal solutions. Every point on AB corresponds to an optimal solution.

Here the objective function has the same slope as the constraint line $5x + 2y = 10$.

IV. Redundancy. A constraint in a given LPP is said to be redundant if the feasible region of the problem is not affected by deleting that constraint.

Example. *Graph the feasible region of the following problem and identify a redundant constraint.*

Maximize $Z = 5x + 6y$

subject to the constraints

$$x + 2y \le 4$$
$$4x + 5y \le 20$$
$$3x + y \le 3$$
$$x, y \ge 0.$$

Sol. The feasible region is shown shaded in the figure. Clearly, the constraint $4x + 5y \le 20$ is redundant as it does not affect the feasible region. If this constraint is dropped, the feasible region remains the same.

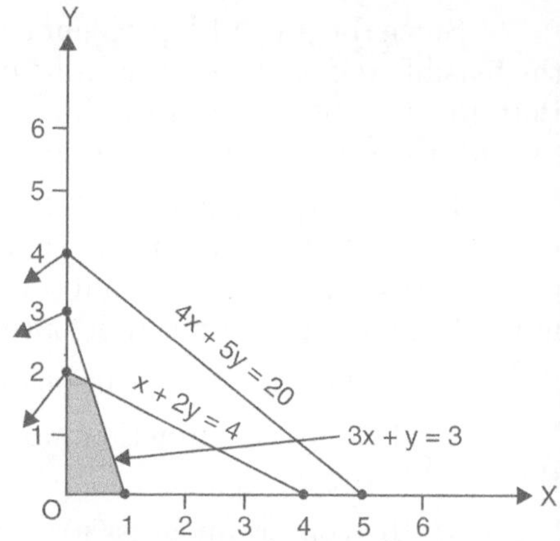

EXERCISE 7.3

Solve the following LP problems graphically (using corner point method):

1. Maximize $Z = 5x_1 + 7x_2$
subject to the constraints
$$x_1 + x_2 \le 4$$
$$3x_1 + 8x_2 \le 24$$
$$10x_1 + 7x_2 \le 35$$
$$x_1, x_2 \ge 0.$$
(M.D.U. 2010)

2. Maximize $Z = 2x + 3y$
subject to the constraints
$$x + y \le 30$$
$$3 \le y \le 12$$
$$x - y \ge 0$$
$$0 \le x \le 20.$$

3. Maximize $Z = 5x_1 + 3x_2$ *(M.D.U. May 2015, Dec. 2015)*
subject to the constraints
$$3x_1 + 5x_2 \le 15$$
$$5x_1 + 2x_2 \le 10$$
$$x_1, x_2 \ge 0.$$

4. Minimize $Z = 8x_1 + 12x_2$
subject to the constraints
$$60x_1 + 30x_2 \ge 240$$
$$30x_1 + 60x_2 \ge 300$$
$$30x_1 + 180x_2 \ge 540$$
$$x_1, x_2 \ge 0.$$

5. Minimize $Z = 3x + 2y$
subject to the constraints
$$5x + y \ge 10$$
$$x + y \ge 6$$
$$x + 4y \ge 12$$
$$x, y \ge 0.$$

6. If x_1, x_2 be real, show that the set

$$S = \left\{ (x_1, x_2) : \left| \begin{array}{c} x_1 + x_2 \le 50 \\ x_1 + 2x_2 \le 80 \\ 2x_1 + x_2 \ge 20 \\ x_1, x_2 \ge 0 \end{array} \right. \right\}$$

is a convex set. Find the extreme points of this set. Hence solve LPP graphically:
Maximize $Z = 4x_1 + 3x_2$ subject to the constraints given in S.

7. (*a*) Minimize $Z = 20x + 10y$

subject to the constraints $x + 2y \le 40$

$$3x + y \ge 30$$

$$4x + 3y \ge 60$$

$$x, y \ge 0.$$

(*b*) Use Graphical method to solve the LPP:

$$\text{Max. } Z = 3x_1 + 4x_2$$

subject to $4x_1 + 2x_2 \le 80$

$$2x_1 + 5x_2 \le 180$$

$$x_1, x_2 \ge 0$$

(*K.U.K. May 2013*)

8. (*a*) Find a geometrical interpretation and solution as well for the following L.P. problem:

Maximize $Z = 3x_1 + 5x_2$

subject to restrictions $x_1 + 2x_2 \le 2000$

$$x_1 + x_2 \le 1500$$

$$x_2 \le 600$$

and $x_1 \ge 0, x_2 \ge 0.$

(*b*) Solve the following L.P.P. graphically:

Maximize $Z = 5x_1 + 3x_2$

subject to $4x_1 + 5x_2 \le 1000$

$$5x_1 + 2x_2 \le 1000$$

$$3x_1 + 8x_2 \le 1200$$

$$x_1, x_2 \ge 0$$

(*K.U.K. 2010, 2013*)

(*c*) Using graphical method, solve the following L.P.P.

Maximize $Z = 2x_1 + 3x_2$ subject to

$$x_1 - x_2 \le 2$$

$$x_1 + x_2 \ge 4$$

$$x_1 \ x_2 \ge 0$$

(*M.D.U. 2011*)

9. (*a*) A housewife wishes to mix two types of foods X and Y in such a way that the vitamin contents of the mixture contain at least 8 units of vitamin A and 11 units of vitamin B. Food X costs ₹ 60 per kg and food Y costs ₹ 80 per kg. Food X contains 3 units per kg of vitamin A and 5 units per kg of vitamin B while food Y contains 4 units per kg of vitamin A and 2 units per kg of vitamin B. Formulate the problem as an LPP to minimise the cost of mixture.

(*b*) A house wife wishes to mix together two kinds of food, X and Y, in such a way that the mixture contains at least 10 units of vitamin A, 12 units of vitamin B and 8 units of vitamin C. The vitamin contents of one kg of food is given below:

	Vitamin A	*Vitamin B*	*Vitamin C*
Food X	1	2	3
Food Y	2	2	1

One kg of food X costs ₹ 6 and one kg of food Y costs ₹ 10. Formulate the above problem as an LPP to find the least cost of the mixture which will produce the diet and solve it.

(*c*) A manufacturer of furniture makes two products: chairs and tables. Processing of these products is done on two machines A and B. A chair requires 2 hours on machine A and 6 hours on machine B. A table requires 5 hours on machine A and no time on machine B. There are 16 hours of time per day available on machine A and 30 hours on machine B. Profit gained by the manufacturer from a chair is ₹ 10 and from a table is ₹ 50. What should be the daily production of the two?

10. A retired person wants to invest an amount of up to ₹ 20,000. His broker recommends investing in two types of bonds A and B, bond A yielding 10% return on the amount invested and bond B yielding 15% return on the amount invested. After some consideration, he decides to invest at least ₹ 5000 in bond A and no more than ₹ 8000 in bond B. He also wants to invest at least as much in bond A as in bond B. What should his broker suggest if he wants to maximize his return on investments? Formulate the above problem as an LPP and solve it.

11. A factory manufactures two types of screws, A and B, each type requiring the use of two machines, an automatic and a hand operated. It takes 4 minutes on the automatic and 6 minutes on hand operated machines to manufacture a package of screws A, while it takes 6 minutes on automatic and 3 minutes on the hand operated machines to manufacture a package of screws B. Each machine is available for at the most 4 hours on any day. The manufacturer can sell a package of screws A at a profit of 70 paise and screws B at a profit of ₹ 1. Assuming that he can sell all the screws he can manufacture, how many packages of each type should the factory owner produce in a day in order to maximize his profit? Determine the maximum profit.

12. A firm manufactures two products X and Y, each requiring the use of three machines M_1, M_2 and M_3. The time required for each product in hours and total time available in hours on each machine are as follows:

Machine	*Product X*	*Product Y*	*Available time (in hours)*
M_1	2	1	70
M_2	1	1	40
M_3	1	3	90

If the profit is ₹ 40 per unit for product X and ₹ 60 per unit for product Y, how many units of each product should be manufactured to maximize profit?

13. A dietician wishes to mix two types of food in such a way that the vitamin contents of the mixture contain at least 8 units of vitamin A and 10 units of vitamin C. Food I contains 2 units/kg of vitamin A and 1 unit/kg of vitamin C while food II contains 1 unit/kg of vitamin A and 2 units/kg of vitamin C. It costs ₹ 5 per kg to purchase food I and ₹ 7 per kg to purchase food II. Determine the minimum cost of such a mixture.

14. (*a*) An oil company has two depots, A and B, with capacities of 7000 *l* and 4000 *l*, respectively. The company is to supply oil to three petrol pumps D, E and F whose requirements are 4500 *l*, 3000 *l* and 3500 *l*, respectively. The distances (in km) between the depots and the petrol pumps are given in the following table:

From / *To*	*Distance (in km)*	
	A	*B*
D	7	3
E	6	4
F	3	2

Assuming that the transportation cost per km is ₹ 1 per litre, how should the delivery be scheduled in order that the transportation cost in minimum?

(*b*) There are two factories located one at place P and the other at place Q. From these locations, a certain commodity is to be delivered to each of the three depots situated at A, B and C. The weekly requirements of the depots are respectively 5, 5 and 4 units of the commodity while the production capacity of the factories at P and Q are respectively 8 and 6 units. The cost of transportation per unit is given below:

To / From	Cost (in ₹)		
	A	B	C
P	160	100	150
Q	100	120	100

How many units should be transported from each factory to each depot in order that the transportation cost is minimum? What will be the minimum transportation cost?

15. A manufacturer of patent medicines is preparing a production plan on medicines A and B. There are sufficient raw materials available to make 20,000 bottles of A and 40,000 bottles of B, but there are only 45,000 bottles into which either of the medicines can be put. Further, it takes 3 hours to prepare enough material to fill 1000 bottles of A and 1 hour to prepare enough material to fill 1000 bottles of B. There are 66 hours available for this operation. The profit is ₹ 8 per bottle for A and ₹ 7 per bottle for B. How should the manufacturer schedule his production in order to maximize his profit?

16. A small manufacturer has employed 5 skilled men and 10 semi-skilled men and makes an article in two qualities, a deluxe model and an ordinary model. The making of a deluxe model requires 2 hours of work by a skilled man and 2 hours of work by a semi-skilled man. The ordinary model requires 1 hour by a skilled man and 3 hours by a semi-skilled man. By union rule, no man may work for more than 8 hours per day. The manufacturer gains ₹ 15 on deluxe model and ₹ 10 on ordinary model. How many of each type should be made in order to maximize his total daily profit?

17. (*a*) Minimize $Z = 3x_1 + 2x_2$

subject to the constraints
$$5x_1 + x_2 \geq 10$$
$$x_1 + x_2 \geq 6$$
$$x_1 + 4x_2 \geq 12$$
$$x_1, x_2 \geq 0.$$

(*b*) Using graphical method, solve the LPP

Minimize $Z = 2x + 3y$

Subject to conditions $x \geq 0,\ y \geq 0,\quad 1 \leq x + 2y \leq 8.$

Also draw the feasible region of the constraints. (*K.U.K. Dec. 2015*)

18. A firm plans to purchase at least 200 quintals of scrap containing high quality metal X and low quality metal Y. It decides that the scrap to be purchased must contain at least 100 quintals of X-metal and not more than 35 quintals of Y-metal. The firm can purchase the scrap from two suppliers (A and B) in unlimited quantities. The percentage of X and Y metals in terms of weight in the scrap supplied by A and B is given below:

Metals	*Supplier A*	*Supplier B*
X	25%	75%
Y	10%	20%

The price of A's scrap is ₹ 200 per quintal and that of B is ₹ 400 per quintal. The firm wants to determine the quantities that it should buy from the two suppliers so that total cost is minimized. Formulate this as an LPP and solve it.

19. G.J. Breweries Ltd. has two bottling plants, one located at 'G' and the other at 'J'. Each plant produces three drinks: whisky, beer and brandy, named A, B and C, respectively. The number of bottles produced per day are as follows:

Drink	*Plant at*	
	G	*J*
Whisky	1,500	1,500
Beer	3,000	1,000
Brandy	2,000	5,000

A market survey indicates that during the month of July, there will be a demand of 20,000 bottles of whisky, 40,000 bottles of beer and 44,000 bottles of brandy. The operating cost per day for plants at G and J are 600 and 400 monetary units. For how many days should each plant be run in July so as to minimize the production cost, while still meeting the market demand? Solve graphically.

20. Verify that the following problems have no feasible solution.

(*i*) Maximize $Z = 15x + 20y$

subject to the constraints
$$x + y \geq 12$$
$$6x + 9y \leq 54$$
$$15x + 10y \leq 90$$
$$x, y \geq 0$$

(*ii*) Maximize $Z = 4x + 3y$

subject to the constraints
$$x + y \leq 3$$
$$2x - y \leq 3$$
$$x \geq 4$$
$$x, y \geq 0$$

21. Verify that the following problems have multiple solutions.

(*i*) Minimize $Z = 12x + 9y$

subject to the constraints
$$3x + 6y \geq 36$$
$$4x + 3y \geq 24$$
$$x + y \leq 15$$
$$x, y \geq 0.$$

(*ii*) Maximize $Z = 6x + 4y$

subject to the constraints
$$x + y \leq 5$$
$$3x + 2y \leq 12$$
$$x, y \geq 0$$

22. Verify that the following problem has an unbounded solution:

Maximize $Z = 20x + 30y$

subject to the constraints
$$5x + 2y \geq 50$$
$$2x + 6y \geq 20$$
$$4x + 3y \geq 60$$
$$x, y \geq 0$$

23. Graph the feasible region of the following problem and identify a redundant constraint.

Minimize $Z = 6x + 10y$

subject of the constraints
$$x \geq 6$$
$$y \geq 2$$
$$2x + y \geq 10$$
$$x, y \geq 0.$$

Answers

1. $x_1 = \frac{8}{5}, x_2 = \frac{12}{5}$; Max. $Z = 24.8$

2. $x = 18,\ y = 12$; Max. $Z = 72$

3. $x_1 = \frac{20}{19}, x_2 = \frac{45}{19}$; Max. $Z = \frac{235}{19}$

4. $x_1 = 2, x_2 = 4$; Min. $Z = 64$

5. $x_1 = 1, x_2 = 5$; Min. $Z = 13$

6. $x_1 = 50, x_2 = 0$; Max. $Z = 200$

7. (*a*) $x = 6, y = 12$; Min. Z = 240 (*b*) $x_1 = 0, x_2 = 40$; Max. 160

8. (*a*) $x_1 = 1000, x_2 = 500$; Max. Z = 5500

(*b*) $x_1 = \frac{3000}{17}, x_2 = \frac{1000}{17}$, Max. Z = 1058.82 (*c*) No maximum

9. (*a*) Minimum cost = ₹ 160

(*b*) 2 kg of food X, 4 kg of food Y; least cost = ₹ 52

(*c*) Number of chairs = 0, Number of tables = 16/5 *i.e.,* 16 in 5 days

10. ₹ 12000 in bond A, ₹ 8,000 in bond B; Maximum return = ₹ 2400

11. Screw A = 30 packages, Screw B = 20 packages; Maximum profit = ₹ 41

12. Product X = 15, Product Y = 25; Maximum profit = ₹ 2100

13. Food I = 2 kg, Food II = 4 kg; Minimum cost = ₹ 38

14. (*a*) From A: 500 *l*, 3000 *l*, 3500 *l* to D, E, F respectively, From B: 4000 *l*, 0 *l*, 0 *l* to D, E, F respectively; Minimum cost = ₹ 44,000

(*b*) From P: 0, 5, 3 units to A, B, C respectively; From Q: 5, 0, 1 units to A, B, C respectively; Minimum transportation cost = ₹ 1550.

15. 10500 bottles of A, 34500 bottles of B; Maximum profit = ₹ 325500

16. Delux model = 10, ordinary model = 20; Maximum profit = ₹ 350

[**Hint.** 5 skilled men cannot work for more than $5 \times 8 = 40$ hours and 10 semi-skilled men cannot work for more than $10 \times 8 = 80$ hours.]

17. (*a*) $x_1 = 1, x_2 = 5$; Min. Z = 13 (*b*) Min. Z = 3/2

18. Supplier A = 100 quintals, supplier B = 100 quintals; Minimum cost = ₹ 60,000

19. Plant at G = 12 days, Plant at J = 4 days; Minimum cost = 8,800 monetary units.

23. $2x + y \geq 10$.

THE SIMPLEX METHOD

7.7. INTRODUCTION

The linear programming problems discussed so far are concerned with two variables, the solution of which can be found out easily by the graphical method. But most real life problems when formulated as an LP model involve more than two variables and many constraints. Thus, there is a need for a method other than the graphical method. The most popular non-graphical method of solving an LPP is called the **simplex method.** This method, developed by George B. Dantzig in 1947, is applicable to any problem that can be formulated in terms of linear objective function subject to a set of linear constraints. There are no theoretical restrictions placed on the number of decision variables or constraints. The development of computers has further made it easy for the simplex method to solve large scale LP problems very quickly.

The concept of simplex method is similar to the graphical method. For LP problems with several variables, the optimal solution lies at a corner point of the many-faced, multi-dimensional figure, called an n-dimensional polyhedron. The simplex method examines the corner points in a systematic manner. It is a computational routine of repeating the same set of steps over and over until an optimal solution is reached. For this reason, it is known as an **iterative method.** As we move from one iteration to the other, the method improves the value of the objective function and achieves optimal solution in a finite number of iterations.

7.8. SOME USEFUL DEFINITIONS *(M.D.U. May 2013; K.U.K. 2013)*

(*i*) **Slack Variable.** A variable added to the left hand side of a less than or equal to constraint to convert the constraint into an equality is called a *slack variable.*

For example, to convert the constraint

$$3x + 2y \leq 18 \quad \ldots(1)$$

into an equation, we add a slack variable s to the left-hand side thereby getting the equality

$$3x + 2y + s = 18$$

Clearly, s must be non-negative, since $s = 18 - (3x + 2y) \geq 0$ [by (1)]

In economic terminology, a slack variable represents unused resource in the form of money, labour hours, time on a machine etc.

(*ii*) **Surplus Variable.** A variable subtracted from the left-hand side of a greater than or equal to constraint to convert the constraint into an equality is called a *surplus variable.* *(M.D.U. May 2013)*

For example, to convert the constraint

$$2x + 3y \geq 30 \quad \ldots(1)$$

into an equation, we subtract a surplus variable s from the left-hand side thereby getting the equality

$$2x + 3y - s = 30$$

Clearly, s must be non-negative, since $s = (2x + 3y) - 30 \geq 0$ [by (1)]

A surplus variable represents the surplus of left-hand side over the right-hand side. It is also called a negative slack variable.

(*iii*) **Basic Solution.** For a system of m simultaneous linear equations in n variables ($n > m$), a solution obtained by setting ($n - m$) variables equal to zero and solving for the remaining m variables for a unique solution is called a **basic solution.** The ($n - m$) variables set equal to zero in any solution are called **non-basic variables.** The other m variables are called **basic variables.**

Total number of basic solution = ${}^nC_m$

(*iv*) **Basic Feasible Solution.** A basic solution which happens to be feasible (*i.e.,* a solution in which each basic variable is non-negative) is called a basic feasible solution.

(*v*) **Degenerate and Non-degenerate Solution.** If one or more of the basic variables in the basic feasible solution are zero, then it is called a **degenerate solution.** If all the variables in the basic feasible solution are positive, then it is called a non-degenerate solution.

Example. *Consider the following LPP:* (M.D.U. Dec. 2014)

Maximize $Z = 5x + 8y$

subject to the constraints $x + y \leq 4$

$$2x + y \leq 6$$

$$x, y \geq 0.$$

Sol. Introducing slack variables s_1, s_2 to convert less than or equal to constraints into equalities, we get

$$x + y + s_1 = 4$$

$$2x + y + s_2 = 6$$

We have two equations ($m = 2$) in four variables ($n = 4$). For a basic solution, we put $n - m = 4 - 2 = 2$ variables equal to zero and solve for the other two. The various possibilities are shown in the following table:

Number of basic solution	*Non-basic variables (each = 0)*	*Basic variables*	*Basic solutions*
1	s_1, s_2	x, y	$x = 2, y = 2$
2	x, s_1	y, s_2	$y = 4, s_2 = 2$
3	x, s_2	y, s_1	$y = 6, s_1 = -2$
4	y, s_1	x, s_2	$x = 4, s_2 = -2$
5	y, s_2	x, s_1	$x = 3, s_1 = 1$
6	x, y	s_1, s_2	$s_1 = 4, s_2 = 6$

There are ${}^nC_m = {}^4C_2 = 6$ basic solutions. Solutions (3) and (4) contain a variable which has a negative value. In the remaining four basic solutions, each variable has a positive value. Therefore solutions (1), (2), (5) and (6) are basic feasible solutions. The four basic feasible solutions (x, y) are:

$$(0, 0), \quad (3, 0), \quad (2, 2) \quad \text{and} \quad (0, 4)$$

which correspond to the corner points O, A, B and C respectively of the feasible region as shown in the figure.

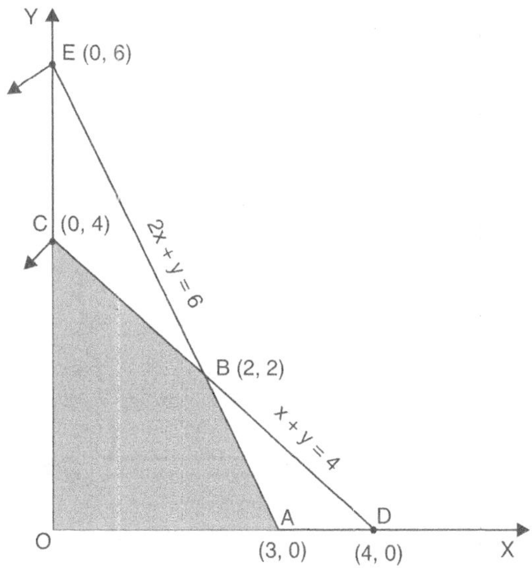

Thus every corner point of the feasible region corresponds to a basic feasible solution and *vice versa*.

Moreover, the basic feasible solutions (0, 0), (3, 0) and (0, 4) are degenerate whereas the basic feasible solution (2, 2) is non-degenerate.

7.9. STANDARD FORM OF AN LPP

The standard form of an LPP should have the following characteristics:

(*i*) Objective function should be of maximization type.

(*ii*) All constraints should be expressed as equations by adding slack or surplus variables, one for each constraint.

(*iii*) The right-hand side of each constraint should be non-negative. If it is negative, then to make it positive, we multiply both sides of the constraint by (– 1), changing ≤ to ≥ and *vice versa*.

(*iv*) All variables are non-negative.

Thus, the standard form of an LPP with n variables and m constraints is:

Maximize $Z = c_1x_1 + c_2x_2 + \cdots\cdots + c_nx_n + 0s_1 + 0s_2 + \cdots\cdots + 0s_m$

subject to the constraints

$$a_{11}x_1 + a_{12}x_2 + \cdots\cdots + a_{1n}x_n + s_1 = b_1$$

$$a_{21}x_1 + a_{22}x_2 + \ldots\ldots + a_{2n}x_n + s_2 = b_2$$

$$\ldots\ldots\ldots\ldots\ldots\ldots\ldots\ldots\ldots\ldots\ldots\ldots$$

$$\ldots\ldots\ldots\ldots\ldots\ldots\ldots\ldots\ldots\ldots\ldots\ldots$$

$$a_{m1}x_1 + a_{m2}x_2 + \ldots\ldots + a_{mn}x_n + s_m = b_m$$

and $$x_1, x_2 \ldots\ldots, x_n, s_1, s_2, \ldots\ldots, s_m \geq 0$$

Remark. Since minimum Z = – maximum Z*, where Z* = – Z, the objective function can always be expressed in the maximization form.

ILLUSTRATIVE EXAMPLES

Example. *Express the following LPP in the standard form:*

$$\textit{Maximize } Z = 2x_1 + 5x_2 + 7x_3$$

subject to the constraints
$$5x_1 - 3x_2 \leq 4$$
$$7x_1 + 6x_2 + 9x_3 \geq 15$$
$$8x_1 + 6x_3 \leq 5$$
$$x_1, x_2, x_3 \geq 0.$$

Sol. Introducing the slack/surplus variables, the given LPP in standard form becomes:

$$\text{Maximize } Z = 2x_1 + 5x_2 + 7x_3 + 0s_1 + 0s_2 + 0s_3$$

subject to the constraints

$$5x_1 - 3x_2 + s_1 = 4$$
$$7x_1 + 6x_2 + 9x_3 - s_2 = 15$$
$$8x_1 + 6x_3 + s_3 = 5$$
$$x_1, x_2, x_3, s_1, s_2, s_3 \geq 0.$$

EXERCISE 7.4

1. Find all the basic solutions of the following system of equations identifying in each case the basic and non-basic variables:
$$2x_1 + x_2 + 4x_3 = 11, \quad 3x_1 + x_2 + 5x_3 = 14.$$
Investigate whether the basic solutions are degenerate basic solutions or not. Hence find the basic feasible solution of the system.

2. Obtain all the basic solutions to the following system of linear equations:
$$x_1 + 2x_2 + x_3 = 4, \quad 2x_1 + x_2 + 5x_3 = 5$$ *(K.U.K. Jan. 2014)*
Which of them are feasible? Point out non-degenerate basic solutions, if any.

3. Show that the following system of linear equations has two degenerate feasible solutions and the non-degenerate basic solution is not feasible.
$$2x_1 + x_2 - x_3 = 2, \quad 3x_1 + 2x_2 + x_3 = 3.$$

4. Find all the basic solutions to the following problem:
Maximize $Z = x_1 + 3x_2 + 3x_3$
subject to the constraints $x_1 + 2x_2 + 3x_3 = 4$
$$2x_1 + 3x_2 + 5x_3 = 7$$
$$x_1, x_2, x_3 \geq 0$$
Which of the basic solutions are
(*i*) non-degenerate basic feasible (*ii*) optimal basic feasible? *(K.U.K. May 2013)*

5. Find an optimal solution to the following LPP by computing all basic solutions and then finding one that maximizes the objective function:

Maximize $Z = 2x_1 + 3x_2 + 4x_3 + 7x_4$

subject to the constraints $2x_1 + 3x_2 - x_3 + 4x_4 = 8$

$$x_1 - 2x_2 + 6x_3 - 7x_4 = -3$$

$$x_1, x_2, x_3, x_4 \geq 0$$

6. Express the following LP problem in the standard form:

Maximize $Z = 3x_1 + 4x_2 + 5x_3$

subject to the constraints $2x_1 + 7x_2 + x_3 \leq 10$

$$5x_1 + 9x_2 + 4x_3 \geq 20$$

$$8x_1 + 15x_3 \leq 30$$

$$x_1, x_2, x_3 \geq 0$$

Answers

1.

Non-basic variables	*Basic variables*	*Basic solutions*	
x_3	x_1, x_2	$x_1 = 3,$	$x_2 = 5$
x_1	x_2, x_3	$x_2 = -1,$	$x_3 = 3$
x_2	x_1, x_3	$x_1 = \frac{1}{2},$	$x_3 = \frac{5}{2}$

First and third are basic feasible solutions which are also non-degenerate basic solutions.

2. $x_1 = 2, \quad x_2 = 1$; feasible, non-degenerate

$x_2 = \dfrac{5}{3}, \quad x_3 = \dfrac{2}{3}$; feasible, non-degenerate

$x_1 = 5, \quad x_3 = -1$; non-feasible.

4. $x_1 = 2, x_2 = 1$; $(x_3 = 0)$ $\quad x_1 = x_3 = 1$; $(x_2 = 0)$ $\quad x_2 = -1, x_3 = 2$ $(x_1 = 0)$

(*i*) First two solutions are non-degenerate basic feasible solutions

(*ii*) First solution is optimal basic feasible and Max. Z = 5.

5. Optimal basic feasible solution is $x_1 = 0, x_2 = 0, x_3 = \dfrac{44}{17}, x_4 = \dfrac{45}{17}$ and Max. $Z = \dfrac{491}{17}$

6. Maximize $Z = 3x_1 + 4x_2 + 5x_3 + 0s_1 + 0s_2 + 0s_3$

subject to the constraints

$$2x_1 + 7x_2 + x_3 + s_1 = 10$$

$$5x_1 + 9x_2 + 4x_3 - s_2 = 20$$

$$8x_1 + 15x_3 + s_3 = 30$$

$$x_1, x_2, x_3, s_1, s_2, s_3 \geq 0$$

7.10. WORKING PROCEDURE OF SIMPLEX METHOD

Assuming the existence of an initial basic feasible solution, the optimal solution to an LPP by simplex method is obtained as follows:

Step 1: To express the LPP in the Standard Form

(*i*) Formulate the mathematical model of the given LPP.

(*ii*) If the objective function is to be minimized, then convert it into a maximization problem by using

$$\text{Min. } Z = -\text{ Max. } (-Z)$$

(*iii*) The right-hand side of each constraint should be non-negative.

(*iv*) Express all constraints as equations by introducing slack/surplus variables, one for each constraint.

(*v*) Restate the given LPP in standard form:

Maximize $Z = c_1x_1 + c_2x_2 + \cdots\cdots + c_nx_n + 0s_1 + 0s_2 + \cdots\cdots + 0s_m$

subject to the constraints $a_{11}x_1 + a_{12}x_2 + \cdots\cdots + a_{1n}x_n + s_1 = b_1$

$$a_{21}x_1 + a_{22}x_2 + \cdots\cdots + a_{2n}x_n + s_2 = b_2$$

$$\cdots\cdots\cdots\cdots\cdots\cdots\cdots\cdots\cdots\cdots$$

$$\cdots\cdots\cdots\cdots\cdots\cdots\cdots\cdots\cdots\cdots$$

$$a_{m1}x_1 + a_{m2}x_2 + \cdots\cdots + a_{mn}x_n + s_m = b_m$$

and $x_1, x_2, \ldots\ldots, x_n, s_1, s_2, \ldots\ldots, s_m \geq 0$

Step 2: To set up the Initial Basic Feasible Solution

Take the m slack/surplus variables $s_1, s_2, \ldots\ldots, s_m$ as the basic variables so that the n given variables $x_1, x_2, \ldots\ldots, x_n$ are non-basic variables. As such, $x_1 = x_2 = \ldots\ldots = x_n = 0$ and $s_1 = b_1, s_2 = b_2, \ldots\ldots, s_m = b_m$.

Since each b_i is non-negative [see Step I (*iii*)], the basic solution is feasible. This basic feasible solution is the starting point of the iterative process. The simplex method then proceeds to solve the LPP by designing and re-designing successively better basic feasible solutions until an optimal solution is obtained.

Step 3: To set up the Initial Simplex Table

The above information is conveniently expressed in the tabular form as shown below. For computational efficiency, the tabular form is better.

		$c_j \rightarrow$	c_1	c_2	......c_n	0	0	 0	Ratio
c_B	*Basic Variables* B	*Solution* $b(= x_B)$	x_1	x_2	*Variables* ...... x_n	s_1	s_2	...... s_m	x_B/x_b
$(c_{B1} =)\ 0$	s_1	$(x_{B1} =) b_1$	a_{11}	a_{12}	...... a_{1n}	1	0	 0	
$(c_{B2} =)\ 0$	s_2	$(x_{B2} =) b_2$	a_{21}	a_{22}	...... a_{2n}	0	1	 0	
⋮	⋮	⋮	⋮	⋮	⋮	⋮	⋮	⋮	
$(c_{Bm} =)\ 0$	s_m	$(x_{Bm} =) b_m$	a_{m1}	a_{m2}	...... a_{mn}	0	0	 1	
$Z = \Sigma c_{Bi}\, x_{Bi}$		$Z_j = \Sigma c_{Bi} a_{ij}$	0	0	 0	0	0	 0	
		$C_j = c_j - Z_j$	$c_1 - Z_1$	$c_1 - Z_2$	...... $c_n - Z_n$	0	0	 0	

The data in the table is interpreted as follows:

1. In the top row labelled c_j, we write the co-efficients of the variables in the objective function. These values will remain the same in subsequent simplex tables.
2. The second row shows the major column headings for the simplex table. These headings remain the same in subsequent simplex tables and apply to the values listed in the first m-rows.
3. In the first column labelled 'c_B' we write the co-efficients of the current basic variables in the objective function. c_{Bi} is the co-efficient of the ith basic variable.
4. In the second column labelled **'Basic Variables'**, we write the basic variables. This column can also be labelled as **'Basis'.**
5. In the column labelled **'Solution'**, we write the current values of the corresponding basic variables. The remaining non-basic variables will always be zero.
6. The **body matrix** under non-basic variables $x_1, x_2, \ldots\ldots, x_n$ consists of the co-efficients of the decision variables in the constraint set.
7. The **identity matrix** under $s_1, s_2, \ldots\ldots, s_m$ represents the co-efficients of the slack variables in the constraint set.
8. To get an entry in the z_j-row under a column, we multiply the entries in that column by the corresponding entries of c_B column and add up the products. Thus the entry in the z_j-row under the column 'x_2' is obtained by multiplying the entries under 'x_2', *viz.*, $a_{12}, a_{22}, \ldots\ldots, a_{m2}$ by the corresponding entries under c_B, *viz.*, $c_{B1}, c_{B2}, \ldots\ldots, c_{Bm}$ and adding, thereby getting

$$c_{B1}a_{11} + c_{B2}a_{22} + \cdots\cdots + c_{Bm}a_{m2}$$

In the initial simplex table, the z_j-row entries will be all equal to zero. The Z entry under the c_B column gives the current value of the objective function. Thus $Z = \Sigma c_{Bi}x_{Bi} = c_{B1}x_{B1} + c_{B2}x_{B2} + \cdots\cdots + c_{Bm}x_{Bm}$. The values of z_j represent the amount by which the value of objective function Z would be decreased (or increased) if one of the variables not included in the solution were brought into the solution.

9. The last row labelled '$C_j = c_j - Z_j$' is called the **index row** or **net evaluation row.**

C_j for any column = (c_j value for that column written at the top of that column) – (Z_j value for that column)

It should be noted that C_j values are meaningful for non-basic variables only. For a basic variable, $C_j = 0$.

The C_j-row represents the net contribution to the objective function that results by introducing one unit of each of the respective column variables. A plus value indicates that a greater contribution can be made by bringing the variable for that column into the solution. A negative value indicates the amount by which contribution would decrease if one unit of the variable for that column were brought into the solution.

Step 4: Test for Optimality. Examine the entries in C_j-row. If all entries in this row are negative or zero, *i.e.*, if $C_j \leq 0$, then the basic feasible solution is optimal. Any positive entry in the row indicates that an improvement in the value of objective function Z is possible and, hence, we proceed to the next step.

Step 5: To Identify the Incoming and Outgoing Variables. If there is a positive entry in the C_j-row, then simplex method shifts from the current basic feasible solution to a better basic feasible solution. For this, we have to replace one current basic variable (called the **outgoing or departing variable)** by a new non-basic variable (called the **incoming or entering variable**).

Determination of Incoming Variable. The column with the largest positive entry in the C_j-row is called the **key** (or **pivot**) **column** (which is shown marked with an arrow ↑). The non-basic variable which will replace a basic variable is the one lying in the key column. Thus the incoming variable is located.

If more than one variable has the same positive largest entry in the C_j-row, then any of these variables may be selected arbitrarily as the incoming variable.

Determination of Outgoing Variable. Divide each entry of the solution column (*i.e.*, x_B-column) by the corresponding positive entry in the key column. These quotients are written in the last column labelled 'Ratio'. The row which corresponds to the smallest non-negative quotient is called the **key** (or **pivot**) **row** (which is shown marked with an arrow →). The departing variable is the corresponding basic variable in this row. The element at the intersection of the key row and key column is called the **key** (or **pivot**) **element.** We place a circle around this element.

If all these ratios are negative or zero, the incoming variable can be made as large as we please without violating the feasibility condition. Hence the problem has an **unbounded solution** and no further iteration is required. *(M.D.U. Dec. 2013)*

Step 6: To set up the new simplex table from the current one.

Drop the outgoing variable and introduce the incoming variable alongwith its associated value under c_B column.

If the key element is 1, then the key row remains the same in the new simplex table.

If the key element is not 1, then to reduce it to 1, divide each element in the key row (including elements in x_B-column) by the key element.

Thus, $$\text{new key row} = \frac{\textbf{old key row}}{\textbf{key element}}$$

Make all other elements of the key column 0 by subtracting suitable multiples of key row from the other rows. In other words, to change the non-key rows, we use the formula:

Number in new non-key row = (Number in old non-key row) – (key column entry)

× (corresponding number in new key-row)

where **'key column entry'** is the entry in this row that is in the key column.

Step 7: Test for Optimality. If there is no positive entry in C_j-row, we have an optimal solution. Otherwise, go to Step 4 and repeat the procedure until all entries in C_j-row are either negative or zero.

ILLUSTRATIVE EXAMPLES

Example 1. *Using simplex method*

Maximize $Z = 2x_1 + 5x_2$

subject to

$$x_1 + 4x_2 \le 24$$
$$3x_1 + x_2 \le 21$$
$$x_1 + x_2 \le 9$$
$$x_1, x_2 \ge 0.$$ (M.D.U. 2012)

Sol. Step 1. To formulate the mathematical model of the problem. We are already given the mathematical model of the LPP. The problem is of maximization type and all b's are positive.

Step 2. To express the problem in standard form. Introducing slack variables s_1, s_2, s_3 (one for each constraint) the problem in standard form is:

Maximize $Z = 2x_1 + 5x_2 + 0s_1 + 0s_2 + 0s_3$

subject to

$$x_1 + 4x_2 + s_1 = 24 \quad ...(1)$$
$$3x_1 + x_2 + s_2 = 21 \quad ...(2)$$
$$x_1 + x_2 + s_3 = 9 \quad ...(3)$$
$$x_1, x_2, s_1, s_2, s_3 \ge 0$$

Step 3. To set up the initial basic feasible solution. Since we have 3 equations in 5 variables, a solution is obtained by setting $5 - 3 = 2$ variables equal to zero and solving for the remaining 3 variables. We start with a basic solution by setting $x_1 = x_2 = 0$. Substituting $x_1 = x_2 = 0$ in (1), (2) and (3), we get the basic solution

$$s_1 = 24, \quad s_2 = 21, \quad s_3 = 9$$

Since s_1, s_2, s_3 are all positive, the basic solution is feasible and non-degenerate.

Thus our initial basic feasible solution is

$$x_1 = 0, \quad x_2 = 0; \quad s_1 = 24, \quad s_2 = 21, \quad s_3 = 9$$

at which Z = 0. This initial basic feasible solution is summarized in the following initial simplex table.

Simplex Table I

c_B	*Basis*	$c_j \rightarrow$	2	5	0	0	0	*Ratio*
		Solution $b(= x_B)$	x_1	x_2	s_1	s_2	s_3	x_B/x_2
0	s_1	24	1	(4)	1	0	0	$\frac{24}{4} = 6 \rightarrow$
0	s_2	21	3	1	0	1	0	$\frac{21}{1} = 21$
0	s_3	9	1	1	0	0	1	$\frac{9}{1} = 9$
Z = 0		Z_j	0	0	0	0	0	
		$C_j = c_j - Z_j$	2	5 ↑	0	0	0	

For writing the body matrix (under x_1, x_2) and the identity matrix (under s_1, s_2, s_3), the left hand sides of equations (1), (2), (3) should be treated as

$$1x_1 + 4x_2 + 1s_1 + 0s_2 + 0s_3$$
$$3x_1 + 1x_2 + 0s_1 + 1s_2 + 0s_3$$
$$1x_1 + 1x_2 + 0s_1 + 0s_2 + 1s_3$$

The Z_j-row entries are all equal to zero in the initial simplex table.

Step 4. To test for optimality. Since some entries in C_j-row are positive, the current solution is not optimal. Therefore, an improvement in the value of objective function Z is possible and we proceed to the next step.

Step 5. To identify the incoming and outgoing variables. The largest positive entry in the C_j-row is 5 and the column in which it appears is headed by x_2. Thus x_2 is the incoming variable and x_2-column is the key column (indicated by ↑).

Dividing each entry of the solution column by the corresponding positive entry in the key column, we find that minimum positive ratio is 6 and it appears in the row headed by s_1. Thus s_1 is the outgoing basic variable and the corresponding row is the key row (indicated by →). The number at the intersection of the key row and the key column is the key number. Thus 4 is the key number. A circle is placed around this number.

Step 6. To set up the new simplex table from the current one. Drop the outgoing variable s_1 from the basis and introduce the incoming variable x_2. The new basis will contain x_2, s_2, s_3 as the basic variables. The co-efficient of x_2 in the objective function is 5. Therefore the entry in c_B column corresponding to the new basic variable x_2 will be 5. Since the key element enclosed in the circle is not 1, divide all elements of the key row by the key element 4 to obtain new values of the elements in this row. Thus the key row

$$0 \quad s_1 \quad 24 \quad 1 \quad 4 \quad 1 \quad 0 \quad 0$$

is replaced by the **new key row**

$$5 \quad x_2 \quad 6 \quad \frac{1}{4} \quad 1 \quad \frac{1}{4} \quad 0 \quad 0$$

Now we make all other elements of the key column (*i.e.*, x_2-column) zero by subtracting suitable multiples of key row from the other rows.

To replace non-key rows, we use the formula:

Number in new non-key row = (Number in old non-key row) – (key column entry) × (corresponding number in new key row)

Here s_2-row and s_3-row (*i.e.,* second and third rows) are non-key rows.

Transformation of R_2. Key column entry in R_2 is 1.

$$\therefore \quad R_2(\text{new}) = R_2(\text{old}) - 1 \times R_1(\text{new})$$

$$21 - 1 \times 6 = 15$$

$$3 - 1 \times \frac{1}{4} = \frac{11}{4}$$

$$1 - 1 \times 1 = 0$$

$$0 - 1 \times \frac{1}{4} = -\frac{1}{4}$$
$$1 - 1 \times 0 = 1$$
$$0 - 1 \times 0 = 0$$

Transformation of R_3. Key column entry in R_3 is 1.

$$\therefore \quad R_3(\text{new}) = R_3(\text{old}) - 1 \times R_1(\text{new})$$

$$9 - 1 \times 6 = 3$$
$$1 - 1 \times \frac{1}{4} = \frac{3}{4}$$
$$1 - 1 \times 1 = 0$$
$$0 - 1 \times \frac{1}{4} = -\frac{1}{4}$$
$$0 - 1 \times 0 = 0$$
$$1 - 1 \times 0 = 1$$

The above information is summarized in the following table:

Simplex Table II

c_B	*Basis*	$c_j \rightarrow$	*2*	*5*	*0*	*0*	*0*	*Ratio*
		Solution $b(= x_B)$	x_1	x_2	s_1	s_2	s_3	x_B/x_1
5	x_2	6	$\frac{1}{4}$	1	$\frac{1}{4}$	0	0	$\frac{6}{(1/4)} = 24$
0	s_2	15	$\frac{11}{4}$	0	$-\frac{1}{4}$	1	0	$\frac{15}{(11/4)} = \frac{60}{11}$
0	s_3	3	$\boxed{\frac{3}{4}}$	0	$-\frac{1}{4}$	0	1	$\frac{3}{(3/4)} = 4 \rightarrow$
Z = 30		Z_j	$\frac{5}{4}$	5	$\frac{5}{4}$	0	0	
		$C_j = c_j - Z_j$	$\frac{3}{4}$ ↑	0	$-\frac{5}{4}$	0	0	

$^*Z = \Sigma c_{Bi} x_{Bi} = 5(6) + 0(15) + 0(3) = 30$

***Entries in Z_j-row.** In the column headed by

x_1	$5\left(\frac{1}{4}\right) + 0\left(\frac{11}{4}\right) + 0\left(\frac{3}{4}\right)$	$= \frac{5}{4}$
x_2	$5(1) + 0(0) + 0(0)$	$= 5$
s_1	$5\left(\frac{1}{4}\right) + 0\left(-\frac{1}{4}\right) + 0\left(-\frac{1}{4}\right)$	$= \frac{5}{4}$
s_2	$5(0) + 0(1) + 0(0)$	$= 0$
s_3	$5(0) + 0(0) + 0(1)$	$= 0$

Step 7. Test for optimality. Since all entries in C_j-row are not negative or zero, the current solution is not optimal. Therefore, an improvement in the value of objective function Z is possible and we repeat steps 5 to 7.

Incoming variable is x_1 and x_1-column is the key-column. s_3 is the outgoing basic variable and s_3-row is the key row. Key number is $\frac{3}{4}$.

The new basis will contain x_2, s_2, x_1 as the basic variables. Dividing all elements of the key row by the key number $\frac{3}{4}$, the new key row is

$$2 \quad x_1 \quad 4 \quad 1 \quad 0 \quad -\frac{1}{3} \quad 0 \quad \frac{4}{3}$$

Transformation of R_1. Key column entry in R_1 is $\frac{1}{4}$

$$\therefore \qquad R_1(\text{new}) = R_1(\text{old}) - \frac{1}{4}R_3(\text{new})$$

$$6 - \frac{1}{4}(4) = 5$$

$$\frac{1}{4} - \frac{1}{4}(1) = 0$$

$$1 - \frac{1}{4}(0) = 1$$

$$\frac{1}{4} - \frac{1}{4}\left(-\frac{1}{3}\right) = \frac{1}{3}$$

$$0 - \frac{1}{4}(0) = 0$$

$$0 - \frac{1}{4}\left(\frac{4}{3}\right) = -\frac{1}{3}$$

Transformation of R_2. Key column entry in R_2 is $\frac{11}{4}$.

$$\therefore \qquad R_2(\text{new}) = R_2(\text{old}) - \frac{11}{4}R_3(\text{new})$$

$$15 - \frac{11}{4}(4) = 4$$

$$\frac{11}{4} - \frac{11}{4}(1) = 0$$

$$0 - \frac{11}{4}(0) = 0$$

$$-\frac{1}{4} - \frac{11}{4}\left(-\frac{1}{3}\right) = \frac{2}{3}$$

$$1 - \frac{11}{4}(0) = 1$$

$$0 - \frac{11}{4}\left(\frac{4}{3}\right) = -\frac{11}{3}$$

The above information is summarized in the following table.

Simplex Table III

		$c_j \rightarrow$	2	5	0	0	0	*Ratio*
c_B	*Basis*	*Solution* $b(=x_B)$	x_1	x_2	s_1	s_2	s_3	
5	x_2	5	0	1	$\frac{1}{3}$	0	$-\frac{1}{3}$	
0	s_2	4	0	0	$\frac{2}{3}$	1	$-\frac{11}{3}$	
2	x_1	4	1	0	$-\frac{1}{3}$	0	$\frac{4}{3}$	
Z = 33		Z_j	2	5	1	0	1	
		$C_j = c_j - Z_j$	0	0	-1	0	-1	

$^*Z = \Sigma\, c_{Bi}x_{Bi} = 5(5) + 0(4) + 2(4) = 33$

As all entries in C_j-row are either negative or zero, the above table gives the optimal basic feasible solution.

$\therefore$ The optimal solution is $x_1 = 4$, $x_2 = 5$ and Max. Z = 33.

Example 2. *Use the simplex method to solve the following LP problem.*

Maximize $Z = 3x_1 + 5x_2 + 4x_3$

subject to

$$2x_1 + 3x_2 \le 8$$
$$2x_2 + 5x_3 \le 10$$
$$3x_1 + 2x_2 + 4x_3 \le 15$$
$$x_1, x_2, x_3 \ge 0$$

(K.U.K. Jan. 2014)

Sol. Step 1. Introducing slack variables s_1, s_2, s_3 (one for each constraint) the problem in standard form is:

Maximize $Z = 3x_1 + 5x_2 + 4x_3 + 0s_1 + 0s_2 + 0s_3$

subject to

$$2x_1 + 3x_2 + s_1 = 8 \quad \text{...(1)}$$
$$2x_2 + 5x_3 + s_2 = 10 \quad \text{...(2)}$$
$$3x_1 + 2x_2 + 4x_3 + s_3 = 15 \quad \text{...(3)}$$
$$x_1, x_2, x_3, s_1, s_2, s_3 \ge 0.$$

Step 2. Since we have 3 equations in 6 variables, a solution is obtained by setting 6 – 3 = 3 variables equal to zero and solving for the remaining 3 variables. We choose initial basic feasible solution as:

$$x_1 = x_2 = x_3 = 0; \quad s_1 = 8, \quad s_2 = 10, \quad s_3 = 15$$

at which Z = 0.

Simplex Table I

c_B	Basis	$c_j \rightarrow$ Solution $b(=x_B)$	3 x_1	5 x_2	4 x_3	0 s_1	0 s_2	0 s_3	Ratio x_B/x_2
0	s_1	8	2	(3)	0	1	0	0	$\frac{8}{3} \rightarrow$
0	s_2	10	0	2	5	0	1	0	$\frac{10}{2} = 5$
0	s_3	15	3	2	4	0	0	1	$\frac{15}{2}$
Z = 0		Z_j	0	0	0	0	0	0	
		$C_j = c_j - Z_j$	3	5 ↑	4	0	0	0	

Step 3. Since some entries in C_j-row are positive, the current solution is not optimal. The incoming variable is x_2 and the outgoing basic variable is s_1. Also 3 is the key number.

Step 4. The co-efficient of x_2 in the objective function is 5. Therefore the entry in c_B column corresponding to the new basic variable x_2 will be 5. Dividing all elements of the key row by the key element 3, the **new key row** is

$$5 \quad x_2 \quad \frac{8}{3} \quad \frac{2}{3} \quad 1 \quad 0 \quad \frac{1}{3} \quad 0 \quad 0$$

Transformation of R_2. Key column entry in R_2 is 2.

$\therefore$ $R_2(\text{new}) = R_2(\text{old}) - 2R_1(\text{new})$

$$10 - 2\left(\frac{8}{3}\right) = \frac{14}{3}$$
$$0 - 2\left(\frac{2}{3}\right) = -\frac{4}{3}$$
$$2 - 2(1) = 0$$
$$5 - 2(0) = 5$$
$$0 - 2\left(\frac{1}{3}\right) = -\frac{2}{3}$$
$$1 - 2(0) = 1$$
$$0 - 2(0) = 0$$

Transformation of R_3. Key column entry in R_3 is 2.

$\therefore$ $R_3(\text{new}) = R_3(\text{old}) - 2R_1(\text{new})$

$$15 - 2\left(\frac{8}{3}\right) = \frac{29}{3}$$
$$3 - 2\left(\frac{2}{3}\right) = \frac{5}{3}$$

$$2 - 2(1) = 0$$
$$4 - 2(0) = 4$$
$$0 - 2\left(\frac{1}{3}\right) = -\frac{2}{3}$$
$$0 - 2(0) = 0$$
$$1 - 2(0) = 1$$

The above information is summarized in the following table.

Simplex Table II

		$c_j \rightarrow$ Solution	3	5	4	0	0	0	Ratio
c_B	Basis	$b(= x_B)$	x_1	x_2	x_3	s_1	s_2	s_3	x_B/x_3
5	x_2	$\frac{8}{3}$	$\frac{2}{3}$	1	0	$\frac{1}{3}$	0	0	—
0	s_2	$\frac{14}{3}$	$-\frac{4}{3}$	0	(5)	$-\frac{2}{3}$	1	0	$\frac{\left(\frac{14}{3}\right)}{5} = \frac{14}{15} \rightarrow$
0	s_3	$\frac{29}{3}$	$\frac{5}{3}$	0	4	$-\frac{2}{3}$	0	1	$\frac{\left(\frac{29}{3}\right)}{4} = \frac{29}{12}$
$Z = \frac{40}{3}$		Z_j	$\frac{10}{3}$	5	0	$\frac{5}{3}$	0	0	
		$C_j = c_j - Z_j$	$-\frac{1}{3}$	0	4 ↑	$-\frac{5}{3}$	0	0	

$$*Z = 5\left(\frac{8}{3}\right) + 0\left(\frac{14}{3}\right) + 0\left(\frac{29}{3}\right) = \frac{40}{3}$$

Since all entries in C_j-row are not negative or zero, the current solution is not optimal. The incoming variable is x_3 and the outgoing basic variable is s_2. Also 5 is the key number.

Step 5. The co-efficient of x_3 in the objective function is 4. Therefore the entry in c_B column corresponding to the new basic variable x_3 will be 4. Dividing all elements of the key row by the key element 5, the **new key row** is

4	x_3	$\frac{14}{15}$	$-\frac{4}{15}$	0	1	$-\frac{2}{15}$	$\frac{1}{5}$	0

Transformation of R_1. Key column entry in R_1 is 0.

$\therefore$ $R_1(\text{new}) = R_1(\text{old}) - 0 \times R_2(\text{new})$
$= R_1(\text{odd})$

Transformation of R_3. Key column entry in R_3 is 4.

$\therefore$ $R_3(\text{new}) = R_3(\text{old}) - 4R_2(\text{new})$

$$\frac{29}{3} - 4\left(\frac{14}{15}\right) = \frac{89}{15}$$
$$\frac{5}{3} - 4\left(-\frac{4}{15}\right) = \frac{41}{15}$$

$$0 - 4(0) = 0$$
$$4 - 4(1) = 0$$
$$-\frac{2}{3} - 4\left(-\frac{2}{15}\right) = -\frac{2}{15}$$
$$0 - 4\left(\frac{1}{5}\right) = -\frac{4}{5}$$
$$1 - 4(0) = 1$$

The above information is summarized in the following table.

Simplex Table III

		$c_j \rightarrow$	3	5	4	0	0	0	Ratio
c_B	Basis	Solution $b(= x_B)$	x_1	x_2	x_3	s_1	s_2	s_3	x_B/x_1
5	x_2	$\frac{8}{3}$	$\frac{2}{3}$	1	0	$\frac{1}{3}$	0	0	$\left(\frac{8}{3}\right)/\left(\frac{2}{3}\right) = 4$
4	x_3	$\frac{14}{15}$	$-\frac{4}{15}$	0	1	$-\frac{2}{15}$	$\frac{1}{5}$	0	—
0	s_3	$\frac{89}{15}$	$\frac{41}{15}$ (circled)	0	0	$-\frac{2}{15}$	$-\frac{4}{5}$	1	$\left(\frac{89}{15}\right)/\left(\frac{41}{15}\right) = \frac{89}{41} \rightarrow$
$Z = \frac{256}{15}$		Z_j	$\frac{34}{15}$	5	4	$\frac{17}{15}$	$\frac{4}{5}$	0	
		$C_j = c_j - Z_j$	$\frac{11}{15}$ ↑	0	0	$-\frac{17}{15}$	$-\frac{4}{5}$	0	

$$*Z = 5\left(\frac{8}{3}\right) + 4\left(\frac{14}{15}\right) + 0\left(\frac{89}{15}\right) = \frac{256}{15}$$

Since all entries in C_j-row are not negative or zero, the current solution is not optimal.

The incoming variable is x_1 and the outgoing basic variable is s_3. Also $\frac{41}{15}$ is the key number.

Step 6. The co-efficient of x_1 in the objective function is 3. Therefore the entry in c_B column corresponding to the new basic variable x_1 will be 3. Dividing all elements of the key row by the key element $\frac{41}{15}$, the **new key row** is

$$3 \qquad x_1 \qquad \frac{89}{41} \qquad 1 \qquad 0 \qquad 0 \qquad -\frac{2}{41} \qquad -\frac{12}{41} \qquad \frac{15}{41}$$

Transformation of R_1. Key column entry in R_1 is $\frac{2}{3}$

$$\therefore \qquad R_1(\text{new}) = R_1(\text{old}) - \frac{2}{3}\, R_3(\text{new})$$

$$\frac{8}{3} - \frac{2}{3}\left(\frac{89}{41}\right) = \frac{50}{41}$$

$$\frac{2}{3} - \frac{2}{3}\,(1) = 0$$

$$1 - \frac{2}{3}\,(0) = 1$$

$$0 - \frac{2}{3}\,(0) = 0$$

$$\frac{1}{3} - \frac{2}{3}\left(-\frac{2}{41}\right) = \frac{15}{41}$$

$$0 - \frac{2}{3}\left(-\frac{12}{41}\right) = \frac{8}{41}$$

$$0 - \frac{2}{3}\left(\frac{15}{41}\right) = -\frac{10}{41}$$

Transformation of R_2. Key column entry in R_2 is $-\dfrac{4}{15}$

$$\therefore \qquad R_2(\text{new}) = R_2(\text{old}) + \frac{4}{15}\,R_3(\text{new})$$

$$\frac{14}{15} + \frac{4}{15}\left(\frac{89}{41}\right) = \frac{62}{41}$$

$$-\frac{4}{15} + \frac{4}{15}\,(1) = 0$$

$$0 + \frac{4}{15}\,(0) = 0$$

$$1 + \frac{4}{15}\,(0) = 1$$

$$-\frac{2}{15} + \frac{4}{15}\left(-\frac{2}{41}\right) = -\frac{6}{41}$$

$$\frac{1}{5} + \frac{4}{15}\left(-\frac{12}{41}\right) = \frac{5}{41}$$

$$0 + \frac{4}{15}\left(\frac{15}{41}\right) = \frac{4}{41}.$$

The above information is summarized in the following table.

Simplex Table IV

		$c_j \rightarrow$ Solution	3	5	4	0	0	0	*Ratio*
c_B	*Basis*	$b(= x_B)$	x_1	x_2	x_3	s_1	s_2	s_3	
5	x_2	$\frac{50}{41}$	0	1	0	$\frac{15}{41}$	$\frac{8}{41}$	$-\frac{10}{41}$	
4	x_3	$\frac{62}{41}$	0	0	1	$-\frac{6}{41}$	$\frac{5}{41}$	$\frac{4}{41}$	
3	x_1	$\frac{89}{41}$	1	0	0	$-\frac{2}{41}$	$-\frac{12}{41}$	$\frac{15}{41}$	
$Z = \frac{765}{41}$		Z_j	3	5	4	$\frac{45}{41}$	$\frac{24}{41}$	$\frac{11}{41}$	
		$C_j = c_j - Z_j$	0	0	0	$-\frac{45}{41}$	$-\frac{24}{41}$	$-\frac{11}{41}$	

$$*Z = 5\left(\frac{50}{41}\right) + 4\left(\frac{62}{41}\right) + 3\left(\frac{89}{41}\right) = \frac{765}{41}$$

Since all entries in C_j-row are either negative or zero, therefore, the optimal solution is reached with

$$x_1 = \frac{89}{41},\ x_2 = \frac{50}{41},\ x_3 = \frac{62}{41} \text{ and Max. } Z = \frac{765}{41}$$

Example 3. *Solve the following linear programming problem by simplex method:*

Maximize $Z = 2x_1 + x_2 - x_3$

subject to

$$x_1 + x_2 \le 1$$
$$x_1 - 2x_2 - x_3 \ge -2$$
$$x_1, x_2, x_3 \ge 0.$$

Sol. Step 1. The problem is of maximization. As the sign of b in the second inequality is negative, it is first converted to positive by multiplying both the sides of the second inequality by (– 1). Thus, the problem can be restated as

Maximize $Z = 2x_1 + x_2 - x_3$

subject to

$$x_1 + x_2 \le 1$$
$$-x_1 + 2x_2 + x_3 \le 2$$
$$x_1, x_2, x_3 \ge 0.$$

Step 2. Introducing slack variables s_1, s_2 (one for each constraint) the problem in standard form is:

Maximize $Z = 2x_1 + x_2 - x_3 + 0s_1 + 0s_2$

subject to

$$x_1 + x_2 + s_1 = 1 \quad \text{...(1)}$$
$$-x_1 + 2x_2 + x_3 + s_2 = 2 \quad \text{...(2)}$$
$$x_1, x_2, x_3, s_1, s_2 \ge 0$$

Step 3. Since we have 2 equations in 5 variables, a solution is obtained by setting 5 – 2 = 3 variables equal to zero and solving for the remaining 2 variables. We choose initial basic feasible solution as:

$$x_1 = x_2 = x_3 = 0;\ s_1 = 1,\ s_2 = 2$$

at which Z = 0

Simplex Table I

		c_j Solution	2	1	– 1	0	0	Ratio
c_B	Basis	$b(= x_B)$	x_1	x_2	x_3	s_1	s_2	x_B/x_1
0	s_1	1	(1)	1	0	1	0	$\frac{1}{1} = 1 \rightarrow$
0	s_2	2	– 1	2	1	0	1	—
Z = 0		Z_j	0	0	0	0	0	
		$C_j = c_j - Z_j$	**2** ↑	1	– 1	0	0	

Step 4. Since some entries in C_j-row are positive, the current solution is not optimal. The incoming variable is x_1 and the outgoing basic variable is s_1. Also 1 is the key number.

Since the key number is 1, new key row is the same as old one.

Transformation of R_2. Key column entry in R_2 is (–1)

∴	R_2(new) = R_2(old) + 1 R_1(new)
	2 + 1(1) = 3
	– 1 + 1(1) = 0
	2 + 1(1) = 3
	1 + 1(0) = 1
	0 + 1(1) = 1
	1 + 1(0) = 1

The above information is summarized in the following table.

Simplex Table II

		$c_j \rightarrow$ Solution	2	1	– 1	0	0	*Ratio*
c_B	*Basis*	$b(=x_B)$	x_1	x_2	x_3	s_1	s_2	
2	x_1	1	1	1	0	1	0	
0	s_2	3	0	3	1	1	1	
Z = 2		Z_j	2	2	0	2	0	
		$C_j = c_j - Z_j$	0	– 1	– 1	– 2	0	

Since all entries in C_j-row are either negative or zero, therefore, the optimal solution is reached with

$$x_1 = 1, \quad x_2 = 0, \quad x_3 = 0 \quad \text{and} \quad \text{Max. } Z = 2$$

Example 4. *Solve the following LPP by simplex method:*

Minimize $Z = x_1 - 3x_2 + 3x_3$

subject to

$$3x_1 - x_2 + 2x_3 \leq 7$$
$$2x_1 + 4x_2 \geq -12$$
$$-4x_1 + 3x_2 + 8x_3 \leq 10$$
$$x_1, x_2, x_3 \geq 0.$$

(M.D.U. 2011, May 2013)

Sol. Step 1. The problem is that of minimization. Converting it into maximization problem by using the relationship

$$\text{Min. } Z = -\text{ Max. } Z^* \qquad \text{where } Z^* = -Z \qquad ...(1)$$

As the sign of b in the second inequality is negative, it is first converted to positive by multiplying both the sides of the second inequality by (– 1). Thus, the problem can be restated as

Maximize $Z^* = -x_1 + 3x_2 - 3x_3$

subject to

$$3x_1 - x_2 + 2x_3 \leq 7$$
$$-2x_1 - 4x_2 \leq 12$$
$$-4x_1 + 3x_2 + 8x_3 \leq 10$$
$$x_1, x_2, x_3 \geq 0.$$

Step 2. Introducing slack variables s_1, s_2, s_3 (one for each constraint), the problem in standard form is:

Maximize $Z^* = -x_1 + 3x_2 - 3x_3 + 0s_1 + 0s_2 + 0s_3$

subject to
$$3x_1 - x_2 + 2x_3 + s_1 = 7$$
$$-2x_1 - 4x_2 + s_2 = 12$$
$$-4x_1 + 3x_2 + 8x_3 + s_3 = 10$$
$$x_1, x_2, x_3, s_1, s_2, s_3 \geq 0.$$

Step 3. Since we have 3-equations in 6 variables, a solution is obtained by setting $6 - 3 = 3$ variables equal to zero and solving for the remaining 3 variables. We choose initial basic feasible solution as:

$$x_1 = x_2 = x_3 = 0;\ s_1 = 7,\ s_2 = 12,\ s_3 = 10 \text{ at which } Z^* = 0$$

Simplex Table I

c_B	*Basis*	$c_j \rightarrow$	-1	3	-3	0	0	0	*Ratio*
		Solution $b(=x_B)$	x_1	x_2	x_3	s_1	s_2	s_3	x_B/x_2
0	s_1	7	3	-1	2	1	0	0	—
0	s_2	12	-2	-4	0	0	1	0	—
0	s_3	10	-4	(3)	8	0	0	1	$\frac{10}{3} \rightarrow$
$Z^* = 0$		Z_j^*	0	0	0	0	0	0	
		$C_j = c_j - Z_j^*$	-1	3 ↑	-3	0	0	0	

Step 4. Since some entries in C_j – row are positive, the current solution is not optimal. The incoming variable is x_2 and the outgoing basic variable is s_3. Also the key number is 3.

Dividing all elements of the key row by the key number 3, the **new key row** is

$$3 \quad x_2 \quad \frac{10}{3} \quad -\frac{4}{3} \quad 1 \quad \frac{8}{3} \quad 0 \quad 0 \quad \frac{1}{3}$$

Transformation of R_1. Key column entry in R_1 is (-1)

$\therefore$ R_1 (new) = R_1(old) + 1 R_3 (new)

$$7 + 1\left(\frac{10}{3}\right) = \frac{31}{3}$$

$$3 + 1\left(-\frac{4}{3}\right) = \frac{5}{3}$$

$$-1 + 1(1) = 0$$

$$2 + 1\left(\frac{8}{3}\right) = \frac{14}{3}$$

$$1 + 1(0) = 1$$
$$0 + 1(0) = 0$$
$$0 + 1\left(\frac{1}{3}\right) = \frac{1}{3}$$

Transformation of R_2. Key column entry in R_2 is (– 4)

$$\therefore \qquad R_2(\text{new}) = R_2(\text{old}) + 4R_3(\text{new})$$

$$12 + 4\left(\frac{10}{3}\right) = \frac{76}{3}$$
$$-2 + 4\left(-\frac{4}{3}\right) = -\frac{22}{3}$$
$$-4 + 4(1) = 0$$
$$0 + 4\left(\frac{8}{3}\right) = \frac{32}{3}$$
$$0 + 4(0) = 0$$
$$1 + 4(0) = 1$$
$$0 + 4\left(\frac{1}{3}\right) = \frac{4}{3}$$

The above information is summarized in the following table.

Simplex Table II

c_B	*Basis*	$c_j \rightarrow$ *Solution* $b(= x_B)$	– 1	3	– 3	0	0	0	*Ratio* x_B/x_1
			x_1	x_2	x_3	s_1	s_2	s_3	
0	s_1	$\frac{31}{3}$	($\frac{5}{3}$)	0	$\frac{14}{3}$	1	0	$\frac{1}{3}$	$\frac{31}{5} \rightarrow$
0	s_2	$\frac{76}{3}$	$-\frac{22}{3}$	0	$\frac{32}{3}$	0	1	$\frac{4}{3}$	—
3	x_2	$\frac{10}{3}$	$-\frac{4}{3}$	1	$\frac{8}{3}$	0	0	$\frac{1}{3}$	—
$Z^* = 10$		Z^*_j	– 4	3	8	0	0	1	
		$C_j = c_j - Z^*_j$	3 ↑	0	– 11	0	0	– 1	

Step 5. Since all entries in C_j-row are not negative or zero, the current solution is not optimal.

The incoming variable is x_1 and the outgoing basic variable is s_1. Also the key number is $\frac{5}{3}$.

Dividing all elements of the key row by the key number $\frac{5}{3}$, the **new key row** is

$$-1 \quad x_1 \quad \frac{31}{5} \quad 1 \quad 0 \quad \frac{14}{5} \quad \frac{3}{5} \quad 0 \quad \frac{1}{5}$$

Transformation of R_2. Key column entry in R_2 is $-\frac{22}{3}$.

$$\therefore \qquad R_2(\text{new}) = R_2(\text{old}) + \frac{22}{3} R_1(\text{new})$$

$$\frac{76}{3} + \frac{22}{3}\left(\frac{31}{5}\right) = \frac{354}{5}$$

$$-\frac{22}{3} + \frac{22}{3}(1) = 0$$

$$0 + \frac{22}{3}(0) = 0$$

$$\frac{32}{3} + \frac{22}{3}\left(\frac{14}{5}\right) = \frac{156}{5}$$

$$0 + \frac{22}{3}\left(\frac{3}{5}\right) = \frac{22}{5}$$

$$1 + \frac{22}{3}(0) = 1$$

$$\frac{4}{3} + \frac{22}{3}\left(\frac{1}{5}\right) = \frac{14}{5}$$

Transformation of R_3. Key column entry in R_3 is $-\frac{4}{3}$

$$\therefore \qquad R_3(\text{new}) = R_3(\text{old}) + \frac{4}{3} R_1(\text{new})$$

$$\frac{10}{3} + \frac{4}{3}\left(\frac{31}{5}\right) = \frac{58}{5}$$

$$-\frac{4}{3} + \frac{4}{3}(1) = 0$$

$$1 + \frac{4}{3}(0) = 1$$

$$\frac{8}{3} + \frac{4}{3}\left(\frac{14}{5}\right) = \frac{32}{5}$$

$$0 + \frac{4}{3}\left(\frac{3}{5}\right) = \frac{4}{5}$$

$$0 + \frac{4}{3}(0) \quad = 0$$

$$\frac{1}{3} + \frac{4}{3}\left(\frac{1}{5}\right) \quad = \frac{3}{5}$$

The above information is summarized in the following table.

Simplex Table III

c_B	Basis	c_j Solution $b(= x_B)$	-1	3	-3	0	0	0	Ratio
			x_1	x_2	x_3	s_1	s_2	s_3	
-1	x_1	$\frac{31}{5}$	1	0	$\frac{14}{5}$	$\frac{3}{5}$	0	$\frac{1}{5}$	
0	s_2	$\frac{354}{5}$	0	0	$\frac{156}{5}$	$\frac{22}{5}$	1	$\frac{14}{5}$	
3	x_2	$\frac{58}{5}$	0	1	$\frac{32}{5}$	$\frac{4}{5}$	0	$\frac{3}{5}$	
$Z^* = \frac{143}{5}$		Z^*_j	-1	3	$\frac{82}{5}$	$\frac{9}{5}$	0	$\frac{8}{5}$	
		$C_j = c_j - Z^*_j$	0	0	$-\frac{97}{5}$	$-\frac{9}{5}$	0	$-\frac{8}{5}$	

Since all entries in C_j-row are either negative or zero, therefore, the optimal solution is reached with

$$x_1 = \frac{31}{5},\ x_2 = \frac{58}{5},\ x_3 = 0 \text{ and Max. } Z^* = \frac{143}{5}$$

Hence, from (1), Min. $Z = -\frac{143}{5}$.

Example 5. *Following data are available for a firm which manufactures three items A, B and C:*

Product	*Time required in hours*		*Profit*
	Assembly	*Finishing*	*(in ₹)*
A	*10*	*2*	*80*
B	*4*	*5*	*60*
C	*5*	*4*	*30*
Firm's Capacity	*2000*	*1009*	

Express the above data in the form of linear programming problem to maximize the profit from the production and solve it by simplex method.

Sol. Step 1. Let x_1, x_2, x_3 denote respectively the number of units of products A, B, C manufactured by the firm.

The mathematical formulation of the problem is

Maximize $Z = 80x_1 + 60x_2 + 30x_3$

subject to the constraints

$$10x_1 + 4x_2 + 5x_3 \leq 2000$$
$$2x_1 + 5x_2 + 4x_3 \leq 1009$$
$$x_1, x_2, x_3 \geq 0$$

Step 2. Introducing slack variables s_1, s_2 (one for each constraint) the problem in standard form is:

Maximize $Z = 80x_1 + 60x_2 + 30x_3 + 0s_1 + 0s_2$

subject to

$$10x_1 + 4x_2 + 5x_3 + s_1 = 2000$$
$$2x_1 + 5x_2 + 4x_3 + s_2 = 1009$$
$$x_1, x_2, x_3, s_1, s_2 \geq 0$$

Step 3. Since we have 2 equations in 5 variables, a solution is obtained by setting 5 – 2 = 3 variables equal to zero and solving for the remaining 2 variables.

An initial basic feasible solution is obtained by setting $x_1 = x_2 = x_3 = 0$ so that $s_1 = 2000$, $s_2 = 1009$.

The initial simplex table corresponding to this solution is given below:

Simplex Table I

c_B	*Basis*	$c_j \rightarrow$ *Solution* $b(= x_B)$	*80*	*60*	*30*	*0*	*0*	*Ratio*
			x_1	x_2	x_3	s_1	s_2	x_B/x_1
0	s_1	2000	(10)	4	5	1	0	200 →
0	s_2	1009	2	5	4	0	1	$\frac{1009}{2}$
Z = 0		Z_j	0	0	0	0	0	
		$C_j = c_j - Z_j$	80 ↑	60	30	0	0	

Step 4. Since C_j-row has some positive entries, the current solution is not optimal.

The incoming variable is x_1 and the outgoing basic variable is s_1. The key number is 10.

Dividing each element in the key row *i.e.*, s_1-row, by the key number 10 the **new key row** is

$$80 \quad x_1 \quad 200 \quad 1 \quad \frac{2}{5} \quad \frac{1}{2} \quad \frac{1}{10} \quad 0$$

Transformation of R_2. Key column entry in R_2 is 2.

∴ $R_2(\text{new}) = R_2(\text{old}) - 2R_1(\text{new})$

$$1009 - 2(200) = 609$$
$$2 - 2(1) = 0$$

$$5 - 2\left(\frac{2}{5}\right) = \frac{21}{5}$$

$$4 - 2\left(\frac{1}{2}\right) = 3$$

$$0 - 2\left(\frac{1}{10}\right) = -\frac{1}{5}$$

$$1 - 2(0) = 1$$

The above information is summarized in the following table.

Simplex Table II

c_B	*Basis*	$c_j \rightarrow$ *Solution* $b(= x_B)$	80	60	30	0	0	*Ratio*
			x_1	x_2	x_3	s_1	s_2	x_B/x_2
80	x_1	200	1	$\frac{2}{5}$	$\frac{1}{2}$	$\frac{1}{10}$	0	500
0	s_2	609	0	$\boxed{\frac{21}{5}}$	3	$-\frac{1}{5}$	1	145 →
Z = 16000		Z_j	80	32	40	8	0	
		$C_j = c_j - Z_j$	0	28 ↑	− 10	− 8	0	

Step 5. Since C_j-row has a positive entry, the current solution is not optimal.

The incoming variable is x_2 and the outgoing basic variable is s_2. The key number is $\frac{21}{5}$.

Dividing each element in the key row by the key number $\frac{21}{5}$, the **new key row** is

60	x_2	145	0	1	$\frac{5}{7}$	$-\frac{1}{21}$	$\frac{5}{21}$

Transformation of R_1. Key column entry in R_1 is $\frac{2}{5}$.

$$\therefore \quad R_1(\text{new}) = R_1(\text{old}) - \frac{2}{5} R_2(\text{new})$$

$$200 - \frac{2}{5}(145) = 142$$

$$1 - \frac{2}{5}(0) = 1$$

$$\frac{2}{5} - \frac{2}{5}(1) = 0$$

$$\frac{1}{2} - \frac{2}{5}\left(\frac{5}{7}\right) = \frac{3}{14}$$

$$\frac{1}{10} - \frac{2}{5}\left(-\frac{1}{21}\right) = \frac{5}{42}$$

$$0 - \frac{2}{5}\left(\frac{5}{21}\right) = -\frac{2}{21}$$

The above information is summarized in the following table.

Simplex Table III

c_B	*Basis*	$c_j \rightarrow$	*80*	*60*	*30*	*0*	*0*	*Ratio*
		Solution $b(= x_B)$	x_1	x_2	x_3	s_1	s_2	
80	x_1	142	1	0	$\frac{3}{14}$	$\frac{5}{42}$	$-\frac{2}{21}$	
60	x_2	145	0	1	$\frac{5}{7}$	$-\frac{1}{21}$	$\frac{5}{21}$	
Z = 20060		Z_j	80	60	60	$\frac{20}{3}$	$\frac{20}{3}$	
		$C_j = c_j - Z_j$	0	0	– 30	$-\frac{20}{3}$	$-\frac{20}{3}$	

Since all entries in C_j-row are either negative or zero, the optimal solution has been obtained with $x_1 = 142$, $x_2 = 145$ and Max. Z = 20060.

7.11. TIE FOR THE INCOMING VARIABLE (Key Column)

At any iteration, if two or more columns have the same largest positive value in C_j-row, a tie for the incoming variable occurs. In order to break this tie, the selection for key column (incoming variable) can be made arbitrarily. However, in order to minimize the number of iterations required to reach the optimal solution, the following rules may be used:

(*i*) If there is a tie between two decision variables, the choice can be made arbitrarily.

(*ii*) If there is a tie between two slack (or surplus) variables, the choice can be made arbitrarily.

(*iii*) If there is a tie between a decision variable and a slack (or surplus) variable, then the decision variable is selected as the incoming variable.

7.12. TIE FOR THE OUTGOING VARIABLE (Key Row)—DEGENERACY

(*M.D.U. Dec. 2013*)

In the simplex method, in order to determine which basic variable to go out of the basis (*i.e.*, to determine the key row), we find the ratio of entries in x_B-column by the corresponding (but positive) entries in the key column and select that variable to go out of the basis for which this ratio is the minimum. Sometimes this minimum ratio is not unique or values of one or more basic variables in the x_B-column are zero. This problem is known as **degeneracy** in linear programming and the basic feasible solution is **degenerate.** It is possible, in a degenerate situation, to obtain a sequence of simplex tables that correspond to basic feasible solutions which give the same value of the objective function. Moreover, we may eventually return to the first simplex table in the sequence. This is called **cycling.** When cycling occurs, it is possible that we may never obtain the optimal value of the objective function. However, this phenomenon seldom occurs in practical problems.

Degeneracy may occur at any iteration of the simplex method. When there is a tie in the minimum ratios, the selection of the key row can be made arbitrarily. However, the number of iterations required to arrive at the optimal solution can be minimized by adopting the following rules.

(*i*) Identify the tied rows.

(*ii*) Divide the co-efficients of slack variables in each tied row by the corresponding positive number of the key column in the row, starting from left to right.

(*iii*) Compare the resulting ratios columnwise from left to right.

(*iv*) Select the row which has the smallest ratio. This row becomes the key row.

If the above ratios fail to break the tie, then find similar ratios for the decision variables.

Thus, the degeneracy is resolved. Simplex method is then continued until an optimal solution is obtained.

Example. *Using simplex method*

Maximize $Z = 5x_1 + 3x_2$

subject to

$$x_1 + x_2 \leq 2$$
$$5x_1 + 2x_2 \leq 10$$
$$3x_1 + 8x_2 \leq 12$$
$$x_1, x_2 \geq 0.$$

(K.U.K. 2010, 2013, May 2013, Dec. 2015)

Sol. Step 1. Introducing slack variables s_1, s_2, s_3 (one for each constraint) the problem in standard form is:

Maximize $Z = 5x_1 + 3x_2 + 0s_1 + 0s_2 + 0s_3$

subject to

$$x_1 + x_2 + s_1 = 2$$
$$5x_1 + 2x_2 + s_2 = 10$$
$$3x_1 + 8x_2 + s_3 = 12$$
$$x_1, x_2, s_1, s_2, s_3 \geq 0.$$

Step 2. Since we have 3 equations in 5 variables, a solution is obtained by setting $5 - 3 = 2$ variables equal to zero and solving for the remaining 3 variables.

Thus an initial basic feasible solution is obtained by setting $x_1 = x_2 = 0$ so that $s_1 = 2$, $s_2 = 10$, $s_3 = 12$.

The initial simplex table corresponding to this solution is given below:

Simplex Table I

		$c_j \rightarrow$ Solution	5	3	0	0	0	Ratio
c_B	Basis	$b(=x_B)$	x_1	x_2	s_1	s_2	s_3	x_B/x_1
0	s_1	2	1	1	1	0	0	2 } Tie
0	s_2	10	(5)	2	0	1	0	2 } Tie
0	s_3	12	3	8	0	0	1	4
Z = 0		Z_j	0	0	0	0	0	
		$C_j = c_j - Z_j$	5 ↑	3	0	0	0	

Since C_j-row has some positive entries, the current solution is not optimal. The incoming variable is x_1.

Step 3. From table I, we observe that there is a tie among the s_1-row and s_2-row. This is an indication of the existence of degeneracy.

To obtain the unique key row, we adop the following procedure:

(*i*) The co-efficients of slack variables s_1 and s_2 are:

Row	*Column*	
	s_1	s_2
s_1	1	0
s_2	0	1

(*ii*) Key column element in s_1-row is 1 and in s_2-row is 5.

Dividing the co-efficients in s_1-row by 1 and in s_2-row by 5, we get the following ratios:

Row	*Column*	
	s_1	s_2
s_1	$\frac{1}{1} = 1$	$\frac{0}{1} = 0$
s_2	$\frac{0}{5} = 0$	$\frac{1}{5}$

(*iii*) Comparing the ratios of Step (*ii*) from left to right columnwise, we find that in the first column, the second row yields the smaller ratio and, therefore, s_2-row becomes the key row and s_2 is the outgoing basic variable. The key number is 5.

Step 4. Dividing each element in the key row by the key number 5, the **new key row** is

$$5 \quad x_1 \quad 2 \quad 1 \quad \frac{2}{5} \quad 0 \quad \frac{1}{5} \quad 0$$

Transformation of R_1. Key column entry in R_1 is 1

$\therefore$ $R_1(\text{new}) = R_1(\text{old}) - 1\ R_2(\text{new})$

$$2 - 1(2) = 0$$
$$1 - 1(1) = 0$$
$$1 - 1\left(\frac{2}{5}\right) = \frac{3}{5}$$
$$1 - 1(0) = 1$$
$$0 - 1\left(\frac{1}{5}\right) = -\frac{1}{5}$$
$$0 - 1(0) = 0$$

Transformation of R_3. Key column entry in R_3 is 3

$\therefore$ $R_3(\text{new}) = R_3(\text{old}) - 3\ R_2(\text{new})$

$$12 - 3(2) = 6$$
$$3 - 3(1) = 0$$

$$8 - 3\left(\frac{2}{5}\right) \quad = \frac{34}{5}$$

$$0 - 3(0) \quad = 0$$

$$0 - 3\left(\frac{1}{5}\right) \quad = -\frac{3}{5}$$

$$1 - 3(0) \quad = 1$$

The above information is summarized in the following table.

Simplex Table II

		$c_j \rightarrow$ Solution	5	3	0	0	0	*Ratio*
c_B	*Basis*	$b(= x_B)$	x_1	x_2	s_1	s_2	s_3	x_B/x_2
0	s_1	0	0	$\frac{3}{5}$ (circled)	1	$-\frac{1}{5}$	0	$0 \rightarrow$
5	x_1	2	1	$\frac{2}{5}$	0	$\frac{1}{5}$	0	5
0	s_3	6	0	$\frac{34}{5}$	0	$-\frac{3}{5}$	1	$\frac{15}{17}$
Z = 10		Z_j	5	2	0	1	0	
		$C_j = c_j - Z_j$	0	1 ↑	0	−1	0	

Since C_j-row has a positive entry, the current solution is not optimal. The incoming variable is x_2 and the outgoing basic variable is s_1. The key number is $\frac{3}{5}$.

Dividing each element in the key row by the key number $\frac{3}{5}$, the **new key row** is

$$3 \quad x_2 \quad 0 \quad 0 \quad 1 \quad \frac{5}{3} \quad -\frac{1}{3} \quad 0$$

Transformation of R_2. Key column entry in R_2 is $\frac{2}{5}$

$\therefore \qquad R_2(\text{new}) = R_2(\text{old}) - \frac{2}{5}\, R_1(\text{new})$

$$2 - \frac{2}{5}(0) \quad = 2$$

$$1 - \frac{2}{5}(0) \quad = 1$$

$$\frac{2}{5} - \frac{2}{5}(1) \quad = 0$$

$$0 - \frac{2}{5}\left(\frac{5}{3}\right) \quad = -\frac{2}{3}$$

$$\frac{1}{5}-\frac{2}{5}\left(-\frac{1}{3}\right)=\frac{1}{3}$$

$$0-\frac{2}{5}(0)=0$$

Transformation of R_3. Key column entry in R_3 is $\frac{34}{5}$

$$\therefore \quad R_3(\text{new}) = R_3(\text{old}) - \frac{34}{5} R_1(\text{new})$$

$$6-\frac{34}{5}(0)=6$$

$$0-\frac{34}{5}(0)=0$$

$$\frac{34}{5}-\frac{34}{5}(1)=0$$

$$0-\frac{34}{5}\left(\frac{5}{3}\right)=-\frac{34}{3}$$

$$-\frac{3}{5}-\frac{34}{5}\left(-\frac{1}{3}\right)=\frac{5}{3}$$

$$1-\frac{34}{5}(0)=1$$

The above information is summarized in the following table.

Simplex Table III

		$c_j \rightarrow$	*5*	*3*	*0*	*0*	*0*	*Ratio*
c_B	*Basis*	*Solution* $b(=x_B)$	x_1	x_2	s_1	s_2	s_3	
3	x_2	0	0	1	$\frac{5}{3}$	$-\frac{1}{3}$	0	
5	x_1	2	1	0	$-\frac{2}{3}$	$\frac{1}{3}$	0	
0	s_3	6	0	0	$-\frac{34}{3}$	$\frac{5}{3}$	1	
Z = 10		Z_j	5	3	$\frac{5}{3}$	$\frac{2}{3}$	0	
		$C_j = c_j - Z_j$	0	0	$-\frac{5}{3}$	$-\frac{2}{3}$	0	

Since all entries in C_j-row are either negative or zero, the optimal solution has been obtained with $x_1 = 2$, $x_2 = 0$ and Max. Z = 10.

EXERCISE 7.5

Using simplex method, solve the following linear programming problems (1–6):

1. (*i*) Maximize $Z = x_1 + 2x_2$

subject to
$$2x_1 + x_2 \le 8$$
$$2x_1 + 3x_2 \le 12$$
$$x_1, x_2 \ge 0$$
(*M.D.U. 2011*)

(*ii*) Maximize $Z = x_1 + 3x_2$

subject to
$$x_1 + 2x_2 \le 10$$
$$0 \le x_1 \le 5$$
$$0 \le x_2 \le 4$$

(*iii*) Maximize $Z = 6x_1 + 4x_2$

subject to
$$x_1 + 2x_2 \le 240$$
$$3x_1 + 2x_2 \le 300$$
$$x_1, x_2 \ge 0$$

(*iv*) Maximize $Z = 4x_1 + 5x_2$

subject to
$$x_1 - 2x_2 \le 2$$
$$2x_1 + x_2 \le 6$$
$$x_1 + 2x_2 \le 5$$
$$-x_1 + x_2 \le 2$$
$$x_1, x_2 \ge 0$$

(*v*) Maximize $Z = 45x_1 + 80x_2$

subject to
$$5x_1 + 20x_2 \le 400$$
$$10x_1 + 15x_2 \le 450$$
$$x_1, x_2 \ge 0$$

(*vi*) Maximize $Z = 6x_1 + 4x_2$

subject to
$$-4x_1 + 5x_2 \le 10$$
$$3x_1 + 2x_2 \le 9$$
$$8x_1 + 3x_2 \le 12$$
$$x_1 \ge 0, x_2 \ge 0$$
(*M.D.U. Dec. 2013*)

2. (*i*) Maximize $Z = 10x_1 + x_2 + 2x_3$

subject to
$$x_1 + x_2 - 2x_3 \le 10$$
$$4x_1 + x_2 + x_3 \le 20$$
$$x_1, x_2, x_3 \ge 0$$

(*ii*) Maximize $Z = 5x_1 + 4x_2 + 3x_3$

subject to
$$3x_1 + 2x_2 + x_3 \le 10$$
$$2x_1 + x_2 + 2x_3 \le 12$$
$$x_1 + x_2 + 3x_3 \le 15$$
$$x_1, x_2, x_3 \ge 0$$

(*iii*) Maximize $Z = x_1 + x_2 + 3x_3$

subject to
$$3x_1 + 2x_2 + x_3 \le 3$$
$$2x_1 + x_2 + 2x_3 \le 2$$
$$x_1, x_2, x_3 \ge 0$$

(*iv*) Maximize $Z = 4x_1 + 3x_2 + 4x_3 + 6x_4$

subject to
$$x_1 + 2x_2 + 2x_3 + 4x_4 \le 80$$
$$2x_1 + 2x_3 + x_4 \le 60$$

$$3x_1 + 3x_2 + x_3 + x_4 \le 80$$
$$x_1, x_2, x_3, x_4 \ge 0$$

3. Maximize $Z = 107x_1 + x_2 + 2x_3$

subject to the constraints,
$$14x_1 + x_2 - 6x_3 + 3x_4 = 7$$
$$16x_1 + \frac{1}{2}x_2 - 6x_3 \le 5$$
$$3x_1 - x_2 - x_3 \le 0$$
$$x_1, x_2, x_3, x_4 \ge 0$$
(M.D.U. 2010)

4. Minimize $Z = x_1 - 3x_2 + 2x_3$

subject to
$$3x_1 - x_2 + 2x_3 \le 7$$
$$-2x_1 + 4x_2 \le 12$$
$$-4x_1 + 3x_2 + 8x_3 \le 10$$
$$x_1, x_2, x_3 \ge 0$$
(K.U.K. 2012; M.D.U. Dec. 2014)

5. (*i*) Maximize $Z = 5x_1 + 6x_2 + x_3$

subject to
$$9x_1 + 3x_2 - 2x_3 \le 5$$
$$4x_1 + 2x_2 - x_3 \le 2$$
$$x_1 - 4x_2 + x_3 \le 3$$
$$x_1, x_2, x_3 \ge 0$$

(*ii*) Maximize $Z = x_1 + 4x_2 - x_3$

subject to
$$-5x_1 + 6x_2 - 2x_3 \le 30$$
$$-x_1 + 3x_2 + 6x_3 \le 12$$
$$x_1, x_2, x_3 \ge 0$$

6. (*i*) Maximize $Z = 3x_1 + 9x_2$

subject to
$$x_1 + 4x_2 \le 8$$
$$x_1 + 2x_2 \le 4$$
$$x_1, x_2 \ge 0$$

(*ii*) Maximize $Z = 50x_1 + 60x_2 + 80x_3$

subject to
$$2x_1 + x_2 + 2x_3 \le 50$$
$$x_1 + 6x_2 + 2x_3 \le 50$$
$$x_1 + 2x_2 + x_3 \le 26$$
$$x_1, x_2, x_3 \ge 0$$

(*iii*) Maximize $Z = x_1 + 2x_2 + x_3$

subject to
$$2x_1 + x_2 - x_3 \le 2$$
$$-2x_1 + x_2 - 5x_3 \ge -6$$
$$4x_1 + x_2 + x_3 \le 6$$
$$x_1, x_2, x_3 \ge 0$$
(M.D.U. May 2015)

7. A firm makes two types of furnitures, chairs and tables. Profits are ₹ 20 per chair and ₹ 30 per table. Both products are processed on three machines M_1, M_2 and M_3. The time required for each product in hours and total time available in hours per week on each machine are as follows:

Machine	*Chair*	*Table*	*Available Time*
M_1	3	3	36
M_2	5	2	50
M_3	2	6	60

How should the manufacturer schedule is production in order to maximize profit?

(Use simplex method)

8. A company makes two kinds of leather belts. Belt A is high quality belt and belt B is of lower quality. The respective profits are ₹ 4 and ₹ 3 per belt. The production of each of type A requires twice as much time as a belt of type B, and if all belts were of type B, the company could make 1000 per day. The supply of leather is sufficient for only 800 belts per day (both A and B combined). Belt A requires a fancy buckle and only 400 per day are available. There are only 700 buckles a day available for belt B.

What should be the daily production of each type of belt? Formulate this problem as an LP model and solve it by simplex method.

9. A farmer has 1000 acres of land on which he can grow corn, wheat or soyabeans. Each acre of corn costs ₹ 100 for preparation, requires 7 man days of work and yields a profit of ₹ 30. An acre of wheat costs ₹ 120 to prepare, requires 10 man days of work and yields a profit of ₹ 40. An acre of soyabeans costs ₹ 70 to prepare, requires 8 man days of work and yields a profit of ₹ 20. If the farmer has ₹ 1,00,000 for preparation and can count on 8000 man days of work, how many acres should be allocated to each crop to maximize profits. Formulate this as an LPP and solve it by simplex method.

10. A manufacturing firm has discontinued production of a certain unprofitable product line. This has created considerable excess production capacity. Management is considering to devote this excess capacity to one or more of three products 1, 2 and 3. The available capacity on machines and the number of machine-hours required for each unit of the respective product, is given below:

Machine Type	*Available Time (in Machine-hours per week)*	*Productivity (in Machine-hours per unit)*		
		Product 1	*Product 2*	*Product 3*
Milling Machine	250	8	2	3
Lathe	150	4	3	0
Grinder	50	2	—	1

The profit per unit would be ₹ 20, ₹ 6 and ₹ 8 respectively for product 1, 2 and 3. Find how much of each product the firm should produce in order to maximize profit?

11. A company can make three different products A, B and C by combining steel and rubber. Product A requires 2 units of steel and 3 units of rubber and can be sold at a profit of ₹ 40 per unit. Product B requires 3 units of steel and 3 units of rubber and can be sold at a profit of ₹ 45 per unit. Product C requires 1 unit of steel and 2 units of rubber and yields ₹ 24 profit per unit. There are 100 units of steel and 120 units of rubber available per day. What should be the daily production of each of the three products in order to maximize profit? Use simplex method.

Answers

1. (*i*) $x_1 = 0, \quad x_2 = 4;$ Max. Z = 8 (*ii*) $x_1 = 2, \quad x_2 = 4;$ Max. Z = 14

(*iii*) $x_1 = 100, \quad x_2 = 0;$ Max. Z = 600 (*iv*) $x_1 = \frac{7}{3}, \quad x_2 = \frac{4}{3};$ Max. Z = 16

(*v*) $x_1 = 24, \quad x_2 = 14;$ Max. Z = 2200 (*vi*) $x_1 = \frac{15}{26}, x_2 = \frac{32}{13};$ Max. Z = $\frac{173}{13}$

2. (*i*) $x_1 = 5, \quad x_2 = x_3 = 0;$ Max. Z = 50 (*ii*) $x_1 = 0, \quad x_2 = 3, \quad x_3 = 4;$ Max. Z = 24

(*iii*) $x_1 = x_2 = 0, \quad x_3 = 1;$ Max. Z = 3

(*iv*) $x_1 = \frac{280}{13}, \quad x_2 = 0, \quad x_3 = \frac{20}{13}, \quad x_4 = \frac{180}{13};$ Max. Z = $\frac{2280}{13}$

3. Unbounded solution. [Here x_4 is a slack variable]

4. $x_1 = 4, \quad x_2 = 5, \quad x_3 = 0;$ Min. Z = – 11

5. (*i*) Unbounded solution (*ii*) Unbounded solution

[**Hint.** When each entry in the key column is negative or zero, Z is unbounded]

6. (*i*) $x_1 = 0, \quad x_2 = 2;$ Max. Z = 18 (*ii*) $x_1 = x_2 = 0, \quad x_3 = 25;$ Max. Z = 2000

(*iii*) $x_1 = 0, \quad x_2 = 4, \quad x_3 = 2;$ Max. Z = 10

7. 3 chairs and 9 tables per week, maximum profit = ₹ 330
[**Hint.** Let x and y be the number of chairs and tables to be produced per week. Then
Maximize $Z = 20x + 30y$
subject to
$$3x + 3y \le 36$$
$$5x + 2y \le 50$$
$$2x + 6y \le 60$$
$$x, y \ge 0]$$

8. Let x_1 and x_2 be the number of belts of type A and B respectively manufactured each day. Then the mathematical LP model is:
Maximize $Z = 4x_1 + 3x_2$
subject to
$$2x_1 + x_2 \le 1000$$
$$x_1 + x_2 \le 800$$
$$x_1 \le 400$$
$$x_2 \le 700$$
$$x_1, x_2 \ge 0$$
Optimal solution is $x_1 = 200$, $x_2 = 600$; max. Z = ₹ 2,600.

9. Let x_1, x_2, x_3 acres of land be allocated to corn, wheat and soyabeans respectively. Then
Maximize $Z = 30x_1 + 40x_2 + 20x_3$
subject to
$$100x_1 + 120x_2 + 70x_3 \le 1{,}00{,}000$$
$$7x_1 + 10x_2 + 8x_3 \le 8000$$
$$x_1 + x_2 + x_3 \le 1000$$
$$x_1, x_2, x_3 \ge 0$$
Optimal solution is $x_1 = 250$, $x_2 = 625$, $x_3 = 0$; max. Z = ₹ 32500.

10. Let x_1, x_2, x_3 be the number of units of product 1, 2 and 3 to be produced per week. Then
Maximize $Z = 20x_1 + 6x_2 + 8x_3$
subject to
$$8x_1 + 2x_2 + 3x_3 \le 250$$
$$4x_1 + 3x_2 \le 150$$
$$2x_1 + x_3 \le 50$$
$$x_1, x_2, x_3 \ge 0$$
Optimal solution: $x_1 = 0$, $x_2 = 50$, $x_3 = 50$; Max. Z = 700. *(DCRUST Murthal, May 2014)*

11. 20 units of A, 20 units of B, 0 units of C with max. profit = ₹ 1700.

7.13. ARTIFICIAL VARIABLES

In the linear programming problems discussed so far, we have the following characteristics:

(*i*) The objective function is of maximization type

(*ii*) All constraints are of ≤ form

(*iii*) Right-hand side of each constraint is positive.

Adding slack variables, all inequality constraints are converted into equalities. Initial solution is found very conveniently by letting the slack variables be the initial basic variables so that each one equals the positive right hand side of its equation.

All linear programming problems donot have the above characteristics. Now we will see what adjustments are required for other forms ($\ge$, $=$ or $b_i < 0$) of linear programming problems.

When the constraints are of ≤ type but some $b_i < 0$, after adding the non-negative slack variable s_i, the initial solution will involve $s_i = b_i < 0$. This solution is not feasible because it violates the non-negativity condition of slack variable.

When the constraints are of ≥ type, after adding the non-negative surplus variable s_i and letting each decision variable equal to zero, we get an initial solution $-s_i = b_i$ or $s_i = -b_i < 0$. This solution is not feasible because it violates the non-negativity condition of surplus variable.

Thus, introduction of slack/surplus variables fails to give a basic solution in many linear programming problems. Such problems are solved by the **artificial variable technique** explained below:

7.14. SIMPLEX METHOD (Minimization Case)

Consider the general linear programming problem:

Minimize $Z = c_1x_1 + c_2x_2 + \ldots + c_nx_n$

subject to the constraints

$$a_{11}x_1 + a_{12}x_2 + \ldots\ldots + a_{1n}x_n \geq b_1$$
$$a_{21}x_1 + a_{22}x_2 + \ldots\ldots + a_{2n}x_n \geq b_2$$
$$\ldots\ldots\ldots\ldots\ldots\ldots\ldots\ldots\ldots\ldots\ldots\ldots$$
$$\ldots\ldots\ldots\ldots\ldots\ldots\ldots\ldots\ldots\ldots\ldots\ldots$$
$$a_{m1}x_1 + a_{m2}x_2 + \cdots\cdots + a_{mn}x_n \geq b_m$$

and $x_1, x_2, \ldots\ldots, x_n \geq 0$

Subtracting surplus variables s_i to convert inequalities into equalities, the above problem reduces to:

Minimize $Z = c_1x_1 + c_2x_2 + \cdots\cdots + c_nx_n + 0s_1 + 0s_2 + \cdots\cdots + 0s_m$

subject to the constraints

$$a_{11}x_1 + a_{12}x_2 + \ldots\ldots + a_{1n}x_n - s_1 = b_1$$
$$a_{21}x_1 + a_{22}x_2 + \ldots\ldots + a_{2n}x_n - s_2 = b_2$$
$$\ldots\ldots\ldots\ldots\ldots\ldots\ldots\ldots\ldots\ldots\ldots\ldots\ldots\ldots$$
$$\ldots\ldots\ldots\ldots\ldots\ldots\ldots\ldots\ldots\ldots\ldots\ldots\ldots\ldots$$
$$a_{m1}x_1 + a_{m2}x_2 + \ldots\ldots + a_{mn}x_n - s_m = b_m$$

and $x_j \geq 0, \quad (j = 1, 2, \ldots, n)$

$s_i \geq 0, \quad (i = 1, 2, \ldots, m)$

An initial basic solution is obtained by assigning zero value to each decision variable *i.e.*, by setting

$$x_1 = x_2 = \ldots\ldots = x_n = 0$$

Thus, we get

$$-s_1 = b_1 \quad \text{or} \quad s_1 = -b_1$$
$$-s_2 = b_2 \quad \text{or} \quad s_2 = -b_2$$
$$\vdots \qquad\qquad\qquad \vdots$$
$$-s_m = b_m \quad \text{or} \quad s_m = -b_m$$

which is not feasible because it violates the non-negativity constraints $s_i \geq 0$.

∴ The simplex algorithm needs modification. We now introduce m new variables $A_1, A_2, \ldots, A_m$ into the system of constraints.

These new variables are called **artificial variables.** The resulting system of equations can now be written as

$$a_{11}x_1 + a_{12}x_2 + \cdots\cdots + a_{1n}x_n - s_1 + A_1 = b_1$$
$$a_{21}x_1 + a_{22}x_2 + \cdots\cdots + a_{2n}x_n - s_2 + A_2 = b_2$$
$$\ldots\ldots\ldots\ldots\ldots\ldots\ldots\ldots\ldots\ldots\ldots\ldots\ldots\ldots\ldots$$
$$\ldots\ldots\ldots\ldots\ldots\ldots\ldots\ldots\ldots\ldots\ldots\ldots\ldots\ldots\ldots$$
$$a_{m1}x_1 + a_{m2}x_2 + \cdots\cdots + a_{mn}x_n - s_m + A_m = b_m$$

where $x_j \geq 0, \quad (j = 1, 2, \ldots, n)$

and $s_i \geq 0, \quad A_i \geq 0, (i = 1, 2, \ldots, m)$

Thus the standard LPP has been reduced to a system of m equations in $(n + 2m)$ variables [n decision variables + m surplus variables + m artificial variables]. An initial basic feasible solution can now be obtained by setting $(n + 2m) - m = n + m$ variables equal to zero (*i.e.*, by setting each decision variable and each surplus variable equal to zero). Thus the initial basic feasible solution is given by $A_1 = b_1$, $A_2 = b_2$, ..., $A_m = b_m$.

This, however, does not constitute a solution to the original system of equations because the two systems are not equivalent. To remove artificial variables from solution, we use

Big-M Method or Method of Penalties

We assign a zero co-efficient to surplus variables and a very large positive co-efficient +M to artificial variable in the objective function.

Therefore, the problem can now be re-written as follows:

Minimize $Z = c_1x_1 + c_2x_2 + \cdots\cdots + c_nx_n + 0s_1 + 0s_2 + \cdots + 0s_m + MA_1 + MA_2 + \ldots + MA_m$

subject to the constraints

$$a_{11}x_1 + a_{12}x_2 + \cdots\cdots + a_{1n}x_n - s_1 + A_1 = b_1$$
$$a_{21}x_1 + a_{22}x_2 + \cdots\cdots + a_{2n}x_n - s_2 + A_2 = b_2$$
$$\ldots\ldots\ldots\ldots\ldots\ldots\ldots\ldots\ldots\ldots\ldots\ldots$$
$$\ldots\ldots\ldots\ldots\ldots\ldots\ldots\ldots\ldots\ldots\ldots\ldots$$
$$a_{m1}x_1 + a_{m2}x_2 + \cdots\cdots + a_{mn}x_n - s_m + A_m = b_m$$

and $\quad x_j \geq 0; (j = 1, 2, \cdots, n)$

$s_i \geq 0, \quad A_i \geq 0; (i = 1, 2, \ldots, m)$

Solve the modified LPP by simplex method.

Test for optimality. Compute the values of $C_j = c_j - Z_j$ in the last row of the simplex table.

(*i*) If all entries in C_j-row are non-negative (*i.e.*, ≥ 0), then the current basic feasible solution is optimal.

(*ii*) If C_k is most negative [*i.e.*, $C_k = \min. \{C_j : C_j < 0\}$] in a column and all entries in this column are negative, then the problem has an unbounded optimal solution.

(*iii*) If C_j has one or more negative values, then select the variable to enter into the basis with the largest negative value, *i.e.*, select x_k when $C_k = \min. \{C_j : C_j < 0\}$.

Determine the key row and key element in the same manner as discussed in the simplex algorithm of the maximization case.

Remarks 1. At any iteration of simplex method, one of the following three cases may arise:

(*i*) There remains no artificial variable in the basis and the optimality condition is satisfied, then the solution is an optimal basic feasible solution to the problem.

(*ii*) There is at least one artificial variable present in the basis with zero value and the co-efficient of M in each C_j value is non-negative, then the given LPP has no solution *i.e.*, the current basic feasible solution is degerate.

(*iii*) There is at least one artificial variable in the basis with positive value and the co-efficient of M in each C_j-value is non-negative, then the given LPP has no optimal basic feasible solution. In this case, the given LPP is said to have a **pseudo optimal basic feasible solution.**

2. The artificial variables are fictitious. They are introduced only for computational purposes and donot have any physical meaning or economic significance.

3. The constraints with $\geq$ inequality sign require a surplus variable and an artificial variable.

The constraints with '=' sign require neither a slack variable nor a surplus variable. Only an artificial variable has to be added to get the initial solution.

4. By assigning a very large per unit penalty to each of the artificial variables in the objective function, these variables can be removed from the simplex table as soon as they become non-basic. Such a penalty is designated as – M for maximization problems and + M for minimization problems where M > 0.

5. *No feasible solution* exists to the problem if all the artificial variables cannot be driven out in the optimum Big-M simplex table. Infeasibility is due to the presence of inconsistent constraints in the formulation of LP problems where the resources are not sufficient to meet the expected demands.

6. In minimization problems, key column is the one with largest negative number in C_j-row.

In maximization problems, key column is the one with largest positive number in C_j-row.

7. It must be noted that all variables, except the artificial variables, once driven out in an iteration, may re-enter in a subsequent iteration. But an artificial variable, once driven out, can never re-enter. Therefore, in the iteration subsequent to the one from which an artificial variable is drawn out, do not compute the column for the particular artificial variable.

ILLUSTRATIVE EXAMPLES

Example 1. *Use the penalty (Big-M) method to solve the following LP problem:*

Minimize $Z = 5x_1 + 3x_2$

subject to the constraints $2x_1 + 4x_2 \leq 12$

$2x_1 + 2x_2 = 10$

$5x_1 + 2x_2 \geq 10$

and $x_1, x_2 \geq 0.$

Sol. Step 1. The first constraint involves $\leq$. We introduce only a slack variable s_1 thereby getting

$$2x_1 + 4x_2 + s_1 = 12$$

The second constraint is strict equality and requires neither a slack variable nor a surplus variable. We add only an artificial variable A_1 thereby getting

$$2x_1 + 2x_2 + A_1 = 10$$

The third constraint involves $\geq$. We introduce a surplus variable s_2 and an artificial variable A_2 thereby getting

$$5x_1 + 2x_2 - s_2 + A_2 = 10$$

$\therefore$ The standard form of the LP problem now becomes:

Minimize $Z = 5x_1 + 3x_2 + 0s_1 + 0s_2 + MA_1 + MA_2$

subject to the constraints $2x_1 + 4x_2 + s_1 = 12$

$2x_1 + 2x_2 + A_1 = 10$

$5x_1 + 2x_2 - s_2 + A_2 = 10$

and $x_1, x_2, s_1, s_2, A_1, A_2 \geq 0$

Step 2. Since we have 3 equations in 6 variables, a solution is obtained by setting 6 – 3 = 3 variables equal to zero and solving for the remaining 3 variables. An initial basic feasible solution is obtained by setting $x_1 = x_2 = s_2 = 0$. Therefore, the initial basic feasible solution is $s_1 = 12$, $A_1 = 10$, $A_2 = 10$ and Z = 10 M + 10 M = 20 M. The initial basic feasible solution is given in the simplex table below.

Simplex Table I

		$c_j \to$	5	3	0	0	M	M	Ratio
c_B	Basis	Solution $b(= x_B)$	x_1	x_2	s_1	s_2	A_1	A_2	x_B/x_1
0	s_1	12	2	4	1	0	0	0	$\frac{12}{2} = 6$
M	A_1	10	2	2	0	0	1	0	$\frac{10}{2} = 5$
M	A_2	10	(5)	2	0	– 1	0	1	$\frac{10}{5} = 2 \to$
Z = 20 M		Z_j	7 M	4 M	0	– M	M	M	
		$C_j = c_j - Z_j$	**5 – 7 M** ↑	3 – 4 M	0	M	0	0	

Some entries in C_j-row being negative, the current solution is not optimal.

Step 3. Largest negative entry in C_j-row is 5 – 7 M which lies in x_1-column. Therefore the incoming variable is x_1. The ratio 2 is minimum in A_2-row, therefore, the outgoing basic variable is A_2. (In further simplex tables, we will not compute the A_2-column). Key element is 5.

New key row is

$$5 \quad x_1 \quad 2 \quad 1 \quad \frac{2}{5} \quad 0 \quad -\frac{1}{5} \quad 0$$

Transformation of R_1. Key column entry in R_1 is 2.

$\therefore$ $R_1(\text{new}) = R_1(\text{old}) - 2\,R_3(\text{new})$

$$12 - 2(2) = 8$$
$$2 - 2(1) = 0$$
$$4 - 2\left(\frac{2}{5}\right) = \frac{16}{5}$$
$$1 - 2(0) = 1$$
$$0 - 2\left(-\frac{1}{5}\right) = \frac{2}{5}$$
$$0 - 2(0) = 0$$

Transformation of R_2. Key column entry in R_2 is 2.

$\therefore$ $R_2(\text{new}) = R_2(\text{old}) - 2\,R_3(\text{new})$

$$10 - 2(2) = 6$$
$$2 - 2(1) = 0$$
$$2 - 2\left(\frac{2}{5}\right) = \frac{6}{5}$$
$$0 - 2(0) = 0$$

$$0-2\left(-\frac{1}{5}\right)=\frac{2}{5}$$

$$1-2(0)=1$$

New simplex table is given below.

Simplex Table II

		$c_j \rightarrow$ Solution	5	3	0	0	M	Ratio
c_B	Basis	$b(=x_B)$	x_1	x_2	s_1	s_2	A_1	x_B/x_2
0	s_1	8	0	$\frac{16}{5}$ (key element)	1	$\frac{2}{5}$	0	$\frac{5}{2}\rightarrow$
M	A_1	6	0	$\frac{6}{5}$	0	$\frac{2}{5}$	1	5
5	x_1	2	1	$\frac{2}{5}$	0	$-\frac{1}{5}$	0	5
$Z=10+6\,M$		Z_j	5	$2+\frac{6\,M}{5}$	0	$-1+\frac{2\,M}{5}$	M	
		$C_j=c_j-Z_j$	0	$1-\frac{6\,M}{5}$ ↑	0	$1-\frac{2\,M}{5}$	0	

Some entries in C_j-row being negative, the current solution is not optimal.

Step 4. Largest negative entry in C_j-row is $1-\frac{6\,M}{5}$ which lies in x_2-column. Therefore the incoming variable is x_2. The ratio $\frac{5}{2}$ is minimum in s_1-row, therefore, the outgoing basic variable is s_1. Key element is $\frac{16}{5}$.

New key row is

3	x_2	$\frac{5}{2}$	0	1	$\frac{5}{16}$	$\frac{1}{8}$	0

Transformation of R_2. Key column entry in R_2 is $\frac{6}{5}$

$$\therefore \quad R_2(\text{new}) = R_2(\text{old}) - \frac{6}{5}\,R_1(\text{new})$$

$$6-\frac{6}{5}\left(\frac{5}{2}\right)=3$$

$$0-\frac{6}{5}(0)=0$$

$$\frac{6}{5}-\frac{6}{5}(1)=0$$

$$0 - \frac{6}{5}\left(\frac{5}{16}\right) = -\frac{3}{8}$$

$$\frac{2}{5} - \frac{6}{5}\left(\frac{1}{8}\right) = \frac{1}{4}$$

$$1 - \frac{6}{5}(0) = 1$$

Transformation of R_3. Key column entry in R_3 is $\frac{2}{5}$.

$$\therefore \qquad R_3(\text{new}) = R_3(\text{old}) - \frac{2}{5} R_1(\text{new})$$

$$2 - \frac{2}{5}\left(\frac{5}{2}\right) = 1$$

$$1 - \frac{2}{5}(0) = 1$$

$$\frac{2}{5} - \frac{2}{5}(1) = 0$$

$$0 - \frac{2}{5}\left(\frac{5}{16}\right) = -\frac{1}{8}$$

$$-\frac{1}{5} - \frac{2}{5}\left(\frac{1}{8}\right) = -\frac{1}{4}$$

$$0 - \frac{2}{5}(0) = 0$$

The new simplex table is given below.

Simplex Table III

		$c_j \rightarrow$ Solution	*5*	*3*	*0*	*0*	*M*	*Ratio*
c_B	*Basis*	$b(=x_B)$	x_1	x_2	s_1	s_2	A_1	x_B/s_2
3	x_2	$\frac{5}{2}$	0	1	$\frac{5}{16}$	$\frac{1}{8}$	0	$\frac{(5/2)}{(1/8)} = 20$
M	A_1	3	0	0	$-\frac{3}{8}$	$\boxed{\frac{1}{4}}$	1	$\frac{3}{\left(\frac{1}{4}\right)} = 12 \rightarrow$
5	x_1	1	1	0	$-\frac{1}{8}$	$-\frac{1}{4}$	0	—
$Z = \frac{25}{2} + 3M$		Z_j	5	3	$\frac{5}{16} - \frac{3M}{8}$	$-\frac{7}{8} + \frac{M}{4}$	M	
		$C_j = c_j - Z_j$	0	0	$-\frac{5}{16} + \frac{3M}{8}$	$\frac{7}{8} - \frac{M}{4}$ ↑	0	

Since one entry in C_j-row is negative, the current solution is not optimal.

Step 5. The only negative entry in C_j-row is $\frac{7}{8} - \frac{M}{4}$ which lies in s_2-column. Therefore the incoming variable is s_2. The ratio 12 is minimum in A_1-row, therefore, the outgoing basic variable is A_1. In further simplex tables, we will not compute the A_1-column. Key element is $\frac{1}{4}$.

New key row is

$$0 \quad s_2 \quad 12 \quad 0 \quad 0 \quad -\frac{3}{2} \quad 1$$

Transformation of R_1. Key column entry in R_1 is $\frac{1}{8}$.

$$\therefore \quad R_1(\text{new}) = R_1(\text{old}) - \frac{1}{8} R_2(\text{new})$$

$$\frac{5}{2} - \frac{1}{8}(12) = 1$$

$$0 - \frac{1}{8}(0) = 0$$

$$1 - \frac{1}{8}(0) = 1$$

$$\frac{5}{16} - \frac{1}{8}\left(-\frac{3}{2}\right) = \frac{1}{2}$$

$$\frac{1}{8} - \frac{1}{8}(1) = 0$$

Transformation of R_3. Key column entry in R_3 is $-\frac{1}{4}$

$$\therefore \quad R_3(\text{new}) = R_3(\text{old}) + \frac{1}{4} R_2(\text{new})$$

$$1 + \frac{1}{4}(12) = 4$$

$$1 + \frac{1}{4}(0) = 1$$

$$0 + \frac{1}{4}(0) = 0$$

$$-\frac{1}{8} + \frac{1}{4}\left(-\frac{3}{2}\right) = -\frac{1}{2}$$

$$-\frac{1}{4} + \frac{1}{4}(1) = 0$$

The new simplex table is given below.

Simplex Table IV

		$c_j \rightarrow$	5	3	0	0	Ratio
c_B	Basis	Solution $b(=x_B)$	x_1	x_2	s_1	s_2	
3	x_2	1	0	1	$\frac{1}{2}$	0	
0	s_2	12	0	0	$-\frac{3}{2}$	1	
5	x_1	4	1	0	$-\frac{1}{2}$	0	
Z = 23		Z_j	5	3	−1	0	
		$C_j = c_j - Z_j$	0	0	1	0	

Since all entries in C_j-row are zero or positive, the current solution is optimal with $x_1 = 4, x_2 = 1, s_1 = 0, s_2 = 12$ and Min. Z = 23.

Example 2. *Use penalty (Big-M) method to solve the following LP problem.*

Maximize $Z = x_1 + 3x_2 - 2x_3$

subject to $-x_1 - 2x_2 - 2x_3 = -6$

$$-x_1 - x_2 + x_3 \leq -2$$

$$x_1, x_2, x_3 \geq 0.$$

Sol. Step 1. The constants on the right hand side of each constraint are negative. Therefore, multiplying both sides of each constraint by − 1, we get

$$x_1 + 2x_2 + 2x_3 = 6$$

$$x_1 + x_2 - x_3 \geq 2$$

Introducing a surplus variable s and two artificial variables A_1 and A_2, we get the standard form of the LP problem as:

Maximize $Z = x_1 + 3x_2 - 2x_3 + 0s - MA_1 - MA_2$

subject to $x_1 + 2x_2 + 2x_3 + A_1 = 6$

$$x_1 + x_2 - x_3 - s + A_2 = 2$$

where $x_1, x_2, x_3, s, A_1, A_2 \geq 0$

and M is a large positive number.

Step 2. Since we have 2 equations in 6 variables, a solution is obtained by setting $6 - 2 = 4$ variables equal to zero and solving for the remaining 2 variables. By setting $x_1 = x_2 = x_3 = s = 0$, the initial basic feasible solution is $A_1 = 6$, $A_2 = 2$ and $Z = -8M$.

The initial basic feasible solution is given in the simplex table below.

Simplex Table I

		$c_j \rightarrow$	*1*	*3*	*– 2*	*0*	*– M*	*– M*	*Ratio*
c_B	*Basis*	*Solution* $b(= x_B)$	x_1	x_2	x_3	s	A_1	A_2	x_B/x_2
– M	A_1	6	1	2	2	0	1	0	6/2 = 3
– M	A_2	2	1	(1)	– 1	– 1	0	1	2/1 = 2→
Z = – 8 M		Z_j	– 2M	– 3M	– M	M	– M	– M	
		$C_j = c_j - Z_j$	1 + 2M	3 + 3M ↑	– 2 + M	– M	0	0	

Some entries in C_j-row being positive, the current solution is not optimal.

Step 3. Largest positive entry in C_j-row is 3 + 3M which lies in x_2-column. Therefore, the incoming variable is x_2. The ratio 2 is minimum in A_2-row, therefore, the outgoing basic variable is A_2. In further simplex tables, we will not compute the A_2-column. Key element is 1.

New key row is

3 x_2 2 1 1 – 1 – 1 0

Transformation of R_1. Key column entry in R_1 is 2.

$\therefore$ $R_1(\text{new}) = R_1(\text{old}) - 2R_2(\text{new})$

$6 - 2(2) = 2$
$1 - 2(1) = -1$
$2 - 2(1) = 0$
$2 - 2(-1) = 4$
$0 - 2(-1) = 2$
$1 - 2(0) = 1$

New simplex table is given below.

Simplex Table II

		$c_j \rightarrow$	*1*	*3*	*– 2*	*0*	*– M*	*Ratio*
c_B	*Basis*	*Solution* $b(= x_B)$	x_1	x_2	x_3	s	A_1	x_B/x_3
– M	A_1	2	– 1	0	(4)	2	1	$\frac{2}{4} = \frac{1}{2}$ →
3	x_2	2	1	1	–1	–1	0	——
Z = 6 – 2 M		Z_j	3 + M	3	– 3 – 4 M	– 3 – 2 M	– M	
		$C_j = c_j - Z_j$	– 2 – M	0	1 + 4 M ↑	3 + 2 M	0	

Some entries in C_j-row being positive, the current solution is not optimal.

Step 4. Largest positive entry in C_j -row is $1 + 4M$ which lies in x_3- column. Therefore, the incoming variable is x_3. The ratio $\frac{1}{2}$ is minimum in A_1 - row, therefore, the outgoing basic variable is A_1. In further simplex tables, we will not compute the A_1-column. Key element is 4.

New key row is

$$-2 \quad x_3 \quad \frac{1}{2} \quad -\frac{1}{4} \quad 0 \quad 1 \quad \frac{1}{2}$$

Transformation of R_2. Key column entry in R_2 is -1.

$\therefore$

$R_2(\text{new}) = R_2(\text{old}) + 1\, R_1(\text{new})$
$2 + 1(\frac{1}{2}) = \frac{5}{2}$
$1 + 1\left(-\frac{1}{4}\right) = \frac{3}{4}$
$1 + 1(0) = 1$
$-1 + 1(1) = 0$
$-1 + 1(\frac{1}{2}) = -\frac{1}{2}$

New simplex table is given below.

Simplex Table III

		$c_j \rightarrow$	*1*	*3*	*– 2*	*0*	*Ratio*
c_B	*Basis*	*Solution* $b(= x_B)$	x_1	x_2	x_3	s	x_B/s
-2	x_3	$\frac{1}{2}$	$-\frac{1}{4}$	0	1	$\boxed{\frac{1}{2}}$	$1 \rightarrow$
3	x_2	$\frac{5}{2}$	$\frac{3}{4}$	1	0	$-\frac{1}{2}$	—
$Z = \frac{13}{2}$		Z_i	$\frac{11}{4}$	3	-2	$-\frac{5}{2}$	
		$C_j = c_j - Z_j$	$-\frac{7}{4}$	0	0	$\frac{5}{2}\uparrow$	

One entry in C_j-row is positive, the current solution is not optimal.

Step 5. Largest positive entry in C_j-row is $\frac{5}{2}$ which lies in s-column. Therefore, the incoming variable is s. Outgoing basic variable is x_3. Key element is $\frac{1}{2}$.

New key row is

$$0 \quad s \quad 1 \quad -\frac{1}{2} \quad 0 \quad 2 \quad 1$$

Transformation of R_2. Key column entry in R_2 is $-\frac{1}{2}$.

$\therefore$ $R_2\text{ (new)} = R_2\text{ (old)} + \frac{1}{2}R_1\text{(new)}$

$$\frac{5}{2}+\frac{1}{2}(1)=3$$
$$\frac{3}{4}+\frac{1}{2}\left(-\frac{1}{2}\right)=\frac{1}{2}$$
$$1+\frac{1}{2}(0)=1$$
$$0+\frac{1}{2}(2)=1$$
$$-\frac{1}{2}+\frac{1}{2}(1)=0$$

New simplex table is given below.

Simplex Table IV

		$c_j \rightarrow$ Solution	*1*	*3*	*−2*	*0*	*Ratio*
c_B	*Basis*	$b(=x_B)$	x_1	x_2	x_3	s	
0	s	1	$-\frac{1}{2}$	0	2	1	
3	x_2	3	$\frac{1}{2}$	1	1	0	
Z = 9		Z_j	$\frac{3}{2}$	3	3	0	
		$C_j = c_j - Z_j$	$-\frac{1}{2}$	0	−5	0	

Since all entries in C_j-row are negative or zero, the optimal solution has been arrived at with $x_1 = 0$, $x_2 = 3$, $x_3 = 0$ and Max. Z = 9.

EXERCISE 7.6

Solve the following LP problems using Big-M method.

1. Minimize $Z = 8x + 12y$

subject to
$$2x + 2y \geq 1$$
$$x + 3y \geq 2$$
$$x, y \geq 0$$

2. Minimize $Z = 5x + 8y$

subject to
$$x + y = 5$$
$$x \leq 4$$
$$y \geq 2$$
$$x, y \geq 0.$$

3. Minimize $Z = 2x_1 + x_2$

subject to

$$3x_1 + x_2 = 3$$
$$4x_1 + 3x_2 \geq 6$$
$$x_1 + 2x_2 \leq 3$$
$$x_1, x_2 \geq 0$$

4. Minimize $z = 3x + 6y$

subject to

$$-x + y \leq 6$$
$$x + y \geq 10$$
$$x, y \geq 0$$

5. Maximize $Z = 2x + y$

subject to

$$x + y \leq 6$$
$$-x + y \geq 4$$
$$x, y \geq 0$$

6. Maximize $Z = -3x + 2y$

subject to

$$x - y \leq 4$$
$$-x + y = 4$$
$$x \geq 6$$
$$x, y \geq 0$$

7. Maximize $Z = 3x + 4y$

subject to

$$x + y \leq 12$$
$$5x + 2y \geq 36$$
$$7x + 4y \geq 14$$
$$x, y \geq 0.$$

8. Minimize $Z = 5x_1 + 4x_2 + 3x_3$

subject to

$$x_1 + x_2 + x_3 \geq 100$$
$$2x_1 + x_2 \geq 50$$
$$x_1, x_2, x_3 \geq 0.$$

9. Minimize $Z = x_1 - x_2 - 3x_3$

subject to

$$x_1 + 2x_2 + x_3 = 4$$
$$x_2 + x_3 = 1$$
$$x_1 + x_2 \leq 6$$
$$x_1, x_2, x_3 \geq 0.$$

10. Minimize $Z = x_1 + 2x_2 + x_3$

subject to

$$x_1 - x_2 - x_3 \leq -1$$
$$6x_1 + 3x_2 + 2x_3 = 12$$
$$x_1, x_2, x_3 \geq 0$$

11. Maximize $Z = 2x_1 + x_2 - x_3$

subject to

$$x_1 + 2x_2 + x_3 \leq 5$$
$$-x_1 + x_2 + x_3 \geq 1$$
$$x_1, x_2, x_3 \geq 0$$

12. Maximize $Z = 3x_1 + 2x_2 + 3x_3$

subject to

$$2x_1 + x_2 + x_3 \leq 2$$
$$3x_1 + 4x_2 + 2x_3 \geq 8$$
$$x_1, x_2, x_3 \geq 0.$$

13. Maximize $Z = x_1 + 2x_2 + 3x_3 - x_4$

subject to

$$x_1 + 2x_2 + 3x_3 = 15$$
$$2x_1 + x_2 + 5x_3 = 20$$
$$x_1 + 2x_2 + x_3 + x_4 = 10$$
$$x_1, x_2, x_3, x_4 \geq 0.$$

Answers

1. $x = 0, y = \frac{2}{3}$; Min. Z = 8
2. $x = 3, y = 2$; Min. Z = 31
3. $x_1 = \frac{3}{5}, x_2 = \frac{6}{5}$; Min. Z = $\frac{12}{5}$
4. $x = 10, y = 0$; Min. Z = 30
5. $x = 1, y = 5$; Max. Z = 7
6. $x = 6, y = 10$; Max. Z = 2
7. $x = 4, y = 8$; Max. Z = 44
8. $x_1 = 25, x_2 = 0, x_3 = 75$; Min. Z = 350
9. $x_1 = 3, x_2 = 0, x_3 = 1$; Min. Z = 0
10. $x_1 = \frac{5}{4}, x_2 = 0, x_3 = \frac{9}{4}$; Min. Z = $\frac{7}{2}$
11. $x_1 = 1, x_2 = 2, x_3 = 0$; Max. Z = 4
12. $x_1 = 0, x_2 = 2, x_3 = 0$; Max. Z = 4
13. $x_1 = x_2 = x_3 = \frac{5}{2}, x_4 = 0$; Max. Z = 15.

7.15. DUALITY CONCEPT

With every linear programming problem, there exists another linear programming problem called its dual since every LPP can be analysed in two different ways without any additional data or information. For example, profit maximization problem can be seen as a problem of cost minimization and cost minimization problem can be viewed as a problem of maximizing the efficiency of using available resources. The original problem is called **primal** and the associated problem is called its **dual**. In general, either problem can be considered as primal and the other its dual.

7.16. FORMULATION OF A DUAL PROBLEM

Suppose the primal LP problem is in the form

Maximize $Z = c_1x_1 + c_2x_2 + \ldots\ldots\ldots\ldots + c_n x_n$

subject to the constraints

$$\begin{aligned} a_{11}x_1 + a_{12}x_2 + \ldots\ldots\ldots\ldots + a_{1n}x_n &\le b_1 \\ a_{21}x_1 + a_{22}x_2 + \ldots\ldots\ldots\ldots + a_{2n}x_n &\le b_2 \\ \ldots\ldots\ldots\ldots\ldots\ldots\ldots\ldots\ldots\ldots\ldots\ldots\ldots\ldots \\ \ldots\ldots\ldots\ldots\ldots\ldots\ldots\ldots\ldots\ldots\ldots\ldots\ldots\ldots \\ a_{m1}x_1 + a_{m2}x_2 + \ldots\ldots\ldots\ldots + a_{mn}x_n &\le b_m \end{aligned}$$

and $x_1, x_2, \ldots\ldots\ldots\ldots, x_n \ge 0$

The corresponding dual problem is

Minimize $Z^* = b_1 y_1 + b_2 y_2 + \ldots\ldots\ldots\ldots + b_m y_m$

subject to the constraints

$$\begin{aligned} a_{11}y_1 + a_{21}y_2 + \ldots\ldots\ldots\ldots + a_{m1}y_m &\ge c_1 \\ a_{12}y_1 + a_{22}y_2 + \ldots\ldots\ldots\ldots + a_{m2}y_m &\ge c_2 \\ \ldots\ldots\ldots\ldots\ldots\ldots\ldots\ldots\ldots\ldots\ldots\ldots\ldots\ldots \\ \ldots\ldots\ldots\ldots\ldots\ldots\ldots\ldots\ldots\ldots\ldots\ldots\ldots\ldots \\ a_{1n}y_1 + a_{2n}y_2 + \ldots\ldots\ldots\ldots + a_{mn}y_m &\ge c_n \end{aligned}$$

and $y_1, y_2, \ldots\ldots\ldots\ldots, y_m \ge 0$

The characteristics of the primal and the dual problem can be stated as follows:

(*i*) The maximization problem in the primal becomes the minimization problem in the dual and *vice versa*.

(*ii*) The $\leq$ type constraints in the primal become $\geq$ type constraints in the dual and *vice versa*.

When the primal variables are unrestricted in sign, corresponding dual constraints are equalities.

(*iii*) The co-efficients $c_1, c_2, \ldots\ldots, c_n$ in the objective function of the primal become the constants $b_1, b_2, \ldots\ldots\ldots, b_m$ in the constraints of the dual and *vice versa*.

(*iv*) If the primal has n variables and m constraints, then the dual has m variables and n constraints. Thus the body matrix of dual is the transpose of body matrix of primal and *vice versa*.

(*v*) A new set of variables appear in the dual.

(*vi*) The variables in both the primal and the dual are non-negative.

(*vii*) The dual of the dual problem is the original primal problem itself.

ILLUSTRATIVE EXAMPLES

Example. *Obtain the dual of*

Maximize $Z = 5x_1 + 3x_2$

subject to

$$x_1 + x_2 \leq 2$$
$$5x_1 + 2x_2 \leq 10$$
$$3x_1 + 8x_2 \leq 12$$
$$x_1, x_2 \geq 0.$$

(M.D.U. 2011)

Sol. (*i*) Primal is of maximization type

Dual will be of 'minimization' type.

(*ii*) Primal has 2 variables x_1, x_2 and 3 constraints.

Dual will have 3 variables y_1, y_2, y_3 and 2 constraints.

(*iii*) Primal has $\leq$ constraints.

Dual will have $\geq$ constraints.

(*iv*) Primal has $c_1 = 5$, $c_2 = 3$ and $b_1 = 2$, $b_2 = 10$, $b_3 = 12$

Dual will have $c_1 = 2$, $c_2 = 10$, $c_3 = 12$ and $b_1 = 5$, $b_2 = 3$

The body matrix in the primal is $\begin{bmatrix} 1 & 1 \\ 5 & 2 \\ 3 & 8 \end{bmatrix}$

The body matrix in the dual will be

$$\begin{bmatrix} 1 & 1 \\ 5 & 2 \\ 3 & 8 \end{bmatrix}' = \begin{bmatrix} 1 & 5 & 3 \\ 1 & 2 & 8 \end{bmatrix}$$

Thus the dual is

Minimize $Z^* = 2y_1 + 10y_2 + 12y_3$

subject to $y_1 + 5y_2 + 3y_3 \geq 5$

$$y_1 + 2y_2 + 8y_3 \geq 3$$

and $$y_1, y_2, y_3 \geq 0$$

EXERCISE 7.7

Find the dual of each of the following problems:

1. Maximize $Z = 20x_1 + 50x_2$
   subject to $5x_1 + 4x_2 \leq 20$
   $3x_1 + 5x_2 \leq 15$
   $x_1, x_2 \geq 0$

2. Maximize $Z = 20x_1 + 30x_2 + 50x_3$
   subject to $2x_1 + 3x_2 + 5x_3 \leq 10$
   $5x_1 + 6x_3 \leq 15$
   $x_1, x_2, x_3 \geq 0$

3. Maximize $Z = 4x_1 + 3x_2$
   subject to $3x_1 - x_2 \leq 2$
   $x_1 + x_2 \geq 1$
   $-4x_1 + x_2 \geq 3$
   $x_1, x_2 \geq 0$

4. Maximize $Z = 2x_1 + 3x_2$
   subject to $x_1 + x_2 \geq 6$
   $2x_1 + x_2 \geq 7$
   $x_1 + 4x_2 \geq 8$
   $x_1, x_2 \geq 0$

5. Minimize $Z = 3x_1 + 5x_1 + x_3$
   subject to $2x_1 + x_2 \geq 10$
   $4x_1 + 3x_2 + 2x_3 \geq 15$
   $x_1, x_2, x_3 \geq 0$

6. Minimize $Z = 6x_1 + 4x_2$
   subject to $-x_1 + x_2 \leq 1$
   $x_1 + x_2 \geq 3$
   $x_1, x_2 \geq 0$

7. Minimize $Z = 3x_1 - 3x_2 + x_3$
   subject to $2x_1 - 3x_2 + x_3 \leq 5$
   $4x_1 - 2x_2 \geq 9$
   $-8x_1 + 4x_2 + 3x_3 = 8$
   $x_1, x_2 \geq 0$ and x_3 is unrestricted. *(K.U.K. Dec. 2015)*

Answers

1. Minimize $Z^* = 20y_1 + 15y_2$
   Subject to $5y_1 + 3y_2 \geq 20$
   $4y_1 + 5y_2 \geq 50$
   $y_1, y_2 \geq 0$

2. Minimize $Z^* = 10y_1 + 15y_2$
   subject to $2y_1 + 5y_2 \geq 20$
   $3y_1 \geq 30$
   $5y_1 + 6y_2 \geq 50$
   $y_1, y_2 \geq 0$

3. Minimize $Z^* = 2y_1 - y_2 - 3y_3$
   subject to $3y_1 - y_2 + 4y_3 \geq 4$
   $-y_1 - y_2 - y_3 \geq 3$
   $y_1, y_2, y_3 \geq 0$

4. Minimize $Z^* = -6y_1 - 7y_2 - 8y_3$
   subject to $-y_1 - 2y_2 - y_3 \geq 2$
   $-y_1 - y_2 - 4y_3 \geq 3$
   $y_1, y_2, y_3 \geq 0$

5. Maximize $Z^* = 10y_1 + 15y_2$
   subject to $2y_1 + 4y_2 \leq 3$
   $y_1 + 3y_2 \leq 5$
   $2y_2 \leq 1$
   $y_1, y_2 \geq 0$

6. Maximize $Z^* = -y_1 + 3y_2$
   subject to $y_1 + y_2 \leq 6$
   $-y_1 + y_2 \leq 4$
   $y_1, y_2 \geq 0$

7. Maximize $Z^* = -5y_1 + 9y_2 + 8y_3$
   subject to $-2y_1 + 4y_2 - 8y_3 \leq 3$
   $3y_1 - 2y_2 + 4y_3 \leq -3$
   $-y_1 + 3y_3 = 1$
   $y_1, y_2 \geq 0$ and y_3 is unrestricted.

7.17. DUALITY PRINCIPLE

If either the primal or the dual problem has an optimal solution then so does the other problem and the optimal value of the primal's objective function is the same as that of its dual.

Example. *Solve the dual problem of the following LPP and hence find Max. Z:*

Maximize $Z = 20x_1 + 30x_2$

subject to
$$3x_1 + 3x_2 \leq 36$$
$$5x_1 + 2x_2 \leq 50$$
$$2x_1 + 6x_2 \leq 60$$
$$x_1, x_2 \geq 0.$$

Sol. The dual of the above LPP is

Minimize $Z^* = 36y_1 + 50y_2 + 60y_3$

subject to
$$3y_1 + 5y_2 + 2y_3 \geq 20$$
$$3y_1 + 2y_2 + 5y_3 \geq 30$$
$$y_1, y_2, y_3 \geq 0$$

Solving this problem by simplex method, it can be verified that the final simplex table is:

		$c_j \rightarrow$ Solution	36	50	60	0	0	M	M
c_B	Basis	$b\ (= y_B)$	y_1	y_2	y_3	s_1	s_2	A_1	A_2
36	y_1	5	1	$\frac{13}{6}$	0	$-\frac{1}{2}$	$\frac{1}{6}$	$\frac{1}{2}$	$-\frac{1}{6}$
60	y_3	$\frac{5}{2}$	0	$-\frac{3}{4}$	1	$\frac{1}{4}$	$-\frac{1}{4}$	$-\frac{1}{4}$	$\frac{1}{4}$
$Z^* = 330$		Z_j^*	36	33	60	-3	-9	3	9
		$C_j = c_j - Z_j^*$	0	17	0	3	9	$M-3$	$M-9$

The optimal solution to the dual problem is

$$y_1 = 5,\ y_2 = 0,\ y_3 = \frac{5}{2} \quad \text{and Min. } Z^* = 330$$

By duality principle = Max. Z = Min. Z* = 330.

EXERCISE 7.8

1. Solve the dual problem of the following LPP and hence find Max. Z.

   Maximize $Z = 3x_1 + 2x_2 + 5x_3$

   subject to
$$x_1 + 2x_2 + x_3 \leq 430$$
$$3x_1 + 2x_3 \leq 460$$
$$x_1 + 4x_2 \leq 420$$
$$x_1, x_2, x_3 \geq 0$$

2. Solve the dual problem of the following LPP and hence find Max. Z.

   Maximize $Z = 7x_1 + 5x_2$

   subject to
$$3x_1 + x_2 \leq 48$$
$$2x_1 + x_2 \leq 40$$
$$x_1, x_2 \geq 0$$

3. Write the dual of the following LPP and find the solution of both dual and primal problem

Maximize $Z = 40x_1 + 25x_2 + 50x_3$

subject to the constraints
$$x_1 + x_2 + x_3 \le 36$$
$$2x_1 + x_2 + 4x_3 \le 60$$
$$2x_1 + 5x_2 + x_3 \le 45$$
$$x_1, x_2, x_3 \ge 0$$

Answers

1. Max. $Z = 1350$ **2.** Max. $Z = 200$

3. Solution of dual is $y_1 = y_2 = 0, y_3 = 10$, Min $Z^* = 1050$

Solution of primal is $x_1 = 20, x_2 = 0, x_3 = 5$, Max. $Z = 1050$

7.18. DUAL SIMPLEX METHOD

Dual simplex method is similar to the regular simplex method which has already been discussed. The only difference lies in the criterion used for selecting the incoming and outgoing variables. In the dual simplex method, we first determine the outgoing variable and then the incoming variable whereas in the regular simplex method, the reverse is done. The regular simplex method starts with a basic feasible but non-optimal solution and works towards optimality, the dual simplex method starts with a basic infeasible but optimal solution and works towards feasibility.

7.19. WORKING PROCEDURE FOR DUAL SIMPLEX METHOD

Setp 1: Check whether the objective function is to be maximized or minimized. If the objective function is to be minimized, then convert it into maximization form.

Setp 2: Convert $\ge$ type constraints, if any, into $\le$ type by multiplying such constraints by -1.

Step 3: Express all constraints as equations by adding slack variables, one for each constraint.

Step 4: Find the initial basic solution and express this information in the form of a table as in regular simplex method.

Step 5: Compute $C_j = c_j - Z_j$

(*a*) If all $C_j \le 0$ and all $b_i \ge 0$, then the solution found above is the optimal basic feasible solution.

(*b*) If all $C_j \le 0$ and at least one $b_i < 0$, then go to the next step.

(*c*) If any $C_j > 0$, the method fails.

Step 6: *Selection of key row and the outgoing variable*

Select the row containing the most negative b_i. This row is the key row and the basic variable heading the key row is the outgoing variable.

Step 7: *Selection of key column and the incoming variable*

(*a*) If all elements in the key row are ≥ 0, then the problem does not have feasible solution.

(*b*) If at least one element in the key row is negative, find the ratios of the corresponding elements of C_j-row to these elements. Ignore the ratios associated with positive or zero elements of the key row. Choose the smallest of these ratios. The corresponding column is the key column and the variable heading the key column is the incoming variable.

Step 8: Mark the key element and make it unity. Perform row operations as in the regular simplex method and repeat iterations until either an optimal feasible solution is attained or there is an indication of non-existence of a feasible solution.

ILLUSTRATIVE EXAMPLES

Example 1. *Using dual simplex method*
Maximize $Z = -3x_1 - x_2$
subject to $x_1 + x_2 \geq 1$, $2x_1 + 3x_2 \geq 2$, $x_1, x_2 \geq 0$.
(M.D.U. May 2014; K.U.K. 2010, Jan. 2013, Dec. 2015)

Sol.

Step 1. The problem is that of maximization.

Step 2. Converting ≥ type constraints into ≤ type, the LPP takes the form
Maximize $Z = -3x_1 - x_2$
subject to $-x_1 - x_2 \leq -1$, $-2x_1 - 3x_2 \leq -2$, $x_1, x_2 \geq 0$

Step 3. Introducing slack variables s_1, s_2 (one for each constraint), the problem in standard form is:

Maximize $Z = -3x_1 - x_2 + 0s_1 + 0s_2$
subject to $-x_1 - x_2 + s_1 = -1$
$-2x_1 - 3x_2 + s_2 = -2$
$x_1, x_2, s_1, s_2 \geq 0$

Step 4. The initial basic solution is given by
$$x_1 = x_2 = 0,\ S_1 = -1,\ S_2 = -2 \text{ at which } Z = 0$$

Dual Simplex Table I

C_B	*Basis*	$C_j \rightarrow$ *Solution* $b\ (= x_B)$	-3	-1	0	0
			x_1	x_2	s_1	s_2
0	S_1	-1	-1	-1	1	0
0	S_2	-2	-2	(-3)	0	1
$Z = 0$		Z_j	0	0	0	0
		$C_j = c_j - Z_j$	-3	-1	0	0

Step 5. Since all $C_j \leq 0$ and all $b_i < 0$, the initial solution is optimal but infeasible. So we proceed further.

Step 6. The most negative b_i (*i.e.*, the b_i which is negative and numerically largest) is $b_2 = -2$, **the second row is the key row and S_2 is the outgoing variable.**

Step 7. The key row has negative entries. The ratios of the elements in C_j-row to the corresponding negative elements of the key row (neglecting ratios corresponding to positive or zero elements of key row) are

$$\frac{-3}{-2} = \frac{3}{2},\ \frac{-1}{-3} = \frac{1}{3}$$

Smallest ratio is $\frac{1}{3}$. **The x_2-column is the key column and x_2 is the incoming variable.**

Step 8. The key element is – 3, shown circled. Drop the outgoing variable s_2 from the basis and introduce the incoming variable x_2. The new basis contains s_1, x_2 as the basic variables. The co-efficient of x_2 in the objective function is – 1. Therefore, the entry in C_B column corresponding to the new basic variable x_2 will be – 1. Since the key element enclosed in the circle is not 1, divide all elements of the key row by the key element – 3 to make it unity and make all other elements in the key column zero. Thus the key row

$$0 \qquad s_2 \qquad -2 \qquad -2 \qquad -3 \qquad 0 \qquad 1$$

is replaced by the new key row

$$-1 \qquad x_2 \qquad \frac{2}{3} \qquad \frac{2}{3} \qquad 1 \qquad 0 \qquad -\frac{1}{3}$$

Now we make all other elements of the key column (*i.e.*, x_2-column) zero.

Transformation of R_1. Key column entry in R_1 is –1

$\therefore$ R_1 (new) = R_1 (old) – (– 1) R_2 (new)

$$-1 + 1\left(\frac{2}{3}\right) = -\frac{1}{3}$$

$$-1 + 1\left(\frac{2}{3}\right) = -\frac{1}{3}$$

$$-1 + 1\,(1) = 0$$

$$1 + 1\,(0) = 1$$

$$0 + 1\left(-\frac{1}{3}\right) = -\frac{1}{3}$$

The above information is summarized in the following table:

Dual Simplex Table II

C_B	*Basis*	c_j *Solution* $b\ (= x_B)$	*–3*	*–1*	*0*	*0*
			x_1	x_2	s_1	s_2
0	s_1	$-\frac{1}{3}$	$-\frac{1}{3}$	0	1	$-\frac{1}{3}$ (circled)
–1	x_2	$\frac{2}{3}$	$\frac{2}{3}$	1	0	$-\frac{1}{3}$
$Z = -\frac{2}{3}$		Z_j	$-\frac{2}{3}$	– 1	0	$\frac{1}{3}$
		$C_j = c_j - Z_j$	$-\frac{7}{3}$	0	0	$-\frac{1}{3}$

Step 9. Test for optimality. Here all $C_j \le 0$ but $b_1 = -\frac{1}{3} < 0$, the current solution is not optimal. Therefore, an improvement in the value of Z is possible and we repeat steps 6 to 8.

The first row is the key row and s_1 is the outgoing variable. The ratios of the elements in C_j-row to the corresponding negative elements of the key row are

$$\frac{-\frac{7}{3}}{-\frac{1}{3}} = 7, \frac{-\frac{1}{3}}{-\frac{1}{3}} = 1$$

The smallest ratio is 1, therefore, s_2-column is the key column and s_2 is the incoming variable. The key element is $-\frac{1}{3}$, shown circled. Drop s_1 and introduce s_2. The new basis contains s_2, x_2 as the basic variables. Dividing all elements of the key row by the key element $-\frac{1}{3}$, the new key row is

0	s_2	1	1	0	– 3	1

Transformation of R_2. Key column entry in R_2 is $-\frac{1}{3}$.

$$\therefore \quad R_2 \text{ (new)} = R_2 \text{ (old)} - \left(-\frac{1}{3}\right) R_1 \text{ (new)}$$

$$\frac{2}{3} + \frac{1}{3}(1) = 1$$

$$\frac{2}{3} + \frac{1}{3}(1) = 1$$

$$1 + \frac{1}{3}(0) = 1$$

$$0 + \frac{1}{3}(-3) = -1$$

$$-\frac{1}{3} + \frac{1}{3}(1) = 0$$

The above information is summarized in the following table:

Dual Simplex Table III

C_B	*Basis*	C_j *Solution* $b\ (= x_B)$	– 3	– 1	0	0
			x_1	x_2	s_1	s_2
0	s_2	1	1	0	–3	1
– 1	x_2	1	1	1	–1	0
Z = – 1		Z_j	–1	–1	1	0
		$C_j = c_j - Z_j$	–2	0	–1	0

Here all $C_j \le 0$ and all $b_i > 0$, the optimal feasible solution has been attained. Thus the optimal feasible solution is

$$x_1 = 0, x_2 = 1, \text{Max. } Z = -1.$$

Example 2. *Using dual simplex method,*

Minimize $Z = x_1 + 2x_2 + 3x_3$

subject to $2x_1 - x_2 + x_3 \geq 4$

$x_1 + x_2 + 2x_3 \leq 8$

$x_2 - x_3 \geq 2$

$x_1, x_2, x_3 \geq 0.$ (K.U.K. May 2013)

Sol.

Step 1. The problem is that of minimization. Converting it into maximization problem by using the relationship

Min. Z = – Max. Z*, where Z* = – Z, we have

Maximize $Z^* = -x_1 - 2x_2 - 3x_3$

subject to $2x_1 - x_2 + x_3 \geq 4$

$x_1 + x_2 + 2x_3 \leq 8$

$x_2 - x_3 \geq 2$

$x_1, x_2, x_3 \geq 0$

Step 2. Converting ≥ type constraints into ≤ type, the L.P.P. takes the form

Maximize $Z^* = -x_1 - 2x_2 - 3x_3$

subject to $-2x_1 + x_2 - x_3 \leq -4$

$x_1 + x_2 + 2x_3 \leq 8$

$-x_2 + x_3 \leq -2$

$x_1, x_2, x_3 \geq 0$

Step 3. Introducing slack variables s_1, s_2, s_3, the problem in standard form is:

Maximize $Z^* = -x_1 - 2x_2 - 3x_3 + 0s_1 + 0s_2 + 0s_3$

subject to $-2x_1 + x_2 - x_3 + s_1 = -4$

$x_1 + x_2 + 2x_3 + s_2 = 8$

$-x_2 + x_3 + s_3 = -2$

$x_1, x_2, x_3, s_1, s_2, s_3 \geq 0$

Step 4. The initial basic solution is given by

$$x_1 = x_2 = x_3 = 0,\ s_1 = -4,\ s_2 = 8,\ s_3 = -2$$

at which Z* = 0

Dual Simplex Table I

C_B	*Basis*	$C_j \rightarrow$ Solution $b\ (= x_B)$	– 1 x_1	– 2 x_2	– 3 x_3	0 s_1	0 s_2	0 s_3
0	s_1	– 4	(– 2)	1	– 1	1	0	0
0	s_2	8	1	1	2	0	1	0
0	s_3	– 2	0	– 1	1	0	0	1
Z* = 0		Z_j^*	0	0	0	0	0	0
		$C_j = c_j - Z_j^*$	– 1	– 2	– 3	0	0	0

Step 5. Since all $C_j \leq 0$ and at least one $b_i < 0$, the initial solution is optimal but infeasible. So, we proceed further.

Step 6. The most negative b_i is $b_1 = -4$, the first row is the key row and s_1 is the outgoing variable.

Step 7. The key row has negative entries. The ratios of the elements in C_j-row to the corresponding negative elements of the key row are

$$\frac{-1}{-2}=\frac{1}{2},\ \frac{-3}{-1}=3$$

Smallest ratio is $\frac{1}{2}$. The x_1-column is the key column and x_1 is the incoming variable.

Step 8. The key element is – 2, shown circled. The new basis contains x_1, s_2, s_3 as the basic variables. Since the key element is not 1, divide all elements of the key row by the key element – 2 to make it unity. Thus, the new key-row is

$$-1 \quad x_1 \quad 2 \quad 1 \quad -\frac{1}{2} \quad \frac{1}{2} \quad -\frac{1}{2} \quad 0 \quad 0$$

Now we make all other elements of the key column zero.

Transformation of R_2. Key column entry in R_2 is 1.

$\therefore$ R_2 (new) = R_2 (old) – 1 R_1 (new)

$$8-1\,(2)=6,\quad 1-1\,(1)=0,\quad 1-1\left(-\frac{1}{2}\right)=\frac{3}{2}$$

$$2-1\left(\frac{1}{2}\right)=\frac{3}{2},\quad 0-1\left(-\frac{1}{2}\right)=\frac{1}{2},\quad 1-1\,(0)=1$$

$$0-1\,(0)=0$$

Transformation of R_3. Key column entry in R_3 is 0.

$\therefore$ R_3 (new) = R_3 (old)

The above information is summarized in the following table:

Dual Simplex Table II

C_B	*Basis*	$C_j \rightarrow$	-1	-2	-3	0	0	0
		Solution $b\ (=x_B)$	x_1	x_2	x_3	s_1	s_2	s_3
– 1	x_1	2	1	$-\frac{1}{2}$	$\frac{1}{2}$	$-\frac{1}{2}$	0	0
0	s_2	6	0	$\frac{3}{2}$	$\frac{3}{2}$	$\frac{1}{2}$	1	0
0	s_3	– 2	0	(– 1)	1	0	0	1
$Z^* = -2$		Z_j^*	– 1	$\frac{1}{2}$	$-\frac{1}{2}$	$\frac{1}{2}$	0	0
		$C_j = c_j - Z_j^*$	0	$-\frac{5}{2}$	$-\frac{5}{2}$	$-\frac{1}{2}$	0	0

Step 9. Here all $C_j \le 0$ but $b_3 = -2 < 0$, the current solution is not optimal. So we repeat steps 6 to 8.

The third row is the key row and s_3 is the outgoing variable. The ratios of the elements in C_j-row to the corresponding negative elements of the key row are

$$\frac{-\frac{5}{2}}{-1}=\frac{5}{2}$$

The x_2-column is the key column and x_2 is the incoming variable. The key element is -1, shown circled. Drop s_3 and introduce x_2. The new basis contains x_1, s_2, x_2 as the basic variables. Dividing all elements of the key row by the key element -1, the new key row is

$$-2 \quad x_2 \quad 2 \quad 0 \quad 1 \quad -1 \quad 0 \quad 0 \quad -1$$

Now we make all other elements of the key column (*i.e.*, x_2-column) zero.

Transformation of R_1. Key column entry in R_1 is $-\frac{1}{2}$.

$$\therefore \quad R_1\text{ (new)} = R_1\text{ (old)} - \left(-\frac{1}{2}\right) R_3\text{ (new)}$$

$$2 + \frac{1}{2}(2) = 3, \quad 1 + \frac{1}{2}(0) = 1, \quad -\frac{1}{2} + \frac{1}{2}(1) = 0$$

$$\frac{1}{2} + \frac{1}{2}(-1) = 0, \quad -\frac{1}{2} + \frac{1}{2}(0) = -\frac{1}{2}, \quad 0 + \frac{1}{2}(0) = 0$$

$$0 + \frac{1}{2}(-1) = -\frac{1}{2}$$

Transformation of R_2. Key column entry in R_2 is $\frac{3}{2}$.

$$\therefore \quad R_2\text{ (new)} = R_2\text{ (old)} - \frac{3}{2} R_3\text{ (new)}$$

$$6 - \frac{3}{2}(2) = 3, \quad 0 - \frac{3}{2}(0) = 0, \quad \frac{3}{2} - \frac{3}{2}(1) = 0$$

$$\frac{3}{2} - \frac{3}{2}(-1) = 3, \quad \frac{1}{2} - \frac{3}{2}(0) = \frac{1}{2}, \quad 1 - \frac{3}{2}(0) = 1$$

$$0 - \frac{3}{2}(-1) = \frac{3}{2}$$

The above information is summarized in the following table:

Dual Simplex Table III

C_B	*Basis*	$C_j \rightarrow$ *Solution* $b\ (= x_B)$	-1 x_1	-2 x_2	-3 x_3	0 s_1	0 s_2	0 s_3
-1	x_1	3	1	0	0	$-\frac{1}{2}$	0	$-\frac{1}{2}$
0	s_2	3	0	0	3	$\frac{1}{2}$	1	$\frac{3}{2}$
-2	x_2	2	0	1	-1	0	0	-1
$Z^* = -7$		Z_j^*	-1	-2	2	$\frac{1}{2}$	0	$\frac{5}{2}$
		$C_j = c_j - Z_j^*$	0	0	-5	$-\frac{1}{2}$	0	$-\frac{5}{2}$

Here all $C_j \leq 0$ and all $b_i > 0$, the optimal feasible solution has been attained. Thus the optimal feasible solution is

$$x_1 = 3, x_2 = 2, x_3 = 0, \text{Max. } Z^* = -7 \quad i.e., \text{Min. } Z = 7.$$

EXERCISE 7.9

Using dual simplex method, solve the following problems:

1. Minimize $Z = 0.7x_1 + 0.5x_2$
   subject to $x_1 \geq 4, x_2 \geq 6, x_1 + 2x_2 \geq 20, 2x_1 + x_2 \geq 18, x_1, x_2 \geq 0$ *(M.D.U. 2010)*
2. Maximize $Z = -2x_1 - x_2$ *(M.D.U. Dec. 2015)*
   subject to $3x_1 + x_2 \geq 3, 4x_1 + 3x_2 \geq 6, x_1 + 2x_2 \geq 3, x_1, x_2 \geq 0$
3. Maximize $Z = -3x_1 - 2x_2$
   subject to $x_1 + x_2 \geq 1, x_1 + x_2 \leq 7, x_1 + 2x_2 \geq 10, x_2 \leq 3, x_1, x_2 \geq 0$
4. Minimize $Z = 2x_1 + x_2$
   subject to $3x_1 + x_2 \geq 3, 4x_1 + 3x_2 \geq 6, x_1 + 2x_2 \leq 3, x_1, x_2 \geq 0$
5. Minimize $Z = 2x_1 + 2x_2 + 4x_3$
   subject to $2x_1 + 3x_2 + 5x_3 \geq 2, 3x_1 + x_2 + 7x_3 \leq 3, x_1 + 4x_2 + 6x_3 \leq 5, x_1, x_2, x_3 \geq 0.$
   (M.D.U. 2011; K.U.K. Jan. 2014)
6. Minimize $Z = 10x_1 + 6x_2 + 2x_3$
   subject to $-x_1 + x_2 + x_3 \geq 1, 3x_1 + x_2 - x_3 \geq 2, x_1, x_2, x_3 \geq 0.$

Answers

1. $x_1 = \frac{16}{3}, x_2 = \frac{22}{3}$, Min. $Z = 7.4$
2. $x_1 = \frac{3}{5}, x_2 = \frac{6}{5}$, Max. $Z = -\frac{12}{5}$
3. $x_1 = 4, x_2 = 3$, Max. $Z = -18$
4. $x_1 = \frac{3}{5}, x_2 = \frac{6}{5}$, Min. $Z = \frac{12}{5}$
5. $x_1 = 0, x_2 = \frac{2}{3}, x_3 = 0$, Min. $Z = \frac{4}{3}$
6. $x_1 = \frac{1}{4}, x_2 = \frac{5}{4}, x_3 = 0$, Min. $Z = \frac{4}{3}$.

❑❑❑

APPENDIX—I

SHORT ANSWER TYPE QUESTIONS

FOURIER SERIES

1. The period of $\sin 4x$ is ....
2. The period of $\cos nx$ is ....
3. The period of $|\sin t|$ is ....
4. The period of $f(x) = \cos x + \frac{1}{2}\cos 2x + \frac{1}{3}\cos 3x$ is ....
5. The period of a constant function is ....
6. The period of $f(x) = \cos\left(\frac{4n\pi x}{3\lambda}\right)$ is ....
7. Fourier expansion of an odd function $f(x)$ in $(-\pi, \pi)$ has only ...... terms.
8. If $f(x)$ is an odd function in $(-\pi, \pi)$, then write the values of a_0, a_n and b_n.
9. Fourier expansion of an even function $f(x)$ in $(-\pi, \pi)$ has only ...... terms.
10. If $f(x)$ is an even function in $(-\pi, \pi)$, then write the values of a_0, a_n and b_n.
11. In the Fourier series expansion of $f(x) = \left(\frac{\pi - x}{2}\right)^2$ in $(0, 2\pi)$, the value of $b_n =$ ....
    (*M.D.U. Dec., 2013*)
12. In the Fourier series expansion of $f(x) = x \sin x$ in $(0, 2\pi)$, the value of $a_n =$ ....
13. Write the Fourier series for $f(x) = x^2$ in $(-\pi, \pi)$.
14. If $f(x) = x^2$ in $(-\pi, \pi)$, then the Fourier series of $f(x)$ contains only ...... terms.
15. In the Fourier series of $f(x) = |x|$ in $(-\pi, \pi)$, the value of $b_n =$ ....
16. Express $f(x) = |x|, -\pi < x < \pi$, as Fourier series.
17. In the Fourier series expansion of $f(x) = x \sin x$ in $(-\pi, \pi)$, the value of $b_n =$ ....
18. In the Fourier series expansion of $f(x) = x$ in $(-\pi, \pi)$, the ...... terms are absent.
19. Expand in a Fourier series the function $f(x) = x$ in the interval $0 < x < 2\pi$.
20. Expand in a Fourier series the function $f(x) = \frac{1}{2}x$ in the interval $(-\pi, \pi)$.
21. In the Fourier series expansion of $f(x) = |\sin x|$ in $(-\pi, \pi)$, the value of $b_n =$ ......
22. The expansion of $f(x) = |\cos x|$ as a Fourier series in $(-\pi, \pi)$ does not contain ...... terms.
23. In the Fourier expansion of $f(x) = x \cos x$ in $(-\pi, \pi)$, the ...... terms are absent.
24. The Fourier series expansion of $f(x) = \sqrt{1 - \cos x}$ in $(-\pi, \pi)$ contains only ...... terms.
25. Is the function $f(x) = \begin{cases} \pi + x, & -\pi < x < 0 \\ \pi - x, & 0 < x < \pi \end{cases}$ even or odd?

26. Is the following function $f(x) = \begin{cases} 1+\dfrac{2x}{\pi}, & -\pi \le x \le 0 \\ 1-\dfrac{2x}{\pi}, & 0 \le x \le \pi \end{cases}$ even or odd?

27. If $f(x) = \begin{cases} -x, & -\pi < x < 0 \\ x, & 0 < x < \pi \end{cases}$, then $f(x)$ is an ...... function in $(-\pi, \pi)$.

28. If $f(t) = \begin{cases} -1, & -1 < t < 0 \\ 1, & 0 < t < 1 \end{cases}$, then $f(t)$ is an ...... function in $(-1, 1)$.

29. Dirichlet conditions for the expansion of a function as a Fourier series in $[a, b]$ are ....

30. If $f(x) = x^4$ in $(-1, 1)$, then the Fourier co-efficient b_n = ...

31. If $f(x)$ is an even function in $(-\pi, \pi)$, then the graph of $f(x)$ is symmetrical about ....

32. If $x = c$ is a point of discontinuity then the Fourier series of $f(x)$ at $x = c$ gives $f(x)$ = ...

33. If $f(x) = \begin{cases} -\pi, & -\pi < x < 0 \\ x, & 0 < x < \pi \end{cases}$, then $f(0)$ = ...

34. If $f(x)$ is a periodic function with period 2T, then the value of Fourier co-efficient a_n = ...

35. The mean value of $f(x) \cos nx$ in $(0, 2\pi)$ = ...

36. If $f(x)$ is an odd function in $(-l, l)$, then the Fourier co-efficient a_n = ...

37. Find the Fourier series to represent $f(x) = x^2 - 2$ when $-2 \le x \le 2$.

38. The half-range cosine series for a function $f(x)$ in the interval $(0, l)$ is ....

39. The half-range sine series for a function $f(x)$ in the interval $(0, \pi)$ is ....

40. The half-range sine series for 1 in $(0, \pi)$ is ....

FOURIER TRANSFORMS

41. Fourier sine integral of $f(x)$ is ....

42. Fourier cosine integral of $f(x)$ is ....

43. Fourier integral of $f(x)$ is ....

44. Fourier cosine transform of $f(x)$ in $(0, \infty)$ is ....

45. The inversion formula for the Fourier cosine transform of $F_c(s)$ is ....

46. Fourier sine transform of $f(x)$ in $(0, \infty)$ is ......

47. The inverse Fourier sine transform of $F_s(s)$ is ....

48. Fourier transform of $f(x)$ is ....

49. The inverse Fourier transform of $F(s)$ is ....

50. The finite Fourier sine transform of $f(x)$ in $0 < x < c$ is defined as ....

51. The inverse finite Fourier sine transform of $F_s(n)$ is given by ....

52. The finite Fourier cosine transform of $f(x)$ in $0 < x < c$ is defined as ....

53. The inverse finite Fourier cosine transform of $F_c(n)$ is given by ....

54. Write the kernel of each of the following transforms:

(*i*) Fourier sine transform (*ii*) Fourier cosine transform

(*iii*) Fourier transform.

55. Find the Fourier sine transform of $e^{-|x|}$.

56. Find the Fourier sine transform of $\frac{1}{x}$.

57. What is the Fourier cosine transform of $\frac{1}{\sqrt{x}}$?

58. Find the Fourier cosine transform of $f(x) = \begin{cases} \cos x, & 0 < x < a \\ 0, & x \geq a \end{cases}$.

59. Find the Fourier transform of $f(x) = \begin{cases} 1, & |x| < a \\ 0, & |x| > a \end{cases}$.

60. Inverse finite Fourier sine transform of $F_s(n) = \frac{1 - \cos n\pi}{n^2\pi^2}$ where $0 \leq x \leq \pi$ is ....

61. Find the Fourier sine transform of $f(x) = \begin{cases} 1, & 0 \leq x < a \\ 0, & x \geq a \end{cases}$.

62. Find the Fourier cosine transform of $f(x) = e^{-ax}$, $a > 0$.

63. Write the finite Fourier cosine transform of $f(x) = 1$ in $(0, \pi)$.

64. If $f(x) = \begin{cases} 1, & 0 < x \leq \pi \\ 0, & x > \pi \end{cases}$, then Fourier cosine integral of $f(x)$ is ....

65. Is Fourier transform a linear operation?

66. If $F_c(s)$ and $G_c(s)$ are the Fourier cosine transforms of $f(x)$ and $g(x)$ respectively, then F_c $[a\, f(x) + b\, g(x)]$ = ..., where a and b are constants.

67. If $F(\lambda)$ is the Fourier transform of $f(x)$, then the Fourier transform of $f(ax)$ is ....

68. If $F_s\,[f(ax)] = \lambda F_s\left(\frac{s}{a}\right)$, then λ = ...

69. If $F(s)$ is the complex Fourier transform of $f(x)$, then $F\,[f(x-a)]$ = ...

70. If $F_s(s)$ is the complex Fourier transform of $f(x)$, then $F\,[f(x)\cos ax]$ = ...

71. If $F_s(s)$ and $F_c(s)$ are Fourier sine and cosine transforms of $f(x)$ respectively, then $F_s\,[f(x)\sin ax]$ = ...

72. If the Fourier transform of e^{-x^2} is $\sqrt{\pi}\, e^{-\frac{s^2}{4}}$, write the Fourier transform of

(*i*) e^{-ax^2} (*ii*) $e^{-x^2}\cos 2x$

73. State convolution theorem for Fourier transforms in words as well as in symbols.

74. Write the relation between Fourier and Laplace transforms.

75. Write the Fourier transform of dirac-delta function.

FUNCTIONS OF A COMPLEX VARIABLE

76. Real part of $e^{5 + \frac{1}{2}i\pi}$ is ....

77. Imaginary part of $e^{(5+3i)^2}$ is ....

78. Real part of e^{e^z} is ....

79. e^z is a periodic function with period ....

80. Find the real and imaginary parts of exp (z^2).

81. Re [Log $(\alpha + i\beta)$] = ..., Im [Log $(\alpha + i\beta)$] = ...

82. Find the general value of log (– 3).

83. Separate Log $(4 + 3i)$ into real and imaginary parts.

84. Prove that $\tan\left(i \log \dfrac{a - ib}{a + ib}\right) = \dfrac{2ab}{a^2 - b^2}$.

85. Is i^i wholly real or purely imaginary?

86. Prove that Log $i^i = -\left(2n + \dfrac{1}{2}\right)\pi$.

87. Imaginary part of log $(1 + i \tan \alpha)$ is ....

88. Period of tanh x is ....

89. Real part of sin z is ....

90. Imaginary part of cos $\bar{z}$ is ....

91. Real part of tan $(x + iy)$ is ....

92. Imaginary part of cot $\bar{z}$ is ....

93. Real part of sec z is ....

94. Imaginary part of cosec $\bar{z}$ is ....

95. Real part of sinh $(x + iy)$ is ....

96. (*a*) Imaginary part of cosh $(x + iy)$ is ....
(*b*) Real part of $(\sin x + i \cos x)^5$ (*K.U.K. 2013*)

97. $\sinh^{-1} x = \log$ (......).

98. $\tanh^{-1} x = \dfrac{1}{2} \log$ (......).

99. If $u + iv = \tan^{-1}(x + iy)$ then $u =$ ....

100. If $c \tan(x + iy) = a + ib$, prove that $\tan 2x = \dfrac{2ca}{c^2 - a^2 - b^2}$.

101. Find tanh x if $5 \sinh x - \cosh x = 5$.

102. Show that $\lim\limits_{z \to 0} \dfrac{(\text{Re}\, z - \text{Im}\, z)^2}{|z|^2}$ does not exist.

103. Show that the function $f(z)$ defined by

$$f(z) = \begin{cases} \dfrac{\text{Re}\,(z^2)}{|z|^2}, & z \neq 0 \\ 0, & z = 0 \end{cases}$$

is not continuous at $z = 0$.

104. Show that the function

$$f(z) = \begin{cases} \dfrac{\text{Im}\,(z)}{|z|}, & z \neq 0 \\ 0, & z = 0 \end{cases}$$

is not continuous at $z = 0$.

105. If $f(z) = u(x, y) + iv(x, y)$ is analytic, then write $f'(z)$ in terms of
(*i*) partial derivatives of u and v, (*ii*) partial derivatives of u only
(*iii*) partial derivatives of v only

106. The C-R equations for $f(z) = u(x, y) + iv(x, y)$ to be analytic are ....

107. Write the polar form of Cauchy-Riemann equations.

108. What is a harmonic function?

109. The real and imaginary parts of an analytic function $f(z) = u + iv$ satisfy the ...... equation.

110. The real and imaginary parts of an analytic function $f(z) = u + iv$ are ...... functions.

111. An analytic function is also called a ...... function or a ...... function.

112. The curves $u(x, y) = c_1$ and $v(x, y) = c_2$ are orthogonal if ....

113. An analytic function with constant modulus is ....

114. Determine a, b, c, d so that the function

$$f(z) = (x^2 + axy + by^2) + i(cx^2 + dxy + y^2)$$

is analytic.

115. Show that $f(z) = z + 2\bar{z}$ is not analytic anywhere in the complex plane.

116. Show that the function $f(z) = z\,|z|$ is not analytic anywhere.

117. Find the only function that is analytic from the following:

(*i*) $\sin z$ (*ii*) $\bar{z}$

(*iii*) Im (z) (*iv*) Re (iz).

118. The function $f(z) = \bar{z}$ is analytic at ....

119. Show that the function $f(z) = |\bar{z}|^2$ is not analytic at any point.

120. If $\phi(x, y) = 2x - x^2 + ay^2$ is a harmonic function, then find the value of a.

121. Is $\phi(x, y) = e^x \cos y$ a harmonic function?

122. Harmonic conjugate of $u = x \cos y$ is ....

123. Harmonic conjugate of $u = x^3 - 3xy^2$ is ....

124. If $u + iv$ is analytic, show that $v - iu$ is also analytic.

125. Are the families of curves $x^2 - y^2 = c_1$ and $xy = c_2$ orthogonal?

126. Find the values of a and b for which the function $f(z) = \cos x(\cosh y + a \sinh y) + i \sin x(\cosh y + b \sinh y)$ is analytic.

127. If $f(z) = u + iv$ is an analytic function and $u = x^2 - y^2 - y$, then its conjugate harmonic function v is given by ....

128. If $f(z) = u + iv$ is an analytic function and $v = y^2 - x^2$, then its conjugate harmonic function u is given by ....

129. The values of 'b' for which the function $e^{bx} \cos 5y$ is harmonic are ....

130. If $f(z)$ and $\overline{f(z)}$ both are analytic, then $f(z)$ is ....

131. The value of $\int_0^{1+i} z^2 dz$ along the line $x = y$ is ....

132. Evaluate $\int_0^{2+i} (\bar{z})^2 dz$ along the line $y = \frac{1}{2}x$.

133. $\oint_C \frac{dz}{z - a} =$ ...... , C: $|z - a| = r$.

134. State Cauchy's integral theorem.

135. State Cauchy's integral formula.

136. Evaluate $\oint_C \frac{5z^2 - 7z + 3}{z - 1} dz$, where C is $|z| = \frac{1}{2}$.

137. Evaluate $\oint_C \frac{dz}{5z-7}$, where C : $|z| = 1$.

138. Evaluate $\oint_C \frac{e^{-z}}{z+1}\,dz$, where C: $|z| = 2$.

139. Evaluate $\oint_C \frac{e^{z}}{(z-2)^3}\,dz$, where C: $|z| = 1$.

140. Evaluate $\oint_C \frac{z^3+z+1}{z^2-3z+2}\,dz$, where C is the ellipse $4x^2 + 9y^2 = 1$.

141. Evaluate $\oint_C \frac{5z^3-7z^2+3z+2}{z-4}\,dz$, where C: $9x^2 + 4y^2 = 36$.

142. Evaluate $\oint_C \frac{\cos z}{z-\pi}\,dz$ where, C: $|z-1| = 3$.

143. The value of $\int_C |z|\,dz$, where C is the contour represented by the straight line from $z = -i$ to $z = i$, is ....

144. The value of $\int_C \frac{z\,dz}{\sin z}$ where C: $|z| = 4$, is ....

POWER SERIES AND CONTOUR INTEGRATION

145. Expand the function $\frac{\sin z}{z-\pi}$ about $z = \pi$.

146. Write Laurent's series for $\frac{1}{z-2}$, $|z| > 2$.

147. Taylor's series expansion of $\log(1+z)$ about $z = 0$ is ....

148. Taylor's series expansion of $\frac{1}{z-2}$ in $|z| < 1$ is ....

149. Show that $z^{-2} = 1 + \sum_{n=1}^{\infty} (n+1)(z+1)^n$ when $|z+1| < 1$.

150. Taylor's series expansion of $\cos z$ about $z = \frac{\pi}{4}$ is ....

151. Expand $\frac{1}{z^2-3z+2}$ in the region $|z| < 1$.

152. Write Laurent's series for $\frac{1}{z^2-3z+2}$ when $|z| > 2$.

153. Write Laurent's series for $\frac{e^z}{(z-1)^2}$ about the singularity $z = 1$.

154. The co-efficient of $(z-a)^{-1}$ in the expansion of $f(z)$ about an isolated singularity $z = a$ is called ....

155. If $f(z)$ has a simple pole at $z = a$, then Res $\{f(z), a\} =$ ....

156. If $f(z)$ has a pole of order m at $z = a$, then Res $\{f(z), a\} =$ ....

157. If $f(z) = \dfrac{\phi(z)}{\psi(z)}$ has a simple pole at $z = a$, then Res $\{f(z), a\} =$ ....

158. Find the poles of the function $f(z) = \dfrac{\tan z}{z}$ inside the circle $|z| = 2$.

159. Find the poles of the function $f(z) = \dfrac{e^z}{\cos \pi z}$ inside the unit circle $|z| = 1$.

160. The poles of $\cot z$ are ....

161. If $f(z)$ has a pole of order 3 at $z = a$, then Res $\{f(z), a\} =$ ....

162. The singularity of $f(z) = \dfrac{z+2}{(z-3)^2}$ is ....

163. Poles of $f(z) = \dfrac{z^2}{(z-2)^3 (z+1)^2}$ are at $z =$ ....

164. Singular points of $\dfrac{\cos(\pi z^2)}{(z-2)(z-3)}$ are ....

165. The poles of $f(z) = \dfrac{z^2+1}{z^3+1}$ are $z =$ ....

166. Residue of $f(z) = \dfrac{1-e^{2z}}{z^4}$ at $z = 0$ is ....

167. Residue of $f(z) = z \cos \dfrac{1}{z}$ at $z = 0$ is ....

168. Residue of $f(z) = \dfrac{\cos z}{z}$ at $z = 0$ is ....

169. Residue of $f(z) = \dfrac{1+e^z}{\sin z + z \cos z}$ at $z = 0$ is ....

170. Residue of $f(z) = e^{\frac{1}{z}}$ at $z = 0$ is ....

171. Residue of $\tan z$ at $z = \dfrac{\pi}{2}$ is ....

172. What is the nature of singularity of $f(z) = \dfrac{z - \sin z}{z^3}$ at $z = 0$?

173. What is the nature of singularity of $f(z) = e^{\frac{1}{z-a}}$ at $z = a$?

PROBABILITY DISTRIBUTIONS

174. A bag contains 50 tickets numbered 1, 2, 3, ..., 50, of which five are drawn at random and arranged in ascending order of magnitude ($x_1 < x_2 < x_3 < x_4 < x_5$). What is the probability that $x_3 = 30$?

175. A card is drawn from a well-shuffled pack of playing cards. What is the probability that it is either a spade or an ace?

176. What is the chance that a leap year should have 53 Sundays?

177. A and B throw alternately with a single die, A having the first throw. The one who first throws ace is to win. What are their respective chances of winning?

178. Three bags A, B, C contain 4 red, 3 black, 2 white; 3 red, 4 black, 4 white; and 5 red, 2 black, 6 white balls respectively. If a bag is selected at random and a ball is drawn from it, find the probability that the ball drawn is red.

179. A card is drawn from an ordinary pack and a gambler bets that it is a spade or an ace. What are the odds against his winning the bet?

180. A speaks truth in 75% cases and B in 80% cases. In what percentage of cases are they likely to contradict each other in stating the same fact?

181. The probabilities of A, B, C solving a problem are $\frac{1}{3}, \frac{2}{7}$ and $\frac{3}{8}$ respectively. If all the three try to solve the problem simultaneously, find the probability that exactly one of them will solve it.

182. A family has two children. What is the probability that both the children are boys given that at least one of them is a boy?

183. A die is thrown twice and the sum of the numbers appearing is observed to be 6. What is the conditional probability that the number 4 has appeared at least once?

184. If $P(A) = 0.8$, $P(B) = 0.5$ and $P(B/A) = 0.4$, find $P(A \cup B)$. *(M.D.U. Dec. 2013)*

185. If $P(A) = \frac{6}{11}$, $P(B) = \frac{5}{11}$ and $P(A \cup B) = \frac{7}{11}$, find $P(B/A)$. *(M.D.U. Dec. 2014)*

186. Given that the two numbers appearing on throwing two dice are different. Find the probability of the event 'the sum of numbers on the dice is 4.'

187. If A and B are events such that $P(A/B) = P(B/A)$, then ....

188. Three cards are drawn successively, without replacement, from a pack of 52 well-shuffled cards. What is the probability that first two cards are kings and the third card is an ace?

189. Prove that if E and F are independent events, then so are the events E′ and F′.

190. A die is thrown thrice. Find the probability of getting an odd number at least once.

191. Bag I contains 3 red and 4 black balls while another Bag II contains 5 red and 6 black balls. One ball is drawn at random and it is found to be red. Find the probability that it was drawn from Bag II.

192. Two cards are drawn successively with replacement from a well-shuffled deck of 52 cards. Write the probability distribution of the number of aces.

193. The probability distribution of a random variable X is

X:	0	1	2	3	4
P(X):	0.1	k	$2k$	$2k$	k,

find k.

194. Find the variance of the number obtained on a throw of an unbiased die.

195. Find the mean number of heads in three tosses of a fair coin.

196. Two dice are thrown simultaneously. If X denotes the number of sixes, find the expectation of X.

197. The mean of the numbers obtained on throwing a die having written 1 on three faces, 2 on two faces and 5 on one face is ....

198. Test the validity of the statement:
Mean and standard deviation of a Binomial Distribution are 10 and 6 respectively.

199. A student obtained the following answer to a certain problem given to him; Mean = 2.4, variance = 3.2 for a binomial distribution. Comment on the result.

200. If the probability of male births is 0.5 find the probability that in a family of 4 children there will be at least one boy.

201. A coin is tossed four times. What is the probability of obtaining more than two heads?

202. A die is thrown six times. If "getting an odd number" is a success, what is the probability of getting at least one success?

203. Suppose that X has a Poisson's distribution. If $P(X = 2) = \frac{2}{3}P(X = 1)$, find $P(X = 0)$.

204. Recurrence formula for binomial distribution is ....

205. For a binomial distribution, mean = ... , standard deviation = ...

206. During war, one ship out of 9 was sunk on an average in making a certain voyage. What is the probability that exactly 3 out of a convoy of 6 ships would arrive safely?

207. The probability of a shooter hitting a target is $\frac{3}{4}$. How many minimum number of times he/she must fire so that the probability of hitting the target at least once is more than 0.99?

208. If X is a Poisson variate such that $P(X = 2) = 9P(X = 4) + 90\,P(X = 6)$, find the S.D.

209. Total area under normal probability curve is ....

210. For a normal distribution, mean, mode and median ....

211. In a normal curve, the maximum ordinate is at the ....

212. The probability density function for the normal distribution, in standard form, is given by ....

Answers

1. $\frac{\pi}{2}$

2. $\frac{2\pi}{n}$

3. π

4. 2π

5. not defined

6. $\frac{3\lambda}{2n}$

7. sine

8. $a_0 = 0,\ a_n = 0,\ b_n = \frac{2}{\pi}\int_0^{\pi} f(x)\sin nx\,dx$

9. cosine

10. $a_0 = \frac{2}{\pi}\int_0^{\pi} f(x)\,dx,\ a_n = \frac{2}{\pi}\int_0^{\pi} f(x)\cos nx\,dx,\ b_n = 0$

11. 0

12. $a_n = \frac{2}{n^2 - 1},\ n \neq 1;\ a_1 = -\frac{1}{2}$

13. $f(x) = \frac{\pi^2}{3} - 4\left(\frac{\cos x}{1^2} - \frac{\cos 2x}{2^2} + \frac{\cos 3x}{3^2} - \frac{\cos 4x}{4^2} + \ldots\ldots\right)$

14. cosine

15. 0

16. $f(x) = \frac{\pi}{2} - \frac{4}{\pi}\left(\cos x + \frac{\cos 3x}{3^2} + \frac{\cos 5x}{5^2} + \ldots\ldots\right)$

17. 0

18. cosine

19. $f(x) = \pi - 2\sum_{n=1}^{\infty} \frac{\sin nx}{n}$

20. $f(x) = \sin x - \frac{\sin 2x}{2} + \frac{\sin 3x}{3} - \frac{\sin 4x}{4} + \ldots\ldots$

21. 0

22. sine

23. cosine

24. cosine

25. even

26. even

27. even

28. odd

29. See Fourier series, 1.4

30. 0

31. y-axis

32. $\frac{1}{2}[f(c-0) + f(c+0)]$

33. $-\frac{\pi}{2}$

34. $\frac{1}{T}\int_c^{c+2T} f(x) \cos \frac{n\pi x}{T}\, dx$

35. $\frac{1}{2} a_n$

36. 0

37. $f(x) = -\frac{2}{3} - \frac{16}{\pi^2}\left(\cos \frac{\pi x}{2} - \frac{1}{4}\cos \pi x + \frac{1}{9}\cos \frac{3\pi x}{2} - \ldots\ldots\right)$

38. $f(x) = \frac{a_0}{2} + \sum_{n=1}^{\infty} a_n \cos \frac{n\pi x}{l}$ where $a_0 = \frac{2}{l}\int_0^l f(x)\, dx$ and $a_n = \frac{2}{l}\int_0^l f(x) \cos \frac{n\pi x}{l}\, dx$

39. $f(x) = \sum_{n=1}^{\infty} b_n \sin nx\, dx$, where $b_n = \frac{2}{\pi}\int_0^{\pi} f(x) \sin nx\, dx$

40. $\frac{4}{\pi}\left(\sin x + \frac{\sin 3x}{3} + \frac{\sin 5x}{5} + \ldots\ldots\right)$

---

41. $f(x) = \frac{2}{\pi}\int_0^{\infty} \sin \lambda x \int_0^{\infty} f(t) \sin \lambda t\, dt\, d\lambda$

42. $f(x) = \frac{2}{\pi}\int_0^{\infty} \cos \lambda x \int_0^{\infty} f(t) \cos \lambda t\, dt\, d\lambda$

43. $f(x) = \frac{1}{\pi}\int_0^{\infty}\int_{-\infty}^{\infty} f(t) \cos \lambda (t - x)\, dt\, d\lambda$

44. $F_c(s) = \int_0^{\infty} f(x) \cos sx\, dx$

45. $f(x) = \frac{2}{\pi}\int_0^{\infty} F_c(s) \cos sx\, ds$

46. $F_s(s) = \int_0^{\infty} f(x) \sin sx\, dx$

47. $f(x) = \frac{2}{\pi}\int_0^{\infty} F_s(s) \sin sx\, ds$

48. $F(s) = \int_{-\infty}^{\infty} f(x)\, e^{isx}\, dx$

49. $f(x) = \frac{1}{2\pi}\int_{-\infty}^{\infty} F(s)\, e^{-isx}\, ds$

50. $F_s(n) = \int_0^{c} f(x) \sin \frac{n\pi x}{c}\, dx$, where n is an integer.

51. $f(x) = \frac{2}{c} \sum_{n=1}^{\infty} F_s(n) \sin \frac{n\pi x}{c}$

52. $F_c(n) = \int_0^c f(x) \cos \frac{n\pi x}{c}\, dx$, where n is an integer.

53. $f(x) = \frac{1}{c} F_c(0) + \frac{2}{c} \sum_{n=1}^{\infty} F_c(n) \cos \frac{n\pi x}{c}$

54. $K(s, x) = \sin sx, \cos sx, e^{isx}$

55. $\frac{s}{1+s^2}$

56. $\frac{\pi}{2}$

57. $\sqrt{\frac{\pi}{2s}}$

58. $\frac{1}{2}\left[\frac{\sin(1+s)a}{1+s} + \frac{\sin(1-s)a}{1-s}\right]$

59. $F\{f(x)\} = \begin{cases} \frac{2\sin as}{s}, & s \neq 0 \\ 2a, & s = 0 \end{cases}$

60. $\frac{2}{\pi^3} \sum_{n=1}^{\infty} \left(\frac{1-\cos n\pi}{n^2}\right) \sin nx$

61. $\frac{1-\cos as}{s}$

62. $\frac{a}{s^2 + a^2}$

63. 0

64. $\frac{2}{\pi} \int_0^{\infty} \cos \lambda x \int_0^{\pi} \cos \lambda t\, dt\, d\lambda$

65. Yes

66. $a F_c(s) + bG_c(s)$

67. $\frac{1}{a} F\left(\frac{\lambda}{a}\right)$

68. $\frac{1}{a}$

69. $e^{isa} F(s)$

70. $\frac{1}{2}[F(s+a) + F(s-a)]$

71. $\frac{1}{2}[F_c(s-a) - F_c(s+a)]$

72. (i) $\sqrt{\frac{\pi}{a}} e^{-\frac{s^2}{4a}}$ (ii) $\frac{\sqrt{\pi}}{2}\left[e^{-\frac{1}{4}(s+2)^2} + e^{-\frac{1}{4}(s-2)^2}\right]$

73. The Fourier transform of the convolution of $f(x)$ and $g(x)$ is equal to the product of the Fourier transforms of $f(x)$ and $g(x)$, *i.e.*,
$F[f(x) * g(x)] = F[f(x)] \,.\, F[g(x)] = F(s)\, G(s)$.

74. If $f(t) = \begin{cases} e^{-xt} g(t), & t > 0 \\ 0, & t < 0 \end{cases}$, then $F[f(t)] = L[g(t)]$

75. $F[\delta(t-a)] = e^{isa}$.

---

76. 0

77. $e^{16} \sin 30$

78. $e^{e^x \cos y} . \cos(e^x \sin y)$

79. $2\pi i$

80. $e^{x^2 - y^2} \cos(2xy)$, $e^{x^2 - y^2} \sin(2xy)$

81. $\frac{1}{2} \log(\alpha^2 + \beta^2), 2n\pi + \tan^{-1} \frac{\beta}{\alpha}$

82. $\log 3 + i(2n + 1)\pi$

83. Real part = log 5, Imaginary part = $2n\pi + \tan^{-1} \frac{3}{4}$

85. Wholly real

87. α

88. πi

89. $\sin x \cosh y$

90. $\sin x \sinh y$

91. $\frac{\sin 2x}{\cos 2x + \cosh 2y}$

92. $\frac{\sinh 2y}{\cosh 2y - \cos 2x}$

93. $\frac{2 \cos x \cosh y}{\cos 2x + \cosh 2y}$

94. $\frac{2 \cos x \sinh y}{\cosh 2y - \cos 2x}$

95. $\sinh x \cos y$

96. (*a*) $\sinh x \sin y$ (*b*) $\sin 5x$

97. $x + \sqrt{x^2 + 1}$

98. $\frac{1+x}{1-x}$

99. $\tan^{-1} \frac{2x}{1 - x^2 - y^2}$

101. $\frac{4}{5}, -\frac{3}{5}$

105. (*i*) $\frac{\partial u}{\partial x} + i \frac{\partial v}{\partial x}$ or $\frac{\partial v}{\partial y} - i \frac{\partial u}{\partial y}$ (*ii*) $\frac{\partial u}{\partial x} - i \frac{\partial u}{\partial y}$

(*iii*) $\frac{\partial v}{\partial y} + i \frac{\partial v}{\partial x}$

106. $\frac{\partial u}{\partial x} = \frac{\partial v}{\partial y}, \frac{\partial u}{\partial y} = -\frac{\partial v}{\partial x}$

107. $\frac{\partial u}{\partial r} = \frac{1}{r} \frac{\partial v}{\partial \theta}, \frac{\partial v}{\partial r} = -\frac{1}{r} \frac{\partial u}{\partial \theta}$

108. Any solution of the Laplace's equation $\frac{\partial^2 \phi}{\partial x^2} + \frac{\partial^2 \phi}{\partial y^2} = 0$ is called a harmonic function.

109. Laplace's

110. Harmonic

111. regular, holomorphic

112. $u + iv$ is an analytic function

113. constant

114. $a = 2, b = -1, c = -1, d = 2$

117. (*i*)

118. no point

120. 1

121. Yes

122. $e^x \sin y + c$

123. $3x^2y - y^3 + c$

125. Yes. [**Hint.** Show that $(x^2 - y^2) + i(xy)$ is analytic.]

126. $a = -1, b = -1$

127. $2xy + x + c$

128. $2xy + c$

129. ± 5

130. constant

131. $\frac{2}{3}(-1 + i)$

132. $\frac{5}{3}(2-i)$

133. $2\pi i$

134. If $f(z)$ is an analytic function and $f'(z)$ is continuous at each point within and on a simple closed curve C, then $\oint_C f(z)\,dz = 0$.

135. If $f(z)$ is analytic within and on a closed curve C and a is any point within C, then $f(a) = \frac{1}{2\pi i}\oint_C \frac{f(z)}{z-a}\,dz$.

136. 0

137. 0

138. $2\pi ie$

139. 0

140. 0

141. 0

142. $-2\pi i$

143. i

144. 0

---

145. $-1+\frac{(z-\pi)^2}{3!}-\frac{(z-\pi)^4}{5!}+\frac{(z-\pi)^6}{7!}-\ldots\ldots$

146. $\frac{1}{z}+\frac{2}{z^2}+\frac{2^2}{z^3}+\frac{2^3}{z^4}+\ldots\ldots=\sum_{n-1}^{\infty}\frac{2^{n-1}}{z^n}$

147. $z-\frac{z^2}{2}+\frac{z^3}{3}-\frac{z^4}{4}+\ldots\ldots$

148. $-\frac{1}{2}\left(1+\frac{z}{2}+\frac{z^2}{4}+\frac{z^3}{8}+\ldots\ldots\right)$

150. $\frac{1}{\sqrt{2}}\left[1-\left(z-\frac{\pi}{4}\right)-\frac{1}{2!}\left(z-\frac{\pi}{4}\right)^2+\frac{1}{3!}\left(z-\frac{\pi}{4}\right)^3+\ldots\ldots\right]$

151. $\frac{1}{2}+\frac{3}{4}z+\frac{7}{8}z^2+\frac{15}{16}z^3+\ldots\ldots$

152. $z^{-2}+3z^{-3}+7z^{-4}+\ldots\ldots$

153. $e\left[(z-1)^{-2}+(z-1)^{-1}+\frac{1}{2!}+\frac{1}{3!}(z-1)+\frac{1}{4!}(z-1)^2+\ldots\ldots\right]$

154. The residue of $f(z)$ at $z = a$.

155. $\lim_{z\to a}(z-a)f(z)$

156. $\frac{1}{(m-1)!}\lim_{z\to a}\frac{d^{m-1}}{dz^{m-1}}[(z-a)^m f(z)]$

157. $\lim_{z\to a}\frac{\phi(z)}{\psi'(z)}$

158. $0, \pm\frac{\pi}{2}$

159. $\pm\frac{1}{2}$

160. $n\pi$, n is an integer

161. $\frac{1}{2!}\lim_{z\to a}\frac{d^2}{dz^2}[(z-a)^3 f(z)]$

162. $z = 3$

163. $2, -1$

164. $z = 2, 3$

165. $-1, \frac{1\pm i\sqrt{3}}{2}$

166. $-\frac{4}{3}$

167. $-\frac{1}{2}$ **168.** 1 **169.** 1

170. 0 **171.** – 1 **172.** removable

173. essential

---

174. $\frac{551}{15134}$ **175.** $\frac{4}{13}$ **176.** $\frac{2}{7}$

177. $\frac{6}{11}, \frac{5}{11}$ **178.** 0.366 **179.** 9 : 4

180. 35% **181.** $\frac{25}{56}$ **182.** $\frac{1}{3}$

183. $\frac{2}{5}$ **184.** 0.98 **185.** $\frac{2}{3}$

186. $\frac{1}{15}$ **187.** P(A) = P(B) **188.** $\frac{2}{5525}$

190. $\frac{7}{8}$ **191.** $\frac{35}{68}$

192.

X:	0	1	2
P(X):	$\frac{144}{169}$	$\frac{24}{169}$	$\frac{1}{169}$

193. 0.15 **194.** $\frac{35}{12}$ **195.** 1.5

196. $\frac{1}{3}$ **197.** 2 **198.** Not valid

199. Not valid **200.** $\frac{15}{16}$ **201.** $\frac{5}{16}$

202. $\frac{63}{64}$ **203.** e^{-4} **204.** $P(r+1) = \frac{n-r}{r+1} \cdot \frac{p}{q} P(r)$

205. $np, \sqrt{npq}$ **206.** $\frac{10240}{9^6}$ **207.** 4

208. 1 **209.** 1 **210.** coincide

211. Mean **212.** $f(z) = \frac{1}{\sqrt{2\pi}} e^{-\frac{1}{2}z^2}$.

APPENDIX—II

Table 1 : NORMAL TABLE

AREAS UNDER THE STANDARD NORMAL

$$\text{CURVE} = \frac{1}{\sqrt{2\pi}} \int_0^z e^{-\frac{z^2}{2}}\, dz$$

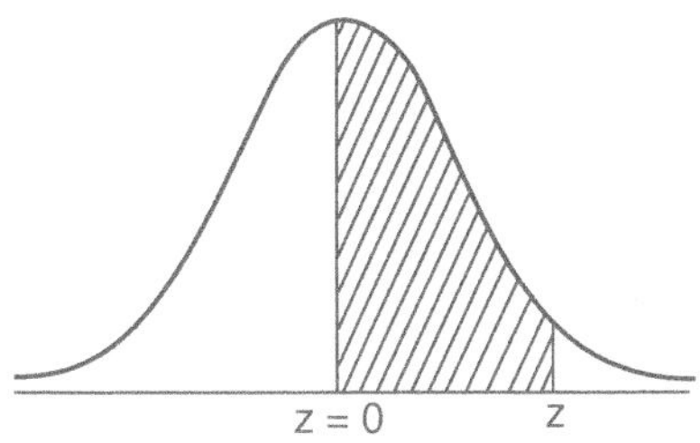

z	.00	.01	.02	.03	.04	.05	.06	.07	.08	.09
0.0	.0000	.0040	.0080	.0120	.0160	.0199	.0239	.0279	.0319	.0359
0.1	.0398	.0438	.0478	.0517	.0557	.0596	.0636	.0675	.0714	.0754
0.2	.0793	.0832	.0871	.0910	.0948	.0987	.1026	.1064	.1103	.1141
0.3	.1179	.1217	.1255	.1293	.1331	.1368	.1406	.1443	.1480	.1517
0.4	.1554	.1591	.1628	.1664	.1700	.1736	.1772	.1808	.1844	.1879
0.5	.1915	.1950	.1985	.2019	.2054	.2088	.2123	.2157	.2190	.2224
0.6	.2257	.2291	.2324	.2357	.2389	.2422	.2454	.2485	.2517	.2549
0.7	.2580	.2611	.2642	.2673	.2704	.2734	.2764	.2794	.2823	.2852
0.8	.2881	.2910	.2939	.2967	.2995	.3023	.3051	.3078	.3106	.3133
0.9	.3159	.3186	.3212	.3238	.3264	.3289	.3315	.3340	.3365	.3389
1.0	.3413	.3438	.3461	.3485	.3508	.3531	.3554	.3577	.3599	.3621
1.1	.3643	.3665	.3686	.3708	.8729	.3749	.3770	.3790	.3810	.3830
1.2	.3849	.3869	.3888	.3907	.3925	.3944	.3962	.3980	.3997	.4015
1.3	.4032	.4049	.4066	.4082	.4099	.4115	.4131	.4147	.4162	.4177
1.4	.4192	.4207	.4222	.4236	.4251	.4255	.4279	.4292	.4306	.4319
1.5	.4332	.4345	.4357	.4370	.4382	.4394	.4406	.4418	.4429	.4441
1.6	.4452	.4463	.4474	.4484	.4495	.4505	.4515	.4525	.4535	.4545
1.7	.4554	.4564	.4573	.4582	.4591	.4599	.4608	.4616	.4625	.4633
1.8	.4641	.4649	.4656	.4664	.4671	.4678	.4686	.4693	.4699	.4706
1.9	.4713	.4719	.4726	.4732	.4738	.4744	.4750	.4756	.4761	.4767
2.0	.4772	.4778	.4783	.4788	.4793	.4798	.4803	.4808	.4812	.4817
2.1	.4821	.4826	.4830	.4834	.4838	.4842	.4846	.4850	.4854	.4857
2.2	.4861	.4864	.4868	.4871	.4875	.4878	.4881	.4884	.4887	.4890
2.3	.4893	.4896	.4898	.4901	.4904	.4906	.4909	.4911	.4913	.4916
2.4	.4918	.4920	.4922	.4925	.4927	.4929	.4931	.4932	.4934	.4936
2.5	.4938	.4940	.4941	.4943	.4945	.4946	.4948	.4949	.4951	.4952
2.6	.4953	.4955	.4956	.4957	.4959	.4930	.4961	.4962	.4963	.4964
2.7	.4965	.4966	.4967	.4968	.4969	.4970	.4971	.4972	.4973	.4974
2.8	.4974	.4975	.4976	.4977	.4977	.4978	.4979	.4979	.4980	.4981
2.9	.4981	.4982	.4982	.4983	.4984	.4984	.4985	.4985	.4986	.4986
3.0	.4987	.4987	.4987	.4988	.4988	.4989	.4999	.4989	.4990	.4990
3.1	.4990	.4991	.4991	.4991	.4992	.4992	.4992	.4992	.4993	.4993

Table 2 : SIGNIFICANT VALUES t_v (α) OF t-DISTRIBUTION (TWO TAIL AREAS) [| t | > $t_v(\alpha)$] = α

d.f.	*Probability (Level of Significance)*					
(v)	*0.50*	*0.10*	*0.05*	*0.02*	*0.01*	*0.001*
1	1.00	6.31	12.71	31.82	63.66	636.62
2	0.82	0.92	4.30	6.97	6.93	31.60
3	0.77	2.32	3.18	4.54	5.84	12.94
4	0.74	2.13	2.78	3.75	4.60	8.61
5	0.73	2.02	2.57	3.37	4.03	6.86
6	0.72	1.94	2.45	3.14	3.71	5.96
7	0.71	1.90	2.37	3.00	3.50	5.41
8	0.71	1.80	2.31	2.90	3.36	5.04
9	0.70	1.83	2.26	2.82	3.25	4.78
10	0.70	1.81	2.23	2.76	3.17	4.59
11	0.70	1.80	2.20	2.72	3.11	4.44
12	0.70	1.78	2.18	2.68	3.06	4.32
13	0.69	1.77	2.16	2.05	3.01	4.22
14	0.69	1.76	2.15	2.62	2.98	4.14
15	0.69	1.75	2.13	2.60	2.95	4.07
16	0.69	1.75	2.12	2.58	2.92	4.02
17	0.69	1.74	2.11	2.57	2.90	3.97
18	0.69	1.73	2.10	2.55	2.88	3.92
19	0.69	1.73	2.09	2.54	2.86	3.88
20	0.69	1.73	2.09	2.53	2.85	3.85
21	0.69	1.72	2.08	2.52	2.83	3.83
22	0.69	1.72	2.07	2.51	2.42	3.79
23	0.69	1.71	2.07	2.50	2.81	3.77
24	0.69	1.71	2.06	2.49	2.80	3.75
25	0.68	1.71	2.06	2.49	2.79	3.73
26	0.68	1.71	2.06	2.48	2.78	3.71
27	0.68	1.70	2.05	2.47	2.77	3.69
28	0.68	1.70	2.05	2.47	2.76	3.67
29	0.68	1.70	2.05	2.46	2.76	3.66
30	0.68	1.70	2.04	2.46	2.75	3.65
∞	0.67	1.65	1.96	2.33	2.58	3.29

Table 3: CHI-SQUARE (χ^2)
Significant Values χ^2 (α) of χ^2 Distribution Right Tail Areas for Given Probability α,
$P = P_r\ (\chi^2 > \chi^2\ (\alpha)) = \alpha$
And is Degrees of Freedom (d.f.)

Degree of freedom (v)	*Probability (Level of Significance)*						
	0 = .99	*0.95*	*0.50*	*0.10*	*0.05*	*0.02*	*0.01*
1	.000157	.00393	.455	2.706	3.841	5.214	6.635
2	.0201	.103	1.386	4.605	5.991	7.824	9.210
3	.115	.352	2.366	6.251	7.815	9.837	11.341
4	.297	.711	3.357	7.779	9.488	11.668	13.277
5	.554	1.145	4.351	9.236	11.070	13.388	15.086
6	.872	2.635	5.348	10.645	12.592	15.033	16.812
7	1.239	2.167	6.346	12.017	14.067	16.622	18.475
8	1.646	2.733	7.344	13.362	15.507	18.168	20.090
9	2.088	3.325	8.343	14.684	16.919	19.679	21.669
10	2.558	3.940	9.340	15.987	18.307	21.161	23.209
11	3.053	4.575	10.341	17.275	19.675	22.618	24.725
12	3.571	5.226	11.340	18.549	21.026	24.054	26.217
13	4.107	5.892	12.340	19.812	22.362	25.472	27.688
14	4.660	6.571	13.339	21.064	23.685	26.873	29.141
15	4.229	7.261	14.339	22.307	24.996	28.259	30.578
16	5.812	7.962	15.338	23.542	26.296	29.633	32.000
17	6.408	8.672	15.338	24.769	27.587	30.995	33.409
18	7.015	9.390	17.338	25.989	28.869	32.346	34.805
19	7.633	10.117	18.338	27.204	30.144	33.687	36.191
20	8.260	10.851	19.337	28.412	31.410	35.020	37.566
21	8.897	11.591	20.337	29.615	32.671	36.343	38.932
22	9.542	12.338	21.337	30.813	33.924	37.659	40.289
23	10.196	13.091	22.337	32.007	35.172	38.968	41.638
24	10.856	13.848	23.337	32.196	36.415	40.270	42.980
25	11.524	14.611	24.337	34.382	37.65	41.566	44.314
26	12.198	15.379	25.336	35.363	38.885	41.856	45.642
27	12.879	16.151	26.336	36.741	40.113	41.140	46.963
28	13.565	16.928	27.336	37.916	41.337	45.419	48.278
29	14.256	17.708	28.336	39.087	42.557	46.693	49.588
30	14.933	18.493	29.336	40.256	43.773	47.962	50.892

Note. For degrees of freedom (v) greater than 30, the quantity $\sqrt{2\chi^2} - \sqrt{2v - 1}$ may be used as a normal variate with unit variance.

APPENDIX—III

SOME IMPORTANT CURVES

Cissoid (Imp.)

$y^2(2a - x) = x^3$

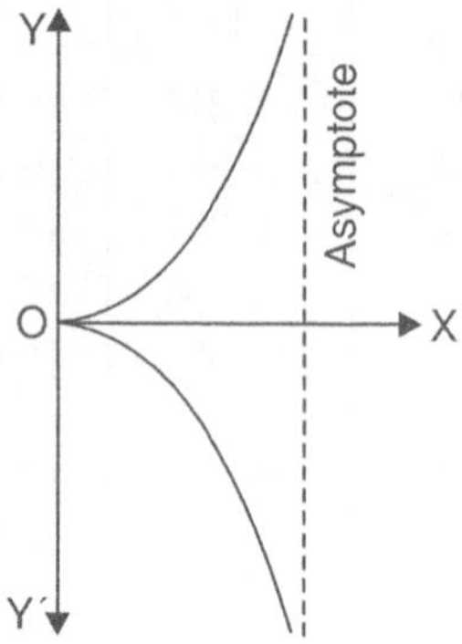

Catenary

$y = a \cosh \frac{x}{a}$

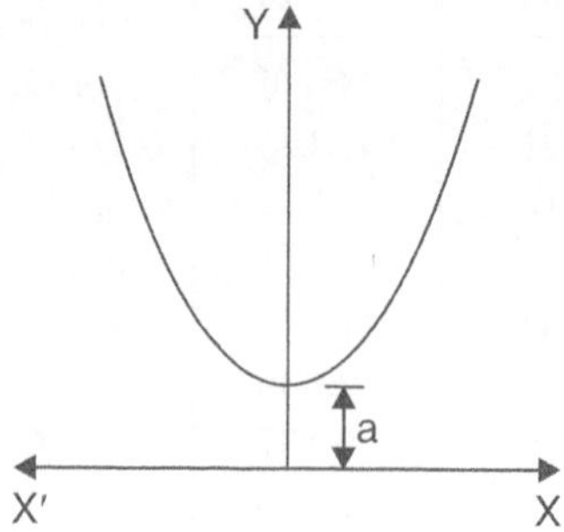

Cubical Parabola

$a^2y = x^3$

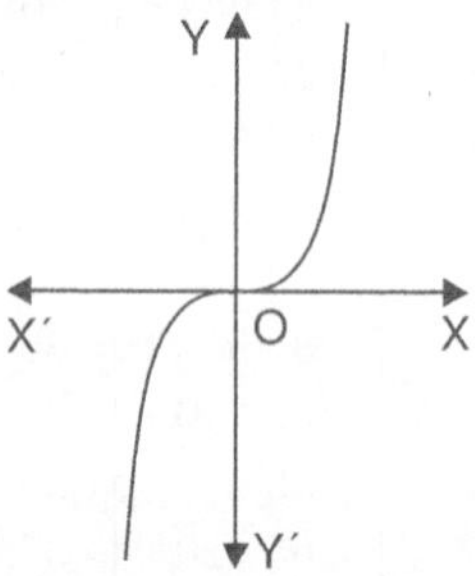

Semicubical Parabola

$ay^2 = x^3$

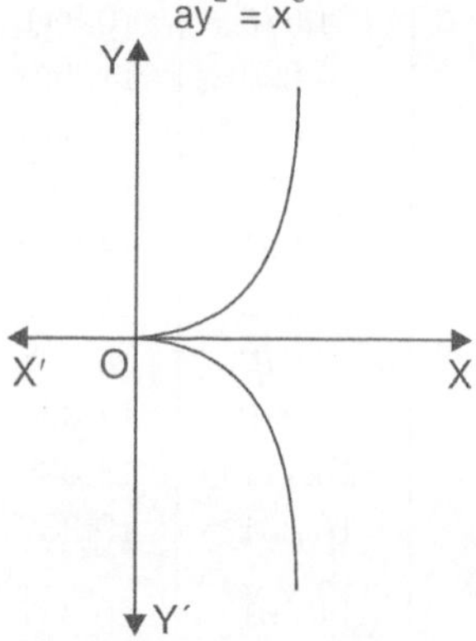

Strophoid

$y^2(a - x) = x^2(a + x)$

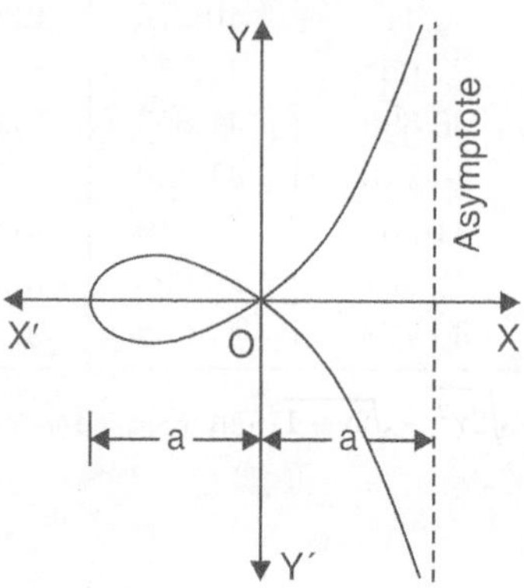

Folium of Descartes

$x^3 + y^3 = 3axy$

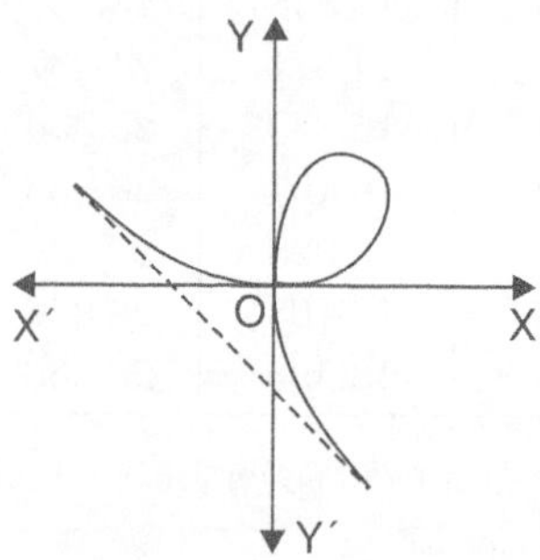

Hypocycloid (Imp.)

$x^{2/3} + y^{2/3} = a^{2/3}$

$x = a\cos^3\theta \quad y = a\sin^3\theta$

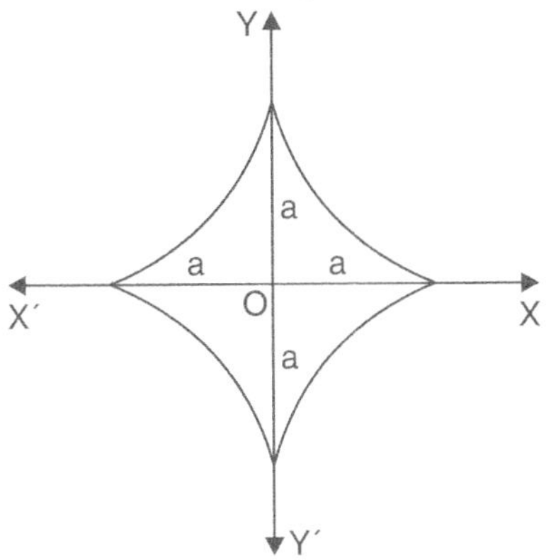

Witon of Agnesi

$xy^2 = 4a^2(2a - x)$

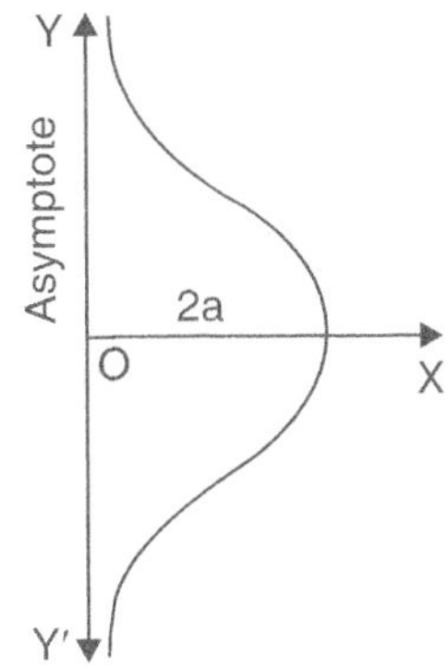

Cardioid (Imp.)

$r = a(1 + \cos\theta)$

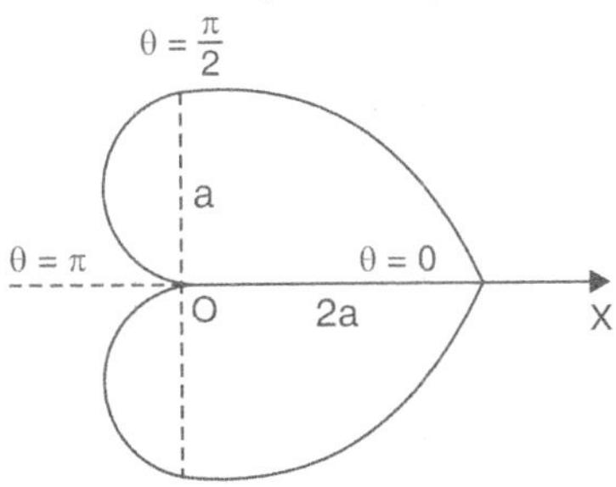

Cardioid (Imp.)

$r = a(1 - \cos\theta)$

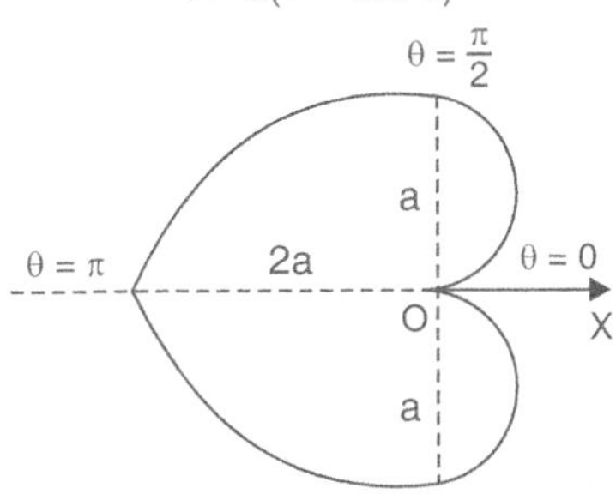

Cycloid (Most Imp.)

$x = a(\theta + \sin\theta)$

$y = a(1 - \cos\theta)$

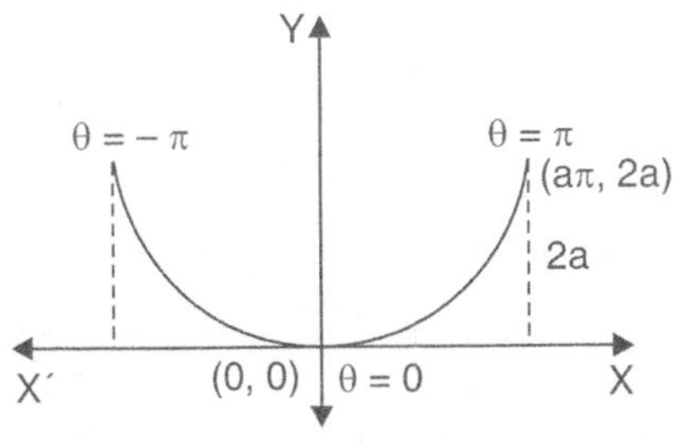

Cycloid

$x = a(\theta - \sin\theta)$

$y = a(1 + \cos\theta)$

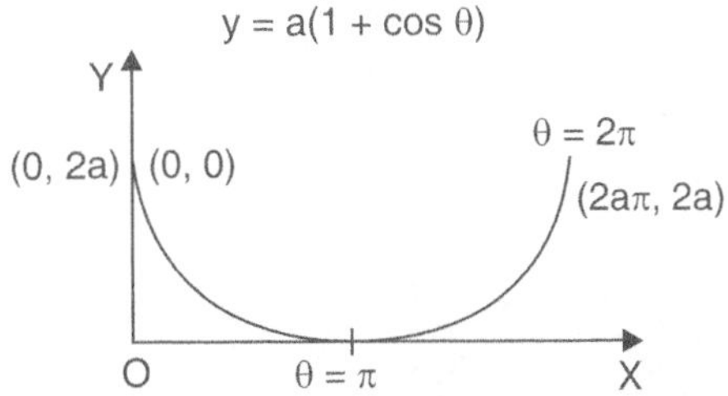

Cycloid (Imp.)
$x = a(\theta - \sin\theta)$
$y = a(1 - \cos\theta)$

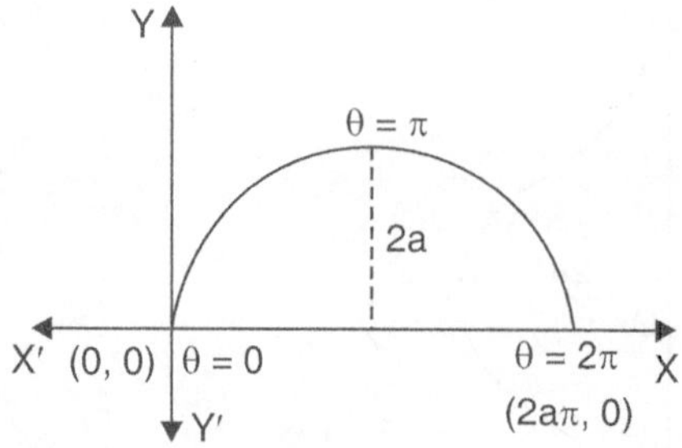

Cycloid
$x = a(\theta + \sin\theta)$
$y = a(1 + \cos\theta)$

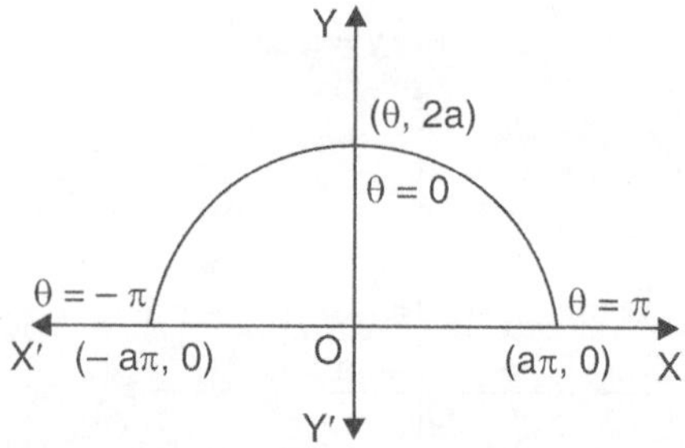

Lemniscate
$r^2 = a^2 \cos 2\theta$

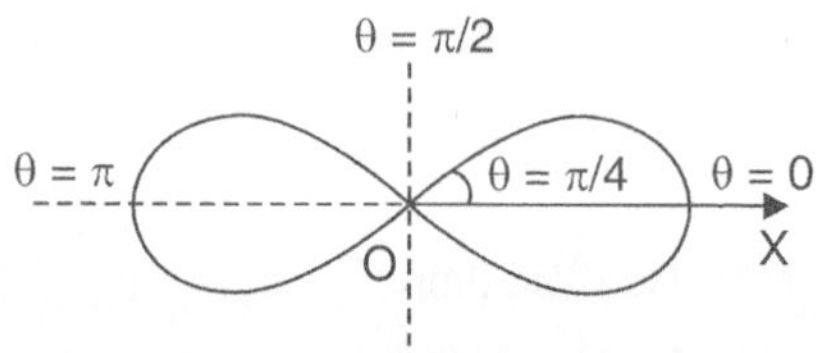

The three leaved rose
$r = a \sin 3\theta$

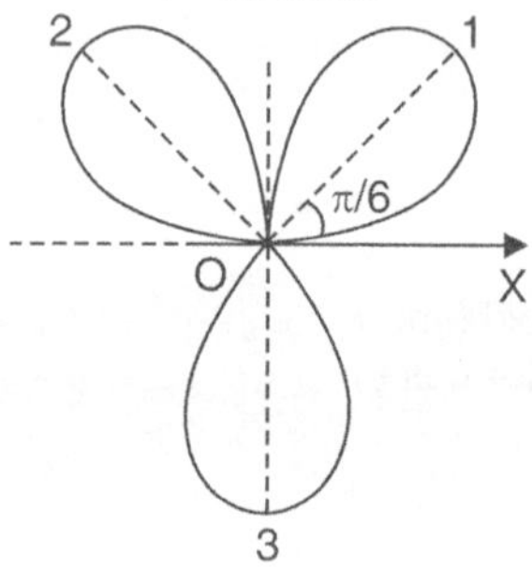

EXAMINATION PAPERS

M.D.U., Rohtak
B.Tech. 3rd Semester (AEIE)
Examination, December, 2015
MATHEMATICS–III
Paper : Math-201-F

Time allowed: 3 hours] [*Maximum Marks:100*

Before answering the question, candidates should ensure that they have been supplied the correct and complete question paper. No complaint in this regard, will be entertained after examination.

Note: Attempt ***five*** questions in total selecting at least ***one*** question from each Section. Question No. 1 is ***compulsory.***

1. (*a*) Find a_n for $f(x) = |x|, -\pi < x < \pi$.

 (*b*) Define Analytic function and hence state necessary and sufficient conditions for a function to be analytic.

 (*c*) Define discrete and continuous probability distribution.

 (*d*) Maximize $Z = 2x + 3y$

 subject to

$$x + y \le 30,$$
$$3 \le y \le 12,$$
$$x - y \ge 0,\ 0 \le x \le 20$$

 by graphical method.

SECTION–A

2. (*a*) Find the Fourier series for $f(x) = \begin{cases} 0, & -\pi \le x \le 0 \\ \sin x, & 0 \le x \le \pi \end{cases}$

 (*b*) Find the half range cosine series for $f(x) = x^2$ in the interval $0 \le x \le \pi$.

3. Solve the equation $\dfrac{\partial u}{\partial t} = \dfrac{\partial^2 u}{\partial x^2}$, $x > 0$, $t > 0$ subject to the conditions:

 (*i*) $u = 0$, when $x = 0$, $t > 0$ (*ii*) $u = \begin{cases} 1, & 0 < x < 1 \\ 0, & x \ge 1 \end{cases}$ when $t = 0$ and

 (*iii*) $u(x, t)$ is bounded.

SECTION–B

4. (*a*) If $a + ib = \tanh\left(v + \dfrac{i\pi}{4}\right)$, prove that $a^2 + b^2 = 1$.

(*b*) If $f(z) = \begin{cases} \dfrac{x^3 y(y - ix)}{x^6 + y^2}, & z \neq 0 \\ 0\ , & z = 0 \end{cases}$

Prove that $\dfrac{f(z) - f(0)}{z} \to 0$ as $z \to 0$ along any radius vector but not as $z \to 0$ in any manner and also that $f(z)$ is not analytic at $z = 0$.

5. (*a*) State and prove Cauchy's integral theorem.

(*b*) Evaluate $\oint_C \dfrac{3z^2 + z}{z^2 - 1}\, dz$, where C is the circle $|\ z - 1\ | = 1$.

SECTION–C

6. (*a*) Find the sum of the residues of the function $f(z) = \dfrac{\sin z}{z \cos z}$ at its pole inside the circle $|\ z\ | = 2$.

(*b*) Evaluate: $\displaystyle\int_0^{2\pi} \frac{d\theta}{2 + \cos\theta}$.

7. (*a*) Prove that mean and variance of the Poisson distribution are equal.

(*b*) A survey was conducted to find the supplies of the consumer durables for the market. It was found that the three major companies A, B and C have market share of 35%, 25% and 40% respectively out of which 2%, 1% and 3% are not up to the satisfaction. A consumer buys a product and is dissatisfied with it. What is the probability that it might be from the company C?

SECTION–D

8. Verify whether Poisson distribution can be assumed from the data given below:

x	0	1	2	3	4
f	109	65	22	3	1

9. Maximize $z = -2x_1 - x_2$

subject to

$$3x_1 + x_2 \geq 3,$$
$$4x_1 + 3x_2 \geq 6,$$
$$x_1 + 2x_2 \geq 3$$
$$x_1, x_2 \geq 0$$

by dual simplex method.

K.U., Kurukshetra
B.Tech. 3rd Semester
BT-3/D-15
MATHEMATICS–III
December, 2015
Paper–Math–201-E

Time allowed : 3 hours] *[Maximum marks : 100*

Note: *Attempt any **five** questions in all, selecting at least **one** question from each unit.*

UNIT–I

1. (*a*) Find the Fourier series expansion of $f(x) = 2x - x^2$ in (0, 3) and hence deduce that

$$\frac{1}{1^2} - \frac{1}{2^2} + \frac{1}{3^2} - \frac{1}{4^2} + \ldots\ldots = \frac{\pi^2}{12}$$ (10)

(*b*) Obtain the Half-range sine series of the function $f(x) = x \sin x$ in $0 < x < \pi$. (10)

2. (*a*) Find the Fourier cosine transform of $f(x) = e^{-ax}$, $a > 0$ and hence find the Fourier sine transform of xe^{-ax}. (10)

(*b*) Solve: $$\frac{\partial u}{\partial t} = k\frac{\partial u^2}{\partial x^2},$$

subject to the conditions $u(x, 0) = 0$, $x \geq 0$

$$\frac{\partial u(0,t)}{\partial x} = -\mu \text{ (constant)}, \; t > 0.$$ (10)

UNIT–II

3. (*a*) Reduce $\tan^{-1}(\cos\theta + i\sin\theta)$ to the form $a + ib$, hence show that

$$\tan^{-1}(e^{i\theta}) = \frac{n\pi}{2} + \frac{\pi}{4} - \frac{i}{2}\log\tan\left(\frac{\pi}{4} - \frac{\theta}{2}\right)$$ (10)

(*b*) Find all values of $(-1)^{1/4}$. (10)

4. (*a*) Determine the analytic function whose real part is given by $e^x((x^2 - y^2)\cos y - 2xy \sin y)$. (10)

(*b*) Under the transformation $w = \dfrac{1}{z}$,

Show that the image of hyperbola $x^2 - y^2 = 1$ is the Lemniscate $\rho^2 = \cos 2\phi$. (10)

UNIT–III

5. (*a*) An insurance company insured 2000 scooter drivers, 4000 car drivers and 6000 truck drivers. The probability of accidents are 0.01, 0.03 and 0.15 respectively. One of the insured persons meet an accident. What is the probability that he is a scooter driver? (10)

(*b*) The frequency distribution of a measurable characteristic varying between 0 and 2 is as under.

$$f(x) = \begin{cases} x^3 & , \quad 0 \le x \le 1 \\ (2-x)^3, & 1 \le x \le 2 \end{cases}$$

Calculate the standard deviation and also the mean deviation about the mean. (10)

6. (*a*) Fit a Binomial distribution to the following data:

x	0	1	2	3	4
f	28	62	46	10	4

(10)

(*b*) Define normal distribution and prove that standard deviation for a normal distribution is approximately 25% more than the mean deviation. (10)

UNIT–IV

7. (*a*) Using graphical method, solve the LPP

Minimize $z = 2x + 3y$

Subject to conditions $x \ge 0, y \ge 0, 1 \le x + 2y \le 8$. Also draw the feasible region of the constraints. (8)

(*b*) Using simplex method, solve the LPP

Maximize $z = 5x_1 + 3x_2$

subject to $x_1 + x_2 \le 2,\ 5x_1 + 2x_2 \le 10,$

$3x_1 + 8x_2 \le 12;$

$x_1, x_2 \ge 0.$ (12)

8. (*a*) Write the duals of the following Problem

Minimize $z = 3x_1 - 3x_2 + x_3$

subject to $2x_1 - 3x_2 + x_3 \le 5,$

$4x_1 - 2x_2 \ge 9,$

$-8x_1 + 4x_2 + 3x_3 = 8,$

$x_1, x_2 \ge 0$ and x_3 is unrestricted. (5)

(*b*) Using dual simplex method,

Maximize $z = -3x_1 - x_2$

subject to $x_1 + x_2 \ge 1,$

$2x_1 + 3x_2 \ge 2,$

$x_1, x_2 \ge 0.$ (15)

DCRUST, Murthal, Sonepat
B.Tech. (Part-Time)
EXAMINATION, May 2014
[1st Semester (Re-appear only)]
CSE-EE, ECE, ME-CE
MATHEMATICS–III
Math-W-101

Time : 3 hours] [*Maximum marks : 100*

Before answering the Question paper candidates should ensure that they have been supplied the correct and complete Question Paper. No complaints in this regard, will be entertained after the examination.

Note: *Attempt any* **five** *questions, selecting at least one from each part. All questions carry equal marks.*

PART–A

1. (*a*) Find a Fourier series to represent $x - x^2$ from $x = -\pi$ to π and hence show that

$$\frac{\pi^2}{12} = \frac{1}{1^2} - \frac{1}{2^2} + \frac{1}{3^2} - \frac{1}{4^2} + \cdots$$

(*b*) If $f(x) = \begin{cases} x, & 0 < x < \pi/2 \\ \pi - x, & \dfrac{\pi}{2} < x < \pi \end{cases}$ then show that $f(x) = \dfrac{4}{\pi}\left[\sin x - \dfrac{1}{3^2}\sin 3x + \dfrac{1}{5^2}\sin 5x \ldots\right]$.

2. (*a*) Using Fourier Transform method, find $f(x)$ from the equation $\int_0^\infty f(x)\cos sx\, dx = e^{-s}$.

(*b*) State and prove convolution theorem on Fourier Transform.

PART–B

3. (*a*) Separate $\tan^{-1}(a + ib)$ into real and imaginary parts.

(*b*) Show that the function $f(z) = \sqrt{|xy|}$ is not regular at the origin, although CR equations are satisfied at this point.

4. (*a*) Find an analytic function $w = u + iv$ given that $v = \dfrac{x}{x^2 + y^2} + \cosh(x)\cos y$.

(*b*) Evaluate $\int_C \dfrac{z - 1}{(z + 1)^2 (z - 2)}\, dz$, where C is given by $|z - i| = 2$.

5. (*a*) Obtain the Taylor's or Laurent's series which represents the function

$$f(z) = \frac{1}{(1+z^2)(z+2)} \text{ when } (i)\ 1 < |z| < 2 \quad (ii)\ |z| > 2$$

(*b*) Using Cauchy Residue Theorem evaluate the integral $\int_0^{2\pi} \frac{d\theta}{2+\cos\theta}$.

PART–C

6. (*a*) Out of 800 families with four children each, how many families would be expected to have

(*i*) 2 boys and 2 girls (*ii*) at least one boy

(*iii*) No girl (*iv*) Atmost two girls.

(*b*) Assuming that the diameters of 1000 brass plugs taken consecutively from a machine form a normal distribution with mean 0.7515 cm and standard deviation 0.0020 cm. How many of plugs are likely to be rejected if the approved diameter is 0.752 ± 0.004 cm?

7. (*a*) In a sample of 100 residents of a colony 60 are found to be wheat eaters and 40 are rice eaters. Can we assume that both food articles are equally popular?

(*b*) A set of five similar coins is tossed 320 times and the result is

No. of Heads	0	1	2	3	4	5
Frequency	6	27	72	112	71	32

Test the hypothesis that the data follow a binomial distribution.

8. Solve the following LPP. using simplex method:

Maximize $Z = 20x_1 + 6x_2 + 8x_3$

subject to the constraints

$$8x_1 + 2x_2 + 3x_3 \le 200$$
$$4x_1 + 3x_2 \le 150$$
$$2x_1 + x_3 \le 50$$
$$x_1, x_2, x_3 \ge 0$$